A HISTORY OF RUSSIA

A HISTORY OF
RUSSIA

FIFTH EDITION

NICHOLAS V. RIASANOVSKY

New York   Oxford
OXFORD UNIVERSITY PRESS
1993

To My Students

Oxford University Press

Oxford   New York   Toronto
Delhi   Bombay   Calcutta   Madras   Karachi
Kuala Lumpur   Singapore   Hong Kong   Tokyo
Nairobi   Dar es Salaam   Cape Town
Melbourne   Auckland   Madrid

and associated companies in
Berlin   Ibadan

Published by Oxford University Press, Inc.,
198 Madison Avenue, New York, New York 10016-4314

Oxford is a registered trademark of Oxford University Press

Library of Congress Cataloging-in-Publication Data
Riasanovsky, Nicholas Valentine, 1923–
A history of Russia / Nicholas V. Riasanovsky. — 5th ed.
p.   cm.
Includes bibliographical references and index.
ISBN 0–19–507462–9
1. Soviet Union — History.   I. Title.
DK40.R5   1993
947—dc20   91–43254

9 8 7

Printed in the United States of America
on acid-free paper

PREFACE TO THE FIRST EDITION

* * * * * * * * * *

For a student of Russian history to write a complete history of Russia is, in a sense, to give an account of his entire intellectual and academic life. And his indebtedness to others is, of course, enormous. I know at least where to begin the listing of my debts: my father, Valentin A. Riasanovsky, made a huge contribution to this *History of Russia* both by his participation in the writing of the book and, still more important, by teaching me Russian history. Next I must mention my teachers of Russian history at Harvard and Oxford, notably the late Professor Michael Karpovich, the late Warden B. H. Sumner, and Professor Sir Isaiah Berlin. A number of colleagues read sections of the manuscript and made very helpful comments. To name only those who read large parts of the work, I thank Professors Gregory Grossman, Richard Herr, and Martin Malia of the University of California at Berkeley, my former teacher Professor Dimitri Obolensky of Oxford University, Professor Richard Pipes of Harvard University, and Professor Charles Jelavich of Indiana University.

I wish, further, to thank the personnel of the Oxford University Press both for great help of every kind and for letting me have things my own way. I am also indebted to several University of California graduate students who served as my research assistants during the years in which this work was written and prepared for publication; in particular, to Mrs. Patricia Grimsted and Mr. Walter Sablinsky, who were largely responsible for the Bibliography and the Index, respectively. Nor will I forget libraries and librarians, especially those in Berkeley. The publication of this volume can be considered a tribute to my wife and my students: my wife, because of her persistent and devoted aid in every stage of the enterprise; my students, because *A History of Russia* developed through teaching them and has its main *raison d'être* in answering their needs.

I would also like gratefully to acknowledge specific contributions of material to my *History of Russia*. The following publishers allowed me to quote at length from the works cited.

Harvard University Press for Merle Fainsod, *How Russia Is Ruled* (Cambridge, 1954), pp. 372–73.

American Committee for Liberation for *News Briefs on Soviet Activities,* Vol. II, No. 3, June 1959.

Houghton Mifflin Company for George Z. F. Bereday, William W.

Brickman, and Gerald H. Read, editors, *The Changing Soviet School* (Boston, 1960), pp. 8–9.

Further, I am deeply grateful to the Rand Corporation and to Harvard University Press for their permission to use Table 51 on page 210 of Abram Bergson, *The Real National Income of Soviet Russia Since 1928* (Cambridge, 1961). A condensed version of that table constitutes an appendix to my history. Professor Bergson not only gave his personal permission to use this material but advised me kindly on this and certain related matters.

Several people have been most generous in lending material for the illustrations. I should like to thank Mr. George R. Hann for making available to me prints of his superb collection of icons: Mrs. Henry Shapiro, who lent photographs taken by her and her husband during recent years spent in Russia; Professor Theodore Von Laue, who took the pictures I have used from our trip to Russia in 1958; Miss Malvina Hoffman, who lent the pictures of Pavlova and Diaghilev; and the Solomon R. Guggenheim Museum, which permitted reproduction of a painting in their collection, *Winter* by Vasily Kandinsky.

As every writer — and reader — in the Russian field knows, there is no completely satisfactory solution to the problems of transliteration and transcription of proper names. I relied on the Library of Congress system, but with certain modifications: notably, I omitted the soft sign, except in the very few cases where it seemed desirable to render it by using *i,* and I used *y* as the ending of family names. A few of these names, such as that of the composer Tchaikovsky, I spelled in the generally accepted Western manner, although this does not agree with the system of transliteration adopted in this book. As to first names, I preferred their English equivalents, although I transliterated the Russian forms of such well-known names as Ivan and used transliterated forms in some other instances as well, as with Vissarion, not Bessarion, Belinsky. The names of the Soviet astronauts are written as spelled in the daily press. I avoided patronymics. In general I tried to utilize English terms and forms where possible. I might have gone too far in that direction; in any case, I feel uneasy about my translation of *kholopy* as "slaves."

As with transliteration, there is no satisfactory solution to constructing an effective bibliography to a general history of a country. I finally decided simply to list the principal relevant works of the scholars mentioned by name in the text. This should enable the interested reader who knows the required languages to pursue further the views of the men in question, and it should provide something of an introduction to the literature on Russian history. The main asset of such a bibliography is that it is manageable. Its chief liability lies in the fact that it encompasses only a fraction of the works on which this volume is based and of necessity omits important authors and studies.

I decided to have as appendixes only the genealogical tables of Russian rulers, which are indispensable for an understanding of the succession to the throne in the eighteenth century and at some other times, and Professor Bergson's estimate of the growth of the gross national product in the U.S.S.R.

Berkeley, California NICHOLAS V. RIASANOVSKY
September 24, 1962

PREFACE TO THE SECOND EDITION

* * * * * * * * * *

THE SECOND EDITION of my *History of Russia* follows in all essentials the first. Still, the passage of time and the continuous development of scholarship resulted in many additions and modifications. In particular, the Soviet period was expanded both to encompass the last six years and to devote a little more attention to certain topics. A dozen additional authors proved important enough to be cited by name in the second edition, and thus enter the bibliography. Numerous other researchers in the field, some of equal importance to me, received no personal citation. In addition to the text and the bibliography, changes were made in the maps and the illustrations. In the appendixes, the table of the U.S.S.R. gross national product was brought up to date and a table of the administrative divisions of the U.S.S.R. was added. Moreover, a new appendix containing a select list of readings in English on Russian history was included in the second edition.

Again, I have very many people to thank. In the first place, I want to thank my students and students throughout the United States who have used my *History* and have thus given it its true test. I have tried to utilize their experience and their opinions. I am also deeply grateful to very numerous colleagues who used *History of Russia* in their courses, or simply read it, and made corrections or comments. While it is not feasible to list all the appropriate names, I must mention at least Professor Gregory Grossman of the University of California at Berkeley, without whom the gross national product table would not have been possible and who, in addition, paid careful attention to the entire section on the Soviet Union, and the Soviet scholar V. B. Vilinbakhov, who has subjected my presentation of the early periods of Russian history to a thorough and searching criticism. Needless to say, as I thank these and other scholars for their help, I must state that they are not responsible for the opinions or the final form of my book. I am further indebted to my research assistants Mrs. Victoria King and Mr. Vladimir Pavloff and, most especially, to my wife.

Berkeley, California
December 19, 1968

NICHOLAS V. RIASANOVSKY

PREFACE TO THE THIRD EDITION

*   *   *   *   *   *   *   *   *   *

No ATTEMPT HAS BEEN MADE in this third edition of *A History of Russia* to alter the character and basic design of previous editions. The passage of time since the completion of the second edition in 1968 has brought us from the occupation of Czechoslovakia to the Twenty-fifth Party Congress in 1976 and the current Tenth Five-Year Plan. Numerous changes have therefore been made in the text as a consequence of recent events and of recent scholarship as well. The bibliography and especially the English reading list have been expanded. The section on the Soviet period has grown slightly in proportion to the whole, although the aim remains to present a single balanced volume.

Many people deserve my special gratitude. Professor Gregory Grossman of the University of California at Berkeley again brought up to date the gross national product table and, moreover, was of invaluable help in up dating the entire Soviet section. Other Berkeley colleagues generously contributed their knowledge and wisdom in regard to subjects which preoccupied me during the preparation of this third edition. Colleagues else where were equally helpful as they used *A History of Russia* as a textbook and informed me of their experience or simply commented on the work. I would like to thank particularly many conscientious reviewers, such as Professor Walter Leitsch of Vienna. Mr. Gerald Surh and Mr. Jacob Picheny proved to be excellent research assistants, who aided me in every way and most notably in the preparation of the English reading list and the index. The mistakes and other deficiencies that remain after all that help are, I am afraid, mine, and, taking into account the scope of the book, they may well be considerable. My most fundamental gratitude goes to my constant helper, my wife, and to the students for whom this textbook was written and who have been using it. May the group of students who recently called me across the continent from Brown University to discuss my *History of Russia* and whose names I do not know accept the thanks I extend to them, as the representatives of students everywhere.

Berkeley, California   NICHOLAS V. RIASANOVSKY
March 12, 1976

PREFACE TO THE FOURTH EDITION

*   *   *   *   *   *   *   *   *   *

T HE DEATH OF LEONID ILICH BREZHNEV on November 10, 1982, and
Iurii Vladimirovich Andropov's prompt succession to the leadership of the
Soviet Union have provided a striking terminal point to this fourth edition
of my *History of Russia*. The new material in the book covers the last
seven years of the Brezhnev regime. It includes also additions and changes
in all previous parts of Russian and Soviet history as well as the updating
of the two bibliographies.

Acknowledging my overall fundamental and grateful indebtedness to the
scholarship in the field, I must record special thanks to my colleagues, particularly Berkeley colleagues, who contributed directly to the preparation
of this edition. Professor Gregory Grossman again updated the population
and gross national product table and, beyond that, offered invaluable help
based on his matchless knowledge of the Soviet economy and of the Soviet
Union in general. Other colleagues, such as Professor George Breslauer,
whose notable book *Khrushchev and Brezhnev as Leaders: Building Authority in Soviet Politics* came out just as Brezhnev died, were also generous with their time and expert advice. For checking, rechecking, typing,
preparing the index, and much else, I was blessed with an excellent research assistant, Mr. Maciej Siekierski, who also contributed his special
knowledge of Poland and Lithuania, and an excellent secretary, Ms. Dorothy Shannon. And, once more, I must emphasize my indebtedness to my
students and my wife: the students have been using *A History of Russia,*
often both enthusiastically and critically, for some twenty years; my debt
to my wife is even more basic as well as of a still longer duration.

Berkeley, California   NICHOLAS V. RIASANOVSKY
September 1983

I ENDED the first four editions of *A History of Russia* with a comment on the contemporary Soviet conundrum. I wrote that the Soviet Union was neither a stable nor a happy country, but that the problem of change, either by revolution or by evolution, was, in its case, an extremely difficult one, which I could not clearly foresee. The last sentences were,

> To conclude, the Soviet system is not likely to last, not likely to change fundamentally by evolution, and not likely to be overthrown by a revolution. History, to be sure, has a way of advancing even when that means leaving historians behind.

Shortly after this assessment appeared in print, an author of a generally kind and even flattering review wrote in exasperation that Professor Riasanovsky, unfortunately, terminated 710 lucid pages with a murky sentence. In response to my critic, I have considered and reconsidered my conclusion throughout the years and with every edition, but always retained it. To be sure, it was not distinguished by perspicacity or precision, but it was the best I could offer. Now, however, I am moving it from the conclusion to the preface. Historians, and all others as well, have been left behind. The first part of my commentary, on the instability and unhappiness in the Soviet Union, needs no elaboration. The second, on the difficulty of change, is something the citizens of the former Soviet Union and even other people in the world are living through day by day.

To be sure, as many friends have advised me, it would be wiser to wait with a new edition of *A History of Russia*. I am not waiting for two reasons: I have always been in favor of writing contemporary history, no matter how contemporary, as well as other kinds, and Oxford University Press has provided me with an excellent determined editor with whom I have been working for many years. Let us hope that the next edition will be lucid in its final as well as its earlier pages. (And, incidentally, that it will bring reliably up to date Tables 5 and 6 of the Appendix, an impossibility at present.)

The next edition may also be richer in historiography. *Glasnost* has been perhaps the most striking substantive change in the Soviet Union in the past few years. It does represent the breaking out from a totalitarian straight jacket so characteristic of Soviet society and culture. It may be irreversible. But so far, because of the shortage of time and other reasons, it has not transformed Soviet historiography. Having participated in the conference, held in Moscow in April 1990, on rewriting Soviet history, having read Soviet publications, and having talked with Soviet historians, I must conclude that the change has been slow. I do not want to minimize the work of such revisionist historians as Evgeny Viktorovich Anisimov, all the more so because to them probably belongs the future, but I have been on the whole impressed and depressed by the difficulty of change. Understandably, if often unfortunately, people who have spent many years or a lifetime at hard work try to retain at least some of their accomplishment rather than sweep it away. Bolder and more important historiographical developments should appear in the coming years.

In the preparation of this new edition, I made the usual additions and changes throughout the manuscript, and considered or introduced at least fifty-seven emendations in Soviet history prior to 1985. If not always minor — the figure of Soviet casualties in the Second World War was raised from 20 million to 27 million, and that is 7 million more dead — they were brief and precise. The last narrative chapter was, of course, written anew, and the "Concluding Remarks" underwent considerable change.

As always, I am deeply indebted to many people: my colleagues at the University of California, Berkeley; other colleagues whom I met at the Wilson Center and the Kennan Institute in Washington, D.C., where I spent the 1989–90 academic year; still other American colleagues elsewhere as well as extremely numerous Soviet scholars and other Soviet visitors. I must emphasize my gratitude to Professor Gregory Grossman, whose help has been, again, invaluable in the treatment of the Soviet economy in the volume and, moreover, whom I consider in general to be our best specialist on the Soviet Union. I am grateful to Nancy Lane and her colleagues at Oxford University Press; to my secretary, Nadine Ghammache; and to my research assistants, Theodore Weeks, John W. Randolph, Jr., and Ilya Vinkovetsky, who had the major responsibility for revising the index. More generally, I am grateful for the continuing response to my *History* abroad as well as in the United States. Since the publication of the fourth American edition, there appeared another and different Italian edition, a French edition, and even a pirated Korean edition in South Korea of the imperial part of my volume (I was told that the earlier part is being prepared for publication). But as usual, in these fluid times, too, my main indebtedness is to my students and my wife.

Berkeley, California   NICHOLAS V. RIASANOVSKY
April 1992

CONTENTS

* * * * * * * * * *

LIST OF MAPS

* * * * * * * * * *

ALL MAPS HAVE BEEN PREPARED BY VAUGHN GRAY.

ILLUSTRATIONS

* * * * * * * * * *

After page 126

Scythian embossed goldwork of the sixth century B.C. (*Leningrad Museum*)
Ancient monuments of Polovtsy (*Sovfoto*)
Icon: St. George and the Dragon (*Sovfoto*)
Icon: The Old Testament Trinity, by A. Rublev, early fifteenth century
    (*Tretiakov Gallery in Moscow*) (*Sovfoto*)
Icon: The Deesis Festival Tier: Entrance into Jerusalem (*Sovfoto*)
St. Nicholas, Bishop of Myra (*Sovfoto*)
Icon: Our Lady of Vladimir (*Sovfoto*)
Ipatiev Monastery in Kostroma (*Howard Sochurek*)
Fourteenth-century wooden church (*Howard Sochurek*)
Preobrazhenskii Cathedral on Volga at Uglich (*Howard Sochurek*)

After page 244

Fresco: Head of St. Peter (*Sovfoto*)
Holy Gates of the Rizpolozhenskii Monastery in Suzdal (*Mrs. Henry Shapiro*)
Preobrazhenskaia Church in Kizhy near Petrozavodsk (*Sovfoto*)
Sixteenth-century view of the city of Moscow (*Bettmann Archive, Inc.*)
Red Square in Moscow, 1844 (*Bettmann Archive, Inc.*)
Church of St. Basil the Blessed, Moscow (*Ewing Galloway*)
Zagorsk (Holy Trinity-St. Sergius Monastery) (*Sovfoto*)
Moscow Kremlin (*Ewing Galloway*)
Moscow State University, on Lenin Hills (*Wide World Photos*)
Hermitage Museum, Leningrad, 1852 (*Sovfoto*)
Simeon Stolpnik Church on Moscow's Kalinin Prospect (*Sovfoto*)

After page 360

Ministries opposite the Winter Palace, Leningrad (*Sovfoto*)
Kazan Cathedral, Leningrad (*Sovfoto*)
Ivan the Terrible and His Son by Repin (*Sovfoto*)
View of Admiralty and St. Isaac's Cathedral, Leningrad (*Sovfoto*)
Petrodvorets (Peterhof), near Leningrad (*author*)
Cossacks of the Zaporozhie by Repin (*Sovfoto*)
St. Dmitrii Cathedral in Vladimir (*Mrs. Henry Shapiro*)
A church in ancient Suzdal (*Mrs. Henry Shapiro*)
Ivan the Terrible (*Sovfoto*)
Catherine the Great (*Sovfoto*)
Peter I, the Great (*Sovfoto*)

Part I: INTRODUCTION

I

* * * * * * * * *

A GEOGRAPHICAL NOTE

Russia! what a marvelous phenomenon on the world scene! Russia —
a distance of ten thousand versts * in length on a straight line from
the virtually central European river, across all of Asia and the Eastern Ocean, down to the remote American lands! A distance of five
thousand versts in width from Persia, one of the southern Asiatic
states, to the end of the inhabited world — to the North Pole. What
state can equal it? Its half? How many states can match its twentieth,
its fiftieth part? . . . Russia — a state which contains all types of
soil, from the warmest to the coldest, from the burning environs of
Erivan to icy Lapland; which abounds in all the products required for
the needs, comforts, and pleasures of life, in accordance with its
present state of development — a whole world, self-sufficient, independent, absolute.

POGODIN

Loe thus I make an ende: none other news to thee
But that the country is too cold, the people beastly bee.
AMBASSADOR GEORGE TURBEVILLE
REPORTING TO ELIZABETH I OF ENGLAND

These poor villages,
This barren nature —
Native land of enduring patience,
The land of the Russian people!

TIUTCHEV

THE RUSSIAN EMPIRE, and more recently the Union of Soviet Socialist
Republics, represents a land mass of over eight and one-half million square
miles, an area larger than the entire North American continent. To quote
the leading Russian encyclopedia: "The Russian empire, stretching in the
main latitudinally, occupies all of eastern Europe and northern Asia, and
its surface constitutes 0.42 of the area of these two continents. The Russian
empire occupies $\frac{1}{22}$ part of the entire globe and approximately $\frac{1}{6}$ part of
its total land surface."

Yet, this enormous territory exhibits considerable homogeneity. Indeed,
homogeneity helps to explain its size. The great bulk of Russia is an immense
plain — at one time the bottom of a huge sea — extending from central and
even western Europe deep into Siberia. Although numerous hills and chains
of hills are scattered on its surface, they are not high enough or sufficiently
concentrated to interfere appreciably with the flow of the mighty plain, the

* A *versta* is not quite two-thirds of a mile, or a little over a kilometer.

3

largest on the entire globe. The Ural mountains themselves, ancient and weather-beaten, constitute no effective barrier between Europe and Asia, which they separate; besides, a broad gap of steppe land remains between the southern tips of the Ural chain and the Caspian and Aral seas. Only in vast northeastern Siberia, beyond the Enisei river, does the elevation rise considerably and hills predominate. But this area, while of a remarkable potential, has so far remained at best on the periphery of Russian history. Impressive mountain ranges are restricted to Russian borders or, at the most, borderlands. They include the Carpathians to the southwest, the high and picturesque Caucasian chain in the south between the Black Sea and the Caspian, and the mighty Pamir, Tien Shan, and Altai ranges further east along the southern border.

Rivers flow slowly through the plain. Most of them carry their waters along a north-south axis and empty either into the Baltic and the Arctic Ocean or into the Black and the Caspian seas. In European Russia, such rivers as the Northern Dvina and the Pechora flow northward, while others, notably the Dniester, the Bug, and the larger Dnieper, Don, and Volga proceed south. The Dnieper and the Don empty into the Black Sea, the Volga into the Caspian. Siberian rivers, the huge Ob and Enisei, as well as the rapid Lena, the Indigirka, and the Kolyma, drain into the Arctic Ocean. The exception is the Amur, which flows eastward, serves during much of its course as the boundary between Russia and China, and empties into the Strait of Tartary. South of Siberia in Russian Central Asia both the Amu Daria and the Syr Daria flow northwestward to the Aral Sea, although the former at one time used to reach the Caspian. These rivers and their tributaries, together with other rivers and lakes, provide Russia with an excellent system of water communication. The low Valdai hills in northwestern European Russia represent a particularly important watershed, for it is there that the Dnieper and the Volga, as well as the Western Dvina and the Lovat, have their sources.

But while Russia abounds in rivers and lakes, it is essentially a landlocked country. By far its longest coastline opens on the icy Arctic Ocean. The neighboring seas include the Baltic and the Black, both of which must pass through narrow straits, away from Russian borders, to connect with broader expanses of water, and the Caspian and the Aral, which are totally isolated. The Aral Sea is also entirely within Russian territory, and it has been listed with such major Russian lakes as Ladoga and Onega in the European part of the country, Balkhash in Central Asia, and the huge and extremely deep Lake Baikal in Siberia. The Russian eastern coastline too is subject to cold and inclement weather, except for the southern section adjacent to the Chinese border.

Latitude and a landlocked condition largely determine Russian climate, which can be best described as severely continental. Northern and even central Russia are on the latitude of Alaska, while the position of southern Russia corresponds more to the position of Canada in the western hemisphere than to that of the United States. The Gulf Stream, which does so much to make the climate of western and northern Europe milder, barely reaches one segment of the northern coastline of Russia. In the absence of interfering mountain ranges, icy winds from the Arctic Ocean sweep across European Russia to the Black Sea. Siberian weather, except in the extreme southeastern corner, is more brutal still. In short, although sections of the Crimean littoral can be described as the Russian Riviera, and although subtropical conditions do prevail in parts of the southern Caucasus, the overwhelming bulk of Russian territory remains subject to a very severe climate. In northern European Russia the soil stays frozen eight months out of twelve. Even the Ukraine is covered by snow three months every year, while the rivers freeze all the way to the Black Sea. Siberia in general and northeastern Siberia in particular belong among the coldest areas in the world. The temperature at Verkhoiansk has been registered at as low as −90° F. Still, in keeping with the continental nature of the climate, when summer finally comes — and it often comes rather suddenly — temperatures soar. Heat waves are common in European Russia and in much of Siberia, not to mention the deserts of Central Asia which spew sand many miles to the west.

Climate determines the vegetation that forms several broad belts extending latitudinally across the country. In the extreme north lies the tundra, a virtually uninhabited frozen waste of swamps, moss, and shrubs covering almost 15 per cent of Russian territory. South of the tundra stretches the taiga, a zone of coniferous forest, merging with and followed by the next zone, that of mixed forest. The two huge forested belts sweep across Russia from its western boundaries to its eastern shoreline and account for over half of its territory. Next comes the steppe, or prairie, occupying southern European Russia and extending into Asia up to the Altai mountains. Finally, the southernmost zone, that of semi-desert and desert, takes up most of Central Asia. Being very wide if considerably shorter than even the steppe belt, it occupies somewhat less than one-fifth of the total area of the country.

One important result of the climate and of this pattern of vegetation in Russia has been a relative dearth of first-rate agricultural land. Only an estimated one million square miles out of an area more than eight times that size are truly rewarding to the tiller of the soil. Other sections of the country suffer from the cold and from insufficient precipitation, which becomes more inadequate as one progresses east. Even the heavy snowfalls add relatively little moisture because of the rapid melting and the quick run-off of water in the spring. In Central Asia farming depends almost entirely on irrigation. The best land in Russia, the excellent black soil of the southern steppe,

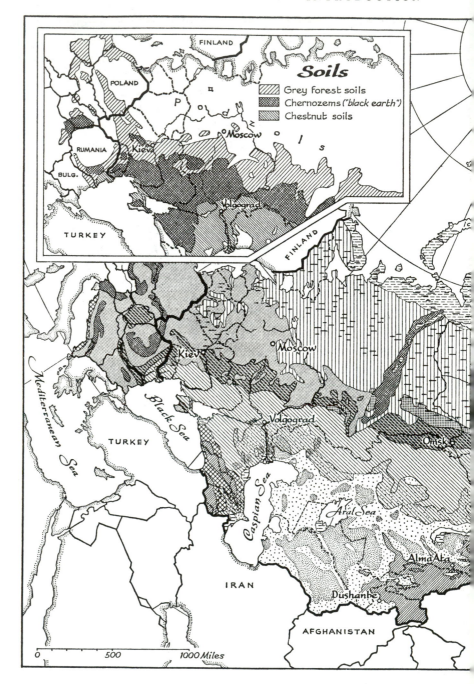

ALASKA

Arctic Ocean

Pacific

Ocean

Iakutsk

Novosibirsk

Irkutsk

MANCHURIA

Vladivostok

MONGOLIA

Vegetation

CHINA

Tundra		Steppe	
Pine and fir forest		Semi-desert	
Larch forest		Desert (loose sand)	
Deciduous mixed forest		Warm temperate broadleaf forest	
Birch forest		Mountain meadow	
Steppe with some trees		Mountain forest	

Peat bog and marsh

offers agricultural conditions comparable to those on the great plains of Canada rather than those in warmer Iowa or Illinois. Russia, on the other hand, is fabulously rich in forests, more so than any other country in the world. And it possesses a great wealth and variety of natural resources, ranging from platinum to oil and from coal to gold. On the whole, however, these resources remained unused and even unexplored for a very long time.

Ever since Herodotus historians have been fascinated by the role of geographic factors in human history. Indeed the father of history referred to the broad sweep of the southern Russian steppe and to the adaptation of the steppe inhabitants, the Scythians, to their natural environment in his explanation of why the mighty Persians could not overcome them. Modern historians of Russia, including such leading Russian scholars as Kliuchevsky and especially his teacher S. Soloviev, as well as such prominent Western writers as Kerner and Sumner, have persistently emphasized the significance of geography for Russian history. Even if we reject the rigid determinism implicit in some of their views and refuse to speculate on such nebulous and precarious topics as the Russian national character and its dependence on the environment — speculations in which Kliuchevsky and others engaged in a fascinating manner — some fundamental points have to be made.

For instance, it appears certain that the growth of the Russian state was affected by the geography of the area: a vast plain with very few natural obstacles to expansion. This setting notably made it easier for the Moscow state to spread across eastern Europe. Beyond the Urals, the Russians advanced all the way to the Pacific, and even to Alaska and California, a progression paralleled only by the great American movement west. As the boundaries of the Russian empire ultimately emerged, they consisted of oceans to the north and east and, in large part, of seas, high mountains, and deserts to the south; only in the west, where the Russians merged with streams of other peoples, did the border seem unrelated to geography. The extremely severe climate contributed to the weakness of the tribes scattered in northern European Russia and of the various inhabitants of Siberia, leading to their utter inability to stem the Russian advance. Whereas the Russians could easily expand, they were well protected from outside attack. Russian distances brought defeat to many, although not all, invaders, from the days of the Persians and the Scythians to those of Napoleon and Hitler.

Occupied territory had to be governed. The problem of administering an enormous area, of holding the parts together, of co-ordinating local activities and efforts remained a staggering task for those in power, whether Ivan the Terrible, Nicholas I, or Stalin. And the variety of peoples on the great plain was bound to make such issues as centralization and federation all the more acute. One can appreciate, if not accept, the opinion of those thinkers, prominent in the Enlightenment and present in other periods, who related the system of government of a country directly to its size and declared despotism to be the natural form of rule in Russia.

The magnificent network of Russian rivers and lakes also left its mark on Russian history. It is sufficient to mention the significance of the Dnieper for Kievan Russia, or of the Volga and its tributaries for the Moscow state. The landlocked position of the country and the search for an access to the waterways of the world made the Russians repeatedly concerned with the Baltic, the Black Sea, and the Straits. Climate and vegetation basically affected the distribution of people in Russia and also their occupations. The poor quality of much agricultural land has led to endemic suffering among Russian peasants and has taxed the ingenuity of tsarist ministers and Khrushchev alike. Russian natural resources, since they began to be developed on a large scale, have added immeasurably to Soviet strength. Both the wealth of Russia and the geographic and climatic obstacles to a utilization of this wealth have perhaps never stood out so sharply as in the course of present efforts to industrialize eastern Siberia.

The location of Russia on its two continents has had a profound impact on Russian history. The southern Russian steppe in particular served for centuries as the highway for Asiatic nomads to burst into Europe. Mongol devastation was for the Russians only the most notable incident in a long series, and it was followed by over two hundred years of Mongol rule. In effect, the steppe frontier, open for centuries, contributed hugely to the militarization of Russian society, a trend reinforced by the generally unprotected and fluid nature of the western border of the country. But proximity to Asiatic lands led also to some less warlike contacts; furthermore, it enabled Russia later in turn to expand grandly in Asia without the need first to rule the high seas. Recently the Eurasian school of historians, represented in the English language especially by Vernadsky, has tried to interpret the entire development of Russia in terms of its unique position in the Old World.

Russian location in Europe may well be regarded as even more important than its connections with Asia. Linked to the West by language, religion, and basic culture, the Russians nevertheless suffered the usual fate of border peoples: invasion from the outside, relative isolation, and retardation. Hence, at least in part, the efforts to catch up, whether by means of Peter the Great's reforms or the Five-Year Plans. Hence also, among other things, the interminable debate concerning the nature and the significance of the relationship between Russia and the West.

As the examples above, which by no means exhaust the subject, indicate, geography does affect history, Russian history included. It has been noted that the influence of certain geographic factors tends to be especially persistent. Thus, while our modern scientific civilization does much to mitigate the impact of climate, a fact brilliantly illustrated in the development of such a northern country as Finland, so far we have not changed mountains into plains or created new seas. Still, it is best to conclude with a reservation: geography may set the stage for history; human beings make history.

RUSSIA BEFORE THE RUSSIANS

> We have only to study more closely than has been done the antiquities
> of South Russia during the period of migrations, i.e., from the fourth
> to the eighth century, to become aware of the uninterrupted evolution
> of Iranian culture in South Russia through these centuries. . . . The
> Slavonic state of Kiev presents the same features . . . because the
> same cultural tradition — I mean the Graeco-Iranian — was the only
> tradition which was known to South Russia for centuries and which
> no German or Mongolian invaders were able to destroy.
>
> ROSTOVTZEFF

> Yes, we are Scythians. Yes, we are Asiatics.
> With slanting and greedy eyes.
>
> BLOK

CONTINUITY is the very stuff of history. Although every historical event
is unique, and every sequence of events, therefore, presents flux and change,
it is the connection of a given present with its past that makes the present
meaningful and enables us to have history. In sociological terms, continuity
is indispensable for group culture, without which each new generation of
human beings would have had to start from scratch.

Non-Slavic Peoples and Cultures

A number of ancient cultures developed in the huge territory that is today
enclosed within the boundaries of the U.S.S.R. Those that flourished in
Transcaucasia and in Central Asia, however, exercised merely a peripheral
influence on Russian history, the areas themselves becoming parts of the
Russian state only in the nineteenth century. As an introduction to Russian
history proper, we must turn to the northern shore of the Black Sea and to
the steppe beyond. These wide expanses remained for centuries on the
border of the ancient world of Greece, Rome, and Byzantium. In fact,
through the Greek colonies which began to appear in southern Russia from
the seventh century before Christ and through commercial and cultural
contacts in general, the peoples of the southern Russian steppe participated
in classical civilization. Herodotus himself, who lived in the fifth century
B.C., spent some time in the Greek colony of Olbia at the mouth of the Bug
river and left us a valuable description of the steppe area and its population.
Herodotus' account and other scattered and scarce contemporary evidence

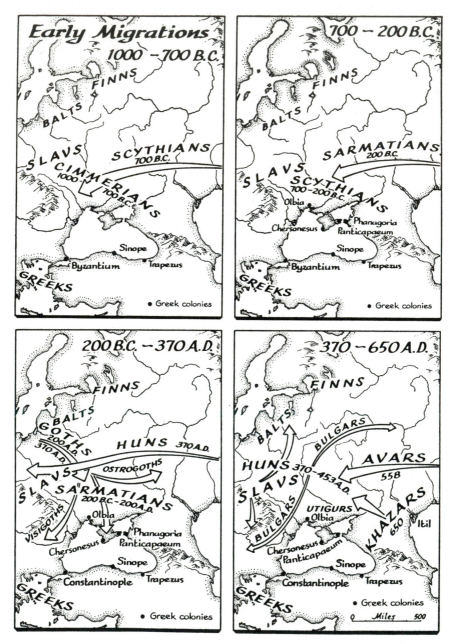

have been greatly augmented by excavations pursued first in tsarist Russia
and subsequently, on an increased scale, in the Soviet Union. At present we
know, at least in broad outline, the historical development of southern
Russia before the establishment of the Kievan state. And we have come to
appreciate the importance of this background for Russian history.

The best-known neolithic culture in southern Russia evolved in the valleys of the Dnieper, the Bug, and the Dniester as early as the fourth millennium before Christ. Its remnants testify to the fact that agriculture was then already entrenched in that area, and also to a struggle between the sedentary tillers of the soil and the invading nomads, a recurrent motif in southern Russian, and later Russian, history. This neolithic people also used domestic animals, engaged in weaving, and had a developed religion. The "pottery of spirals and meander" links it not only to the southern part of Central Europe, but also and especially, as Rostovtzeff insisted, to Asia Minor, although a precise connection is difficult to establish. At about the same time a culture utilizing metal developed in the Kuban valley north of the Caucasian range, contemporaneously with similar cultures in Egypt and Mesopotamia. Its artifacts of copper, gold, and silver, found in numerous burial mounds, testify to the skill and taste of its artisans. While the bronze age in southern Russia is relatively little known and poorly represented, that of iron coincided with, and apparently resulted from, new waves of invasion and the establishment of the first historic peoples in the southern Russian steppe.

The Cimmerians, about whom our information is very meager, are usually considered to be the earliest such people, again in large part thanks to Herodotus. They belonged to the Thracian subdivision of the Indo-European language family and ruled southern Russia from roughly 1000 B.C. to 700 B.C. At one time their dominion extended deep into the Caucasus. Recent historians have generally assumed that the Cimmerians represented the upper crust in southern Russia, while the bulk of the population consisted of indigenous elements who continued the steady development of culture on the northern shore of the Black Sea. The ruling group was to change several times during the subsequent centuries without destroying this fundamental cultural continuity.

The Scythians followed the Cimmerians, defeating them and destroying their state. The new invaders, who came from Central Asia, spoke an Iranian tongue and belonged thus to the Indo-European language family, although they apparently also included Mongol elements. They ruled southern Russia from the seventh to the end of the third century B.C. The Scythian sway extended, according to a contemporary, Herodotus, from the Danube to the Don and from the northern shore of the Black Sea inland for a distance traveled in the course of a twenty-day journey. At its greatest extent, the Scythian state stretched south of the Danube on its western flank and across the Caucasus and into Asia Minor on its eastern.

The Scythians were typical nomads: they lived in tentlike carriages dragged by oxen and counted their riches by the number of horses, which also served them as food. In war they formed excellent light cavalry, utilizing the saddle and fighting with bows and arrows and short swords. Their military tactics based on mobility and evasion proved so successful that even their great Iranian rivals, the mighty Persians, could not defeat them in their home territory. The Scythians established a strong military state in southern Russia and for over three centuries gave a considerable degree of stability to that area. Indigenous culture continued to develop, enriched by new contacts and opportunities. In particular, in spite of the nomadic nature of the Scythians themselves, agriculture went on flourishing in the steppe north of the Black Sea. Herodotus who, in accordance with the general practice, referred to the entire population of the area as Scythian, distinguished, among other groups, not only "the royal Scythians," but also "the Scythian ploughmen."

The Scythians were finally defeated and replaced in southern Russia by the Sarmatians, another wave of Iranian-speaking nomads from Central Asia. The Sarmatian social organization and culture were akin to the Scythian, although some striking differences have been noted. Thus, while both peoples fought typically as cavalry, the Sarmatians used stirrups and armor, lances, and long swords in contrast to the light equipment of the Scythians. What is more important is that they apparently had little difficulty in adapting themselves to their new position as rulers of southern Russia and in fitting into the economy and the culture of the area. The famous Greek geographer Strabo, writing in the first century A.D., mentions this continuity and in particular observes that the great east-west trade route through the southern Russian steppe remained open under the Sarmatians. The Sarmatians were divided into several tribes of which the Alans, it would seem, led in numbers and power. The Ossetians of today, a people living in the central Caucasus, are direct descendants of the Alans. The Sarmatian rule in southern Russia lasted from the end of the third century B.C. to the beginning of the third century A.D.

It was during the Scytho-Sarmatian period that the Graeco-Iranian culture developed on the northern shore of the Black Sea and in the Russian steppe. The Iranian element was represented in the first place by the Scythians and the Sarmatians themselves. They established large and lasting military states which provided the basic pattern of political organization for the area. They brought with them their languages, their customs, their religion emphasizing war, an original style in decorative art known as the Scythian animal style, and generally vigorous and varied art and craftsmanship, especially in metalwork. The enormously rich Greek civilization came to the area primarily through Greek colonies. These colonies began as fishing enterprises and grew into major commercial centers and flourishing communities. They included the already mentioned Olbia, founded as early as the middle of the seventh century B.C., Chersonesus in the Crimea near present-day Sevastopol, Tanais at the mouth of the Don, and Panticapaeum and Phanagoria on either side of the Strait of Kerch, which links the Sea of Azov to the Black Sea and separates the Crimea and the Caucasus. The

Greeks engaged in varied trade, but especially significant was their impor-
tation of southern Russian grain into the Hellenic world. The settlements
near the Strait of Kerch, enjoying a particularly favorable position for trade
and defense, formed the nucleus of the Bosporan kingdom which was to
have a long and dramatic history. That kingdom as well as other Greek
centers in southern Russia fell in the first century before Christ under the
sway of Mithridates the Great of Pontus and, after his ultimate defeat by
the Romans, of Rome. Even after a retrenchment of the Roman Empire and
its eventual collapse, some former Greek colonies on the northern shore of
the Black Sea, such as Chersonesus, had another revival as outposts of the
Byzantine Empire.

The Sarmatian rule in the steppe north of the Black Sea was shattered
side by side. It has been noted that the Scythians and the Sarmatians made
no sustained effort to destroy Greek colonies in southern Russia, choosing
instead to maintain vigorous trade relations and other contacts with them.
Intermarriage, Hellenization of Iranians, and Iranization of Greeks proceeded apace. The resulting cultural and at times political synthesis was
such that the two elements became inextricably intertwined. As Rostovtzeff
explains in regard to the Bosporan kingdom, a prize example of this sym-
biosis: "It is a matter of great interest to trace the development of the new
community. A loosely knit confederation of cities and tribes in its beginning, it became gradually a political body of dual nature. The ruler of this
body was for the Greeks an elected magistrate, for the natives a king ruling
by divine right." Today one can readily appreciate some of the sweep and
the glory of the ancient Graeco-Iranian culture in southern Russia after
visiting the appropriate rooms of the Hermitage or of the historical museum
in Moscow.

The Sarmatian rule in the steppe north of the Black Sea was shattered
by the Goths. These Germanic invaders came from the north, originally
from the Baltic area, reaching out in a southeasterly direction. In southern
Russia they split into the Visigoths and the Ostrogoths, and the latter eventually established under Hermanric a great state stretching from the Black
Sea to the Baltic. But the Gothic period in Russia, dated usually from
A.D. 200 to A.D. 370, ended abruptly with the appearance of new intruders
from Asia, the Huns. Furthermore, while the Goths proved themselves
to be fine soldiers and sailors, their general cultural level lagged considerably
behind the culture of southern Russia, to which they had little to contribute.

The Huns, who descended upon the Goths around A.D. 370, came in a
mass migration by the classic steppe road from Central Asia to southern
Russia. A remarkably mixed group when they appeared in European history, the Huns were, on best evidence, a Turkic-speaking people supported
by large Mongol and Ugrian contingents. Later, as they swept into central
and even western Europe, they also brought with them different Germanic

and Iranian elements which they had overwhelmed and picked up on the way. Although one of the most primitive peoples to come to southern Russia, the Huns had sufficient drive and military prowess to conquer that area and, indeed, to play a key role in the so-called period of great migrations in Europe. Even after their defeat in the battle of Châlons, deep in France, in 451, they invaded Italy and, according to tradition, spared Rome only because of the influence of Pope Leo I on their leader, Attila. But with the sudden death of Attila in 453 the poorly organized Hunnic state crumbled. Its successors included the large horde of the Bulgars and the smaller ones of the Utigurs and the Kutrigurs.

The next human wave to break into southern Russia consisted again of an Asiatic, Mongol- and Turkic-speaking, and relatively primitive people, the Avars. Their invasion is dated A.D. 558, and their state lasted for about a century in Russia and for over two and a half centuries altogether, at the end of which time it dissolved rapidly and virtually without trace, a common fate of fluid, politically rudimentary, and culturally weak nomadic empires. At the height of their power, the Avars ruled the entire area from eastern Russia to the Danubian plain, where they had their capital and where they remained after they had lost control in Russia. Avar armies threatened Byzantium, and they also waged major, although unsuccessful, wars against Charlemagne and his empire.

In the seventh century A.D. a new force emerged in southern Russia, to be more exact, on the lower Volga, in the northern Caucasus, and the southeastern Russian steppe in general: the Khazar state. The impact of the Khazars split the Bulgars sharply in two: one group definitely settled in the Balkans to dissolve in the Slavic mass and give its name to present-day Bulgaria; the other retreated to the northeast, eventually establishing a state at the confluence of the Volga and the Kama, with the town of Great Bulgar as its capital. The Utigurs and the Kutrigurs retrenched to the lands along the Sea of Azov and the mouth of the Don.

Although the Khazars were still another Turkic-speaking people from Asia, their historical role proved to be quite different from that of the Huns or of the Avars. To begin with, they fought bitter wars against the Arabs and served as a bulwark against the spread of Islam into Europe. When their own state assumed form in southeastern European Russia, it became notable for its commerce, its international connections, and the tolerance and enlightenment of its laws. Although a semi-nomadic people themselves, the Khazars promoted the building of towns, such as their capital of Itil — not far from the mouth of the Volga — Samandar, Sarkil, and certain others. The location at the crossroads of two continents proved to be of fundamental importance for the Khazar economy. In the words of a recent historian of the Khazars, Dunlop: "The prosperity of Khazaria evidently depended less on the resources of the country than on its favorable position across important trade-routes." The Khazar revenue, consequently, came especially from commercial imposts as well as from the tribute which increased as the Khazar rule expanded westward on the Russian plain. Pagans, Moslems, Christians, and Jews mingled in Khazaria, where all enjoyed considerable freedom and autonomy to live under their own laws. In the eighth and ninth centuries the Khazars themselves embraced Judaism, or at least their ruler, who bore the title of khakan, and the upper class did, thus adding another exceptional chapter to their unusual history. The Khazars have also been cited as one of the first peoples to institute a permanent paid armed force. The development of Khazaria, with its close links to the Arabic and Byzantine worlds, as well as to some other civilizations, its far-flung trade connections, and its general cosmopolitanism, well represents one line of political, economic, and cultural evolution on the great Russian plain at the time of the emergence of the Kievan state. It may be added that, while the Khazars were outstanding in commercial development, varied commercial intercourse on a large scale also grew further north, in the country of the Volga Bulgars.

The East Slavs

Cultures on the northern shore of the Black Sea and in the southern Russian steppe, from the neolithic period to the time of the Khazars, form an essential part of the background of Kievan Russia. Yet it is true too that the people of the Kievan state who came to be known as Russians were not Scythians, Greeks, or Khazars, much as they might have been influenced in one way or another by these and other predecessors and neighbors; they were East Slavs. Therefore, East Slavs also demand our attention. The term itself is linguistic, as our better classifications of ancient peoples usually are. It refers to a group speaking the Eastern variety of Slavic. With time, three distinct East Slavic languages developed: Great Russian, often called simply Russian, Ukrainian, and White Russian or Belorussian. Other branches of the Slavic languages are the West Slavic, including Polish and Czech, and the South Slavic, represented, for instance, by Serbo-Croatian and Bulgarian. The Slavic languages, in turn, form a subdivision of the Indo-European language family which includes most of the tongues spoken today in Europe and some used in Asia. To be more precise, in addition to the Slavic this family contains the Teutonic, Romance, Hellenic, Baltic, Celtic, Iranian, Indic, Armenian, and Thraco-Illyrian subfamilies of languages. The Cimmerians, it might be recalled, belonged apparently to the Thraco-Illyrian subfamily, the Scythians and the Sarmatians to the Iranian, and the Goths to the Teutonic or Germanic, while the Greeks are, of course, the great representatives of the Hellenic. Early Russian history was also influenced by other Indo-European peoples, such as the Baltic Lithuanians, as well as by some non-Indo-Europeans, notably by different Turkic tribes — some of which have already been mentioned — the Mongols, and Finno-Ugrian elements.

Languages are organically and intrinsically related within the same subfamily and also within the same family. By contrast, no fundamental connection, as distinct from chance borrowing, has been firmly established between languages in different families, for example, the Indo-European and the Ural-Altaic. In fact, there is even an opinion that speech originated on our planet in a number of separate places, division thus being the rule in the linguistic world from the very beginning. To explain the relatedness of the languages within a family and the much closer relationship of the languages of the same subfamily, scholars have postulated an original language and homeland for each family — such as for all Indo-European peoples whence they spread across Europe and parts of Asia —and later languages and homelands for different linguistic subfamilies before further separation and differentiation. Within the framework of this theory, the Slavs have usually been assigned a common homeland in the general area of the valley of the Vistula and the northern slopes of the Carpathians. Their split has been dated, by Shakhmatov and others, in the sixth century A.D., and the settlement by the East Slavs of the great plain of European Russia in the seventh, the eighth, and the ninth. In reconstructing Slavic migrations, allowance has frequently been made for the fact that the East Slavic languages are closer to the South Slavic than those of either of these branches to the West Slavic ones. It should be emphasized that in relying on original languages and their homelands one is dealing with languages, not races. The categories listed above are all linguistic, not racial, and do not necessarily correspond to any physical traits. Besides, intermarriage, conquest, imitation, as well as some other factors, have repeatedly changed the number and composition of those speaking a given language. Today, for instance, English is the native tongue of African-Americans as well as of Yorkshiremen. An entire people can lose a language and adopt a new one. Invaders have often been absorbed by the indigenous population, as in the case of the Turkic Bulgars in the Balkans. Other invaders have been able to overwhelm and incorporate native peoples. Thus some historians explain the Germanic expansion in eastern Europe by a Germanization, not an extermination, of different Slavic and Lithuanian tribes. There are also such puzzling cases as the language of the Lapps in the far north of Scandinavia and Russia: it is a Finno-Ugrian tongue, but, in the opinion of certain specialists, it appears to be superimposed on a radically different linguistic structure.

Recent scholarship has subjected the theory of original languages and homelands to a searching criticism. At present few specialists speak with any confidence about the historical homeland of the Indo-Europeans, and some reject it even as a theoretical concept. More important for students of

Russian history, the Slavic homeland has also been thoroughly questioned. The revaluation has been largely instigated by discoveries of the presence of the Slavs at a much earlier time and over a much larger area in Russia than had been traditionally supposed. To meet new evidence, some scholars have redefined the original Slavic homeland to include parts of Russia. Others have postulated an earlier dispersal of the Slavs, some suggesting that it proceeded in several waves to explain both their ancient presence on the Russian plain and their later migration thither. Still others have given up the Slavic homeland altogether. While recent work concerning Slavic prehistory has produced many new facts, it has lacked a convincing general theory to replace that which has been found wanting.

The first extant written references to the Slavs belong to the classical writers early in our era, including Pliny the Elder and Tacitus. Important later accounts include those of the sixth century produced by the Byzantine historian Procopius and the Gothic Jordanes. The terms most frequently used to designate the Slavs were "Venedi" and "Antes," with the latter coming to mean the East Slavs — although "Antes" has also been given other interpretations, such as pre-Slavic Iranian inhabitants of southern Russia or Goths. Soviet archaeologists insist that Slavic settlements in parts of Russia, notably in the Don area, date at least from the middle of the first millennium B.C. It is now assumed by some historians that the Slavs composed a significant part, perhaps the bulk, of the population of southern and central Russia from the time of the Scythians. For instance, they may be hidden under various designations used by Herodotus, such as "Scythian ploughmen." It is known that the East Slavs fought against the Goths, were swept westward with the Huns, and were conquered by the Avars; certain East Slavic tribes were paying tribute to the Khazars at the dawn of Kievan history. At that time, according to our main written source, the Kievan *Primary Chronicle* of the early twelfth century, the East Slavs were divided into twelve tribes located on the broad expanses of the Russian plain, from the Black Sea, the Danube, and the Carpathian mountains, across the Ukraine, and beyond, northward to the Novgorod territory and eastward toward the Volga. Their neighbors included, in addition to some of the peoples already mentioned, Finnic elements scattered throughout northern and eastern Russia and Lithuanian tribes to the west.

By the ninth century A.D. East Slavic economy, society, and culture had already experienced a considerable development. Agriculture was well and widely established among the East Slavs. Other important occupations included fishing, hunting, apiculture, cattle-raising, weaving, and pottery-making, as well as other arts and crafts, such as carpentry. The East Slavs had known the use of iron for centuries. They had also been engaging in varied and far-flung commerce. They possessed a remarkable number of towns; even Tikhomirov's count of them, some 238, is not complete.

Certain of these towns, such as Novgorod, Smolensk, and Kiev, a town belonging to the tribe of the Poliane, were to have long and important histories. Very little is known about the political organization of the East Slavs. There exist, however, a few scattered references to the rulers of the Antes and of some of the component tribes: for example, Jordanes's mention of Bozh, a prince of the Antes at the time of the Gothic wars; and the statement of Masudi, an Arabian writer, concerning Madzhak, apparently a prince of the East Slavic tribe of the Duleby in the Avar period.

PART II: KIEVAN RUSSIA

THE ESTABLISHMENT OF THE KIEVAN STATE

They accordingly went overseas to the Varangian Russes.

THE PRIMARY CHRONICLE

THE PROBLEM of the origin of the first Russian state, that of Kiev, is exceedingly complex and controversial. No other chapter of Russian history presents the same number and variety of difficulties. Yet the modern student of the subject, although he can by no means produce all the answers, should at least be able to avoid the cruder mistakes and oversimplifications of the past.

The first comprehensive, scholarly effort to explain the appearance of the Kievan state was made in the eighteenth century in terms of the so-called Norman theory. As formulated by Bayer, Schlözer, and others, this view stressed the role of the vikings from Scandinavia — that is, Norsemen, or, to follow the established usage in Russian historiography, Normans — in giving Russia government, cohesion, and, in large part, even culture. The Norman period of Russian history was thus postulated as the foundation for its subsequent evolution. In the course of some two hundred years the Norman theory has been developed, modified, and changed by many prominent scholars. Other specialists, however, opposed it virtually from the very beginning, offering instead a dazzling variety of possibilities. More recently Soviet historians have turned violently against it, and it remained largely out of bounds for Soviet scholarship until 1985 and *glasnost*.

In estimating the value of the Norman theory it is important to appreciate its drastic limitations in the field of culture. The original assertion of the Norman influence on Russia was made before the early history of southern Russia, outlined in the preceding chapter, had been discovered. With our present knowledge of that history there is no need to bring in the Norsemen to account for Kievan society and culture. What is more, Scandinavia itself, located in the far north, lay at that time much farther from cultural centers and crosscurrents than did the valley of the Dnieper. Not surprisingly, once the Kievan state emerged, its culture developed more richly and rapidly than that of its northern neighbor; whether we consider written literature and written law or coin stamping, we have to register their appearance in Kievan Russia a considerable time before their arrival in Scandinavia.

Detailed investigations of Scandinavian elements in Russian culture serve to emphasize their relative insignificance. Norman words in the Russian language, formerly supposed to be numerous, number actually only six or seven. Old Russian terms pertaining to navigation were often Greek, those dealing with trade, Oriental or native Slavic, but not Scandinavian. Written literature in Kiev preceded written literature in Scandinavia, and it experienced clear Byzantine and Bulgarian rather than Nordic influences; under these circumstances, persistent efforts to link it to the Scandinavian epic fail to carry conviction. Claims of Norman contributions to Russian law have suffered a fiasco: while at one time scholars believed in the Scandinavian foundation of Russian jurisprudence, it has in fact proved impossible to trace elements of Kievan law back to Norman prototypes. Similarly, there is no sound evidence for Norman influence on Kievan paganism: Perun, the god of thunder and the chief deity of the East Slavic pantheon, far from being a copy of Thor, was described as the supreme divinity of the Antes by Procopius in the sixth century; a linguistic analysis of the names of East Slavic gods reveals a variety of cultural connections, but none of them with Scandinavia. Other assertions of Norman cultural influences, for instance, on the organization of the Kievan court or on Russian dress, tend to be vague and inconclusive, especially when compared to the massive impact of Byzantium and the tangible effects of some Oriental cultures on Russia.

But, while the importance of Scandinavian culture for Russian culture no longer represents a major historical issue, the role of the Normans in the establishment of the Kievan state itself remains highly controversial. The question of the origin of the Kievan state is very closely connected with a group, tribe, or people known as the Rus, and it is also from the Rus that we derive the later name of the Russians. Almost everything connected with the Rus has become a subject of major controversy in Russian historiography. Under the year A.D. 862 the *Primary Chronicle* tells briefly about the arrival of the Rus following an invitation from the quarreling Slavic tribes of the Sloveni and the Krivichi and some Finnish tribes:

> They accordingly went overseas to the Varangian Russes: these particular Varangians were known as Russes, just as some are called Swedes, and others Normans, Angles, and Goths, for they were thus named. The Chuds, the Slavs and the Krivichians then said to the people of Rus, "Our whole land is great and rich, but there is no order in it. Come to rule and reign over us!" They thus selected three brothers, with their kinsfolk, who took with them all the Russes and migrated. The oldest, Rurik, located himself in Novgorod; the second, Sineus, in Byeloozero; and the third, Truvor, in Izborsk. On account of these Varangians, the district of Novgorod became known as the land of the Rus. The present inhabitants

of Novgorod are descended from the Varangian race, but aforetime they were Slavs.*

The proponents of the Norman theory accepted the *Chronicle* verbatim, with the understanding that the Rus were a Scandinavian tribe or group, and proceeded to identify the *Rus-Ros-Rhos* of other sources with the Scandinavians. However, before long grave complications arose. A group called Rus could not be found in Scandinavia itself and were utterly unknown in the West. Although the *Chronicle* referred to Novgorod, *Rus* became identified with the Kievan state, and the very name came to designate the southern Russian state as distinct from the north, Novgorod included. Still more important was the discovery that the Rus had been known to some Byzantine and Oriental writers before A.D. 862 and was evidently located in southern Russia. Finally, the *Primary Chronicle* itself came to be suspected and underwent a searching criticism.

As one of their first tasks, the supporters of the Norman view set out to find the Scandinavian origin of the name *Rus*. Their search, from the time of Schlözer to the present, has had mixed success at best. A number of derivations had to be abandoned. The deduction of *Rus* from the Finnish word for the Swedes, *Ruotsi*, developed by Thomsen and upheld by Stender-Petersen and others, seems linguistically acceptable, but it has been criticized as extremely complicated and unlikely on historical grounds.

Because they considered the Rus a Scandinavian group, the proponents of the Norman theory proceeded to interpret all references to the Rus in Norman terms. Under the year A.D. 839 a Western source, *The Bertinian Annals,* tells about the Rus ambassadors who came to Ingelheim through Constantinople and who were men of Khakan-Rus, but who turned out to be Swedes. Some scholars even concluded that the ambassadors must have come all the way from Sweden, and they read *khakan* to mean *Haakon*. But the Russian khakanate was probably located in southern Russia, and the title of khakan suggests Khazar rather than Norman influence. The early date made certain other scholars advance the hypothetical arrival of the Scandinavian Rus into Russia from A.D. 862 to "approximately A.D. 840." A slight change in the original chronology also enabled these specialists to regard as Scandinavian the Rus who staged an attack on Constantinople in A.D. 860 and who were described on that occasion by Patriarch Photius.

In the tenth century Bishop Liutprand of Cremona referred to the *Rusios* in his description of the neighbors of the Byzantine Empire. A controversy

* I am using the standard English translation of the *Primary Chronicle* by Professor S. Cross (*The Russian Primary Chronicle, Laurentian Text*. Cambridge, Mass., 1930), although I am not entirely satisfied with it either in general or in this particular instance.

still continues as to whether Liutprand described his *Rusios* as Normans or merely as a northern people. Also in the tenth century the Byzantine emperor and scholar Constantine Porphyrogenitus gave the names of seven Dnieper rapids "in Slavic" and "in Russian." The "Russian" names, or at least most of them, can best be explained from Scandinavian languages. This evidence of "the language of the Rus" is rather baffling: there is no other mention of any Scandinavian tongue of the Rus; on the contrary, the *Chronicle* itself states that the Slavic and the Russian languages are one. The supporters of the Norman theory were quick to point to the Scandinavian names of the first Russian princes and of many of their followers listed in the treaties between Kievan Russia and Byzantium. Their opponents challenged their derivation of some of the names and stressed the fact that the treaties were written in Greek and in Slavic and that the Rus swore by Slavic gods.

Certain Arabic authors also mention and sometimes discuss and describe the Rus, but their statements have also been variously interpreted by different scholars. In general the Rus of the Arabic writers are a numerous people rather than a viking detachment, "a tribe of the Slavs" according to Ibn-Khurdadhbih. The Rus had many towns, and its ruler bore the title of *khakan*. True, the Rus are often contrasted with the Slavs. The contrast, however, may refer simply to the difference between the Kievan Slavs and other Slavs to the north. Some of the customs of the Rus, described in Arabic sources, seem to be definitely Slavic rather than Norman: such are the posthumous marriage of bachelors and the suicide of wives following the death of their husbands. The Rus known to the Arabs lived most probably somewhere in southern Russia. Although Arabic writers refer primarily to the ninth century, the widespread and well-established relations of the Rus with the East at that time suggest an acquaintance of long standing.

Other evidence, it has been argued, also points to an early existence of the Rus in southern Russia. To mention only some of the disputed issues, the Rus, reportedly, attacked Surozh in the Crimea early in the ninth century and Amastris on the southern shore of the Black Sea between A.D. 820 and 842. Vernadsky derives the name of *Rus* from the Alanic tribe of the Roxolans. Other scholars have turned to topographic terms, ranging from the ancient word for Volga, *Rha,* to Slavic names for different rivers. An ingenious compromise hypothesis postulates both a Scandinavian and a southern derivation of *Rus-Ros* and the merger of the two.

The proponents of the Norman view have reacted in a number of ways to assertions of the antiquity of the Rus and their intrinsic connection with southern Russia. Sometimes they denied or challenged the evidence. Vasiliev, for instance, refused to recognize the early attacks of the Rus on Surozh and Amastris. The first he classified as apocryphal, the second as referring in fact to the well-known campaign of Igor in A.D. 941. Other specialists, in order to account for all the events at the dawn of Russian history and to connect them with the Scandinavian north, have postulated more than one separate Scandinavian Rus, bringing, rather arbitrarily, some of them from Denmark and others from Sweden. Their extremely complex and unverified schemes serve little purpose, unless one is to assume that the Rus could be nothing but Scandinavians. For example, Vernadsky in his reconstruction of early Russian history conveyed one group of Normans to the shores of the Black Sea as early as A.D. 740. Vernadsky's reasoning unfortunately is highly speculative and generally not at all convincing. By contrast, recently many scholars have considered the Normans as merely one element in the composition of the Rus linked fundamentally to southern Russia and its inhabitants.

The *Primary Chronicle* itself, a central source for the Norman theory, has been thoroughly analyzed and criticized by Shakhmatov and other specialists. This criticism threw new light on the obvious inadequacies of its narrative and revealed further failings in it. The suspiciously peaceful establishment of Riurik and his brothers in northern Russia was related to similar Anglo-Saxon and other stories, in particular to a passage in Widukind's *Res gestae saxonicae,* to indicate, in the opinion of some scholars, the mythical character of the entire "invitation of the Varangians." Oleg's capture of Kiev in the name of Riurik's son Igor in A.D. 882, the starting point of Kievan history according to the *Chronicle,* also raised many issues. In particular it was noted that, due to considerations of age, Igor could hardly have been Riurik's son, and that no Kievan sources anterior to the *Primary Chronicle,* that is, until the early twelfth century, knew of Riurik, tracing instead the ancestry of Kievan princes only to Igor. Moreover, the *Chronicle* as a whole is no longer regarded as a naïve factual narrative, but rather as a work written from a distinct point of view and possibly for definite dynastic purposes, such as providing desirable personal or territorial connections for the Kievan ruling family. On the other hand, the proponents of the Norman theory argue plausibly that the *Chronicle* remains our best source concerning the origin of the Russian state, and that its story, although incorrect in many details, does on the whole faithfully reflect real events.

To sum up, the Norman theory can no longer be held in anything like its original scope. Most significantly, there is no reason to assert a fundamental Scandinavian influence on Kievan culture. But the supporters of the theory stand on a much firmer ground when they rely on archaeological, philological and other evidence to substantiate the presence of the Normans in Russia in the ninth century. In particular the names of the first princes, to and excluding Sviatoslav, as well as the names of many of their followers in the treaties with Byzantium, make the majority of scholars outside the Soviet Union today consider the first Russian dynasty and its immediate retinue as Scandinavian. Yet, even if we accept this view, it remains danger-ous to postulate grand Norman designs for eastern Europe, or to interpret the role of the vikings on the Russian plain by analogy with their much better known activities in Normandy or in Sicily. A historian can go beyond his evidence only at his own peril.

In any case, whether through internal evolution, outside intervention, or some peculiar combination of the two, the Kievan state did arise in the Dnieper area toward the end of the ninth century.

I V

* * * * * * * * * *

KIEVAN RUSSIA: A POLITICAL OUTLINE

In that city, in the city of Kiev. . . .
THE FIRST LINE OF AN EPIC POEM

Kievan political history can be conveniently divided into three periods. The first starts with Oleg's semi-legendary occupation of the city on the Dnieper in 882 and continues until 972 or 980. During that initial century of Kievan history, Kievan princes brought the different East Slavic tribes under their sway, exploiting successfully the position of Kiev on the famous road "from the Varangians to the Greeks" — that is, from the Scandinavian, Baltic, and Russian north of Europe to Constantinople — as well as other connections with the inhabitants both of the forest and the steppe, and building up their domain into a major European state. At the end of the century Prince Sviatoslav even engaged in a series of far-reaching campaigns and conquests, defeated a variety of enemies, and threatened the *status quo* in the Balkans and the Byzantine Empire itself.

The failure of Sviatoslav's more ambitious plans as well as a gradual consolidation of the Kievan state in European Russia marks the transition to the next period of Kievan history, when Kievan Russia attained in most respects its greatest development, prosperity, stability, and success. This second period was occupied almost entirely by the reigns of two remarkable princes, Saint Vladimir and Iaroslav the Wise, and it ended with the death of the latter in 1054. While the Kievan rulers from Oleg through Sviatoslav established Kievan Russia as an important state, it was early in the time of Vladimir that a new element of enormous significance entered the life and culture of Kiev: Christianity. The new Christian civilization of Kievan Russia produced impressive results as early as the first half of the eleventh century, adding literary and artistic attainment to the political power and high economic development characteristic of the age.

The third and last period of Kievan history, that of the decline and fall, is the most difficult one to define chronologically. It may be said to begin with the passing of Iaroslav the Wise in 1054, but there is no consensus about the point at which foreign invasions, civil wars, and the general diminution in the significance of Kiev brought the Kievan era of Russian history to a close. Vladimir Monomakh, who reigned from 1113 to 1125, has often been considered the last effective Kievan ruler, and the same has been said of his son, Mstislav, who reigned from 1125 to 1132. Other

29

historians indicate as the terminal point, for example, the capture and the sacking of Kiev in 1169 by Prince Andrew Bogoliubskii of Suzdal and his decision to remain in the northeast rather than move to the city on the Dnieper. As the ultimate date of Kievan history, 1240 also has a certain claim: in that year Kiev, already a shadow of its former self in importance, was thoroughly destroyed by the Mongols, who established their dominion over conquered Russia.

The Rise of the Kievan State

Oleg, the first historical ruler of Kiev, remains in most respects an obscure figure. According to the *Primary Chronicle* he was a Varangian, a relative of Igor, who occupied Kiev in 882 and died in 913. Assisted by his retainers, the *druzhina,* Oleg spread his rule from the territory of the Poliane to the areas of several neighboring East Slavic tribes. Some record of a subsequent bitter opposition of the Drevliane to this expansion has come down to our time; certain other tribes, it would seem, submitted with less struggle. Tribute became the main mark and form of their allegiance to Kiev. Still other tribes might have acted simply as associates of Oleg and his successor Igor in their various enterprises, without recognizing the supreme authority of Kiev. Toward the end of his life Oleg had gathered a sufficient force to undertake in 907 a successful campaign against Byzantium. Russian chronicles exaggerate Oleg's success and tell, among other things, the story of how he nailed his shield to the gates of Constantinople. Byzantine sources are strangely silent on the subject of Oleg's campaign. Yet some Russian victories seem probable, for in 911 Oleg obtained from Byzantium an extremely advantageous trade treaty.

Oleg's successor, Prince Igor, ruled Kievan Russia from 913 until his death in 945. Our knowledge of him comes from Greek and Latin, in addition to Russian, sources, and he stands out, by contrast with the semi-legendary Oleg, as a fully historical person. Igor had to fight the Drevliane as well as to maintain and spread Kievan authority in other East Slavic lands. That authority remained rather precarious, so that each new prince was forced to repeat in large part the work of his predecessor. In 941 Igor engaged in a major campaign against Constantinople and devastated its suburbs, but his fleet suffered defeat by the Byzantine navy which used the celebrated "Greek fire." * The war was finally terminated by the treaty of 944, the provisions of which were rather less favorable to the Russians than those of the preceding agreement of 911. In 943 the Russians campaigned successfully in the distant transcaspian provinces of

* The Greek fire was an incendiary compound projected through copper pipes by Byzantine sailors to set on fire the ships of their opponents. Its exact composition remains unknown.

Persia. Igor was killed by the Drevliane in 945 while collecting tribute in their land.

Oleg's and Igor's treaties with Byzantium deserve special attention. Their carefully worded and remarkably detailed provisions dealt with the sojourn of the Russians in Constantinople, Russian trade with its inhabitants, and the relations between the two states in general. It may be noted that the Russians in Constantinople were subject to their own courts, but that, on the other hand, they were free to enter Byzantine service.

While their relations with Byzantium increased the prestige and the profits of the Russians, the inhabitants of the steppe continued to threaten the young Kievan state. In addition to the relatively stabilized and civilized Khazars, more primitive peoples pressed westward. At the dawn of Kievan history, the Magyars, a nomadic horde speaking a Finno-Ugrian language and associated for a long time with the Khazar state, moved from the southern Russian steppe to enter, at the end of the ninth century, the Pannonian plain and lay the foundations for Hungary. But they were replaced and indeed in part pushed out of southern Russia by the next wave from the east, rather primitive and ferocious Turkic nomads, the Pechenegs or Patzinaks. The approach of the Pechenegs is mentioned in the *Chronicle* under the year 915; and they began to carry out constant assaults on the Kievan state in the second half of the tenth century, after the decline of the Khazars.

Igor's sudden death left his widow Olga in charge of the Kievan state, for their son Sviatoslav was still a boy. Olga rose to the occasion, ruling the land from 945 to about 962 and becoming the first famous woman in Russian history as well as a saint of the Orthodox church. The information concerning Olga describes her harsh punishment of the Drevliane and her persistent efforts to strengthen Kievan authority among other East Slavic tribes. It tells also of her conversion to Christianity, possibly in 954 or 955, and her journey to Constantinople in 957. There she was received by the emperor Constantine Porphyrogenitus, who left us an account of her visit. But the conversion of Olga did not mean a conversion of her people, nor indeed of her son Sviatoslav.

The ten years of Sviatoslav's rule of Kievan Russia, 962 to 972, which marked the culmination of the first period of Kievan history in the course of which the new state obtained a definite form and role on the east European plain, have been trenchantly called "the great adventure." If successful, the adventure might have given Russian history a new center and a different course. Even with their ultimate failure, Sviatoslav's daring campaigns and designs left their imprint all the way from Constantinople to the Volga and the Caspian Sea. Sviatoslav stands out in history as a classic warrior-prince, simple, severe, indefatigable, brave, sharing with his men uncounted hardships as well as continuous battles. He has been likened to the cossack hetmans and to the viking captains as well as to leaders in other military traditions, and the cossack, if not the viking, comparison has a point: Sviatoslav's appearance, dress, and manner of life all remind us of the steppe. In the words of the *Primary Chronicle:* "Upon his expeditions he carried with him neither wagons nor kettles, and boiled no meat, but cut off small strips of horseflesh, game, or beef, and ate it after roasting it on the coals. Nor did he have a tent, but he spread out a piece of saddle cloth under him, and set his saddle under his head."

In 964 Sviatoslav started out on a great eastern campaign. First he subjugated the East Slavic tribe of the Viatichi, who had continued to pay tribute to the Khazars rather than to Kiev. Next he descended to the mouth of the river Oka bringing the surrounding Finnic-speaking tribes under his authority. From the mouth of the Oka he proceeded down the Volga, attacked the Volga Bulgars, and sacked their capital, the Great Bulgar. But instead of developing his campaign against the Bulgars, he resumed in 965 his advance down the Volga toward the Khazar state, subduing Finnic and Turkic tribes on the way. Sviatoslav's war against the Khazars had a sweeping scope and impressive results: the Russians smashed the Khazar army, captured and sacked the Khazar capital, Itil, reached the Caspian and advancing along its western shore seized the key fortress of Samandar. Next, turning west, they defeated the Alans and some other peoples of the northern Caucasus, came to the mouth of the Don and stormed the Khazar fortress of Sarkil, which dominated that area. The Khazars, although their state lasted for another half century, never recovered from these staggering blows. Sviatoslav returned to Kiev in 967. His remarkable eastern campaign, which led to the defeat of the Volga Bulgars and the Khazars, completed the unification of the East Slavs around Kiev, attaching to it both the Viatichi and other groups to the southeast, notably in the Don area. Also, it brought under Russian control the entire flow of the Volga, and thus the great Volga-Caspian Sea trade route — a more ancient and perhaps more important north-south communication artery than the Dnieper way itself — whereas formerly the Russians had held only the upper reaches of the Volga. Yet the magnificent victory over the Khazars had its reverse side; it weakened decisively their effectiveness as a buffer against other Asiatic peoples, in particular the Pechenegs.

In 968 Sviatoslav became involved in another major undertaking. On the invitation of the Byzantine emperor Nicephorus Phocas, he led a large army into the Balkans to attack the Bulgarian state in the Danubian valley. Once more the Russians achieved notable military successes, capturing the capital of the Bulgarians and taking prisoner their ruler Boris, although they had to interrupt the campaign to defeat the Pechenegs, who in 969 in the absence of Sviatoslav and his troops had besieged Kiev. Sviatoslav, who thus came to control the territory from the Volga to the Danubian plain, apparently liked the Balkan lands especially well. According to the *Chronicle,* he declared: "I do not care to remain in Kiev, but should prefer to live in Pereiaslavets on the Danube, since that is the center of my realm, where all riches are concentrated: gold, silks, wine, and various fruits from Greece, silver and horses from Hungary and Bohemia, and from Russia furs, wax, honey, and slaves." One can only speculate on the possible implications of such a change of capital for Russian history.

But the Byzantine state, ruled from 969 by the famous military leader Emperor John Tzimisces, had become fully aware of the new danger. As Sviatoslav would not leave the Balkans, a bitter war ensued. In his characteristic manner the Russian prince rapidly crossed the Balkan mountains and invaded the Byzantine Empire, capturing Philippopolis and threatening Adrianople and Constantinople itself. However, John Tzimisces managed in the nick of time to restore his position in Asia, which had been threatened by both a foreign war and a rebellion, and to shift his main effort to the Balkans. He counterattacked, crossing in his turn the Balkan range and capturing Great Preslav, the Bulgarian capital. The Russian army, its lines of communication endangered, had to retreat to the fortress of Dorostolon on the Danube — present-day Dristra or Silistria — which, after a hard-fought battle, John Tzimisces placed under siege. Following more desperate fighting, in July 971 Sviatoslav was finally reduced to making peace with Byzantium on condition of abandoning the Balkans, as well as the Crimea, and promising not to challenge the Byzantine Empire in the future. On his way back to Russia, with a small retinue, he was intercepted and killed by the Pechenegs. Tradition has it that the Pecheneg khan had a drinking cup made out of Sviatoslav's skull. The great adventure had come to its end. Sviatoslav's Balkan wars attract attention not only because of the issues involved but also because of the sizes of the contending armies and because of their place in military history; Byzantine sources indicate that Sviatoslav fought at the head of 60,000 troops of whom 22,000 remained when peace was concluded.

After the death of his mother Olga in 969, Sviatoslav, constantly away with the army, entrusted the administration of the Kiev area to his elder son Iaropolk, dispatched the second son Oleg to govern the territory of the Drevliane, and sent the third, the young Vladimir, with an older relative to manage Novgorod. A civil war among the brothers followed Sviatoslav's death. At first Iaropolk had the upper hand, Oleg perishing in the struggle and Vladimir escaping abroad. But in two years Vladimir returned and with foreign mercenaries and local support defeated and killed Iaropolk. About 980 he became the ruler of the entire Kievan realm.

Kiev at the Zenith

Vladimir, who reigned until 1015, continued in most respects the policies of his predecessors. Among the East Slavs, he reaffirmed the authority of the Kievan state which had been badly shaken during the years of civil war. He recovered Galician towns from Poland and, further to the north, subdued the warlike Baltic tribe of the Iatviags, extending his domain in that area to the Baltic Sea. Vladimir also made a major and generally successful effort to contain the Pechenegs. He built fortresses and towns, brought settlers into the frontier districts, and managed to push the steppe border to two days, rather than a single day, of travel time from Kiev.

However, Vladimir's great fame rests on his relations with Byzantium and, most especially, on his adoption of Christianity, which proved to be of immense significance and long outlasted the specific political and cultural circumstances that led to the step. Interest in Christianity was not unprecedented among the Russians. In fact, there may even have been a Russian diocese of the Byzantine Church as early as 867, although not all scholars agree on this inference from a particular tantalizing passage in an early document. Whether or not an early Christian Rus existed on the shores of the Sea of Azov, Kiev itself certainly experienced Christian influences before the time of Vladimir. A Christian church existed in Kiev in the reign of Igor, and we know that Olga, Vladimir's grandmother, became a Christian; Vladimir's brother Iaropolk has also been described as favorably inclined to Christianity. But it should be emphasized that Olga's conversion did not affect the pagan faith of her subjects and, furthermore, that, in the first part of the reign of Vladimir, Kievan Russia experienced a strong pagan revival. Vladimir's turnabout and the resulting "baptism of Russia" were accompanied by an intricate series of developments that has been given different explications and interpretations by scholars: Vladimir's military aid to Emperor Basil II of Byzantium, the siege and capture by the Russians of the Byzantine outpost of Chersonesus in the Crimea, and Vladimir's marriage to Anne, Basil II's sister. Whatever the exact import and motivation of these and certain other events, the Kievan Russians formally accepted Christianity from Constantinople in or around 988 and probably in or near Kiev, although some historians prefer Chersonesus.

The conversion of Kievan Russia to Christianity fits into a broad historical pattern. At about the same time similar conversions from paganism were taking place among some of the Baltic Slavs, and in Poland, Hungary, Denmark, and Norway. Christendom in effect was spreading rapidly across all of Europe, with only a few remote peoples, such as the Lithuanians, holding out. Nevertheless, it can well be argued that Vladimir's decision represented a real and extremely important choice. The legendary account

of how the Russians selected their religion, spurning Islam because it prohibited alcohol — for "drink is the joy of the Russian" — and Judaism because it expressed the beliefs of a defeated people without a state, and opting for Byzantine liturgy and faith, contains a larger meaning: Russia did lie at cultural crossroads, and it had contacts not only with Byzantium and other Christian neighbors but also with the Moslem state of the Volga

Bulgars and other more distant Moslems to the southeast as well as with the Jewish Khazars. In other words, Vladimir and his associates chose to become the Eastern flank of Christendom rather than an extension into Europe of non-Christian civilizations. In doing so, they opened wide the gates for the highly developed Byzantine culture to enter their land. Kievan literature, art, law, manners, and customs experienced a fundamental impact of Byzantium. The most obvious result of the conversion was the appearance in Kievan Russia of the Christian Church itself, a new and extremely important institution which was to play a role similar to that of the Church in other parts of medieval Europe. But Christianity, as already indicated, remained by no means confined to the Church, permeating instead Kievan society and culture, a subject to which we shall return in later chapters. In politics too it gave the Kievan prince and state a stronger ideological basis, urging the unity of the country and at the same time emphasizing its links with Byzantium and with the Christian world as a whole. Dvornik, Obolensky, Meyendorff, and many other scholars have given us a rich picture of the Byzantine heritage and of the Russian borrowing from it.

It must be kept in mind that Christianity came to Russia from Byzantium, not from Rome. Although at the time this distinction did not have its later significance and although the break between the Eastern and the Western Churches occurred only in 1054, the Russian allegiance to Byzantium determined or helped to determine much of the subsequent history of the country. It meant that Russia remained outside the Roman Catholic Church, and this in turn not only deprived Russia of what that Church itself had to offer, but also contributed in a major way to the relative isolation of Russia from the rest of Europe and its Latin civilization. It helped notably to inspire Russian suspicions of the West and the tragic enmity between the Russians and the Poles. On the other side, one can well argue that Vladimir's turn to Constantinople represented the richest and the most rewarding spiritual, cultural, and political choice that he could make at the time. Even the absence of Latinism and the emphasis on local languages had its advantages: it brought religion, in the form of a readily understandable Slavic rite, close to the people and gave a powerful impetus to the development of a national culture. In addition to being remembered as a mighty and successful ruler, Vladimir was canonized by the Church as the baptizer of the Russians, "equal to the apostles."

Vladimir's death in 1015 led to another civil war. Several of Vladimir's sons who had served in different parts of the realm as their father's lieutenants and had acquired local support became involved in the struggle. The eldest among them, Sviatopolk, triumphed over several rivals and profited from strong Polish aid, only to be finally defeated in 1019 by another son Iaroslav, who resumed the conflict from his base in Novgorod.

Sviatopolk's traditional appelation in Russian history can be roughly translated as "the Damned," and his listed crimes — true or false, for Iaroslav was the ultimate victor — include the assassination of three of his brothers, Sviatoslav, Boris, and Gleb. The latter two became saints of the Orthodox Church.

Prince Iaroslav, known in history as Iaroslav the Wise, ruled in Kiev from 1019 until his death in 1054. His reign has been generally acclaimed as the high point of Kievan development and success. Yet, especially in its first part, it was fraught with danger, and the needs of the state continued to demand strenuous exertion from the prince and his subjects. Civil war did not end with Iaroslav's occupation of Kiev. In fact, he had to flee it and ultimately, by an agreement of 1026, divide the realm with his brother Mstislav the Brave, prince of Tmutorokan, a principality situated in the area where the Kuban flows into the Sea of Azov and the Black Sea: Iaroslav kept Kiev and authority over the lands west of the Dnieper; Mstislav secured as his domain the territory east of it, with the center in Chernigov. Only after the death of Mstislav in 1036 did Iaroslav become the ruler of the entire Kievan state, and even then the Polotsk district retained a separate prince. Besides fighting for his throne, Iaroslav had to suppress a whole series of local rebellions, ranging from a militant pagan revival in the Suzdal area to the uprisings of various Finnish and Lithuanian tribes.

Iaroslav's foreign wars included a successful effort in 1031 to recover from Poland the southwestern section which that country obtained in return for supporting Sviatopolk, and an unsuccessful campaign against Byzantium some twelve years later which proved to be the last in the long sequence of Russian military undertakings against Constantinople. But especial significance attaches to Iaroslav's struggle with the attacking Pechenegs in 1037: the decisive Russian victory broke the might of the invaders and led to a quarter-century of relative peace on the steppe frontier, until the arrival from the east of new enemies, the Polovtsy.

At the time of Iaroslav the prestige of the Kievan state stood at its zenith; the state itself stretched from the Baltic to the Black Sea and from the mouth of the Oka river to the Carpathian mountains, and the Kievan ruling family enjoyed close connections with many other reigning houses of Europe. Himself the husband of a Swedish princess, Iaroslav obtained the hands of three European princesses for three of his sons and married his three daughters to the kings of France, Hungary, and Norway; one of his sisters became the wife of the Polish king, another the wife of a Byzantine prince. Iaroslav offered asylum to exiled rulers and princes, such as the princes who fled from England and Hungary and St. Olaf, the king of Norway, with his son, and his cousin Harold Hardrada. It should be added that while the links with the rest of Europe were particularly numerous in the reign of Iaroslav, they were in general a rather common occurrence in Kievan Russia. Following Baumgarten, Vernadsky has calculated, for instance, that six Kievan matrimonial alliances were established with Hungary, five with Bohemia, some fifteen with Poland, and at least eleven with Germany, or, to be more precise on the last point, at least six Russian princes had German wives, while "two German marquises, one count, one landgrave, and one emperor had Russian wives."

Iaroslav's great fame, however, rests more on his actions at home than on his activities in foreign relations. His name stands connected with an impressive religious revival, and with Kievan law, education, architecture, and art. Church affairs of the reign present certain very intricate puzzles to the historian. For some reason Kievan sources, and most importantly the *Primary Chronicle,* virtually omit Russian ecclesiastical history from the conversion in 988 to 1037, and, furthermore, give the impression that the years around the latter date, at the time of Iaroslav, produced a new departure in Russian Christianity, marked by such a strange act as the consecration in 1039 of a Kievan church which had been erected by Vladimir. In search of an explanation, Priselkov suggested that until 1037 the Russian Church was linked to the Bulgarian archbishopric of Ochrid rather than to Byzantium. Some specialists proposed that the Church at Kiev turned from Constantinople to Rome or simply took an independent and disobedient stand vis-à-vis Constantinople. A more recent interpretation, by Stokes, shifted the emphasis from international ecclesiastical politics to the internal history of the Kievan state and argued that the change under Iaroslav consisted in the transfer of the religious center of Russia, the seat of the metropolitan, from its original location in the city of Pereiaslavl, east of the Dnieper, to Kiev. At least until further evidence, it seems best to assume that Russia remained under the jurisdiction of the Byzantine Church and also had its own metropolitan, whether in Kiev or Pereiaslavl, from the time of the conversion. Whatever the interpretation of its pre-1037 development, Iaroslav did leave an impact on the Russian Church, changing or confirming its organization, having an able and educated Russian, Hilarion, serve as the first native metropolitan, and building and supporting churches and monasteries on a large scale. He has usually been credited with a major role in the dissemination and consolidation of Christianity in Russia.

Iaroslav the Wise has the reputation also of a lawgiver, for he has generally been considered responsible for the first Russian legal code, *The Russian Justice,* an invaluable source for our knowledge of Kievan society and life. And he played a significant role in Kievan culture by such measures as his patronage of artists and architects and the establishment of a large school and a library in Kiev.

The Decline and Fall of the Kievan State

Before his death Iaroslav assigned separate princedoms to his sons: Iziaslav, the eldest, received the Kiev and Novgorod areas; Sviatoslav, the second, the area centered on Chernigov; Vsevolod, the third, Pereiaslavl; Viacheslav, the fourth, Smolensk; and Igor, the fifth, Vladimir-in-Volynia — always with their surrounding territories. The princes, apparently, were expected to co-operate and to hold Kievan Russia together. Moreover, it would seem that when a vacancy occurred, they were to move up step by step, with the position in Kiev the summit. Some such moves did in fact take place, but the system — if indeed it can be called a system — quickly bogged down: Iaroslav's arrangement, based quite possibly on old clan concepts and relations still present in the ruling family, worked to break the natural link between a prince and his state, and it excluded sons from succession in favor of their uncles, their late father's brothers. Besides, with a constant increase in the number of princes, precise calculations of appropriate appointments became extremely difficult. At their meeting in Liubech in 1097 the princes agreed that the practice of succession from father to son should prevail. Yet the principle of rotation from brother to brother remained linked for a long time to the most important seat of all, that of the Grand Prince in Kiev.

The reigns of Iziaslav, Sviatoslav, and Vsevolod, the last of whom died in 1093, as well as that of Iziaslav's son Sviatopolk, who succeeded Vsevolod and ruled until his death in 1113, present a frightening record of virtually constant civil wars which failed to resolve with any degree of permanence the problem of political power in Kievan Russia. At the same time the Kievan state had to face a new major enemy, the Polovtsy, or the Cumans as they are known to Western authors. This latest wave of Turkic invaders from Asia had defeated the Pechenegs, pushing them toward the Danube, and had occupied the southeastern steppe. They attacked Kievan territory for the first time in 1061, and after that initial assault became a persistent threat to the security and even existence of Kievan Russia and a constant drain on its resources.

Although hard beset, the Kievan state had one more revival, under an outstanding ruler, Vladimir Monomakh. A son of Grand Prince Vsevolod, Vladimir Monomakh became prominent in the political life of the country long before he formally assumed the highest authority: he acted with and for his father in many matters and he took the lead at princely conferences, such as those of 1097 and 1100 to settle internecine disputes or that of 1103 to concert action in defense of the steppe border. Also, he played a major role in the actual fighting against the Polovtsy, obtaining perhaps his greatest victory over them, in 1111 at Salnitsa, before his elevation to the Kievan seat. As Grand Prince, that is, from 1113 until his death in 1125, Vladimir Monomakh fought virtually all the time. He waged war in Livonia, Finland, the land of the Volga Bulgars, and the Danubian area, repulsing the Poles and the Hungarians among others; but above all he campaigned against the Polovtsy. His remarkable *Testament* speaks of a grand total of eighty-three major campaigns and also of the killing of two hundred Polovetsian princes; according to tradition, Polovetsian mothers used to scare their children with his name. Vladimir Monomakh distinguished himself as an effective and indefatigable organizer and administrator, a builder, for instance, possibly, of the town of Vladimir in the northeast on the river Kliazma, which was to become in two generations the seat of the grand prince, and also as a writer of note. Of special interest is his social legislation intended to help the poor, in particular the debtors.

Vladimir Monomakh was succeeded by his able and energetic son Mstislav (ruled 1125–32) and after him by another son, Iaropolk, who reigned until his death in 1139. But before long the Kievan seat became again the object of bitter contention and civil war which often followed the classic Kievan pattern of a struggle between uncles and nephews. In 1169 one of the contenders, Prince Andrew, or Andrei, Bogoliubskii of the northeastern principalities of Rostov and Suzdal, not only stormed and sacked Kiev but, after his victory in the civil war, transferred the capital to his favorite city of Vladimir. Andrew Bogoliubskii's action both represented the personal preference of the new grand prince and reflected a striking decline in importance of the city on the Dnieper. Kiev was sacked again in 1203. Finally, it suffered virtually complete destruction in 1240, at the hands of the Mongols.

The Fall of Kiev: The Reasons

The decline and collapse of Kievan Russia have been ascribed to a number of factors; but there is considerable controversy about the precise nature of these factors and no consensus concerning their relative weight. The most comprehensive general view, held by Soviet historians as a group and by some others, emphasizes the loose nature of the Kievan state and its evolution in the direction of further decentralization and feudalism. In fact, certain specialists raised the question of whether Kievan Russia could be called a state at all. Aside from this extreme opinion, it has been generally recognized that the Kievan state, very far from resembling its modern counterparts, represented in a sense a federation or association of a number of areas which could be effectively held together only for limited periods of time and by exceptionally able rulers. Huge distances and poor communications made the issue of centralization especially acute. Moreover, it is argued that Russia, as well as Europe in general, evolved toward natural economy, particularism, and feudalism. Therefore, the relatively slender unifying bonds dissolved, and Russia emerged as an aggregate of ten or twelve separate areas. We shall return to this view when we discuss the question of feudalism in Russia, and on other occasions.

Soviet historians, as well as some other specialists, have also pointed to social conflicts as a factor in the decline of Kiev. They refer in particular to the gradual enserfment of the peasants by the landlords and to the worsening position of the urban poor, as indicated by events at the time of Vladimir Monomakh. Slavery, which Kievan Russia inherited from earlier societies, has also been cited as an element of weakness.

Another essentially economic explanation of the fall of Kievan Russia stresses trade, or rather the destruction of trade. In its crude form it argues that the Kievan state arose on the great commercial route "from the Varangians to the Greeks," lived by it, and perished when it was cut. In a more limited and generally accepted version, the worsening of the Kievan position in international trade has been presented as one major factor in the decline of Kiev. The city on the Dnieper suffered from the change in trade routes which began in the eleventh century and resulted, largely through the activities of Italian merchants in the Mediterranean, in the establishment of closer connections between western and central Europe on the one hand and Byzantium and Asia Minor on the other, and a bypassing of Kiev. It was adversely affected by the Crusades, and in particular by the sacking of Constantinople by the Crusaders in 1204, as well as by the decline of the Caliphate of Bagdad. The fact that certain Russian towns and areas, such as Smolensk and especially Novgorod, profited by the rearrangement of the commercial map of Europe and the rise of Italian and German cities only tended to make Kievan control over them less secure. Finally, Kiev experienced tremendous difficulty, and ultimately failed, in protecting from the steppe peoples the commercial line across the southern steppe to the Black Sea.

In addition to the economic and social analyses, one can turn to the political. A number of historians have placed much stress on the failure of the Kievan system of government which they consider a major, possibly decisive, cause of the collapse of Kievan Russia, rather than merely a reflection of more fundamental economic and social difficulties. There is a consensus that the Kievan princely political system did not function well, but no agreement as to the exact nature of that system. Of the two main interpretations, one considers it simply to be confusion worse confounded and a rule of force without broad agreement on principle, while the other gives full credence and weight to the practice of joint clan rule and of brother to brother rotation with such further provisions as the equation of the claims of the elder son of a prince to those of his father's third brother, his third uncle. In any case, the system did collapse in constant disputes and endemic internal strife. Pogodin calculated that of the 170 years following the death of Iaroslav the Wise 80 witnessed civil war. Kievan princes have also been blamed for various faults and deficiencies and in particular for being too militant and adventurous and often lacking the more solid attributes of rulers. On this point it would seem, however, that their qualities in general were well suited to the age.

Towns added further complications to princely rule and princely relations. Towns in Kievan Russia had existed before princely authority appeared, and they represented, so to speak, a more fundamental level of political organization. As princely disputes increased and princely power declined, the towns proceeded to play an increasingly significant role in Kievan politics, especially in determining what prince would rule in a given town and area. The later evolution of Novgorod represents an extreme case of this Kievan political tendency.

At least one other factor must be mentioned: foreign pressure. While it can well be argued that Kievan economics, social relations, and politics all led to the collapse of the state, the fall of Kiev can also — perhaps paradoxically — be explained primarily in terms of outside aggression. For Kiev had to fight countless exhausting wars on many fronts, but above all in the southeast against the inhabitants of the steppe. The Pechenegs replaced the Khazars, and the Polovtsy the Pechenegs, but the fighting continued. After the Polovtsy and the Kievan Russians virtually knocked each other out, the Mongols came to give the *coup de grâce*. In contrast to the wars of medieval Europe, these wars were waged on a mass scale with tremendous effort and destruction. It might be added that during the centuries of Kievan history the steppe had crept up on the forest, and deforestation has been cited as one development weakening the military defenses of Kiev. There exists an epic Russian tale about the destruction of the Russian land. It tells of the *bogatyri,* the mighty warriors of Kievan Russia, meeting the invaders head on. The bogatyri fought very hard; indeed they split their foes in two with the blows of their swords. But then each half would become whole, and the enemies kept pressing in ever-increasing numbers until finally they overwhelmed the Russians.

V

* * * * * * * * * *

KIEVAN RUSSIA: ECONOMICS, SOCIETY, INSTITUTIONS

> . . . merry-go-round, moving harmoniously and melodiously, full of
> joy. . . . This spirit permeates, this form marks everything that comes
> from Russia; such is our song itself, such is its tune, such is the organ-
> ization of our Land.
>
> K. AKSAKOV

> The decisive factor in the process of feudalization proved to be the
> emergence of private ownership in land and the expropriation of the
> small farmer, who was turned into a feudal "tenant" of privately
> owned land, and his exploitation by economic or extra-economic com-
> pulsion.
>
> LIASHCHENKO

THE TRADITIONAL VIEW of Kievan economy stresses the role of trade. Its classic document is an account of the activities of the Rus composed by the tenth-century Byzantine emperor and scholar Constantine Porphyrogenitus. Every November, writes Constantine Porphyrogenitus, the Kievan princes and their retainers went on a tour of the territories of different tributary Slavic tribes and lived on the fat of those lands during the winter. In April, after the ice on the Dnieper had broken, they returned, with the tribute, down the river to Kiev. In the meantime, Slavs, subject to the Rus, would fell trees, build boats, and in the spring, when rivers became navi-gable, take them to Kiev and sell them to the prince and his retinue. Having outfitted and loaded the boats, the Rus next moved down the Dnieper to Vitichev where they waited for more boats carrying goods from Novgorod, Smolensk, Liubech, Chernigov, and Vyshgorod to join them. Finally, the entire expedition proceeded down the Dnieper toward the Black Sea and Constantinople.

Kliuchevsky and other historians have expounded how this brief Byzan-tine narrative summarizes some of the most essential characteristics of Kievan Russia, and even, so to speak, its life cycle. The main concern of the prince and his retainers was to gather tribute from subject territories, either, as described above, by visiting the different parts of the realm during the winter — a process called *poliudie* in Russian — or by having the tribute brought to them — *povoz*. The tribute in kind, which the prince obtained in his capacity as ruler and which consisted in particular of such items as furs, wax, and honey, formed the foundation of the commercial undertakings of the Rus. Slaves constituted another major commodity: the

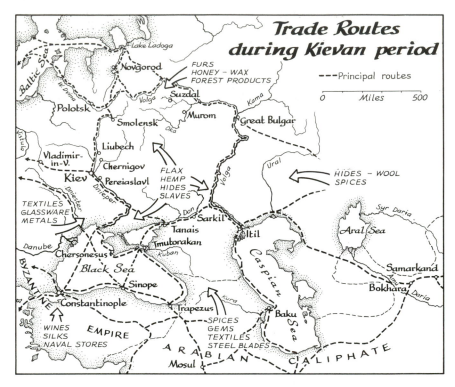

Trade Routes during Kievan period

---Principal routes

FURS
HONEY – WAX
FOREST PRODUCTS

FLAX
HEMP
HIDES
SLAVES

HIDES – WOOL
SPICES

TEXTILES
GLASSWARE
METALS

WINES
SILKS
NAVAL STORES

SPICES
GEMS
TEXTILES
STEEL BLADES

continuous expansion of the Kievan state connected with repeated wars enabled the prince constantly to acquire human chattels for foreign markets. The Kievan ruler thus acted as a merchant-prince on a grand scale. His retainers, the druzhina, emulated him as best they could: they helped him gather tribute in winter, and received their share of it, which they took for sale abroad with the great summer expedition of the Rus. Many other merchants from different parts of Kievan Russia with their merchandise joined the princely train to secure protection on the way and support for their interests at the end of the journey. The gathering of tribute, the construction of boats and their sale each spring near Kiev, the organization of the commercial convoy, and finally the expedition itself linked the entire population of the Dnieper basin, and even of Kievan Russia in the large, and constituted the indispensable economic foundation of the Kievan state. With regularity, coins from Byzantium or Bagdad found their way to the banks of the Oka or the Volkhov rivers.

Constantine Porphyrogenitus' account, it is further argued, explains also the foreign policy of the Rus which followed logically from their economic interests. The rulers in Kiev strove to gain foreign markets and to protect the lifelines of trade leading to those markets. The Kievan state depended above all on the great north-south commercial route "from the Varangians to the Greeks" which formed its main economic and political axis, and it perished with the blocking of this route. The famous Russian campaigns against Constantinople, in 860, under Oleg in 907, under Igor in 941 and 944, under Sviatoslav in 970, and in the reign of Iaroslav the Wise in 1043, demonstrate in an especially striking manner this synthesis between trade and foreign policy. Typically, wars began over such incidents as attacks on Russian merchants in Constantinople and ended with trade pacts. All the Russo-Byzantine treaties which have come down to us exhibit a commercial character. Furthermore, their provisions dealing with trade are both extremely detailed and juridically highly developed, constituting in fact an engaging chapter in the history of international relations and international law. Russian commercial interests, it may well be noted, obtained various advantages from these agreements; and they were considered in Constantinople not as private enterprise but as trade missions of the allied Kievan court.

Full evidence for a history of Kievan commerce goes, of course, far beyond Constantine Porphyrogenitus' narration and even beyond the significant story of Russo-Byzantine relations. Its main points include trade routes and activities in southern Russia prior to the formation of the Kievan state, a subject expertly treated by Rostovtzeff and some other specialists. Attention must also be drawn to the widespread commercial enterprises of the East Slavs themselves long before the time of Oleg, as well as to the fact that at the dawn of Kievan history they already possessed many towns. Saveliev, for instance, estimates that the trade of the East Slavs with Oriental countries, which extended to the borders of China, dates at least from the seventh century A.D. Some Russian weights and measures were borrowed from the east, notably from Mesopotamia, while others came originally from Rome. Similarly, to the west at an early date the East Slavs established trade relations with their closer neighbors and also with some more remote European countries, like Scandinavia. With the flowering of the Kievan state, Russian trade continued to grow, and on an impressive scale. Its complexity and high degree of development find strong reflection, for example, in the eleventh-century legal code, *The Russian Justice*.

Whereas the traditional estimate of Kievan economy stresses commerce, a different interpretation emphasizing agriculture has more recently risen into prominence. Grekov was the ablest exponent of this view, and his work has been continued by other Soviet historians. These scholars carefully delineate the early origin of agriculture in Russia and its great complexity and extent prior to as well as after the establishment of the Kievan state. In point of time, as mentioned earlier, agriculture in southern Russia goes back to the Scythian ploughmen and even to a neolithic civilization of the fourth millennium before Christ. The past of the East Slavs also testifies to their ancient and fundamental link with agriculture. For example, linguistic data indicate that from deep antiquity they were acquainted with various kinds of grains, vegetables, and agricultural tools and implements. Their pagan religion contained the cults of mother earth and the sun, and their different beliefs and rites connected with the agricultural cycle survived in certain aspects of the worship of the Virgin and of Saints Elijah, George, and Nicholas, among others. The East Slavic calendar had its months named after the tasks which an agricultural society living in a forest found it necessary to perform: the month when trees are cut down, the month when they dry, the month when burned trees turn to ashes, and so on. Archaeological finds similarly demonstrate the great antiquity and pervasiveness of agriculture among the East Slavs; in particular they include metallic agricultural implements and an enormous amount of various grains, often preserved in separate buildings.

Written sources offer further support of the case. "Products of the earth" were mentioned as early as the sixth century in a reference to the Antes. Slavic flax was reported on Central Asiatic markets in the ninth century, where it came to be known as "Russian silk." Kievan writings illustrate the central position of agriculture in Kievan life. Bread emerges as the principal food of men, oats of horses. Bread and water represent the basic ration, much bread is associated with abundance, while a drought means a calamity. It should be noted that the Kievan Russians knew the difference between winter grain and spring grain. The *Russian Justice,* for all its concern with trade, also laid extremely heavy penalties for moving field boundaries. Tribute and taxes too, while sometimes paid in furs, were more generally connected to the "plough" as the basic unit, which probably referred to a certain amount of cultivated land.

Grekov and other Soviet historians argue further that this fundamental role of agriculture in Kievan economy determined the social character of the prince and his druzhina and indeed the class structure of Kievan society. They emphasize the connections of the prince and his retainers with the land as shown in references to elaborate princely households, the spread of princely and druzhina estates throughout Kievan territory, and nicknames associated with the land. They consider that Kievan Russia was developing into a fully feudal society, in the definition of which they stress the prevalence of manorial economy.

It can readily be seen that the evidence supporting the significance of trade in Kievan Russia and the evidence urging the importance of agriculture supplement, rather than cancel, each other. Both occupations, then, must be recognized as highly characteristic of the country. But the interrelationship of the two does present certain difficulties. One view holds that the bulk of population supported itself by agriculture, whereas the prince and the upper class were mainly interested in trade. Other specialists stress the evolution in time, suggesting that, while Constantine Porphyrogenitus' account may be a valid guide for the middle of the tenth century, subsequent Kievan development tipped the scales increasingly in favor of agriculture. Furthermore, there is no consensus on the social structure of Kievan Russia which is intimately related to this complicated economic picture.

Kievan exports, as has already been mentioned in the case of Byzantium, consisted primarily of raw materials, in particular furs, wax, and honey, and also, during the earlier part of Kievan history, of slaves. Other items for sale included flax, hemp, tow, burlap, hops, sheepskin, and hides. In return the Kievan Russians purchased such luxury goods as wines, silk fabrics, and objects of art from Byzantium, and spices, precious stones, and various fine fabrics from the Orient. Byzantium also supplied naval stores, while Damask blades and superior horses came from the east. From the west the Kievan Russians imported certain manufactured goods, for instance textiles and glassware, as well as some metals and other items, such as Hungarian horses. Russian merchants went abroad in many directions and foreign traders came in large numbers to Russia, where they established themselves, sometimes as separate communities, in Kiev, Novgorod, Smolensk, Suzdal, and other centers. The newcomers included Germans, Greeks, Armenians, Jews, Volga Bulgars, merchants from the Caucasus, and representatives of still other nationalities. Russian traders themselves were often organized in associations similar to Western guilds, not to mention less formal groupings. Financial transactions and commercial activity in general enjoyed a high development. It should be added that, in addition to exchange for direct consumption, the Kievan Russians engaged in transit trade on a large scale.

Internal trade, although less spectacular than foreign commerce, likewise dated from time immemorial and satisfied important needs. Kiev, Novgorod, and other leading towns served as its main centers, but it also spread widely throughout the land. Some of this domestic trade stemmed from the division of the country into the steppe and the forest, the grain-producing south and the grain-consuming north — a fact of profound significance throughout Russian history — and the resulting prerequisites for exchange.

Commerce led to a wide circulation of money. Originally furs were used as currency in the north and cattle in the south. But, beginning with the reign of St. Vladimir, Kievan minting began with, in particular, silver bars and coins. Foreign money too accumulated in considerable quantities in Kievan Russia.

Agriculture developed both in the steppe and in the forest. In the steppe it acquired an extensive, rather than intensive, character, the peasant cultivating new, good, and easily available land as his old field became less productive. In the forest a more complex process evolved. The trees had to be cut down — a process called *podseka* — and the ground prepared for sowing. Moreover, when the soil became exhausted, a new field could be obtained only after further hard work. Therefore, the *perelog* practice emerged: the cultivator utilized one part of his land and left the other fallow, alternating the two after a number of years. Eventually a regular two-field system grew out of the perelog, with the land divided into annually rotated halves. Toward the end of the Kievan period the three-field system appeared, marking a further important improvement in agriculture and a major increase in the intensity of cultivation: the holding came to be divided into three parts, one of which was sown under a spring grain crop, harvested in the autumn, another under a so-called winter grain crop, sown in the autumn and harvested in the summer, while the third was left fallow; the three parts were rotated in sequence each year. Agricultural implements improved with time; the East Slavs used a wooden plough as early as the eighth and even the seventh century A.D. Wheat formed the bulk of the produce in the south; rye, also barley and oats, in the north. With the evolution of the Kievan state, princes, boyars, and monasteries developed large-scale agriculture. It may be noted in this connection that, in the opinion of some scholars, private ownership of land in Kievan Russia should be dated from the eleventh century at the earliest, while, on indirect evidence, other specialists ascribe the origins of this institution to the tenth or the ninth centuries, and even to a still more distant past.

The East Slavs and later the Kievan Russians engaged in many other occupations as well. Cattle raising has existed since very ancient times in the steppe of southern Russia, and a Byzantine author of the sixth century A.D. wrote about the great number and variety of cattle possessed by the Antes. Forest environment on the other hand led to the acquisition of such skills as carpentry and woodworking in general, as well as apiculture, and the forests also served as enormous game preserves. Hunting for furs, hides, and meat, together with fishing in the many rivers and lakes, developed long before the formation of the state on the Dnieper and continued to be important in Kievan Russia. The Kievan people mined metal, primarily iron, and extracted salt. Their other industries included pottery, metalwork, furriery, tanning, preparation of textiles, and building in stone, not to mention many less widespread arts and crafts practiced at times with a consummate artistry. Rybakov and some other investigators have recently shed much light on this interesting aspect of Kievan life.

Kievan Society

Vernadsky's well-known and perhaps high estimate has placed the population of Kievan Russia in the twelfth century at seven or eight million.

At the top stood the prince and the ever-increasing princely family with its numerous branches, followed by the retainers of the prince, the druzhina. The latter, divided according to their importance and function into the senior and the junior druzhina, together with the local aristocracy formed the upper class of the country, known in the *Russian Justice* and other documents of the time as the *muzhi*. With the evolution of the Kievan state the retainers of the prince and the regional nobility fused into a single group which was to play for centuries an important role in Russian history under the name of the *boyars*. After the muzhi came the *liudi* who can be generally described as the Kievan middle class. Because of the great number and significance of towns in Kievan Russia, this class had considerable relative weight, more than its counterparts in other European countries at the time or in Russia in later periods, even though apparently it diminished with the decline of the state.

The bulk of the population, the so-called *smerdy,* remained agricultural and rural. Kievan peasants, or at least the great majority of them, seem to have been free men at the dawn of Kievan history, and free peasantry remained an important element throughout the evolution of the Kievan state, although bondage gradually increased. Indeed several kinds of bondsmen emerged, their dependence often resulting from their inability to repay the landlord's loan which they had needed to establish or re-establish their economy in troubled times. The slaves occupied the bottom of the social pyramid. It may be added that the principal taxes in Kiev were levied on the "plough" or the "smoke," meaning a household, and were gathered only in the countryside and apparently exclusively from the peasants.

A special group consisted of people connected with the Church, both the clergy who married and had families and the monks and nuns, together with others serving the huge ecclesiastical establishment in many different capacities. The Church operated hospitals and hostels, dispensed charity, and engaged in education, to mention only some of its activities, in addition to performing the fundamental religious functions. Still another classification, that of the *izgoi,* encompassed various displaced social elements, such as freed slaves.

Soviet historians — and, for different reasons, Pavlov-Silvansky and a few other early scholars — consider the evolution of Kievan society in terms of the establishment of a full-fledged feudalism. But the prevalence of money economy in Kievan Russia, the importance of towns and trade, the unrestricted rather than feudal attitude to landed property, the limited and delegated authority of the local magnates, as well as certain other factors, indicate serious weaknesses of any such view and suggest that the issue of feudalism in Russia can be more profitably discussed when dealing with a later period of Russian history.

Kievan Institutions

The chief Kievan political institutions were the office of prince, the *duma* or council of the boyars, and the *veche* or town assembly, which have been linked, respectively, to the autocratic or monarchic, aristocratic, and democratic aspects of the Kievan state. While princes in Kievan Russia proliferated, the one in Kiev retained a special position. From the twelfth century he carried the title of the great, or grand, prince. Princely tasks included military leadership, the rendering of justice, and administration. In war the prince could rely first of all on his own druzhina, and after that on the regiments of important towns, and even, in case of need, on a mass levy. Kievan military history, as has already been mentioned, proved to be unusually rich, and the organization and experience of Kievan armies left a legacy for later ages.

In both justice and administration the prince occupied the key position. Yet he had to work with elected as well as his own appointed officials and in general co-ordinate his efforts with the local elements. To repeat a point made earlier, princely government came relatively late and had to be superimposed on rather well-developed local institutions, notably so in towns. The customary law of the Kievan Russians, known to us best through the *Russian Justice,* a code associated with Iaroslav the Wise, indicates a relatively high development of Kievan society, especially in the fields of trade and finance. It has also attracted attention for the remarkable mildness of its punishments, including a reliance on fines in preference to the death penalty. Canon law came with Christianity from Byzantium. In addition to the direct taxes on the "smoke" and the "plough," state revenue accumulated from judicial fees and fines, as well as from tariffs and other imposts on commerce.

The boyar duma developed, it would seem, from consultations and joint work of the prince and his immediate retinue, the senior druzhina. It expanded with the evolution of Kievan Russia, reflecting the rise of the boyar class and also such developments as the conversion of Russia to Christianity, for the higher clergy found a place in the duma. While it would be quite incorrect to consider the boyar duma as analogous to a parliament — although it might be compared to its immediate predecessor, the *curia regis* — or even to claim for it a definite legal limitation of princely power, it remained an extremely important institution in its customary capacity as the constant adviser and collaborator of the prince. We know of a few occasions when the senior druzhina refused to follow the prince because he had failed to consult it.

Finally, the democratic element in the Kievan state found a certain expression in the veche or town meeting similar to the assemblies of freemen in the barbarian kingdoms of the West. All heads of households could participate in these gatherings, held usually in the market place and called to decide such basic issues as war and peace, emergency legislation, and conflicts with the prince or between princes. The frequently unruly veche practice of decision by unanimity, can be described as an application of direct democracy, ignoring such principles as representation and majority rule. The veche derived from prehistoric times and thus preceded princely authority with which it never became fully co-ordinated. In the Kievan period, the veche in Kiev itself played an especially significant role, but there were other vecha in action all over Russia. In fact, the most far-reaching development of this institution was to occur a little later in Novgorod.

The economic and social development of Kievan Russia, and in particular its institutions, deserve study not only in themselves but also as the heritage of the subsequent periods of Russian history. For example, we shall time and again be concerned with the prince, the duma, and the veche as they evolved differently under changing circumstances in various parts of what used to be the Kievan state.

VI

* * * * * * * * * *

KIEVAN RUSSIA: RELIGION AND CULTURE

> Old customs and beliefs have left but the slightest trace in the documents of the earlier period, and no systematic attempt to record the national epic was made until the middle of the nineteenth century. Moreover, it is generally admitted that the survival of folklore has suffered important modifications in the course of time. Under these conditions any attempt to present a comprehensive survey of Russian cultural developments previous to the seventeenth century meets with insurmountable obstacles and is necessarily incomplete and one-sided. The sources have preserved merely the Christian literature, while the bulk of the national epic has been irretrievably lost. . . . The early literary efforts of native origin were hardly more than slavish imitations of the Byzantine patterns.
>
> FLORINSKY

> Yet, Kievan Russia, like the golden days of childhood, was never dimmed in the memory of the Russian nation. In the pure fountain of her literary works anyone who wills can quench his religious thirst; in her venerable authors he can find his guide through the complexities of the modern world. Kievan Christianity has the same value for the Russian religious mind as Pushkin for the artistic sense: that of a standard, a golden measure, a royal way.
>
> FEDOTOV

THE KIEVAN RUSSIANS, as we have seen, had two religions in succession: paganism and Christianity. The heathen faith of the East Slavs included a deification of the forces of nature, animism in general, and a worship of ancestral spirits. Of the many gods, Perun, the deity of thunder and lightning, claimed special respect. East Slavic paganism lacked elaborate organization or institutional development. Vladimir's efforts to strengthen it proved to be short-lived, and the conversion to Christianity came quickly and relatively painlessly, although we know of some instances of the use of force by the government, and of certain rebellions. But the effectiveness of the baptism of Russia represents a more controversial matter. Some historians, including Golubinsky and other Church historians, have declared that the new religion for centuries retained only a superficial hold on the masses, which remained stubbornly heathen in their true convictions and daily practices, incorporating many of their old superstitions into Christianity. Some scholars speak of *dvoeverie,* meaning a double faith, a term used originally by such religious leaders of the time as St. Theodosius to designate this troublesome phenomenon.

52

Kievan Christianity presents its own problems to the historian. Rich in content and relatively well known, it revealed the tremendous impact of its Byzantine origin and model as well as changes to fit Russian circumstances. The resulting product has been both unduly praised as an organically Russian and generally superior type of Christianity and excessively blamed for its superficiality and derivative nature. In drawing a balance it should be made clear that in certain important respects Kievan Christianity could not even copy that of Byzantium, let alone surpass it. Thus theology and philosophy found little ground on which to grow in Kievan Russia and produced no major fruits. In fact, Kievan religious writings in general closely followed their Byzantine originals and made a minimal independent contribution to the Christian heritage. Mysticism too remained alien to Kievan soil. Yet in another sense Kievan Christianity did grow and develop on its own. It represented, after all, the religion of an entire, newly baptized people with its special attitudes, demands, and ethical and esthetic traditions. This Russification, so to speak, of Byzantine Christianity became gradually apparent in the emergence of Kievan saints, in the creative growth of church architecture and art, in the daily life of the Kievan Orthodox Church, and in its total influence on Russian society and culture.

Kievan saints, who, it might be added, were sometimes canonized with considerable delay and over pronounced opposition from Byzantium, which was apparently unwilling to accord too much luster to the young Russian Church, included, of course, Vladimir the baptizer of Russia, Olga the first Christian ruler of Kiev, and certain princes and religious leaders. Of these princes, Boris and Gleb deserve special notice as reflecting both Kievan politics and in a sense — in their lives and canonization — Kievan mentality. As mentioned before, the brothers, sons of St. Vladimir and his Bulgarian wife, were murdered, allegedly, by their half-brother Sviatopolk, in the fratricidal struggles preceding Iaroslav the Wise's accession to power. They were elevated to sainthood as innocent victims of civil war, but also, at least in the case of Boris, because they preferred death to active participation in the deplorable conflict. St. Anthony, who lived approximately from 982 to 1073, and St. Theodosius, who died in 1074, stand out among the canonized churchmen. Both were monks and both are associated with the establishment of monasticism in Russia and with the creation and organization of the Monastery of the Caves near Kiev. Yet they possessed unlike personalities, represented dissimilar religious types, and left different impacts on Russian Christianity. Anthony, who took his monastic vows on Mount Athos, and whose very name recalled that of the founder of all monasticism, St. Anthony the Great, followed the classic path of asceticism and struggle for the salvation of one's soul. His disciple, Theodosius, while extremely ascetic in his own life, made his major contribution in developing the monastic community and in stressing the social ideal of service to the needy, be they princes who required advice or the hungry poor. The advice, if need be, could become an admonition or even a denunciation. A number of St. Theodosius' writings on different subjects have been preserved. Following the lead and the organizational pattern of the Monastery of the Caves near Kiev, monasteries spread throughout the land, although in Kievan Russia, in contrast to later periods of Russian history, they clustered in and near towns.

At the end of the Kievan period the Russian Church, headed by the metropolitan in Kiev, encompassed sixteen dioceses, a doubling from St. Vladimir's original eight. Two of them had the status of archbishoprics. The Russian metropolitan and Church remained under the jurisdiction of the patriarch of Constantinople. In the days of Kiev only two metropolitans are known to have been Russians, Hilarion in the eleventh century and Clement in the twelfth; especially at first, many bishops also came from Byzantium. The link with Byzantium contributed to the strength and independence of the Russian Church in its relations with the State. But in general the period witnessed a remarkable co-operation, rather than conflict, between Church and State.

As already mentioned, the Church in Kievan Russia obtained vast holdings of land and pre-empted such fields as charity, healing the sick, and sheltering travelers, in addition to its specifically religious functions. Canon law extended not only to those connected with the ecclesiastical establishment but, especially on issues of morality and proper religious observance, to the people at large. The Church also occupied a central position, as we shall see, in Kievan education, literature, and the arts. The over-all impact of religion on Kievan society and life is much more difficult to determine. Kievan Christianity has been described, often in glowing terms, as peculiarly associated with a certain joyousness and affirmation of man and his works; as possessing a powerful cosmic sense and emphasizing the transfiguration of the entire universe, perhaps under the influence of the closeness to nature of the pagan East Slavs; or as expressing in particular the kenotic element in Christianity, that is, the belief in the humble Christ and His sacrifice, in contrast to the Byzantine stress on God the Father, the ruler of heaven and earth. Whatever the validity of these and other similar evaluations of Kievan Christianity — and they seem to contain some truth in spite of the complexity of the issues involved and the limited and at times biased nature of our sources — Christian principles did affect life in Kievan Russia. Their influence can be richly illustrated from Kievan literature and especially its ethical norms, such as the striking concept of the good prince which emerges from Vladimir Monomakh's *Testament,* the constant emphasis on almsgiving in the writings of the period, and the sweeping endorsement of Christian standards of behavior.

Language and Literature

The language of the Russians too was affected by their conversion to Christianity. The emergence among the Russians of a written language, using the Cyrillic alphabet, has been associated with the baptism of the country, the writing itself having been originally devised by St. Cyril and St. Methodius, the apostles to the Slavs, in the second half of the ninth century for the benefit of the Moravians. More precisely, the dominant view today is that St. Cyril invented the older Glagolithic alphabet and that the Cyrillic was a somewhat later development carried out by one of his disciples, probably in Bulgaria. While there exists some evidence, notably in the early treaties with Byzantium and in the fact that these treaties were translated into Slavic, that the Russians had been acquainted with writing before 988, the conversion firmly and permanently established the written language in Russia. To repeat, the liturgy itself, as well as the lesser services of the Church and its other activities, were conducted in Church Slavonic, readily understandable to the people, not in Greek, nor in Latin as in the West. A written literature based on the religious observances grew quickly and before long embraced other fields as well. The language of this Kievan written literature has traditionally been considered to be the same as Church Slavonic, a literary language based on an eastern South Slavic dialect which became the tongue of Slavic Christianity. Recently, however, certain scholars, and especially Obnorsky, have advanced the highly questionable argument that the basic written, as well as spoken, language of Kievan society had been and remained essentially Russian, although it experienced strong Church Slavonic influences. Perhaps it would be best to say that many written works of the Kievan period were written in Church Slavonic, others in Russian — Old Church Slavonic and Old Russian, to be more exact — and still others in a mixture or blend of both. In any case, the Kievan Russians possessed a rather rich and well-developed literary language; one comparison of an eleventh-century Russian translation with the original Byzantine chronicle indicates that the Russian version had the exact equivalents of eighty per cent of the Greek vocabulary. The conversion to Christianity had meant not only an influx of Greek terms, dominant in the sphere of religion and present in many other areas, but also certain borrowings from the Balkan Slavs, notably the Bulgarians, who had accepted Christianity earlier and who helped its dissemination in Russia.

Kievan literature consisted of two sharply different categories: oral creations, and written works linked to particular authors. Although it is highly probable that the great bulk of Kievan folklore has been lost, enough remains to demonstrate its richness and variety. That folklore had developed largely in the immemorial past, and it expanded further to incorporate Kievan experiences. It has been noted, for example, that different Russian wedding songs reflect several distinct stages of social relations: marriage by kidnapping, marriage by purchase, and marriage by consent. Funeral dirges too go very far back in expressing the attitude of the East Slavs toward death. These and other kinds of Russian folk songs often possess outstanding lyrical and generally artistic qualities that have received recognition throughout the world. Kievan folklore also included sayings, proverbs, riddles, and fairy tales of different kinds.

But special interest attaches to the epic poems, the famous *byliny*. They represent one of the several great epic cycles of Western literature, comparable in many ways to the Homeric epic of the Greeks, or to the Serbian epic. The byliny narrate the activities of the bogatyri, the mighty warriors of ancient Russia, who can be divided into two categories: a few senior bogatyri and the more numerous junior ones. Members of the first group, concerning whom little information remains, belong to hoary antiquity, overlap with or even become part of mythology, and seem often to be associated with forces or phenomena of nature. The junior Kievan bogatyri, about whom we possess some four hundred epic songs, reflect Kievan history much better, although their deeds too usually belong to the realm of the fantastic and the miraculous. Typically, they form the entourage of St. Vladimir, at whose court many byliny begin and end, and they fight the deadly enemies of the Russian land. The Khazars, with their Hebrew faith, may appear in the guise of the legendary Zhidovin, the Jew; or Tugor Khan of the Polovtsy may become the dragon Tugarin. The junior bogatyri express the peculiarly Kievan mixture of a certain kind of knighthood, Christianity, and the unremitting struggle against the steppe peoples.

Ilia of Murom, Dobrynia Nikitych, and Alesha Popovich stand out as the favorite heroes of the epic. Ilia of Murom, the mightiest of them and in many respects the most interesting, is depicted as an invalid peasant who only at the age of thirty-three after a miraculous cure started on his great career of defending Kievan Russia against its enemies: his tremendous military exploits do not deprive him of a high moral sense and indeed combine with an unwillingness to fight, except as a last resort. If Ilia of Murom represents the rural masses of Kiev, Dobrynia Nikitych belongs clearly to the upper stratum: his bearing and manners strike a different note than those of the peasant warrior, and in fact he, more than other bogatyri, has links to an actual historical figure, an uncle and associate of St. Vladimir. Alesha Popovich, as the patronymic indicates, comes from the clerical class; his characteristics include bragging, greediness, and a certain shrewdness that often enables him to defeat his opponents by means other than valor. In addition to the great Kievan cycle, we know some Novgorod byliny that will be mentioned later in a discussion of that city-state and a few stray epic poems not fitting into any cycle, as well as the artistically much less valuable historical songs of the Moscow period.

Kievan written literature, as already noted, developed in close association with the conversion of the Russians to Christianity. It contained Church service books, collections of Old Testament narratives, canonical and apocryphal, known as *Palaea* after the Greek word for Old Testament, sermons and other didactic works, hymns, and lives of saints. Among the more prominent pieces one might mention the hymns composed by St. Cyril of Turov; a collection of the lives of the saints of the Monastery of the Caves near Kiev, the so-called *Paterikon;* and the writings of Hilarion, a metropolitan in the reign of Iaroslav the Wise and a leading Kievan intellectual, who has been described by Fedotov as "the best theologian and preacher of all ancient Russia, the Muscovite period included." Hilarion's best-known work, a sermon *On Law and Grace,* begins with a skillful comparison of the law of Moses and the grace of Christ, the Old and the New Testaments, and proceeds to a rhetorical account of the baptism of Russia and a paean of praise to St. Vladimir, the baptizer. It has often been cited as a fine expression of the joyously affirmative spirit of Kievan Christianity.

The chronicles of the period deserve special notice. Although frequently written by monks and reflecting the strong Christian assumptions of Kievan civilization, they belong more with the historical than the religious literature. These early Russian chronicles have been praised by specialists for their historical sense, realism, and richness of detail. They indicate clearly the major problems of Kievan Russia, such as the struggle against the peoples of the steppe and the issue of princely succession. Still more important, they have passed on to us the specific facts of the history of the period. The greatest value attaches to the *Primary Chronicle* — to which we have already made many references — associated especially with two Kievan monks, Nestor and Sylvester, and dating from around 1111. The earliest extant copies of it are the fourteenth-century Laurentian and the fifteenth-century Hypatian. The *Primary Chronicle* forms the basis of all later general Russian chronicles. Regional chronicles, such as those of Novgorod or Vladimir, a number of which survive, also flourished in Kievan Russia.

The secular literature of Kievan Russia included a variety of works ranging from Vladimir Monomakh's remarkable *Testament* to the most famous product of all, *The Lay of the Host of Igor.* The *Lay,* a poetic account of the unsuccessful Russian campaign against the Polovtsy in 1185, written in verse or rhythmic prose, has evoked much admiration and considerable controversy. Although one view, championed by Mazon, more recently Zimin, and some other scholars, holds it to be a modern forgery, the *Lay* has been accepted by Jakobson and most specialists as a genuine, if in cer-

58   KIEVAN RUSSIA

tain respects unique, expression of Kievan genius. Its unknown author apparently had a detailed knowledge of the events that he described, as well as a great poetic talent. The narrative shifts from the campaign and the decisive battle of one of the local Russian princes, Igor and his associates, to Kiev where Grand Prince Sviatoslav learns of the disaster, and to Putivl where Igor's wife Iaroslavna speaks her justly celebrated lament for her lost husband. The story concludes with Igor's escape from his captors and the joy of his return to Russia. The *Lay* is written in magnificent language which reproduces in haunting sounds the clang of battle or the rustle of the steppe; and it also deserves praise for its impressive imagery, its lyricism, the striking treatment of nature — in a sense animate and close to man — and the vividness, power, and passion with which it tells its tale.

Architecture and Other Arts

If Kievan literature divides naturally into the oral or popular and the written, Kievan architecture can be classified on a somewhat parallel basis as wooden or stone. Wooden architecture, like folk poetry, stems from the prehistoric past of the East Slavs. Stone architecture and written literature were both associated with the conversion to Christianity, and both experienced a fundamental Byzantine influence. Yet they should by no means be dismissed for this reason as merely derivative, for, already in the days of Kiev, they had developed creatively in their new environment and produced valuable results. Borrowing, to be sure, forms the very core of cultural history.

Because wood is highly combustible, no wooden structures survive from the Kievan period, but some two dozen of the stone churches of that age have come down to our times. Typically they follow their Byzantine models in their basic form, that of a cross composed of squares or rectangles, and in many other characteristics. But from the beginning they also incorporate such Russian attributes as the preference for several and even many cupolas and, especially in the north, thick walls, small windows, and steep roofs to withstand the inclement weather. The architects of the great churches of the Kievan age came from Byzantium and from other areas of Byzantine or partly Byzantine culture, such as the Slavic lands in the Balkans and certain sections of the Caucasus, but they also included native Russians.

The Cathedral of St. Sophia in Kiev, built in 1037 and the years following, has generally been considered the most splendid surviving monument of Kievan architecture. Modeled after a church in Constantinople and erected by Greek architects, it follows the form of a cross made of squares, with five apses on the eastern or sanctuary side, five naves, and thirteen cupolas. The sumptuous interior of the cathedral contains columns of porphyry, marble, and alabaster, as well as mosaics, frescoes, and other decoration. In Novgorod another majestic and luxurious Cathedral of St. Sophia — a favorite Byzantine dedication of churches to Christ as Wisdom — built by Greeks around 1052, became the center of the life of that city and territory. But still more outstanding from the artistic point of view, according to Grabar, was the St. George Cathedral of the St. George Monastery near Novgorod. Erected by a Russian master, Peter, in 1119–30, this building with its three apses, three cupolas, and unornamented walls of white stone produces an unforgettable impression of grace, majesty, and simplicity.

The architecture of the Kievan period achieved especially striking results in the twelfth and the first half of the thirteenth century in the eastern part of the country, the Vladimir-Suzdal area, which became at that time also the political center of Russia. The churches of that region illustrate well the blending of the native tradition with the Romanesque style of the West together with certain Caucasian and, of course, Byzantine influences. The best remaining examples include the two cathedrals in Vladimir, that of the Assumption of Our Lady, which later became the prototype for the cathedral by the same name in the Moscow Kremlin, and that of St. Dmitrii; the Cathedral of St. George in Iuriev Polskii, with its marked native characteristics; and the church of the Intercession of Our Lady on the Nerl river, near Vladimir, which has often been cited as the highest achievement of ancient Russian architecture. Built in 1166–71 and representing a rectangle with three apses and a single cupola, it has attracted unstinting praise for harmony of design and grace of form and decoration.

Other forms of art also flourished in Kievan Russia, especially in connection with the churches. Mosaics and frescoes richly adorned St. Sophia in Kiev and other cathedrals and churches in the land. Icon-painting too came to Russia with Christianity from Byzantium. Although the Byzantine tradition dominated all these branches of art, and although many masters practicing in Russia came from Byzantium or the Balkans, a Russian school began gradually to emerge. It was to have a great future, especially in icon-painting, in which St. Alipii of the Monastery of the Caves and other Kievan pioneers started what has often been considered the most remarkable artistic development in Russian history. Fine Kievan work in illumination and miniatures in general, as well as in different decorative arts, has also come down to our time. By contrast, because of the negative attitude of the Eastern Church, sculpture proper was banned from the churches, the Russians and other Orthodox peoples being limited to miniature and relief sculpture. Reliefs, however, did develop, reaching the high point in the Cathedral of St. Dmitrii in Vladimir, which has more than a thousand relief pieces, and in the cathedral in Iuriev Polskii. Popular entertainment, combining music and elementary theater, was provided by traveling performers, the *skomorokhi,* whom the church tried continuously to suppress as immoral and as remnants of paganism.

ucation. Concluding Remarks

The scope and level of education in Kievan Russia remain controversial subjects, beclouded by unmeasured praise and excessive blame. On the positive side, it seems obvious that the Kievan culture outlined above could not have developed without an educated layer of society. Moreover, as Kliuchevsky, Chizhevsky, and others have emphasized, Kievan sources, such as the *Primary Chronicle* and Vladimir Monomakh's *Testament,* express a very high regard for learning. As to specific information, we have scattered reports of schools in Kiev and other towns, of monasteries fostering learning and the arts, and of princes who knew foreign languages, collected books, patronized scholars, and generally supported education and culture. Beyond that, recent Soviet discoveries centering on Novgorod indicate a considerable spread of literacy among artisans and other broad layers of townspeople, and even to some extent among the peasants in the countryside. Still it would appear that the bulk of the Kievan population, in particular the rural masses, remained illiterate and ignorant.

Even a brief account of Kievan culture indicates the variety of foreign influences which it experienced and their importance for its evolution. First and foremost stands Byzantium, but it should not obscure other significant contributions. The complexity of the Kievan cultural heritage would become even more apparent had we time to discuss, for example, the links between the Kievan and the Iranian epic, the musical scales of the East Slavs and of certain Turkic tribes, or the development of ornamentation in Kiev with its Scythian, Byzantine, and Islamic motifs. In general, these influences stimulated, rather than stifled, native growth — or even made it possible. Kievan Russia had the good fortune of being situated on the crossroads, not the periphery, of culture.

Perhaps too much emphasis has been placed on the destruction of Kievan civilization and the loss of its unique qualities. True, Kievan Russia, like other societies, went down never to reappear. But it left a rich legacy of social and political institutions, of religion, language, and culture that we shall meet again and again as we study the history of the Russians in the long centuries that followed their brilliant debut on the world scene.

PART III: APPANAGE RUSSIA

V I I

* * * * * * * * * *

APPANAGE RUSSIA: INTRODUCTION

> The grass bends in sorrow, and the tree is bowed down to earth by
> woe. For already, brethren, a cheerless season has set in: already our
> strength has been swallowed up by the wilderness. . . . Victory of
> the princes over the infidels is gone, for now brother said to brother:
> "This is mine, and that is mine also," and the princes began to say of
> little things, "Lo! this is a great matter," and to forge discord against
> themselves. And on all sides the infidels were victoriously invading the
> Russian land.
>
> "THE LAY OF THE HOST OF IGOR"
> (S. CROSS'S TRANSLATION)

THE KIEVAN LEGACY stood the Russians in good stead. It included, as has
already been noted, a uniform religion, a common language and literature,
and, with numerous regional and local modifications, common arts and culture in general. It embraced a similarly rich heritage in the economic, social,
and political fields. While the metropolitan in Kiev headed the Church of the
entire realm, the grand prince, also in Kiev, occupied the seat of the temporal power of the state. Both offices outlived by centuries the society which
had created them and both remained of major significance in Russian history, in spite of a shift in their locale and competition for preference among
different branches of the huge princely clan. In a like manner the concept
of one common "Russian land," so dear to Kievan writers and preachers,
stayed in the Russian consciousness. These bonds of unity proved to be of
decisive importance in the age of division and defeat which followed the
collapse of the Kievan state, in particular during the dark first hundred
years following the Mongol conquest, that is, approximately from the middle of the thirteenth to the middle of the fourteenth century. In that period
the persistence of these bonds ensured the survival of the Russians as a
major people, thus making possible their future historical role. The powerful
Moscow state which finally emerged on the east European plain looked,
and often was, strikingly different from its Kievan predecessor. Yet, for the
historian in any case, Muscovite Russia remains linked to Kievan Russia in
many essential, as well as less essential, ways. And it affirmed and treasured
at least a part of its Kievan inheritance.

The twin terrors of Kievan Russia, internal division and invasion from
abroad, prevailed in the age which followed the collapse of the Kievan state.

63

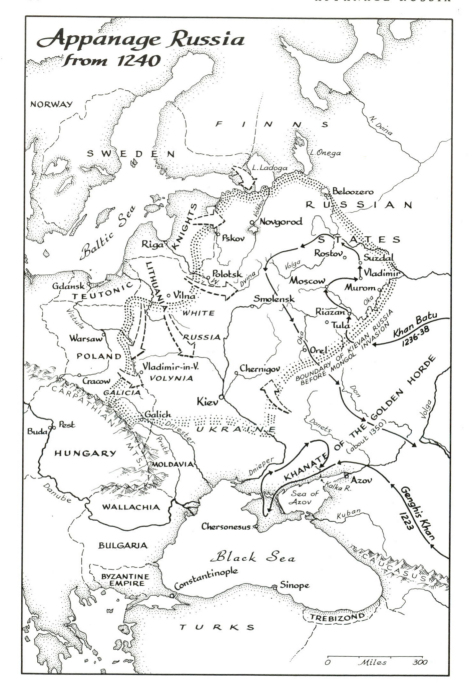

Appanage Russia from 1240

The new period has been named after the *udel,* or appanage, the separate holding of an individual prince. And indeed appanages proliferated at that time. Typically, in his will a ruler would divide his principality among his sons, thus creating with a single act several new political entities. Subdivision followed upon subdivision, destroying the tenuous political unity of the land. As legal historians have emphasized, private law came to the fore at the expense of public law. The political life of the period corresponded to — some would say was determined by — the economic, which was dominated by agriculture and local consumption. Much Kievan trade, and in general a part of the variety and richness of the economy of Kievan Russia, disappeared.

The parceling of Russia in the appanage period combined with population shifts, a political, social, and economic regrouping, and even the emergence of new peoples. These processes began long before the final fall of Kiev, on the whole developing gradually. But their total impact on Russian history may well be considered revolutionary. As the struggle against the inhabitants of the steppe became more exhausting and as the fortunes of Kiev declined, migrants moved from the south to the southwest, the west, the north, and especially the northeast. The final terrible Mongol devastation of Kiev itself and southern Russia only helped to emphasize this development. The areas which gained in relative importance included Galicia and Volynia in the southwest, the Smolensk and Polotsk territories in the west, Novgorod with its huge holdings in the north, as well as the principalities of the northeast, notably Rostov, Suzdal, Vladimir, and eventually Moscow. Population movements led to a colonization of vast lands in the north and northeast of European Russia, although there too the continuity with the Kievan period persisted, for the new expansion radiated from such old Kievan centers as Novgorod, Rostov, and Suzdal.

Of special significance was the linguistic and ethnic differentiation of the Kievan Russians into three peoples: the Great Russians, usually referred to simply as Russians, the Ukrainians, and the Belorussians or White Russians. While certain differences among these groups go far back, the ultimate split was in part caused by the collapse of the Kievan state and the subsequent history of its population, in particular by the fact that southwestern and western Russia, where the Ukrainian and the White Russian nationalities grew, experienced Lithuanian and Polish rule and influences, whereas virtually the entire territory of the Great Russians remained out of their reach.

Appanage Russia was characterized not only by internal division and differentiation but also by external weakness and, indeed, conquest. The Mongol domination over the Russians lasted from 1240 to 1380 or even 1480 depending on whether we include the period of a more or less nominal Mongol rule. But divided Russia became subject to aggression from numerous other quarters as well. As already mentioned, the western and south-western parts of the country fell to the Lithuanians — whose state as we shall see represented in a sense a successor state to that of Kiev — and eventually fell to the Poles. Novgorod to the north had to fight constant wars against the German Knights, the Swedes, and the Norwegians, in addition to the Lithuanians. With the collapse of the Kievan state and the Mongol conquest, Russia lost its important international position, even though a few principalities, such as Novgorod, acted vigorously on the diplomatic stage. In general, in contrast to the earlier history of the country, a relative isolation from the rest of Europe became characteristic of appanage Russia, cut off from many former outside contacts and immersed in local problems and feuds. Isolation, together with political, social, and economic parochialism, led to stagnation and even regression, which can be seen in the political thought, the law, and most, although not all, fields of culture of the period.

The equilibrium of appanage Russia proved to be unstable. Russian economy would not permanently remain at the dead level of local agriculture. Politically, the weak appanage principalities constituted easy prey for the outside aggressor or even for the more able and ambitious in their own midst. Thus Lithuania and Poland obtained the western part of the country. In the rest, several states contended for leadership until the final victory of Moscow over its rivals. The successful Muscovite "gathering of Russia" marked the end of the appanage period and the dawn of a new age. Together with political unification, came economic revival and steady, if slow, cultural progress, the entire development reversing the basic trends of the preceding centuries. The terminal date of the appanage period has been variously set at the accession to the Muscovite throne of Ivan III in 1462, or Basil III in 1505, or Ivan IV, the Terrible, in 1533. For certain reasons of convenience, we shall adopt the last date.

V I I I

* * * * * * * * * *

THE MONGOLS AND RUSSIA

> The churches of God they devastated, and in the holy altars they shed
> much blood. And no one in the town remained alive: all died equally
> and drank the single cup of death. There was no one here to moan,
> or cry — neither father and mother over children, nor children over
> father and mother, neither brother over brother, nor relatives over
> relatives — but all lay together dead. And all this occurred to us for
> our sins.
>
> "THE TALE OF THE RAVAGE OF RIAZAN BY BATU"

> And how could the Mongol influence on Russian life be considerable,
> when the Mongols lived far off, did not mix with the Russians, and
> came to Russia only to gather tribute or as an army, brought in for
> the most part by Russian princes for the princes' own purposes?
> . . . Therefore we can proceed to consider the internal life of Russian
> society in the thirteenth century without paying attention to the fact
> of the Mongol yoke. . . .
>
> PLATONOV

> A convenient method of gauging the extent of Mongol influence on
> Russia is to compare the Russian state and society of the pre-Mongol
> period with those of the post-Mongol era, and in particular to con-
> trast the spirit and institutions of Muscovite Russia with those of
> Russia of the Kievan age. . . . The picture changed completely after
> the Mongol period.
>
> VERNADSKY

THE MONGOLS — or Tatars as they are called in Russian sources * — came
upon the Russians like a bolt from the blue. They appeared suddenly in
1223 in southeastern Russia and smashed the Russians and the Polovtsy in
a battle near the river Kalka, only to vanish into the steppe. But they returned to conquer Russia, in 1237–40, and impose their long rule over it.

Unknown to the Russians, Mongolian-speaking tribes had lived for centuries in the general area of present-day Mongolia, and in the adjoining parts
of Manchuria and Siberia. The Chinese, who watched their northern neigh-

---

* "Tatars" referred originally to a Mongol tribe. But, with the expansion of the
Mongol state, the Tatars of the Russian sources were mostly Turkic, rather than
Mongol, linguistically and ethnically. I am using "Mongol" throughout in preference
to "Tatar."

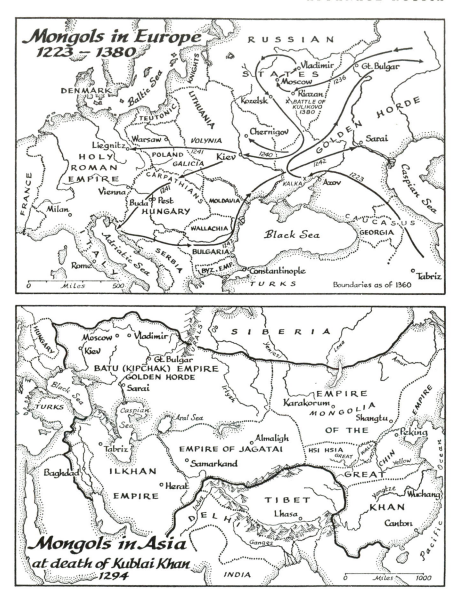

bors closely, left us informative accounts of the Mongols. To quote one
Chinese author:

> . . . they are preoccupied exclusively with their flocks, they roam and
> they possess neither towns, nor walls, neither writing, nor books; they
> conclude all agreements orally. From childhood they practice riding and
> shooting arrows . . . and thus they acquire courage necessary for pillage
> and war. As long as they hope for success, they move back and forth; when
> there is no hope, a timely flight is not considered reprehensible. Religious

rites and legal institutions they know not. . . . They all feed on the meat of the animals which they kill . . . and they dress in their hides and furs. The strongest among them grab the fattest pieces; the old men, on the other hand, eat and drink what is left. They respect only the bravest; old age and feebleness are held in contempt.

While excellent fighters and warlike, the Mongols generally directed their efforts to fratricidal strife among the many tribes, their rivalries skillfully fanned by the Chinese. Only an extraordinary leader managed to unite the Mongols and suddenly transform them into a power of world significance. Temuchin, born probably in 1155 or 1162 and a son of a tribal chief, finally in 1206 after many years of desperate struggle became the head of all the Mongols with the title of Jenghiz Khan. One of the decisively important figures in history, Jenghiz Khan remains something of an enigma. It has been suggested that he was inspired by an urge to avenge the treasonable poisoning of his father and the subsequent humiliation of his family. With time, Jenghiz Khan apparently came to believe in his sweeping divine mission to re-establish justice on earth, and as in the case of some other great leaders, he seems to have had an unshakable conviction in the righteousness of his cause. The new Mongol ruler joined to this determination and sense of mission a remarkable intelligence and outstanding military, diplomatic, and administrative ability.

After uniting the Mongols, Jenghiz Khan subdued other neighboring tribes, and then in 1211 invaded the independent Chin empire in northern China, piercing the Great Wall. What followed has been described as the conquest, in five years, of one hundred million people by one hundred thousand soldiers. The western campaigns of Jenghiz Khan and his generals proved to be still more notable. In spite of bitter resistance, the Mongols smashed the Moslem states of Central Asia and reached the Caucasus. It was through Caucasian passes that they staged a raid into southern Russia to defeat the Russians and the Polovtsy on the river Kalka in 1223. Jenghiz Khan died in 1227. Before his death he had made provisions for succession, dividing the empire among four sons, although its substantial unity was to be preserved by the leadership of one of them with the title of "great khan," a position which fell to the third son, Ugedey. Jenghiz Khan's successors continued his sweeping conquests and spread Mongol rule to Turkestan, Armenia, Georgia, and other parts of the Caucasus, the state of the Volga Bulgars, Russia, Persia, Mesopotamia, Syria, Korea, and all of China. At the time of Kublai Khan, the founder of the Yuan dynasty in China who ruled as Great Khan from 1259 to 1294, Mongol dominion stretched from Poland and the Balkans to the Pacific and from the Arctic Ocean to Turkey, the Persian Gulf, and the southern borders of China. Moreover, the Mongols had penetrated deep into Central Europe, defeating the Poles, the Germans, and the Hungarians in the process.

The remarkable success of Mongol armies can no longer be ascribed, as in the past, to overwhelming numbers. It stemmed rather from the effective strategy of the Mongols, their excellence as highly mobile cavalry, their endurance, and their disciplined and co-ordinated manner of fighting assisted by an organization which in certain ways resembled a modern general staff. These assets acquired particular importance because the military forces of the invaded countries, especially in Europe, were frequently cumbersome, undisciplined and unco-ordinated. Espionage, terrorism, and superior siege equipment, borrowed from China and other lands, have also been cited as factors contributing to the amazing spread of Mongol rule. The Mongols held occupied territories with the aid of such devices as newly built roads, a courier system, and a crude census for purposes of taxation.

Batu, a grandson of Jenghiz Khan and a nephew of Ugedey, who succeeded his father Juchi to the greater part of Juchi's empire, directed the Mongol invasion of Europe. He had some 150,000 or 200,000 troops at his disposal and the veteran Subudey to serve as his chief general. The Mongols crossed the Urals in 1236 to attack first the Volga Bulgars. After that, in 1237, they struck at the Russian eastern principality of Riazan, coming unexpectedly from the north. In the Mongol strategy, the conquest of Russia served to secure their flank for a further major invasion of Europe. The Russian princes proved to be disunited and totally unprepared. Characteristically, many of them stayed to protect their own appanages rather than come to the aid of invaded principalities or make any joint effort. Following the defeat of a Russian army, the town of Riazan was besieged and captured after five days of bitter fighting and its entire population massacred. Next, in the winter of 1237/38, the Mongols attacked the Suzdal territory with its capital of Vladimir, the seat of the grand prince. The sequence of desperate fighting and massacre recurred on a larger scale and at many towns, the grand prince himself and his army perishing in the decisive battle near the river Sit. Thus, in a matter of several months, the Mongols succeeded in conquering the strongest section of the country. Furthermore, they attained their objectives by means of a winter campaign, the Mongol cavalry moving with great speed on frozen rivers — the only successful winter invasion of Russia in history. But a spring thaw that made the terrain virtually impassable forced the Mongols to abandon their advance on Novgorod and retreat to the southern steppe. They spent the next year and a half in preparation for a great campaign as well as in devastating and conquering some additional Russian territories, notably that of Chernigov.

The Mongol assault of 1240, continued in 1241 and the first part of 1242, aimed at more than Russia. In fact, it had been preceded by an order to the king of Hungary to submit to the Mongol rule. The Mongols began by invading the Kievan area proper. Overcoming the stubborn defenders, they took Kiev by storm, exterminated the population, and leveled the city. The same fate befell other towns of the area, whose inhabitants either died or became slaves. After Kiev, the Mongols swept through the southwestern principalities of Galicia and Volynia, laying everything waste. Poland and Hungary came next. One Mongol army defeated the Poles and the Germans, the most important battle taking place at Liegnitz in Silesia in 1241, while another army smashed the Hungarians. Undeterred by the Carpathian mountains, the Mongols occupied the Hungarian plain; their advance guard reached the Adriatic. Whereas campaigning in central Europe presented certain problems to the Mongols, particularly the need to reduce fortresses, many historians believe that only the death of Great Khan Ugedey saved a number of European countries. Concerned with internal Mongol politics, his nephew Batu decided to retrench; and in the spring of 1242 he withdrew his armies to the southern steppe, subjugating Bulgaria, Moldavia, and Wallachia on the way back. Although the Mongols thus retreated to the east, all of Russia, including the northwestern part which escaped direct conquest, remained under their sway.

Batu established his headquarters in the lower Volga area in what became the town of Old Sarai and the capital of the domain known as the Golden Horde. The Golden Horde constituted first a part of the Mongol empire and later, as the central ties weakened, an independent state. A department in Old Sarai, headed by a *daruga,* handled Russian affairs. Mongol dominion over Russia meant that the Russian rulers recognized the Mongol overlordship, that the Mongols, initially the great khan in Mongolia and subsequently the potentate of the Golden Horde, invested the Russian grand prince with his office, and that to be so invested the Russian prince had to journey to the Mongol headquarters and pay humble obeisance to his suzerain. Further, it meant that the Mongols collected tribute from the Russians, at first by means of their own agents and afterwards through the intermediacy of Russian princes. Also, the Russians occasionally had to send military detachments for the Mongol army. We know of several such levies and of Russians serving in the Mongol forces as far away from their homeland as China.

In general, although the Mongols interfered little in Russian life, they maintained an effective control over Russia for almost a century and a half, from 1240 to 1380. In 1380 the prince of Moscow Dmitrii succeeded in defeating the Mongols in a major battle on the field of Kulikovo. Although the Mongols managed to stage a comeback, their invincibility had been destroyed and their rule greatly weakened. Still, another century passed before the Mongol yoke was finally overthrown. Only in 1480 Ivan III of Moscow renounced his, and Russian, allegiance to the khan, and the Mongols failed to challenge his action seriously. Later yet, Russia expanded to absorb the successor states to the Golden Horde: the khanate of Kazan in 1552, of Astrakhan in 1556, and, at long last, that of Crimea in 1783.

The Role of the Mongols in Russian History

Thus, the Mongol rule over the Russians lasted, with a greater or a lesser degree of effectiveness, fôr almost 250 years. There exists, however, no consensus among specialists concerning the role of the Mongols in Russian history. Traditionally Russian historians have paid little attention to the Mongols and their impact on Russia; nevertheless, some of them did stress the destructive and generally negative influence of the Mongol invasion and subjugation. Others virtually dismissed the entire matter as of minor significance in the historical development of their country. While a few earlier scholars held radically different views, a thorough reconsideration of the problem of the Mongols and Russia occurred only in the twentieth century among Russian émigré intellectuals. A new, so-called Eurasian, school proclaimed the fundamental affiliation of Russia with parts of Asia and brought the Mongol period of Russian history to the center of interest. What is more, the Eurasian school interpreted the Mongol impact largely in positive and creative terms. Their views, particularly as expressed in Vernadsky's historical works, have attracted considerable attention.

The destructive and generally negative influence of the Mongols on the course of Russian history has been amply documented. To begin with, the Mongol invasion itself brought wholesale devastation and massacre to Russia. The sources, both Russian and non-Russian, tell, for instance, of a complete extermination of population in such towns as Riazan, Torzhok, and Kozelsk, while in others those who survived the carnage became slaves. A Mongol chronicle states that Batu and his lieutenants destroyed the towns of the Russians and killed or captured all their inhabitants. A papal legate and famous traveler, Archbishop Plano Carpini, who crossed southern Russia in 1245–46 on his way to Mongolia, wrote as follows concerning the Mongol invasion of Russia:

> . . . they went against Russia and enacted a great massacre in the Russian land, they destroyed towns and fortresses and killed people, they besieged Kiev which had been the capital of Russia, and after a long siege they took it and killed the inhabitants of the city; for this reason, when we passed through that land, we found lying in the field countless heads and bones of dead people; for this city had been extremely large and very populous, whereas now it has been reduced to nothing: barely two hundred houses stand there, and those people are held in the harshest slavery.

These and other similar contemporary accounts seem to give a convincing picture of the devastation of the Mongol invasion even if we allow for possible exaggeration.

The Mongol occupation of the southern Russian steppe deprived the

Russians for centuries of much of the best land and contributed to the shift of population, economic activity, and political power to the northeast. It also did much to cut Russia off from Byzantium and in part from the West, and to accentuate the relative isolation of the country typical of the time. It has been suggested that, but for the Mongols, Russia might well have participated in such epochal European developments as the Renaissance and the Reformation. The financial exactions of the Mongols laid a heavy burden on the Russians precisely when their impoverished and dislocated economy was least prepared to bear it. Rebellions against the Mongol taxes led to new repressions and penalties. The entire period, and especially the decades immediately following the Mongol invasion, acquired the character of a grim struggle for survival, with the advanced and elaborate Kievan style of life and ethical and cultural standards in rapid decline. We learn of new cruel punishments established by law, of illiterate princes, of an inability to erect the dome of a stone cathedral, and of other clear signs of cultural regression. Indeed, certain historians have estimated that the Mongol invasion and domination of Russia retarded the development of the country by some 150 or 200 years.

Constructive, positive contributions of the Mongols to Russian history appear, by contrast, very limited. A number of Mongolian words in the fields of administration and finance have entered the Russian language, indicating a degree of influence. For example, the term *iarlyk,* which means in modern Russian a trademark or a customs stamp, comes from a Mongol word signifying a written order of the khan, especially the khan's grant of privileges; similarly the Russian words *denga,* meaning coin, and *dengi,* money, derive from Mongolian. The Mongols did take a census of the Russian population. They have also been credited with affecting the evolution of Russian military forces and tactics, notably as applied to the cavalry. Yet even these restricted Mongol influences have to be qualified. The financial measures of the Mongols together with the census and the Mongol roads added something to the process of centralization in Russia. Yet these taxes had as their aim an exaction of the greatest possible tribute and as such proved to be neither beneficial to the people nor lasting. The invaders replaced the old "smoke" and "plough" taxes with the cruder and simpler head tax, which did not at all take into account one's ability to pay. This innovation disappeared when Russian princes, as intermediaries, took over from the Mongol tax collectors. Thinking simply in terms of pecuniary profit, the Mongols often acted with little wisdom: they sold the position of grand prince to the highest bidder and in the end failed to check in time the rise of Moscow. Rampant corruption further vitiated the financial policy of the Mongols. As to military matters, where the invaders did excel, the fact remains that Russian armies and tactics of the appanage period, based on foot soldiers, evolved directly from those of Kiev, not from the Mongol cavalry. That cavalry, however, was to influence later Muscovite gentry horse formations.

Similarly, the Mongols deserve only limited credit for bringing to Russia the postal service or the practice of keeping women in seclusion in a separate part of the house. A real postal system came to Russia as late as the seventeenth century, and from the West; the Mongols merely resorted to the Kievan practice of obligating the local population to supply horses, carriages, boats, and other aids to communication for the use of officials, although they did implement this practice widely and bequeath several words in the field of transportation to the Russians. The seclusion of women was practiced only in the upper class in Russia; it probably reflected the general insecurity of the time to which the Mongols contributed their part rather than the simple borrowing of a custom from the Mongols. The Mongols themselves, it might be added, acquired this practice late in their history when they adopted the Moslem faith and some customs of conquered peoples.

Turning to the more far-reaching claims made, especially by scholars of the Eurasian school, on behalf of the Mongols and their impact on Russia, one has to proceed with caution. Although numerous and varied, Eurasian arguments usually center on the political role of the Mongols. Typically they present the Muscovite tsar and the Muscovite state as successors to the Mongol khan and the Golden Horde, and emphasize the influence of the Mongols in transforming weak and divided appanage Russia into a powerful, disciplined, and monolithic autocracy. Institutions, legal norms, and the psychology of Muscovite Russia have all been described as a legacy of Jenghiz Khan.

Yet these claims can hardly stand analysis. As already mentioned, the Mongols kept apart from the Russians, limiting their interest in their unwilling subjects to a few items, notably the exaction of tribute. Religion posed a formidable barrier between the two peoples, both at first when the Mongols were still pagan and later when the Golden Horde became Moslem. The Mongols, to repeat a point, were perfectly willing to leave the Russians to their own ways; indeed, they patronized the Orthodox Church.

Perhaps a still greater significance attaches to the fact that the Mongol and the Russian societies bore little resemblance to each other. The Mongols remained nomads in the clan stage of development. Their institutions and laws could in no wise be adopted by a much more complex agricultural society. A comparison of Mongol law, the code of Jenghiz Khan, to the Pskov *Sudebnik,* an example of Russian law of the appanage age, makes the difference abundantly clear. Even the increasing harshness of Russian criminal law of the period should probably be attributed to the conditions of the time rather than to borrowing from the Mongols. Mongol influence on Russia could not parallel the impact of the Arabs on the West, because, to quote Pushkin, the Mongols were "Arabs without Aristotle and algebra" — or other cultural assets.

The Eurasian argument also tends to misrepresent the nature of the Mongol states. Far from having been particularly well organized, efficient or lasting, they turned out to be relatively unstable and short-lived. Thus, in 1260 Kublai Khan built Peking and in 1280 he completed the conquest of southern China, but in 1368 the Mongol dynasty was driven out of China; the Mongol dynasty in Persia lasted only from 1256 to 1344; and the Mongol Central Asiatic state with its capital in Bukhara existed from 1242 until its destruction by Tamerlane in 1370. In the Russian case the dates are rather similar, but the Mongols never established their own dynasty in the country, acting instead merely as overlords of the Russian princes. While the Mongol states lasted, they continued on the whole to be rent by dissensions and wars and to suffer from arbitrariness, corruption, and misrule in general. Not only did the Mongols fail to contribute a superior statecraft, but they had to borrow virtually everything from alphabets to advisers from the conquered peoples to enable their states to exist. As one of these advisers remarked, an empire could be won on horseback, but not ruled from the saddle. True, cruelty, lawlessness, and at times anarchy, in that period characterized also the life of many peoples other than the Mongols, the Russians included. But at least most of these peoples managed eventually to surmount their difficulties and organize effective and lasting states. Not so the Mongols, who, after their sudden and stunning performance on the world scene, receded to the steppe, clan life, and the internecine warfare of Mongolia.

When the Muscovite state emerged, its leaders looked to Byzantium for their high model, and to Kievan Russia for their historical and still meaningful heritage. As to the Mongols, a single attitude toward them pervades all Russian literature: they were a scourge of God sent upon the Russians for their sins. Historians too, whether they studied the growth of serfdom, the rise of the gentry, or the nature of princely power in Muscovite Russia, established significant connections with the Russian past and Russian conditions, not with Mongolia. Even for purposes of analogy, European countries stood much closer to Russia than Mongol states. In fact, from the Atlantic to the Urals absolute monarchies were in the process of replacing feudal division. Therefore, Vernadsky's affirming the importance of the Mongol impact by contrasting Muscovite with Kievan Russia appears to miss the point. There existed many other reasons for changes in Russia; and, needless to say, other countries changed during those centuries without contact with the Mongols.

It is tempting, thus, to return to the older view and to consider the Mongols as of little significance in Russian history. On the other hand, their destructive impact deserves attention. And they, no doubt, contributed something to the general harshness of the age and to the burdensome and exacting nature of the centralizing Muscovite state which emerged out of this painful background. Mongol pressure on Russia and its resources continued after the end of the yoke itself, for one of the authentic legacies of Jenghiz Khan proved to be the successor states to the Golden Horde which kept southeastern Russia under a virtual state of siege and repeatedly taxed the efforts of the entire country.

LORD NOVGOROD THE GREAT

> The Italian municipalities had, in earlier days, given signal proof of
> that force which transforms the city into the state.
>
> BURCKHARDT

> The men of Novgorod showed *Knyaz* * Vsevolod the road. "We do
> not want thee, go whither thou wilt." He went to his father, into
> Russia.
>
> "THE CHRONICLE OF NOVGOROD"
> (R. MICHELL'S AND N. FORBES'S TRANSLATION)

N<small>OVGOROD</small> or, to use its formal name, Lord Novgorod the Great stands out
as one of the most impressive and important states of appanage Russia.
When Kievan might and authority declined and economic and political
weight shifted, Novgorod rose as the capital of northern Russia as well as
the greatest trading center and, indeed, the leading city of the entire country.
Located in a lake area, in the northwestern corner of European Russia, and
serving throughout the appanage period as a great Russian bulwark against
the West, it came to rule enormous lands, stretching east to the Urals and
north to the coast line. Yet, for the historian, the unusual political system of
the principality of Novgorod and its general style of life and culture possess
even greater interest than its size, wealth, and power.

The Historical Evolution of Novgorod

Novgorod was founded not later than the eighth century of our era —
recent excavations and research emphasize its antiquity and its connection
with the Baltic Slavs — and, according to the *Primary Chronicle,* it was
to Novgorod that Riurik came in 862 at the dawn of Russian history.
During the hegemony of Kiev, Novgorod retained a position of high importance. In particular, it served as the northern base of the celebrated
trade route "from the Varangians to the Greeks," and also as a center of
trade between the East and the West by means of the Volga river. The city
seems to have remained outside the regular Kievan princely system of
succession from brother to brother. Instead, it was often ruled by sons of
the grand princes of Kiev who, not infrequently, themselves later ascended
the Kievan throne; although some persons not closely related to the grand

* *Knyaz* means "prince."

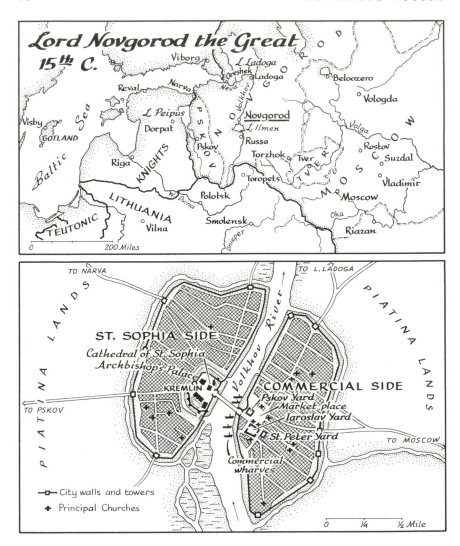

prince also governed in Novgorod on occasion. St. Vladimir, Iaroslav the
Wise, and Vladimir Monomakh's son Mstislav all were at some time
princes of Novgorod. Iaroslav the Wise in particular came to be closely
linked to Novgorod where he ruled for a number of years before his
accession to the Kievan throne; even the *Russian Justice* has been considered by many scholars as belonging to the Novgorodian period of his
activities. And Novgorod repeatedly offered valuable support to the larger
ambitions and claims of its princes, for example, to the same Iaroslav the
Wise in his bitter struggle with Sviatopolk for the Kievan seat.

The evolution of authority and power within Novgorod proved to be
even more significant than the interventions of the Novgorodians on behalf

of their favorite princes. While we know of a few earlier instances when Novgorod refused to accept the prince allotted to the city — in one case advising that the appointee should come only if he had two heads — it is with the famous expulsion of a ruler in 1136 that the Novgorodians embarked upon their peculiar political course. After that date the prince of Novgorod became in essence a hired official of the city with strictly circumscribed authority and prerogatives. His position resembled that of the *podestà* in Italian city-states, and it made some historians refer to Novgorod as a "commercial republic." In 1156 Novgorod obtained virtual independence in religious administration too by seizing the right to elect its own archbishop. To be exact, under the new system the Novgorodian veche selected three candidates for the position of archbishop; next, one of the three was chosen by lot to fill the high office; and, finally, he was elevated to his new ecclesiastical rank by the head of the Russian Church, the metropolitan.

The emergence of Novgorod as an independent principality formed a part of the general process of collapse of the Kievan state accompanied by the appearance of competing regional entities which were frequently mutually hostile. For Novgorod the great rivals were the potentates of the northeast, notably the princes of Suzdal, who controlled the upper reaches of the Volga and thus the Volga trade artery and who — the most important point — could cut the grain supply of Novgorod. Moreover, for centuries vast and distant lands in northeastern Russia remained in contention between the city of Novgorod and the princes of the northeast, at times owing allegiance to both. In 1216 the Novgorodians, led by the dashing prince Mstislav of Toropets, scored a decisive victory over their rivals at Lipitsa. But, although Novgorod also acquitted itself well in subsequent struggles, the troublesome issues remained to be resolved finally only with the destruction of the independence of Novgorod and its absorption into the Muscovite state.

Novgorod's defense of Russian lands from foreign invasions, stemming from its location in the northwestern corner of Russia, might well have had a greater historical significance than its wars against other Russian principalities. The most celebrated chapter of this defense is linked to the name of Prince Alexander, known as Alexander Nevskii, that is, of the Neva, for his victory over the Swedes on the banks of that river. Alexander became the prince of Novgorod and later the grand prince of Russia at a particularly difficult time in the history of his country. Born in 1219 and dying in 1263, Alexander had to face the Mongol invasion and the imposition of the Mongol yoke on Russia, and he also was forced to deal with major assaults on Russia from Europe. These assaults came from the Swedes and the Teutonic Knights, while neighboring Finnish and especially strong Lithuanian tribes applied additional pressure. The German attack was the most ominous: it represented a continuation and extension of the long-term German drive eastward which had already resulted in the Germanization or extermination of many Baltic Slavic and western Lithuanian tribes and which had spread to the Estonian, Latvian, and Lithuanian neighbors of Russia. A forcible conversion of all these peoples to Roman Catholicism, as well as their subjugation and Germanization, constituted the aims of the Teutonic Knights who had begun as a crusading order in the Holy Land and later transferred their activities to the Baltic area.

In the year in which Kiev fell to the Mongols, 1240, Alexander seized the initiative and led the Novgorodians to a victory over the advancing Swedes on the banks of the Neva River. The chronicles tell us that Alexander himself wounded the Swedish commander Birger, who barely escaped capture. In the meantime the Teutonic Knights had begun their systematic attack on northwestern Russian lands in 1239, and they succeeded in 1241 in capturing Pskov. Having defeated the Swedes and settled some differences with the Novgorodians, Alexander Nevskii turned against the new invaders. In short order he managed to drive them back and free Pskov. What is more, he carried warfare into enemy territory. The crucial battle took place on April 5, 1242, on the ice of Lake Chud, or Peipus, in Estonia. It became known in Russian historical tradition as "the massacre on the ice" and has been celebrated in song and story — more recently in Prokofiev's music and Eisenstein's brilliant film *Alexander Nevskii*. The massed force of mailclad and heavily armed German knights and their Finnish allies struck like an enormous battering ram at the Russian lines; the lines sagged but held long enough for Alexander Nevskii to make an enveloping movement with a part of his troops and assail an enemy flank; a complete rout of the Teutonic Knights followed, the spring ice breaking under them to aid their destruction.

Alexander Nevskii's victories were important, but they represented only a single sequence in the continuous struggle of Novgorod against its western and northwestern foes. Two Soviet specialists have calculated that between 1142 and 1446 Novgorod fought the Swedes twenty-six times, the German knights eleven times, the Lithuanians fourteen times, and the Norwegians five times. The German knights then included the Livonian and the Teutonic orders, which merged in 1237.

Relations with the Mongols took a different turn. Although the Mongol invasion failed to reach Novgorod, the principality together with other Russian lands submitted to the khan. In fact, the great warrior Alexander Nevskii himself instituted this policy of co-operation with the Mongols, becoming a favorite of the khan and thus the grand prince of Russia from 1252 until his death in 1263. Alexander Nevskii acted as he did because of a simple and sound reason: he considered resistance to the Mongols hopeless. And it was especially because of his humble submission to the khan and his consequent ability to preserve the principality of Novgorod as well as some other Russian lands from ruin that the Orthodox Church canonized Alexander Nevskii.

Throughout the appanage period Novgorod remained one of the most important Russian principalities. It played a significant role in the rivalry between Moscow and Tver as well as in the struggle between Moscow and Lithuania. As Moscow successfully gathered other Russian lands, the position of Novgorod became increasingly difficult. Finally in 1471 the city surrendered to Ivan III of Moscow. Trouble followed several years later and in 1478 the Muscovites severely suppressed all opposition, exiling many people, and incorporated the city organically into the Moscow state.

Novgorod: Institutions and Way of Life

Novgorod was an impressive city. Its population at the time of its independence numbered more than 30,000. Its location on the river Volkhov in a lake district assisted commerce and communication and supported strong defense. The Volkhov flows from Lake Ilmen to Lake Ladoga, opening the way to the Baltic Sea and trade centers beyond. This complex of waterways represented the northern section of the famed commercial route "from the Varangians to the Greeks," and it also connected well with the Volga and trade routes going east. As to defense, its location and the skill of the Novgorodians made the city virtually inaccessible to the enemy, at least during much of the year. Novgorod reportedly possessed sturdy wooden walls with towers of stone, although recently a fourteenth-century stone wall was discovered. It found further protection in defensive perimeters constructed roughly two and a half, seven, and twelve miles from the city. These defensive lines frequently had monasteries as strong points, and they skillfully utilized the difficult terrain. In particular, the Novgorodians were excellent hydraulic engineers and knew how to divert water against an advancing enemy.

Like other medieval towns, Novgorod suffered from crowding because everyone wanted to dwell within the walls. The rich families and their servants lived in large houses built in solid blocks and the poorer inhabitants used whatever area they could obtain. The Volkhov divided the city into two halves: the commercial side, where the main market was located, and that of St. Sophia. On the St. Sophia side stood, of course, the cathedral itself as well as the ancient kremlin, or citadel, of the city. The Novgorodians enjoyed the advantages of fire protection, streets ingeniously paved with wood, and a wooden water pipe system, the principles of which they had learned from Byzantium.

Local initiative, organization, and autonomy constituted the distinguishing traits of Novgorod. Several block houses in the city composed a street which already had the status of a self-governing unit with its own elected elder. Several streets formed a *sotnia,* that is, a hundred. Hundreds in their turn combined into quarters, or *kontsy,* which totaled five. Each *konets* enjoyed far-reaching autonomy: not only did it govern itself through its own veche and officials, but it also possessed separately a part of the *piatina* lands, a large area outside the city limits and subject to Novgorod. The piatina holdings of a particular konets usually radiated from its city boundary. It should be added that distant Novgorodian territories did not belong to the piatina lands and were managed by the city as a whole. Also, because of the autonomy of the kontsy, formal Novgorodian documents had to be confirmed at times with as many as eight seals: one for each of the five kontsy and three for central authorities.

The chief central official remained the prince, who commanded the army and played a major role in justice and in administration. However, after the popular revolution and the expulsion of 1136, the veche proceeded to impose severe and minute restrictions on his power and activities. We have the precise terms of a number of such contracts between princes and the city, the earliest concluded with Alexander Nevskii's brother Iaroslav in 1265. As in most of these contracts, the prince promised to follow ancient Novgorodian custom in his government, to appoint only Novgorodians as administrators of the city's lands, not to dismiss officials without court action, and not to hold court without the *posadnik,* an elected official, or his delegate to represent the city. He had to establish his headquarters outside the city limits; he and his druzhina could not own land in Novgorod or trade with the Germans; his remuneration as well as his rights to hunt and to fish were all regulated in great detail. Thus, although in the course of time the grand prince of Moscow or at least a member of the Muscovite ruling family came to hold the office of prince in Novgorod, his power there remained quite limited.

The posadnik and the *tysiatskii,* elected by the veche, shared executive duties with the prince and if need be, especially the posadnik, protected the interests of the city from the prince. The posadnik served as the prince's main associate and assistant, who took charge of the administration and the army in the prince's absence. The tysiatskii, or chiliarch, had apparently at least two important functions: he commanded the town regiment or thousand — hence probably his name — and he settled commercial disputes. He has sometimes been regarded as a representative of the common people of Novgorod. The archbishop of Novgorod must also be mentioned. In addition to performing the highest ecclesiastical functions in the principality, he continuously played a leading role in political affairs, presiding over the Council of Notables, advising secular authorities, reconciling antagonistic factions, and sometimes heading Novgorodian embassies abroad.

Truly outstanding was the power of the Novgorodian veche, or town council, which usually met in the main market place. As we have seen, it invited and dismissed the prince, elected the posadnik and the tysiatskii, and determined the selection of the archbishop by electing three candidates for that position. It decided the issues of war and peace, mobilized the army, proclaimed laws, raised taxes, and acted in general as the supreme authority in Novgorod. A permanent chancellery was attached to it. The veche could be called together by the prince, an official, the people, or even a single person, through ringing the veche bell. One might add parenthetically that the removal of the bell by the Muscovites symbolized the end of the independence of Novgorod and of its peculiar constitution. The veche, composed as usual of all free householders, did settle many important matters, but it also frequently bogged down in violent factional quarrels promoted by its practices of direct democracy and unanimity of decision. The Novgorodians won respect as independent and self-reliant people who managed their own affairs. Yet the archbishop made many solemn appearances at the veche in a desperate effort to restore some semblance of order; and a legend grew up that the statue of the pagan god Perun, dumped into the river when the Novgorodians became Christian, reappeared briefly to leave a stick with which the townspeople have belabored one another ever since.

The Council of Notables also rose into prominence in Novgorodian politics, both because the veche could not conduct day-to-day business efficiently and, still more fundamentally, as a reflection of the actual distribution of wealth and power in the principality. Presided over by the archbishop, it included a considerable number of influential boyars, notably present and past holders of the offices of posadnik and tysiatskii, as well as heads of the kontsy and of the hundreds. The Council elaborated the legislative measures discussed or enacted by the veche and could often control the course of Novgorodian politics. It effectively represented the wealthy, so to speak aristocratic, element in the principality.

The judicial system of Novgorod deserves special mention. It exhibited a remarkable degree of elaboration, organization, and complexity, as well as high juridical and humanitarian standards. The prince, the posadnik, the tysiatskii, and the archbishop, all had their particular courts. A system of jurymen, *dokladchiki,* functioned in the high court presided over by the posadnik; the jurymen, ten in number, consisted of one boyar and one commoner from each of the five kontsy. Novgorodian jurisprudence also resorted frequently to mediation: the contending persons were asked to nominate two mediators, and only when the four failed to reach an agreement did court action follow. Judicial combat, after a solemn kissing of the cross, was used to reach the right decision in certain dubious cases. There seem to have been instances of such combat even between women.

Novgorodian punishments remained characteristically mild. Although the death penalty was not unknown, they consisted especially of fines and, on particularly grave occasions, of banishment with the loss of property and possessions which could be pillaged at will by the populace. In contrast to the general practices of the time, torture occupied little, if any, place in the Novgorodian judicial process. Much evidence reflects the high regard for human life characteristic of Novgorod; the *Novgorodian Chronicle* at times refers to a great slaughter when it speaks of the killing of several persons.

Novgorod stood out as a great trading state. It exploited the enormous wealth of northern Russian forests, principally in furs, but also in wax and honey, for export to foreign markets, and it served, as already mentioned, as an intermediary point on extensive trade routes going in several directions. Manufactured goods, certain metals, and other items, such as herring, wine, and beer, were typical imports. Novgorod traded on a large scale with the island of Gotland and with the ports of the Baltic coast line, but its merchandise also reached England, Flanders, and other distant lands. Many merchants, especially from Gotland and Germany, came to Novgorod where they enjoyed autonomy and a privileged position. Yet, the Novgorodians themselves engaged for a long time in active trade — a point which some scholars failed to appreciate. They went to foreign lands and, on the basis of reciprocal treaties, established Novgorodian commercial communities abroad, as attested by the two Russian churches on the island of Gotland and other evidence. It was in the second half of the thirteenth century, with the beginnings of the Hanseatic commercial league of northern European cities and the growth of its special commercial ships vastly superior to the rather simple boats of the Novgorodians, that Novgorod gradually shifted to a strictly passive role in trade.

While merchants, especially prosperous merchants engaged in foreign trade, constituted a very important element in Novgorod, Soviet research has emphasized the significance of landed wealth, together with the close links between the two upper-class groups. In any case, social differentiation in Novgorod increased with time, leading to political antagonisms, reminiscent again of Italian cities and their conflicts between the rich and the poor, the *populo grosso* and the *populo minuto*. Apparently in the fourteenth and fifteenth centuries Novgorod became increasingly an oligarchy, with a few powerful families virtually controlling high offices. Thus, during most of the fourteenth century the posadniki came from only two families.

At the time when social tensions inside Novgorod increased, the city also found it more difficult to hold its sprawling lands together. The huge Novgorodian territories fell roughly into two groups: the piatina area and the more distant semi-colonial possessions in the sparsely populated far north and east. In line with Novgorodian political practice, piatina towns, with their surrounding countryside, received some self-government, although their posadniki and tysiatskie were appointed from Novgorod rather than elected. Gradually decentralization increased, Viatka, in fact, becoming independent in the late twelfth century and Pskov in the middle of the fourteenth. In addition, as has been noted, Novgorod had to struggle continuously for the security and allegiance of many of its territories against the princes of the northeast, who came to be ably represented by the powerful and successful Muscovite rulers.

Moscow finally destroyed Novgorod. The outcome of their conflict had been in a sense predetermined by the fact that Novgorod, in spite of its swollen size, had remained essentially a city-state. Not surprisingly, many historians consider the twelfth and the thirteenth centuries as the golden age of Novgorod, although the principality gained additional territory in the fourteenth and fifteenth centuries. Devoted to its highly specific and particularistic interests, it flourished in the appanage period when it stood out because of its wealth and its strength and when it could utilize the rivalries of its neighbors. Furthermore, by controlling its prince it had escaped subdivision into new appanages. But it proved unable to compete with Moscow in uniting the Russian people. As Moscow gathered Russian lands and as its last serious rival, the Lithuanian state, came to be linked increasingly to Poland and Catholicism, Novgorod lost its freedom of maneuver. Moscow's absorption of the city and its huge holdings in northern Russia represented the same kind of historical logic — with much less bloodshed — that led to the incorporation of southern France into the French state. Social conflicts made their contribution to this end when class differences and antagonism grew in Novgorod. It seems that in the decisive struggle with Moscow the poor of Novgorod preferred Ivan III to their own oligarchical government with its Lithuanian orientation.

Novgorodian culture too developed in an impressive manner. The city had the good fortune to escape Mongol devastation. In contrast to other appanage principalities, it contained sufficient wealth to continue Kievan cultural traditions on a grand scale. And it benefited from its rich contacts with the West. While Russian culture in the appanage period will be discussed in a later chapter, it is appropriate to note here that Novgorod became famous for its church architecture and its icon-painting, as well as for its vigorous and varied literature.

Moreover, it was in Novgorod that a Soviet search uncovered some seven hundred so-called birchbark documents, usually succinct businesslike notes or messages, which suggest a considerable spread of literacy among the general population of that city and area. Novgorodian literature embraced the writings of such archbishops of the city as Moses and Basil, travelogues, in particular accounts of visits to the Holy Land, and extremely useful chronicles, together with an oral tradition which included a special cycle of byliny. The oldest surviving Russian, that is, Church Slavonic, manuscript, the illuminated so-called *Ostromirovo Gospel* of 1056–57, comes from Novgorod. Indeed, as is frequently the case, the culture of Novgorod survived the political downfall of the city to exercise a considerable influence on Moscow and on Russia in general.

Specialists have cited certain characteristics of Novgorodian culture as reflecting the peculiar nature and history of that city-state. The *Chronicle of Novgorod* and other Novgorodian writings express a strong and constant attachment to the city, its streets, buildings, and affairs. Moreover, the whole general tone of Novgorodian literature has been described as strikingly realistic, pragmatic, and businesslike, even when dealing with religious issues. For example, Archbishop Basil adduced the following arguments, among others, to prove that paradise was located on earth rather than in heaven or in imagination: four terrestrial rivers flow from paradise, one of which, the Nile, Basil described with some relish; St. Macarius lived near paradise; St. Efrosimius even visited paradise and brought back to his abbot three apples, while St. Agapius took some bread there; two Novgorodian boats once reached the paradise mountain as they sailed in a distant sea. Together with realism and practicality went energy and bustle, manifested, for example, in constant building — about one hundred stone churches were erected in the city in the last two centuries of its independence. Visitors described the Novgorodians as an extremely vigorous and active people, whose women were equal to men and prominent in the affairs of the city.

The heroes of Novgorodian literature also reflect the life of the city. The main protagonists of the Novgorodian cycle of byliny included the extraordinary businessman and traveler, merchant Sadko, and the irrepressible and irresponsible young giant Basil Buslaev, whose bloody forays against his neighbors could be checked only by his mother. Buslaev's death illustrates well his behavior: given the choice by a skull to jump in one direction and live or to jump in another and perish, he naturally chose the second and cracked his head. Buslaev has been cited as a genuine representative of the free adventurers of Novgorod, who did so much to spread the sway of their city over enormous lands populated both by Russians and by Finnic-speaking and other tribes.

The history of Novgorod, remarkable in itself, attracts further attention as one variant in the evolution of the lands of Kievan Russia after the decline of Kiev. While it is usual to emphasize the peculiar qualities of Novgorod, it is important to realize also that these qualities stemmed directly from the Kievan — and to some extent pre-Kievan — period and represented, sometimes in an accentuated manner, certain salient Kievan characteristics. The urban life and culture of Novgorod, the important position of its middle class, its commerce, and its close contacts with the outside world all link Novgorod to the mainstream of Kievan history. The veche too, of course, had had a significant role in Kievan life and politics. In emphasizing further its authority and functions, the Novgorodians developed one element of the political synthesis of Kievan Russia, the democratic, at the expense of two others, the autocratic and the aristocratic, which, as we shall see, found a more fertile soil in other parts of the country.

Pskov

The democratic political evolution characteristic of Novgorod occurred also in a few other places, especially in another northwestern Russian town, Pskov. Long subject to Novgorod, this extreme Russian outpost became in 1348 a small independent principality with a territory of some 250 by 75 miles. Pskov had a prince whose powers were even more restricted than those of the prince of Novgorod and a veche which in some ways exceeded that of the larger town in importance. Notably, the Pskovian veche, in addition to its other functions, acted as a court for serious crimes. The town had two elected posadniki as well as the elders of the kontsy, but no tysiatskii; and it was subdivided, much like Novgorod, into streets and kontsy. A council of elders also operated in Pskov.

Being much smaller than Novgorod, Pskov experienced less social differentiation and social tension. It has been generally described as more compact, democratic, and peaceful in its inner life than its "big brother." On the other hand, this "little brother" — a title given to Pskov by Novgorod at one point — participated fully in the high development of urban life and culture typical of Novgorod. In fact, Pskovian architects obtained wide renown, while the legal code issued by the Pskovian veche, the celebrated *Sudebnik* of 1397, with supplements until about 1467 — mentioned earlier in contrasting the Russians and the Mongols — represents a most impressive compendium of highly developed Russian medieval law.

Pskov's relations with Moscow differed from those of Novgorod. Never a rival of the Muscovite state, Pskov, on the contrary, constantly needed its help against attacks from the west. Thus it fell naturally and rather peacefully under the influence of Moscow. Yet when the Muscovite state finally incorporated Pskov around 1511, the town, after suffering deportations, lost its special institutions, all of its independence, and in the face of Muscovite taxes and regulations, its commercial and middle-class way of life.

In spite of brilliance and many successes, the historical development of Novgorod and Pskov proved to be, in the long run, abortive.

THE SOUTHWEST AND THE NORTHEAST

> At the end of the twelfth century the Russian land has no effective
> political unity; on the contrary, it possesses several important centers,
> the evolutions of which, up to a certain point, follow different di-
> rections and assume diverse appearances.
>
> MIAKOTIN

WHILE THE HISTORY of Novgorod represented one important variation
on the Kievan theme, two others were provided by the evolutions of the
southwestern and the northeastern Russian lands. As in the case of Novgorod, these areas formed parts of Kievan Russia and participated fully
in its life and culture. In fact, the southwest played an especially important
role in maintaining close links between the Russians of the Kievan period
and the inhabitants of eastern and central Europe; whereas the northeast
gradually replaced Kiev itself as the political and economic center of the
Russian state and also made major contributions to culture, for instance,
through its brilliant school of architecture which we discussed earlier. With
the collapse of the Kievan state and the breakdown of unity among the Russians, the two areas went their separate ways. Like the development of
Novgorod, their independent evolutions stressed certain elements in the
Kievan heritage and minimized others to produce strikingly different, yet
intrinsically related, societies.

The Southwest

The territory inhabited by the Russians directly west and southwest of
the Kiev area was divided into Volynia and Galicia. The larger land,
Volynia, sweeps in a broad belt, west of Kiev, from the foothills of the
Carpathian Mountains into White Russia. The smaller, Galicia, which is
located along the northern slopes of the Carpathians, irrigated by such
rivers as the Prut and the Dniester, and bordered by Hungary and Poland,
represented the furthest southwestern extension of the Kievan state. During
the Kievan period the Russian southwest attracted attention by its international trade, its cities, such as Vladimir-in-Volynia and Galich, as well
as many others, and in general by its active participation in the life and
culture of the times. Vladimir-in-Volynia, it may be remembered, ranked
high as a princely seat, while the entire area was considered among the more
desirable sections of the state. The culture of Volynia and Galicia formed

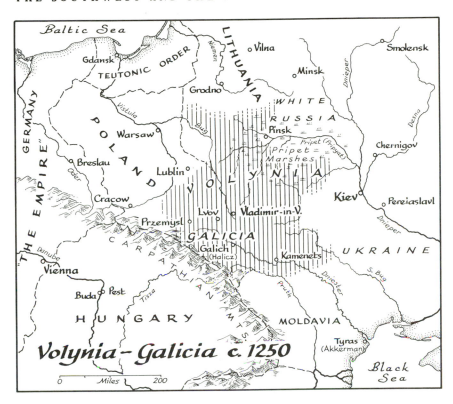

Volynia – Galicia c. 1250

an integral part of Kievan culture, but it experienced particularly strong foreign, especially Western, influences. The two lands played their part in the warfare of the period; Galicia became repeatedly a battleground for the Russians and the Poles.

As Kiev declined, the southwest and several other areas rose in importance. In the second half of the twelfth century Galicia had one of its ablest and most famous rulers, prince Iaroslav Osmomysl, whose obscure appellation has been taken by some scholars to mean "of eight minds" and to denote his wisdom, and whose power was treated with great respect in the *Lay of the Host of Igor*. After Iaroslav Osmomysl's death in 1187, Andrew, king of Hungary, made an abortive effort to reign in the principality, which was followed by the rule of Iaroslav's son Vladimir who died in 1197. After Vladimir, Galicia obtained a strong and celebrated prince, Roman of Volynia, who united the two southwestern Russian lands and also extended his sway to Kiev itself. Roman campaigned successfully against the Hungarians, the Poles, the Lithuanians, and the Polovtsy. Byzantium sought his alliance, while Pope Innocent III offered him a royal crown, which Roman declined. The chronicle of Galicia and Volynia, a work of high literary merit noted for its vivid language, pictured Roman as follows: "he threw himself against the pagans like a lion, he raged like a lynx, he brought destruction like a crocodile, and he swept over their land like an eagle, brave he was like an aurochs." Roman died in a Polish ambush in 1205, leaving behind two small sons, the elder, Daniel, aged four.

After Roman's death, Galicia experienced extremely troubled times marked by a rapid succession of rulers, by civil wars, and by Hungarian and Polish intervention. In contrast, Volynia had a more fortunate history, and from 1221 to 1264 it was ruled by Roman's able son Daniel. Following his complete victory in Volynia, which required a number of years, Daniel turned to Galicia and, by about 1238, brought it under his own and his brother's jurisdiction. Daniel also achieved fame as a creator of cities, such as Lvov, which to an extent replaced Kiev as an emporium of East-West trade, a patron of learning and the arts, and in general as a builder and organizer of the Russian southwest. His rule witnessed, in a sense, the culmination of the *rapprochement* between Russia and the West. In 1253 Daniel accepted a king's crown from the pope — the only such instance in Russian history — while his son Roman married into the Austrian reigning house. Daniel's work, however, received a shattering blow from the Mongol invasion. The Mongols laid waste Galicia and Volynia, and the Russians of the southwest, together with their compatriots elsewhere, had to submit to the overlordship of the khan.

Following the death of Daniel in 1264 and of his worthy son and successor Leo in 1301, who had had more trouble with the Mongols, Volynia and Galicia began to decline. Their decline lasted for almost a century and was interrupted by several rallies, but they were finally absorbed by neighboring states. Volynia gradually became part of the Lithuanian state which will be discussed in a later chapter. Galicia experienced intermittently Polish and Hungarian rule until the final Polish success in 1387. Galicia's political allegiance to Poland contributed greatly to a spread of Catholicism and Polish culture and social influences in the southwestern Russian principality, at least among its upper classes. Over a period of time, Galicia lost in many respects its character as one of the Kievan Russian lands.

The internal development of Volynia and Galicia reflected the exceptional growth and power of the boyars. Ancient and well-established on fertile soil and in prosperous towns, the landed proprietors of the southwest often arrogated to themselves the right to invite and depose princes, and they played the leading role in countless political struggles and intrigues. In a most extraordinary development, one of the boyars, a certain Vladislav, even occupied briefly the princely seat of Galicia in 1210, the only occasion in ancient Russia when a princely seat was held by anyone other than a member of a princely family. Vladimirsky-Budanov and other specialists have noted such remarkable activities of Galician boyars as their direct administration of parts of the principality, in disregard of the prince, and their withdrawal *in corpore* from the princedom in 1226 in their dispute with Prince Mstislav. By contrast with the authority of the boyars, princely authority in Galicia and Volynia represented a later, more superficial, and highly circumscribed phenomenon. Only exceptionally strong rulers, such as Iaroslav Osmomysl, could control the boyars. The veche in Galicia and Volynia, while it did play a role in politics and at least occasionally supported the prince against the boyars, could not consistently curb their power. It should be noted that the rise of the boyars in southwest Russia resembled in many respects the development of the landlord class in adjacent Poland and Hungary.

The Northeast

The northeast, like the southwest, formed an integral part of the Kievan state. Its leading towns, Rostov and Suzdal and some others, belonged with the oldest in Russia. Its princes, deriving from Vladimir Monomakh, participated effectively in twelfth-century Kievan politics. In fact, as we have seen, when Kiev and the Kievan area declined, the political center of the state shifted to the northeast, to the so-called princedom of Vladimir-Suzdal which covered large territories in the central and eastern parts of European Russia. It was a ruler of this principality, Andrew Bogoliubskii, who sacked Kiev in 1169 and, having won the office of the grand prince, transferred its seat to his favorite town of Vladimir in the northeast. His father, the first independent prince of Suzdal and a son of Vladimir Monomakh, the celebrated Iurii Dolgorukii, that is, George of the Long Arm, had already won the grand princedom, but had kept it in Kiev; with Andrew, it shifted definitively to the northeast. Although Andrew Bogoliubskii fell victim to a conspiracy in 1174, his achievements of building up his principality and of emphasizing the authority of the princes of Suzdal in their own territory and in Russia, remained. His work was resumed in 1176 by Andrew's brother Vsevolod, known as Vsevolod III, because he was the third Russian grand prince with that name, or Vsevolod of the Large Nest because of his big family. Vsevolod ruled until his death in 1212 and continued to build towns, fortresses, and churches, to suppress opposition, and to administer the land effectively. At the same time, as grand prince, he made his authority felt all over Russia.

It will be remembered that the Mongol invasion dealt a staggering blow to the Russian northeast. The grand prince at the time, Iurii, a son of Vsevolod III, fell in battle, the Russian armies were smashed, and virtually the entire land was laid waste. Yet, after the Golden Horde established its rule in Russia, the northeastern principalities had some advantages. In contrast to the steppe of the south, they remained outside the zone directly

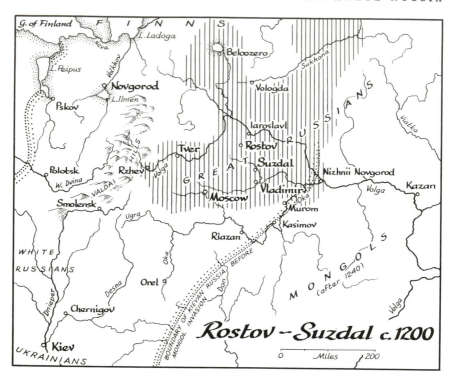

Rostov – Suzdal c.1200

occupied by the Mongols and on the whole could slowly rebuild and develop. A certain distance from the invaders, it might be added, gave them an advantage not only over the old Kievan south, but also over the southeastern principality of Riazan, which evolved along lines parallel to the evolution of the northeast, but experienced greater Mongol pressure. Moreover, the seat of the grand prince stayed in the northeast with the descendants of Vladimir Monomakh. To be more exact, after the death in 1263 of Alexander Nevskii, who, as mentioned earlier, had managed to stabilize relations with the Mongols, the office of the grand prince went successively to his brothers Iaroslav of Tver and Basil of Kostroma and to his sons Dmitrii and Andrew. Following the death of Andrew in 1304, Michael of Tver, Iaroslav's son and Alexander Nevskii's nephew, ruled as grand prince until he was killed by the Mongols at the court of the Golden Horde in 1319. Michael was succeeded by his rival, a grandson of Alexander Nevskii, Iurii, or George, who became the first prince of Moscow to assume the office of grand prince.

But, while the position of the grand prince, with its location in the northeast and the complicated Kievan practice of princely succession, continued as a symbol of Russian unity, in other respects division prevailed. Appanages multiplied as princes divided their holdings among their sons. On the death of Vsevolod III, the Vladimir-Suzdal princedom had already split into five principalities which proceeded to divide further. Ultimately some princes inherited tiny territories, while still others could not be provided for and had to find service with more fortunate members of the family. In the continuous shifting of political boundaries, four leading principalities emerged in the northeast in the first half of the fourteenth century: the princedoms of Vladimir, Rostov, Tver, and Moscow. A proliferation of appanages, characteristic of the northeast, occurred also in the western lands and in the southeastern principality of Riazan, in fact, everywhere in Russia, except in Novgorod which knew how to control its princes.

Whereas the evolution of Novgorod emphasized the role of the veche, and the evolution of Galicia and Volynia that of the boyars, the prince prevailed in the northeast. Although, as already mentioned, Rostov, Suzdal, and some other towns and areas of the northeast formed integral and important parts of Kievan Russia, they generally lay, in contrast to the southwest, in a wilderness of forests with no definite boundaries and hence with great possibilities of expansion to the north and the east. That expansion took place in the late Kievan and especially the appanage periods. This celebrated "colonization" of new lands was considered by S. Soloviev, Kliuchevsky, and some other specialists to have been decisive for subsequent Russian history. The princes played a major role in the expansion by providing economic support, protection, and social organization for the colonists. In the new pioneer society there existed little in the nature of vested interests or established institutions to challenge princely authority. It may be noted that Andrew Bogoliubskii had already transferred his capital from ancient Suzdal to the new town of Vladimir and that his chief political opponents were the boyars from the older sections of his realm. The Mongol invasion and other wars and disasters of the time also contributed to the growth of princely authority, for they shattered the established economic and social order and left it to the prince to rebuild and reorganize devastated territory. The increasing particularism and dependence on local economy, together with the proliferation of appanages, meant that the prince often acted simply as the proprietor of his principality, entering into every detail of its life and worrying little about the distinction between public and private law. With the passage of years, the role of the prince in the northeast came to bear little resemblance to that of the princes in Novgorod or in Galicia.

Kliuchevsky and other Russian historians seem to overstate the case when they select the evolution of the northeast as the authentic Russian development and the true continuation of Kievan history. It would seem better to consider Novgorod, the southwest, and the northeast, all as fully Kievan and as accentuating in their later independent growth certain aspects of the mixed and complicated Kievan society and system: the democratic veche, the aristocratic boyar rule, or the autocratic prince; the city or the countryside; trade or agriculture; contacts with the West or proximity to Asia. Nor should other Russian areas — not included in our brief discussion — such as those of Smolensk, Chernigov, or Riazan, be denied their full share of Kievan inheritance. The more catholic point of view would not minimize the significance of the northeast in Russian history. It was in the northeast, together with the Novgorodian north and certain other adjacent lands, that the Great Russian ethnic type developed, as distinct from the Ukrainian and the White Russian. The conditions of its emergence, all characteristic of the northeast, included the breakdown of Kievan unity and the existence of a more primitive style of life in a forest wilderness inhabited also by Finnic-speaking tribes. And it was a northeastern principality, Moscow, which rose to gather the Russian lands and initiate a new epoch in Russian history.

THE RISE OF MOSCOW

. . . we can imagine the attitude towards the princedom of Moscow
and its prince which developed amidst the northern Russian population. . . . 1) The senior Grand Prince of Moscow came to be regarded as a model ruler-manager, the establisher of peace in the land
and of civil order, and the princedom of Moscow as the starting point
of a new system of social relations, the first fruit of which was precisely the establishment of a greater degree of internal peace and
external security. 2) The senior Grand Prince of Moscow came to
be regarded as the leader of the Russian people in its struggle against
foreign enemies, and Moscow as the instrument of the first popular
successes over infidel Lithuania and the heathen "devourers of raw
flesh," the Mongols. 3) Finally, in the Moscow prince northern Russia
became accustomed to see the eldest son of the Russian church, the
closest friend and collaborator of the chief Russian hierarch; and it
came to consider Moscow as a city on which rests a special blessing
of the greatest saint of the Russian land, and to which are linked the
religious-moral interests of the entire Orthodox Russian people. Such
significance was achieved, by the middle of the fifteenth century, by
the appanage princeling from the banks of the Moscow River, who,
a century and a half earlier, had acted as a minor plunderer, lying
around a corner in ambush for his neighbors.

KLIUCHEVSKY

The unification of Great Russia took place through a destruction of
all local, independent political forces, in favor of the single authority
of the Grand Prince. But these forces, doomed by historical circumstances, were the bearers of "antiquity and tradition," of the customary-
legal foundations of Great Russian life. Their fall weakened its firm
traditions. To create a new system of life on the ruins of the old
became a task of the authority of the Grand Prince which sought not
only unity, but also complete freedom in ordering the forces and the
resources of the land. The single rule of Moscow led to Muscovite
autocracy.

PRESNIAKOV

THE NAME MOSCOW first appears in a chronicle under the year 1147, when
Iurii Dolgorukii, a prince of Suzdal mentioned in the preceding chapter,
sent an invitation to his ally Prince Sviatoslav of the eastern Ukrainian
principality of Novgorod-Seversk: "Come to me, brother, to Moscow."
And in Moscow, Iurii feasted Sviatoslav. Under the year 1156, the chron-
icler notes that Grand Prince Iurii Dolgorukii "laid the foundations of the
town of Moscow," meaning — as on other such occasions — that he built
the city wall. Moscow as a town is mentioned next under 1177 when Gleb,

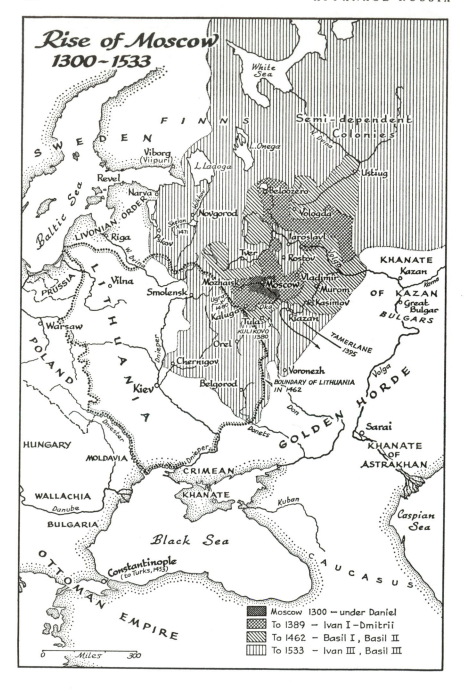

Rise of Moscow 1300 ~ 1533

White Sea

SWEDEN

FINNS

Semi-dependent Colonies

S. Dvina

L. Onega

Viborg (Viipuri)

L. Ladoga

Ustiug

Revel

Volkhov

Beloozero

Narva

Novgorod

Vologda

Shelon 1471

Baltic Sea

LIVONIAN ORDER

Riga

W. Dvina

Pskov

Tver

Yaroslavl

Rostov

KHANATE

Kazan

Kama

PRUSSIA

LITHUANIA

Vilna

Smolensk

Mozhaisk

Moscow

Vladimir

Murom

OF KAZAN

Great Bulgar

BULGARS

Ugra 1480

Kaluga

Oka

Kasimov

Warsaw

Tula

Riazan

POLAND

KULIKOVO 1380

Orel

TAMERLANE 1395

Dnieper

Chernigov

Voronezh

Volga

Kiev

Belgorod

BOUNDARY OF LITHUANIA IN 1462

Dniester

Don

Donets

GOLDEN HORDE

Sarai

HUNGARY

MOLDAVIA

Dnieper

CRIMEAN

KHANATE OF ASTRAKHAN

WALLACHIA

KHANATE

Kuban

Caspian Sea

Danube

BULGARIA

Black Sea

CAUCASUS

OTTOMAN EMPIRE

Constantinople (to Turks, 1453)

0   Miles   300

	Moscow 1300 — under Daniel
	To 1389 — Ivan I–Dmitrii
	To 1462 — Basil I, Basil II
	To 1533 — Ivan III, Basil III

Prince of Riazan, "came upon Moscow and burned the entire town and the villages." It would seem, then, that Moscow originated as a princely village or settlement prior to 1147, and that about the middle of the twelfth century it became a walled center, that is, a town. Moscow was located in Suzdal territory, close to the borders of the principalities of Novgorod-Seversk and Riazan.

The Rise of Moscow to the Reign of Ivan III

We know little of the early Muscovite princes, who changed frequently and apparently considered their small and insignificant appanage merely as a stepping stone to a better position, although one might mention at least one Vladimir who was one of the younger sons of Vsevolod III and probably the first prince of Moscow in the early thirteenth century, and another Vladimir who perished when Moscow was destroyed by the Mongols in 1237. It was with Daniel, the youngest son of Alexander Nevskii, who became the ruler of Moscow in the second half of the thirteenth century that Moscow acquired a separate family of princes who stayed in their appanage and devoted themselves to its development. Daniel concentrated his efforts both on building up his small principality and on extending it along the flow of the Moscow river, of which he controlled originally only the middle course. Daniel succeeded in seizing the mouth of the river and its lower course from one of the Riazan princes; he also had the good fortune of inheriting an appanage from a childless ruler.

Daniel's son Iurii, or George, who succeeded him in 1303, attacked another neighbor, the prince of Mozhaisk, and by annexing his territory finally established Muscovite control over the entire flow of the Moscow river. After that he turned to a much more ambitious undertaking: a struggle with Grand Prince Michael of Tver for leadership in Russia. The rivalry between Moscow and Tver was to continue for almost two centuries, determine in large part which principality would unite the Russian people, and also add much drama and violence to the appanage period. In 1317 or 1318 Iurii married a sister of the khan of the Golden Horde, the bride having become Orthodox, and received from the khan the appointment as grand prince. During the resulting campaign against Tver, the Muscovite army suffered a crushing defeat, and, although Iurii escaped, his wife fell prisoner. When she died in captivity, Iurii accused Michael of poisoning her. The Tver prince had to appear at the court of the Golden Horde, where he was judged, condemned, and executed. In consequence, Iurii was reaffirmed in 1319 as grand prince. Yet by 1322 the khan had made Michael's eldest son, Dmitrii, grand prince. Iurii accepted this decision, but apparently continued his intrigues, traveling in 1324 to the Golden Horde. There, in 1325, he was met and dispatched on the spot by Dmitrii, who was in turn killed by the Mongols. Dmitrii's younger brother, Alexander of Tver, became grand prince. However, he too soon ran into trouble with the Mongols. In 1327 a punitive Mongol expedition, aided by Muscovite troops, devastated Tver, although Alexander escaped to Pskov and eventually to Lithuania. In 1337 Alexander was allowed to return as prince of Tver, but in 1338 he was ordered to appear at the court of the Golden Horde and was there executed.

Following the devastation of Tver and Alexander's flight, Iurii's younger brother Ivan Kalita, Prince of Moscow, obtained the position of grand prince, which he held from 1328, or according to another opinion from 1332, until his death in 1341. Ivan Kalita means "John the Moneybag," and Ivan I remains the prototype of provident Moscow princes with their financial and administrative talents. Always careful to cultivate the Golden Horde, he not only retained the office of grand prince, but also received the commission of gathering tribute for the khan from other Russian princes. He used his increasing revenue to purchase more land: both entire appanages from bankrupt rulers and separate villages. The princedom of Vladimir, which he held as grand prince, he simply added to his own principality, keeping the capital in Moscow. He ransomed Russian prisoners from the Mongols to settle them on Muscovite lands. All in all, Ivan Kalita managed to increase the territory of his princedom severalfold.

It was also in Ivan Kalita's reign that Moscow became the religious capital of Russia. After the collapse of Kiev, and in line with the general breakup of unity in the land, no ecclesiastical center immediately emerged to replace Kiev, "the cradle of Christianity in Russia." In 1326 the head of the Russian Church, Metropolitan Peter, died while staying in Moscow. He came to be worshipped as a saint and canonized, his shrine bringing a measure of sanctity to Moscow. Moreover, in 1328 Ivan Kalita persuaded Peter's successor, Theognost, to settle in Moscow. From that time on, the metropolitans "of Kiev and all Russia" — a title which they retained until the mid-fifteenth century — added immeasurably to the importance and prestige of the upstart principality and its rulers. Indeed, the presence of the metropolitan not only made Moscow the spiritual center of Russia, but, as we shall see, it also proved time and again to be helpful to the princedom in diverse material matters.

Following the passing of Ivan Kalita in 1341, his son Simeon, surnamed the Proud, was confirmed as grand prince by the khan of the Golden Horde. Simeon's appellation, his references to himself as prince "of all Russia," and his entire bearing indicated the new significance of Moscow. In addition to emphasizing his authority over other Russian rulers, Simeon the Proud continued his predecessor's work of enlarging the Muscovite domain proper. He died in 1353 at the age of thirty-six, apparently of the plague which had been devastating most of Europe. In his testament Simeon the Proud urged his heirs to obey a remarkable Russian cleric, Alexis, who was to become one of the most celebrated Muscovite metropolitans.

Alexis, in fact, proceeded to play a leading role in the affairs of the Muscovite state both during the reign of Simeon the Proud's weak brother and successor, Ivan the Meek, which lasted from 1353 to 1359, and during the minority of Ivan's son Grand Prince Dmitrii. Besides overseeing the management of affairs in Moscow and treating with other Russian princes, the metropolitan traveled repeatedly to the Golden Horde to deal with the Mongols. Alexis's wise leadership of Church and State contributed to his enshrinement as one of the leading figures in the Muscovite pantheon of saints. During Ivan II's reign, beginning with 1357, civil strife erupted in the Golden Horde: no less than twenty rulers were to change in bloody struggle in the next twenty years. Yet, if Mongol power declined, that of Lithuania, led by Olgerd, grew; and the Moscow princes had to turn increasing attention to the defense of their western frontier.

Ivan the Meek's death resulted in a contest for the office of grand prince, with Prince Dmitrii of Suzdal and Ivan's nine-year-old son Dmitrii as the protagonists. In a sense, the new crisis represented a revival of old Kievan political strife between "uncles'" and "nephews": Dmitrii of Suzdal, who, as well as Dmitrii of Moscow, was descended directly from Vsevolod III, was a generation older than the Muscovite prince and claimed seniority over him. Rapidly changing Mongol authorities endorsed both candidates. The rally of the people of Moscow behind their boy-ruler and the principle of direct succession from father to son carried the day: Dmitrii of Suzdal abandoned his headquarters in Vladimir without a fight, and Ivan the Meek's son became firmly established as the Russian grand prince. The Kievan system of succession failed to find sufficient support in the northeast.

Grand Prince Dmitrii, known as Dmitrii Donskoi, that is, of the Don, after his celebrated victory over the Mongols near that river, reigned in Moscow for three decades until his death in 1389. The early part of his reign, with Metropolitan Alexis playing a major role in the government, saw a continuing growth of Muscovite territory, while in Moscow itself in 1367 stone walls replaced wooden walls in the Kremlin. It also witnessed a bitter struggle against Tver supported by Lithuania. Indeed Prince Michael of Tver obtained from the Golden Horde the title of grand prince and, together with the Lithuanians, tried to destroy his Muscovite rival. Twice, in 1368 and 1372, Olgerd of Lithuania reached Moscow and devastated its environs, although he could not capture the fortified town itself. Dmitrii managed to blunt the Lithuanian offensive and make peace with Lithuania, after which he defeated Tver and made Michael recognize him as grand prince. Muscovite troops also scored victories over Riazan and over the Volga Bulgars, who paid tribute to the Golden Horde.

But Dmitrii's fame rests on his victorious war with the Golden Horde itself. As Moscow grew and as civil strife swept through the Golden Horde, the Mongol hegemony in Russia experienced its first serious challenge since the time of the invasion. We have seen that Dmitrii had successfully defied the Mongol decision to make Michael of Tver grand prince and had defeated the Volga Bulgars, whose principality was a vassal state of the Golden Horde. A series of incidents and clashes involving the Russians and the Mongols culminated, in 1378, in Dmitrii's victory over a Mongol army on the banks of the Vozha river. Clearly the Mongols had either to reassert their mastery over Moscow or give up their dominion in Russia. A period of relative stability in the Golden Horde enabled the Mongol military leader and strong man, Mamai, to mount a major effort against Dmitrii.

The Mongols made an alliance with Lithuania, and Mamai set out with some 200,000 troops to meet in the upper Don area with forces of Grand Prince Jagiello of Lithuania for a joint invasion of Muscovite lands. Dmitrii, however, decided to seize the initiative and crossed the Don with an army of about 150,000 men, seeking to engage the Mongols before the Lithuanians arrived. The decisive battle, known as the battle of Kulikovo field, was fought on the eighth of September 1380 where the Nepriadva river flows into the Don, on a hilly terrain intersected by streams which the Russians selected to limit the effectiveness of the Mongol cavalry. The terrain was such that the Mongols could not simply envelop Russian positions, but had to break through them. Fighting of desperate ferocity — Dmitrii himself, according to one source, was knocked unconscious in combat and found after the battle in a pile of dead bodies — ended in a complete rout of Mamai's army when the last Russian reserve came out of ambush in a forest upon the exhausted and unsuspecting Mongols. Jagiello, whose Lithuanian forces failed to reach Kulikovo by some two days, chose not to fight Dmitrii alone and turned back. The great victory of the Russians laid to rest the belief in Mongol invincibility. What is more, the new victor of the Don rose suddenly as the champion of all the Russians against the hated Mongol oppressors. While certain important Russian rulers failed to support Dmitrii, and those of Riazan even negotiated with the Mongols, some twenty princes rallied against the common enemy in an undertaking blessed by the Church and bearing some marks of a crusade. The logic of events pointed beyond the developments of 1380 to a new role in Russian history for both the principality and prince of Moscow.

Nevertheless, the years following the great victory at Kulikovo saw a reversal of its results. In fact, only two years later, in 1382, the Mongols came back, led this time by the able Khan Tokhtamysh. While the surprised Dmitrii was in the north gathering an army, they besieged Moscow and, after assaults failed, managed to enter the city by a ruse: Tokhtamysh swore that he had decided to stop the fighting and that he and his small party wanted to be allowed within the walls merely to satisfy their curiosity; once inside, the Mongols charged their hosts and, by seizing a gate, obtained reinforcements and hence control of Moscow, which they sacked and burned. Although Tokhtamysh retreated, with an enormous booty, rather than face Dmitrii's army, the capital and many of the lands of the principality were desolated and its resources virtually exhausted. Dmitrii, therefore, had to accept the overlordship of the Mongol khan, who in return confirmed him as the Russian grand prince. Still, after Kulikovo, the Mongol grip on Russia lacked its former firmness. Dmitrii Donskoi spent the last years of his reign in strengthening his authority among Russian princes, especially those of Tver and Riazan, and in assisting the rebuilding and economic recovery of his lands.

When Dmitrii Donskoi died in 1389 at the age of thirty-nine, his son Vasilii, or Basil, became grand prince without challenge either in Russia or in the Golden Horde. Basil I's long reign, from 1389 until his death in 1425, deserves attention for a number of reasons. The cautious and intelligent ruler continued very successfully the traditional policy of the Muscovite princes of enlarging their own principality and of making its welfare their first concern. Thus, Basil I acquired several new appanages as well as a number of individual towns with their surrounding areas. Also he waged a continuous struggle against Lithuania for western Russian lands. Although the warlike Grand Prince Vitovt of Lithuania scored some victories over his Russian son-in-law, Basil's persistent efforts led to a military and political deadlock in much of the contested area. It might be noted that, after the conclusion of a treaty with Lithuania in 1408, a number of appanage princes in the western borderlands switched their allegiance from Lithuania to Moscow.

Relations with the East presented as many problems as relations with the West. In 1395 Moscow barely escaped invasion by the army of one of the greatest conquerors of history, Tamerlane, who had spread his rule through the Middle East and the Caucasus and in 1391 had smashed Tokhtamysh. Tamerlane's forces actually devastated Riazan and advanced upon Moscow, only to turn back to the steppe before reaching the Oka river. Around 1400 Muscovite troops laid waste the land of the Volga Bulgars, capturing their capital Great Bulgar and other towns. In 1408 the Golden Horde, pretending to be staging a campaign against Lithuania, suddenly mounted a major assault on Moscow to punish Basil I for not paying tribute and for generally disobeying and disregarding his overlord. The Mongols devastated the principality, although they could not capture the city of Moscow itself. In the later part of his reign, Basil I, preoccupied by his struggle with Lithuania and Tver, maintained good relations with the khan and sent him "gifts."

The death of Basil I in 1425 led to the only war of succession in the history of the principality of Moscow. The protagonists in the protracted struggle were Basil I's son Basil II, who succeeded his father at the age of ten, and Basil II's uncle Prince Iurii, who died in 1434 but whose cause was taken over by his sons, Basil the Squint-eyed and Dmitrii Shemiaka. Prince Iurii claimed seniority over his nephew, and he represented, in some sense, a feudal reaction against the growing power of the grand princes of Moscow and their centralizing activities. By 1448, after several reversals of fortune and much bloodshed and cruelty — which included the blinding of both Basil the Squint-eyed and of Basil II himself, henceforth known as Basil the Blind — the Muscovite prince had prevailed. Dmitrii Shemiaka's final rebellion was suppressed in 1450. Indeed, having obtained sufficient support from the boyars and the people of Moscow, Basil II managed, although at a very heavy cost, not only to defeat his rivals but also to expand his principality at the expense of Basil the Squint-eyed and Dmitrii Shemiaka and also of some other appanage princes.

Relations with the Mongols continued to be turbulent as the Golden Horde began to break up and Moscow asserted its independence. In 1445 Basil II was badly wounded and captured in a battle with dissident Mongol leaders, although soon he regained his freedom for a large ransom. The year 1452 marked a new development: a Mongol prince of the ruling family accepted Russian suzerainty when the princedom of Kasimov was established. Basil II had taken into his service Mongol nobles with their followers fleeing from the Golden Horde, and he rewarded one of them, Kasim, a descendant of Jenghiz Khan, with the principality for his important assistance in the struggle against Dmitrii Shemiaka. The creation of this Mongol princedom subject to the grand prince of Moscow was only one indication of the decline of Mongol power. Still more significant was the division of the vast lands held directly by the Golden Horde, with the Crimean khanate separating itself in 1430, that of Kazan in 1436, and that of Astrakhan in 1466 during the reign of Basil II's successor, Ivan III. In 1475 the Crimean state recognized Ottoman suzerainty, with Turkish troops occupying several key positions on the northern shore of the Black Sea. Of course, the khans of the Golden Horde tried to stem the tide and, among other things, to bring their Russian vassal back to obedience. Khan Ahmad directed three campaigns against Moscow, in 1451, 1455, and 1461, but failed to obtain decisive results. For practical purposes, Moscow can be considered as independent of the Mongols after 1452 at least, although the formal and final abrogation of the yoke came only in 1480. In fact, Vernadsky regards the establishment of the principality of Kasimov as a decisive turning point in the relations between the forest and the steppe and thus in what is, to him, the basic rhythm of Russian history.

Basil II's long reign from 1425 to 1462 also witnessed important events in Europe which were to influence Russian history profoundly, although they did not carry an immediate political impact like that implicit in the break-up of the Golden Horde. At the Council of Florence in 1439, with Byzantium struggling against the Turks for its existence and hoping to obtain help from the West, the Greek clergy signed an abortive agreement with Rome, recognizing papal supremacy. The Russian metropolitan, Isidore, a Greek, participated in the Council of Florence and, upon his return to Moscow, proclaimed its results during a solemn service and read a prayer for the pope. After the service he was arrested on orders of the grand prince and imprisoned in a monastery, from which he escaped before long to the West. A council of Russian bishops in 1443 condemned the Church union, deposed Isidore, and elected Archbishop Jonas metropolitan. The administrative dependence of the Russian Church on the Byzantine came to an end. Furthermore, many Russians remained suspicious of the Greeks even after they repudiated the very short-lived Union of Florence. Then in 1453 Constantinople fell to the Turks, who proceeded to acquire complete control of the Balkan peninsula and of what used to be the Byzantine Empire. As we know, it was with Byzantium and the Balkan Slavs that ancient Russia had its most important religious and cultural ties, in the appanage period as well as in the days of Kiev. The success of the Turks contributed greatly to a weakening of these ties and, therefore, to a more complete isolation of Russia. As we shall see, it also strengthened Muscovite xenophobia and self-importance and various teachings based on these attitudes. It should be noted that this boost to Muscovite parochialism occurred at the very time when the northeastern Russian princedom was being transformed into a major state that was bound to play an important role in international relations and was in need of Western knowledge.

The Reigns of Ivan III and Basil III

The long reign of Ivan III, which extended from 1462 to 1505, has generally been considered, together with the following reign of Basil III, as the termination of the appanage period and the beginning of a new age in Russian history, that of Muscovite Russia. These two reigns provide a fitting climax to the story of the rise of Moscow. Ivan III's predecessors had already increased the territory of their principality from less than 600 square miles at the time of Ivan Kalita to 15,000 toward the end of Basil II's reign. But it remained for Ivan III to absorb such old rivals as Novgorod and Tver and to establish virtually a single rule in what used to be appanage Russia. Also, it was Ivan III who, as the conclusion to the developments described earlier in this chapter, successfully asserted full Russian independence from the Mongols. And it was in his reign that the position and authority of the grand prince of Moscow, continuing their long-term rise, acquired attributes of majesty and formality unknown in the appanage period. Ivan III, also called Ivan the Great, suited his important role well: while sources differ concerning certain traits of his character, the general impression remains of a mighty figure combining practical abilities of an appanage prince with unusual statesmanship and vision. Although only twenty-two years old at the time of Basil II's death, the new grand prince was fully prepared to succeed him, having already acted for several years as his blind father's chief assistant and even co-ruler.

Under Ivan III "the gathering of Russia" proceeded apace. The following catalogue of events might give some indication of the nature and diversity of the process. In 1463 — or about a decade later according to Cherepnin — Ivan III purchased the patrimony of the appanage princes of Iaroslavl, and in 1474 the remaining half of the town of Rostov. In 1472 he inherited an appanage, the town of Dmitrov, from his childless brother Iurii; and in the same year he conquered the distant northeastern land of Perm, inhabited by a Finnic-speaking people and formerly under the vague suzerainty of Novgorod. In 1481 the Muscovite grand prince obtained another appanage after the death of another brother, Andrew the Little. In 1485 he forced Prince Michael of Vereia to bequeath to him Michael's principality, bypassing Michael's son, who had chosen to serve Lithuania. In 1489 he annexed Viatka, a northern veche-ruled state founded by emigrants from Novgorod. And in 1493 Ivan III seized the town of Uglich from his brother, Andrew the Big, and imprisoned Andrew for failing to carry out his instructions to march with an army to the Oka river, against the Mongols. Around 1500 the Muscovite grand prince inherited, from Prince Ivan of Riazan, half of his principality and was appointed warden of the other half bequeathed to Ivan of Riazan's young son.

Ivan III's most famous acquisitions, however, were Novgorod and Tver. Novgorod, which we discussed in an earlier chapter, collapsed because of both the Muscovite preponderance of strength and its own internal weaknesses. After the treaty of 1456 imposed by Basil II on Novgorod, the boyar party in the city — led by the Boretsky family which included Martha the celebrated widow of a posadnik — turned to Lithuania as its last hope. The common people of Novgorod, on the other hand, apparently had little liking either for Lithuania or for their own boyars. In the crucial campaign of 1471 Novgorodian troops made a poor showing, the archbishop's regiment refusing outright to fight against the grand prince of Moscow. After winning the decisive battle fought on the banks of the Shelon river, Ivan III had the Novgorodians at his mercy. They had to promise allegiance to the grand prince and his son, pay a large indemnity, and cede to Moscow some of their lands. The new arrangement, which meant a thorough defeat and humiliation of Novgorod but left its system and position essentially intact, could not be expected to last. And indeed the authorities of Novgorod soon refused to recognize Ivan III as their sovereign and tried again to obtain help from Lithuania. In 1478 the angry grand prince undertook his second campaign against Novgorod; because Lithuanian help failed to materialize and the Novgorodians split among themselves, the city finally surrendered without a battle to the besieging Muscovite army. This time Ivan III executed some of his opponents as traitors, exiled others, and transferred a considerable number of Novgorodian boyar families to other parts of the country. He declared, as quoted in a chronicle: "The veche bell in my patrimony, in Novgorod, shall not be, a posadnik there shall not be, and I will rule the entire state." The veche, the offices of the posadnik and the tysiatskii, and in effect the entire Novgorodian system were accordingly abolished; even the veche bell was carted away. Further large-scale deportations took place in 1489, and Novgorod became an integral part of the Muscovite state.

Tver's turn came next; and the principality offered even less resistance than Novgorod. Another Tver prince named Michael also tried to obtain Lithuanian help against the expanding might of Moscow, signing an agreement in 1483 with Casimir IV of Lithuania and Poland. But when Ivan III marched on Tver, Michael repudiated the agreement and declared himself an obedient "younger brother" of the Muscovite ruler. Yet in 1485 he tried to resume relations with Lithuania; his messages to Casimir IV were intercepted and his plans discovered by Moscow. Thereupon, Ivan III promptly besieged Tver. Michael's support among his own followers collapsed, and he escaped to Lithuania, while the town surrendered without battle to the Muscovite army. When Michael died in Lithuania he left no heir, and in this manner ended the greatest rival family to the princes of Moscow. In contrast to Novgorod, the incorporation of Tver, which was a northeastern principality, presented no special problems to Muscovite authorities. The sum of Ivan III's acquisitions, large and small, meant that very few Russian appanages remained to be gathered, and as a rule even these few, such as Pskov or the last half of Riazan, survived because of their co-operation with the grand princes of Moscow.

Ivan III's ambitions were not limited to the remaining Russian appanages. The grand prince of Moscow considered himself the rightful heir to all the former Kievan lands, which in his opinion constituted his lawful patrimony. Ivan III made his view of the matter quite clear in foreign relations, and at home he similarly emphasized his position as the sole ruler of the whole country. In 1493 he assumed the title of Sovereign — gosudar in Russian — of All Russia. Ivan III's claim to the entire inheritance of the Kievan state represented above all else, a challenge to Lithuania which, following the collapse of Kiev, had extended its dominion over vast western and southwestern Russian territories. The Princedom of Lithuania, called by some the Lithuanian-Russian Princedom, which we shall discuss in a later chapter, arose in large part as a successor to Kiev: on the outcome of the struggle between Moscow on one side and Lithuania and Poland on the other depended the final settlement of the Kievan estate.

After Ivan III acquired Novgorod and Tver, a number of appanage princes in the Upper Oka area, a border region between Lithuania and Moscow, switched their allegiance from their Lithuanian overlord to him. Lithuania failed to reverse their decision by force and had to accept the change in an agreement in 1494. But new defections of princes to Moscow, this time further south, led to war again in 1500. The Russians won the crucial battle on the banks of the Vedrosha river, capturing the Lithuanian commander, artillery, and supplies. By the peace treaty of 1503, the Lithuanians recognized as belonging to the grand prince of Moscow those territories that his armies had occupied. Ivan III thus obtained parts of the Smolensk and the Polotsk areas and much of Chernigov-Seversk, a huge land in southern and central European Russia based on the old principality of Chernigov. Another peace treaty in 1503 ended the war which Moscow had effectively waged to defend the principality of Pskov against the Livonian Order. All in all, Ivan III's successes in other Russian states and in foreign wars enormously increased his domain.

The grand prince's growing power and prestige led him logically to a final break with the Mongols. This definitive lifting of the Mongol yoke, however, represented something of an anticlimax compared to the catastrophe of the Mongol invasion or the epic battle of Kulikovo. Ivan III became grand prince without being confirmed by the khan and, following the practice of his father Basil II, he limited his allegiance to the Golden Horde to the sending of "presents" instead of the regular tribute, finally discontinuing even those. Mongol punitive expeditions in 1465 and 1472 were checked in the border areas of the Muscovite state. Finally in 1480, after Ivan III publicly renounced any allegiance to the Golden Horde, Khan Ahmad decided on an all-out effort against the disobedient Russians. He made an alliance with Casimir IV of Lithuania and Poland and invaded Muscovite territory. Ivan III, in turn, obtained the support of Mengli-Geray, the Crimean khan, and disposed his forces so as to block the Mongol advance and above all to guard river crossings. The main Mongol and Muscovite armies reached the opposite banks of the Ugra river and remained there facing each other. The Mongols had failed to cross the river before the Muscovites arrived, and they did not receive the expected Lithuanian and Polish help because these countries had to concentrate on beating back the Crimean Tartars who had made a large raid into Lithuania. Strangely enough, when the river froze, making it possible for the cavalry of the

Golden Horde to advance, and the Russians began to retreat, the Mongols suddenly broke camp and rushed back into the steppe. Apparently they were frightened by an attack on their home base of Sarai that was staged by a Russian and Tartar detachment. In any case, Khan Ahmad's effort to restore his authority in Russia collapsed. Shortly after, he was killed during strife in the Golden Horde, and around 1500 the Horde itself fell under the blows of the Crimean Tartars.

Another important event in Ivan III's reign was his marriage in 1472 to a Byzantine princess, Sophia, or Zoe, Paleologue. The marital alliance between the grand prince of Moscow and a niece of the last Byzantine emperor, Constantine XI, who had perished on the walls of Constantinople in the final Turkish assault, was sponsored by the Vatican in the hope of bringing Russia under the sway of the pope and of establishing a broad front against the Turks. These expectations failed utterly, yet for other reasons the marriage represented a notable occurrence. Specifically, it fitted well into the general trend of elevating the position of the Muscovite ruler. Ivan III added the Byzantine two-headed eagle to his own family's St. George, and he developed a complicated court ceremonial on the Byzantine model. He also proceeded to use the high titles of *tsar* and *autocrat* and to institute the ceremony of coronation as a solemn church rite. While *autocrat* as used in Moscow originally referred to the complete independence of the Muscovite sovereign from any overlord, and thus to the termination of the Mongol yoke, the word itself — although translated into the Russian — and the attendant concept of power and majesty were Greek, just as *tsar* stemmed from the Roman, and hence Byzantine, *caesar*. Ivan III also engaged in an impressive building program in Moscow, inviting craftsmen from many countries to serve him. In 1497 he promulgated for his entire land a code of law which counted the *Russian Justice* and the Pskov *Sudebnik* among its main sources. It may be added that legends and doctrines emphasizing the prestige of Moscow and its ruler grew mainly in Ivan III's reign, and in that of his successor. They included the stories of the bringing of Christianity to Russia by St. Andrew the apostle, the descent of the Muscovite princes from the Roman emperors and the significance of the regalia of Constantine Monomakh, and even the rather well-developed doctrine of Moscow the Third Rome. Apparently, the Muscovite ruler took the attitude of a distant superior toward his collaborators, especially after his Byzantine marriage. Or, at least, so the boyars complained for years to come.

Although Ivan III asserted his importance and role as the successor to the Kievan princes, he refused to be drawn into broader schemes or sacrifice any of his independence. Thus he declined papal suggestions of a union with Rome and of a possible re-establishment, in the person of the Muscovite ruler, of a Christian emperor in Constantinople. And when the Holy

Roman Emperor offered him a kingly crown, he answered as follows: "We pray God that He let us and our children always remain, as we are now, the lords of our land; as to being appointed, just as we had never desired it, so we do not desire it now." Ivan III has been called the first national Russian sovereign.

Ivan III was succeeded by his son Basil III, who ruled from 1505 to 1533. The new reign in many ways continued and completed the old. Basil III annexed virtually all remaining appanages, such as Pskov, obtained in 1511, and the remaining part of Riazan, which joined the Muscovite state in 1517, as well as the principalities of Starodub, Chernigov-Seversk, and the upper Oka area. The Muscovite ruler fought Lithuania, staging three campaigns aimed at Smolensk before that town was finally captured in 1514; the treaty of 1522 confirmed Russian gains. Continuing Ivan III's policy, he exercised pressure on the khanate of Kazan, advancing the Russian borders in that direction and supporting a pro-Russian party which acted as one of the two main contending political factions in the turbulent life of the city and the state. Profiting from the new standing of Muscovite Russia, Basil III had diplomatic relations with the Holy Roman Empire — the ambassador of which, Sigismund von Herberstein, left an important account of Russia, *Rerum moscovitarum commentarii* — with the papacy, with the celebrated Turkish sultan Suleiman I, the Magnificent, and even with the founder of the great Mogul empire in India, Babar. Ironically, in the case of this last potentate, of whom next to nothing was known in Moscow, the Russians behaved with extreme caution not to pay excessive honors to his empire and thus to demean the prestige of their ruler. Invitations to foreigners to enter Russian service continued. It was in the reigns of Ivan III and Basil III that a whole foreign settlement, the so-called German suburb, appeared in Moscow.

In home affairs too Basil III continued the work of his father. He sternly ruled the boyars and members of former appanage princely families who had become simply servitors of Moscow. In contrast to the practice of centuries, but in line with Ivan III's policy, the abandonment of Muscovite service in favor of some other power — which in effect came to mean Lithuania — was judged as treason. At the same time the obligations imposed by Moscow increased. These and other issues connected with the transition from appanages to centralized rule were to become tragically prominent in the following reign.

Incidentally, it was Basil III who forbade his merchants to attend the Kazan fair and established instead a fair first in Vasilsursk and soon after near the monastery of St. Macarius where the Vetluga flows into the Volga; the new fair was transferred in 1817 to Nizhnii Novgorod to become the most famous and important annual event of its kind in modern Russia.

Why Moscow Succeeded

The rise of Moscow was a fundamental development in Russian history. The ultimate success of the northeastern principality meant the end of the appanage period and the establishment of a centralized state, and the particular character of Muscovite government and society affected the evolution of Russia for centuries to come. Yet, while the role of Moscow proved to be in the end overwhelming, its ability to attain this role long remained subject to doubt and thus its success needs a thorough explanation. Moscow, after all, began with very little and for a long time could not be compared to such flourishing principalities as Novgorod or Galicia. Even in its own area, the northeast, it started as a junior not only to old centers like Rostov and Suzdal but also to Vladimir, and it defeated Tver in a long struggle which it appeared several times to have lost. Written sources, on their part, indicate the surprise of contemporaries at the unexpected emergence of Moscow. In explaining the rise of Moscow, historians have emphasized several factors, or rather groups of factors, many of which have already become apparent in our brief narrative.

First, attention may be given to the doctrine of geographical causation which represents both one of the basic and one of the earliest explanations offered, having already been fully developed by S. Soloviev. It stresses the decisive importance of the location of Moscow for the later expansion of the Muscovite state and includes several lines of argument. Moscow lay at the crossing of three roads. The most important was the way from Kiev and the entire declining south to the growing northeast. In fact, Moscow has been described as the first stopping and settling point in the northeast. But it also profited from movements in other directions, including the reverse. Thus, it seems, immigrants came to Moscow after the Mongol devastation of the lands further to the northeast. Moscow is also situated on a bend of the Moscow river, which flows from the northwest to the southeast into the Oka, the largest western tributary of the Volga. To speak more broadly of water communications which span and unite European Russia, Moscow had the rare fortune of being located near the headwaters of four major rivers: the Oka, the Volga, the Don, and the Dnieper. This offered marvelous opportunities for expansion across the flowing plain, especially as there were no mountains or other natural obstacles to hem in the young principality.

In another sense too Moscow benefited from a central position. It stood in the midst of lands inhabited by the Russian, and especially the Great Russian, people, which, so the argument runs, provided a proper setting for a natural growth in all directions. In fact, some specialists have tried to estimate precisely how close to the geographic center of the Russian people Moscow was situated, noting also such circumstances as its proximity to the line dividing the two main dialects of the Great Russian language. Central location within Russia, to make an additional point, cushioned Moscow from outside invaders. Thus, for example, it was Novgorod, not Moscow, that continuously had to meet enemies from the northwest, while in the southeast Riazan absorbed the first blows, a most helpful situation in the case of Tamerlane's invasion and on some other occasions. All in all, the considerable significance of the location of Moscow for the expansion of the Muscovite state cannot be denied, although this geographic factor certainly is not the only one and indeed has generally been assigned less relative weight by recent scholars.

The economic argument is linked in part to the geographic. The Moscow river served as an important trade artery, and as the Muscovite principality expanded along its waterways it profited by and in turn helped to promote increasing economic intercourse. Soviet historians in particular have treated the expansion of Moscow largely in terms of the growth of a common market. Another economic approach emphasizes the success of the Muscovite princes in developing agriculture in their domains and supporting colonization. These princes, it is asserted, clearly outdistanced their rivals in obtaining peasants to settle on their lands, their energetic activities ranging from various inducements to free farmers to the purchase of prisoners from the Mongols. As a further advantage, they managed to maintain in their realm a relative peace and security highly beneficial to economic life.

The last view introduces another key factor in the problem of the Muscovite rise: the role of the rulers of Moscow. Moscow has generally been considered fortunate in its princes, and in a number of ways. Sheer luck constituted a part of the picture. For several generations the princes of Moscow, like the Capetian kings who united France, had the advantage of continuous male succession without interruption or conflict. In particular, for a long time the sons of the princes of Moscow were lucky not to have uncles competing for the Muscovite seat. When the classic struggle between "the uncles" and "the nephews" finally erupted in the reign of Basil II, direct succession from father to son possessed sufficient standing and support in the principality of Moscow to overcome the challenge. The princedom has also been considered fortunate because its early rulers, descending from the youngest son of Alexander Nevskii and thus representing a junior princely branch, found it expedient to devote themselves to their small appanage instead of neglecting it for more ambitious undertakings elsewhere.

It is generally believed that the policies of the Muscovite princes made a major and massive contribution to the rise of Moscow. From Ivan Kalita to Ivan III and Basil III these rulers stood out as "the gatherers of the

Russian land," as skillful landlords, managers, and businessmen, as well as warriors and diplomats. They all acted effectively even though, for a long time, on a petty scale. Kliuchevsky distinguishes five main Muscovite methods of obtaining territory: purchase, armed seizure, diplomatic seizure with the aid of the Golden Horde, service agreements with appanage princes, and the settlement by Muscovite population of the lands beyond the Volga. The relative prosperity, good government, peace, and order prevalent in the Muscovite principality attracted increasingly not only peasants but also, a fact of great importance, boyars, as well as members of other classes, to the growing grand princedom.

To be sure, not every policy of the Muscovite rulers contributed to the rise of Moscow. For example, they followed the practice of the appanage period in dividing their principality among their sons. Yet in this respect too they gained by comparison with other princedoms. In the Muscovite practice the eldest son of a grand prince received a comparatively larger share of the inheritance, and his share grew relatively, as well as absolutely, with time. Thus, Dmitrii Donskoi left his eldest son one-third of his total possessions, Basil II left his eldest one-half, and Ivan III left his eldest three-fourths. Furthermore, the eldest son became, of course, grand prince and thus had a stronger position in relation to his brothers than was the case with other appanage rulers. Gradually the right to coin money and to negotiate with foreign powers came to be restricted to the grand prince.

The development of the Muscovite state followed the pattern mentioned earlier in our general discussion of the northeast: in a relatively primitive society and a generally fluid and shifting situation, the prince became increasingly important as organizer and owner as well as ruler — with little distinction among his various capacities — while other elements of the Kievan political system declined and even atrophied. We know, for instance, that Basil Veliaminov, the last Muscovite tysiatskii, died in 1374 and that thenceforth that office was abolished. The Muscovite "gathering of Russia," while it was certainly a remarkable achievement, also reflected the trend of the time. The very extent of the division of Russia in the appanage period paved the way for the reverse process, because most principalities proved to be too small and weak to offer effective resistance to a centralizing force. After Moscow triumphed in the northeast, in the old principality of Vladimir-Suzdal, it had to deal with only two other major Russian lands, those of Novgorod and of Riazan, the rest having already been absorbed by the expanding Lithuanian-Russian state.

To appreciate better the success of the princes of Moscow, it is necessary to give special attention to one aspect of their policy: relations with the Mongols. In their dealings with the Golden Horde, the Muscovite rulers managed to eat the proverbial cake and to have it too. The key to their remarkable performance lay in good timing. For a long time, while the Mongols retained their strength, the princes of Moscow demonstrated complete obedience to the khans, and indeed eager co-operation with them. In this manner they became established as grand princes after helping the Mongols to devastate the more impatient and heroic Tver and some other Russian lands to their own advantage. In addition, they collected tribute for the Mongols, thus acquiring some financial and, indirectly, judicial authority over other Russian princes. "The gathering of the Russian land" was also greatly facilitated by this connection with the Golden Horde: Liubavsky and other historians have stressed the fact that the khans handed over to the Muscovite princes entire appanages which were unable to pay their tribute, while, for that reason, rulers of other principalities preferred to sell their lands directly to Moscow in order to save something for themselves. But, as the Golden Horde declined and the Muscovite power rose, it was a grand prince of Moscow, Dmitrii Donskoi, who led the Russian forces against the Mongol oppressors on the field of Kulikovo. The victory of Kulikovo and the final lifting of the Mongol yoke by Ivan III represented milestones in the rise of the princedom of Moscow from a northeastern appanage principality to a national Russian state.

Yet another major factor in that rise was the role of the Church. To estimate its significance one should bear in mind the strongly religious character of the age, which was similar to the Middle Ages in the West. Moscow became the seat of the metropolitan and thus the religious capital of Russia in 1326 or 1328, long before it could claim any effective political domination over most of the country. It became, further, the city of St. Alexis and especially St. Sergius, whose monastery, the Holy Trinity-St. Sergius Monastery north of Moscow, was a fountainhead of a broad monastic movement and quickly became a most important religious center, rivaled in all Russian history only by the Monastery of the Caves near Kiev. Religious leadership, very valuable in itself, also affected politics. St. Alexis, as we saw, acted as one of the most important statesmen of the princedom of Moscow; and the metropolitans in general, linked to Moscow and at least dimly conscious of broader Russian interests, favored the Muscovite "gathering of Russia." Their greatest service to this cause consisted probably in their frequent intervention in princely quarrels and struggles, through advice, admonition, and occasionally even excommunication; this intervention was usually in favor of Moscow.

Judgments of the nature and import of the rise of Moscow are even more controversial than descriptions and explanations of that process. Most prerevolutionary Russian historians praised it as a great and necessary achievement of the princes of Moscow and of the Russian people, who had to unite to survive outside aggression and to play their part in history. Soviet historians have come to share the same view. On the other hand, some Russian doubters, for example, Presniakov, together with many scholars in other traditions, such as the Polish, the Lithuanian, or the nationalist Ukrainian, have argued on the other side: they have emphasized in particular that the vaunted "gathering of Russia" consisted, above all, in a skillful aggression by the Muscovite princes against both Russians, such as the inhabitants of Novgorod and Pskov, and eventually various non-Russian nationalities, which deprived them of their liberties, subjugating everyone to Muscovite despotism. As is frequently the case in major historical controversies, both schools are substantially correct, stressing as they do different aspects of the same complicated phenomenon. Without necessarily taking sides on this or other related issues, we shall appreciate a little better the complexity and the problems of the period after devoting some attention to the economic, social, and cultural life of appanage Russia.

APPANAGE RUSSIA: ECONOMICS, SOCIETY, INSTITUTIONS

> Thus our medieval boyardom in its fundamental characteristics of
> territorial rule; the dependence of the peasants, with the right of de-
> parture; manorial jurisdiction, limited by communal administration;
> and economic organization, characterized by the insignificance of the
> lord's own economy: in all these characteristics our boyardom rep-
> resents an institution of the same nature with the feudal seigniory,
> just as our medieval rural commune represents, as has been demon-
> strated above, an institution of the same essence with the commune
> of the German Mark.
>
> PAVLOV-SILVANSKY

> . . . the "service people" was the name of the class of population ob-
> ligated to provide service (court, military, civil) and making use, in
> return, on the basis of a conditional right, of private landholdings.
> The basis for a separate existence of this class is provided not by its
> rights, but by its obligations to the state. These obligations are varied,
> and the members of this class have no corporate unity.
>
> VLADIMIRSKY-BUDANOV

> Here, of course, you have in fact the process of a certain *feudalization*
> of simpler state arrangements in their interaction and mutual limi-
> tation.*
>
> STRUVE

Whereas the controversy continues concerning the relative weight of
commerce and agriculture in Kievan Russia, scholars agree that tilling the
soil represented the main occupation of the appanage period. Rye, wheat,
barley, millet, oats, and a few other crops continued to be the staples of
Russian agriculture. The centuries from the fall of Kiev to the unification
of the country under Moscow saw a prevalence of local, agrarian economy,
an economic parochialism corresponding to political division. Furthermore,
with the decline of the south and the Mongol invasion, the Russians lost
much of their best land and had to establish or develop agriculture in forested areas and under severe climatic conditions. Mongol exactions further
strained the meager Russian economy. In Liubavsky's words: "A huge
parasite attached itself to the popular organism of northeastern Russia; the
parasite sucked the juices of the organism, chronically drained its life
forces, and from time to time produced great perturbations in it."

---

* Italics in the original. Struve's statement refers to a particular development
during the period, but I think that it can also stand fairly as the author's general
judgment on the issue of feudalism in Russia.

The role of trade in appanage Russia is more difficult to determine. While it retained great importance in such lands as Galicia, not to mention the city and the principality of Novgorod, its position in the northeast, and notably in the princedom of Moscow, needs further study. True, the Moscow river served as a trade route from the very beginning of Moscow's history, and the town also profited commercially from its excellent location on the waterways of Russia in a more general sense. Soviet historians stress the ancient Volga trade artery, made more usable by firm Mongol control of an enormous territory to the east and the southeast; and, as already indicated, they also link closely the expansion of the Muscovite principality to the growth of a common market. In addition to the Volga, the Don became a major commercial route, with Genoese and Venetian colonies appearing on the Black Sea. Around 1475, however, the Turks established a firm hold on that sea, eliminating the Italians. The Russians continued to export such items as furs and wax and to import a wide variety of products, including textiles, wines, silverware, objects of gold, and other luxuries. Yet, although the inhabitants of northeastern Russia in the appanage period did retain some important commercial connections with the outside world and establish others, and although internal trade did grow in the area with the rise of Moscow, agricultural economy for local consumption remained dominant. Commercial interests and the middle class in general had remarkably little weight in the history of the Muscovite state.

Other leading occupations of the period were hunting, fishing, cattle raising, and apiculture, as well as numerous arts and crafts. Carpentry was especially well developed, while tannery, weaving, work in metal, and some other skills found a wide application in providing for the basic needs of the people. Certain luxurious and artistic crafts sharply declined, largely because of the poverty characteristic of the age, but they survived in some places, principally in Novgorod; with the rise of Moscow, the new capital gradually became their center.

The Question of Russian Feudalism

The question of the social structure of appanage Russia is closely tied to the issue of feudalism in Russian history. Traditionally, specialists have considered the development of Russia as significantly different from that of other European countries, one of the points of contrast being precisely the absence of feudalism in the Russian past. Only at the beginning of this century did Pavlov-Silvansky offer a brilliant and reasonably full analysis of ancient Russia supporting the conclusion that Russia too had experienced a feudal stage. Pavlov-Silvansky's thesis became an object of heated controversy in the years preceding the First World War. After the Revolution, Soviet historians proceeded to define "feudal" in extremely broad terms and to apply this concept to the development of Russia all the way from the days of Kiev to the second half of the nineteenth century. Outside the Soviet Union, a number of scholars, while disagreeing with Pavlov-Silvansky on important points, nevertheless accepted at least a few feudal characteristics as applicable to medieval Russia.

Pavlov-Silvansky argued that three traits defined feudalism and that all three were present in appanage Russia: division of the country into independent and semi-independent landholdings, the seigniories; inclusion of these landholdings into a single system by means of a hierarchy of vassal relationships; and the conditional quality of the possession of a fief. Russia was indeed divided into numerous independent principalities and privileged boyar holdings, that is, seigniories. As in western Europe, the vassal hierarchy was linked to the land: the *votchina,* which was an inherited estate, corresponded to the seigniory; the *pomestie,* which was an estate granted on condition of service, to the benefice. Pavlov-Silvansky, it should be noted, believed that the pomestiia, characteristic of the Muscovite period of Russian history, already represented a significant category of landholding in the appanage age. The barons, counts, dukes, and kings of the West found their counterparts in the boyars, service princes, appanage princes, and grand princes of medieval Russia. Boyar service, especially military service, based on free contract, provided the foundation for the hierarchy of vassal relationships. Special ceremonies, comparable to those in the West, marked the assumption and the termination of this service. Appanage Russia knew such institutions as feudal patronage, commendation — personal or with the land — and the granting of immunity to the landlords, that is, of the right to govern, judge, and tax their peasants without interference from higher authority. Vassals of vassals appeared, so that one can also speak of sub-infeudation in Russia.

Pavlov-Silvansky's opponents, however, have presented strong arguments on their side. They have stressed the fact that throughout the appanage period Russian landlords acquired their estates through inheritance, not as compensation for service, thus retaining the right to serve whom they pleased. The estate of an appanage landlord usually remained under the jurisdiction of the ruler in whose territory it was located, no matter whom the landlord served. Furthermore, numerous institutions and even entire aspects of Western feudalism either never developed at all in Russia, or, at best, failed to grow there beyond a rudimentary stage. Such was the case, for example, with the extremely complicated Western hierarchies of vassals, with feudal military service, or with the entire phenomenon of chivalry. Even the position of the peasants and their relationship with the landlords differed markedly in the East and in the West, for serfdom became firmly established in Russia only after the appanage period.

In sum, it would seem that a precise definition of feudalism, with proper attention to its legal characteristics, would not be applicable to Russian society. Yet, on the other hand, many developments in Russia, whether we think of the division of power and authority in the appanage period, the economy of large landed estates, or even the later pomestie system of state service, bear important resemblances to the feudal West. As already indicated, Russian social forms often appear to be rudimentary, or at least simpler and cruder, versions of Western models. Therefore, a number of scholars speak of the social organization of medieval Russia as incipient or undeveloped feudalism. That feudalism proved to be particularly weak when faced with the rising power of the grand princes and, especially, of the autocratic tsars.

Soviet historians require an additional note. Starting from the Marxist emphasis on similarities in the development of different societies and basing their periodization on economic factors, they offer an extremely broad definition of feudalism in terms of manorial economy, disregarding the usual stress on the distribution of power and legal authority. Thus, they consider Russia as feudal from the later Kievan period to the second half of the nineteenth century. The Soviet approach, it may be readily seen, does little to differentiate between the appanage period of Russian history and the preceding and succeeding epochs.

Appanage Society and Institutions

The social structure of appanage Russia represented, of course, a continuation and a further evolution of the society of the Kievan period, with no sharp break between the two. The princes occupied the highest rung on the social ladder. The already huge Kievan princely family proliferated and differentiated further during the centuries which followed the collapse of a unitary state. The appanage period naturally proved to be the heyday of princes and princelings, ranging from grand princes to rulers of tiny principalities and even to princes who had nothing to rule and were forced to find service with their relatives. It might be added that in addition to the grand princes "of Moscow and all Russia," grand princes emerged in several other regional centers, notably Tver and Riazan, where the lesser members of a particular branch of the princely family paid a certain homage to their more powerful elder. The expansion of Moscow ended this anarchy of princes, and with it the appanage period.

Next came the boyars, followed by the less aristocratic "free servants" of a prince who performed a similar function. The boyars and the free servants made contracts with their prince, and they were at liberty to leave him and seek another master. The boyars had their own retinues, sometimes quite numerous. For instance, in 1332 a boyar with a following of 1,700 persons entered the service of the grand prince of Moscow, while shortly after his arrival another boyar with a retinue of 1,300 left it. As already emphasized, members of the upper classes of appanage Russia were landlords. They acted as virtual rulers of their large estates, levying taxes and administering justice, although it is worth noting that, as Moscow rose, the immunities which they received to govern their lands no longer extended to jurisdiction in cases of major crimes. Votchiny, that is, hereditary landholdings, prevailed in the appanage period. However, with the rise of Moscow, the pomestie, that is, an estate granted by a prince to a servitor during the term of his personal service, became common. The earliest extant reference to a pomestie goes back to Ivan Kalita's testament, but the pomestie system developed on a large scale only in the fifteenth and subsequent centuries. We shall meet it again when we discuss Muscovite Russia.

Traders, artisans, and the middle class as a whole experienced a decline during the appanage period. Except in Novgorod and a few other centers, members of that layer of society were relatively few in number and politically ineffective.

Peasants constituted the bulk of the population. It is generally believed that their position worsened during the centuries which followed the collapse of the Kievan state. Political division, invasions, and general insecurity increased the peasant's dependence on the landlord and consequently his bondage, thus accelerating a trend which had already become pronounced in the days of Kiev. While serfdom remained incomplete even at the end of the appanage period — for the peasant could still leave his master once a year, around St. George's day in late autumn, provided his accounts had been settled — it grew in a variety of forms. Principal peasant obligations were of two types: the as yet relatively little developed *barshchina,* or corvée, that is, work for the landlord, and *obrok,* or quitrent, that is, payment to the landlord in kind or in money. It should be noted, however, that many peasants, especially in the north, had no private landlords, a fortunate situation for them, even though they bore increasingly heavy obligations to the state.

The slaves, *kholopy,* of the Kievan period continued to play a significant role in the Russian economy, performing all kinds of tasks in the manorial households and estates. In fact, a small upper group of kholopy occupied important positions as managers and administrators on the estates. Indeed Diakonov suggested that in the Muscovite principality, as in France, court functionaries and their counterparts in most noble households were originally slaves, who were later replaced by the most prominent among the free servitors.

In the period which followed the fall of Kiev, the Church in Russia maintained and developed its strong and privileged position. In a time of division it profited from the best and the most widespread organization in the country, and it enjoyed the benevolence of the khans and the protection of Russian, especially Muscovite, princes. Ecclesiastical lands received exemptions from taxation and sweeping immunities; also, as in the West — although this is a controversial point — they probably proved to be more attractive to the peasants than other estates because of their relative peace, good management, and stability. The Church, or rather individual monasteries and monks, often led the Russian penetration into the northeastern wilderness. Disciples of St. Sergius alone founded more than thirty monasteries on or beyond the frontier of settlement. But the greatest addition to ecclesiastical possessions came from continuous donations, in particular the bequeathing of estates or parts of estates in return for prayers for one's soul, a practice similar to the granting of land in free alms to the Catholic Church in the feudal West. It has been estimated that at the end of the appanage period the Church in Russia owned over 25 per cent of all cultivated land in the country. As we shall see, these enormous ecclesiastical, particularly monastic, holdings created major problems both for the religious conscience and for the state.

The unification of Russia under Moscow meant a victory for a northeastern political system, characterized by the dominant position of the prince. Princes, of course, played a major part in the appanage period. It was during that time that they acted largely as managers and even proprietors of their principalities, as illustrated in the celebrated princely wills and testaments which deal indiscriminately with villages and winter coats. Princely activities became more and more petty; public rights and interests became almost indistinguishable from private. With the rise of Moscow, the process was reversed. The rulers "of Moscow and all Russia" gained in importance until, at about the time of Ivan III, they instituted a new era of autocratic tsardom. Yet, for all their exalted majesty, the tsars retained much from their northeastern princeling ancestry, combining in a formidable manner sweeping authority with petty despotism and public goals with proprietary instincts. Their power proved to be all the more dangerous because it went virtually unopposed. After the absorption of Novgorod, Pskov, and Viatka, the veche disappeared from Russian politics. The third element of the Kievan system of government, the boyar duma, it is true, continued to exist side by side with the princes and with the tsars. However, as will be indicated in later chapters, the duma in Muscovite Russia supported rather than effectively circumscribed the authority of the ruler. The evolution of Russia in the appanage period led to autocracy.

XIII

* * * * * * * * * *

APPANAGE RUSSIA: RELIGION AND CULTURE

> The Mongol yoke, which dealt a heavy blow to the manufactures of the Russian people in general, could not but be reflected, in a most grievous manner, in the artistic production and technique closely related to manufacturing. . . . The second half of the thirteenth and the entire fourteenth century were an epoch "of oppression of the life of the people, of despair among the leaders, of an impoverishment of the land, of a decline of trades and crafts, of a disappearance of many technical skills."
>
> BAGALEI

> If we consider nothing but its literature, the period that extends from the Tatar invasion to the unification of Russia by Ivan III of Moscow may be called a Dark Age. Its literature is either a more or less impoverished reminiscence of Kievan traditions or an unoriginal imitation of South Slavonic models. But here more than ever it is necessary to bear in mind that literature does not give the true measure of Old Russian culture. The fourteenth and fifteenth centuries, the Dark Age of literature, were at the same time the Golden Age of Russian religious painting.
>
> MIRSKY

> The Russian icon was the most significant artistic phenomenon of ancient Russia, the fundamental and preponderant means, and at the same time a gift, of its religious life. In its historical origin and formation the icon was an expression of the highest artistic tradition, while in its development it represented a remarkable phenomenon of artistic craftsmanship.
>
> KONDAKOV

THE RELIGION AND CULTURE of appanage Russia, like its economic and social development, stemmed directly from the Kievan period. The hard centuries which followed the collapse of a unitary state witnessed, however, a certain retardation, and even regression, in many fields of culture. Impoverishment and relative isolation had an especially adverse effect on education in general and on such costly and difficult pursuits as large-scale building in stone and certain luxury arts and crafts. Literature too seemed to have lost much of its former artistry and *élan*. Yet this decline in many areas of activity coincided with probably the highest achievements of Russian creative genius in a few fields which included wooden architecture and, especially, icon painting.

Religion in appanage Russia reflected, in its turn, the strong and weak points, the achievements and failings of the period, as it continued to occupy a central position in the life and culture of the people. In an age of division, the unity and organization of the Church stood out in striking manner. In the early fifteenth century the Orthodox Church in Russia had, in addition to the metropolitan in Moscow, fifteen bishops, of whom three, those of Novgorod, Rostov, and Suzdal, had the title of archbishop. In 1448, after suspicions of the Greek clergy had been aroused in Russia by the Council of Florence, Jonas became metropolitan without the confirmation of the patriarch of Constantinople, thus breaking the old Russian allegiance to the Byzantine See and inaugurating the autocephalous, in effect independent, period in the history of the Russian Church. Administrative unity within the Russian Church, however, finally proved impossible to preserve. The growing division of the land and the people between Moscow and Lithuania resulted in the establishment, in Kiev, of a separate Orthodox metropolitanate for the Lithuanian state, the final break with Moscow coming in 1458.

As we know, the Church, with its enormous holdings and its privileged position, played a major role in the economic and political life of appanage Russia, influencing almost every important development of the period, from the rise of Moscow to the colonization of the northeastern wilderness. But the exact impact of the Church in its own religious and spiritual sphere remains difficult to determine. It has been frequently, and on the whole convincingly, argued that the ritualistic and esthetic sides of Christianity prevailed in medieval Russia, finding their fullest expression in the liturgy and other Church services, some of which became extremely long and elaborate. Fasting, celebrating religious holidays, and generally observing the Church calendar provided further occasions for the ritualism of the Russian people, while icon painting and church architecture served as additional paths in their search for beauty. Still, the ethical and social import of Russian Christianity should not be underestimated in this period any more than during the hegemony of Kiev. Many specialists credit the teaching of the Church with the frequent manumission of slaves by individual masters, realized often by means of a provision in last wills and testaments. And, in a general sense, Christian standards of behavior remained at least the ideal of the Russian people.

Saints continued to reflect the problems and aspirations of the Russians. Figures of the appanage period who became canonized ranged from princes, such as Alexander Nevskii, and ecclesiastical statesmen exemplified by Metropolitan Alexis, to obscure hermits. But the strongest impression on the Russian religious consciousness was made by St. Sergius of Radonezh. St. Sergius, who died in 1392 at the age of about seventy-eight, began as a monk in a forest wilderness and ended as the recognized spiritual leader of Russia. His blessing apparently added strength to Grand Prince Dmitrii and the Russian army for the daring enterprise of Kulikovo, and his word could on occasion stop princely quarrels. Although he refused to be metropolitan, he became in effect the moral head of the Russian Church. As already mentioned, the monastery which St. Sergius founded north of Moscow and which came to be known as the Holy Trinity-St. Sergius Monastery, became one of the greatest religious and cultural centers of the country and the fountainhead of a powerful monastic movement. For centuries after the death of St. Sergius tens and hundreds of thousands of pilgrims continued to come annually from all over Russia to his burial place in one of the churches in the monastery. They still come. As in the case of many other saints, the chief explanation of the influence of St. Sergius lies in his ability to give a certain reality to the concepts of humility, kindness, brotherhood, and love which remain both beliefs and hopes of the Christians. It might be added that St. Sergius tried constantly to help all who needed his help and that he stressed work and learning as well as religious contemplation and observance.

The disciples of St. Sergius, as already mentioned, spread the Christian religion to vast areas in northern Russia, founding scores of monasteries. St. Stephen of Perm, the most distinguished of the followers of St. Sergius, brought Christianity to the Finnic-speaking tribes of the Zyriane: he learned their tongue and created a written language for them, utilizing their decorative designs as a basis for letters. Thus, following the Orthodox tradition, the Zyriane could worship God in their native language.

In medieval Russia, as in medieval Europe as a whole, intellectual life centered on religious problems, although their ramifications often encompassed other areas of human activity. While, in the main, Russia stayed outside the rationalist and reforming currents which developed in Western Christendom, it did not remain totally unaffected by them. Significantly, Russian religious movements stressing rationalism and radical reform emerged in western parts of the country and especially in Novgorod. As early as 1311 a Church council condemned the heresy of a certain Novgorodian priest who denounced monasticism. In the second half of the fourteenth century, in Novgorod, the teaching of the so-called *strigolniki* acquired prominence. These radical sectarians, quite similar to the evangelical Christians in the West, denied the authority of the Church and its hierarchy, as well as all sacraments except baptism, and wanted to return to the time of the apostles; an extreme faction within the movement even renounced Christ and sought to limit religious observances to prayer to God the Father. It might be noted that the protest began apparently over the issue of fees for the sacraments, and that the dissidents came rapidly to adhere to increasingly radical views. All persuasion failed, but violent repression by the population and authorities in Novgorod and Pskov, together with disagreements among the strigolniki, led to the disappearance of the sect in the early fifteenth century.

Later in the century, however, new heretics appeared, known as the

Judaizers. Their radical religious movement has been linked to the arrival in Novgorod in 1470 of a Jew Zechariah, or Skharia, and to the spread of his doctrines. The Judaizers in effect accepted the Old Testament, but rejected the New, considering Christ a prophet rather than the Messiah. Consequently they also denounced the Church. Through the transfer of two Novgorodian priests to Moscow, the movement obtained a foothold in the court circles of the capital. Joseph of Volok, an abbot of Volokolamsk, led the ecclesiastical attack on the heretics. They were condemned by the Church council of 1504, and Ivan III, finally ceding to the wishes of the dominant Church party, cruelly suppressed the Judaizers, having their leaders burned at the stake.

Controversies within the Russian Orthodox Church at the time had an even greater historical significance than did challenges to the Church from the outside. The most important and celebrated dispute of the age pitted the "possessors" against the "non-possessors," with Joseph of Volok again occupying a central position as the outstanding leader of the first-named faction. Joseph of Volok and the possessors believed in a close union of an autocratic ruler and a rich and powerful Church. The prince, or tsar, was the natural protector of the Church with all its lands and privileges. In return, he deserved complete ecclesiastical support, his authority extending not only to all secular matters but also to Church administration. The possessors emphasized, too, a formal and ritualistic approach to religion, the sanctity of Church services, rituals, practices, and teachings, and a violent and complete suppression of all dissent.

The non-possessors, who because of their origin in the monasteries of the northeast, have sometimes been called the "elders from beyond the Volga," had as their chief spokesman Nil Sorskii — or Nilus of Sora — a man of striking spiritual qualities. The non-possessors, as their name indicates, objected to ecclesiastical wealth and in particular to monastic landholding. They insisted that the monks should in fact carry out their vows, that they must be poor, must work for their living, and must remain truly "dead to the world." The Church and the State should be independent of each other; most especially, the State, which belonged to a lower order of reality, had no right to interfere in religious matters. The non-possessors stressed contemplation and the inner spiritual light, together with a striving for moral perfection, as against ecclesiastical formalism and ritualism. Furthermore, by contrast with the possessors, they differentiated in the teaching of the Church among Holy Writ, tradition, and human custom, considering only Holy Writ — that is, God's commandments — as completely binding. The rest could be criticized and changed. But even those who challenged the foundations of the Church were to be met with persuasion, never with force.

The Church council of 1503 decided in favor of the possessors. Joseph of Volok and his associates cited Byzantine examples in support of their position and also argued, in practical terms, the necessity for the Church to have a large and rich establishment in order to perform its different functions, including the exercise of charity on a large scale. Their views, especially on relations of Church and State, suited on the whole the rising absolutism of Moscow, although it seems plausible that Ivan III sympathized with the non-possessors in the hope of acquiring monastic lands. After Joseph of Volok died in 1515, subsequently to be proclaimed a saint, other high clerics continued his work, notably Daniel, who became metropolitan in 1521. At the councils of 1524 and 1531, and even as late as 1554–55, some of Nil Sorskii's chief followers were declared to be heretics. Nil Sorskii himself, however, was canonized.

In explaining the controversy between the possessors and the non-possessors, many scholars, including Soviet historians as a group, have emphasized that the possessors championed the rise of the authority of the Muscovite rulers and the interests of those elements in Russian society which favored this rise. The non-possessors, on the other hand, with their high social connections, reflected the aristocratic opposition to centralization. In a different context, that of the history of the Orthodox Church, the non-possessors may be considered to have derived from the mystical and contemplative tradition of Eastern monasticism, especially as practiced on Mount Athos. However, in a still broader sense, the possessors and the non-possessors expressed two recurrent attitudes that devoted Christians have taken toward things of this world, burdened as they have been by an incompatibility between the temporal and the eternal standards and goals of behavior. The non-possessors, thus, resemble the Franciscans in the West as well as other religious groups that have tried hard to be in, and yet not of, this world. And even after all the sixteenth-century councils they remained an important part of the Russian Church as an attitude and a point of view.

Such essentially secular intellectual issues of the period as that of the position and power of the ruler often acquired a religious coloring. The problem of authority, its character and its limitations, became paramount as Moscow rose to "gather Russia" and as its princes turned into autocratic tsars. As already mentioned, a number of legends and doctrines appeared to justify and buttress these new developments. For example, one tale about the princes of Vladimir, which originated, apparently, in the first quarter of the sixteenth century, related how Vladimir Monomakh of Kiev, the celebrated ancestor of the Muscovite princes, received from his maternal grandfather, the Byzantine emperor Constantine Monomakh, certain regalia of his high office: a headdress which came to be known as "the hat of Monomakh" and some other items of formal attire. Still more grandly, the princes of Moscow came to be connected to the Roman emperors. According to the new genealogy, Augustus, a sovereign of Rome and the world, in his old age divided his possessions among his relatives, placing his brother Prus as ruler on the banks of the Vistula. Riurik was a fourteenth-generation descendant of this Prus, St. Vladimir a fourth-generation descendant of Riurik, and Vladimir Monomakh a fourth-generation descendant of St. Vladimir. Concurrently with this revision of the genealogy of the princes of Moscow, Christianity in Russia was antedated and St. Andrew, the apostle, was proclaimed its true originator.

But the most interesting doctrine — and one that has received divergent interpretations from scholars — was that of Moscow as the Third Rome. Its originator, an abbot from Pskov named Philotheus or Filofei, wrote a letter to Basil III in 1510 which described three Romes: the Church of Old Rome, which fell because of a heresy, the Church of Constantinople brought down by the infidels, and finally the Church in Basil III's own tsardom which, like the sun, was to illumine the entire world — further-more, after two Romes had fallen, Moscow the Third Rome would stand permanently, for there was to be no fourth. Some scholars have stressed the political aspects of this doctrine, and recently it has even been re-peatedly cited as evidence of a secular Russian imperialism and aggression. It is, therefore, necessary to emphasize that Philotheus thought, in the first place, of Churches, not States, and that he was concerned with the pres-ervation of the true faith, not political expansion. And, in any case, the Muscovite rulers in their foreign policy never endorsed the view of Moscow as the Third Rome, remaining, as already mentioned, quite uninterested in the possibility of a Byzantine inheritance, while at the same time deter-mined to recover the inheritance of the princes of Kiev.

Literature and the Arts

The literature of the appanage period has generally been rated rather low. This judgment applies with full force only to the extant written works, although the oral, folkloristic tradition too, while it continued to be rich and varied, failed to produce tales equal in artistry to the Kievan byliny. As a qualification it might be added that, in the opinion of certain scholars, surviving material is insufficient to enable us to form a definitive view of the scope and quality of appanage literature.

The Mongol conquest of Russia gave rise to a number of factual narra-tives as well as semi-legendary and legendary stories. These dwelt on the bitter fighting, the horror, and the devastation of the invasion and inter-preted the events as divine punishment for the Russians' sins. The best artis-tic accounts of the catastrophe can be read in the series dealing with the Mon-gol ravage of Riazan and in the *Lay of the Destruction of the Russian Land,* written early in the appanage period about the middle of the thirteenth century, of which only the beginning has survived. The victory of Kulikovo in turn found reflection in literature. Thus the *Story of the Massacre of Mamai,* written with considerable artistry some twenty years after the event, tells about the departure of Prince Dmitrii from Moscow, the grief of his wife, the visit of the prince to the blessed Sergius of Radonezh, the eve of the battle, and the battle itself. Another well-known account of Kulikovo, the *Zadonshchina* composed at the end of the fifteenth century, has little literary merit and is a clumsy imitation of the *Lay of the Host of Igor.* The expansion of Moscow, as seen from the other side, inspired the *Tale about the Capture of Pskov,* written by a sorrowing patriot of that city. Chronicles in Novgorod and elsewhere continued to give detailed and consecutive information about developments in their localities.

Accounts of the outside world can be found in the sizeable travel literature of the period. Foremost in this category stands Athanasius Nikitin's cele- brated *Wanderings beyond the Three Seas,* a narrative of this Tver mer- chant's journey to Persia, Turkey, and India from 1466 to 1472. Particular value attaches to the excellent description of India, which Nikitin saw some twenty-five years before Vasco da Gama. Other interesting records of travel during the period include those of a Novgorodian named Stephen to the Holy Land in 1350, of Metropolitan Pimen to Constantinople in 1389, and of a monk Zosima to Constantinople, Mount Athos, and Jeru- salem in 1420 and also two accounts of journeys to the Council of Florence.

Church literature, including sermons, continued to be produced on what must have been a considerable scale. Hagiography deserves special notice. Lives of saints composed in the thirteenth and fourteenth centuries, for example, of Abraham of Smolensk, Alexander Nevskii, Michael of Cherni- gov, and Metropolitan Peter, are characterized by simplicity and biographi- cal detail. Unfortunately for the historian, a new style, artificial, pompous, and opposed to realistic description, came to the fore with the fifteenth century. This style came from the southern Slavs and was introduced by such writers as Cyprian in his life of St. Peter the Metropolitan, and Epiphanius the Wise, who dealt with St. Sergius of Radonezh and St. Stephen of Perm. The southern Slavs, it should be added, exercised a strong influence on appanage literature and thought, as for example in the formulation of the doctrine of Moscow as the Third Rome.

In contrast to literature, architecture has frequently been considered one of the glories of the appanage period in spite of the fact that the age wit- nessed relatively little building in stone. Russian wooden architecture, to say the least, represents a remarkable achievement. Although it dates, without doubt, from the Kievan and the pre-Kievan eras, no buildings survive from those early times. It is only with the appanage and the Mus- covite periods that we can trace the consecutive development of this archi- tecture and study its monuments.

A *klet* or *srub,* a rectangular structure of stacked beams, each some

Scythian embossed goldwork of the sixth century B.C.

Ancient monuments on the graves of the Polovtsy.

St. George and the Dragon. Novgorod School. Early 15th century.

The Old Testament Trinity. A. Rublev, early fifteenth century.

Icon from the Deesis Festival tier: Entrance into Jerusalem. Novgorod School, about 1475.

St. Nicholas, Bishop of Myra, Moscow School. 14th century.

Our Lady of Vladimir. Moscow School. End of 15th century.

Cathedral and cemetery at Ipatiev Monastery in Kostroma.

Fourteenth-century wooden church displayed at Ipatiev Monastery in Kostroma.

Preobrazhenskii Cathedral on Volga river at Uglich.

twenty or twenty-five feet long, constituted the basis of ancient Russian wooden architecture. The walls were usually eight or nine feet high. A steep, two-slope roof offered protection and prevented an accumulation of snow, while moss and later hemp helped to plug cracks and holes. At first the floors were earthen, later wooden floors were constructed. A klet represented the living quarters of a family. Another, usually smaller, klet housed livestock and supplies. Generally the two were linked by a third small structure, a passageway, which also contained the door to the outside. A peasant household thus consisted of three separate, although connected, units. As the owner became more prosperous, or as his sons started families of their own, additional kleti were built and linked to the old ones, the ensemble growing, somewhat haphazardly, as a conglomeration of distinct, yet joined, structures.

After the Russians accepted Christianity, they adapted their wooden architecture to the Byzantine canons of church building. The three required parts of a church were erected as follows: the sanctuary, always on the eastern side, consisted of a small klet; the main section of the church, where the congregation stood, was built as a large double klet, one on top of the other; finally, another small klet on the western side constituted the *pritvor,* or separate entrance hall, where originally catechumens waited for the moment to enter the church proper. The high two-slope roof of the large klet was crowned with a small cupola topped by a cross. Churches of this simple ancient type can be seen on old icons, and a few of them in northern Russia — built, however, in the seventeenth century — have come down to our times.

Various developments in church architecture followed. Frequently a special basement klet was constructed under each of the three kleti constituting the church proper, which was thus raised to a second-floor level while its main part acquired a three-story elevation. The basement could be used for storage; a high outside staircase and porch were built to secure entrance to the church. The sanctuary sometimes assumed the form not of a quadrangle, but of a polygon, for instance, an octagon. The roofs of the churches became steeper and steeper, until many of them resembled wedges. In contrast to the Byzantine tradition of building churches with one or five cupolas, the Russians, whether they worked in stone or in wood, early demonstrated a liking for more cupolas. It might be noted that St. Sophia in Kiev had thirteen cupolas, and another Kievan church, that of the Tithe, had twenty-five. Numerous wooden churches also possessed many cupolas, including a remarkable one with seventeen and another with twenty-one.

The Russians not only translated Byzantine stone church architecture into another medium, wood, but they also developed it further in a creative and varied manner. Especially original and striking were the so-called tent, or pyramidal, churches, of which some from the late sixteenth and the seventeenth centuries have escaped destruction. In the tent churches the main part of the church was a high octagon — although occasionally it had six or twelve sides — which provided the foundation for a very high pyramidal, sometimes conical, roof, capped by a small cupola and a cross. The elevation of these roofs ranged from 125 to well over 200 feet. The roofs of the altar and the pritvor were, by contrast, usually low. To quote Grabar, perhaps the most distinguished historian of Russian architecture and art, concerning tent churches:

> Marvelously strict, almost severe, in their majestic simplicity are these giants, grown into the earth, as if one with it. . . . The idea of the eternity and immensity of the church of Christ is expressed here with unbelievable power and utmost simplicity. The simplicity of outline has attained in them the highest artistic beauty, and every line speaks for itself, because it is not forced, not contrived, but absolutely necessary and logically inevitable.

Weidle has written of undeveloped Gothic in Russia, an approach not unrelated to the general concept of undeveloped Russian feudalism.

By contrast, architecture in stone, as already indicated, experienced a decline in the appanage period, although stone churches continued to be built in Novgorod and in lesser numbers in some other centers. To illustrate regression, historians have often cited the inability of Russian architects in the 1470's to erect a new Cathedral of the Assumption, the patron church of Moscow, using the Cathedral of the Assumption in Vladimir as their model. Yet this incident also marked the turning point, for Ivan III invited foreign specialists to Moscow and initiated stone building on a large scale. The most important result of the revival of stone architecture was the construction of the heart of the Kremlin in Moscow, a fitting symbol of the new authority, power, and wealth of the Muscovite rulers.

Beginning in 1474, Ivan III sent a special agent to Venice and repeatedly invited Italian architects and other masters to come to work for him in Moscow. The volunteers included a famous architect, mathematician, and engineer, Aristotle Fieravanti, together with such prominent builders as Marco Ruffo, Pietro Solario, and Alevisio. Fieravanti, who lived in Russia from 1475 to 1479, erected the Cathedral of the Assumption in the Kremlin on the Vladimir model, but with some differences. In 1490 architects from Pskov constructed in the same courtyard the Cathedral of the Annunciation, a square building with four inside pillars, three altar apses, five cupolas, and interesting decorations. It reflected the dominant influence of Vladimir architecture, but also borrowed elements from the tradition of Novgorod and Pskov and from wooden architecture. Next, still working on the Kremlin courtyard, Ivan III ordered the construction of a new Cathedral of the Archangel in place of the old one, just as he had done earlier with the

Cathedral of the Assumption. Alevisio accomplished this task between 1505 and 1509, following the plan of the Cathedral of the Assumption, but adding such distinct traits as Italian decoration of the façade. The three cathedrals of the Annunciation, the Assumption, and the Archangel Michael became, so to speak, the sacred heart of the Kremlin and served, among other functions, respectively as the place for the wedding, the coronation, and the burial of the rulers of Russia.

Stone palaces also began to appear. As with the cathedrals, probably the greatest interest attaches to the palace in the Kremlin in Moscow. It was constructed by Ruffo, Solario, Alevisio and other Italian architects, but following the canons of Russian wooden architecture: the palace was a conglomerate of separate parts, not a single building. Indeed stone structures often replaced the earlier wooden ones piecemeal. Italian architects also rebuilt walls and erected towers in the Kremlin, while Alevisio surrounded it with a moat by joining the waters of the rivers Moscow and Neglinnaia. Soviet specialists insist that the Muscovite Kremlin became the greatest citadel of its kind in Europe. They also stress the point that its architecture made use of the existing terrain, by contrast with the Italian tradition which required leveling and preparation of a site for building. But we shall return to the Kremlin when we deal with Muscovite Russia.

More than architecture, icon painting has frequently been considered the medieval Russian art par excellence, the greatest and most authentic expression of the spirituality and the creative genius of the Russians of the appanage period. As we have seen, icon painting came to Russia with Christianity from Byzantium. However, apparently quite early the Russians proceeded to modify their Byzantine heritage and to develop the rudiments of an original style. In the centuries which followed the collapse of the Kievan state several magnificent Russian schools of icon painting came into their own. To understand their role in the life and culture of the Russians, one should appreciate the importance of icons to a believer who finds in them a direct link with the other world and, in effect, a materialization of that other world. If, on one hand, icons might suggest superstition and even idolatry, they represent, on the other, one of the most radical and powerful attempts to grapple with such fundamental Christian doctrines as the incarnation and the transfiguration of the universe. And, in the appanage period, pictorial representation provided otherwise unobtainable information and education for the illiterate masses.

The first original Russian school of icon painting appeared in Suzdal at the end of the thirteenth century, flourished in the fourteenth, and merged early in the fifteenth with the Muscovite school. Like the architecture of Suzdal, the icons are characterized by elegance, grace, and fine taste, and can also be distinguished, according to Grabar, by "a general tone, which is always cool, silvery, in contrast to Novgorodian painting which inevitably

tends towards the warm, the yellowish, the golden." The famous icon of
Saints Boris and Gleb and that of Archangel Michael on a silver background provide excellent examples of the icon painting of Suzdal.

"The warm, the yellowish, the golden" Novgorodian school deserves
further notice because of its monumentality and generally bright colors.
The icons are often in the grand style, large in size, massive in composition,
and full of figures and action. "The Praying Novgorodians" and "The
Miracle of Our Lady," also known as "The Battle between the Men of
Suzdal and the Novgorodians," illustrate the above-mentioned points. The
Novgorodian school reached its highest development around the middle of
the fifteenth century, and its influence continued after the fall of the city.

In the second half of the fourteenth century a distinct school formed in
and around Moscow. Soon it came to be led by the most celebrated icon
painter of all times, Andrew Rublev, who lived approximately from 1370
to 1430. The few extant works known to be Rublev's, especially his masterpiece, a representation of the so-called Old Testament Holy Trinity, demonstrate exquisite drawing, composition, rhythm, harmony, and lyricism.
Muratov, stressing the influence of St. Sergius on the artist, describes
Rublev's chef d'oeuvre as follows:

> This masterpiece is imbued with a suave and mystical spirituality. The
> composition is simple and harmonious; following its own rhythm, free
> from any emphasis or heaviness, it obeys a movement clearly discernible
> and yet hardly noticeable. The impression of harmony, peace, light and
> integrity which this icon produces, is a revelation of the spirit of St. Sergius.

Dionysus, who was active in the first decade of the sixteenth century, stood
out as the greatest continuer of the traditions of Rublev and the Muscovite
school. Contemporaries mentioned his name immediately after Rublev's,
and his few remaining creations support this high esteem. The icons of
Dionysus are distinguished by a marvelous grace, especially in the de-
lineation of figures, and by a certain perfection and polish. For subjects
he often chose the Virgin Mary, the protectress of the city of Moscow,
and the Holy Family. It should be noted that the works of Rublev and
Dionysus set the high standard of icon paintings not only in Russia, but
also generally in the Orthodox East.

In addition to the icons, some very valuable frescoes have come down
to us from the appanage period. Located in old churches, they include
works possibly of Rublev and certainly of Dionysus and his followers. The
art of the miniature also continued to develop, achieving a high degree
of excellence in the fifteenth century. The so-called Khitrovo Gospels of
the beginning of the fifteenth century and some other manuscripts contained excellent illustrations and illumination. By contrast with all these
forms of painting, sculpture was stifled because the Orthodox Church continued its ban on statuary, although, contrary to a popular misconception,

even large-scale sculpture was not unknown in ancient Russia. Miniature sculpture, which was permitted, developed in a remarkable manner. Cutting figures one inch and less in height, Russian artists managed to represent saints, scenes from the Gospels, and even trees, hills, and buildings as background. The most famous practitioner of this difficult art was the monk Ambrosius, whose work is linked to the Holy Trinity-St. Sergius Monastery. In spite of general poverty, certain artistic crafts, especially embroidery, also developed brilliantly in the appanage period.

Education

In the appanage period, education was in eclipse. As already indicated, the Mongol devastation and the relative isolation and poverty characteristic of the age led to a diminution in culture and learning. The decline of Russian towns played an especially significant role in this process, because Kievan culture had been essentially urban. Studying documents of the appanage period, we find mention of illiterate princes, and we note repeated complaints on the part of the higher clergy of the ignorance of priests. The masses of people, of course, received no education at all, although a certain slight qualification of that statement might be in order on the basis of the already-mentioned Novgorodian birchbark documents. Yet some learning and skills did remain to support the cultural development outlined in this chapter. They were preserved and promoted largely by the monasteries — as happened earlier and under comparable conditions in the West — not only by the great Holy Trinity-St. Sergius Monastery north of Moscow, but also by such distant ones as that of St. Cyril on the White Lake or the Solovetskii on the White Sea. The first century after the Mongol invasion seems to have been the nadir. With the rise of Moscow, education and learning in Russia likewise began a painful ascent.

X I V

* * * * * * * * * *

THE LITHUANIAN-RUSSIAN STATE

> And one more trait distinguishing the grand princedom of Lithuania from its origin revealed itself. This state from the very beginning was not simply Lithuanian, but Lithuanian-Russian.
>
> LIUBAVSKY

> Lithuania's expansion, almost unique in its rapid success, thus proved beyond the real forces of the Lithuanians alone and of a dynasty which in spite of the unusual qualities of many of its members was too divided by the petty rivalries of its various branches to guarantee a joint action under one chief. . . . The comparatively small group of ethnic Lithuanians would have been the main victim, but the whole of East Central Europe would have suffered from a chaotic situation amidst German, Muscovite, and possibly Tartar interference. . . . A union of Poland with Lithuania and her Ruthenian lands, added to those already connected with Poland, could indeed create a new great power, comprising a large and crucial section of East Central Europe and strong enough to check both German and Muscovite advance. The amazing success of a plan which would seem almost fantastic was a turning point in the history not only of that region but also of Europe.
>
> HALECKI

Whereas by the reign of Basil III the Muscovite rulers had managed to bring a large part of the former territory of the Kievan state under their authority, another large part of the Kievan inheritance remained in the possession of the grand princes of Lithuania. In effect, the history of the western Russian lands was linked for centuries to the social systems and fortunes of Lithuania and Poland.

The Evolution of the Lithuanian State

The Lithuanians, whose language belongs to the Baltic subfamily of the Indo-European family, appeared late on the historical scene, although for a very long time they had inhabited the forests of the Baltic region. It was apparently the pressure of the Teutonic Knights — the same who attacked Novgorod — that finally forced a number of Lithuanian tribes into a semblance of unity under the leadership of Mindovg, or Mindaugas, whose rule is dated approximately 1240–63. Mindovg accepted Christianity and received a crown from Pope Innocent IV only to sever his Western connections and relapse into paganism. A period of internal strife and

132

The Lithuanian-Russian State after c.1300

SWEDEN

NOVGOROD

Ladoga

Novgorod

Baltic Sea

KNIGHTS LIVONIA

Pskov

Riga

Volga

Tver

Danzig

TEUTONIC

Grodno

Vilna

Polotsk

Vitebsk

Smolensk

Kaluga

Oka

Orel

TANNENBERG
1410

Minsk

WHITE
RUSSIA

POLAND

Warsaw

KINGDOM OF POLAND
(after Union of Lublin, 1569)

Lublin

Vladimir-in-V.

VOLYNIA

Pripyat

Desna

Chernigov

Kiev

L I T H U A N I A

Dnieper

Worskla

1399

Cracow

Lvov

GALICIA

CARPATHIAN MTS.

HUNGARY

S. Bug

Dniester

Pruth

MONGOLS

Kievan Russia about 1100
Lithuania about 1300
Lithuania about 1360
× Battle sites

0   Miles   200

MOLDAVIA

WALLACHIA

Danube

Black
Sea

CRIMEA

Vistula

Bug

rapidly changing rulers followed his assassination. However, toward the
end of the thirteenth century Viten, or Vytenis, managed to unite the
Lithuanians again. He ruled as grand prince from 1295 to 1316, acted
energetically at home and in foreign relations, and perished fighting the
Teutonic Knights.

Viten's brother Gedymin (Gediminas), who reigned from 1316 to 1341,
has been called the true founder of the Lithuanian state. He completed the
unification of the Lithuanian tribes and strove hard to organize his possessions into a viable political unit. What is more, he extended his dominion

to the southeast. Some Russian territories, notably in the Polotsk area, had already become parts of the Lithuanian principality under Mindovg; with Gedymin, that principality began a massive expansion into Russia. Vilna—Vilnius in Lithuanian — became the capital of the growing state.

Gedymin's famous son Olgerd, or Algirdas, who died in 1377, carried the work of his father much further. Assisted by his valiant brother Keistut, or Kestutis, who undertook the heavy task of blocking the formidable Teutonic Order in the west, Olgerd expanded eastward with a stunning rapidity. The Russian lands which he brought under his authority included, among others, those of Volynia, Kiev, and Chernigov, and a large part of Smolensk. In the process, he defeated the Polish effort to win Volynia and fought successfully against the Mongols. Lithuanian sway spread from the Baltic to the Black Sea. Indeed, Olgerd wanted to rule all of Russia. Three times he campaigned against the Muscovite state, and twice he besieged Moscow itself, although he failed to capture it or to force the issue otherwise.

The sweeping Lithuanian expansion into Russia has more explanations than one. Obviously, internal division and foreign invasions had made the Russian power of resistance extremely low. But it should also be noted that the attacks of the Lithuanians could not be compared in destruction and brutality to the invasions of the Mongols or the Teutonic Knights, and that their domination, in a sense, did not represent foreign rule for the Russians. Indeed, many historians speak, on good evidence, of a Lithuanian-Russian state. Population statistics help to illustrate the situation: it has been estimated that, after the expansion of the Lithuanian state virtually to the Black Sea, two-thirds or even three-fourths and more of its people were Russians. Also, very little social displacement took place: the towns retained their Russian character; the Russian boyars and the Orthodox Church kept their high positions and extensive privileges; Russian princes continued to rule in different appanages next to Lithuanian princes, all subject to the Lithuanian grand prince; and intermarriage between the two aristocracies was quite common. Perhaps as important as the superior numbers of the Russian element was the fact that the Lithuanians, on their part, had little to offer and much to learn. Coming from a still pagan and relatively isolated and culturally backward area, the ruling circles of Lithuania eagerly accepted the culture of Kievan Russia. The Lithuanian army, administration, legal system, and finance were organized on the Russian pattern, and Russian became the official language of the new state. As Platonov insisted in the case of Grand Prince Olgerd of Lithuania: "In relation to different nationalities, it can be said that Olgerd's entire sympathy and attention concentrated on the Russian nationality. By his opinions, habits, and family connections, Olgerd belonged to the Russian nationality and served as its representative in Lithuania." Not surprisingly, then, the

Lithuanian state could well be considered as another variation on the Kievan theme and an heir to Kiev, rather than a foreign body imposed upon Russia. And this made its rivalry with Moscow, the other successful heir, all the more fundamental and significant.

However, shortly after Olgerd's death a new major element entered the situation: a link between Lithuania and Poland. In 1386, following the dynastic agreement of Krewo of 1385, Olgerd's son and successor Jagiello, or Jogaila — who reigned from 1377 to 1434 — married Queen Jadwiga of Poland. Because the Polish Piast ruling family had no male members left, Jagiello became the legitimate sovereign of both states, with the Polish name of Wladyslaw II. The states remained distinct, and the union personal. In fact, in 1392 Jagiello had to recognize his cousin, Keistut's son Vitovt, or Vytautas, as a separate, although vassal, grand prince of Lithuania, an arrangement extended in 1413 to subsequent rulers of the two states. Yet both positions came to be occupied by the same man again when, in 1447, Casimir IV ascended the Polish throne without relinquishing his position as grand prince of Lithuania. Whether with the same or different rulers, Poland exercised a major and increasing influence on Lithuania after 1385.

The late fourteenth and early fifteenth century was a remarkable period in the history of the Lithuanian state. Within the decade from 1387 to 1396, Moldavia, Wallachia, and Bessarabia accepted Lithuanian suzerainty. Vitovt's rule, which lasted from 1392 to 1430, witnessed the greatest extension of the Lithuanian domain, with still more alluring possibilities in sight, as Lithuania continued to challenge Moscow for supremacy on the great Russian plain. In addition, in 1410 Vitovt personally led his army in the crucial battle of Tannenberg, or Grünwald, where the joint forces of Poland and Lithuania crushed the Teutonic Knights, thus finally eliminating this deadly threat to both Slav and Lithuanian. The Lithuanian prince's great defeat came in 1399, when his major campaign against the Mongols met disaster at their hands. Some historians believe that had Vitovt won rather than lost on the banks of the Vorskla, he could then have asserted his will successfully against both Moscow and Poland and given a different direction to eastern European history.

Jagiello's marriage, in the last analysis, proved more important for Lithuania than Vitovt's wars. It marked the beginning of a Polonization of the country. Significantly, in order to marry Jadwiga, Jagiello forsook Orthodoxy for Roman Catholicism. Moreover, he had his pagan Lithuanians converted to Catholicism. The clergy, naturally, came to Lithuania from Poland, and the Church became a great stronghold of Polish influence. It has been noted, for instance, that three of the first four bishops of Vilnius were definitely Poles, and that the Poles constituted the majority in the Vilnius chapter even at the end of the fifteenth century. Education followed religion: the first schools were either cathedral or monastic schools, and their teachers were mainly members of the clergy. To obtain higher education, unavailable at home, the Lithuanians went to the great Polish university at Cracow, which provided the much-needed training for the Lithuanian elite. Russian historians, who stress the cultural impact of the Russians on the Lithuanians, often fail to appreciate the powerful attraction of the glorious Polish culture of the late Middle Ages and the Renaissance. Naturally the Lithuanians were dazzled by what Poland had to offer. Naturally too Polish specialists, ranging from architects and artists to diplomats, appeared in Lithuania. Even Polish colonists came. But, to return to the Church, its influence extended, of course, beyond religion proper, education, and culture, to society, economics, and politics. Church estates grew, and they remained exempt from general taxation. The bishops sat in the council of the grand prince, while many clerics, highly esteemed for their education, engaged in the conduct of state business.

Polonization was the most extensive at the court and among the upper classes. Poland, with its sweeping privileges and freedom for the gentry, proved to be extremely attractive to Lithuanian landlords. Indeed, many western Russian landlords as well were Polonized, to complicate further the involved ethnic and cultural pattern of the area and contribute another element for future conflicts. Polish language and Polish customs and attitudes, stressing the independence and honor of the gentry, came gradually to dominate Lithuanian life. For example, in 1413 forty-seven Polish noble families established special relations with the same number of Lithuanian aristocratic families, each Polish family offering its coat of arms to its Lithuanian counterpart. It should also be emphasized that between 1386, that is, the marriage of Jagiello and Jadwiga and the beginning of a close relationship between Lithuania and Poland, and 1569, the year of the Union of Lublin, the Lithuanian upper classes underwent a considerable change: in general their evolution favored the development of a numerous gentry, similar to the Polish *szlachta,* while the relative importance of the great landed magnates declined.

The Union of Lublin

Over a period of time, the principality of Lithuania came into the Polish cultural and political sphere and thus ceased to be a successor state to Kiev. The Union of Lublin, which bound Poland and Lithuania firmly together, represented, one can argue, a logical culmination of the historical evolution of the Lithuanian princedom. Still, its accomplishment required a major and persistent effort on the part of the Poles. In fact, in spite of

Polish pressure and a sympathetic attitude toward Poland on the part of their own petty gentry, the Lithuanian magnates managed to block an effective union even as late as the Lublin meeting itself in 1569. Only when Sigismund II, or Sigismund Augustus, of Poland proceeded to seize large Russian territories from Lithuania and incorporate them into his own kingdom, did the Lithuanians accept Polish proposals. The Union of Lublin provided for a merger of the two states: they were to have a common sovereign and a common diet, although they retained separate laws, administrations, treasuries, and even armies. Notwithstanding an explicit recognition of equality between Lithuania and Poland and a grant of vast autonomy to the Lithuanians, the new arrangement meant a decisive Polish victory. To begin with, Poland kept the Russian lands that it had just annexed from Lithuania and that constituted the entire southern section of the principality and over a third of its total territory, including some of the richest areas. Because each county sent two representatives to the common diet and because there were many more counties in Poland than in Lithuania, the Poles outnumbered the Lithuanians in the diet by a ratio of three to one. Perhaps still more important, under conditions of union Polish influences of almost every sort were bound to spread further in Lithuania, assuring for Poland the position of the senior partner in the new commonwealth.

Constituting as it does a crucial event in the histories of several peoples, the Union of Lublin has received sharply divergent evaluations and interpretations. Polish historians in general consider it very favorably, emphasizing the diffusion of high Polish culture as well as the political and other successes resulting from the Polish-Lithuanian association. Further, they stress that the large new political entity in eastern Europe resulted from agreement, not conquest, and occasionally they even suggest it as a model for the future. Lithuanian historians, by contrast, complain that their country did not receive a fair break from Poland, which used every means to dominate its neighbor. The Russians show special concern with the fate of the Russian population: Poland's seizure of the Kiev, Volynia, and other southern areas of the Lithuanian principality in 1569 meant that their Orthodox Russian people found themselves no longer in a state which continued their traditions and to which they had become accustomed, but under foreign rule, Polish and Catholic. Besides, whatever the Polish system promised to the gentry, it had nothing but oppression for the peasants. This note of tragedy is prominent in nationalist Ukrainian historiography. For the Ukrainians, the transfer of the bulk of their land to Polish rule — the Poles had obtained Galicia earlier — marked the beginning of a new chapter in the trials and tribulations of the Ukrainian people and also set the stage for a heroic struggle for independence. In any case, for good or evil, the Union of Lublin terminated the independent history of the Lithuanian principality.

The Lithuanian State and Russian History

From the standpoint of Russian history, the Lithuanian, or Lithuanian-Russian, princedom presents particular interest as the great, unsuccessful rival of Moscow for the unification of the country. Liubavsky and other specialists have provided thoughtful explanations of why Vilna lost where Moscow won. A fundamental cause, in their opinion, was the contrast in the evolutions of central authority in the two states. Whereas princely absolutism developed in Moscow, the position of the Lithuanian rulers became progressively weaker rather than stronger. Limited by the interests of powerful boyars and largely self-governing towns, the grand princes of Lithuania turned into elected, constitutional monarchs who granted ever-increasing rights and privileges to their subjects: first they came to depend on the sanction of their aristocratic council; after the statutes of 1529 and 1566 they also needed the approval of the entire gentry gathered in a diet. Thus, as the Muscovite autocracy reached an unprecedented high in the reign of Ivan the Terrible, the authority of the Lithuanian grand princes sank to a new low. Whereas the Muscovite rulers strove, successfully on the whole, to build up a great central administration and to control the life of the country, those of Lithuania increasingly relied on, or resigned themselves to, the administration of local officials and the landlord class in general. In the showdown, the Muscovite system proved to be the stronger.

Important causes, of course, lay behind the contrasting evolutions of the two states. To refer to our earlier analysis, the princedom of Moscow arose in a relatively primitive and pioneer northeast, where rulers managed to acquire a dominant position in a fluid and expanding society. The Lithuanian principality, on the other hand, as it emerged from the Baltic forests, came to include primarily old and well-established Kievan lands. It encompassed much of the Russian southwest, and its economic, social, and political development reflected the southwestern pattern, which we discussed in a preceding chapter and which was characterized by the great power of the boyars as against the prince. Detailed studies indicate that in the princedom of Lithuania the same noble families frequently occupied the same land in the seventeenth as in the sixteenth or fifteenth centuries, that at times they were extremely rich, even granting loans to the state, and that the votchina landholding remained dominant, while the pomestie system played a secondary role. The rulers found this entrenched landed aristocracy, as well as, to a lesser degree, the old and prosperous towns, too much to contend with and had to accept restrictions on princely power. The Lithuanian connection with Poland contributed to the same end. Poland served as a model of an elective monarchy with sweeping privileges for the gentry; in fact, it presented an entire gentry culture and way of life. While the social and political structure of Lithuania evolved out of its own past, Polish influences supported the rise of the gentry, supplying it with theoretical justifications and legal sanctions. Lithuania in contrast to mono-lithic Moscow, always had to deal with different peoples and cultures and formed a federal, not a unitary, state. In the end, as already indicated, it became a junior partner to Poland rather than a serious contender for the Kievan succession.

The Lithuanian-Russian princedom also attracts the attention of his-torians of Russia because of its role in the linguistic and ethnic division of the Russians into the Great Russians, often called simply Russians, the Ukrainians, and the White Russians or Belorussians, and its particular importance for the last two groups. While the roots of the differentiation extend far back, one can speculate that events would have taken a different shape if the Russians had preserved their political unity in the Kievan state. As it actually happened, the Great Russians came to be associated with the Muscovite realm, the Ukrainians and the White Russians with Lithuania and Poland. Political separation tended to promote cultural differences, although all started with the same Kievan heritage. Francis Skorina, a scholar from Polotsk, who, early in the sixteenth century, translated the Bible and also published other works in Prague and in Vilna, has fre-quently been cited as the founder of a distinct southwestern Russian literary language and, in particular, as a forerunner of Belorussian litera-ture. The Russian Orthodox Church too, as we know, finally split admin-istratively, with a separate metropolitan established in Kiev to head the Orthodox in the Lithuanian state. The division of the Russians into the Great Russians, the Ukrainians, and the Belorussians, reinforced by cen-turies of separation, became a major factor in subsequent Russian history.

PART IV: MUSCOVITE RUSSIA

THE REIGNS OF IVAN THE TERRIBLE, 1533–84, AND OF THEODORE, 1584–98

> There is nothing more unjust than to deny that there was a principle at stake in Ivan's struggle with the boyars or to see in this struggle only political stagnation. Whether Ivan IV was himself the initiator or not — most probably he was not — yet this "oprichnina" was an attempt, a hundred and fifty years before Peter's time, to found a personal autocracy like the Petrine monarchy. . . . Just as the "reforms" had been the work of a coalition of the bourgeoisie and the boyars, the coup of 1564 was carried out by a coalition of the townsmen and the petty vassals.
>
> POKROVSKY

> The new system which he [Ivan the Terrible] set up was madness, but the madness of a genius.
>
> PARES

WITH THE REIGN of Ivan IV, the Terrible, the appanage period became definitely a thing of the past and Muscovite absolutism came fully into its own. Ivan IV was the first Muscovite ruler to be crowned tsar, to have this action approved by the Eastern patriarchs, and to use the title regularly and officially both in governing his land and in conducting foreign relations. In calling himself also "autocrat" he emphasized his complete power at home as well as the fact that he was a sovereign, not a dependent, monarch. Nevertheless, it was Ivan the Terrible's actions, rather than his titles or ideas, that offered a stunning demonstration of the new arbitrary might of the Muscovite, and now Russian, ruler. Indeed, Ivan the Terrible remains the classic Russian tyrant in spite of such successors as Peter the Great, Paul I, and Nicholas I.

Ivan the Terrible's Childhood and the First Part of His Rule

Ivan IV was only three years old in 1533 when his father, Basil III, died, leaving the government of Russia to his wife — Ivan's mother Helen, of the Glinsky family — and the boyar duma. The new regent acted in a haughty and arbitrary manner, disregarding the boyars and relying first on her uncle, the experienced Prince Michael Glinsky, and after his death on her lover, the youthful Prince Telepnev-Obolensky. In 1538 she died suddenly, possibly of poison. Boyar rule — if this phrase can be used to

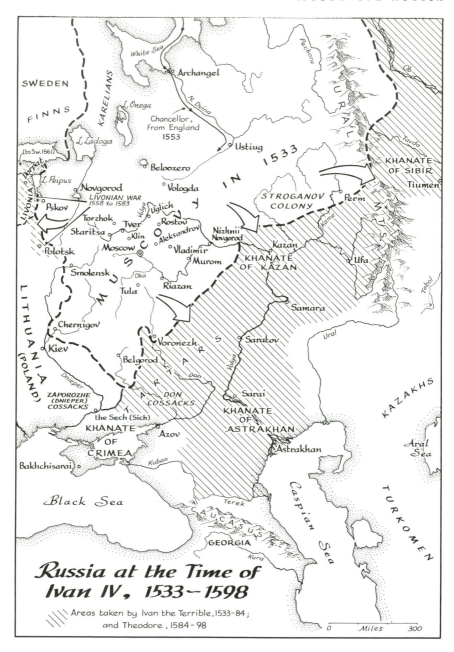

Russia at the Time of
Ivan IV, 1533–1598

Areas taken by Ivan the Terrible, 1533–84;
and Theodore, 1584–98

0   Miles   300

characterize the strife and misrule which ensued — followed her demise.
To quote one brief summary of the developments:

> The regency was disputed between two princely houses, the Shuiskys and
> the Belskys. Thrice the power changed hands and twice the Metropolitans
> themselves were forcibly changed during the struggle, one of them, Joseph,

being done to death. The Shuiskys prevailed, and three successive members of this family held power in turn. Their use of it was entirely selfish, dictated not even by class interests but simply by those of family and favour.

Imprisonments, exiles, executions, and murders proliferated.

All evidence indicates that Ivan IV was a sensitive, intelligent, and precocious boy. He learned to read early and read everything that he could find, especially Muscovite Church literature. He became of necessity painfully aware of the struggle and intrigues around him and also of the ambivalence of his own position. The same boyars who formally paid obeisance to him as autocrat and treated him with utmost respect on ceremonial occasions, neglected, insulted, and injured him in private life. In fact, they deprived him at will of his favorite servants and companions and ran the palace, as well as Russia, as they pleased. Bitterness and cruelty, expressed, for instance, in his torture of animals, became fundamental traits of the young ruler's character.

At the age of thirteen Ivan IV suddenly turned on Andrew Shuisky, who was arrested and dispatched by the tsar's servants. The autocrat entered into his inheritance. The year 1547 is commonly considered the introduction to Ivan IV's effective reign. In that year, at the age of sixteen, he decided to be crowned, not as grand prince, but as tsar, paying minute attention to details in planning the ceremony in order to make it as majestic and awe-inspiring as possible. In the same year Ivan IV married Anastasia of the popular Romanov boyar family: again, he acted with great seriousness and deliberation in selecting Anastasia from a special list of eligible young Russian ladies after he had considered and dismissed the alternative of a foreign marital alliance. The marriage turned out to be a very happy one. Still in the same year, a great fire, followed by a riot, swept Moscow. As the city burned, and even the belfry of Ivan the Great in the Kremlin collapsed, crazed mobs killed an uncle of the tsar and imperiled the tsar's own life before being dispersed. The tsar himself experienced one of the psychological crises which were periodically to mark his explosive reign. He apparently believed the disaster to be a punishment for his sins: he repented publicly in Red Square and promised to rule in the interests of the people.

What followed has traditionally been described as the first, the good, half of Ivan IV's rule. The young tsar, beneficially influenced by his kind and attractive wife, worked with a small group of able and enlightened advisers, the Chosen Council, which included Metropolitan Macarius, a priest named Sylvester, and a court official of relatively low origin, Alexis Adashev. In 1549 he called together the first full *zemskii sobor,* an institution similar to a gathering of the representatives of estates in other European countries, which will be discussed in a later chapter. While our knowledge of the assembly of 1549 remains fragmentary, it seems that Ivan IV solicited and received its approval for his projected reforms, notably for a new code of law and for changes in local government, and that he also used that occasion to hear complaints and learn opinions of his subjects concerning various matters.

In 1551 a great Church council, known as the Council of a Hundred Chapters, took place. Its decrees did much to regulate the position of the Church in relation to the state and society as well as to regulate ecclesiastical affairs proper. Significantly, the Church lost the right to acquire more land without the tsar's explicit permission, a regulation which could not, however, be effectively put into practice. In general, Metropolitan Macarius and his associates accomplished a great deal in tightening and perfecting the organization of the Church in the sprawling, but now firmly united, Russian state. One interesting aspect of this process was their incorporation of different regional Russian saints — with a number of new canonizations in 1547 and 1549 — into a single Church calendar.

Ivan the Terrible also presented to the Church council his new legal code, the *Sudebnik* of 1550, and the local government reform, and received its approval. Both measures became law. The institution of a novel scheme of local government deserves special attention as one of the more daring attempts in Russian history to resolve this perennially difficult problem. The new system aimed at the elimination of corruption and oppression on the part of centrally appointed officials by means of popular participation in local affairs. Various localities had already received permission to elect their own judicial authorities to deal, drastically if need be, with crime. Now, in areas whose population guaranteed a certain amount of dues to the treasury, other locally elected officials replaced the centrally appointed governors. And even where the governors remained, the people could elect assessors to check closely on their activities and, indeed, impeach them when necessary. But we shall return to the Muscovite system of government in a later chapter.

In 1556 Ivan IV established general regulations for military service of the gentry. While this service had existed for a long time, it remained without comprehensive organization or standardization until the new rules set a definite relationship between the size of the estate and the number of warriors and horses the landlord had to produce on demand. It should be noted that by the middle of the sixteenth century the distinction between the hereditary votchina and the pomestie, granted for service, had largely disappeared: in particular, it had become impossible to remain a landlord, hereditary or otherwise, without owing service to the tsar. In 1550 and thereabout Ivan the Terrible and his advisors also engaged in an army reform, which included new emphasis on artillery and engineering as well as development of the southern defense line. Moreover, the first permanent, regular regiments, known because of their chief weapon as the *streltsy* or musketeers, were added to the Russian army.

The military improvements came none too soon, for in the 1550's the Muscovite state was already engaging in a series of wars. Most important, a new phase appeared in the struggle against the peoples of the steppe. After Ivan IV became tsar, just as in the time of his predecessors, Russia remained subject to constant large-scale raids by a number of Tartar armies, particularly from the khanates of Kazan, Astrakhan, and the Crimea. These repeated invasions in search of booty and slaves cost the Muscovite state dearly, because of the havoc and devastation which they wrought and the immense burden of guarding the huge southeastern frontier. Certain developments in the early years of Ivan the Terrible's reign indicated that the Tartars were increasing their strength and improving their co-ordination. In 1551, however, the Russians began an offensive against the nearest Tartar enemy, the khanate of Kazan, conquering some of its vassal tribes and building the fortress of Sviiazhsk near Kazan itself. But as soon as the great campaign against Kazan opened in 1552, the Crimean Tartars, assisted by some Turkish janissaries and artillery, invaded the Muscovite territory, aiming for Moscow itself. Only after they had been checked and had withdrawn to the southern steppe could the Russians resume their advance on Kazan. The tsar's troops surrounded the city by land and water, and after a siege of six weeks stormed it successfully, using powder to blow up some of the fortifications. The Russian heroes of the bitter fighting included commanders Prince Michael Vorotynsky and Prince Andrew Kurbsky, who led the first detachment to break into the city. It took another five years to establish Russian rule over the entire territory of the khanate of Kazan.

Following the conquest of Kazan on the middle Volga, the Russians turned their attention to the mouth of the river, to Astrakhan. They seized it first in 1554 and installed their candidate there as khan. After this vassal khan established contacts with the Crimea, the Russians seized Astrakhan once more in 1556, at which time the khanate was annexed to the Muscovite state. Thus of the three chief Tartar enemies of Russia, only the Crimean state remained, with its Ottoman suzerain looming behind it. Crimean forces invaded the tsar's domain in 1554, 1557, and 1558, but were beaten back each time. On the last occasion the Russians counterattacked deep into the southern steppe, penetrating the Crimean peninsula itself.

Another major war was waged at the opposite end of the Russian state, in the northwest, against the Livonian Order. It started in 1558 over the issue of Russian access and expansion to the Baltic beyond the small hold on the coastline at the mouth of the Neva. The first phase of this war, to 1563, brought striking successes to the Muscovite armies. In 1558 alone they captured some twenty Livonian strongholds, including the greatest of them, the town of Dorpat, originally built by Iaroslav the Wise and named Iuriev. In 1561 the Livonian Order was disbanded, its territories were secularized, and its last master, Gotthard Kettler, became the hereditary Duke of Courland and a vassal of the Polish king. Yet the resulting Polish-Lithuanian offensive failed, and the Russian forces seized Polotsk from Lithuania in 1563.

Ivan IV and his assistants had many interests in the outside world other than war. As early as 1547 the Muscovite government sent an agent, the Saxon Slitte, to western Europe to invite specialists to serve the tsar. Eventually over one hundred and twenty doctors, teachers, artists, and different technicians and craftsmen from Germany accepted the Russian invitation. But when they reached Lübeck, authorities of the Hanseatic League and of the Livonian Order refused to let them through, with the result that only a few of their number ultimately came to Russia on their own. In 1553 an English captain, Richard Chancellor, in search of a new route to the East through the Arctic Ocean, reached the Russian White Sea shore near the mouth of the Northern Dvina. He went on to visit Moscow and establish direct relations between England and Russia. The agreement of 1555 gave the English great commercial advantages in the Muscovite state, for they were to pay no dues and could maintain a separate organization under the jurisdiction of their own chief factor. Arkhangelsk — Archangel in English — on the Northern Dvina became their port of entry. Ivan IV valued his English connection highly. Characteristically, the first Russian mission to England returned with some specialists in medicine and mining.

The Second Part of Ivan the Terrible's Rule

However, in spite of improvements at home and successes abroad, the "good" period of Ivan the Terrible's rule came gradually to its end. The change in the Muscovite government involved the tsar's break with the Chosen Council and his violent turning against many of his advisers and their associates and afterwards, as his suspicion and rage expanded, against the boyars as a whole. His personal despotism became extreme. Furthermore, Ivan the Terrible's assault on the boyars, bringing with it changes in the administrative mechanism of the state and a reign of terror, came to dominate, and to a considerable extent shatter, Russian political life, society, and economy.

In a sense, a conflict between the tsar and the boyars followed logically from preceding history. As Muscovite absolutism rose to its heights with Ivan the Terrible, the boyar class, constantly growing with the expansion of Moscow, represented one of the few possible checks on the sovereign's power. Furthermore, the boyars remained partly linked to the old appanage order, which the Muscovite rulers had striven hard and successfully to destroy. The size and composition of the Muscovite boyardom reflected the rapid growth of the state. While in the first half of the fifteenth century some forty boyar families served the Muscovite ruler, in the first half of the sixteenth the number of the families had increased to over two hundred. The Muscovite boyars included descendants of former Russian or Lithuanian grand princes, descendants of former appanage princes, members of old Muscovite boyar families, and, finally, members of boyar families from other parts of Russia who had transferred their service to Moscow. The first two groups, the so-called service princes, possessed the greatest influence and prestige and also the strongest links with the past: they remained at least to some extent rulers in their own localities even after they became servitors in Moscow. The power of the Muscovite boyars, however, should not be overestimated. They showed little initiative and lacked solidarity and organization. In fact, they constantly engaged in petty squabbles and intrigues against one another, a deplorable situation well illustrated during the early years of Ivan the Terrible's reign. The Muscovite system of appointments, the notorious *mestnichestvo,* based on a hierarchical ranking of boyar families, as well as of the individual members within a given family, added to the boyar disunity.

Ivan the Terrible's attitude toward his advisers and the boyars as a whole changed over a period of years under the strong impact, it would seem, of certain events. In 1553 the tsar fell gravely ill and believed himself to be on his deathbed. He asked the boyars to swear allegiance to his infant son Dmitrii, but met opposition even from some of his closest associates, such as Sylvester, not to mention a considerable number of boyars: they apparently resented the merely boyar, not princely, family of Ivan the Terrible's wife, were afraid of more misfortunes for the Muscovite state during another reign of a minor, and favored Ivan the Terrible's cousin, Prince Vladimir of Staritsa, as tsar. Although the oath to Dmitrii was finally sworn, Ivan the Terrible never forgot this troubling experience. Shortly afterwards some boyars were caught planning to escape to Lithuania. New tensions resulted from the Livonian War. In fact it led to the break between the tsar and his advisers, Sylvester and Adashev, who disapproved of the proposed offensive in the Baltic area, preferring an assault against the Crimean Tartars.

In 1560 Ivan the Terrible's young and beloved wife Anastasia died suddenly. Convinced that Sylvester and Adashev had participated in a plot to poison her, the tsar had them condemned in extraordinary judicial proceedings, in the course of which they were not allowed to appear to state their case. The priest was apparently exiled to a distant monastery; the layman thrown into jail where he died. Before long Ivan the Terrible's wrath descended upon everyone connected with the Chosen Council. Adashev's and Sylvester's relatives, associates, and friends perished without trial. Two princes lost their lives merely because they expressed disapproval of the tsar's behavior. At this turn of events, a number of boyars fled to Lithuania. The escapees included a famous commander and associate of the tsar, Prince Andrew Kurbsky, who spent the rest of his life organizing forces and coalitions against his former sovereign. Kurbsky is best known, however, for the remarkable letters which he exchanged with Ivan the Terrible in 1564–79 and which will demand our attention when we deal with the political thought of Muscovite Russia.

In late 1564 Ivan IV suddenly abandoned Moscow for the small town of Aleksandrov some sixty miles away. A month later two letters, addressed to the metropolitan, arrived from the tsar. In them Ivan IV expressed his desire to retire from the throne and denounced the boyars and the clergy. Yet, in the letter to be read to the masses, he emphasized that he had no complaints against the common people. In confusion and consternation, the boyars and the people of Moscow begged the tsar to return and rule over them. Ivan the Terrible did return in February 1565, after his two conditions had been accepted: the creation of a special institution and subdivision in the Muscovite state, known as the *oprichnina* — from the word *oprich,* that is, *apart, beside* — to be managed entirely at the tsar's own discretion; and an endorsement of the tsar's right to punish evil-doers and traitors as he would see fit, executing them when necessary and confiscating their possessions. After the tsar returned to Moscow, it became apparent to those who knew him that he had experienced another shattering psychological crisis, for his eyes were dim and his hair and beard almost gone.

The oprichnina acquired more than one meaning. It came to stand for a separate jurisdiction within Russia which consisted originally of some twenty towns with their countryside, several special sections scattered throughout the state, and a part of Moscow where Ivan the Terrible built a new palace. Eventually it extended to well over a third of the Muscovite realm. The tsar set up a separate state administration for the oprichnina, paralleling the one in existence which was retained for the rest of the country, now known as the *zemshchina.* Much later there was even established a new and nominal ruler, a baptized Tartar prince Simeon, to whom Ivan the Terrible pretended to render homage. Our knowledge of the structure and functioning of the oprichnina administration remains fairly limited. Platonov suggested that after the reform of 1564 the state had actually one set of institutions, but two sets of officials. In any case, new men under the direct control of Ivan the Terrible ran the oprichina, whereas the zemshchina stayed within the purview of the boyar duma and old officialdom. In fact, many landlords in the territory of the oprichnina were transferred elsewhere, while their lands were granted to the new servitors of the tsar. The term *oprichnina* also came to designate especially this new corps of servants to Ivan the Terrible — called *oprichniki* — who are described sometimes today as gendarmes or political police. The oprichniki, dressed in black and riding black horses, numbered at first one thousand and later as many as six thousand. Their purpose was to destroy those whom the tsar considered to be his enemies.

A reign of terror followed. Boyars and other people linked to Prince Kurbsky, who had escaped to Lithuania, fell first. The tsar's cousin, Prince Vladimir of Staritsa, perished in his turn, together with his relatives, friends, and associates. The circle of suspects and victims kept widening: not only more and more boyars, but also their families, relatives, friends, and even servants and peasants were swept away in the purge. The estates of the victims and the villages of their peasants were confiscated by the state, and often plundered or simply burned. Ivan the Terrible brooked no contradiction. Metropolitan Philip, who dared remonstrate with the tsar, was thrown into jail and killed there by the oprichniki. Entire towns, such as Torzhok, Klin, and, especially, in 1570, Novgorod, suffered utter devastation and ruin. It looked as if a civil war were raging in the Muscovite state, but a peculiar civil war, for the attackers met no resistance. It might be added that the wave of extermination engulfed some of the leading oprichniki themselves. In 1572 Ivan the Terrible declared the oprichnina abolished, although division of the state into two parts lasted at least until 1575.

Following the death of his first wife, Ivan the Terrible appeared to have lost his emotional balance. His six subsequent wives never exercised the same beneficial influence on him as had Anastasia. The tsar was increasingly given to feelings of persecution and outbreaks of wild rage. He saw traitors everywhere. After the oprichnina began its work, Ivan the Terrible's life became part of a nightmare which he had brought into being. With Maliuta Skuratov and other oprichniki the sovereign personally participated in the investigations and the horrid tortures and executions. Weirdly he alternated dissolution and utmost cruelty with repentance, and blasphemy with prayer. Some contemporary accounts of the events defy imagination. In 1581, in a fit of violence, Ivan the Terrible struck his son and heir Ivan with a pointed staff and mortally wounded him. It has been said that from that time on he knew no peace at all. The tsar died in March 1584, a Soviet autopsy of his body indicating poisoning.

While the oprichnina was raging inside Russia, enemies pressed from the outside. Although the Crimean Tartars failed to take Astrakhan in 1569, in 1571 Khan Davlet-Geray led them to Moscow itself. Unable to seize the Kremlin, they burned much of the city. They withdrew from the Muscovite state only after laying waste a large area and capturing an enormous booty and 100,000 prisoners. Famine and plague added to the horror of the Tartar devastation. The following year, however, a new invasion by the Crimean Tartars met disaster at the hands of a Russian army.

The Muscovite unpreparedness for the Crimean Tartars resulted largely from the increasing demands of the Livonian War. Begun by Ivan the Terrible in 1558 and prosecuted with great success for a number of years, this major enterprise, too, started to turn against the Russians. In his effort to expand in the Baltic area, the tsar found himself opposed by a united Lithuania and Poland after 1569, and also by Sweden. After the death of Sigismund II in 1572, Poland had experienced several turbulent years: two elections to the Polish throne involved many interests and intrigues, with the Hapsburgs making a determined bid to secure the crown, and Ivan the Terrible himself promoted as a candidate by another party; also, the successful competitor, Henry of Valois, elected king in 1573, left the country the following year to succeed his deceased brother on the French throne. The situation changed after the election in 1575 of the Hungarian Prince of Transylvania, Stephen Bathory, as King of Poland. The new ruler brought stability and enhanced his reputation as an excellent general. In 1578 the Poles started an offensive in southern Livonia. The following year they captured Polotsk and Velikie Luki, although, in exceptionally bitter combat, they failed to take Pskov. On their side, in 1578, the Swedes smashed a Russian army at Wenden. By the treaties of 1582 with Poland and 1583 with Sweden, Russia had to renounce all it had gained during the first part of the war and even cede several additional towns to Sweden. Thus, after some twenty-five years of fighting, Ivan the Terrible's move to the Baltic failed dismally. The Muscovite state lay prostrate from the internal ravages of the oprichnina and continuous foreign war.

In concluding the story of Ivan the Terrible, mention should be made of one more development, in the last years of his reign, pregnant with consequences for subsequent Russian history: Ermak's so-called conquest of Siberia. Even prior to the Mongol invasion the Novgorodians had penetrated beyond the Urals. The Russians used northern routes to enter Siberia by both land and sea and, by the middle of the sixteenth century, had already reached the mouth of the Enisei. In the sixteenth century the Stroganov family developed large-scale industries, including the extracting of salt and the procurement of fish and furs, in northeastern European Russia, especially in the Ustiug area. After the conquest of Kazan, the Stroganovs obtained from the government large holdings in the wild upper Kama region, where they maintained garrisons and imported colonists. The local native tribes' resistance to the Russians was encouraged by their nominal suzerain, the so-called khan of Sibir, or Siberia, beyond the Urals. In 1582 the Stroganovs sent an expedition against the Siberian khanate. It consisted of perhaps 1650 cossacks and other volunteers, led by a cossack commander,

Ermak. Greatly outnumbered, but making good use of their better organization, firearms, and daring, the Russians defeated the natives in repeated engagements and seized the headquarters of the Siberian Khan Kuchum. Ivan the Terrible appreciated the importance of this unexpected conquest, accepted the new territories into his realm, and sent reinforcements. Although Ermak perished in the struggle in 1584 before help arrived and although the conquest of the Siberian khanate had to be repeated, the Stroganov expedition marked in effect the beginning of the establishment of Russian control in western Siberia. Tiumen, a fortified town, was built there in 1586, and another fortified town, Tobolsk, was built in 1587 and subsequently became an important administrative center.

Explanations

The eventful and tragic reign of Ivan the Terrible has received different evaluations and interpretations. In general, the judgments of historians have fallen into two categories: an emphasis on the tsar's pathological character, indeed madness, and an explanation of his actions on the basis of fundamental Muscovite needs and problems, and thus in terms of a larger purpose on his part. Personal denunciation of Ivan the Terrible, together with the division of his reign into the first, good, half, when the tsar listened to his advisers, and the second, bad, half, when he became a bloodthirsty tyrant, derives from the accounts of Andrew Kurbsky, as well as, to a lesser extent, of some other contemporaries. Karamzin adopted this view in his extremely influential history of the Russian state, and it has been accepted by many later scholars.

The view stressing political, social, and economic reasons for the events of Ivan IV's reign has also had numerous adherents. Platonov did particularly valuable work in elucidating the nature of the oprichnina and the reasons for its establishment. He argued that the Chosen Council had indeed ruled Russia, representing a usurpation of power by the boyars. Ivan the Terrible's struggle against it and against the boyars as a whole marked one of the most important developments in the evolution of the centralized Russian monarchy. Moreover, the tsar waged this struggle with foresight and intelligence. Platonov pointed out that the lands taken into the oprichnina, in particular in central Russia, included many estates of the descendants of former appanage princes and princelings who in their hereditary possessions had retained the prestige and largely the authority of rulers, including the rights to judge and collect taxes. Their transfer to other lands where they had no special standing or power and their replacement with reliable new men, together with the wholesale suppression of the boyar opposition, ensured the tsar's victory over the remnants of the old order.

Henceforth, the boyars were to be their monarch's obedient servants both in the duma and in their assigned military and administrative posts. In addition, the oprichnina territory contained important commercial centers and routes, notably the new trade artery from Archangel to central Russia. Platonov saw in this arrangement Ivan the Terrible's effort to satisfy the financial needs of the oprichnina; some Marxist historians have offered it as evidence of a new class alignment. Furthermore, the oprichnina gave the tsar an opportunity to bypass the mestnichestvo system and to bring to the fore servicemen from among the gentry, most of whom remained in important government work even after the country had returned to normalcy. And it provided an effective police corps to fight opposition and treason. The bitterness and the cruelty of the struggle stemmed likewise from more basic reasons than the tsar's character. In fact, in this respect too Ivan the Terrible's reign provided a close parallel to those of Louis XI in France or Henry VIII in England, who similarly suppressed their aristocracies. Platonov added that the tsar began with relatively mild measures and turned to severe punishments only after the boyar opposition continued.

Marxist historians have developed an analysis of Ivan IV's reign in terms of the class struggle. Pokrovsky and others have interpreted the reforms of 1564 as a shift from boyar control of the government to an alliance between the crown and the service gentry and merchants, to whom the tsar turned at the zemskii sobor of 1566 on the issue of the Livonian War and on other occasions. In fact, Ivan IV tried to establish, long before Peter the Great, an effective personal autocracy. Other Soviet scholars, especially Wipper, have placed heavy emphasis on the reality of treason in the reign of Ivan the Terrible and the need to combat it. In general, Soviet historians have gradually come to stress the progressive nature of Ivan IV's rule in Russia as well as the tsar's able championing of Russian national interests against foreign foes, although recently Makovsky restated emphatically the negative view of the reign. The Soviet cinema versions of the reign of Ivan the Terrible reflect some of the major characteristics and problems of the shifting Soviet interpretations of the tsar and the period. It might be added that the Soviet evaluation of Ivan IV has, apparently, interesting points of contact with the image which the brilliant and restless tsar left with the Russian people. It seems that his popular epithet *Groznyi* — usually rendered ambiguously and inadequately in English as "Terrible" — implied admiration rather than censure and referred to his might, perhaps in connection with the victory over the khanate of Kazan or other successes. On occasion the epithet was also applied to Ivan III in this sense.

Yet, after all the able and valuable rational explanations of Ivan the Terrible's actions in the broad setting of Russian history, grave doubts remain. Even if the boyars, or at least their upper layer, constituted an element linked to the appanage past and opposed to the Muscovite centralization, we have very little evidence to indicate that they were organized, aggressive, or otherwise presented a serious threat to the throne. Probably, given time, their position would have declined further, eliminating any need for drastic action. The story of the oprichnina is that of civil massacre, not civil war. Also, even Platonov failed to provide objective reasons for many of Ivan IV's measures, such as his setting up Simeon as the Russian ruler to whom Ivan himself paid obeisance — although it should be added that some other historians tried to find rational explanations where Platonov admitted defeat. Most important, the pathological element in the tsar's behavior cannot be denied. People of such character have brought about many private tragedies. Ivan the Terrible, however, was not just a private person but the absolute ruler of a huge state.

The Reign of Theodore

The reign of Ivan IV's eldest surviving son Theodore, or Fedor, 1584–98, gave Russia a measure of peace. Physically weak and extremely limited in intelligence and ability, but well meaning as well as very religious, the new tsar relied entirely on his advisers. Fortunately, these advisers, especially Boris Godunov, performed their task fairly well.

An important and extraordinary event of the reign consisted in the establishment of a patriarchate in Russia in 1589. Largely as a result of Boris Godunov's skillful diplomacy, the Russians managed to obtain the consent of the patriarch of Constantinople, Jeremiah, to elevating the head of the Russian Church to the rank of patriarch, the highest in the Orthodox world. Later all Eastern patriarchs agreed to this step, although with some reluctance. Boris Godunov's friend, Metropolitan Job, became the first Muscovite patriarch. The new importance of the Russian Church led to an upgrading and enlargement of its hierarchy through the appointment of a number of new metropolitans, archbishops, and bishops. This strengthening of the organization of the Church proved to be significant in the Time of Troubles.

Foreign relations in the course of the reign included Theodore's unsuccessful candidacy to the Polish throne, following Stephen Bathory's death in 1586, and a successful war against Sweden, which ended in 1595 with the return to the Muscovite state of the towns and territory near the Gulf of Finland which had been ceded by the treaty of 1583. The pre-Livonian War frontier was thus re-established. In 1586 an Orthodox Georgian kingdom in Transcaucasia, beset by Moslems, begged to be accepted as a vassal of the Russian tsar. While Georgia lay too far away for more than a nominal, transitory connection to be established in the sixteenth century, the request pointed to one direction of later Russian expansion.

Theodore's reign also witnessed, in 1591, the death of Prince Dmitrii of Uglich in a setting which made it one of the most famous detective stories of Russian history. Nine-and-a-half-year-old Dmitrii, the tsar's brother and the only other remaining male member of the ruling family, died, his throat slit, in the courtyard of his residence in Uglich. The populace rioted, accused the child's guardians of murder, and killed them. An official investigating commission, headed by Prince Basil Shuisky, declared that Dmitrii had been playing with a knife and had injured himself fatally while in an epileptic fit. Many contemporaries and later historians concluded that Dmitrii had been murdered on orders of Boris Godunov who had determined to become tsar himself. Platonov, however, argued persuasively against this view: as a son of Ivan the Terrible's seventh wife — while canonically only three were allowed — Dmitrii's rights to the throne were highly dubious; the tsar, still in his thirties, could well have a son or sons of his own; Boris Godunov would have staged the murder much more skillfully, without immediate leads to his agents and associates. More recently Vernadsky established that no first-hand evidence of an assassination exists at all, although accusations of murder arose immediately following Prince Dmitrii's apparently accidental death. But, whereas scholars may well remain satisfied with Platonov's and Vernadsky's explanation, the general public will, no doubt, prefer the older version, enshrined in Pushkin's play and Musorgsky's opera, *Boris Godunov*.

Even if Boris Godunov did not murder Dmitrii, he made every other effort to secure power. Coming from a Mongol gentry family which had been converted to Orthodoxy and Russified, himself virtually illiterate, Boris Godunov showed uncanny intelligence and abilities in palace intrigue, diplomacy, and statecraft. He capitalized also on his proximity to Tsar Theodore, who was married to Boris's sister, Irene. In the course of several years Boris Godunov managed to defeat his rivals at court and become the effective ruler of Russia in about 1588. In addition to power and enormous private wealth, Boris Godunov obtained exceptional outward signs of his high position: a most impressive and ever-growing official title; the formal right to conduct foreign relations on behalf of the Muscovite state; and a separate court, imitating that of the tsar, where foreign ambassadors had to present themselves after they had paid their respects to Theodore. When the tsar died in 1598, without an heir, Boris Godunov stood ready and waiting to ascend the throne. His reign, however, was to be not so much a successful consummation of his ambition as a prelude to the Time of Troubles.

XVI

* * * * * * * * * *

THE TIME OF TROUBLES, 1598–1613

> O God, save thy people, and bless thine heritage . . . , preserve this
> city and this holy Temple, and every city and land from pestilence,
> famine, earthquake, flood, fire, the sword, the invasion of enemies,
> and from civil war. . . .
>
> AN ORTHODOX PRAYER

THE TIME OF TROUBLES — *Smutnoe Vremia,* in Russian — refers to a particularly turbulent, confusing, and painful segment of Russian history at the beginning of the seventeenth century, or, roughly, from Boris Godunov's accession to the Muscovite throne in 1598 to the election of Michael as tsar and the establishment of the Romanov dynasty in Russia in 1613. Following the greatest student of the Time of Troubles, Platonov, we may subdivide those years into three consecutive segments on the basis of the paramount issues at stake: the dynastic, the social, and the national. This classification immediately suggests the complexity of the subject.

The dynastic aspect stemmed from the fact that with the passing of Tsar Theodore the Muscovite ruling family died out. For the first time in Muscovite history there remained no natural successor to the throne. The problem of succession was exacerbated because there existed no law of succession in the Muscovite state, because a number of claimants appeared, because Russians looked in different directions for a new ruler, and because, apparently, they placed a very high premium on some link with the extinct dynasty, which opened the way to fantastic intrigues and impersonations.

While the dynastic issue emerged through the accidental absence of an heir, the national issue resulted largely from the centuries-old Russian struggle in the west and in the north. Poland, and to a lesser extent Sweden, felt compelled to take advantage of the sudden Russian weakness. The complex involvement of Poland, especially, in the Time of Troubles reflected some of the key problems and possibilities in the history of eastern Europe.

But it is the social element that demands our main attention. For it was the social disorganization, strife, and virtual collapse that made the dynastic issue so critical and opened the Muscovite state to foreign intrigues and invasions. The Time of Troubles can be understood only as the end product of the rise of the Muscovite state with its attendant dislocations and tensions. It has often been said that Russian history, by comparison with the histories of western European countries, has represented a cruder or simpler process, in particular that Russian social structure has exhibited a certain lack of complexity and differentiation. While this approach must be treated circumspectly, it must not be dismissed. We noted earlier that it might be appropriate to describe appanage Russia in terms of an incipient or undeveloped feudalism. The rise of Moscow meant a further drastic simplification of Russian social relations.

To expand and to defend its growing territory, the Muscovite state relied on service people, that is, on men who fought its battles and also performed the administrative and other work for the government. The service people — eventually known as the service gentry, or simply gentry — were supported by their estates. In this manner, the pomestie, an estate granted for service, became basic to the Muscovite social order. After the acquisition of Novgorod, in its continuing search for land suitable for pomestiia, the Muscovite government confiscated most of the holdings of the Novgorodian boyars and even half of those of the Novgorodian Church. Hereditary landlords too, it will be remembered, found themselves obligated to serve the state. The rapid Muscovite expansion and the continuous wars on all frontiers, except the north and northeast, taxed the resources of the government and the people to the breaking point. Muscovite authorities made frantic efforts to obtain more service gentry. "Needing men fit for military service, in addition to the old class of its servitors, free and bonded, nobles and commoners, the government selects the necessary men and establishes on pomestiia people from everywhere, from all the layers of Muscovite society in which there existed elements answering the military requirements." Thus, for example, small landholders in the areas of Novgorod and Pskov and an ever-increasing number of Mongols, some of whom had not even been converted to Christianity, became members of the Muscovite service gentry.

When Moscow succeeded in the "gathering of Russia" and the appanages disappeared, the princes and boyars failed to make a strong stand against Muscovite centralization and absolutism. Many of them, indeed, were slaughtered, without offering resistance, by Ivan the Terrible. But the relatively easy victory of the Muscovite despots over the old upper classes left problems in its wake. Notably, it has been argued that the Muscovite government displaced the appanage ruling elements all too rapidly, more rapidly than it could provide effective substitutes. The resulting weakening of the political and social framework contributed its share to the Time of Troubles. And so did the boyar reaction following the decline in the tsar's authority after Boris Godunov's death.

As the Muscovite state expanded, centralizing and standardizing administration and institutions and subjugating the interests of other classes to those of the service gentry, towns also suffered. They became administrative and military centers at the expense of local self-government, commercial elements, and the middle class as a whole. This transformation occurred most strikingly in Novgorod and Pskov, but similar changes affected many other towns as well.

Most important, however, was a deterioration in the position of the peasants, who constituted the great bulk of the people. They, of course, provided the labor force on the estates of the service gentry, and, therefore, were affected immediately and directly by the rise of that class. Specifically, the growth of the service gentry meant that more and more state lands and peasants fell into gentry hands through the pomestie system. Gentry landlords, themselves straining to perform burdensome state obligations, squeezed what they could from the peasants. Furthermore, the ravages of the oprichnina brought outright disaster to the already overtaxed peasant economy of much of central Russia. Famine, which appeared in the second half of Ivan the Terrible's reign, was to return in the frightful years of 1601–3.

Many peasants tried to escape. The Russian conquest of the khanates of Kazan and Astrakhan opened up fertile lands to the southeast, and at first the government encouraged migration to consolidate the Russian hold on the area. But this policy could not be reconciled with the interests of the service gentry, whose peasants had to be prevented from fleeing if their masters were to retain the ability to serve the state. Therefore, in the last quarter of the sixteenth century, Muscovite authorities made an especially determined effort to secure and guarantee the labor force of the gentry. Legal migration ceased. The state also tried to curb Church landholding, and especially to prevent the transfer of any gentry land to the Church. Furthermore, serfdom as such finally became fully established in Russia. While the long-term process of the growth of serfdom will be discussed later, it should be mentioned here that the government's dedication to the interests of the service gentry at least contributed to it.

Hard-pressed economically and increasingly deprived of their rights, the peasants continued to flee to the borderlands in spite of all prohibitions. The shattering impact of the oprichnina provided another stimulus for the growth of that restless, dislocated, and dissatisfied lower-class element which played such a significant role during the Time of Troubles. Moreover, some fugitive peasants became cossacks. The cossacks, first mentioned in the chronicles in 1444, represented free or virtually free societies of warlike adventurers that began to emerge along distant borders and in areas of overlapping jurisdictions and uncertain control. Combining military organization and skill, the spirit of adventure, and a hatred of the Muscovite political and social system, and linked socially to the broad masses, the cossacks were to act as another major and explosive element in the Time of Troubles.

Dissatisfied elements in the Russian state included also a number of conquered peoples and tribes, especially in the Volga basin. The gentry itself, while a privileged class, had many complaints against the exacting government. Finally, it should be emphasized that conditions and problems varied in the different parts of the huge Muscovite state, and that the Time of Troubles included local as much as national developments. The Russian north, for example, had no problem of defense and very few gentry or serfs. Since a brief general account can pay only the scantest attention to these local variations, the interested student must be referred to more specialized literature, particularly to the writings of Platonov.

The Reign of Boris Godunov and the Dynastic Phase of the Time of Troubles

With the passing of Theodore, the Muscovite dynasty died out and a new tsar had to be found. While it is generally believed that Boris Godunov remained in control of the situation, he formally ascended the throne only after being elected by a specially convened zemskii sobor and implored by the patriarch, the clergy, and the people to accept the crown. He proved to be, or rather continued to be, an intelligent and able ruler. Interested in learning from the West, Boris Godunov even thought of establishing a university in Moscow, but abandoned this idea because of the opposition of the clergy. He did, however, send eighteen young men to study abroad. In foreign policy, Boris Godunov maintained peaceful relations with other countries and promoted trade, concluding commercial treaties with England and with the Hansa.

But, in spite of the efforts of the ruler, Boris Godunov's brief reign, 1598–1605, witnessed tragic events. In 1601 drought and famine brought disaster to the people. The crops failed again in 1602 and also, to a considerable extent, in 1603. Famine reached catastrophic proportions; epidemics followed. Although the government tried to feed the population of Moscow free of charge, direct supplies to other towns, and find employment for the destitute, its measures availed little against the calamity. It has been estimated that more than 100,000 people perished in the capital alone. Starving people devoured grass, bark, cadavers of animals and, on occasion, even other human beings. Large bands of desperate men that roamed and looted the countryside and sometimes gave battle to regular troops appeared and became a characteristic phenomenon of the Time of Troubles.

At this point rumors to the effect that Boris Godunov was a criminal and a usurper and that Russia was being punished for his sins began to spread. It was alleged that he had plotted to assassinate Prince Dmitrii; it was alleged further that in reality another boy had been murdered, that the prince has escaped and would return to claim his rightful inheritance. The claimant soon appeared in person. Many historians believe that False

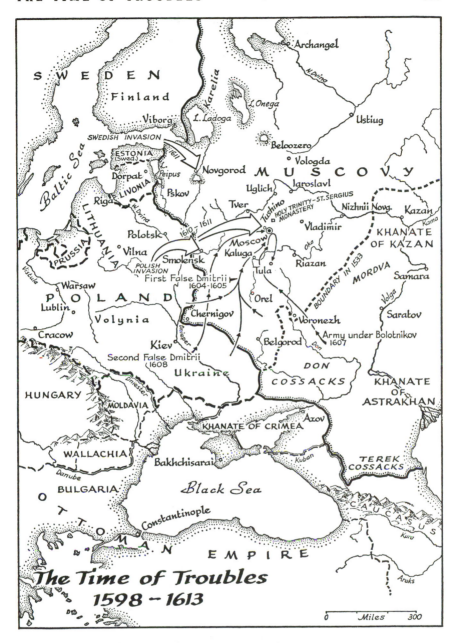

The Time of Troubles
1598 – 1613

Dmitrii was in fact a certain Gregory Otrepiev, a young man of service class origin, who had become a monk and then left his monastery. Very possibly he believed himself to be the true Prince Dmitrii. Apparently he lived in Moscow in 1601 and early 1602, but escaped to the cossacks when authorities became interested in his assertions and decided to arrest him. Next he appeared in Lithuania, where he reiterated his claim to be Ivan the Terrible's son Prince Dmitrii. While the Polish government gave him no official recognition, he obtained support from the Jesuits and from certain Lithuanian and Polish aristocrats. He also fell in love with the daughter of a Polish aristocrat, the beautiful Marina Mniszech. The Jesuits received from him the promise to champion Catholicism in Russia. The role of the Muscovite boyars in the rise of False Dmitrii remains less clear. Yet, in spite of the paucity and frequent absence of evidence, many scholars have become convinced that important boyar circles secretly supported False Dmitrii in order to destroy Boris Godunov. Indeed, the entire False Dmitrii episode has been described as a boyar stratagem. Boris Godunov, on his part, in an effort to defend his position, turned violently against the boyars around the throne, instituting in 1601 a veritable purge of them. In October 1604, False Dmitrii invaded Russia at the head of some 1,500 cossacks, Polish soldiers of fortune, and other adventurers.

Most surprisingly, the foolhardy enterprise succeeded. False Dmitrii's manifestoes proclaiming him to be the true tsar had their effect, in spite of Boris Godunov's attempts to confirm that Prince Dmitrii was dead and to brand the pretender as an impostor and a criminal by such means as his excommunication from the Church and the testimony of Gregory Otrepiev's uncle. Much of southern Russia, including such large centers as Chernigov, welcomed False Dmitrii; in a number of places authorities and population wavered in their stand, but failed to offer firm resistance. Dissatisfaction and unrest within the Muscovite state proved to be more valuable to the pretender's cause than Polish and Lithuanian aid. False Dmitrii's motley forces suffered repeated defeats, but regrouped and reappeared. Still, False Dmitrii probably owed his victory to a stroke of luck: in April 1605, when the military odds against the pretender appeared overwhelming, Boris Godunov suddenly died. Shortly after his death his commander, Theodore Basmanov, went over to False Dmitrii's side, Boris Godunov's wife and his young son and successor Theodore were deposed and murdered in Moscow, and on June 20, 1605, False Dmitrii entered the capital in triumph.

The people rejoiced at what they believed to be the miraculous return of the true tsar to ascend his ancestral throne. On the eve of the riots that overthrew the Godunovs, Basil Shuisky himself had already publicly reversed his testimony and claimed that in Uglich Prince Dmitrii had escaped the assassins, who killed another boy instead. In July 1605, Prince Dmitrii's mother, who had become a nun under the name of Martha, was brought to identify her alleged long-lost child: in the course of a tender meeting she proclaimed him her own. Followers of False Dmitrii, such as Theodore Basmanov, succeeded the supporters of Godunov around the throne. A Greek cleric, Ignatius, who had been among the first to side with the pretender, replaced Boris Godunov's friend Job as patriarch. The new tsar returned from disgrace, prison, or exile the boyars who had suffered during the last years of his predecessor's reign. Those regaining favor included Philaret, formerly Theodore, Romanov, the abbot of a northern monastery whom Boris Godunov had forced to take holy orders and exiled. Philaret became the metropolitan in Rostov.

False Dmitrii has been described as an unprepossessing figure with no waistline, arms of unequal length, red hair that habitually stood up, a large wart on his face, a big ugly nose, and an expression both unsympathetic and melancholy. His qualities, however, included undeniable courage and considerable intelligence and ability. He refused to be anyone's puppet, and in particular failed to honor his promises concerning the introduction of Catholicism into Russia. Instead of acting on these promises, he propounded the grandiose project of driving the Turks out of Europe.

Their new ruler's manners upset the Muscovites. False Dmitrii repeatedly failed to observe the established traditions and etiquette. He would not attend church services, and did not take a nap in the afternoon, but instead wandered on his own in the city, dressed as a Pole. The Polish entourage of the tsar proved still more disturbing: these Poles, loud and prominent, generally despised the Russians, who in turn suspected and hated them as enemies and heretics. But the main argument against False Dmitrii, in the opinion of Platonov and many other specialists, rested simply in the fact that he had already served his purpose. The boyars had utilized him successfully against the Godunovs and now made arrangements to dispose of him in his turn.

It would seem that almost immediately after False Dmitrii's victory Basil Shuisky and his brothers began to spread rumors to the effect that the new tsar was, after all, an impostor. Caught and condemned to death, they were instead exiled and, after several months, entirely pardoned by the clement tsar — a sure sign in the opinion of some specialists that False Dmitrii believed himself to be the true heir to the throne. The next important event of the reign, the tsar's marriage, served to increase tensions. In November 1605 in Cracow, False Dmitrii became engaged to Marina Mniszech. The tsar's proxy for the ritual, Athanasius Vlasiev, surprised those in attendance by refusing to answer the ceremonial question as to whether the tsar had promised to marry anyone else, on the ground that he had no instructions on the subject. Marina came to Moscow on May 2, 1606, and the wedding was celebrated on May 8. Marina, however, remained a Catholic, and she brought with her another large group of Poles. Arguments and clashes between the Poles and the Russians increased.

Having prepared the ground, Prince Basil Shuisky, Prince Basil Golitsyn, and other boyars on the night of May 26 led into Moscow a very large military detachment stationed nearby. Their coup began under the slogan of saving the tsar from the Poles, but as it progressed the tsar himself was denounced as an impostor. The defenders of the palace were overwhelmed.

False Dmitrii tried to escape, but was handed over to the rebels and death by a guard of the streltsy, apparently after they had been persuaded by the mother of Prince Dmitrii of Uglich, the nun Martha, that their tsar was an impostor. Theodore Basmanov and two or three thousand other Russians and Poles perished. The Patriarch Ignatius was deposed.

Both the Godunovs and their rival had thus disappeared from the scene. Prince Basil Shuisky became the next tsar with no greater sanction than the wishes of his party and the endorsing shouts of a Muscovite crowd. The new ruler made certain revealing promises: he would not execute anyone without the decision of the boyar duma; innocent members of a family would not suffer because of a guilty relative; denunciations would not be given credence without a careful investigation; and false informers would be punished. Although historians who see in Basil Shuisky's declaration an effective limitation of autocracy seem to overstate the case, the tsar's assurances did reflect his ties to the boyars as well as the efforts of the latter to obtain minimal guarantees against the kind of persecution practiced by such rulers as Ivan the Terrible and Boris Godunov. Moreover, it appears that the boyars acquired a certain freedom under the new monarch and often behaved willfully and disobediently in their relations with him.

The government tried its best to assure the people that False Dmitrii had been an impostor who had won the throne by magic and had forced the nun Martha and others to recognize him as the authentic prince. The body of False Dmitrii was exposed in Red Square and then burned, and the ashes were fired from a cannon in the direction of Poland. In addition to this, and to Basil Shuisky's and Martha's denunciations of False Dmitrii, another novel attempt at persuasion was made: in June 1606 Prince Dmitrii of Uglich was canonized and his remains were brought to Moscow.

The Social Phase

Basil Shuisky's elevation to tsardom may be said to mark the transition in the Time of Troubles from the dynastic to the social phase. Not that dynastic issues lost their importance: in fact, the contest for the throne remained a basic aspect of the Time of Troubles to the end. But the social conflict became dominant. We have already seen how social discontent assisted False Dmitrii and how mobs in Moscow were significant in the struggle for the seat of power. With the deposition and murder of False Dmitrii, authority in the land was further weakened, whereas the forces of discontent and rebellion grew in size and strength. Indeed, the Russians had seen four tsars — Boris and Theodore Godunov, False Dmitrii, and Basil Shuisky — within thirteen and a half months, and the once firm government control and leadership had collapsed in intrigue, civil war, murder, and general weakness. Then too, whatever advantages the changes brought to the boyars, the masses had gained nothing, and their dissatisfaction grew. In effect, Basil Shuisky's unfortunate reign, 1606–10, had no popular sanction and very little popular support, representing as it did merely the victory of a boyar clique.

Opposition to the government and outright rebellion took many forms. An enemy of Basil Shuisky, Prince Gregory Shakhovskoy, and others roused southern Russian cities against the tsar. Disorder swept towns on the Volga, and in Astrakhan in the far southeast the governor, Prince Ivan Khvorostinin, turned against Basil Shuisky. Similarly in other places local authorities refused to obey the new ruler. The political picture in the Muscovite state became one of extreme disorganization, with countless local variations and complications. Rumors persisted that False Dmitrii had escaped death, and people rallied to his mere name. Serfs and slaves started numerous and often large uprisings against their landlords and the state. On occasion they joined with native tribes, such as the Finnic-speaking Mordva, who on their part also sought to overturn the oppressive political and social system of Muscovite Russia.

The rebellion in the south, led by Shakhovskoy and by Bolotnikov, presented the gravest threat to the government and in fact to the entire established order. Ivan Bolotnikov was a remarkable person who was thrown into prominence by the social turmoil of the Time of Troubles: a slave, and a captive of the Tatars and the Turks from whom he escaped, he rallied the lower classes — the serfs, peasants, slaves, fugitives, and vagabonds — in a war against authority and property. Bolotnikov's manifestoes clearly indicate the importance of the social issue, not simply of the identity of the ruler, as a cause of this rebellion. The masses were to fight for their own interests, not for those of the boyars. In October 1606, the southern armies came to the gates of Moscow, where, however, they were checked by government forces commanded by the tsar's brilliant young nephew, Prince Michael Skopin-Shuisky. Perhaps inevitably, the rebels split. The gentry armies of Riazan, led by the Liapunov brothers, Procopius and Zachary, and those of Tula, led by Philip Pashkov, broke with the social rebel Bolotnikov and even in large part went over to Basil Shuisky's side. The tsar also received other reinforcements. In 1607 a huge government army invested the rebels in Tula and, after a bitter four-month siege and a partial flooding of the town, forced them to surrender. Shakhovskoy was exiled to the north; Bolotnikov was also exiled and, shortly afterwards, dispatched.

It should be noted that Shakhovskoy and Bolotnikov claimed to act in the name of Tsar Dmitrii, although they had no such personage in their camp. Later they did acquire a different pretender, False Peter, who claimed to be Tsar Theodore's son, born allegedly in 1592, although this son never existed. False Peter was hanged after the capture of Tula. As order collapsed and disorganization spread, more and more pretenders appeared. The cossacks in particular produced them in large numbers and with different names, claiming in that strange manner, it would seem, a certain legal sanction for their bands and movements. But it was another False Dmitrii, the second, who became a national figure. Although he emerged in August, 1607, shortly before the fall of Tula, and thus too late to join Shakhovskoy and Bolotnikov, he soon became a center of attraction in his own right.

The new False Dmitrii, who claimed to be Prince Dmitrii of Uglich and also the Tsar Dmitrii who defeated the Godunovs and was deposed by a conspiracy of the boyars, resembled neither. In contrast to the first pretender, he certainly realized that he was an impostor, and his lieutenants also had no illusions on that score. Nothing is known for certain about the second False Dmitrii's identity and background. The earliest mention in the sources locates him in a Lithuanian border town, in jail. Yet, in spite of these unpromising beginnings, the new pretender quickly gathered many supporters. After the defeat of Shakhovskoy and Bolotnikov he became the focal point for forces of social discontent and unrest. He attracted a very large following of cossacks, soldiers of fortune, and adventurers, especially from Poland and Lithuania, including several famous Polish commanders. Marina Mniszech recognized him as her husband and later bore him a son; the nun Martha declared him her child.

Basil Shuisky made the grave mistake of underestimating his new enemy and of not acting with vigor in time. In the spring of 1608 the second False Dmitrii defeated a government army under the command of one of the tsar's brothers, Prince Dmitrii Shuisky, and approached Moscow. He established his headquarters in a nearby large village called Tushino — hence his historical appellation, "The Felon of Tushino." Prince Michael Skopin-Shuisky again prevented the capture of the capital, but he could not defeat or dislodge the pretender. A peculiar situation arose: in Tushino the second False Dmitrii organized his own court, a boyar duma, and an administration, parallel to those in Moscow; he collected taxes, granted lands, titles, and other rewards, judged, and punished. Southern Russia and a number of cities in the north recognized his authority. Moscow and Tushino, so close to each other, maintained a constant clandestine intercourse. Many Russians switched sides; some families served both rulers at the same time. The second False Dmitrii suffered a setback, however, when his forces tried to capture the well-fortified Holy Trinity-St. Sergius Monastery, one of the gateways to northern Russia. A garrison of 1,500 men, reinforced later by another 900, withstood for sixteen months the siege of a force numbering up to 30,000 troops. Also, the Felon of Tushino's rule in those northern

Russian cities which had recognized his authority proved to be ephemeral once they had a taste of his agents and measures.

In his desperate plight, Basil Shuisky finally, in February 1609, made an agreement with Sweden, obtaining the aid of a detachment of Swedish troops 6,000 strong, commanded by Jakob De la Gardie, in return for abandoning all claims to Livonia, ceding a border district, and promising eternal alliance against Poland. Throughout the rest of the year and early in 1610, Prince Michael Skopin-Shuisky, assisted by the Swedes, cleared northern Russia of the Felon of Tushino's troops and bands, lifted the siege of the Holy Trinity-St. Sergius Monastery, and finally relieved Moscow of its rival Tushino neighbor. The pretender and a part of his following fled to Kaluga. After his departure, and before the entire camp disbanded, the Russian gentry in Tushino asked King Sigismund III of Poland to let his son Wladyslaw, a youth of about fifteen, become the Russian tsar on certain conditions.

Sigismund III granted the request and signed an agreement in February 1610 with Russian emissaries from Tushino, who by that time had ceased to represent any organized body in Russia. The Polish king had become deeply involved in Russian affairs in the autumn of 1609, when he declared war on the Muscovite state on the ground of its anti-Polish alliance with Sweden. His advance into Russia, however, had been checked by a heroic defense of Smolensk. It would seem that from the beginning of his intervention Sigismund III intended to play for high stakes and obtain the most from the disintegration of Russia: his main goal was to become himself ruler of Russia as well as Poland. The invitation to Wladyslaw, however, gave him an added opportunity to participate in Muscovite affairs.

In March 1610 the successful and popular Prince Michael Skopin-Shuisky triumphantly entered Moscow at the head of his army. But his triumph did not last long. In early May he died suddenly, although he was only about twenty-four years old. Rumor had it that he had been poisoned by Dmitrii Shuisky's wife, who wanted to assure the throne to her husband after the death of childless Tsar Basil. New disasters soon followed. The Polish commander, Stanislaw Zolkiewski, defeated Dmitrii Shuisky when the latter tried to relieve Smolensk, and marched on Moscow. In the area occupied by Polish troops, the population swore allegiance to Wladyslaw. At this turn of events, the Felon of Tushino too advanced again on Moscow, establishing himself once more near the capital. In July 1610 Basil Shuisky finally lost his throne: he was deposed by an assembly of Muscovite clergy, boyars, gentry, and common people, and forced to become a monk. The boyar duma in the persons of seven boyars, with Prince Theodore Mstislavsky as the senior member, took over the government, or what there was left of it. The interregnum was to last from 1610 to 1613.

The National Phase

The national phase of the Time of Troubles began after Sweden, and especially Poland, became involved in Russian affairs. Wladyslaw's candidacy to the Muscovite throne, supported by various groups in Russia, tended to deepen and complicate the national issue. The eventual great rally of the Russians found its main inspiration in their determination to save the country from the foreign and heretical Poles. The increasing prominence of the national and religious struggle also explains the important role of the Church during the last years of the Time of Troubles. Yet, needless to say, dynastic and social issues retained their significance during those years. In fact any neat classification of the elements which, together, produced the fantastically complicated Time of Troubles is of necessity arbitrary and artificial.

The condition of the country prevented the calling of a zemskii sobor. Yet some decision had to be taken, and urgently. At the gathering of Muscovite boyars, clergy, and ranking service gentry opinions differed. Those proposed for the throne included Prince Basil Golitsyn, and a boy, Michael Romanov, Metropolitan Philaret's son; however, the candidacy of the Polish prince Wladyslaw, which found backing especially among the boyars, prevailed. Probably Wladyslaw profited from a general lack of enthusiasm for another boyar tsar. But, more importantly, he was one of the only two strong and active candidates in the field, the other being the Felon of Tushino who was supported by the lower classes in Russia and probably in Moscow itself. In late August 1610, the Muscovites reached an agreement with the Polish commander Zolkiewski concerning the invitation to Wladyslaw to rule Russia; Russian conditions, which stressed that Wladyslaw was to become Orthodox, resembled in most respects those offered to the Polish prince earlier by the Tushino group, although they acquired a boyar, rather than gentry, coloring. Ten days later Moscow swore allegiance to Wladyslaw. An impressive embassy headed by Prince Basil Golitsyn, Metropolitan Philaret, and other dignitaries departed for Sigismund III's headquarters near Smolensk to confirm the new arrangement with the Polish king. The Felon of Tushino fled again to Kaluga, while Zolkiewski's troops entered Moscow.

At this point, when the Muscovite state appeared finally to be settling its affairs and obtaining a firm government, another reversal occurred: unexpectedly Sigismund III rejected the Russian offer. He objected especially to the conversion of Wladyslaw to Orthodoxy and to the lifting of the siege of Smolensk. But — beyond these and other specified issues — his real intention was to become the Russian ruler himself and without conditions. No agreement could be reached. Finally, contrary to international usage,

Sigismund III arrested the Russian representatives, except those few who endorsed his claims, and sent them to Poland where they were to remain for nine years. Then he proceeded openly to develop his campaign to win the Russian throne by arms, diplomacy, and propaganda.

The autumn of 1610 saw the Muscovite state in utterly desperate straits. The Poles were again enemies of the Russians, and they held Moscow as well as a large area in the western part of the country. The Swedes had declared war on the Russians after Moscow had sworn allegiance to Wladyslaw. They advanced in the north, threatened Novgorod, and before long claimed the Muscovite throne for their own candidate, Prince Philip. With the collapse of Wladyslaw's candidacy, the Felon of Tushino again increased his following, much of eastern Russia turning to him for leadership. Innumerable bands of lawless men were roaming and devastating the land. Yet — as if to illustrate the Russian proverb "there is no evil, but that it brings some good" — at least the issues gradually became clearer. Sigismund III's rejection of the arrangement to put Wladyslaw on the Russian throne eliminated one major alternative for the Russians. More important still, Swedish and especially Polish aggression led to a national rally. Moreover, the cause of Russian unity received an unexpected and mighty boost in December 1610 when the Felon of Tushino was killed by one of his men in a settlement of personal accounts.

In the absence of a tsar and because of the impotence of the boyar duma and other branches of government in Polish-occupied Moscow, the Church headed the rally. Patriarch Hermogen in Moscow declared the Russians released from allegiance to Wladyslaw; and through trusted emissaries he sent manifestoes to other towns, urging them to organize an army and liberate the capital. The patriarch's appeals had a strongly religious as well as national character, for the Poles were Catholic, and Hermogen feared especially the extension of the Uniate jurisdiction to Muscovite territories — a subject to be discussed later when we deal with the Ukraine. Other clerics and laymen joined the patriarch in trying to arouse the people. The first response came from Riazan, where Procopius Liapunov formed an army of gentry, peasants, certain remnants of Skopin-Shuisky's troops and other elements. As Liapunov's army marched on Moscow in early 1611, it was joined by other forces, including even former troops of the Felon of Tushino who came from Kaluga, notably a mixed group commanded by Prince Dmitrii Trubetskoy, and the cossacks led by Ivan Zarutsky. It should be noted that this so-called first national army, headed by Procopius Liapunov, Trubetskoy, and Zarutsky, acted also as the government of the Muscovite state. In particular, it contained a council of representatives who concerned themselves with state legislation and policy as well as with the more immediate demands of the campaign.

The Poles, who had but a small garrison in Moscow, retreated under pressure, burned most of the city, and entrenched themselves principally in the Kremlin. The large Russian army appeared to be in control of the situation. But once more social antagonisms asserted themselves. The cossacks, furious because certain legislative measures in the interest of the gentry were passed, especially on the subject of land, fugitive serfs, and cossack brigandage, and also possibly believing a false document manufactured by the Poles, killed Procopius Liapunov in July 1611. Deprived of its leader and unwilling to co-operate with the cossacks, the gentry army disbanded. The men of Trubetskoy and Zarutsky, on the other hand, stayed around Moscow to continue the siege and seized the government machinery of the defunct first national army. In June 1611 the main Polish army finally captured Smolensk, the population of the town having been reduced from 80,000 to 8,000 in the course of the siege. In July the Swedes took Novgorod by a stratagem. And in Pskov, a new pretender appeared, sometimes called the third False Dmitrii. In Kaluga Marina Mniszech and her son by the Felon of Tushino, known as the Little Felon, constituted another center of attraction for dissatisfied elements.

Yet the Russians did not collapse under all these blows; instead they staged another rally. They profited from a certain lack of energy and initiative on the part of their enemies: instead of advancing with a large army, Sigismund III sent merely a cavalry detachment to the relief of the Poles in Moscow, and that detachment was blocked by the cossacks; the Swedes, after the capture of Novgorod, appeared to rest on their laurels. Still, the magnitude of the Russian recovery should not be underestimated. Stimulated again by the appeals of Patriarch Hermogen, of Abbot Dionysus of the Holy Trinity-St. Sergius Monastery, and of others, the new liberation movement began in the town of Nizhnii Novgorod, present-day Gorky, on the Volga. It found a remarkable leader in Kuzma Minin, a butcher by trade, who combined exalted patriotism and the ability to inspire others with level-headedness and organizational and other practical talents. The people of Nizhnii Novgorod donated a third of their possessions to the cause and, together with other northeastern towns, soon organized a large army that was entrusted to a veteran warrior, Prince Dmitrii Pozharsky. Minin became its quartermaster and treasurer. The entire movement marked a religious, as well as a national, revival, accompanied by fasting and prayer. The second national army, just like its predecessor, acted as the government of the Muscovite state as well as its military force. It too apparently contained an assembly of representatives from different localities, something in the nature of a traveling zemskii sobor.

In early September 1612, the second national army reached Moscow and besieged the Poles. The cossacks blockading the city remained passive;

eventually one part of them joined Minin and Pozharsky, while another, with Zarutsky, went to the borderlands to continue their rebellion. In early November the Russians stormed Moscow and, after bitter fighting, captured Polish positions in the heart of the city, in particular in the Kremlin. Moscow was free at last of the enemy. All Polish efforts, finally led by Sigismund III himself, to come to the aid of the Polish garrison in Moscow failed.

The first aim of the victors was to elect a tsar and thus establish a firm, legitimate government in Russia and end the Time of Troubles. The specially called zemskii sobor which met for that purpose in the beginning of 1613 consisted of 500 to perhaps 700 members, although only 277 signatures have come down to us on the final document. It included the clergy, the boyars, the gentry, the townspeople, and even some representatives of peasants, almost certainly of the state peasants of northern Russia rather than of serfs. Twelve of the signatures belonged to peasants. While we have no records of the assembly and very little information about its deliberations, we know that the number of possible candidates for tsar was first reduced by the decision to exclude foreigners. From a half dozen or more Russians mentioned, the assembly selected Michael Romanov to be tsar, and the Romanov family ruled Russia for over 300 years, from 1613 to 1917.

Historians have adduced a number of reasons for this choice. Through Ivan the Terrible's marriage to Anastasia Romanova, Michael Romanov was related to the old dynasty. The family enjoyed popularity with the masses. In particular, the people remembered Anastasia, Ivan the Terrible's good first wife, and her brother, Nikita Romanov, who dared defend some of the victims of the violent tsar. Metropolitan Philaret, Nikita's son and Michael's father, who was a prisoner of the Poles at the time of the zemskii sobor, added to the advantageous position of the family. In particular, Miliukov and others have stressed that he stood closer to the Tushino camp and had much better relations with the cossacks than other boyars. Michael's youth too counted in his favor: only sixteen years old, he had not been compromised by serving the Poles or the pretenders, and he generally remained free of the extremely complicated and painful entanglements of the Time of Troubles. Michael Romanov also gained stature as Patriarch Hermogen's choice, although the patriarch himself did not live to see the election, having perished as a prisoner of the Poles shortly before the liberation of Moscow.

Thus, in February 1613, the zemskii sobor decided in favor of Michael Romanov. Next, special emissaries were dispatched to different parts of the Muscovite state to sound local opinion. When they reported the people's strong endorsement of the decision, Michael Romanov was elected to rule Russia as tsar, and the title was to pass on to his future descendants. It took additional time to persuade his mother and him to accept the offer.

Finally, Michael Romanov was crowned tsar on July 21, 1613. In Platonov's words: "According to the general notion, God himself had selected the sovereign, and the entire Russian land exulted and rejoiced."

The Nature and Results of the Time of Troubles

Platonov's authoritative evaluation of the Time of Troubles contains several major points: the explosive crisis which Russia experienced represented the culmination and the overcoming of a dangerous disease, or perhaps several diseases. It ended with a decisive triumph over Polish intervention, over the aristocratic reaction inside Russia, over the cossacks and anarchy. The result meant a national victory for Russia and a social victory for its stable classes, that is, the service gentry, the townspeople, and the state peasants of the north. The state gained in strength, and the entire experience, which included popular participation in and indeed rescue of the government, contributed greatly to the growth of national sentiment and to a recognition of public, as against private, rights and duties by sovereign and subject alike.

Many other historians, both before and after Platonov, noted positive results of the Time of Troubles. S. Soloviev, for example, claimed that it marked the victory in Russia, at long last, of the concept of state over that of family and clan. The Slavophiles — whom we shall consider when we discuss Russian thought in the nineteenth century — were probably the most enthusiastic of all: to them the Time of Troubles represented a revelation of the greatness of the Russian people, who survived the hardest trials and tribulations, overcame all enemies, saved their faith and country, and re-established the monarchy.

Critical opinions too have not been lacking. Kliuchevsky, for one, stressed the social struggle, the abandonment of the tradition of patient suffering by the masses, and the legacy of devastation and discord which pointed to the great popular rebellions of later years. He also emphasized the peculiar role and importance of the pretenders which demonstrated the political immaturity of the Russians. Michael Romanov himself could be considered a successful pretender, for his main asset lay in his link with the extinct dynasty. It might be added that Basil Shuisky, for his part, pointed out in his manifestoes that he belonged to an even older branch of the princely house of Suzdal and Kiev than the former Muscovite rulers and thus possessed every claim to legitimacy.

Soviet historians have devoted considerable attention to the Time of Troubles, which they often characterize as a period of peasant revolts and foreign intervention. They have concentrated on the class struggle exemplified by Bolotnikov's rebellion, on the role of the poorer classes generally, and sometimes on the role of the non-Russian nationalities. In contrast to

Platonov they have favored the revolutionary not the "stable" elements. Among the weaknesses of Soviet interpretations has been an underestimation of the significance of the Church.

In conclusion, we may glance at the Muscovite government and society as they emerged from the Time of Troubles. In spite of everything that happened between 1598 and 1613, autocracy survived essentially unimpaired. In fact, at the end of it all, autocracy must have appeared more than ever the only legitimate form of government and the only certain guarantee of peace and security. Centralization, too, increased in the wake of social disorganization. In particular, local self-government that had developed in Ivan the Terrible's reign did not outlast the Time of Troubles. The Church, on its side, gained authority and prestige as the great champion of the interests of the country and the people and the most effective organization in the land that had survived the collapse of the secular order.

The service gentry also won. We know something about the aspirations of that class from such documents as the invitation to ascend the Muscovite throne sent to Wladyslaw by the service gentry in Tushino. The conditions of the offer included full protection of the Orthodox Church in Russia and freedom of religion, for Wladyslaw was a Catholic; rule with the help of the boyar duma and the zemskii sobor; no punishment without trial in court; the preservation and extension of the rights of the clergy, the service gentry, and to a degree the merchants; the rewarding of servitors according to merit; the right to study abroad; and at the same time a prohibition of serfs leaving their masters and a guarantee that slaves would not be freed. This attempt by the Tushino gentry to establish a government failed, but, in a broader sense, the Muscovite gentry succeeded in defending its interests during the Time of Troubles and in preserving and in part re-establishing a political and social order in which it occupied the central position. The Muscovite system, based on a centralizing autocracy and the service gentry, thus surmounted the great crisis and challenge of the Time of Troubles and continued to develop in the seventeenth century as it had in the sixteenth. It is this fundamental continuity that makes it difficult to find any lasting results of the Time of Troubles, anything beyond Platonov's "disease overcome."

The losers included, on one hand, the boyars and, on the other, the common people. The boyars attained their greatest power in the reign of Basil Shuisky and the period immediately following his deposition. Yet this power lacked popular support and failed to last. In the end, autocracy returned with its former authority, while the boyars, many of their families further decimated during the Time of Troubles, had to become unequivocally servants of the tsar. The desires of the boyars found expression in the remarkably mild "conditions" associated with the accession of Basil Shuisky, that is in his promise not to purge the boyars arbitrarily, and in the Muscovite invitation to Wladyslaw, which changed the earlier Tushino stipulations to exclude promotion according to merit and the right to study abroad and insisted that foreigners must not be brought in over the heads of the Muscovite princely and boyar families.

The common people also suffered a defeat. They, and especially the serfs, slaves, fugitives, vagabonds, and uprooted, together with the cossacks, fought for Bolotnikov, for the various pretenders, and also in countless lesser armies and bands. Although they left little written material behind them, their basic demand seems clear enough: a complete overturn, a destruction of the oppressive Muscovite social and economic order. But the order survived. The decades which followed the Time of Troubles saw a final and complete establishment of serfdom in Russia and in general a further subjugation of the working masses to the interests of the victorious service gentry.

The legacy of the Time of Troubles, good and bad, was the point of departure for the reign of Michael Romanov.

THE REIGNS OF MICHAEL, 1613–45, ALEXIS, 1645–76, AND THEODORE, 1676–82

> The seventeenth century cannot be separated either from the preceding or the succeeding epoch. It is the continuation and the result of the past just as it is the preparation for the future. It is essentially an age of transition, which lays the groundwork, and rapidly, for the reforms of Peter.
>
> MILIUKOV

In Kostomarov's words, "Few examples can be found in history when a new sovereign ascended the throne in conditions so extremely sad as those in which Mikhail Fedorovich, a minor, was elected." And indeed Michael Romanov assumed power over a devastated country with the capital itself, as well as a number of other towns, burned down. The treasury was empty, and financial collapse of the state appeared complete. In Astrakhan, Zarutsky, who had Marina Mniszech and the Little Felon in his camp, rallied the cossacks and other malcontents, continuing the story of pretenders and social rebellion so characteristic of the Time of Troubles. Many roaming bands, some of them several thousand strong, continued looting the land. Moreover, Muscovy remained at war with Poland and Sweden, which had seized respectively Smolensk and Novgorod as well as other Russian territory and promoted their own candidates to the Muscovite throne, Prince Wladyslaw and Prince Philip.

Under the circumstances, the sixteen-year-old tsar asked the zemskii sobor not to disband, but to stay in Moscow and help him rule. The zemskii sobor, while its personnel changed several times, in fact participated in the government of Russia throughout the first decade of the new reign. Platonov and others have pointed to the naturalness of this alliance of the "stable" classes of the Muscovite society with the monarchy which they had established. Michael worked very closely also with the boyar duma. Some historians even believe that at his accession he had given the duma certain promises limiting autocracy — an interesting supposition that has not been corroborated by the evidence. The tsar's advisers, few of whom showed ability, at first included especially members of the Saltykov family, relatives on his mother's side. In 1619, however, Michael's father, Metropolitan Philaret, returned from imprisonment in Poland, was made patriarch, and became the most important man in the state. In addition to his ecclesiastical dignities, Philaret received the title of Great Sovereign, with the result that the country had two great sovereigns and documents were issued in the names of both. But Philaret's real power lay in his ability and experience and especially in his forceful character that enabled him to dominate his rather weak son. Philaret died in 1633, almost eighty years old.

In 1613 and the years following, the most pressing problems were those of internal disorder, foreign invasion, and financial collapse. Within some three years the government had dealt effectively with the disorder, in spite of new rebellions. Authorities made certain concessions to the cossacks and amnestied all bandits, provided they would enroll in the army to fight the Swedes. Then they proceeded to destroy the remaining opponents, group by group. The especially dangerous enemies, Zarutsky, the Little Felon, and Marina Mniszech, were defeated in Astrakhan and captured in 1614. The first two were executed, while Marina Mniszech died in prison.

Everything considered, Tsar Michael's government could also claim success in checking foreign aggression and stabilizing international relations, although at a price. Sweden, with its new king Gustavus II, or Gustavus Adolphus, occupied elsewhere in Europe, concluded peace in Stolbovo in 1617. According to the agreement, the Swedes returned Novgorod and adjacent areas of northern Russia, but kept the strip of territory on the Gulf of Finland, thus pushing the Russians further from the sea. In addition, Sweden received twenty thousand rubles. The Poles had greater ambitions; however, an understanding was attained after Wladyslaw's campaign of 1617–18 reached but failed to capture Moscow. By the truce of Deulino of 1618, which was to last for fourteen years, Poland kept Smolensk and certain other gains in western Russia. It was by the terms of this agreement that Russian prisoners, including Philaret, were allowed to return home. At the termination of the treaty in 1632, hostilities were resumed. But in 1634 peace was made: Poland again kept its gains in western Russia and, besides, received twenty thousand rubles, while Wladyslaw finally withdrew his claims to the Muscovite throne.

During Michael's reign important events also occurred south of the Muscovite borders. In 1637 Don cossacks, on their own, seized the distant Turkish fortress of Azov by the sea of the same name. In 1641 a huge Turkish army and navy returned, but in the course of an epic siege of four months could not dislodge the intruders. Having beaten back the Turks, the cossacks offered Azov to Tsar Michael. Acceptance meant war with Turkey. At the especially convened zemskii sobor of 1642 the delegates of the service class opted for war, but those of the merchants and the townspeople argued that financial stringency precluded large-scale military action. The tsar endorsed the latter opinion, and the cossacks had to abandon

Azov. In the Azov area, as in the area of the Gulf of Finland, the next Russian effort was to be led by Peter the Great. '

Financial stability proved to be more difficult to attain than security at home or peace abroad. Miliukov and others have pointed out that the catastrophic financial situation of the Muscovite state resulted from its overextension, from the fact that its needs and requirements tended to exceed the economic capacity of the people. The Time of Troubles caused a further depletion and disorganization. In a desperate effort to obtain money, Tsar Michael's government tried a variety of measures: collection of arrears, new taxes, and loans, including successive loans of three, sixteen, and forty thousand rubles from the Stroganovs. In 1614 an extraordinary levy of "the fifth money" in towns, and of corresponding sums in the countryside, was enacted. While specialists dispute whether this impost represented one fifth of one's possessions or one fifth of one's income, its Draconian nature is obvious. On two later occasions the government made a similar collection of "the tenth money." On the whole, enough funds were obtained for the state to carry on its activities; but at the end of Michael's reign, as in the beginning, the financial situation remained desperate. Finances were to plague the tsar's successors with further crises.

The Reigns of Alexis and Theodore

Michael died in 1645 at the age of forty-eight, and his only son Alexis or Aleksei, a youth of sixteen, succeeded him as tsar. Known as *Tishaishii,* the Quietest One, in spite of his outbursts of anger and general impulsiveness, Alexis left a favorable impression with many contemporaries, as well as with subsequent historians. In his brilliant reconstruction of the tsar's character Kliuchevsky called Alexis "the kindest man, a glorious Russian soul" and presented him both as the epitome of Muscovite culture and as one of the pioneers of the new Russian interest in the West. Even if we allow for a certain exaggeration and stylization in Kliuchevsky's celebrated analysis, there remains the image of an attractive person, remarkably sensitive and considerate in his relations with other people, an absolute ruler who was not at all a despot. Alexis had been brought up in the Muscovite religious tradition, and he continued to be a dedicated and well-informed churchgoer and to observe fasts and rituals throughout his life. At the same time he developed an interest in the West and Western culture, including architecture and also the theatre, which was an innovation for Russia. The tsar liked to write and left behind him many fascinating letters.

Alexis's long reign, 1645–76, was by no means quiet. Old crises and problems persisted and some new ones appeared. In addition, the tsar was a weak ruler, although an attractive person, and especially at first depended

very heavily on relatives and other advisers, who often failed him. The
boyar Boris Morozov, Alexis's Western-oriented tutor who married a sister
of Alexis's wife, and Prince Elijah Miloslavsky, Alexis's father-in-law, became especially prominent after the accession of the new sovereign. Morozov acted with intelligence and ability, but his efforts to replenish the treasury by such means as an increase in the salt tax and the sale of the hitherto
forbidden tobacco, to which the Church objected, antagonized the masses.
Also, some of his protégés and appointees robbed the people. Narrow
selfishness, greed, and corruption characterized the behavior of Miloslavsky
and his clique. In May 1648 the exasperated inhabitants of Moscow staged
a large rebellion, killing a number of officials and forcing the tsar to execute
some of the worst offenders, although both Morozov and Miloslavsky
escaped with their lives. Shortly afterwards rebellions swept through several
other towns, including Novgorod and especially Pskov.

Later in the reign, when the government was still in desperate straits
financially, it attempted to improve matters by debasing the coinage. The
debasing of silver with copper, begun in 1656, proved to be no more successful than similar efforts in other countries: it led to inflation, a further
financial dislocation, and the huge "copper coin riot" of 1662. But the
greatest rebellion of the reign, headed by Stenka, or Stepan, Razin and long
remembered by the people in song and story, occurred in 1670–71. It bore
striking similarities to the lower-class uprisings of the Time of Troubles.
Razin, a commander of a band of Don cossacks, first attracted attention as
a daring freebooter who raided Persia and other lands along the Caspian
Sea and along the lower Volga. In the spring of 1670, he started out with
his band on a more ambitious undertaking, moving up the Volga and everywhere proclaiming freedom from officials and landlords. In town after town
along the river members of the upper classes were massacred, while the
soldiers and the common people welcomed Razin. Razin's emissaries had
similar success in widespread areas in the hinterland. Native tribes as well
as the Russian masses proved eager to overthrow the established order. The
rebel army reached Simbirsk and grew to some 20,000 men. Yet its poor
organization and discipline gave the victory to the regular Muscovite troops,
which included several regiments trained in the Western manner. Razin and
some followers escaped to the Don. But the following spring, in 1671, he
was seized by cossack authorities and handed over to Muscovite officials to
be publicly executed. Several months later Astrakhan, the last center of the
rebellion, surrendered.

In addition to suppressing uprisings, the government took steps to improve administration and justice in order to assuage popular discontent. Of
major importance was the introduction of a new legal code, the *Ulozhenie*
of 1649. Approved in principle by the especially convened zemskii sobor
of 1648 and produced by a commission elected by the sobor, the new code

provided the first systematization of Muscovite laws since 1550. It marked a great improvement over its predecessors and was not to be superseded until 1835.

The extension of Muscovite jurisdiction to the Ukraine in 1654, represented an event of still greater and more lasting significance. As we remember, that land after 1569 found itself under Polish, rather than Lithuanian, control. Association with Poland meant increasing pressure of the Polish social order — based on the exclusive privileges of the gentry and servitude of the masses — as well as pressure of Catholicism on the Orthodox Ukrainian people. The religious issue became more intense after 1596. That year marked the Union of Brest and the establishment of the so-called Uniate Church, that is, a Church linked to Rome but retaining the Eastern ritual, the Slavonic language in its services, and its other practices and customs. Although the Orthodox community split violently on the subject of union, each side anathemizing the other, the Polish government chose to proceed as if the union had been entirely successful and the Uniate Church had replaced the Orthodox in the eastern part of the realm. Yet, in fact, although most Orthodox bishops in the Polish state favored the union, the majority of the Orthodox people did not. Two churches, therefore, competed in the Ukraine: the Uniate, promoted by the government but often lacking other support, and the Orthodox, opposed and sometimes persecuted by authorities but supported by the masses. Lay Orthodox brotherhoods and a small, diminishing, but influential group of Orthodox landed magnates helped the Church of the people.

The cossacks also entered the fray. Around the middle of the sixteenth century the Dnieper cossacks, the most celebrated of all cossack "hosts," had established their headquarters, the *Sech — Sich* in Ukrainian — on an island in the Dnieper beyond the cataracts. They proceeded to stage unbelievably daring raids in all directions, but especially against the Crimean Tartars and Turkey — as described in detail by Hrushevsky and other Ukrainian historians. The cossacks developed a peculiar society, both military and democratic, for their offices were elective and a general gathering of all cossacks made the most important decisions. The Polish government faced difficulties in trying to control the cossacks. Stephen Bathory and his successors allowed them very considerable autonomy, but also established a definite organization for the "host" and introduced the category of registered, that is, officially recognized, cossacks to whom both autonomy and the new organization applied. All other cossacks were to be treated simply as peasants. The Polish policy had some success in that it helped to develop economic and social ties between the cossack upper stratum and the Polish gentry. Yet the same well-established cossacks retained ethnic and, especially, religious links with the Ukrainian people. The ambivalent position of the registered cossacks, particularly of their commanders, repeatedly affected their behavior. An example is the case of the hetman, that is, the chief commander, Peter Sagaidachny, or Sahaidachny, who did so much to strengthen and protect the Orthodox Church in the Ukraine, but in many other matters supported the policies of the Polish government. Nevertheless, as the struggle in the Ukraine deepened, the cossacks sided on the whole with the people. And if the hetmans and registered cossacks, who after the expansion in 1625 numbered six thousand men, obtained certain advantages from their association with Poland and found themselves often with divided loyalties, the unrecognized cossacks, who were several times more numerous, as well as the peasants, saw in Poland only serfdom and Catholicism and had no reason to waver.

From 1624 to 1638 a series of cossack and peasant rebellions swept the Ukraine. Only with great exertion and after several defeats did the Polish army and government at last prevail. The ruthless Polish pacification managed to force obedience for no longer than a decade. In 1648 the Ukrainians rose again under an able leader Bogdan, or Bohdan, Khmelnitsky in what has been called the Ukrainian War of Liberation. After some brilliant successes, achieved with the aid of the Crimean Tartars, and two abortive agreements with Poland, the Ukrainians turned again to Moscow. Earlier, in 1625, 1649, and 1651, the Muscovite government had failed to respond to the Ukrainian request, which, if acceded to, would have meant war against Poland. However, the zemskii sobor of 1653 urged Tsar Alexis to take under his sovereign authority Hetman Bogdan Khmelnitsky and his entire army "with their towns and lands." Both sides thus moved toward union.

The final step was taken in Pereiaslavl in January 1654. A *rada,* or assembly, of the army and the land considered the alternatives open to the Ukraine — subjection to Poland, a transfer of allegiance to Turkey, or a transfer of allegiance to Muscovy — and decided in favor of the Orthodox tsar. After that, the Ukrainians swore allegiance to the tsar. A boyar, Basil Buturlin, represented Tsar Alexis at the assembly of Pereiaslavl. It would seem that, contrary to the opinion of many Ukrainian historians, the new arrangement represented unconditional Ukrainian acceptance of the authority of Moscow. The political realities of the time, with the Ukrainians, not the Muscovite government, pressing for union, the political practice of the Muscovite state, and the specific circumstances of the union all lead to this conclusion. It should be noted, on the other hand, that in subsequent decades and centuries the Ukrainians acquired good reasons to complain of the Russian government, which eventually abrogated entirely the considerable autonomy granted to the Ukrainians after they had sworn allegiance to the Muscovite tsar, and which imposed, or helped to impose, upon them many heavy burdens and restrictions, including serfdom and measures meant to arrest the development of Ukrainian literary language and culture. After the union, the Ukrainians proceeded to play a very important part in Muscovite government and culture, for they were of the same religion as the Great Russians and very close to them ethnically, but were more familiar with the West. In particular, many Ukrainians distinguished themselves as leading supporters of the reforms of Peter the Great and his successors.

The war between the Muscovite state and Poland, which with Swedish intervention at one point threatened complete disaster to Poland, ended in 1667 with the Treaty of Andrusovo, which was negotiated on the Russian side by one of Alexis's ablest assistants, Athanasius Ordyn-Nashchokin. The Dnieper became the boundary between the two states, with the Ukraine on the left bank being ceded to Moscow and the right-bank Ukraine remaining under Poland. Kiev, on the right bank, was an exception, for it was to be left for two years under Muscovite rule. Actually Kiev stayed under Moscow beyond the assigned term, as did Smolensk, granted to the tsar for thirteen and a half years; and the treaty of 1686 confirmed the permanent Russian possession of the cities. The Muscovite state also fought an inconclusive war against Sweden that ended in 1661 and managed to defend its new possessions in the Ukraine in a long struggle with Turkey that lasted until 1681. In Ukrainian history the period following the Union of Pereiaslav, Bogdan Khmelnitsky's death in 1657, and the Treaty of Andrusovo is vividly described as "the Ruin," and its complexities rival those of the Russian Time of Troubles. Divided both physically and in orientation and allegiance, the Ukrainians followed a number of competing leaders who usually, in one way or another, played off Poland against Moscow; Hetman Peter Doroshenko even paid allegiance to Turkey. Constant and frequently fratricidal warfare decimated the people and exhausted the land. Yet the Muscovite hold on the left-bank Ukraine remained, and the arrangement of 1654 acquired increasing importance with the passage of time.

Significant events in the second half of Alexis's reign include the ecclesiastical reform undertaken by Patriarch Nikon and the resulting major split in the Russian Orthodox Church. Nikon himself certainly deserves notice. Of peasant origin, intelligent, and possessing an extremely strong and domineering character, he attracted the favorable attention of the tsar, distinguished himself as metropolitan in Novgorod, and, in 1652, became patriarch. The strong-willed cleric proceeded to exercise a powerful personal influence on the younger and softer monarch. Alexis even gave Nikon the title of Great Sovereign, thus repeating the quite exceptional honor bestowed upon Patriarch Philaret by his son, Tsar Michael. The new patriarch, expressing a viewpoint common in the Catholic West, but not in the Orthodox world, claimed that the church was superior to the state and endeavoured to assert his authority over the sovereign's. Charged with papism, he answered characteristically: "And why not respect the pope for that which is good." Nikon pushed his power and position too far. In 1658

Alexis quarreled with his exacting colleague and mentor. Finally, the Church council of 1666–67, in which Eastern patriarchs participated, deposed and defrocked Nikon. The former Great Sovereign ended his days in exile in a distant monastery.

The measures of Patriarch Nikon that had the most lasting importance concerned a reform of Church books and practices that resulted in a permanent cleavage among the Russian believers. While this entire subject, the fascinating issue of the Old Belief, will be considered when we discuss religion in Muscovite Russia, it might be mentioned here that the same ecclesiastical council of 1666–67 that condemned Nikon entirely upheld his reform. The last decade of Tsar Alexis's reign passed in religious strife and persecution.

Alexis's successor Theodore, his son by his first wife, became tsar at the age of fourteen and died when he was twenty. He was a sickly and undistinguished person, whose education, it is interesting to note, included not only Russian and Church Slavonic, but also Latin and Polish taught by a learned theologian and writer, Simeon of Polotsk. Theodore's brief reign, 1676–82, has been noted for the abolition of mestnichestvo. It was in 1682 that this extremely cumbersome and defective system of service appointments at last disappeared, making it easier later for Peter the Great to reform and govern the state. The mestnichestvo records were burned.

* * * * * * * * * *

MUSCOVITE RUSSIA: ECONOMICS, SOCIETY, INSTITUTIONS

The debate concerns the issue as to whether the peasants had been tied to their masters prior to the *Ulozhenie*. As we already had reason to learn from the above, the gentry and the lower servitors did not ask for the repeal of St. George's Day. They, as well as the peasants, knew that it had been repealed, even if temporarily. The peasants hoped for the restoration of their ancient right and indubitably wanted that to happen; the landlords neither wanted it, nor thought it likely to occur. The *Ulozhenie* put an end to the hopes of the peasants and fully met the demands of the gentry and the lower servitors, not directly, however, but indirectly, by means of the recognition of the time-tested practice of forbidden years, which was not to be repealed.

GREKOV

The *zemskie sobory* in the Muscovite state represent a form of popular participation in the discussion and decision of some of the most important questions of legislation and government. But what form of participation it is, how it arose and developed — these problems have led to no agreement in historical literature.

DIAKONOV

One of the most spectacular aspects of Russian history is the unique, enormous, and continuous expansion of Russia.

LANTZEFF

To QUOTE LIASHCHENKO, and in effect the entire Marxist school of historians: "The agrarian order and rural economy again serve as a key to the understanding of all economic and social relationships within the feudal economy and society of the Moscow state during the fifteenth to the seventeenth centuries." And while the term *feudal* in this passage exemplifies the peculiar Soviet usage mentioned in an earlier chapter, Liashchenko is essentially correct in emphasizing the importance of agriculture for Muscovite Russia.

Rye, wheat, oats, barley, and millet constituted the basic crops. Agricultural technique continued the practices of the appanage period, which actually lasted far into modern times. The implements included wooden or iron ploughs, harrows, scythes, and sickles. Oxen and horses provided draft power and manure served as fertilizer. Cattle-raising, vegetable-gardening, and, particularly in the west, the growing of more specialized crops such as flax and hemp, as well as hunting, fishing, and apiculture, constituted some other important occupations of the people. Many scholars

Industry and Agriculture — 17th Century

Industry before the reign of Peter the Great

▲ Iron ore mining   🔨 Hand-operated blast furnaces   Glassmaking

■ Iron foundries & smelters   ♦ Copper smelters and ◊ processing   Shipbuilding

● Metal processing   🧂 Salt extracting   Ⅱ Textiles   △ Potash mills

✹ Fur-trade centers   ◎ Principal Fairs

have noted a crisis in Muscovite rural economy, especially pronounced in the second half of the sixteenth century, and ascribed it both to the general difficulties of transition from appanages to a centralized state based on gentry service and exploitation of peasants and to Ivan the Terrible's oprichnina.

Trade, crafts, and manufacturing grew, although slowly, with the expansion and development of the Muscovite state. Russia continued to sell raw materials to other countries, and its foreign trade received a boost from the newly established relations with the English and the Dutch. The Russians, however, lacked a merchant marine, and their role in the exchange remained passive. Domestic trade increased, especially after the Time of Troubles, and profited from a rather enlightened new commercial code promulgated in 1667. The mining of metal and manufacturing had to provide, first of all, for the needs of the army and the treasury. Industrial enterprises belonged either to the state or to private owners; among the latter were the Stroganov family which engaged in various undertakings, especially in extracting salt, and the Morozovs, so prominent in Alexis's reign, who developed a huge business in potash. Foreign entrepreneurs and specialists played a leading role in the growth of Muscovite mining and manufacturing, and we shall return to them when we discuss Western influences on Muscovy. As a result of intensified and more varied economic activity, regional differentiation increased. For example, metalwork developed in the Urals, the town of Tula, and Moscow, while the salt enterprises centered principally in the northeast.

Serfdom. Muscovite Society

Serfdom was the mainstay of Muscovite agriculture. Serf labor supported the gentry and thus the entire structure of the state. As we saw earlier, certain types of peasant bondage originated in the days of Kiev, and had undergone centuries of evolution before the times of Ivan the Terrible and Tsar Alexis. Originally, it would seem, peasant dependence on the landlords began through contracts: in return for a loan of money, grain, or agricultural tools, the peasant would promise to pay dues, the quitrent or *obrok,* to the landlord and perform work, the corvée or *barshchina,* for him. Although made for a period ranging from one to ten years, the agreements tended to continue, for the peasant could rarely pay off his obligations. Indeed his annual contributions to the landlord's economy often constituted merely interest on the loan. Invasions, civil wars, droughts, epidemics, and other disasters, so frequent in Russian history in the period from the fall of Kiev to the rise of Moscow, increased peasant dependence and bondage. Gradually it became possible for the peasant to leave his master only once a year, around St. George's day in late autumn, provided, of course, his debts had been paid.

All these developments that laid the foundations for full-fledged serfdom — which were discussed in previous chapters — preceded the Muscovite period proper. Yet the contributions which the Muscovite system itself made to serfdom should not be underestimated. The new pomestie agriculture meant that bondage spread rapidly as lands with peasants were granted by the tsar to his gentry servitors. It is worth noting that serfdom predominated in southern, southeastern, and, in large part, western Russia, but not in the huge northern territories which faced no enemy and needed no gentry officers. The government continued to promote the interests of the gentry, in particular by its efforts to limit or eliminate peasant transfer and to stop peasant flights. While it is now generally agreed that no law directly establishing serfdom was ever issued, certain legislative acts contributed to that end. In particular the government proclaimed forbidden years, that is, years when the peasants could not move — or, more realistically, be moved by those who paid their obligations — even around St. George's day. We know, for example, of such legislation in regard to many categories of peasants in 1601 and 1602. Also, the government proceeded to lengthen the period of time after which a fugitive serf could no longer be returned to his master: from five years at the end of the sixteenth century to an indefinite term, as we find it in the *Ulozhenie* of 1649. Further, in 1607 and other years, the state legislated penalties for harboring fugitive serfs; while the first census, taken from 1550 to 1580, as well as later ones, also helped the growth of serfdom by providing a record of peasant residence and by listing children of serfs in the same category as their parents.

With the *Ulozhenie* of 1649, serfdom can be considered as fully established in the Muscovite state. The new code disregarded the once important distinction between old settlers and new peasants, considering as serfs all tillers of soil on private holdings, and their progeny; it eliminated, as already indicated, any statute of limitations for fugitives; and it imposed heavy penalties for harboring them. Although a few highly special exceptions remained, the *Ulozhenie* in essence assumed the principle "once a serf always a serf" and gave full satisfaction to the gentry. Vladimirsky-Budanov and others have argued convincingly that after 1649 the government continued to consider the serfs its responsible subjects rather than merely gentry property; nevertheless, in fact their position in relation to their masters deteriorated rapidly. Their obligations undefined, the serfs were at the mercy of the landlords who came to exercise increasing judicial and police authority on their estates. By the end of the century, the buying, selling, and willing of serfs had developed; that is, they were treated virtually as slaves.

Serfdom in Russia had a number of striking characteristics. It has been observed that serfdom commenced and ended first in western Europe, and that the time lag increases as we consider areas further east. Thus in Russia, and also Poland, it appeared and disappeared last. Serfdom in Russia appeared simultaneously with a centralized monarchy not with any kind of feudalism. It resulted from two major factors: the old and growing economic dependence of the peasant on the landlord, and the activity of the Muscovite government in support of the gentry. Pre-revolutionary Russian historians, with some notable exceptions, emphasized the first element; Soviet scholars have paid particular attention to the second, as has an American specialist, Hellie, in a recent reconsideration of the issue.

Lower classes in Muscovite Russia included slaves and state peasants as well as serfs. Slaves continued to play a significant role in large households and on large estates. More people joined this category during the disturbances and disasters of the late sixteenth and early seventeenth centuries by selling themselves into slavery. With the growth and final triumph of serfdom, the distinction between slaves and serfs became less and less pronounced. State peasants, that is, peasants who owed their obligations to the state rather than to a private landlord, constituted the bulk of the population in the north and the northeast. Although they were regulated by the state, and although their obligations increased with the development of the Muscovite tsardom, their position was far superior to that of the serfs.

The townspeople, or middle classes, consisted of merchants, subdivided into several hierarchical groups, and artisans. For reasons of fiscal control, trade was strictly regulated as to its location and nature. In general, the government levied the greater part of its taxes in the towns. Also, the merchants had to serve the tsar in state finance and state commerce. The latter included the monopoly of foreign trade and of certain products sold at home, such as wine and tobacco, as well as the greatest single interest in the fur trade and other interests. As the *Ulozhenie* of 1649 and other evidence indicate, the merchants and artisans, as well as the serfs and peasants, tended to become a closed caste, with sons following the occupation of their fathers.

Landlords can be considered the upper class of Muscovite Russia. They ranged from extremely rich and influential boyars to penniless servitors of the tsar who frequently could not meet their service obligations. Yet, as already indicated, with the growth of the pomestie system and the uniform extension and standardization of state service, differences diminished in importance and the landlords gradually coalesced into a fairly homogeneous class of service gentry.

The history of the mestnichestvo illustrates well the peculiar adjustment of ancient Russian princely and boyar families to Muscovite state service, as well as the eventual discarding of the arrangements they cherished in favor of uniformity, efficiency, and merit. The mestnichestvo may be described as the system of state appointments in which the position of a given person had to correspond to the standing of his family and to his own place in the family; nobody who ranked lower on the mestnichestvo scale could be appointed above him. The resulting cumbersomeness, inefficiency, and complication can easily be imagined. For example, the system led to deplorable rigidity in the assignment of military commands. A Muscovite army consisted of five segments or regiments: the big or main regiment, the right arm or wing, the left arm or wing, the forward regiment or advance guard, and the security regiment or rear guard. In the honor of command, the main regiment came first, followed by the right wing, the advance guard and the rear guard which were considered equal, and finally the left wing. The refined calculations involved in awarding these appointments in accordance with the mestnichestvo had nothing to do with military ability. Moreover, the system made it extremely difficult in any case for a man of talent who did not belong to a leading aristocratic family to receive an important command. True, the government proclaimed certain campaigns exempt from the mestnichestvo, and on other occasions it kept high-ranking but unintelligent boyars in Moscow "for advice," while entrusting the direction in the field to abler hands. But these measures proved to be at best palliatives. The same encumbrance hindered the operation of the state machine in civil matters.

The mestnichestvo dated formally from 1475, when boyar families in the Muscovite service were entered into the state genealogical book and all appointments began to be listed in special registers which became indispensable for subsequent assignments. The boyars valued their own and their families' "honor" and "just position" extremely highly, all the more so because any occasional downgrading would be added to the permanent record. The history of Muscovite government often resembled one long squabble among boyars over "honor" and appointments, with some of them dramatically determined to eat sitting on the floor, rather than at a position at the table which they considered below their rank. Even Ivan the Terrible, who dealt so violently with the boyars, failed to abrogate the mestnichestvo. It disappeared at last, as already mentioned, a full century later in 1682 to allow greater simplicity and uniformity in the service and more reward for merit in the interests of Muscovite absolutism and gentry.

Muscovite Institutions

Muscovite tsars developed the emphasis on autocracy that was begun by Muscovite grand princes. They truthfully claimed to be absolute rulers of perhaps ten to fifteen million subjects. Yet they did not exercise their high authority alone: the boyar duma persisted as their constant companion, and a new important state institution, the zemskii sobor, appeared. Both the boyar duma and the zemskii sobor deserve attention for a number of reasons, not the least of which stems from their interesting and suggestive resemblances to Western institutions.

The boyar duma of the Muscovite tsars represented, of course, a continuation of the boyar duma of the Muscovite grand princes. However, in the conditions of a new age, it gradually underwent certain changes. Thus although it still included the great boyars, an increasing portion of the membership were less aristocratic people brought in by the tsar, a bureaucratic element so to speak. The duma membership grew, to cite Diakonov's figures, from 30 under Boris Godunov to 59 under Alexis and 167 under

Theodore. Large size interfered with work in spite of the creation of various special committees. The boyar duma met very frequently, usually daily, and could be considered as continually in session. It dealt with virtually every kind of state business. Kliuchevsky and others have demonstrated convincingly that the boyar duma was essentially an advisory body and that it did not limit autocracy. Indeed service in the Muscovite boyar duma might well be regarded as one of the many obligations imposed by the state. But, on the other hand, the ever-present boyar duma formed in effect an integral part of the supreme authority of the land rather than merely a government department or agency. The celebrated Muscovite formula for state decisions, "the sovereign directed and the boyars assented," reminds one strongly of the English legal phrase "King in Council," while the boyar duma itself bears resemblance to royal councils in different European monarchies. The boyar duma assumed the directing authority in the absence of the tsar from Moscow or in case of an interregnum, such as that which followed the deposition of Basil Shuisky.

The nature of the zemskie sobory and their relationship to the Muscovite autocracy present even more complicated problems than does the boyar duma. Again, one should bear in mind that Muscovite political practice showed little evidence of the clear disjunctions of modern political theory and that it was based on custom, not written constitutions. The zemskie sobory, as we had occasion to see earlier, were essentially sporadic gatherings convened by the tsar when he wanted to discuss and decide a particularly important issue "with all the land." Fortunately for the students of the zemskie sobory, they had much in common with certain Western institutions and especially with the so-called Estates General. In fact, their chief characteristic, in the opinion of most scholars, consisted precisely in their inclusion of at least three estates: the clergy, the boyars, and the gentry servitors of the tsar. These were usually supplemented by the townspeople and, on at least one occasion, in 1613, by the peasants. The representation was by estates. Sometimes, as in the West, the estates would first meet separately, for instance, in the boyar duma or a Church council, and afterwards present their opinion to the entire zemskii sobor. The numbers of the participants in the different zemskie sobory varied from about two hundred to perhaps five hundred or more in 1613, with the service gentry invariably strongly in evidence.

The assembly of 1471, called by Ivan III before his campaign against Novgorod, has usually been listed as a "forerunner" of the zemskie sobory. The first full-fledged zemskie sobory occurred in the reign of Ivan the Terrible, in 1549, 1566, 1575, and possibily 1580, and dealt with such important matters as the tsar's program of reforms and the Livonian War. Immediately after Ivan the Terrible's death, in 1584, another zemskii sobor confirmed his son Theodore as tsar, a step possibly suggested by the fact that Ivan the Terrible had left no testament and no formal law of succession existed in Muscovite Russia. In 1598 a zemskii sobor offered the throne to Boris Godunov. The celebrated zemskii sobor of 1613, which we discussed earlier, elected Michael Romanov and his successors to rule Russia. As we know, at the time of Tsar Michael the zemskie sobory reached the peak of their activity: they met almost continually during the first decade of the reign; later, in 1632–34, 1636–37, and 1642, they convened to tackle the issue of special taxes to continue war against Poland and the problem of the Crimea, Azov, and relations with Turkey. In 1645 a zemskii sobor confirmed Alexis's accession to the throne, while during his reign one zemskii sobor dealt with the *Ulozhenie* of 1649, another in 1650 with the disturbances in Pskov, and still another in 1651–53 with the Ukrainian problem. Many historians add to the list of zemskie sobory the gathering or gatherings of 1681–82 connected with the abolition of the mestnichestvo and the accession of a new ruler. Unknown zemskie sobory may yet be uncovered; recently a Soviet historian claimed to have discovered one in 1575. But, in any case, the zemskie sobory belonged clearly to Muscovite Russia, and the period of their activity corresponded roughly to its chronological boundaries. They found no place in Peter the Great's reformed empire.

The key controversial issue in the literature on the zemskie sobory has been the scope of their authority and their exact position in the Muscovite order of things. Kliuchevsky and some other leading specialists have shown that the zemskie sobory aided and supported the policies of the tsars, but did not limit their power. The question of restricting the sovereign's authority never arose at their gatherings. Moreover, at least in the sixteenth century, the members were appointed by the government rather than elected. Although in the Time of Troubles, with the collapse of the central government and an interregnum, the elective principle appeared and a zemskii sobor emerged as the highest authority in the country, it proved only too eager to hand over full power to a new tsar. In the seventeenth as in the sixteenth century, membership in a zemskii sobor continued to represent obligation and service to the sovereign, rather than rights or privileges against the crown. At most the participants could state their grievances and petition for redress; the monarch retained full power of decision and action.

A different view of the situation has been emphasized by Tikhomirov and other Soviet historians, as well as by certain Western scholars such as Keep. They point out that the zemskie sobory, after all, dealt with most important matters, and often dealt with them decisively: the succession to the throne, war and peace, major financial measures. The most famous zemskii sobor, that of 1613 which led Russia out of the Time of Troubles and established the Romanov dynasty on the throne, deservedly received great attention. It should also be noted that during a large part of Michael's reign no subsidy was levied or benevolence extorted without the consent of zemskie sobory; thus they had a hand on the purse strings, if they did not actually control state finances. Many edicts carried the characteristic sentence: "By the desire of the sovereign and all the land." Again, such epoch-making decisions as the extension of the tsar's jurisdiction to the Ukraine depended on the opinion of a zemskii sobor. Besides, particularly in the seventeenth century, with the elective principle persisting after the Time of Troubles and asserting itself in the composition of several of the zemskie sobory, these assemblies acted by no means simply as rubber stamps for the tsars. For example, it has been argued that the *Ulozhenie* of 1649 represented the decision and initiative of a zemskii sobor that it forced on the government. In fact, the argument proceeds, the tsars and their advisers in the second half of the seventeenth century began to convene the zemskie sobory less and less frequently precisely because of their possible threat to the position of the monarch. The assertion of tsarist absolutism in Russia against the zemskie sobory corresponded to parallel developments in a number of other European countries, such as France, where the Estates General did not meet between 1614 and 1789, and England, where the seventeenth century witnessed a great struggle between the Stuarts and Parliament. But, whether the story of the zemskie sobory resembles its Western counterparts only faintly or rather closely, the net result in Russian social conditions consisted in arrested evolution at best and in the continuing sway of autocracy.

The expansion of the Muscovite state brought with it centralization and standardization, whether sudden or gradual. First the *Sudebniki* of 1497 and 1550 and later the *Ulozhenie* of 1649 became the law of the entire land. In the course of time uncounted legal peculiarities and local practices of appanage Russia disappeared, as did such foreign imports as the so-called Magdeburg Law, German in origin, that was granted to western Russian towns by their Lithuanian and Polish rulers. This interesting law — although oligarchical in nature and often applied in a selective manner, for instance, with discrimination against the Orthodox — had effectively supported the self-government of towns in Poland and Lithuania. Autocracy and legal and administrative centralization in Muscovite Russia were to help immeasurably Peter the Great's far-reaching reforms.

The central administration of Muscovite Russia represented a rather haphazard growth of different departments and bureaus. In the seventeenth century these agencies, which came to be known as the *prikazy* — singular *prikaz* — already numbered about fifty. Many prikazy developed from the simpler offices and functions at the court of Muscovite rulers; others, for example the prikaz dealing with the pomestiia and the one concerned with Siberia, reflected new activities or acquisitions of the state. The authority of a prikaz extended over a certain type of affairs, such as foreign policy in the case of the ambassadorial prikaz; certain categories of people, such as the slaves and the streltsy; or a certain area, such as Siberia and the former khanates of Kazan and Astrakhan. Overlapping and confusion increased with time, although some scholars see in the unwieldly Muscovite arrangement the wise intention to maintain mutual supervision and checks. Bureaucracy continued to proliferate on both the central and the local levels.

Local government constituted one of the weakest parts of the Muscovite political system. The problem, of course, became enormous as the state grew to gigantic size. As a ruler of Moscow acquired new territories, he sent his representatives, the *namestniki* and *volosteli,* to administer them. The appointments, known as *kormleniia,* that is, feedings, were considered personal awards as well as public acts. The officials exercised virtually full powers and at the same time enriched themselves at the expense of the people, a practice which could not be effectively stopped by customary and later written restrictions on the amount of goods and services which the population had to provide for its administrators.

However, as already mentioned, local self-government developed in the sixteenth century, with earlier measures leading up to Ivan the Terrible's legislation of 1555. In addition to the locally elected judicial and police officials — the so-called *gubnye* officials — who were already functioning to combat crime, the enactments of that year provided for local *zemstvo* institutions concerned with finance, administration, and justice. Where the population guaranteed a certain amount of dues to the treasury, locally-elected town administrators — *gorodovye prikazchiki* — replaced centrally appointed officials; and even where the latter remained, the population could elect assessors to check closely on their activities and, indeed, impeach them when necessary. Unfortunately, although both earlier historians and such contemporary scholars as Nosov have shown the considerable development and broad competence of the institutions of local self-government in sixteenth-century Muscovy, these institutions did not last. After the Time of Troubles self-government appeared no more, and the state relied mainly on its military governors, the *voevody*. The failure of local self-government, which was also to plague Peter the Great and his successors, points again to a deficiency in social stratification, independence, initiative, and education in old Russia.

The Eastward Expansion. Concluding Remarks

The expansion of the Muscovite state brought under the scepter of the tsar not only ancient Russian lands but also colonial territories to the east and southeast. The advance continued after the conquest of the khanates of Kazan and Astrakhan. It has been estimated that between 1610 and 1640

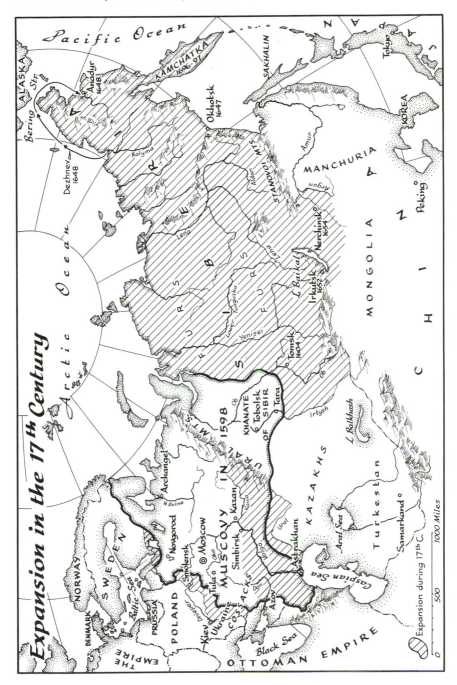

Expansion in the 17th Century

alone the Russian military line and colonists moved three hundred miles further into the southern steppe, under conditions of continuous struggle with the Crimean Tartars and other nomads. But the most spectacular expansion occurred in the direction of the more open east, where, in the course of the same three decades, the Russians advanced three thousand miles from the Ob river to the Pacific, exploring and conquering, if not really settling, gigantic Siberia.

In sweep and grandeur the Russian penetration into Siberia resembles the exploration of Africa, or, to find a closer parallel, the American advance westward. To mention a few highlights, in 1639 a cossack, Ivan Moskvitianin, at the head of a small group of men, reached the Pacific. In 1648 Semen Dezhnev, another cossack, and his followers sailed in five boats, of which three survived, from the mouth of the Kolyma river, around the northeastern tip of Siberia, and through the strait that was later to be named in honor of Bering. Dezhnev's report, incidentally, attracted no attention at the time and was rediscovered in a Siberian archive only in 1736. Other remarkable explorations during the seventeenth century included expeditions in the Amur river basin and the penetration of the Kamchatka peninsula in 1696 and the years immediately following. In the Amur area the Russians finally reached and clashed with China. The settlement of Nerchinsk in 1689 established the boundary between the two countries along the Argun and Gorbitsa rivers and the Stanovoi mountain range. This settlement lasted until 1858.

Furs presented the main attraction in Siberia, where sable, ermine, beaver, and other valuable fur-bearing animals abounded. It should be emphasized that furs constituted an extremely important item in Muscovite finance and foreign trade. In fact, as mentioned earlier, the government acted as the principal dealer in furs. As Russian rule spread among the thinly scattered natives in Siberia, they were required to pay the *iasak,* a tax in furs, to their new sovereign. Also the central authorities expended great effort — needless to say, not always successful — to limit the private acquisition of furs by the administrators in Siberia, so that the state treasury would not suffer. In general, although precise calculation remains difficult, the annexation of Siberia was a highly profitable undertaking for the Muscovite state.

The Siberian prikaz in Moscow had charge of that enormous land. Its jurisdiction, however, overlapped with the jurisdiction of several other institutions, not the least of which was the Church, which established an archbishopric in Siberia in 1621. The system, in typical Muscovite manner, provided some mutual supervision and checks, which were especially important in this distant, primitive, and fantastically large territory. Still, both the voevody and lesser administrators exercised great power and often proved difficult to control from Moscow.

As Lantzeff and others have demonstrated, the policy of the Muscovite state in Siberia, as well as that of the Church, can be considered enlightened. The natives were not to be forcibly baptized. On the other hand, if they became Orthodox, they were treated thenceforth as Russians — a condition which, among other things, excused them from paying the *iasak* and thus might have given the government second thoughts about the desirability of conversion. The government also tried to extend a paternalistic care to both natives and Russian settlers and made an effort to learn and, if possible, to redress their grievances. It encouraged colonists and tried from an early time to develop local agriculture, a perennially difficult problem in Siberia. But Moscow was very far away, whereas the local situation encouraged extreme exploitation and cruelty on the part of officials and other Russians. Often government edicts and instructions had little relation to the harsh reality of Siberia. Still, Siberian life was not all dark. Of most importance is the fact that, with very few gentry and endless spaces for the fugitive, Siberia escaped serfdom. As Siberian society developed, profiting from an assimilation of natives — for intermarriage was common — as well as from migration from European Russia, it came to represent a freer and more democratic social system than the one across the Urals and to exhibit certain qualities of sturdiness and independence often associated with the American frontier.

In concluding our brief survey of Muscovite government and society, it may be appropriate to point out again the enormous effort which the creation and maintenance of the centralized Russian monarchy demanded. In fact, the main tradition of pre-revolutionary Russian historiography placed extremely heavy emphasis on the state: autocracy, gentry service, obligations and restrictions imposed on other classes, serfdom itself, as well as other major characteristics of Muscovy, all fitted into the picture of a great people mobilizing its resources to defend its existence and assert its independence. Soviet historians, however, have shifted the focus of attention to class interests and the class struggle, presenting the history of Muscovite Russia above all in terms of a victory of the gentry over the peasants, not of a national rally. Both interpretations have much to recommend them.

MUSCOVITE RUSSIA: RELIGION AND CULTURE

The Emperor was seated upon an Imperiall Throne, with Pillars of silver and gold, which stood 3 or 4 stepps high, an Imperiall Crowne upon his Head, his Scepter in his right hand and his Globe in his left. And so he sate without any motion that I could perceave, till such time as I had repeated all the King my Masters titles and his owne, and given him greeting in his Majesties name. And then he stood up, and with a very gratious aspect, asked me how his Loving Brother the King of England did, to which when I had made him Answer, he sate downe agayne. Then the Lord Chancellor who stood upon a strada close by me with a high furred Capp upon his head: told me that the great Lord and Emperor of all Russia did very Lovingly receave that Present which stood all this while before the Emperor, and likewise his Majesties Letters which I had presented; then he looke upon a Paper which he had in his hand and said with a loud voyce: "Simon Digby, The great Lord and Emperor of all Russia askes you how you do, and desires you to come neere unto him to kiss his Hand." The first stepp I made towards him upon the state: there stood foure Noble men in Cloth of silver Roabes, with Polates in their hands advanced over me as if they would have knocked me on the head; under which I went, and having stepped up one stepp upon the Emperors throne, it was as much as I could do to reache his Hand, which when I had kissed, I retired unto the Place when I had my first Posture. . . . As I was to goe out of the roome, I observed betwixt 20ty and 30ty great Princes and Councellors of State, sitting upon the left hand of the Emperor, who were all in long Roabes of Cloth of gold, imbrodered with Pearles and Precious Stones, and high Capps either of Sables or Black Foxe about three quarters of a yard high upon their heads. To them, at my going out of the Doore, I bowed myself and they all rose up and putt of their Capps unto me.

SIMON DIGBY TO SIR JOHN COKE

O you Teachers of Christendom! Rome fell away long ago and lies prostrate, and the Poles fell in the like ruin with her, being to the end the enemies of the Christian. And among you orthodoxy is of mongrel breed; and no wonder — if by the violence of the Turkish Mahound you have become impotent, and henceforth it is you who should come to us to learn. By the gift of God among us there is autocracy; till the time of Nikon, the apostate, in our Russia under our pious princes and tsars the orthodox faith was pure and undefiled, and in the Church was no sedition.

AVVAKUM (J. HARRISON'S AND H. MIRRLEES'S TRANSLATION)

Muscovy appeared strange to foreigners. Visitors from the West, such as Guy de Miege, secretary to the embassy sent to Alexis by Charles II of England, as well as many others, described it as something of a magic world: weird, sumptuous, colorful, unlike anything they had ever seen, and utterly barbarian. The church of St. Basil the Blessed, one might add, continues to produce a similar impression on many European and American visitors. Foreign emissaries noticed the rich costumes, especially the furs, the striking grey beards, the elaborate court ceremonial, the lavish banquets and the tremendous drinking. They added, however, that the state dinners, with their endless courses, proved deficient in plates and silver and that the wise grey beards as a rule said nothing. Of more importance were the fundamental characteristics of Muscovy that the visitors quickly discovered: the enormous power and authority of the tsar and the extreme centralization which required that even insignificant matters be referred for decision to high officials. Other interesting facts were reported; however, to sum up, what they saw was an intricate, cohesive, and well-organized society, but one which they found uncongenial and very odd. Indeed, we find references to the effect that Turkey stood closer to the West than Muscovy and sincere doubts as to whether the Muscovites were really Christians.

The view of Muscovy as a strange world apart, a view shared by foreign travelers with such diverse later groups as the Slavophiles and certain Polish historians, contains some truth. Muscovite Russia existed in relative isolation by contrast, for example, with Kievan Russia. Moreover, it developed a distinctive culture based on religion and ritualism and assumed a tone of self-righteousness and suspicion toward any outside influence. This peculiar and parochial culture, it must be added, apparently had a great hold on the people. But the case should not be overstated. In reality the main elements of Muscovite culture — religion, language, law, and others — served as links to the outside world. In terms of time, too, Muscovy represented not simply a self-contained culture, but the transition from appanage Russia to the Russian Empire. And, after all, it was the Muscovites themselves, led by Peter the Great, who transformed their country and culture — the fairy land and at times the nightmare of Western travelers — into one of the great states of modern Europe.

Religion and Church. The Schism

Religion occupied a central position in Muscovite Russia and reflected the principal aspects and problems of Muscovite development: the growth and consolidation of the state; ritualism and conservatism; parochialism and the belonging to a larger world; ignorant, self-satisfied pride and the recognition of the need for reform. As already mentioned, the expansion and strengthening of the Muscovite state found a parallel in the evolution of the Church in Muscovy. The Church councils of 1547, 1549, 1551, and 1554 strove to improve ecclesiastical organization and practices and eliminate various abuses. In 1547 twenty-two Russians were canonized, and in 1549 seventeen more. The resulting consolidated national pantheon of saints represented a religious counterpart to the political unification. The

Hundred-Chapter Council of 1551 dealt, as its name indicates, with many matters in the life of the Church. The council of 1554 condemned certain Russian heretics and heresies which had roots either in Protestantism or in the teachings of the non-possessors. None of them, it might be noted, gained popular support.

The rising stature of the Russian Church at a time when many other Orthodox Churches, including the patriarchate of Constantinople itself, fell under the sway of the Moslem Turks increased Muscovite confidence and pride. References to the holy Russian land, to Holy Russia, date from the second half of the sixteenth century. In 1589, as we know, Muscovy obtained its own patriarch. Some later incumbents of this position, such as Hermogen, Philaret, and Nikon, were to play different but major roles in Russian history. The upgrading of numerous Muscovite sees after the establishment of the patriarchate was followed by a further expansion of the Church when the Ukraine, which included the ancient metropolitanate of Kiev and several other dioceses, joined Moscow in 1654. It should be added that the Church, especially the monasteries, enjoyed enormous wealth in land and other possessions in spite of the repeated efforts of the government to curb its holdings and particularly to prevent its encroachments on the gentry.

The great split or schism in the seventeenth century — raskol in Russian — revealed serious weaknesses in the apparently mighty and monolithic Muscovite Church. Over a long period of time, errors in translation from the Greek and other mistakes had crept into some Muscovite religious texts and rituals. Tsar Michael had already established a commission to study the matter and make the necessary corrections. Some visiting Orthodox dignitaries also urged reform. But in the face of general ignorance, inertia, and opposition little was done until Nikon became patriarch in 1652. The new head of the Church proceeded to act in his usual determined manner which before long became a drastic manner. The reign of Tsar Alexis was witnessing a religious and moral revival in the Church, an effort to improve the performance of the clergy and to attach a higher spiritual tone and greater decorum to various ecclesiastical functions. Yet, once Nikon introduced the issue of corrections, many leaders of this revival, such as Stephen Vonifatiev, Ivan Neronov, and the celebrated Archpriest Avvakum, or Habakkuk, turned against him. In 1653 they accused him of heresy.

To defeat the opposition, the patriarch proceeded to obtain the highest possible authority and support for his reforms: in 1654 a Russian Church council endorsed the verification of all religious texts; next, in response to inquiries from the Russian Church, the patriarch of Constantinople called a council that added its sanction to Nikon's reforms; a monk was sent to bring five hundred religious texts from Mount Athos and the Orthodox East, while many others arrived from the patriarchs of Antioch and Alexandria; a committee of learned Kievan monks and Greeks was set up to do the collating and correcting; another Russian Church council in 1656 also supported Nikon's undertaking. Nikon widened the scope of the reform to include the ritual in addition to texts, introducing in particular the sign of the cross in the Greek manner with three rather than two fingers. But the patriarch's opponents refused to accept all the high authorities brought to bear against them and stood simply on the Muscovite precedent — to keep everything as their fathers and grandfathers had it. They found encouragement in Nikon's break with the tsar in 1658 and in the ineffectiveness of the cleric who replaced him at the head of the Church. To settle matters once and for all, a Russian Church council was held in 1666 and another Church council, attended by the patriarchs of Alexandria and Antioch, who also represented those of Constantinople and Jerusalem, convened later that year and continued in 1667, in Moscow. This great council, which deposed Nikon for his bid for supreme political power, considered the issue of his reforms, listened to the dissenters, and in the end completely endorsed the changes. The opponents had to submit or defy the Church openly.

It is remarkable that, although no dogmatic or doctrinal differences were involved, priests and laymen in considerable numbers refused to obey ecclesiastical authorities, even though the latter received the full support of the state. The raskol began in earnest. The Old Believers or Old Ritualists — *starovery* or *staroobriadtsy* — rejected the new sign of the cross, the corrected spelling of the name of Jesus, the tripling instead of the doubling of the "Hallelujah," and other similar emendations, and hence rejected the Church. Persecution of the Old Believers was soon widespread. Avvakum himself — whose stunning autobiography represents the greatest document of Old Belief and one of the great documents of human faith — perished at the stake in 1682. The Solovetskii Monastery in the far north had to be captured by a siege that lasted from 1668 to 1676. Apocalyptic views prevailed among the early Old Believers, who saw in the Church reform the end of the world, and in Nikon the Antichrist. It has been estimated that between 1672 and 1691 over twenty thousand of them burned themselves alive in thirty-seven known communal conflagrations.

Yet, surprisingly, the Old Belief survived. Reorganized in the eighteenth century by a number of able leaders, especially by the Denisov brothers, Andrew and Simeon, it claimed the allegiance of millions of Russians up to the Revolution of 1917 and after. It exists today. With no canonical foundation and no independent theology to speak of, the Old Belief divided again and again, but it never disappeared. The main cleavage came to be between the *popovtsy* and the *bespopovtsy,* those who had priests and those who had none. For, although the Old Believers refused to change a tittle in the texts or the least detail in the ritual, they soon found themselves without priests and thus without the liturgy, without most of the sacraments, and in general without the very core of traditional religious life: bishops were required for elevation to the priesthood, and no bishops joined the Old Belief. Some dissenters, the popovtsy, bent all their efforts to obtain priests by every possible means, for instance, by enticing them away from the established Church. The priestless, on the other hand, accepted the catastrophic logic of their situation and tried to organize their religious life along different lines. It is from the priestless Old Believers that most Russian sects derive. But all this takes us well beyond the Muscovite period of Russian history.

The raskol constituted the only major schism in the history of the Orthodox Church in Russia. It was in an important sense the opposite of the Reformation: in the West, Christians turned against their ecclesiastical authorities because they wanted changes; in Russia believers revolted because they refused to accept even minor modifications of the traditional religious usage. Many scholars have tried to explain the strange phenomenon of the raskol. Thus Shchapov and numerous others have stressed the social composition of the Old Believers and the social and economic reasons for their rebellion. The dissenters were originally and continued to be mostly well-established peasants and traders. Their action could, therefore, be interpreted as a protest against gentry domination and the entire oppressive Muscovite system. More immediately, they reacted against the increased ecclesiastical centralization under Nikon which led to the appointment of priests — formerly they had been elected in northern parishes — and to the loss of parish autonomy and democracy. In addition to being democrats — so certain historians have claimed — the Old Believers expressed the entrepreneurial and business acumen of the Russian people. Over a period of time they made a remarkable record for themselves in commerce. Some parallels have even been drawn with the Calvinists in the West. As to the other side, the drive for reform has been ascribed, in addition to the obvious reason, to the influence of the more learned Ukrainian clergy, and to the desire of the Muscovite Church and state to adapt their practices to include the Ukrainians and the White Russians, with a further view, according to S. Zenkovsky, to a possible expansion to the Balkans and Constantinople.

Even more rewarding as an explanation of the raskol has been the emphasis on the ritualism and formalism of Muscovite culture. The Old Believers were, characteristically, Great Russians, that is, Muscovite Russians and not, for example, Ukrainians. To them the perfectly correct form and the untainted tradition in religion could not be compromised. This, and their arrogant but sincere belief in the superiority of the Muscovite Church and its practices, go far to explain the rebellion. The reformers exhibited a similar formalism. In spite of the advice of such high authorities as the patriarch of Constantinople, Nikon and his followers refused to allow any local practice or insignificant variation to remain, thus on their part, too, confusing the letter with the spirit. As we have noted, the Russian Church had developed especially in the direction of religious ceremony, ritualism, and formalism, which for the believers served as a great unifying bond and a tangible basis for their daily life. It has been estimated, for instance, that the tsar often spent five hours or more a day in church. Even visiting Orthodox hierarchs complained of the length of Russian services. The appearance of the Old Belief, as well as the excessively narrow and violent reaction to it, indicated that in Muscovy religious content in certain respects lagged behind religious form. The raskol can thus be considered a tribute to the hold that Muscovite culture had on the people, and, as time made apparent, to its staying power. It also marked the dead end of that culture.

Miliukov and others have argued that, because of the split, the Russian Church lost its most devoted and active members and, in effect, its vitality: those who had the courage of their convictions joined the Old Belief; the cowardly and the listless remained in the establishment. Even if we allow for the exaggeration implicit in this view and note further that many of the most ignorant and fanatical must also have joined the dissenters, the loss remains great. It certainly made it easier for Peter the Great to treat the Church in a high-handed manner.

Muscovite Thought and Literature

In addition to the issue of the true faith, the issue of the proper form of government preoccupied certain Muscovite minds. It concerned essentially the nature and the new role of autocracy, and discussion of it continued the intellectual trend clearly observable in the reigns of Ivan III and Basil III. Such publicists as Ivan Peresvetov, who wrote in the middle of the sixteenth century, upheld the new power and authority of the tsar, while the events of the Time of Troubles provided variations on this theme of proper government and seemed to offer to the Russians unwanted political experience. The most famous debate on the subject took place between Ivan the Terrible and Prince Andrew Kurbsky in two letters from the tsar and five from the fugitive nobleman, written between 1564 and 1579. The sovereign's brilliant letters strike the reader by the sweep of their assertions and their grandiose tone. Ivan the Terrible believed in the divine foundation of autocracy, and he declared that, even if he were a tyrant, Kurbsky's only alternative, as a Christian and a faithful subject, remained patient suffering. The prince, on his part, proved to be stronger in his criticism of the tsar's conduct and in personal invective than in political theory. Yet his views, too, represented a system of belief: they harked back to an earlier

order of things, when no great gulf separated the ruler from his chief lieutenants, and when an aristocrat enjoyed more freedom and more respect
than Ivan IV wanted to allow.

In foreign relations, as in domestic matters, Ivan the Terrible and other
tsars reiterated the glory of autocracy and demanded full respect for it.
They considered the Polish kings degraded because the latter had been put
on their throne by others, and thus could not be regarded as hereditary or
rooted rulers. They asked why Swedish monarchs treated their advisers as
companions. Or, to quote the frequently mentioned bitter letter of Ivan
the Terrible to Elizabeth of England, written in 1570: "We had thought
that you were sovereign in your state and ruled yourself, and that you saw
to your sovereign honor and to the interests of the country. But it turns out
that in your land people rule besides you, and not only people, but trading
peasants. . . ."

Passing on to the subject of Muscovite literature as a whole, one should
note the development of the "chancellery language," based on the Muscovite spoken idiom, in which official documents were written, and also the
gradual penetration of popular language into literature in place of the bookish Slavonic-Russian. Avvakum's autobiography, written in the racy spoken
idiom, was a milestone in Russian literature. Religious writings continued
and indeed flourished, especially in the seventeenth century. They included
hagiography and, in particular, menologia, that is calendars with the lives
of saints arranged under the dates of their respective feasts, the most important of which was compiled by Metropolitan Macarius. They also included theological and polemical works, sermons, and other items. After the
Ukraine joined Muscovy, the more learned and less isolated Ukrainian
clerics began to play a leading role in a Russian literary revival.

The *Domostroi,* or "house manager," constituted one of the most noteworthy works of Muscovite Russia. Attributed to Sylvester and dating in
its original version from about 1556, it intends in sixty-three didactic chapters to instruct the head of a Muscovite family and its other members how
properly to run their household and lead their lives. The *Domostroi* teachings reflect the ritualism, piety, severity, and patriarchal nature of Muscovite society. Some commentators have noted in horror that the author, or
more likely authors, write in the same peremptory manner about the venera-
tion of the Holy Trinity and about the preservation of mushrooms. Possibly
the most often cited directive reads:

> Punish your son in his youth, and he will give you a quiet old age, and
> restfulness to your soul. Weaken not beating the boy, for he will not die
> from your striking him with the rod, but will be in better health: for while
> you strike his body, you save his soul from death. If you love your son,
> punish him frequently, that you may rejoice later.

If the *Domostroi,* with its remarkable ritualism, formalism, and emphasis on the preservation of appearances, is considered by some to be a kind of Muscovite *summa,* other events in literature, especially in the seventeenth century, pointed in new directions. Gradually the lay literature of the West spread in Russia. Coming through Poland, the Ukraine, the Balkans, and sometimes more directly, the stories assumed a romantic, didactic, or satirical character and were usually full of adventure, which the religious writings of ancient Russia as a rule lacked. Often, through the vehicle of such recurrent themes as the tales of the seven wise men or of Tristan and Isolde, the stories acquainted Muscovites with the world of knighthood, courtly love, and other concepts and practices unknown in the realm of the tsars. Soon, Russian tales following Western models made their appearance: for instance, stories about Savva Grudtsin, who sold his soul to the devil, and about the rogue Frol Skobeev. Numbers of these tales enjoyed great popularity.

Syllabic versification also came from the West, from the Latin and Polish languages, largely through the efforts of Simeon of Polotsk, who died in 1680. It remained the dominant form in Russian poetry until the middle of the eighteenth century. After some productions of plays arranged by private individuals, Tsar Alexis established a court theater in 1672 under the direction of a German pastor, Johann Gregory. Before long, a few Russian plays enriched the repertoire which was devoted primarily to biblical subjects.

The traditional oral literature of the people continued to thrive throughout the Muscovite period. Tales and songs commemorated such significant events as the capture of Kazan, the penetration into Siberia, or Stenka Razin's rebellion. The byliny retained their popularity. Pilgrims and beggars composed religious poems at venerated shrines. The skomorokhi went on entertaining the people, in spite of all prohibitions. All in all it seems quite unfair to characterize Muscovite culture as silent, as has sometimes been done, all the more so because it is probable that many writings of the period have been lost. On the other hand, Muscovite literary life does appear meager by comparison with the riches of its contemporary West. Nor did it measure up, in the opinion of specialists, to Muscovite architecture and other arts.

The Arts

In architecture, as well as in literature and in culture as a whole, no divide rises between the appanage and the Muscovite periods of Russian history. Building in both wood and stone flourished in the sixteenth and seventeenth centuries. As described earlier, wooden houses of the boyars and mansions of the rulers — the so-called *khoromy* — were remarkable conglomerations of independent units which usually lacked symmetry but compensated for it by the abundance and variety of parts. Outstanding examples of this type of building included the khoromy of the Stroganovs in Solvychegodsk and the summer palace of the tsars in the village of Kolomenskoe near Moscow. Furthermore, it was especially during the Muscovite age that the principles of Russian wooden architecture, with its reliance on small independent structural units and its favorite geometric forms, found a rich expression also in the stone medium, notably in churches.

The church of St. Basil the Blessed at one end of Red Square, outside the Kremlin wall, provides the most striking illustration of this wooden type of construction in stone. Built in 1555–60 by two architects from Pskov, Barma and Posnik, it has never ceased to dazzle visitors and to excite the imagination. This church, known originally as the Cathedral of the Intercession of the Virgin, consists in fact of nine separate churches on a common foundation. All nine have the form of tall octagons — a narrower octagon on top of a broader one in each case — and the central church, around which the other eight are situated, is covered by a tent roof. Striking and different cupolas further emphasize the variety and independence of the parts of the church. Bright colors and abundant decorations contribute their share to the powerful, if somewhat bizarre, impression. While the church of St. Basil the Blessed and its predecessor, the church in the village of Diakovo that consisted of five churches, seem strange and unsymmetric to Western eyes, they succeed, in the opinion of many specialists, in combining their separate units into one magnificent whole.

In the Moscow Kremlin itself the construction went on, although the most important work had already been done in the reigns of Ivan III and Basil III. The Golden Gate arose in the first half of the seventeenth century, and as late as 1670–90 towers in the Kremlin wall were topped with roofs, usually in the Russian tent style, while within the walls palaces and churches continued to grow. In addition to the kremlin in Moscow, the beautiful kremlin of ancient Rostov, built mainly in the seventeenth century, and parts of kremlins in a score of other Russian cities have come down to our time.

In the second half of the seventeenth century the baroque style reached Moscovy through the Ukraine and quickly gained popularity, developing into the so-called Muscovite, or Naryshkin, baroque — the last name referring to the boyar family which sponsored it. It has been said that the Russians found baroque especially congenial because of their love of decoration. The church built in 1693 in the village of Fili, now part of Moscow, provides an interesting example of Russian baroque.

The great Russian tradition of icon painting continued during the sixteenth and seventeenth centuries but then was effectively terminated. Two prominent new schools emerged: the Stroganov school and the school of the tsar's icon-painters. The first, supported by the great merchant family of the northeast, was active approximately from 1580 to 1630. Its characteristics included bright backgrounds, rich colors, elaborate and minute design, and a penchant for decorative elements and gold, for instance gold contours. In fact, the Stroganov icons tended to become miniatures, "lovely and highly precious objects, if no longer great works of art" in the words of one critic. Procopius Chirin, who later joined the tsar's icon-painters and even became Tsar Michael's favorite artist, was an outstanding member of the Stroganov group.

The tsar's icon-painters dominated the scene in the second half of the seventeenth century. They found patronage in the so-called *Oruzheinaia Palata* headed by an able and enlightened boyar, Bogdan Khitrovo. The Oruzheinaia Palata began early in the sixteenth century as an arsenal, but, to quote Voyce: "It became successively a technical, scientific, pedagogical, and art institute, and contained shops and studios of icon and portrait painting, gold and silversmith work, keeping at the same time its original purpose — the manufacture of arms." The tsar's icon-painters developed a monumental style and reflected the influence of the West with its knowledge of perspective and anatomy. Simon Ushakov, who lived approximately from 1626 to 1686, was the school's celebrated master. We can still admire his skillful composition and precise execution in such icons as that of Christ the Ruler of the World painted for the cathedral of the Novodevichii Convent in Moscow.

Although Russian icon painting in the Muscovite period produced notable works and although its prestige and influence in the entire Orthodox world then reached its height, the school of the tsar's icon-painters marked the end of a long road. Ushakov himself has been praised for his remarkable ability to combine Byzantine and Western elements in his art, and the same can be said more modestly of his companions. Before long, the West swept over the East. Secular painting, including portrait painting, had already become popular in Muscovite Russia. After Peter the Great's reforms, art in Russia, as well as all of Russian culture, joined the Western world. Icon painting, of course, continued to exist, and on a very large scale, but as a craft rather than a highly creative and leading art.

Fresco painting and illumination also prospered in Muscovy. In fact, the second half of the seventeenth century saw a great flowering of fresco painting, which centered in Iaroslavl and spread to other towns in the Volga area. The gigantic scope and the fine quality of the work can best be studied in two churches in Iaroslavl: the church of the Prophet Elijah painted by Gurii Nikitin, Sila Savin, and their thirteen associates, and that of St. John the Baptist, where Dmitrii Grigoriev and fifteen other men painted the frescoes. The frescoes in the last-named church, which were created in

1694–95 and contain approximately 4,200 figures, represent the greatest effort of its kind in the world. Illumination also flourished, as evidenced, for instance, by the 1,269 miniatures — another 710 spaces remained blank — of the huge first volume of a sixteenth-century Russian chronicle of the world. In Muscovite frescoes and miniatures, as in icons, Western influences became increasingly apparent. By the end of the seventeenth century all ancient Russian graphic art was being rapidly replaced by the modern art of the West. It might be added in passing that in many other highly-skilled arts and crafts, such as carving, enamel, ceramics, and work with jewelry and precious metals, Muscovite Russia also left a rich legacy.

Education

Education in pre-Petrine Russia remains a controversial subject. Estimates of Muscovite enlightenment have ranged from an emphasis on well-nigh total illiteracy and ignorance to assertions that there existed in the realm of the tsars a widespread ability to read, write, and understand Church teachings and practices. The highly skeptical views of Miliukov and other critics appear on the whole rather convincing. Still, in this case, as in so many others, one has to strive for a balanced judgment. The Muscovite culture that we have discussed in this chapter could not have existed without some enlightenment. The enormous Muscovite state, and in particular its numerous bureaucracy, required, as a minimum, some education of officials. More speculative, although not necessarily fantastic, is Vladimirsky-Budanov's suggestion that Muscovites, like later Old Believers, generally could read and had thorough knowledge of their religious books. Finally, we do possess considerable direct evidence of education in Muscovite Russia.

Some education remained and developed in towns, in the many monasteries, and among the clergy generally. While much of it must have been of an extremely elementary character, more advanced schools appeared in the seventeenth century, especially after the acquisition of the Ukraine by Muscovy. In Kiev in the Ukraine, which was more open to the West, and where Orthodoxy had to defend itself against Catholicism, Metropolitan Peter Mogila, or Mohila, founded an Academy modeled on Jesuit colleges in 1631. In Moscow in 1648–49, a boyar Theodore Rtishchev built a monastery and invited some thirty Kievan monks to teach Slavonic, Latin, Greek, rhetoric, philosophy, and other disciplines. In 1666 Simeon of Polotsk established a school where he taught Latin and the humanities. After his death the school was re-established by his student, Sylvester Medvedev. In 1683 a school that offered Greek was opened in conjunction with a printing office and eventually contained up to two hundred and thirty students. Later in the 1680's the Medvedev and the printing press schools combined to form the Slavonic-Greek-Latin Academy, headed by learned Greek monks, the Lichud brothers, Ioannicius and Sofronius. As planned, the Academy was to protect the faith and to control knowledge as well as disseminate it. While Kiev and Moscow clearly stood out as centers of Russian enlightenment, some relatively advanced teaching also went on in such places as the Holy Trinity-St. Sergius Monastery and the cities of Novgorod and Kharkov.

The Muscovite school curriculum resembled closely, at corresponding levels, that of medieval Europe. In particular, it included almost no study of science and technology. Of the humanities, history fared best. In the sixteenth and, especially, the seventeenth centuries Russian textbooks in such fields as arithmetic, history, and grammar, dictionaries, and even elementary encyclopedias made their appearance, and toward the end of the period Sylvester Medvedev compiled the first Russian bibliography.

Western Influences. The Beginnings of Self-Criticism

Even if we make full allowance for Muscovite enlightenment, the fact remains that in a great many ways Muscovy lagged behind the West. Russia experienced no Renaissance and no Reformation, and it took no part in the maritime discoveries and the scientific and technological advances of the early modern period. Deficiencies became most apparent in war and in such practical matters as medicine and mining. They extended, however, into virtually every field. It should be noted that the Muscovite government showed a continuous and increasing interest in the West and in the many things that it had to offer. Muscovite society too, in spite of all the parochialism and prejudice, began gradually to learn from "the heretics."

Diplomacy constituted one obvious contact between the Muscovite state and other European countries. Although we traced the highlights of Russian foreign relations in preceding chapters, we should note here that these relations repeatedly included distant lands, such as England and Holland, as well as neighbors like Poland and Sweden, and that they dealt with many matters. For instance, an English merchant, Sir John Merrick, helped to negotiate the Treaty of Stolbovo between Sweden and Russia. Or, less happily, after the execution of Charles I, Tsar Alexis restricted English traders to Archangel, and he helped the king's son, later Charles II, with money and grain. Diplomatic correspondence published by Konovalov in the *Oxford Slavonic Papers* illustrates well the variety of issues encompassed in Anglo-Russian relations.

Many foreigners came to Muscovy and stayed. The number continued to increase after the first large influx in the reign of Ivan III. At the end of the sixteenth century foreigners in Muscovite service could be counted in hundreds, and even thousands if we include Poles, Lithuanians, and

Ukrainians, while the foreign section of the tsar's army consisted of 2,500 men. The Time of Troubles reduced these numbers, but with the reign of Michael the influx of foreigners resumed. In 1652 Tsar Alexis assigned them a northeastern suburb of Moscow, the so-called *Nemetskaia Sloboda,* or German Suburb. Incidentally, the Russian word for *German, nemets,* derived from the Russian for *dumb, nemoi,* came to mean all Europeans except Slavs and Latins. A visitor in the sixteen-seventies estimated that about eighteen thousand foreigners lived in Muscovy, mostly in the capital, but also in Archangel and other commercial centers, and in mining areas.

The importance of the foreign community, in particular for the economic development of the country, far exceeded its numbers. In addition to handling Russia's foreign trade, the newcomers began to establish a variety of manufactures and industries. Sir John Merrick, already mentioned as a diplomat, concentrated on producing hemp and tow. Andrew Vinius, a Dutchman, organized the industrial processing of iron ore and built the first modern ironworks in Muscovy. A Swede established a glass factory near Moscow. Others manufactured such items as gunpowder and paper. Second-generation foreigners often proved particularly adept at advancing both the economy of Russia and their own fortunes. Foreigners also acted as military experts, physicians, and other specialists.

Slowly the Russians turned to Western ways. In addition to reading and even writing secular stories, constructing baroque buildings, and painting portraits, as indicated above, they began to eat salad and asparagus, to snuff and smoke tobacco in spite of all the prohibitions, and to cultivate roses. Western clothing gained in popularity; some audacious persons also trimmed their hair and beards. In 1664 the postal service appeared, based on a Western model. And in the reign of Tsar Theodore a proposal was advanced to deal with the poor "according to the new European manner."

The stage was set for Peter the Great. In conclusion, however, it might be added that the reformer's wholesale condemnation of the existing order, although highly unusual, also had certain precedents in the Muscovite past. Not to mention the religious jeremiads, the secular writers often complained that there was no justice in the land even when praising the Muscovite form of government, as in the case of Peresvetov. More radical critics included Prince Ivan Khvorostinin, who died in 1625 and has been described as the first Russian free-thinker, George Križanič, and Gregory Kotoshikhin. Križanič, a Croatian and a Catholic priest, spent eighteen years in the realm of the tsars, from 1659 to 1677, and wrote there some nine books on religious, philosophical, linguistic, and political subjects. He combined an extremely high regard for Russia as the natural leader and savior of Slavdom with a sweeping condemnation of its glaring defects and, above all, its abysmal ignorance. Križanič's writings were apparently known to the Russian ruling circles. Kotoshikhin, an official in the foreign office, escaped to Sweden in 1664 after some personal trouble. There — before being executed in 1667 for the murder of his landlord — he wrote a sweeping denunciation of his native land. Kotoshikhin emphasized Muscovite pride, deceit, and, again, the isolation and ignorance of the people. As it turned out, the system that he condemned did not long outlast him.

PART V: IMPERIAL RUSSIA

X X

* * * * * * * * * *

THE REIGN OF PETER THE GREAT, 1682–1725

> Now an academician, now a hero,
> Now a seafarer, now a carpenter,
> He, with an all-encompassing soul,
> Was on the throne an eternal worker.
>
> PUSHKIN

> If we consider the matter thoroughly, then, in justice, we must be
> called not *Russians,* but *Petrovians.* . . . Russia should be called
> *Petrovia,* and we *Petrovians.* . . .
>
> KANKRIN

PETER THE GREAT'S reign began a new epoch in Russian history, known variously as the Imperial Age because of the new designation of ruler and land, the St. Petersburg Era because of the new capital, or the All-Russian Period because the state came to include more and more peoples other than the Great Russians, that is, the old Muscovites. The epoch lasted for approximately two centuries and ended abruptly in 1917. Although the chronological boundaries of Imperial Russia are clearly marked — by contrast, for instance, with those of appanage Russia — the beginning of Peter the Great's reign itself can be variously dated. The reformer, who died on February 8, 1725, attained supreme power in several stages, and with reversals of fortune: in 1682 as a boy of ten he was proclaimed at first tsar and later that same year co-tsar with his elder half-brother Ivan; in 1689 he, or rather his family and party, regained effective control of the government; in 1694 Peter's mother died and he started to rule in fact as well as in name; finally in 1696 Ivan died, leaving Peter the only and absolute sovereign of Muscovy. Therefore, before turning to the celebrated reformer and his activities, we must consider a number of years during which Peter's authority remained at best nominal.

Russian History from 1682 to 1694

Tsar Alexis had been married twice, to Mary Miloslavskaia from 1648 to 1669, and to Nathalie Naryshkina from 1671 until his death in 1676. He had thirteen children by his first wife, but of the sons only two, Theodore and Ivan, both of them sickly, survived their father. Peter, strong and healthy, was born on June 9, 1672, about a year after the tsar's second marriage. Theodore, as we know, succeeded Alexis and died without an

213

heir in 1682. In the absence of a law of succession, the two boyar families, the Miloslavskys and Naryshkins, competed for the throne. The Naryshkins gained an early victory: supported by the patriarch, a majority in the boyar duma, and a gathering of the gentry, Peter was proclaimed tsar in April 1682. Because of his youth, his mother became regent, while her relatives and friends secured leading positions in the state. However, as early as May, the Miloslavsky party, led by Alexis's able and strong-willed daughter Sophia, Peter's half-sister, inspired a rebellion of the regiments of the streltsy, or musketeers, concentrated in Moscow. Leading members of the Naryshkin clique were murdered — Peter witnessed some of these murders — and the Miloslavskys seized power. At the request of the streltsy, the boyar duma declared Ivan senior tsar, allowed Peter to be junior tsar, and, a little later, made Sophia regent. It might be added that the streltsy, strongly influenced by the Old Belief, proceeded to put more pressure on the government and cause further trouble, but in vain: the new regent managed to punish the leaders and control the regiments.

From 1682 to 1689 Sophia and her associates governed Muscovy, with Ivan V incapable of ruling and Peter I, together with the entire Naryshkin party, kept away from state affairs. Prince Basil Golitsyn, the regent's favorite, played a particularly important role. An enlightened and humane person who spoke several foreign languages and arranged his own home and life in the Western manner, Golitsyn cherished vast projects of improvement and reform including the abolition of serfdom and education on a large scale. He did liberalize the Muscovite penal code, even if he failed to implement his more ambitious schemes. Golitsyn's greatest success came in 1686 when Russia and Poland signed a treaty of "eternal peace" that confirmed the Russian gains of the preceding decades, including the acquisition of Kiev. Yet the same treaty set the stage for the war against the Crimean Tartars, who were backed by Turkey. This war proved disastrous to Muscovite arms. In 1687 and again in 1689 Golitsyn led a Muscovite army into the steppe only to suffer heavy losses and defeat as the lack of water and the huge distances exhausted his troops, while the Tartars set the grass on fire. Golitsyn's military fiasco, together with other accumulating tensions, led to Sophia's downfall.

As Peter grew older, his position as a tsar without authority became increasingly invidious. Sophia, on her part, realized the insecurity of her office and desired to become ruler in her own right. In 1689 Theodore Shaklovity, appointed by Sophia to command the streltsy, apparently tried to incite his troops to stage another coup, put the regent on the throne, and destroy her opponents. Although the streltsy failed to act, a denouement resulted. Frightened by the report of a plot, Peter escaped in the dead of night from the village of Preobrazhenskoe, near Moscow, where he had

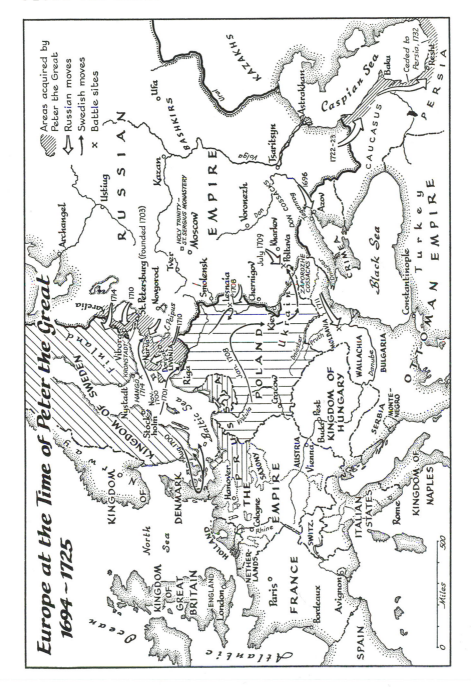

Europe at the Time of Peter the Great
1694~1725

been living, to the Holy Trinity-St. Sergius Monastery. In the critical days
that followed, the patriarch, many boyars and gentry, the military units
trained in the Western manner and commanded by General Patrick Gordon,
and even several regiments of the streltsy, rallied behind Peter. Many
others wavered, but did not back Sophia. In the end the sister capitulated
to the brother without a fight and was sent to live in a convent. Shaklovity
and two of his aides were executed; several other officers and boyars, including Basil Golitsyn, suffered exile. Thus, in August 1689, Peter won
acknowledgment as the effective ruler of Russia, although Ivan retained
his position as co-tsar. Still, at seventeen, Peter showed no desire to take
personal charge of affairs. Instead the government fell into the hands of his
mother Nathalie and her associates, notably her brother, the boyar Leo
Naryshkin, Patriarch Joachim, and, after his death in 1690, Patriarch
Hadrian. The years 1689–94 witnessed the last flowering of Muscovite
religiosity, ritualism, parochialism, and suspicion of everything foreign —
it was even forbidden to train troops in the Western manner. But in 1694
Nathalie died, and Peter I finally assumed the direction of the state at the
age of twenty-two.

Peter the Great: His Character, Childhood, and Youth

The impression that Peter I commonly made on his contemporaries
was one of enormous strength and energy. Almost seven feet tall and
powerfully built, the tsar possessed astonishing physical strength and vigor.
Moreover, he appeared to be in a constant state of restless activity, taking
on himself tasks normally done by several men. Few Russians could keep
up with their monarch in his many occupations. Indeed, as he walked with
rapid giant strides, they had to run even to continue conversation. In addition to his extraordinary physical attributes, Peter I exhibited some remarkable qualities of mind and character. The tsar had an insatiable intellectual
curiosity coupled with an amazing ability to learn. He proceeded to participate personally in all kinds of state affairs, technical and special as well
as general, becoming deeply involved in diplomacy, administration, justice,
finance, commerce, industry, education, and practically everything else besides. In his reforms the tsar invariably valued expert advice, but he was
also generally independent in thought and did not hesitate to adapt projects to circumstances. Peter I also developed into an accomplished military and naval commander. He studied the professions of soldier and sailor
from the bottom up, serving first in the ranks and learning the use of each
weapon before promoting himself to his first post as an officer. The monarch
attained the rank of full general after the victory of Poltava and of full
admiral after the successful conclusion of the Great Northern War. In

addition, the sovereign found time to learn some twenty different trades and prided himself on his ability to make almost anything, from a ship to a pair of shoes. With his own hands he pulled the teeth of his courtiers and cut off their beards. Characteristically, he wanted to be everywhere and see everything for himself, traveling indefatigably around his vast state as no Muscovite monarch had ever done. In a still more unprecedented manner he went twice to the West to learn, in 1697–98 and in 1717. Peter I's mind can best be described as active and practical, able quickly to grasp problems and devise solutions, if not to construct theories.

As to character, the tsar impressed those around him by his energy, unbending will, determination, and dedication. He recovered quickly from even the worst defeats and considered every obstacle as an invitation to further exertion and achievement. Less attractive, but at times equally imposing, traits included a violent temper, crudeness, and frequent cruelty. The sovereign could be an executioner, as well as a dentist, and his drunken, amorous, and blasphemous pastimes exceeded the measure of the rough times in which he lived. Yet Peter the Great must not be confused with Ivan the Terrible, whom he, incidentally, admired. The reformer never lost himself in the paranoid world of megalomania and delusions of persecution, and he even refused to identify himself with the state. To mention one significant detail, when reforming the army, Peter I crossed out "the interests of His Tsarist Majesty" as the object of military devotion and substituted "the interests of the state." Consistently he made every effort to serve his country, to bring to it change and enlightenment. As the sovereign wrote in the last month of his life, in connection with dispatching Vitus Bering's first expedition: "Having ensured the security of the state against the enemy, it is requisite to endeavor to win glory for it by means of the arts and sciences." Or, to support Peter the Great's emphasis on education with another quotation — and one especially appropriate in a textbook — "For learning is good and fundamental, and as it were the root, the seed, and first principle of all that is good and useful in church and state."

Although a precocious child, Peter received no systematic education, barely being taught to read and write. Instead, from a very early age he began to pick things up on his own and pursue a variety of interests. He devoted himself in particular to war games with a mixed assortment of playmates. These games, surprisingly enough, developed over a period of years into a serious military undertaking and resulted in the formation of the first two regiments of the guards, the *Preobrazhenskii* — for Peter lived in the village of Preobrazhenskoe — and the *Semenovskii*, named after a nearby village. Similarly, the young tsar showed an early interest in the navy. At first he built small vessels, but as early as 1694 he established a dockyard in Archangel and constructed a large ship there all by himself. For information and instruction Peter went to the foreign quarter in Moscow. There he learned from a variety of specialists what he wanted to know most about military and naval matters, geometry and the erection of fortifications. There too, in a busy, informal, and unrestrained atmosphere, the tsar apparently felt much more at ease than in the conservative, tradition-bound palace environment, which he never accepted as his own. The smoking, drinking, love-making, rough good humor, and conglomeration of tongues, first discovered in the foreign quarter in Moscow, became an enduring part of Peter the Great's life. The determined attempt of Peter's mother to make him mend his ways by marrying him to Eudoxia Lopukhina in 1689 failed completely to accomplish the desired purpose.

Peter's Assistants

After Peter took over the conduct of state affairs and began to reform Muscovy, he found few collaborators. His own family, the court circles, and the boyar duma overwhelmingly opposed change. Because he discovered little support at the top of the state structure, and also because he never attached much importance to origin or rank, the sovereign proceeded to obtain assistants wherever possible. Before long an extremely mixed but on the whole able group emerged. To quote Kliuchevsky's colorful summary:

> Peter gathered the necessary men everywhere, without worrying about rank and origin, and they came to him from different directions and all possible conditions: one arrived as a cabin-boy on a Portuguese ship, as was the case of the chief of police of the new capital, de Vière; another had shepherded swine in Lithuania, as it was rumored about the first Procurator-General of the Senate, Iaguzhinsky; a third had worked as a clerk in a small store, as in the instance of Vice-Chancellor Shafirov; a fourth had been a Russian house serf, as in the case of the Vice-Governor of Archangel, the inventor of stamped paper, Kurbatov; a fifth, i.e., Ostermann, was a son of a Westphalian pastor. And all these men, together with Prince Menshikov, who, the story went, had once sold pies in the streets of Moscow, met in Peter's society with the remnants of the Russian boyar nobility.

Among foreigners, the tsar had the valuable aid of some of his old friends, such as Patrick Gordon and the Swiss, Francis Lefort, who played a prominent role until his early death in 1699. Later such able newcomers from Germany as the diplomat, Andrew Ostermann, and the military expert, Burkhard Münnich, joined the sovereign's entourage. Some of his numerous foreign assistants, for example, the Scot James Bruce who helped with the artillery, mining, the navy and other matters, had been born in Russia and belonged to the second generation of foreign settlers in Muscovy.

Russian assistants to Peter ranged over the entire social gamut. Alexander Menshikov, Paul Iaguzhinsky, Peter Shafirov, and Alexis Kurbatov, among others, came from the lower classes. A large group belonged to the service gentry, of whom only two examples are the chief admiral of the reign, Theodore Apraksin, and Chancellor Gabriel Golovkin. Even old aristocratic families contributed a number of important figures, such as Field Marshal Count Boris Sheremetev and Senator Prince Jacob Dolgoruky. The Church too, although generally opposed to reform, supplied some able clerics who furthered the work of Peter the Great. The place of honor among them belongs to Archbishop Theophanes, or Feofan, Prokopovich, who, like many other promoters of change in Russia, came from the Ukraine. Of all the "fledglings of Peter's nest" — to use Pushkin's expression — Menshikov acquired the greatest prominence and power. This son of a corporal or groom, who reportedly was once a pie vendor, came closest to being the sovereign's alter ego and participating in the entire range of his activity. Beginning as the boy tsar's orderly in the Preobrazhenskii regiment, Menshikov rose to be Generalissimo, Prince in Russia, and Prince of the Holy Roman Empire, to mention only his most outstanding titles. Vain and thoroughly corrupt, as well as able and energetic, he constituted a permanent target for investigations and court proceedings and repeatedly suffered summary punishment from Peter the Great's cudgel, but somehow managed to maintain his position.

The First Years of Peter's Rule

War against Turkey was the first major action of Peter I after he took the government of Russia into his own hands in 1694, following the death of his mother. In fighting Turkey, the protector of the Crimean Tartars and the power controlling the Black Sea and its southern Russian shore, the new monarch followed in the steps of his predecessors. However, before long it became apparent that he managed his affairs differently. The war began in 1695, and the first Russian campaign against Azov failed: supplied by sea, the fortress remained impregnable to the Muscovite army. Then, in one winter, the tsar built a fleet in Voronezh on the Don river. He worked indefatigably himself, as well as ordering and urging others, and utilized to the best advantage the knowledge of all available foreign specialists along with his own previously acquired knowledge. By displaying his tremendous energy everywhere, Peter the Great brought thirty sea-going vessels and about a thousand transport barges to Azov in May

1696. Some of the Russian fleet, it might be noted, had been built as far away as Moscow and assembled in Voronezh. This time besieged by sea as well as by land, the Turks surrendered Azov in July.

With a view toward a further struggle against Turkey and a continuing augmentation and modernization of the Russian armed forces, the tsar next sent fifty young men to study, above all shipbuilding and navigation, in Holland, Italy, and England. Peter dispatched groups of Russians to study abroad several more times in his reign. After the students returned, the sovereign often examined them personally. In addition to experts, the tsar needed allies to prosecute war against Turkey. The desire to form a mighty coalition against the Ottoman Empire, and an intense interest in the West, prompted Peter to organize a large embassy to visit a number of European countries and — a most unusual act for a Muscovite ruler — to travel with the embassy.

Headed by Lefort, the party of about 250 men set out in March 1697. The sovereign journeyed incognito under the name of Peter Mikhailov. His identity, however, remained no secret to the rulers and officials of the countries he visited or to the crowds which frequently gathered around him. The tsar engaged in a number of important talks on diplomatic and other state matters. But, above all, he tried to learn as much as possible from the West. He seemed most concerned with navigation, but he also tried to absorb other technical skills and crafts, together with the ways and manners and, in fact, the entire life of Europe as he saw it. As the so-called Grand Embassy progressed across the continent and as Peter Mikhailov also took trips of his own, most notably to the British Isles, he obtained some first-hand knowledge of the Baltic provinces of Sweden, Prussia, and certain other German states, and of Holland, England, and the Hapsburg Empire. From Vienna the tsar intended to go to Italy, but instead he rushed back to Moscow at news of a rebellion of the streltsy. Altogether Peter the Great spent eighteen months abroad in 1697–98. At that time over 750 foreigners, especially Dutchmen, were recruited to serve in Russia. Again in 1702 and at other times, the tsar invited Europeans of every nationality — except Jews, whom he considered parasitic — to come to his realm, promising to subsidize passage, provide advantageous employment, and assure religious tolerance and separate law courts.

The streltsy had already caused trouble to Peter and suffered punishment on the eve of the tsar's journey to the West — in fact delaying the journey. Although the new conspiracy that was aimed at deposing Peter and putting Sophia in power had been effectively dealt with before the sovereign's return, the tsar acted with exceptional violence and severity. After investigation and torture more than a thousand streltsy were executed, and their mangled bodies were exposed to the public as a salutary lesson. Sophia was forced to become a nun, and the same fate befell Peter's wife, Eudoxia, who had sympathized with the rebels.

If the gruesome death of the streltsy symbolized the destruction of the old order, many signs indicated the coming of the new. After he returned from the West, the tsar began to demand that beards be cut and foreign dress be worn by courtiers, officials, and the military. With the beginning of the new century, the sovereign changed the Russian calendar: henceforth years were to be counted from the birth of Christ, not the creation of the world, and they were to commence on the first of January, not the first of September. More important, Peter the Great rapidly proceeded to reorganize his army according to the Western pattern.

The Great Northern War

The Grand Embassy failed to further Peter the Great's designs against Turkey. But, although European powers proved unresponsive to the proposal of a major war with the Ottomans, other political opportunities emerged. Before long Peter joined the military alliance against Sweden organized by Augustus II, ruler of Saxony and Poland. Augustus II, in turn, had been influenced by Johann Reinhold Patkul, an émigré Livonian nobleman who bore a personal grudge against the Swedish crown. The interests of the allies, Denmark, Russia, and Poland-Saxony — although, to be exact, Saxony began the war without Poland — clashed with those of Sweden, which after its extremely successful participation in the Thirty Years' War had acquired a dominant position on the Baltic and in the Baltic area. The time to strike appeared ripe, for Charles XII, a mere youth of fifteen, had ascended the Swedish throne in 1697. While Peter I concentrated on concluding the long-drawn-out peace negotiations with Turkey, Augustus II declared war on Sweden in January 1700, and several months later Denmark followed his example. On July 14 the Russo-Turkish treaty was finally signed in Constantinople: the Russians obtained Azov and Taganrog as well as the right to maintain a resident minister in Turkey. On August 19, ten days after Peter the Great learned of the conclusion of the treaty with the Porte and the day after he officially announced it, he declared war on Sweden. Thus Russia entered what came to be known as the Great Northern War.

Immediately the Russians found themselves in a much more difficult situation than they had expected. Charles XII turned out to be something of a military genius. With utmost daring he crossed the straits and carried the fight to the heart of Denmark, quickly forcing the Danes to surrender. Unknown to Peter, the peace treaty of Travendal marking the Danish defeat and abandonment of the struggle was concluded on the very day on which Russia entered the war. Having disposed of Denmark, the Swedish king promptly attacked the new enemy. Transporting his troops across the

Baltic to Livonia, on November 30, 1700, he suddenly assaulted the main Russian army that was besieging the fortress of Narva. In spite of the very heavy numerical odds against them the Swedes routed the Russian forces, killing or capturing some ten thousand troops and forcing the remaining thirty thousand to abandon their artillery and retreat in haste. The prisoners included ten generals and dozens of officers. In the words of a recent historian summarizing the Russian performance at Narva: "The old-fashioned cavalry and irregulars took to flight without fighting. The new infantry levies proved 'nothing more than undisciplined militia,' the foreign officers incompetent and unreliable. Only the two guards and one other foot regiment showed up well."

It was believed by some at the time and has been argued by others since that after Narva Charles XII should have concentrated on knocking Russia out of the war and that by acting in a prompt and determined manner he could have accomplished this purpose. Instead, the Swedish king for years underestimated and neglected his Muscovite opponent. After lifting the Saxon siege of Riga in the summer of 1701, he transferred the main hostilities to Poland, considering Augustus II his most dangerous enemy. Again Swedish arms achieved notable successes, but for about six years they could not force a decision. In the meantime, Peter made utmost use of the respite he received. Acting with his characteristic energy, the tsar had a new army and artillery ready within a year after the debacle of Narva. Conscription, administration, finance, and everything else had to be strained to the limit and adapted to the demands of war, but the sovereign did not swerve from his set purpose. The melting of church bells to make cannons has remained an abiding symbol of that enormous war effort.

Peter I used his reconstructed military forces in two ways: he sent help to Augustus II, and he began a systematic advance in Livonia and Estonia, which Charles XII had left with little protection. Already in 1701 and 1702 Sheremetev at the head of a large army devastated these provinces, twice defeating weak Swedish forces, and the Russians began to establish themselves firmly on the Gulf of Finland. The year 1703 marked the founding of St. Petersburg near the mouth of the Neva. The following year Peter the Great built the island fortress of Kronstadt to protect his future capital, while the Russian troops captured the ancient city of Dorpat, or Iuriev, in Estonia and the stronghold of Narva itself. The tsar rapidly constructed a navy on the Baltic, his southern fleet being useless in the northern war, and the new ships participated effectively in amphibious and naval operations.

But time finally ran out for Augustus II. Brought to bay in his own Saxony, he had to sign the Treaty of Altranstädt with Charles XII in late September, 1706: by its terms Augustus II abdicated the Polish crown in favor of pro-Swedish Stanislaw Leszczynski and, of course, withdrew from the war. Peter the Great was thus left alone to face one of the most feared armies and one of the most successful generals of Europe. Patkul, incidentally, was handed over to the Swedes by Augustus II and executed. The Swedish king began his decisive campaign against Russia in January 1708, crossing the Vistula with a force of almost fifty thousand men and advancing in the direction of Moscow.

Peter I's position was further endangered by the need to suppress rebellions provoked both by the exactions of the Russian government and by opposition to the tsar's reforms. In the summer of 1705 a monk and one of the streltsy started a successful uprising in Astrakhan aimed against the upper classes and the foreign influence. It was even rumored in Astrakhan that all Russian girls would be forced to wed Germans, a threat which led to the hasty conclusion of many marriages. The town was recaptured by Sheremetev only in March 1706, after bitter fighting. In 1707 Conrad Bulavin, a leader of the Don cossacks, led a major rebellion in the Don area. Provoked by the government's determination to hunt down fugitives and also influenced by the Old Belief, Bulavin's movement followed the pattern of the great social uprisings of the past. At its height, the rebellion spread over a large area of southern Russia, including dozens of towns, and the rebel army numbered perhaps as many as one hundred thousand men. As usual in such uprisings, however, this huge force lacked organization and discipline. Government troops managed to defeat the rebels decisively a year or so before the war with Sweden reached its climax in the summer of 1709. Still another rebellion, that of the Turkic Bashkirs who opposed the Russian disruption of their way of life as well as the heavy exactions of the state, erupted in the middle Volga area in 1705 and was not finally put down until 1711.

Some historians believe that Charles XII would have won the war had he pressed his offensive in 1708 against Moscow. Instead he swerved south and entered the Ukraine. The Swedish king wanted to rest and strengthen his army in a rich land untouched by the fighting before resuming the offensive, and he counted heavily on Hetman Ivan Mazepa who had secretly turned against his sovereign. His calculations failed: Mazepa could bring only some two thousand cossacks to the Swedish side — with a few thousand more joining later — while a general lack of sympathy for the Swedes together with Menshikov's energetic and rapid countermeasures assured the loyalty of the Ukraine to Peter the Great. Also, Charles XII's move south made it easier for a Russian force led by the tsar to intercept and smash Swedish reinforcements of fifteen thousand men on October 9, 1708, at Lesnaia. What is more, at Lesnaia the Russians captured the huge supply train which was being brought to Charles. Largely isolated from the people, far from home bases, short of supplies, and unable to advance their cause militarily or diplomatically, the Swedish army spent a dismal, cold winter in 1708–09 in the Ukraine. Yet Charles XII would not retreat. The hour of decision struck in the middle of the following summer when the main Russian army finally came to the rescue of the small fortress of Poltava besieged by the Swedes, and the enemies met in the open field.

The Swedish army was destroyed on July 8, 1709, in the battle of Poltava. The Swedes, numbering only from 22,000 to 28,000 as against over 40,000 Russians, and vastly inferior in artillery, put up a tremendous fight before their lines broke. Most of them, including the generals, eventually surrendered either on the field or several days later near the Dnieper which they could not cross. Charles XII and Mazepa did escape to Turkish territory. Whereas in retrospect the outcome of Poltava occasions no surprise, it bears remembering that a few years earlier the Swedes had won at Narva against much greater odds and that Charles XII had acquired a reputation as an invincible commander. But, in contrast to the debacle at Narva, Russian generalship, discipline, fighting spirit, and efficiency all splendidly passed the test of Poltava. Peter the Great, who had himself led his men in the thick of battle and been lucky to survive the day, appreciated to the full the importance of the outcome. And indeed he had excellent reasons to celebrate the victory and to thank his captive Swedish "teachers" for their most useful "lessons."

Yet not long after Poltava the fortunes of Peter I and his state reached perhaps their lowest point. Instigated by France, as well as by Charles XII, Turkey, which had so far abstained from participation in the hostilities, declared war on Russia in 1710. Peter acted rashly, underestimating the enemy and relying heavily on the problematical support of the vassal Ottoman principalities of Moldavia and Wallachia and of Christian subjects of the sultan elsewhere, notably in Serbia and Montenegro. In July 1711, the tsar found himself at the head of an inadequate army in need of ammunition and supplies and surrounded by vastly superior Turkish forces near the Pruth river. Argument persists to this day as to why the Turks did not make more of their overwhelming advantage. Suggested answers have ranged from the weariness and losses of the Turkish troops to skillful Russian diplomacy and even bribery. In any case Peter the Great signed a peace treaty, according to which he abandoned his southern fleet, returned Azov and other gains of 1700 to the Turks, promised not to intervene in Poland, and guaranteed to Charles XII safe passage to Sweden. But, at the price of renouncing acquisitions to the south, he was enabled to extricate himself from a catastrophic situation and retain a dominant hand in the Great Northern War.

That war, decided in effect in 1709, dragged on for many more years. After Poltava, the tsar transferred his main effort to the Baltic, seizing

Viborg — or Viipuri — Riga, and Reval in 1710. St. Petersburg could be considered secure at last. The debacle of Charles XII in the Ukraine led to a revival of the coalition against him. Saxony, Poland, Denmark, Prussia, and Hanover joined Russia against Sweden. In new circumstances, Peter the Great developed his military operations along two chief lines: Russian troops helped the allies in their campaigns on the southern shore of the Baltic, while other forces continued the advance in the eastern Baltic area. Thus in 1713–14 the tsar occupied most of Finland. The new Russian navy became ever more active, scoring a victory under Peter's direct command over the Swedish fleet off Hangö in 1714.

It may be worth noting that the sudden rise of Russia came as something of a shock to other European countries, straining relations, for example, between Great Britain and Russia. It also led to considerable fear and worried speculations about the intentions and future steps of the northern giant; this was reflected later in such forgeries as the purported testament of Peter the Great which expressed his, and Russia's, aim to conquer the world. In 1717 the tsar traveled to Paris, and, although he failed to obtain any diplomatic results beyond the French promise not to help Sweden, once more he saw and learned much. In December 1718, Charles XII was killed in a minor military engagement in Norway. His sister Ulrika Eleonora and later her husband Frederick I succeeded to the Swedish throne. Unable to reverse the course of the war and, indeed, increasingly threatened, for Peter the Great proceeded to send expeditions into Sweden proper in 1719–21, the Swedes finally admitted defeat and made peace. In 1720–21, by the Treaties of Stockholm, Frederick I reached settlements with Saxony, Poland, Denmark, Prussia, and Hanover, abandoning some islands and territory south of the Baltic, mostly in favor of Prussia. And on August 30, 1721, Sweden concluded the Treaty of Nystadt with Russia.

By the provisions of the Treaty of Nystadt Russia acquired Livonia, Estonia, Ingermanland, part of Karelia, and certain islands, although it returned the bulk of Finland and paid two million rix-dollars. In effect it obtained the so-called Baltic provinces which were to become, after the Treaty of Versailles, the independent states of Estonia and Latvia and later the corresponding Soviet republics, and also obtained southeastern Finnish borderlands located strategically next to St. Petersburg and the Gulf of Finland. The capture and retention of the fortress of Viborg in particular gave Russia virtual control of the Gulf. At a solemn celebration of the peace settlement the Senate prevailed upon Peter I to accept the titles of "Great," "Father of the Fatherland," and "Emperor." In this manner Russia formally became an empire, and one can say that the imperial period of Russian history was officially inaugurated, even though some European powers took their time in recognizing the new title of the

Russian ruler: only Prussia and the Netherlands did so immediately, Sweden in 1723, Austria and Great Britain in 1742, France and Spain as late as 1745.

In modern European history the Great Northern War was one of the important wars and Poltava one of the decisive battles. The Russian victory over Sweden and the resulting Treaty of Nystadt meant that Russia became firmly established on the Baltic, acquiring its essential "window into Europe," and that in fact it replaced Sweden as the dominant power in the north of the continent. Moreover, Russia not only humiliated Sweden but also won a preponderant position vis-à-vis its ancient rival Poland, became directly involved in German affairs — a relationship which included marital alliances arranged by the tsar for his and his half-brother Ivan V's daughters — and generally stepped forth as a major European power. The Great Northern War, and the War of the Spanish Succession fought at the same time, can be regarded as successful efforts to change the results of the Thirty Years' War and to curb the two chief victors of that conflict, Sweden and France. The settlement in the north, it might be added, turned out to be more durable than that in the west. Indeed, because of the relative sizes, resources, and numbers of inhabitants of Russia and Sweden, Peter the Great's defeat of Charles XII proved to be irreversible.

Foreign Relations: Some Other Matters

Although the Great Northern War lasted for most of Peter's reign and although it had first claim on Russian efforts and resources, the tsar never forgot Turkey or the rest of Asia either. We have noted the two wars that he fought against the Ottomans, the first successful and the second unsuccessful in the midst of the hostilities with Sweden. After Nystadt, the emperor turned south once more, or rather southeast. In 1722–23 he fought Persia successfully, in spite of great difficulties of climate and communication, to obtain a foothold on the western and southern shores of the Caspian sea. This foothold was relinquished by Russia in 1732, shortly after Peter's death.

Earlier the tsar had shown a considerable interest in Central Asia, its geography, peoples — particularly the Kazakhs — and routes, and especially in the possibility of large-scale trade with India. Whereas most of the Russian contacts with Central Asia were peaceful, a tragic exception occurred in 1717 when a considerable force commanded by Prince Alexander Bekovich-Cherkassky was tricked and massacred by the supposedly friendly khan of Khiva. Peter the Great ordered young men to learn Turkish, Tartar, and Persian, assigning them for this purpose to appropriate diplomatic missions. He even established classes in Japanese, utilizing the services of a castaway from that hermit island empire. The tsar sent a mission to Mongolia and maintained diplomatic and commercial relations with China, which resulted in the negotiation of the Treaty of Kiakhta shortly after his death, and in the permanent establishment of an important mission of the Russian Orthodox Church in Peking. He initiated the scholarly study of Siberia; and, indeed, the emperor's interest extended even to the island of Madagascar!

The Reforming of Russia: Introductory Remarks

In regard to internal affairs during the reign of Peter the Great, we find that scholars have taken two extreme and opposite approaches. On the one hand, the tsar's reforming of Russia has been presented as a series, or rather a jumble, of disconnected *ad hoc* measures necessitated by the exigencies of the moment, especially by the pressure of the Great Northern War. Contrariwise, the same activity has been depicted as the execution of a comprehensive, radically new, and well-integrated program. In a number of ways, the first view seems closer to the facts. As Kliuchevsky pointed out, only a single year in Peter the Great's whole reign, 1724, passed entirely without war, while no more than another thirteen peaceful months could be added for the entire period. Connected to the enormous strain of war was the inadequacy of the Muscovite financial system, which was over-burdened and in a state of virtual collapse even before Peter the Great made vastly increased demands upon it. The problem for the state became simply to survive, and survival exacted a heavy price. Under Peter the Great the population of Russia might have declined. Miliukov, who made a brilliant analysis of Petrine fiscal structure and economy, and other scholars of his persuasion have shown how military considerations repeatedly led to financial measures, and in turn to edicts aiming to stimulate Russian commerce and industry, to changes in the administrative system without whose improvement these and other edicts proved ineffective, to attempts to foster education in whose absence a modern administration could not function, and on and on. It has further been argued, on the whole convincingly, that in any case Peter the Great was not a theoretician or planner, but an intensely energetic and practical man of affairs.

Yet a balanced judgment has to allow something to the opposite point of view as well. Although Peter the Great was preoccupied during most of his reign with the Great Northern War and although he had to sacrifice much else to its successful prosecution, his reforming of Russia was by no means limited to hectic measures to bolster the war effort. In fact, he wanted to Westernize and modernize all of the Russian government, society, life, and culture, and even if his efforts fell far short of this stupendous goal, failed to dovetail, and left huge gaps, the basic pattern emerges, nevertheless, with sufficient clarity. Countries of the West served as the emperor's model. We shall see, however, when we turn to specific legislation, that Peter did not merely copy from the West, but tried to adapt Western institutions to Russian needs and possibilities. The very number and variety of European states and societies offered the Russian ruler a rich initial choice. It should be added that with time Peter the Great became more interested in general issues and broader patterns. Also, while the reformer was no theoretician, he had the makings of a visionary. With characteristic grandeur and optimism he saw ahead the image of a modern, powerful, prosperous, and educated country, and it was to the realization of that image that he dedicated his life. Both the needs of the moment and longer-range aims must therefore be considered in evaluating Peter the Great's reforms. Other fundamental questions to be asked about them include their relationship to the Russian past, their borrowing from the West — and, concurrently, their modification of Western models — their impact on Russia, and their durability.

The Army and the Navy

Military reforms stemmed most directly from the war. In that field Peter the Great's measures must be regarded as radical, successful, and lasting, as well as imitative of the West; and he has rightly been considered the founder of the modern Russian army. The emperor's predecessors had large armies, but these were poorly organized, technically deficient, and generally of low quality. They assembled for campaigns and disbanded when the campaign ended. Only gradually did "regular" regiments, with Western officers and technicians, begin to appear. Even the streltsy, founded by Ivan the Terrible and expanded to contain twenty-two regiments of about a thousand men each, represented a doubtful asset. Stationed mainly in Moscow, they engaged in various trades and crafts and constituted at best a semi-professional force. Moreover, as mentioned earlier, the streltsy became a factor in Muscovite politics, staged uprisings, and were severely punished and then disbanded by Peter the Great. The reformer instituted general conscription and reorganized and modernized the army. The gentry, of course, had been subject to personal military service ever since the formation of the Muscovite state. Under Peter the Great this obligation came to be much more effectively and, above all, continuously enforced. Except for the unfit and those given civil assignments, the members of the gentry were to remain with their regiments for life. Other classes, with the exception of the clergy and members of the merchant guilds, who were needed elsewhere, fell under the draft. Large numbers were conscripted, especially in the early years of the Great Northern War. In 1715 the Senate established the norm of one draftee from every seventy-five serf households. Probably the same norm operated in the case of the state peasants, while additional recruits were obtained from the townspeople. All were to be separated from their families and occupations and to serve for life, a term which was reduced to twenty-five years only in the last decade of the eighteenth century.

Having obtained a large body of men, Peter I went on to transform them into a modern army. He personally introduced a new and up-to-date military manual, became proficient with every weapon, and learned to command units from the smallest to the largest. He insisted that each draftee, aristocrat and serf alike, similarly work his way from the bottom up, advancing exactly as fast and as far as his merit would warrant. Important changes in the military establishment included the creation of the elite regiments of the guards, and of numerous other regular regiments, the adoption of the flintlock and the bayonet, and an enormous improvement in artillery. By the time of Poltava, Russia was producing most of its own flintlocks. The Russian army was the first to use the bayonet in attack — a weapon originally designed for defense against the charging enemy. As to artillery, Peter the Great developed both the heavy siege artillery, which proved very effective in 1704 in the Russian capture of Narva, and, by about 1707, light artillery, which participated in battles alongside the infantry and the cavalry. The Russian victory over Sweden demonstrated the brilliant success of the tsar's military reforms. At the time of Peter the Great's death the Russian army numbered 210,000 regular troops and 100,000 cossacks who retained their own organization.

The select regiments of guards, however, were not only the elite of Peter's army; they had, so to speak, grown up with the emperor, and contained many of his most devoted and enthusiastic supporters. Especially in the second half of his reign, Peter the Great frequently used officers and non-commissioned officers of the guards for special assignments, bypassing the usual administrative channels. Often endowed with summary powers, which might include the right to bring a transgressing governor or other high official back in chains, they were sent to speed up the collection of taxes or the gathering of recruits, to improve the functioning of the judiciary or to investigate alleged administrative corruption and abuses. Operating outside the regular bureaucratic structure, these emissaries could be considered as extensions of the ruler's own person. Later emperors, such as Alexander I and Nicholas I, continued Peter the Great's novel practice on a large scale, relying on special, and usually military, agents to obtain immediate results in various matters and in general to supervise the workings of the government apparatus.

To an even greater extent than the army, the modern Russian navy was the creation of Peter the Great. One can fairly say it was one of his passions. He began from scratch — with one vessel of an obsolete type, to be exact — and left to his successor 48 major warships and 787 minor and auxiliary craft, serviced by 28,000 men. He also bequeathed to those who followed him the first Russian shipbuilding industry and, of course, the Baltic ports and coastline. Moreover, the navy, built on the British model, had already won high regard by defeating the Swedish fleet. The British considered the Russian vessels comparable to the best British ships in the same class, and the British government became so worried by the sudden rise of the Russian navy that in 1719 it recalled its men from the Russian service. Incidentally, in connection with shipbuilding the emperor introduced forestry regulations in Russia; however, they proved virtually unenforceable.

Administrative Reforms: Central Government, Local Government, the Church

Although mainly occupied with military matters, Peter reformed the central and local government in Russia as well as Church administration and finance, and he also effected important changes in Russian society, economy, and culture. Peter I ascended the throne as Muscovite tsar and autocrat — although, to be sure, until Ivan V's death in 1696 the country had two tsars and autocrats — and he proved to be one of the most powerful and impressive absolute rulers of his age, or any age. Yet comparisons with Ivan the Terrible or other Muscovite predecessors can be misleading. Whatever the views of earlier tsars concerning the nature and extent of their authority — and that is a complicated matter — Peter the Great believed in enlightened despotism as preached and to an extent practiced in Europe during the so-called Age of Reason. He borrowed his definition of autocracy and of the relationship between the ruler and his subjects from Sweden, not from the Muscovite tradition. The very title of *emperor* carried different connotations and associations than that of *tsar*. In contrast to Ivan the Terrible, Peter the Great had the highest regard for law, and he considered himself the first servant of the state. Yet, again in accord with his general outlook, he had no use for the boyar duma, or the zemskii sobor, and treated the Church in a much more high-handed manner than his predecessors had. Thus the reformer largely escaped the vague, but nevertheless real, traditional hindrances to absolute power in Muscovy. It was the discarding of the old and the creation of the new governing institutions that made the change in the nature of the Russian state explicit and obvious.

In 1711, before leaving on his campaign against Turkey, Peter the Great published two orders which created the Governing Senate. The Senate was founded as the highest state institution to supervise all judicial, financial, and administrative affairs. Originally established only for the time of the monarch's absence, it became a permanent body after his return. The number of senators was first set at nine and in 1712 increased to ten. A special high official, the Ober-Procurator, served as the link between the sovereign and the Senate and acted, in the emperor's own words, as "the sovereign's eye." Without his signature no Senate decision could go into effect; any disagreements between the Ober-Procurator and the Senate were to be settled by the monarch. Certain other officials and a chancellery were also attached to the Senate. While it underwent many subsequent changes, the Senate became one of the most important institutions of imperial Russia, especially in administration and law.

In 1717 and the years immediately following, Peter the Great established *collegia,* or colleges, in place of the old, numerous, overlapping, and unwieldy prikazy. The new agencies, comparable to the later ministries, were originally nine in number: the colleges of foreign affairs, war, navy, state expenses, state income, justice, financial inspection and control, commerce, and manufacturing. Later three colleges were added to deal with mining, estates, and town organization. Each college consisted of a president, a vice-president, four councilors, four assessors, a procurator, a secretary, and a chancellery. At first a qualified foreigner was included in every college, but as a rule not as president. At that time collegiate administration had found considerable favor and application in Europe. Peter the Great was especially influenced by the example of Sweden and also, possibly, by Leibniz's advice. It was argued that government by boards assured a greater variety and interplay of opinion, since decisions depended on the majority vote, not on the will of an individual, and that it contributed to a strictly legal and proper handling of state affairs. More bluntly, the emperor remarked that he did not have enough trustworthy assistants to put in full charge of the different branches of the executive and had, therefore, to rely on groups of men, who would keep check on one another. The colleges lasted for almost a century before they were replaced by ministries in the reign of Alexander I. Some prikazy, however, lingered on, and the old system went out of existence only gradually.

Local government also underwent reform. In 1699 towns were reorganized to facilitate taxation and obtain more revenue for the state. This system, run for the government by merchants, took little into account except finance and stemmed from Muscovite practices rather than Western influences. In 1720–21, on the other hand, Peter the Great introduced a thorough municipal reform along advanced European lines. Based on the elective principle and intended to stimulate the initiative and activity of the townspeople, the ambitious scheme failed to be translated into practice because of local inertia and ignorance.

Provincial reform provided probably the outstanding example of a major reforming effort of Peter's come to naught. Again, changes began in a somewhat haphazard manner, largely under the pressure of war and a desperate search for money. After the reform of 1708 the country was divided into huge *gubernii,* or governments, eight, ten, and finally eleven in number. But with the legislation of 1719 a fully-developed and extremely far-reaching scheme appeared. Fifty provinces, each headed by a *voevoda,* became the main administrative units. They were subdivided into *uezdy* administered by commissars. The commissars, as well as a council of from two to four members attached to the voevoda, were to be elected by the local gentry from their midst. All officials received salaries and the old Muscovite practice of *kormleniia* — "feedings" — went out of existence. Peter the Great went beyond his Swedish model in charging provincial bodies with responsibility for local health, education, and economic development. And it deserves special notice that the reform of 1719 introduced into Russia a separation of administrative and judicial power. But all this proved to be premature and unrealistic. Local initiative could not be aroused, nor suitable officials found. The separation of administration and justice disappeared by about 1727, while some other ambitious aspects of the reform never came into more than paper existence. In the case of local government, Peter the Great's sweeping thought could find little or no application in Russian life.

The reign witnessed a strengthening of government control in certain borderlands. After the suppression of Bulavin's great revolt, the emperor tightened his grip on the Don area, and that territory came to be more closely linked to the rest of Russia. The cossacks, however, did retain a distinct administration, military organization, and way of life until the very end of the Russian empire and even into the Soviet period — as readers of the novels of Sholokhov realize. Similarly, after Mazepa's defection to Charles XII in the Ukraine, the government proceeded to tie that land, too, more closely to the rest of the empire. For example, an interesting order in 1714 emphasized the desirability of mixing the Ukrainians and the Russians and of bringing Russian officials into the Ukraine, buttressing its argument with references to successful English policies vis-à-vis Scotland, Wales, and Ireland.

The change in the organization of the Church paralleled Peter the Great's reform of the government. When the reactionary patriarch Hadrian died in 1700, the tsar kept his seat vacant, and the Church was administered for over two decades by a mere *locum tenens,* the very able moderate supporter of reform Metropolitan Stephen Iavorsky. Finally in 1721, the so-called Spiritual Reglament, apparently written mainly by Archbishop Theophanes Prokopovich, established a new organization of the Church. The Holy Synod, consisting of ten, later twelve, clerics, replaced the patriarch. A lay official, the Ober-Procurator of the Holy

Synod, was appointed to see that that body carried on its work in a perfectly legal and correct manner. Although the new arrangement fell under the conciliar principle widespread in the Orthodox Church and although it received approval from the Eastern patriarchs, the reform belonged — as much as did Peter the Great's other reforms — to Western, not Muscovite or Byzantine, tradition. In particular, it tried to reproduce the relationship between Church and state in the Lutheran countries of northern Europe. Although it did not make Russia Byzantine as some writers assert, nor even caesaropapist — for the emperor did not acquire any authority in questions of faith — it did enable the government to exercise effective control over Church organization, possessions, and policies. If Muscovy had two supreme leaders, the tsar and the patriarch, only the tsar remained in the St. Petersburg era. The Holy Synod and the domination of the Church by the government lasted until 1917.

Peter the Great's other measures in the religious domain were similarly conditioned by his general outlook. He considered monks to be shirkers and wastrels and undertook steps to limit ecclesiastical possessions and eventually to control ecclesiastical wealth. On the other hand, he tried to strengthen and broaden Church schools and improve the lot of the impoverished secular clergy. As one might expect, the reformer exhibited more tolerance toward those of other denominations than had his Muscovite predecessors, on the whole preferring Protestants to Catholics. In 1721 the Holy Synod permitted intermarriage between the Orthodox and Western Christians. The emperor apparently felt no religious animosity toward the Old Believers and favored tolerance toward them. They, however, proved to be bitter opponents of his program of reform. Therefore, the relaxation in the treatment of the Old Believers early in the reign gave way to new restrictions and penalties, such as special taxation.

An evaluation of the total impact of Peter the Great's administrative reforms presents certain difficulties. These reforms copied and adapted Western models, trying to import into Russia the best institutions and practices to be found anywhere in Europe. Efforts to delimit clearly the authority of every agency, to separate powers and functions, to standardize procedure, and to spell out each detail could well be considered revolutionary from the old Muscovite point of view. On the surface at least the new system seemed to bear a greater resemblance to Sweden or the German states than to the realm of the good Tsar Alexis. The very names of the new institutions and the offices and technical terms associated with them testified to a flood of Western influences and a break with the Muscovite past. Yet reality differed significantly from this appearance. Even where reforms survived — and sometimes, as in the case of the local government, they did not — the change turned out to be not nearly as profound as the emperor had intended. Statutes, prescriptions, and precise rules looked well on paper; in actuality in the main cities and especially in the enormous expanses of provincial Russia, everything depended as of old on the initiative, ability, and behavior of officials. The kormleniia could be abolished, but not the all-pervasive bribery and corruption. Personal and largely arbitrary rule remained, in sum, the foundation of Russian administration; all the more so because despite the reformer's frantic efforts the new system, which was much too complicated to be discussed here with anything approaching completeness, lacked integration, co-ordination, and cohesion. In fact a few scholars, such as Platonov, have argued that the administrative order established by Peter the Great proved to be more disjointed and disorganized than that of Muscovite Russia.

Financial and Social Measures

The difficulty of transforming Russian reality into something new and Western becomes even more evident when we consider Peter the Great's social legslation and his overall influence on Russian society. Before turning to this topic, however, we must mention briefly the emperor's financial policies, for they played an important and continuous part in his plans and actions.

Peter the Great found himself constantly in dire need of money, and at times the need was utterly desperate. The only recourse was to squeeze still more out of the Russian masses, who were already overburdened and strained almost to the breaking point. According to one calculation, the revenue the government managed to exact in 1702 was twice, and in 1724 five and a half times, the revenue obtained in 1680. In the process it taxed almost everything, including beehives, mills, fisheries, beards, and bath houses; and it also extended the state monopoly to new items. For example, stamped paper, necessary for legal transactions, became an additional source of revenue for the state, and so did oak coffins. In fact, finding or concocting new ways to augment government funds developed into a peculiar kind of occupation in the course of the reign. Another and perhaps more significant change was in the main form of direct taxation; in 1718 Peter the Great introduced the head, or poll, tax in place of the household tax and the tax on cultivated land.

One purpose of the head tax was to catch shirkers who combined households or failed to till their land. It was levied on the entire lower class of population and it represented a heavy assessment — considerably heavier than the taxes that it replaced. Set at seventy or eighty kopecks per serf and at one ruble twenty kopecks for each state peasant and non-exempt townsman, the new tax had to be paid in money. From 1718 to 1722 a census, a so-called revision, of the population subject to the head tax took place. On private estates, serfs and those slaves who tilled the soil were registered first. Next came orders to add to the lists household slaves and all dependent people not on the land, and finally even vagrants of every sort. Each person registered during the revision had to pay the same set head tax; on estates, the landlords were held responsible for the prompt flow of money to the treasury. A number of scholars have stressed that Peter the Great's tax legislation thus led to the final elimination of the ancient difference between serf and slave, and the merging of the landlords' peasants into one bonded mass. Legally the mass consisted of serfs, not slaves. In actuality, as already indicated, the arbitrary power of the landlord and the weakness of the peasant made Russian serfdom differ little from slavery. After the revision the serfs were allowed to leave the estate only with their master's written permission, a measure which marked the beginning of a passport system. The head tax, it might be added, proved to be one of the emperor's lasting innovations.

On the whole Peter the Great had to accept and did accept Russian society as it was, with serfdom and the economic and social dominance of the gentry. The emperor, however, made a tremendous effort to bend that society to serve his purposes: the successful prosecution of war, Westernization, and reform. Above all, the government needed money and men. The head tax presents an excellent example of an important social measure passed for financial reasons. But whereas the head tax affected the lower classes, other social groups also found themselves subject to the insatiable demands of the tireless emperor. For example, the merchants, the few professional people, and other middle class elements, who were all exempt from the head tax, had to work harder than ever before to discharge their obligations to the state in the economic domain and other fields of activity.

However, the emperor insisted on service especially in the case of the gentry. State service, of course, constituted an ancient obligation of that class. But, as we have already seen in dealing with the army, under Peter the Great it became a more regular and continuous as well as much heavier obligation. Every member of the gentry was required to serve from about the age of sixteen to the end of his days, and the sovereign himself often gave an examination to boys as young as fourteen or even ten and assigned them to schools and careers. After an inspection, held usually in Moscow, the gentry youths were divided roughly two-thirds to one-third between the military and the civilian branches of service. Peter the Great insisted that in the civilian offices as in the regiments or aboard ships all novices must start at the bottom and advance only according to their merit. In 1722 he promulgated the Table of Ranks, which listed in hierarchic order the fourteen ranks, from the fourteenth to the first, to be attained in the parallel services — military, civil, and court. The Table, with its impressive ranks borrowed from abroad, served as the foundation of the imperial Russian bureaucracy and lasted, with modifications, until 1917. The emperor opened advancement in service to all. Entrance into service brought personal nobility, while those of non-gentry origin who attained the eighth rank in the civil service or the twelfth in the military became hereditary members of the gentry. He also began to grant titles of nobility, including "prince," for extraordinary achievements, and later emperors continued this practice.

Peter the Great's handling of the gentry represented something of a tour de force, and it proved successful to the extent that the emperor did obtain a great deal of service from that class. But the reformer's successors could not maintain his drastic policies. In fact, we shall see how in the course of the eighteenth century the gentry gradually escaped from its service obligations. At the same time entry into that class became more difficult, so that Peter the Great's effort to open the road to all talents was somewhat diminished. It might be added that some of the emperor's social legislation failed virtually from the start. Thus, for example, in 1714, in opposition to the established Russian practice of dividing land among sons, the reformer issued a law of inheritance according to which the entire estate had to go to one son only — by choice, and to the elder son if no choice had been made — the others thus being forced to exist, as in the case of the British nobility, solely by service. But this law turned out to be extremely difficult to enforce even during Peter the Great's reign, and it was repealed as early as 1731.

The Development of the National Economy

The development of the national economy constituted another aim of the reformer and another field for his tireless activity. Again, the emperor thought first of war and its immediate demands. But, in addition, from about 1710 he strove to develop industries not related to military needs, to increase Russian exports, and in general to endow the country with a more varied and active economy. Peter the Great made every effort to stimulate private enterprise, but he also acted on a large scale directly through the state. Ideologically the emperor adhered to mercantilism, popular in Europe at the time, with its emphasis on the role of the government, a favorable balance of trade, and the protection of home industries as reflected in the Russian tariff of 1724. One account gives the figure of 200 manufacturing establishments founded in Peter the Great's reign — 86 by the state and 114 by private individuals and companies — to add to the 21 in existence in Russia by 1695; another account mentions 250 such establishments in operation at the time of the emperor's death. The greatest development occurred in metallurgy, mining, and textiles. In effect, the emperor created the Russian textile industry, while he developed mining and metallurgy impressively from very modest beginnings, establishing them, notably, in the Urals. He promoted many other industries as well, including the production of china and glass.

To facilitate trade Peter the Great built canals and began the construction of a merchant marine. For instance, a canal was built between 1703 and 1709 to connect the Neva with the Volga. Indeed, the Volga-Don canal itself, finally completed by the Soviet government after the Second World War, had been one of the reformer's projects. In the course of Peter the Great's reign Russian foreign trade increased fourfold, although it continued to be handled in the main by foreign rather than Russian merchants. On the whole, although some of the emperor's economic undertakings failed and many exacted a heavy price, Peter the Great exercised a major and creative influence on the development of the Russian economy. Later periods built on his accomplishments — there was no turning back.

Education and Culture

There could be no turning back in culture either. In a sense Peter the Great's educational and cultural reforms proved to be the most lasting of all, for they pushed Russia firmly and irrevocably in the direction of the West. While these measures will be discussed in more detail in the chapter dealing with Russian culture in the eighteenth century, it should be pointed out here that they fitted well into the general pattern of the emperor's activity. Utilitarian in his approach, the sovereign stressed the necessity of at least a minimum education for service; and he also encouraged schools that would produce specialists, such as the School of Mathematics and Navigation established in 1701. His broader plans included compulsory education for the gentry — which could not be translated into practice at the time — and the creation of the Academy of Sciences to develop, guide, and crown learning in Russia. This academy did come into existence a few months after the reformer's death. Throughout his life Peter the Great showed a burning interest in science and technology as well as some interest in other areas of knowledge.

In bringing the civilization of the West to his native land, the emperor tried to introduce Western dress, manners, and usages, often by fiat and against strong opposition. The shaving of beards is a celebrated and abiding symbol of the reign. While the government demanded it "for the glory and comeliness of the state and the military profession" — to quote from Sumner's excellent little book on Peter the Great — the traditionalists objected on the ground that shaving impaired the image of God in men and made the Russians look like such objectionable beings as Lutherans, Poles, Kalmyks, Tartars, cats, dogs, and monkeys. Similarly it was argued that the already-mentioned calendar reform stole time from God and that the new simplified civil script should not be allowed to replace Church Slavonic. The *assemblées* or big society parties that women attended, who hitherto had been secluded, also aroused a storm. Yet by the end of Peter's reign members of the civil service, army, and navy, of the upper classes, and to some extent even of the middle classes, particularly in the two leading cities, were shaven and wore foreign dress. Other Western innovations also generally succeeded in winning more adherents with time. It might be added in passing that the criticism frequently levied at Peter the Great that he split Russian society in two appears to miss the point. The reformer had no choice, for he could not bring Western culture to all of his subjects at the same time. The gap between the Westernized segment of the population and the masses had to be bridged by his successors, if at all.

The Problem of Succession

The conflict between old Muscovy and new imperial Russia was played out in the sovereign's own family. Both Peter the Great's mother and his first wife Eudoxia, whom he forced to become a nun in 1698, belonged to the unreformed. In 1690 Eudoxia gave Peter a son, Alexis. The boy lived with his mother until her seclusion and later with aunts, in the old Muscovite palace. The emperor had little time for his son and never established rapport with him. Instead Alexis became the hope of the opponents of the new order and their rallying point. In 1711 Peter the Great married Alexis to a German princess. In 1712 the emperor himself married for the second time, taking as his wife a Lithuanian woman of low origin named Catherine, whom he had found in Menshikov's household, with whom he had been living happily for a few years, and by whom he had had children. It might be added that, because of her understanding and energy, Catherine proved on the whole to be a good companion to the emperor, whom she accompanied even in his campaigns. In 1715 Alexis's wife died after giving birth to a son, Peter.

At that point Peter the Great demanded that Alexis either endorse Peter's reforms and become a worthy successor to his father or renounce his rights to the throne. With characteristically passive resistance, Alexis agreed to give up his rights. Soon after that, in 1716, when Peter the Great, then in Denmark, called for his son, Alexis used the opportunity to escape to Austria and ask the protection of Emperor Charles VI, who had married a sister of Alexis's late wife. The following year, however, Peter the Great's emissary persuaded Alexis to return to Russia. He arrived in Moscow in 1718 and received pardon from his father on condition that he renounce his rights to the throne and name those who had urged him to escape. The last point led to an investigation, which, although it failed to discover an actual plot against the emperor, brought to light a great deal of opposition to and hatred of the new order, as well as some scandals. The pardon of Alexis was withdrawn as a result of the investigation and a trial set. Over a hundred high dignitaries of the state acted as the special court that condemned Alexis to death. But before the execution could be carried out Alexis expired in the fortress of Peter and Paul in the summer of 1718, probably from shock and also torture used during the questioning. Nine of his associates were executed, nine sentenced to hard labor, while many others received milder punishments.

Peter the Great's several sons born to Catherine died at an early age. Possible heirs, therefore, included the emperor's grandson Peter, the emperor's daughters and those of his half-brother Tsar Ivan V, and the emperor's wife Catherine. In 1722 Peter the Great passed a law of succession which disregarded the principle of hereditary seniority and proclaimed instead that the sovereign could appoint his successor. Once more position was to be determined by merit! But the emperor never used his new law. His powerful organism worn out by disease, strain, and an irregular life, he died on February 8, 1725, without designating a successor to his gloriously victorious, multinational, modernizing, and exhausted empire.

Evaluations of Peter the Great

Peter the Great hit Muscovy with a tremendous impact. To many of his contemporaries he appeared as either a virtually superhuman hero or the Antichrist. It was the person of the emperor that drove Russia forward in war and reform and inspired the greatest effort and utmost devotion. It was also against Peter the Great that the streltsy, the Bashkirs, the inhabitants of Astrakhan, and the motley followers of Bulavin staged their rebellions, while uncounted others, Old Believers and Orthodox, fled to the borderlands and into the forests to escape his reach. Rumor spread and legends grew that the reformer was not a son of Tsar Alexis, but a foreigner who substituted himself for the true tsar during the latter's journey abroad, that he was an imposter, a usurper, indeed the Antichrist. Peter himself contributed much to this polarization of opinion. He too saw things in black and white, hating old Muscovy and believing himself to be the creator of a new Russia. Intolerance, violence, and compulsion became the distinguishing traits of the new regime, and St. Petersburg — built in the extreme northwestern corner of the country, in almost inaccessible swamps at a cost in lives far exceeding that of Poltava — became its fitting symbol. The emperor's very size, strength, energy, and temperament intensified his popular image.

So the matter stood for about one hundred and fifty years, or roughly until the third quarter of the nineteenth century. Peter the Great was revered and eulogized by the liberals, who envisaged him as a champion of light against darkness, and also by the imperial government and its ideologists, for, after all, that government was the first emperor's creature. Those who hated the reformer and his work included, in addition to the Old Believers and some other members of the inarticulate masses, such quixotic romantic intellectuals as the Slavophiles, who fancied to have discovered in pre-Petrine Russia the true principles and way of life of their people and who regarded the emperor as a supreme perverter and destroyer. It took a sensitive writer like Pushkin to draw a balance, emphasizing the necessity and the greatness of Peter's reforms and state, while at the same time lamenting their human cost. And Pushkin too was, in fact, overwhelmed by the image of Peter the Great.

Finally, with the work of S. Soloviev, himself a great admirer of the reformer, and other nineteenth-century historians the picture began gradually to change. Scholarly investigations of the last hundred years, together with large-scale publication of materials on the reformer's reign, undertaken by a number of men from Golikov to Bogoslovsky, have established beyond question many close connections between Peter the Great and the Muscovite past. Entire major aspects of the reformer's reign, for example, foreign policy and social relations and legislation, testified to a remarkable continuity with the preceding period. Even the reformer's desire to curb and control ecclesiastical landholding had excellent Muscovite precedents. The central issue itself, the process of Westernization, had begun long before the reformer and had gathered momentum rapidly in the seventeenth century. In the words of a modern scholar, Peter the Great simply marked Russia's transition from an unconscious to a conscious following of her historical path.

Although in the perspective of Russian history Peter the Great appears human rather than superhuman, the reformer is still of enormous importance. Quite possibly Russia was destined to be Westernized, but Peter the Great cannot be denied the role of the chief executor of this fate. At the very least the emperor's reign brought a tremendous speeding up of the irreversible process of Westernization, and it established state policy and control, where formerly individual choice and chance prevailed.

Since Peter the Great was practical, and a utilitarian, it may be better to conclude this discussion on a more mundane note than historical destiny. Long ago Pogodin, a historian, a Right-wing intellectual, and one of the many admirers of the emperor, wrote:

> Yes, Peter the Great did much for Russia. One looks and one does not believe it, one keeps adding and one cannot reach the sum. We cannot open our eyes, cannot make a move, cannot turn in any direction without

encountering him everywhere, at home, in the streets, in church, in school, in court, in the regiment, at a promenade — it is always he, always he, every day, every minute, at every step!

We wake up. What day is it today? January 1, 1841 — Peter the Great ordered us to count years from the birth of Christ; Peter the Great ordered us to count the months from January.

It is time to dress — our clothing is made according to the fashion established by Peter the First, our uniform according to his model. The cloth is woven in a factory which he created; the wool is shorn from the sheep which he started to raise.

A book strikes our eyes — Peter the Great introduced this script and himself cut out the letters. You begin to read it — this language became a written language, a literary language, at the time of Peter the First, superseding the earlier church language.

Newspapers are brought in — Peter the Great introduced them.

You must buy different things — they all, from the silk neckerchief to the sole of your shoe, will remind you of Peter the Great; some were ordered by him, others were brought into use or improved by him, carried on his ships, into his harbors, on his canals, on his roads.

At dinner, all the courses, from salted herring, through potatoes which he ordered grown, to wine made from grapes which he began to cultivate, will speak to you of Peter the Great.

After dinner you drive out for a visit — this is an *assemblée* of Peter the Great. You meet the ladies there — they were admitted into masculine company by order of Peter the Great.

Let us go to the university — the first secular school was founded by Peter the Great.

You receive a rank — according to Peter the Great's Table of Ranks.

The rank gives me gentry status — Peter the Great so arranged it.

I must file a complaint — Peter the Great prescribed its form. It will be received — in front of Peter the Great's mirror of justice. It will be acted upon — on the basis of the General Reglament.

You decide to travel abroad — following the example of Peter the Great; you will be received well — Peter the Great placed Russia among the European states and began to instill respect for her; and so on, and so on, and so on.

XXI

* * * * * * * * * *

RUSSIAN HISTORY FROM PETER THE GREAT TO CATHERINE THE GREAT: THE REIGNS OF CATHERINE I, 1725–27, PETER II, 1727–30, ANNE, 1730–40, IVAN VI, 1740–41, ELIZABETH, 1741–62, AND PETER III, 1762

> The period between the death of Peter the Great and the accession of Catherine the Great, 1725 to 1762, has been considered by some historians as an era of shallowness, confusion, and decay, whereas others attribute to it much of Russia's spiritual growth and political advancement. The truth seems to lie on both sides. Rapid and violent changes, as under Peter, were discontinued, but slowly the process of Westernization went on, gaining in depth and leading to a better proportion between the ambitions and the actual potentialities of the country.
>
> KIRCHNER

> With the second quarter of the eighteenth century a new period of Russian social history begins.
>
> KIZEVETTER

RUSSIAN HISTORY from the death of Peter the Great to the accession of Catherine the Great has been comparatively neglected. Moreover, the treatments available turn out not infrequently to be superficial in nature and derisive in tone. Sandwiched between two celebrated reigns, this period — "when lovers ruled Russia," to quote one writer — offers little to impress, dazzle, or inspire. Rather it appears to be taken up with a continuous struggle of unfit candidates for the crown, with the constant rise and fall of their equally deplorable favorites, with court intrigues of every sort, with Biren's police terror, Elizabeth's absorption in French fashions, and Peter III's imbecility. Florinsky's description of the age, although verging on caricature, has its points. In the course of thirty-seven years Russia had, sardonic commentators remark, six autocrats: three women, a boy of twelve, an infant, and a mental weakling.

But the tragicomedy at the top should not be allowed to obscure important developments which affected the country at large. Westernization continued to spread to more people and broader areas of Russian life. Foreign relations followed the Petrine pattern, bringing Russia into an ever-closer relationship with other European powers. And the gentry made a successful bid to escape service and increase their advantages.

242

Catherine I. Peter II

When the first emperor died without naming his successor, several candidates for the throne emerged. The dominant two were Peter, Alexis's son and Peter the Great's grandson, and Catherine, Peter the Great's second wife. The deceased sovereign's daughters, Anne and Elizabeth, and his nieces, daughters of his half-brother Tsar Ivan V, Catherine and Anne, appeared as more remote possibilities at the time, although before very long two of them were to rule Russia, while descendants of the other two also occupied the throne. Peter was the only direct male heir and thus the logical successor to his grandfather. He had the support of the old nobility, including several of their number prominent in the first emperor's reign, and probably the support of the masses. Catherine, who had been crowned empress in a special ceremony in 1724 — in the opinion of some, a clear indication of Peter the Great's intentions with regard to succession — possessed the backing of "the new men," such as Iaguzhinsky and especially Menshikov, who had risen with the reforms and dreaded everything connected with Peter's son Alexis and old Muscovy. The Preobrazhenskii and Semenovskii guard regiments decided the issue by demonstrating in favor of the empress. Opposition to her collapsed, and the dignitaries of the state proclaimed Catherine the sovereign of Russia, "according to the desire of Peter the Great." The guards, as we shall see, were subsequently to play a decisive role in determining who ruled Russia on more than one occasion.

Catherine's reign, during which Menshikov played the leading role in the government, lasted only two years and three months. The empress's most important act was probably the creation, in February 1726, of the Supreme Secret Council to deal with "matters of exceptional significance." The six members of the council, Menshikov and five others, became in effect constant advisers and in a sense associates of the monarch, a departure from Peter the Great's administrative organization and practice. Catherine I died in 1727, having appointed young Peter to succeed her and nominated as regent the Supreme Secret Council, to which Anne and Elizabeth, her daughters and the new ruler's aunts, were added.

Peter II, not yet twelve when he became emperor, fell into the hands of Menshikov, who even transferred the monarch from the palace to his residence and betrothed him to his daughter. But Peter II did not like Menshikov; he placed his confidence in young Prince Ivan Dolgoruky. The Dolgoruky family used this opportunity to have Menshikov arrested. The once all-powerful favorite and the closest assistant of Peter the Great died some two years later in exile in northern Siberia, and the Dolgorukys replaced him at the court and in the government. Two members of that family sat in the Supreme Secret Council, and late in 1729 the engagement of Peter II to a princess Dolgorukaia was officially announced. But again the picture changed suddenly and drastically. Early in 1730, before the marriage could take place and when Peter II was not quite fifteen years old, he died of smallpox.

Anne. Ivan VI

The young emperor had designated no successor. Moreover, with his death the male line of the Romanovs came to an end. In the disturbed and complicated deliberations which ensued, the advice of Prince Dmitrii Golitsyn to offer the throne to Anne, daughter of Ivan V and childless widow of the Duke of Courland, prevailed in the Supreme Secret Council and with other state dignitaries. Anne appeared to be weak and innocuous, and thus likely to leave power in the hands of the aristocratic clique. Moreover, the Supreme Secret Council, acting on its own, invited Anne to reign only under certain rigid and highly restrictive conditions. The would-be empress had to promise not to marry and not to appoint a successor. The Supreme Secret Council was to retain a membership of eight and to control state affairs: the new sovereign could not without its approval declare war or make peace, levy taxes or commit state funds, grant or confiscate estates, or appoint anyone to a rank higher than that of colonel. The guards as well as all other armed forces were to be under the jurisdiction of the Supreme Secret Council, not of the empress. These drastic conditions, which had no precedent in Russian history, stood poles apart from Peter the Great's view of the position and function of the monarch and his translation of this view into practice. But Anne, who had very little to lose, accepted the limitations, thus establishing constitutional rule in Russia.

Russian constitutionalism, however, proved to be extremely short-lived. Because the Supreme Secret Council had acted in its narrow and exclusive interest, tension ran high among the gentry. Some critics spoke and wrote of extending political advantages to the entire gentry, while others simply denounced the proceedings. Anne utilized a demonstration by the guards and other members of the gentry, shortly after her arrival, to tear up the conditions she had accepted, asserting that she had thought them to represent the desires of her subjects, whereas they turned out to be the stratagem of a selfish cabal. And she abolished the Supreme Secret Council. Autocracy came back into its own.

Empress Anne's ten-year reign left a bitter memory. Traditionally, it has been presented as a period of cruel and stupid rule by individual Ger-

Head of St. Peter of Alexandria, from the fresco in the Church of the Savior on the Nereditsa, Novgorod, 1197.

Holy Gates of the Rizpolozhenskii Monastery in Suzdal, seventeenth century.

Preobrazhenskaia Church, 1714, in Kizhy near Petrozavodsk. It has 22 cupolas.

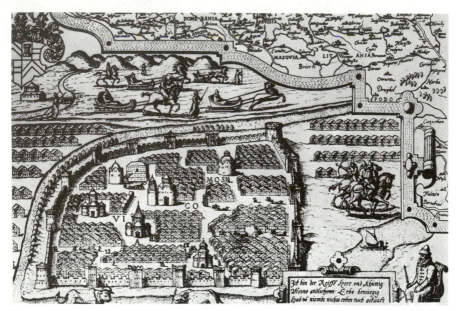

Sixteenth-century view of the city of Moscow.

Red Square in Moscow, 1844.

Church of St. Basil the Blessed, Moscow, 1555–60.

Zagorsk, with Holy Trinity–St. Sergius Monastery, Kremlin, and Assumption Cathedral. Fifteenth and sixteenth centuries and later.

Moscow Kremlin

Contemporary Russian architecture: Moscow State University, on Lenin Hills.

Hermitage Museum, Leningrad, 1852.

The seventeenth-century Simeon Stolpnik Church on Moscow's Kalinin Prospect in front of a new housing project.

mans and even "the German party" in Russia. And while this interpretation should not be overdone — for, after all, the 1730's, in foreign policy, in social legislation, and in other major respects constituted an integral part of the Russian evolution in the eighteenth century rather than anything specifically German — it remains true that Anne brought with her from Courland a band of favorites, and that in general she patronized Germans as well as other foreigners and distrusted the Russian nobility. The sovereign herself proved largely unfit and thoroughly disinclined to manage state affairs. Certain departments, such as the foreign office with Ostermann at the head and the army with Münnich, profited from able German leadership of the Petrine vintage; but many new favorites had no qualifications for their positions, acted simply in their personal interest, and buttressed their remarkable ignorance of Russia with their disdain of everything Russian. Ernst-Johann Biren, or Biron, the empress's lover from Courland, acquired the highest honors and emoluments and became the most hated figure and symbol of the reign. *Bironovshchina* — that is, Bironism — refers especially to the police persecution and political terror during the reign, which led to the execution of several thousand people and to the exile of some twenty or thirty thousand to Siberia. Although many of the victims were Old Believers and even common criminals rather than political opponents, and although the cruelty of Biren and his associates perhaps should not be considered exceptional for the age, the persecutions excited the popular imagination and made the reign compare unfavorably, for example, with the rule of Elizabeth which was to follow it. It might be added that after the abolition of the Supreme Secret Council Anne did not restore the Senate to its former importance as the superior governing institution, but proceeded to rely on a cabinet of two or three members to take charge of state affairs.

Anne died in the autumn of 1740. Shortly before her death she had nominated a two-month-old infant, Ivan, to be her successor on the throne. Ivan was a great-grandson of Ivan V and a grandson of Anne's elder sister, Catherine, who in 1716 had married the Duke of Mecklenburg, Charles Leopold. A daughter from this marriage, Anna Leopoldovna, became the wife of Duke Anthony Ulric of Brunswick-Bevern-Lüneburg. The new emperor was the child of Anna Leopoldovna and Anthony Ulric. But, although both of his parents resided at the Russian court, Empress Anne appointed Biren as regent. The arrangement failed to last. First, within a month Biren was overthrown by Münnich and Anna Leopoldovna became regent. Then, in another year, late in 1741, Ivan VI, Anna Leopoldovna, and the entire "German party" were tumbled from authority and power. This last coup was executed by the guards led by Peter the Great's daughter Elizabeth, who then ascended the throne as Empress Elizabeth of Russia.

Elizabeth. Peter III

Just as Anne and her reign have been excessively blamed in Russian historiography, Elizabeth has received more than her fair share of praise. Kind, young, handsome, and charming, the new monarch symbolized to many contemporaries and later commentators the end of a scandalous "foreign" domination in Russia and even, to an extent, a return to the glorious days of Peter the Great, an association which the empress herself stressed as much as she could. But in truth there was little resemblance between the indolent, easy-going, and disorganized, although by no means stupid, daughter and the fantastically energetic, active, and forceful father. Although the cabinet was abolished, the Senate was restored to its former importance, and certain other administrative changes were made that also harked back to the reign of Peter the Great, the spirit and vigor of the celebrated reformer could not be recaptured, nor in fact was a serious attempt made to recapture them. Moreover, the social and economic evolution of the country continued under Elizabeth as under Anne: neither ruler made a strong personal impress on it. Even Elizabeth's abolition of capital punishment, enlightened and commendable in itself as well as strikingly different from the practices of Anne's government, pales into insignificance when compared to the enormous, persistent, and in fact growing evil of serfdom.

Favorites continued to occupy the stage, although their identity changed, and the new group proved on the whole more attractive than the one sponsored by Empress Anne. Alexis Razumovsky, who may have been morganatically married to Elizabeth, was closest to the monarch. His rise to eminence represents an earlier and truer version of Andersen's tale of the princess and the swineherd. He was a simple cossack in origin, who tended the village flock in his native Ukraine. Because of his magnificent voice, the future favorite was brought to the court as a singer. Elizabeth fell in love with him, and her attachment lasted until her death. Yet, while Alexis Razumovsky became a very close associate and perhaps even the husband of the empress, his impact on state affairs remains difficult to discern. Indeed one historian, in a rather typical evaluation, dismisses the favorite as follows:

> He became the bearer of all Russian decorations, a General Field Marshal, and he was raised to the position of Count of the Holy Roman Empire. He was very imperious, even lived in the palace. But he was distinguished by an honest, noble, and lazy character. He had little influence on the governing of the country; he constantly dodged state affairs. He did much good in the Ukraine and in Russia, and in his tastes and habits he remained more a simple Ukrainian than a Russian lord. In the history

of the Russian court he was a remarkable individual, in the history of the state — an entirely insignificant actor.

Alexis Razumovsky's younger brother, Cyril, received a good education abroad and occupied such important offices as those of President of the Academy of Sciences, Field Marshal, and Hetman of the Ukraine.

The Shuvalovs, the brothers Peter and Alexander and their cousin Ivan, displayed more energy than the Razumovskys. Ivan Shuvalov, the empress's favorite, left behind him an almost unique reputation for integrity and kindness, for refusing honors and rewards, and for selfless service in several capacities, especially in promoting enlightenment in Russia. The University of Moscow, which he founded, remains his lasting monument. Peter Shuvalov was made Count by the empress — a title which Ivan Shuvalov refused — and used his strong position at the court to have a hand in every kind of state business, in particular in financial and economic matters and in the military establishment. Able, but shamelessly corrupt and cynical, Peter Shuvalov contributed much to the ruinous financial policy of the reign and has been credited with saying that debased coinage would be less of a load to carry and that the tax on vodka suited a time of distress because people would then want to get drunk. Elizabeth's own extravagance, which included the building of the extremely expensive Winter Palace and the acquisition of, reportedly, fifteen thousand dresses, added greatly to the financial crisis. A French milliner finally refused further credit to the Russian empress! Of much more importance is the fact that the financial chaos, together with the fundamental and overwhelming burden of serfdom, led to the flight and uprisings of peasants that became characteristic of the age. Alexander Shuvalov, the third prominent member of that family, served as the head of the security police. Other close associates of Elizabeth included her old friend Chancellor Count Michael Vorontsov and Count Alexis Bestuzhev-Riumin who specialized in foreign policy. The replacement of Germans by Russians in the imperial entourage under Elizabeth had some connection with the increasing interest of the Russian court and educated public in French society and culture and their declining concern with the German states.

The German orientation, however, came back with a vengeance, if only briefly, in the reign of Peter III. When Elizabeth died in late 1761 or early 1762 — depending on whether we use the Old or the New Style — Peter, Duke of Holstein-Gottorp, who had been nominated by the empress as her successor as early as 1742, became Emperor Peter III. The new ruler was a son of Elizabeth's older sister, Anne — therefore a grandson of Peter the Great — and of Charles Frederick, Duke of Holstein-Gottorp. Having lost his mother in infancy and his father when a boy, Peter was brought up first with the view of succeeding to the Swedish throne, for his father was a son of Charles XII's sister. After Elizabeth's decision, he was educated to succeed to the throne of the Romanovs. Although he lived in Russia from the age of fourteen, Peter III never adjusted to his new country. Extremely limited mentally, as well as crude and violent in his behavior, he continued to fear and despise Russia and the Russians while he held up Prussia and in particular Frederick II as his ideal. His reign of several months, best remembered in the long run for the law abolishing the compulsory state service of the gentry, impressed many of his contemporaries as a violent attack on everything Russian and a deliberate sacrifice of Russian interests to those of Prussia. While not given to political persecution and in fact willing to sign a law abolishing the security police, the new emperor threatened to disband the guards, and even demanded that icons be withdrawn from churches and that Russian priests dress like Lutheran pastors, both of which orders the Holy Synod did not dare execute. In foreign policy Peter III's admiration for Frederick the Great led to the withdrawal of Russia from the Seven Years' War, an act which probably saved Prussia from a crushing defeat and deprived Russia of great potential gains. Indeed, the Russian emperor refused to accept even what Frederick the Great was willing to give him for withdrawing and proceeded to make an alliance with the Prussian king.

While Peter III rapidly made enemies, his wife Catherine, who had married him in 1745 and who was originally a princess of the small German principality of Anhalt-Zerbst, behaved with far greater intelligence and understanding. Isolated and threatened by her boorish husband, who had a series of love affairs and wanted to marry one of his favorites, she adapted herself to her difficult environment, learned much about the government and the country, and found supporters. In mid-summer 1762 Catherine profited from the general dissatisfaction with Peter III to lead the guards in another palace revolution. The emperor was easily deposed and shortly after killed, very possibly by one of the leaders of the insurrection, Alexis Orlov, in a drunken argument. Catherine became empress, bypassing her son Paul, born in 1754 during her marriage with Peter III, who was proclaimed merely heir to the throne. Although the coup of 1762 appeared to be simply another one in a protracted sequence of overturns characteristic of Russian history in the eighteenth century, and although Catherine's chances of securing her power seemed, if anything, less promising than those of a number of her immediate predecessors, in fact her initial success meant the beginning of a long and celebrated reign. That reign will form the subject of another chapter.

The Gains of the Gentry and the Growth of Serfdom

While rulers changed rapidly and favorites constantly rose and fell in Russia between 1725 and 1762, basic social processes went on in a continuous and consistent manner. Most important was the growth of the power and standing of the gentry together with its complementary process, a further deterioration in the position of the serfs. As we know, Peter the Great's insistence that only one son inherit his father's estate could hardly be enforced even in the reformer's reign and was formally repealed in 1731. Empress Anne began giving away state lands to her gentry supporters on a large scale, the peasants on the lands becoming serfs, and Elizabeth enthusiastically continued the practice. These grants were no longer connected to service obligations.

In 1731 Empress Anne opened a cadet school for the gentry in St. Petersburg. The graduates of this school could become officers without serving in the lower ranks, a privilege directly opposed to Peter the Great's intentions and practice. As the century progressed the gentry came to rely increasingly on such cadet schools for both education and advancement in service. Also to their advantage was the Gentry Bank that was established by Empress Elizabeth in St. Petersburg, with a branch in Moscow, to supply the landlords with credit at a moderate rate of interest. The gentry became increasingly class-conscious and exclusive. An order of 1746 forbade all but the gentry to acquire "men and peasants with and without land." In 1758 the members of other classes who owned serfs were required to sell them. A Senate decision of 1756 affirmed that only those who proved their gentry origin could be entered into gentry registers, while decisions in the years 1758–60 in effect eliminated the opportunity to obtain hereditary gentry status through state service, thus destroying another one of Peter the Great's characteristic arrangements. At the same time "personal," or non-hereditary, members of the class came to be rigidly restricted in their gentry rights.

The most significant evolution took place in regard to the service obligations of the gentry to the state. In 1736 this service, hitherto termless, was limited to twenty-five years — the gentry themselves had asked for twenty years — with a further provision exempting one son from service so he could manage the estates. Immediately following the publication of the law and in subsequent decades, many members of the gentry left service to return to their landholdings. Moreover, some landlords managed to be entered in regimental books from the age of eight or ten to complete the twenty-five-year period of service relatively early in their lives. Finally, on March 1, 1762 — February 18, Old Style — in the reign of Peter III, compulsory gentry service was abolished. Henceforth members of the gentry could serve the state, or not serve it, at will, and they could even serve foreign governments abroad instead, if they so desired. The edict also urged upon the gentry the importance of education and proper care of their estates.

The law of 1762 has attracted much attention from historians. To many older scholars, exemplified by Kliuchevsky, it undermined the basic structure of Russian society, in which everyone served: the serfs served the landlords, the landlords served the state. In equity the repeal of compulsory gentry service should have been followed promptly by the emancipation of the serfs. Yet — again to cite Kliuchevsky — although the abolition of serfdom did take place on the following day, the nineteenth of February, that day came ninety-nine years later. The serfs themselves, it would seem, shared the feeling that an injustice had been committed, for the demand for freedom of the peasants, to follow the freedom of the gentry, became a recurrent motif of their uprisings. By contrast, some specialists, such as V. Leontovich and Malia, have emphasized the positive results of the law of 1762: it represented the acquisition of an essential independence from the state by at least one class of Russian society, and thus the first crucial step taken by Russia on the road to liberalism; besides, it contributed to the growth of a rich gentry culture and, beyond that, to the emergence of the intelligentsia.

As the gentry rose, the serfs sank to a greater depth of misery. In the reign of Peter II they were already prohibited from volunteering for military service and thus escaping their condition. By a series of laws under Empress Anne peasants were forbidden to buy real estate or mills, establish factories, or become parties to government leases and contracts. Later, in the time of Elizabeth, serfs were ordered to obtain their master's permission before assuming financial obligations. Especially following the law of 1731, landlords acquired increasing financial control over their serfs, for whose taxes they were held responsible. After 1736 serfs had to receive the permission of their masters before they could leave for temporary employment elsewhere. Landlords obtained further the right to transfer serfs from one estate to another and, by one of Elizabeth's laws, even to exile delinquent serfs to Siberia and to fetch them back, while the government included these exiles in the number of recruits required from a given estate. The criminal code of 1754 listed serfs only under the heading of property of the gentry. Russian serfdom, although never quite the same as slavery and in the Russian case not concerned with race or ethnicity, came to approximate it closely, as demonstrated in the works of Kolchin and other scholars.

The Foreign Policy of Russia from Peter to Catherine

Russian foreign policy from Peter the Great to Catherine the Great followed certain clearly established lines. The first emperor, as we know, brought Russia forcefully into the community of European nations as a major power that was concerned with the affairs of the continent at large, not, as formerly, merely with the activities of its neighbors, such as Turkey, Poland, and Sweden. From the time of Peter the Great, permanent —

rather than only occasional — representatives were exchanged between Russia and other leading European states. Ostermann in the years immediately following the death of Peter the Great, and Bestuzhev-Riumin in the time of Elizabeth, together with lesser officials and diplomats, followed generally in the steps of Peter.

As Karpovich, to mention one historian, has pointed out, Russian foreign policy from 1726 to 1762, and immediately before and after that period, approached what has been called the checkerboard system: Russia was to a considerable degree an enemy of its neighbors and a friend of its neighbors' neighbors, with other relations affected by this basic pattern. France, for example, consistently remained an antagonist of Russia, because in its struggle for the mastery of the continent it relied on Turkey, Poland, and Sweden to envelop and weaken its arch-enemy, the Hapsburgs. France had maintained an alliance with Turkey from 1526, in the days of Suleiman the Magnificent; with Poland from 1573, when Henry of Valois was elected to the Polish throne; and with Sweden from the time of the Thirty Years' War in the early seventeenth century. Russia, of course, had repeatedly fought against the three eastern European allies of France.

Austria, ruled by the Hapsburgs, stood out, by contrast, as the most reliable Russian ally. The two states shared hostility toward France, and, more importantly for Russia, also toward Turkey and Sweden, which, beginning with its major intervention in the Thirty Years' War, acted repeatedly in Germany against the interests of the Hapsburgs. In Poland also both Russia and Austria found themselves opposed to the French party. The first formal alliance between the two eastern European monarchies was signed in 1726, and it remained, with certain exceptions, a cornerstone of Russian foreign policy until the Crimean War in mid-nineteenth century.

Prussia, the other leading German power, represented a threat to Russia rather than a potential ally. Prussia's rise to great power rank under Frederick the Great after 1740, together with Russia's rise under Peter the Great which had just preceded it, upset the political equilibrium in Europe. Bestuzhev-Riumin was one of the first continental statesmen to point to the Prussian menace. He worried especially about the Russian position on the Baltic, called Frederick the Great "the sudden prince," and spoke in a typically eighteenth-century doctrinaire manner of Russia's "natural friends," Austria and Great Britain, and its "natural enemies," France and Prussia. The hostile Russian attitude toward Prussia lasted, with some interruptions, until the time of Catherine the Great and the partitions of Poland which satisfied both monarchies and brought them together.

In the period under consideration, Great Britain could well be called a "natural friend" of Russia. After the scare occasioned by the achievements of Peter the Great and his navy, no serious conflicts arose between the two until the last part of the century. On the contrary, Great Britain valued Russia both as a counterweight to France and as a trade partner from which it obtained raw materials, including naval stores, in exchange for manufactured goods. Thus it is no surprise that Russia concluded its first modern commercial treaty with Great Britain.

In line with its interests and alliances, Russia participated in five wars between 1725 and 1762. In 1733–35 Russia and Austria fought against France in the War of the Polish Succession, which resulted in the defeat of the French candidate Stanislaw Leszczynski and the coronation of Augustus II's son as Augustus III of Poland. In 1736–39 Russia, again allied to Austria, waged a war against Turkey who was supported by France. Münnich and other Russian commanders scored remarkable victories over the Ottoman forces. However, because of Austrian defeats and French mediation, Russia, after losing approximately 100,000 men, gained very little according to the provisions of the Treaty of Belgrade: a section of the steppe between the Donets and the Bug, and the right to retain Azov, captured during the war, on condition of razing its fortifications and promising not to build a fleet on the Black Sea. In 1741–43, Russia, supported by Austria, fought Sweden, who was supported by France. Sweden started the war to seek revenge, but was defeated, and by the Treaty of Åbo ceded some additional Finnish territory to Russia.

In its new role as a great power Russia became involved also in wars fought away from its borders over issues not immediately related to Russian interests. Thus in 1746–48 she participated in the last stages of the War of the Austrian Succession, begun in 1740 when Frederick the Great seized Silesia from Austria. That conflict saw Bestuzhev-Riumin's theory of alliances come true: Russia joined Austria and Great Britain against Prussia and France. The Russian part in this war proved to be, however, entirely inconsequential.

A much greater importance must be attached to the Russian intervention in the Seven Years' War, 1756–63, fought again largely over Silesia. The conflict was preceded by the celebrated diplomatic revolution of 1756 which saw France ally itself with its traditional enemy Austria, while Prussia turned for support to Great Britain. In the war Russia joined Austria, France, Sweden, and Saxony against Prussia, Great Britain, and Hanover. Yet it should be noted that Russia never declared war on Great Britain and that she found it natural to aid Austria against Prussia, so that in the case of the empire of the tsars the diplomatic revolution had a rather narrow meaning. Russian armies participated in great battles, such as those of Zorndorf and Kunersdorf, and in 1760 Russian troops even briefly held Berlin. Moreover, Russia and its allies managed to drive Prussia to the brink of collapse. Only the death of Empress Elizabeth early in 1762, and the accession to the throne of Peter III, who admired Frederick the Great, saved the Prussian king. Russia withdrew without any compensation from the war and made an alliance with Prussia, which in turn was discontinued when Catherine the Great replaced Peter III.

Although Russian foreign policy between 1725 and 1762 has been severely criticized for its cost in men and money, its meddling in European affairs which had no immediate bearing on Russia, and its alleged sacrifice of national interests to those either of Austria or of the "German party" at home, these criticisms on the whole are not convincing. In its new role Russia could hardly disengage itself from major European affairs and conflicts. In general Russian diplomats successfully pursued the interests of their country, and the wars themselves brought notable gains, for example, the strengthening of the Russian position in Poland and the defeat of the Swedish challenge, even though Peter III did write off in a fantastic manner the opportunities produced by the Seven Years' War. Catherine the Great would continue the basic policies of her predecessors. Militarily the Russians acquitted themselves well. The Russian army, reorganized, improved, and tempered in the wars, scored its first major victories against Turkey in 1736–39, and played its first major part in the heart of Europe in the course of the Seven Years' War. Such famous commanders as Peter Rumiantsev and Alexander Suvorov began their careers in this interim period between two celebrated reigns.

X X I I

* * * * * * * * * *

THE REIGNS OF CATHERINE THE GREAT, 1762–96, AND PAUL, 1796–1801

> Long live the adorable Catherine!
>
> VOLTAIRE

> What interest, therefore, could the young German princess take in that *magnum ignotum,* that people, inarticulate, poor, semi-barbarous, which concealed itself in villages, behind the snow, behind bad roads, and only appeared in the streets of St. Petersburg like a foreign outcast, with its persecuted beard, and prohibited dress — tolerated only through contempt.
>
> HERZEN

> Of the three celebrated "Philosophic Despots" of the eighteenth century Catherine the Great could boast of the most astonishing career.
>
> GOOCH

CATHERINE THE GREAT was thirty-three years old when she ascended the Russian throne. She had acquired considerable education and experience. Born a princess in the petty German principality of Anhalt-Zerbst, the future empress of Russia grew up in modest but cultured surroundings. The court in Anhalt-Zerbst, like many other European courts in the eighteenth century, was strongly influenced by French culture, and Catherine started reading French books in childhood. In 1744, at the age of fifteen, she came to Russia to marry Peter of Holstein-Gottorp and prepare herself to be the wife of a Russian sovereign.

The years from 1744 to 1762 were hard on Catherine. Peter proved to be a miserable husband, while the German princess's position at the imperial court could be fairly described as isolated and even precarious. To add to Catherine's difficulties, her mother was discovered to be Frederick the Great's agent and had to leave Russia. Yet the future empress accomplished much more than merely surviving at court. In addition to becoming Orthodox in order to marry Peter, she proceeded to learn Russian language and literature well and to obtain some knowledge of her new country. Simultaneously she turned to the writings of the *philosophes,* Voltaire, Montesquieu, and others, for which she had been prepared by her earlier grounding in French literature. As we shall see, Catherine the Great's

254

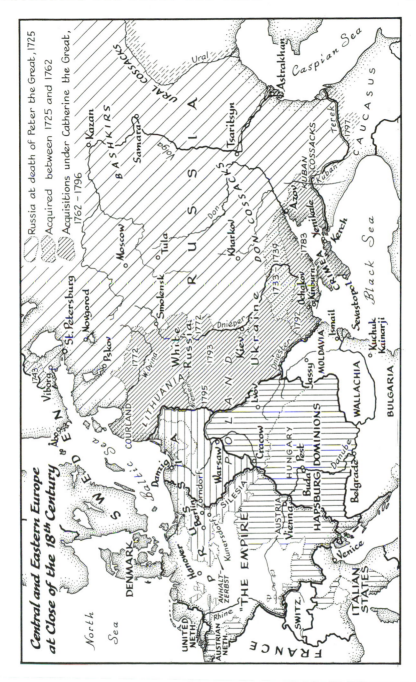

Central and Eastern Europe at Close of the 18th Century

interest in the Enlightenment was to constitute an important aspect of her reign. The young princess adapted herself skillfully to the new environment, made friends, and won a measure of affection and popularity in court circles. While simulating innocence and submissiveness, she participated in political intrigues and plots, carefully covering up her tracks, however, until she led the successful coup in mid-summer 1762, which brought deposition and death to her husband and made her Empress Catherine II.

Catherine the Great's personality and character impressed many of her contemporaries as well as later commentators. A woman quite out of the ordinary, the empress possessed high intelligence, a natural ability to administer and govern, a remarkable practical sense, energy to spare, and an iron will. Along with her determination went courage and optimism: Catherine believed that she could prevail over all obstacles, and more often than not events proved her right. Self-control, skill in discussion and propaganda, and a clever handling of men and circumstances to serve her ends were additional assets of that unusual monarch. The empress herself asserted that it was ambition that sustained her. The historian can agree, provided that ambition is understood broadly, that is, not merely as a desire to snatch the crown, or attain glory by success in war, or gain the admiration of the *philosophes,* but as a constant, urgent drive to excel in everything and bring everything under one's control. For the first time since Peter the Great, Russia acquired a sovereign who worked day and night, paying personal attention to all kinds of matters, great and small.

Yet, together with her formidable virtues, Catherine the Great had certain weaknesses. Indeed the two were intrinsically combined. Determination easily became ruthlessness, ambition fed vanity just as vanity fed ambition, skill in propaganda would not stop short of asserting lies. Above all, the empress was a supreme egoist. As with most true egoists, she had few beliefs or standards of value outside of herself and her own overpowering wishes. Even Catherine II's admirers sometimes noticed that she lacked something, call it charity, mercy, or human sympathy, and, incidentally, that she looked her best in masculine attire. It was also observed that the sovereign took up every issue with the same unflagging drive and earnestness, be it Pugachev's rebellion or correspondence with Voltaire, the partitions of Poland or her latest article for a periodical. Restless ambition served as the only common denominator in her many activities, and, apparently, the only thing that mattered. Similarly, in spite of Catherine's enormous display of enlightened views and sentiments and of her adherence to the principles of the Age of Reason, it remains extremely difficult to tell what the empress actually believed, or whether she believed anything. In fact, the true relationship of Catherine the Great to the Enlightenment constitutes one of the most controversial subjects in the historiography of her reign.

Catherine the Great's notorious love affairs also reflected her peculiar personality: grasping, restless, determined, and somehow, in spite of all passion and sentimentality, essentially cold and unable to establish a happy private life. It has been asserted that her first lover was forced on Catherine, so that she would have a son and Russia an heir, and that Paul resulted from that liaison rather than from the marriage to Peter. In any case, Catherine soon took matters into her own hands. The empress had twenty-one known lovers, the last after she had turned sixty. The favorites included Gregory Orlov, an officer of the guards who proved instrumental in elevating Catherine the Great to the throne and whose brother may have killed Peter III; Stanislaw Poniatowski, a Polish nobleman whom the empress made King of Poland; and, most important, Gregory Potemkin. Potemkin came to occupy a unique position both in the Russian government, to the extent that he can be considered the foremost statesman of the reign, and in the empress's private life. Some specialists believe he married her; he certainly continued to be influential with her after the rise of other favorites. One description of the unusual ménage says: "From 1776 to 1789 her favorites succeeded one another almost every year, and they were confirmed in their position, as a court poet would be, by Potemkin himself, who, after he had lost the empress's heart, remained for thirteen years the manager of her male harem."

The First Years of the Reign. The Legislative Commission

Catherine II had to behave carefully during her first years on the throne. Brought to power by a palace revolution and without a legal title to the crown, the empress had the enthusiastic support of guardsmen such as the Orlov brothers, but otherwise little backing. Elder statesmen looked at her with some suspicion. There persisted the possibility that another turn of fortune would make her son Paul sovereign and demote Catherine to the position of regent or even eliminate her altogether. A different danger struck in July 1764, when a young officer, Basil Mirovich, tried to liberate Ivan VI from his confinement in the Schlüsselburg fortress. The attempt failed, and in the course of it Ivan VI — who, apparently, because of isolation since early childhood, had never grown up mentally and emotionally but remained virtually subhuman — was killed by his guards, who carried out emergency instructions of long standing. The depressing impact of the incident on Russian society was heightened by the execution of Mirovich, an event all the more striking because Elizabeth had avoided executions. Catherine also ran into a certain amount of trouble when, in

1763–64, she completed the long process of divesting the Church of its huge real estate by secularizing Church lands. This reform, which we shall discuss briefly in a later chapter, evoked a violent protest on the part of Metropolitan Arsenii of Rostov, who did not stop short of excommunicating those connected with the new policy. Fortunately for the empress, other hierarchs failed to support Metropolitan Arsenii, and, after two trials, the empress had him defrocked and imprisoned for life.

Gradually Catherine II consolidated her position. She distributed honors and rewards on a large scale, in particular state lands with peasants, who thus became serfs. She traveled widely all over Russia, reviving Peter the Great's practice, both to learn more about the country and to win popularity. She selected her advisers carefully and well. Time itself worked for the empress: with the passage of years memories of the coup of 1762 faded, and the very fact that Catherine II continued to occupy the throne gave the reign a certain legitimacy. In late 1766 she felt ready to introduce into Russia important changes based on the precepts of the Enlightenment, and for that purpose she called the Legislative Commission.

The aim of the Commission was to codify laws, a task last accomplished in 1649, before the Westernization of the country. Moreover, Catherine the Great believed that the work of the Commission would go a long way toward rationalizing and modernizing Russian law and life. Although the empress had certainly no desire to grant her subjects a constitution, and although her propaganda greatly exaggerated the radical nature of her intentions, the *Nakaz,* or *Instruction,* which she prepared for the Legislative Commission, was in fact, even in its final attenuated version, a strikingly liberal document. Composed by Catherine the Great herself over a period of eighteen months, the *Instruction* found its inspiration in the thought of the Enlightenment, particularly in the writings of Montesquieu and the jurist Beccaria. Montesquieu, whose *Spirit of the Laws* the empress referred to as her prayer book, served as the chief guide in political theory. Yet it should be noted that the willful sovereign adapted rather than copied the French philosopher: she paid lip service to his ideas, but either left them conveniently vague or changed them drastically in application to Russian reality; for example, Montesquieu's celebrated admiration of the division of powers in England into the executive, the legislative, and the judicial became an administrative arrangement meant to improve the functioning of Russian autocracy. The empress continued to believe that autocracy was the only feasible form of government for holding enormous Russia together. And in fairness it should be added that here, too, she had some support in the thought of the Enlightenment. Serfdom, on the other hand, she was willing to condemn in theory, although again she largely avoided the issue: the final draft of the *Instruction* contained merely a pious wish that masters would not abuse their serfs. As to the influence of Beccaria, Catherine the Great could afford to follow his views more closely, as they were expressed in his treatise *Crimes and Punishments,* and she did. Thus the *Instruction* denounced capital punishment — which had already been stopped in Russia by Elizabeth — as well as torture, argued for crime prevention, and in general was abreast of advanced Western thought in criminology. On the whole, the liberalism of the *Instruction* produced a strong impression in a number of European countries, and led to its being banned in France.

The Legislative Commission, which opened deliberations in the summer of 1767, consisted of 564 deputies, 28 appointed and 536 elected. The appointees represented the state institutions, such as the Senate. The elected deputies comprised delegates from different segments of the population of the empire: 161 from the landed gentry, 208 from the townspeople, 79 from the state peasants, and 88 from the cossacks and national minorities. Yet this numerous gathering — an "all-Russian ethnographic exhibition," to quote Kliuchevsky — excluded large bodies of the Russian people: the serfs, obviously, but also, in line with the secular tendency of the Enlightenment, the clerical class, although the Holy Synod was represented by a single appointed deputy. Delegates received written instructions or mandates from their electorates, including the state peasants, who, together with the cossacks and national minorities, supplied over a thousand such sets of instructions. Taken together, the instructions of 1767 offer the historian insight into the Russian society of the second half of the eighteenth century comparable to that obtainable for France in the famous *cahiers* of 1789. Kizevetter and other scholars have emphasized the following well-nigh universal characteristics of the instructions: a practical character; a definite acceptance of the existing regime; a desire for decentralization; complaints of unbearable financial demands and, in particular, requests to lower the taxes; and a wish to delineate clearly the rights and the obligations of all the classes of society.

The Legislative Commission met for a year and a half, holding 203 sessions; in addition, special committees were set up to prepare the ground for dealing with particular issues. But all this effort came to naught. The commission proved unwieldy, not enough preliminary work had been done, often there seemed to be little connection between the French philosophy of the empress's *Instruction* and Russian reality. Most important, however, the members of the commission split along class lines. For example, gentry delegates argued with merchant representatives over serf ownership and rights to engage in trade and industry. More ominously, gentry deputies clashed with those of the peasant class on the crucial issue of serfdom. No doubt Catherine the Great quickly realized the potential danger of such confrontations. The outbreak of war against Turkey in 1768 provided a good occasion for disbanding the Legislative Commission. Some committees continued to meet for several more years until the Pugachev rebellion, but again without producing any practical results. Still, the abortive convocation of the Commission served some purpose: it gave Catherine the Great considerable information about the country and influenced both the general course of her subsequent policy and certain particular reforms.

Pugachev's Rebellion

Social antagonisms which simmered in the Legislative Commission exploded in the Pugachev rebellion. That great uprising followed the pattern of earlier lower-class insurrections, such as the ones led by Bolotnikov, Razin, and Bulavin, which strove to destroy the established order. A simple Don cossack, a veteran of several wars and a deserter, Emelian Pugachev capitalized on the grievances of the Ural cossacks to lead them in revolt against the authorities in the autumn of 1773. Before long the movement spread up and down the Ural river and also westward to the Volga basin. At its height the rebellion encompassed a huge territory in eastern European Russia, engulfing such important cities as Kazan and posing a threat to Moscow itself.

Pugachev profited from the fact that Russia was engaged at the time in a major war against Turkey, that few troops were stationed in the eastern part of the country, and that many local officials, as well as, to some extent, the central government itself, panicked when they belatedly realized the immediacy and extent of the danger. Yet his most important advantages stemmed from the nature and the injustice of the Russian social system. The local uprising of the Ural cossacks became a mass rebellion. Crowds of serfs, workers in the Ural mines and factories, Old Believers, Bashkirs, Tartars, and certain other minority peoples, joined Pugachev's original cossack following. Indeed, some specialists believe that Pugachev should have shown more daring and marched directly on Moscow in the heart of the serf area. As Pushkin's A Captain's Daughter illustrates, few, except officials, officers, and landlords, tried to stem the tide.

Pugachev acted in the grand manner. He proclaimed himself Emperor Peter III, alleging that he had fortunately escaped the plot of his wife Catherine; and he established a kind of imperial court in imitation of the one in St. Petersburg. He announced the extermination of officials and landlords, and freedom from serfdom, taxation, and military service for the people. Pugachev and his followers organized an active chancellery and engaged in systematic propaganda. Also, the leaders of the rebellion arranged elections for a new administration in the territory that they held, and they tried to form a semblance of a regular army with a central staff and an artillery, for which Ural metal workers supplied some of the guns.

Although the extent and organization of the Pugachev uprising deservedly attract attention, it still suffered from the usual defects of such movements: a lack of preparation, co-ordination, and leadership. Small army detachments, when well commanded, could defeat peasant hordes. After government victories and severe reprisals, the raging sea of rebellion would vanish almost as rapidly as it had appeared. In late 1774, following the defeat of his troops and his escape back to the Ural area, Pugachev was handed over by his own men to the government forces. He was brought to Moscow, tried, and executed in an especially cruel manner. The great uprising had run its course.

The Pugachev rebellion served to point out again, forcefully and tragically, the chasm between French philosophy and Russian reality. Catherine the Great had in any case allied herself with the gentry from the time of the palace coup which gave her the throne, and it is highly doubtful that she had ever seriously intended to act against any essential interests of the landlords. The sharp division of her reign into the early liberal years and a later period of conservatism and reaction appears none too convincing. Still, the enormous shock of the revolt, following the milder one of the collapse of the work of the Legislative Commission because of social antagonisms, made the alliance between the crown and the gentry very close, explicit, and even militant. In the conditions of eighteenth-century Russia and as a logical result of the policies followed by the Russian government, the two had to sink or swim together. Yet Catherine the Great was too intelligent to become simply a reactionary. She intended instead to combine oppression and coercion with a measure of reform and a great deal of propaganda.

Reforms. The Gentry and the Serfs

The new system of local government introduced by Catherine the Great in 1775 was closely related to the Pugachev rebellion, although it also represented an attempt to bolster this perennially weak aspect of the administration and organization of Russia. Frightened by the collapse of authority at the time of the revolt, the empress meant to strengthen government in the provinces by means of decentralization, a clear distribution of powers and functions, and local gentry participation. She divided some fifteen major administrative units, through which the country was governed at the time, to make a total of fifty units by the end of her reign. Each of these gubernii — "governments" or "provinces" — was subdivided into some ten uezdy, or districts. Every province contained about 300,000 inhabitants and every district about 30,000, while historical and regional considerations were completely disregarded in the drawing of the boundaries.

An appointed governor at the head of the administration of each province was assisted by a complicated network of institutions and officials. Catherine the Great tried — not too successfully — to separate the legislative, executive, and judicial functions, without, of course, impairing her autocracy or ultimate control from St. Petersburg. Local gentry participated in local administration, and were urged to display initiative and energy in supporting the new system. The judicial branch too was organized, quite explicitly, on a class basis, with different courts and procedures for different estates. Catherine the Great's reform of local government was apparently influenced by the example of England, more particularly by Blackstone's views on the matter, and also by the example of the Baltic provinces. The arrangement that she introduced lasted until the fundamental reform of 1864.

Catherine the Great's scheme of local government fitted well into her program of cooperating with and strengthening the landlords. Other measures contributing to the same purpose included the granting of corporate representation and other privileges to the gentry. The incorporation of the gentry began in earnest with the formation of district gentry societies in 1766–67, developed further through the legislation of 1775 concerning local government, and reached its full development in the Charter to the Nobility of 1785. The Charter represented the highwater mark of the position and privileges of the Russian gentry. It recognized the gentry of each district and province as a legal body headed by an elected district or provincial Marshal of the Nobility. The incorporated gentry of a province could petition the monarch directly in connection with issues which aroused its concern, a right denied the rest of the population. Moreover, the Charter confirmed the earlier privileges and exemptions of the landlords and added certain new ones to give them a most advantageous and distinguished status. Members of the gentry remained free from obligations of personal service and taxation, and they became exempt from corporal punishment. They could lose their gentry standing, estates, or life only by court decision. The property rights of the landlords reached a new high; members of the gentry were recognized as full owners of their estates, without any restriction on the sale or exploitation of land, forests, or mineral resources; in case of forfeiture for crime, an estate remained within the family. Indeed some scholars speak — exaggeratedly to be sure — of Catherine the Great's introducing the modern concept of private property into Russia. Also in 1785, the empress granted a largely ineffective charter to towns which provided for a quite limited city government controlled by rich merchants.

As earlier, a rise in the position of the gentry meant an extension and strengthening of serfdom, a development which characterized Catherine the Great's entire reign. Serfdom spread to new areas, and in particular to the Ukraine. Although Catherine's government in essence confirmed an already-existing system in the Ukraine, it does bear the responsibility for helping to legalize serfdom in the Ukraine and for, so to speak, standardizing that evil throughout the empire. A series of laws, fiscal in nature, issued in 1763–83, forbade Ukrainian peasants to leave an estate without the landlord's permission and in general directed them "to remain in their place and calling." Catherine the Great personally extended serfdom on a large scale by her frequent and huge grants of state lands and peasants to her favorites, beginning with the leaders of the coup of 1762. The total number of peasants who thus became serfs has been variously estimated, but it was in the order of several hundred thousand working males — the usual way of counting peasants in imperial Russia — and well over a million persons. The census of 1794–96 indicated this growth of serfdom, with the serfs constituting 53.1% of all peasants and 49% of the entire population of the country. As to the power of the masters over their serfs, little could be added, but the government nevertheless tried its best: it became easier for the landlords to sentence their peasants to hard labor in Siberia, and they were empowered to fetch the peasants back at will; the serfs were forbidden, under a threat of harsh punishment, to petition the empress or the government for redress against the landlords. Catherine the Great also instituted firmer control over the cossacks, abolishing the famed Sech on the Dnieper in 1775 and limiting the autonomy of the Don and the Ural "hosts." Some of the Dnieper cossacks were transferred to the Kuban river to establish a cossack force in the plain north of the Caucasian mountains.

Other government measures relating to land and people included a huge survey of boundaries and titles — an important step in legalizing and confirming landholdings — the above-mentioned final secularization of vast Church estates with some two million peasants who became subject to the so-called College of Economy, and a program of colonization. Colonists were sought abroad, often on very generous conditions and at great cost, to populate territories newly won from Turkey and other areas, because serfdom and government regulations drastically restricted the mobility of the Russian people. Elizabeth had already established Serbian communities in Russia. Catherine the Great sponsored many more colonies of foreigners, especially of Germans along the Volga and in southern Russia.

Catherine II's efforts to promote the development of industry, trade, and also education and culture in Russia will be treated in appropriate chapters. Briefly, in economic life the empress turned in certain respects from rigid mercantilism to the newly popular ideas of free enterprise and trade. In culture she cut a broad swath. A friend of the *philosophes,* one

who corresponded with Voltaire and arranged for Diderot to visit Russia
— unprofitably, as it turned out — a writer and critic in her own right, and
a determined intellectual, Catherine the Great had plans and projects for
everything, from general education to satirical reviews. Indeed, she considered it her main mission to civilize Russia. For this reason, too, she
established a Medical Collegium in 1763, founded hospitals, led the way
in the struggle against infectious diseases, and decreed that Russia be
equipped to produce its own medicines and surgical instruments. And,
again in the interests of civilization, the empress pioneered in introducing
some feeble measures to help the underprivileged, for example, widows and
orphans.

Foreign Affairs: Introductory Remarks

In spite of her preoccupation with internal affairs, Catherine the Great
paid unflagging attention to foreign policy. Success and glory could be
attained by diplomacy as well as by enlightened reform at home, in war perhaps even more than in peace. Assisted by such statesmen as Nikita Panin
and Potemkin and such generals as Rumiantsev and Suvorov, the empress
scored triumph after triumph on the international stage, resulting in a
major extension of the boundaries of the empire, the addition of millions
of subjects, and Russia's rise to a new importance and eminence in Europe.
However, Catherine the Great's foreign policy was by no means a novel
departure. New ideas did appear: for example, Panin's early doctrine of a
northern accord or alliance of all leading northern European states to
counterbalance Austria, France, and Spain; and Potemkin's celebrated
"Greek project," which we shall discuss in its proper place. But, in
fact, these ideas proved ephemeral, and Russia continued on her old
course. As Russian historians like to put it, Peter the Great had solved one
of the three fundamental problems of Russian foreign relations: the
Swedish. Catherine the Great settled the other two: the Turkish and the
Polish. In addition to these key issues, the famous empress dealt with many
other questions, ranging from another Swedish war to the League of
Armed Neutrality and the need to face the shocking reality of the French
Revolution.

In foreign affairs, important events of the reign clustered in two brief
segments of time. The years 1768–74 witnessed the First Turkish War,
together with the first partition of Poland in 1772. Between 1787 and 1795
Russia participated in the Second Turkish War, 1787–92, a war against
Sweden, 1788–90, and the second and third partitions of Poland, 1793
and 1795. It was also during that time, of course, that Catherine the
Great became increasingly hostile to the French Revolution. Fortunately
for the empress, Great Britain was immersed in a conflict with its North

American colonies during the latter part of the First Turkish War, while during the second crucial sequence of years all powers had to shift their attention to revolutionary France.

Russia and Turkey

In their struggle against Turkey the Russians aimed to reach the Black Sea and thus attain what could be considered their natural southern boundary as well as recover fertile lands lost to Asiatic invaders since the days of the Kievan state. The Crimean Tartars, successors to the Golden Horde in that area, had recognized the suzerainty of the Sultan of Turkey. In pushing south Catherine the Great followed the time-honored example of Muscovite tsars and such imperial predecessors as Peter the Great and Anne. The First Turkish War, 1768–74, was fought both on land and, more unusual for Russia, on sea. A Russian army commanded by Rumiantsev advanced into Bessarabia and the Balkans, scoring impressive victories over large Turkish forces and appealing to the Christians to rise against their masters; another Russian army invaded and eventually captured the Crimea. A Russian fleet under Alexis Orlov sailed from the Baltic to Turkish waters and sank the Ottoman navy in the Bay of Chesme on July 6, 1770; however, it did not dare to try to force the Straits. After Alexis Orlov's expedition Russia maintained for a considerable period of time a direct interest in the Mediterranean — witness Paul's efforts at the end of the century to gain Malta and the Ionian islands — and gave up its attempt to obtain a permanent foothold there only in the reign of Alexander I, under British pressure. In spite of the fact that the Russian drive into the Balkans had bogged down, Turkey was ready in the summer of 1774 to make peace.

By the Treaty of Kuchuk Kainarji, Russia received the strategic points of Kinburn, Yenikale and Kerch in and near the Crimea as well as part of the Black Sea coast, west and east of the peninsula, reaching almost to the foot of the Caucasian range and including Azov. The Crimean Tartars were proclaimed independent, although they recognized the sultan as caliph, that is, the religious leader of Islam. Russia obtained the right of free commercial navigation in Turkish waters, including permission to send merchantmen through the Straits. Moldavia and Wallachia were returned to Turkey, but they were to be leniently ruled, and Russia reserved the prerogative to intervene on their behalf. Also, Russia acquired the right to build an Orthodox church in Constantinople, while the Turks promised to protect Christian churches and to accept Russian representations in behalf of the new church to be built in the capital. The provisions of the treaty relating to Christians and Christian worship became the basis of many subsequent Russian claims in regard to Turkey.

Although the First Turkish War in Catherine the Great's reign marked the first decisive defeat of Turkey by Russia and although the Treaty of Kuchuk Kainarji reflected the Russian victory, Russian aims had received only partial satisfaction. Some of the northern littoral of the Black Sea remained Turkish, while the Crimea became independent. From the Ottoman point of view, the war was a disaster which could only be remedied by exaction of revenge and by restoration of Turkey's former position by force of arms. The unstable political situation in the Crimea added to the tension. In 1783 Russia moved in to annex the Crimea, causing many Crimean Tartars to flee to the sultan's domain. By 1785 Russia had built a sizable fleet in the Black Sea, with its main base in Sevastopol. At the same time Potemkin made a great effort to populate and develop the newly-won southern lands. The display which Potemkin put up for Catherine the Great, Emperor Joseph II of Austria, and the Polish king Stanislaw Poniatowski when they visited the area in early 1787 gave rise to the expression "Potemkin villages," i.e., pieces of stage décor which passed at a distance for real buildings and communities. Actually, without minimizing Potemkin's showmanship, recent studies by Soloveytchik and others indicate that progress in southern Russia had proved to be real enough.

At that time Potemkin and Catherine the Great nursed very far-reaching aims which came to be known as "the Greek project." Roughly speaking, the project involved conquering the Ottomans, or at least their European possessions, and establishing — re-establishing the sponsors of the project insisted — a great Christian empire centered on Constantinople. Catherine the Great had her second grandson named Constantine, entrusted him to a Greek nurse, and ordered medals struck with a reproduction of St. Sophia! Austria finally agreed to allow the project after receiving assurance that the new empire would be entirely separate from Russia and after an offer of compensations in the Balkans and other advantages. Yet, like many other overly ambitious schemes, the Greek project proved to be ephemeral. Neither it nor its chief promoter Potemkin survived the Second Turkish War.

Turkey declared war on Russia in 1787 after the Russians rejected an ultimatum demanding that they evacuate the Crimea and the northern Black Sea littoral. The Porte enjoyed the sympathy of several major European powers, especially Great Britain which almost entered the war in 1791, and before long Sweden gave active support by attacking Russia. Catherine the Great had Austria as her military ally. The Second Turkish War, 1787–92, was confined to land action. Russian troops led by Suvorov scored a series of brilliant victories over Turkish forces, notably in 1790 when Suvorov stormed and won the supposedly impregnable fortress of Ismail. Incidentally, it was Michael Kutuzov, the hero of 1812, who first broke into Ismail. At the end of the war, Suvorov was marching on

Constantinople. By the Treaty of Jassy, signed on January 9, 1792, Russia gained the fortress of Ochakov and the Black Sea shore up to the Dniester River, while Turkey recognized Russian annexation of the Crimea. Russia had reached what appeared to be her natural boundaries in the south; the Turkish problem could be considered essentially solved.

The Partitioning of Poland

Catherine the Great's Polish policy turned out to be as impressive as her relations with Turkey. In a sense the partitioning of Poland, an important European state, represented a greater tour de force than the capture of a huge segment of a largely uninhabited steppe from the Ottomans. But, whereas the settlement with Turkey proved definitive and, as many scholars have insisted, logical and natural, the same could not be asserted by any stretch of the imagination in the case of Poland. Indeed, the partitioning of that country left Russia and Europe with a constant source of pain and conflict.

It has often been said, and with some reason, that Poland was ready for partitioning in the second half of the eighteenth century. Decentralization and weakening of central power in that country rapidly gathered momentum from about the middle of the seventeenth century. Elected kings proved increasingly unable to control their unruly subjects. The only other seat of central authority, the *sejm*, or diet, failed almost entirely to function. Composed of instructed delegates from provincial diets, the sejm in its procedure resembled a diplomatic congress more than a national legislature. The objection of a single deputy, the notorious *liberum veto*, would defeat a given measure and, in addition, dissolve the sejm and abrogate all legislation which it had passed prior to the dissolution. Between 1652 and 1674, for example, forty-eight of the fifty-five diets were so dissolved, almost one-third of them by the veto of a single deputy. The only traditional recourse when the sejm was dissolved consisted in proclaiming a confederation, that is, a gathering of the adherents of a given position; a confederation could no longer be obstructed by a *liberum veto*, and it tried to impose its views by force. The Polish political system has been described as "anarchy tempered by civil war."

The weakness of the Polish government acquired additional significance because that government had to face many grave problems. The Polish king ruled over Poles, Lithuanians, White Russians, Ukrainians, and Jews, not to mention smaller ethnic groups, over Catholics, Orthodox, Protestants, and subjects of Hebrew faith. He had to contend with an extremely strong and independent gentry, which, composing some eight per cent of the population, arrogated to itself all "Polish liberties," while keeping the bulk of the people, the peasants, in the worst condition of serfdom imaginable.

And he had to deal with powerful and greedy neighbors who surrounded Poland on three sides.

The last point deserves emphasis, because, after all, Poland did not partition itself: it was divided by three mighty aggressors. In fact, in the eighteenth century, Polish society experienced an intellectual and cultural revival which began to spread to politics. Given time, Poland might well have successfully reformed itself. But its neighbors were determined that it would not have the time. It was the misfortune of Poland that precisely when its political future began to look more promising, Catherine the Great finally agreed to a plan of partition of the kind which Prussia and Austria had been advancing from the days of Peter the Great.

The last king of Poland — and Catherine the Great's former lover — Stanislaw Poniatowski, who reigned from 1764 to 1795, tried to introduce certain reforms but failed to obtain firm support from Russia and Prussia, which countries had agreed in 1764 to co-operate in Polish affairs. In 1766–68 the allies reopened the issue of the dissidents, that is, the Orthodox and the Protestants, and forced the Polish government to grant them equal rights with the Catholics. That concession in turn led to violent protest within Poland, the formation of the Confederation of Bar, and civil war, with France lending some support to the Confederation, and Turkey even using the pretext of defending "Polish liberties" to declare war on Russia. Eventually Russian troops subdued the Confederates, and the first partition of Poland came in 1772.

That unusual attempt to solve the Polish problem stemmed in large part from complicated considerations of power politics: Russia had been so successful in the Turkish War that Austria was alarmed for its position; Frederick the Great of Prussia proposed the partition of a part of Poland as a way to satisfy Catherine the Great's expansionist ambitions and at the same time to provide compensation for Austria — which in effect had taken the initiative in 1769 by seizing and "re-incorporating" certain Polish border areas — as well as to obtain for Prussia certain long-coveted Polish lands which separated Prussian dominions. By the first partition of Poland Russia obtained White Russian and Latvian Lithuania to the Dvina and the Dnieper rivers with some 1,300,000 inhabitants; Austria received so-called Galicia, consisting of Red Russia, with the city of Lemberg, or Lvov, of a part of western Podolia, and of southern Little Poland, with a total population of 2,650,000; Prussia took the so-called Royal, or Polish, Prussia, except Danzig and Thorn. Although moderate in size and containing only 580,000 people, the Prussian acquisition represented the most valuable gain of the three from the political, military, and financial points of view. In all, Poland lost about one-third of her territory and more than a third of her population.

This disaster spurred the Poles finally to enact basic reforms. Changes

Poland 1662-1667 and Partitions of Poland

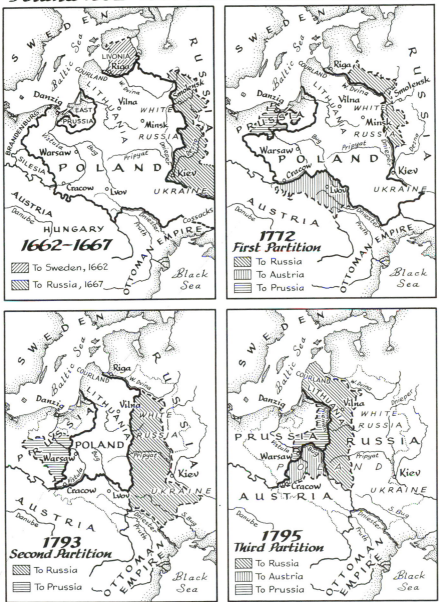

began in 1773 and culminated in the activities of the celebrated Four Years' Diet of 1788–92 and in the constitution of May 3, 1791. The monarchy was to become hereditary, and the king obtained effective executive power; legislative authority was vested in a two-chamber diet with the lower chamber in a dominant position; the *liberum veto* disappeared in favor of majority rule; the diet included representatives of the middle class;

a cabinet of ministers, organized along modern lines, was created and made responsible to the diet. The Polish reform party profited from the benevolent attitude of Prussia, which hoped apparently to obtain further concessions from new Poland; Russia and Austria were again preoccupied with a Turkish war. But the May constitution brought matters to a head. While Prussia and Austria accepted it, Russia instigated the organization of the Confederation of Targowica in defense of the old order in May 1792. When the Russian army entered Poland on the invitation of the Confedera- ion, the Prussians reversed themselves and joined the invaders. The second partition of Poland followed in January 1793. This time Russia took more of Lithuania and most of the western Ukraine with a total of 3,000,000 inhabitants; Prussia seized Danzig, Thorn, and Great Poland with a com- bined population of 1,000,000; Austria did not participate. In addition, Russia obtained the right to move its troops into what remained of Poland and control its foreign policy.

The Poles responded in March 1794 with a great national uprising led by Thaddeus Kosciuszko. But, in spite of their courage, their fight was hope- less. The Poles were crushed by the Russians, commanded by Suvorov, and the Prussians. Austria rejoined her allies to carry out the third partition of Poland in October 1795. By its provisions, Russia acquired the re- mainder of Lithuania and the Ukraine, with 1,200,000 inhabitants, as well as the Duchy of Courland, where Russian influence had predominated from the time of Empress Anne; Prussia took Mazovia, including Warsaw, with 1,000,000 people; Austria appropriated the rest of Little Poland, with Cracow, and another 1,500,000 inhabitants. Poland ceased to exist as an independent state.

The partitioning of Poland brought tragedy to the Poles. Its impact on the successful aggressors is more difficult to assess. As Lord and other historians have shown in detail, Prussia, Russia, and Austria scored a re- markable, virtually unprecedented, diplomatic and military coup. They dismembered and totally destroyed a large European state, eliminating an old enemy, rival, and source of conflicts, while at the same time adding greatly to their own lands, resources, and populations. Eastern Europe fell under their complete control, with France deprived of her old ally. Significantly, after the division of Poland, the three east European mon- archies for a long time co-operated closely on the international scene — partners in crime, if you will. Yet, even some of the *philosophes* praised at least the first partition of Poland, calling it "a triumph of rationality." But the Poles thought differently and never accepted the dismemberment. As a result, Poland, Polish rights, and Polish boundaries remained an un- resolved problem, or series of problems, for Europe and the world all the way to the Oder-Neisse line of today. For imperial Russia, the partitioning of Poland resulted in, among other things, the Polish support of Napoleon in 1812 and the great rebellions of 1831 and 1863.

Russian scholars like to emphasize that the Russian case contrasted sharply with those of Prussia and Austria: in the three partitions of Poland Russia took old Russian lands, once part of the Kievan state, populated principally by Orthodox Ukrainians and White Russians, whereas the two German powers grabbed ethnically and historically Polish territory; the Russians, therefore, came as liberators, the Prussians and the Austrians as oppressors. If Catherine the Great deserved blame, it was not for her own acquisitions, but for allowing Prussia and Austria to expand at the expense of the Poles. Much can be said for this point of view, for it states the facts of the dismemberment correctly; yet at least two caveats seem in order. The brutal Russian policy toward Poland had to allow for the interests of other aggressors and indeed led to further repartitioning, with Warsaw and the very heart of the divided country linked to Russia in 1815. Also, Catherine the Great herself cared little about the faith or the ethnic origins of her new subjects. She thought simply in terms of power politics, position, and prestige — everything to the greater glory of Russia, and of course, to her own greater glory. After suppressing the Confederation of Bar, Russian troops also suppressed a desperate uprising of Ukrainian peasants against their Polish and Polonized landlords. These landlords continued to dominate and exploit the masses quite as effectively after the partitions as before them. In fact, some Ukrainian historians have complained that the oppression increased, because the strong Russian government maintained law and order more successfully than had the weak Polish authorities.

Foreign Policy: Certain Other Matters

Catherine the Great's foreign policy encompassed a wide range of activities and interests in addition to the relations with Turkey and Poland. Important developments included the Russian role in the League of Armed Neutrality, a war against Sweden, and the empress's reaction to the French Revolution. Russia advanced the doctrine of armed neutrality at sea in 1780 to protect the commerce of non-combatant states against arbitrary actions of the British who were engaged in a struggle with their American colonies. Several other European countries supported Russian proposals which eventually became part of international maritime law. Russia and her partners in the League insisted that neutral ships could pass freely from port to port and along the coast to combatants, that enemy goods in neutral ships, except contraband, were not subject to seizure, and that to be legal a blockade had to be enforced, rather than merely proclaimed.

Sweden, as already mentioned, attacked Russia in 1788, when the Russian armies were engaged in fighting Turkey. The Swedes repeatedly threatened St. Petersburg; however, the war proved inconclusive. The Treaty of Werälä signed in August 1790, merely confirmed the pre-war boundary. Denmark, allied with Russia, participated in the hostilities against Sweden.

The French Revolution made a strong impression on Catherine the Great. At first she tried to minimize the import of the events in France and to dissociate them from the main course of European history and the Enlightenment. But, as the Revolution became more radical, the empress reacted with bitterness and hostility. At home she turned against critical intellectuals and indeed against much of the cultural climate that she herself had striven so hard to create. In respect to revolutionary France, she became more and more antagonistic and broke off relations in 1793 after the execution of Louis XVI. Of course, she also used the confusion and disarrangement produced in Europe by the French Revolution to carry out the second and third partitions of Poland without interference. Some historians believe that only the empress's sudden death prevented her from joining a military coalition against France.

Evaluations of Catherine the Great

Much has been written for and against Catherine the Great. The sovereign's admirers have included many intellectuals, from eighteenth-century *philosophes* led by Voltaire to Sidney Hook, who not long ago proclaimed her an outstanding example of the hero who makes history. The empress has received praise from numerous historians, in particular specialists in the cultural development, foreign relations, and expansion of Russia, including such judicious scholars as B. Nolde and Isabel de Madariaga. A few, for instance V. Leontovich, also commended her policy toward the gentry, in which they saw the indispensable first step in the direction of liberalism — rights, privileges, and advantages had to be acquired first by the top social group, and only after that could they percolate downward.

The critics of Catherine the Great, who have included many pre-revolutionary Russian historians as well as the Soviet scholars as a group, have concentrated overwhelmingly on the empress's social policy and the social conditions during her reign. Above all, they have castigated the reign as the zenith of serfdom in Russia. For this reason many of them would deny that Catherine II, in spite of her display and championing of culture, was an enlightened despot in the sense in which this term would apply to Emperor Joseph II of Austria, who did care for the masses. Even though very few social historians have ascribed personally to the empress a fundamental influence on the evolution of Russian society, they have been repelled by the contrast between her professedly progressive views and her support of serfdom, as well as by the ease and thoroughness of her accommodation to that great evil. Her immediate successors, Paul and Alexander I, showed different attitudes.

But whatever judgment we make of the empress — and it should be clear that the views mentioned above rarely clash directly, covering as they do different aspects of Catherine the Great's activity — we must recognize the importance of her reign. In foreign policy, with the acquisition of southern Russia and the partitioning of Poland; in internal affairs, with the development of serfdom and of the gentry position and privileges; and in culture, with striking progress in Westernization, the time of Catherine the Great marked a culmination of earlier trends and set the stage for Russian history in the nineteenth century. But before we turn to Russia in the nineteenth century, we have to consider the reign of Paul and some broad aspects of the evolution of Russia from Peter the Great to Alexander I.

The Reign of Paul

Emperor Paul was forty-two years old when he ascended the throne. In the course of the decades during which his mother had kept him away from power, he came to hate her, her favorites, her advisers, and everything she stood for. Reversing Catherine the Great's decisions and undoing her work was, therefore, one salient trait of Paul's brief reign, 1796–1801. Another stemmed directly from his character and can best be described as petty tyranny. Highly suspicious, irritable, and given to frequent outbreaks of rage, the emperor promoted and demoted his assistants with dazzling rapidity and often for no apparent reason. He changed the drill and the uniforms of the Russian army, himself entering into the minutest details; imperial military reviews inspired terror in the participants. Paul generously freed from prison and exile those punished by Catherine the Great, including liberal and radical intellectuals and leaders of the Polish rebellion such as Kosciuszko. But their places were quickly taken by others who had in some manner displeased the sovereign, and the number of the victims kept mounting. Above all, the emperor insisted on his autocratic power and majesty even in small things like dancing at a palace festival and saluting. As Paul reportedly informed the French ambassador, the only important person in Russia was the one speaking to the emperor, and only while he was so speaking. With the same concept of the majesty of the Russian monarchy in mind, and also reacting, no doubt, to his own long and painful wait for the crown, Paul changed the law of succession to the Russian throne at the time of his coronation in 1797: primogeniture in the male line replaced Peter the Great's provision of free selection by the reigning monarch. Russia finally acquired a strictly legal and stable system of succession to the throne.

The emperor's views and attitudes found reflection in his treatment of the crucial problem of serfdom and the gentry. On the one hand Paul continued Catherine the Great's support and promotion of serfdom by spreading it to extreme southern Russia, so-called New Russia, in 1797, and by distributing state lands and peasants to his favorites at an even faster rate than had his mother. Also, he harshly suppressed all peasant disturbances and tolerated no disobedience or protest on the part of the lower classes. Yet, on the other hand, Paul did not share his mother's confidence in and liking for the gentry. For this reason he tried for the first time to regulate and limit the obligations of the serfs to their masters by proclaiming in 1797 that they should work three days a week for their landlords and three days for themselves, with Sunday sanctified as a day of rest. Although Paul's new law was not, and possibly could not be, enforced, it did represent a turning point in the attitude of the Russian government toward serfdom. From that time on limitation and, eventually, abolition of serfdom became real issues of state policy. The emperor gave further expression to his displeasure with the gentry through such measures as the restoration of corporal punishment for members of that class as well as for the townspeople, and through increased reliance on the bureaucracy in preference to the gentry in local self-government and in general administration.

It was in the field of foreign policy and especially of war that Paul's reign left its most lasting memory. Just before her death, Catherine the Great had come close to joining an anti-French coalition. Paul began with a declaration of the Russian desire for peace, but before long he too, provoked by French victories and certain mistakes of tact on the part of France, turned to the enemies of the revolutionary government. Russia entered the war against France as a member of the so-called Second Coalition, organized in large measure by Paul and composed of Russia, Great Britain, Austria, Naples, Portugal, and Turkey. In the campaigns that followed, a Russian fleet under the command of Theodore Ushakov sailed through the Straits, seized the Ionian Islands from the French, and established there a Russian-controlled republic under the protectorate of Turkey. Russian influence extended even further west in the Mediterranean, for Paul had accepted his election as the grand master of the Knights of Malta and thus ruler of that strategic island.

The main theater of operations, however, remained on land. Russian troops joined allied armies in the Low Countries and in Switzerland, but their most effective intervention took place in northern Italy. There a force of 18,000 Russians and 44,000 Austrians led by Suvorov drove out the French in the course of five months in 1798–99, winning three major battles and about a dozen lesser engagements and capturing some 25 fortresses and approximately 80,000 prisoners. Suvorov wanted to invade France. Instead, because of defeats on other fronts and the change of plans in the allied high command, he had to retreat in 1799–1800 to southern Germany through the Swiss Alps held by a French force. His successfully managing the retreat has been considered one of the great feats of military history. On the whole, Suvorov, who died very shortly after the Swiss campaign at the age of seventy, is regarded as the ablest military commander Russia ever produced — and this is a high honor. The qualities of this eccentric and unpredictable general included heavy reliance on speed and thrust and remarkable psychological rapport with his soldiers.

Disgusted with Austria and also with Great Britain, which failed to support Russian troops adequately in the Netherlands, Paul abandoned the coalition. In fact, in 1800 he switched sides and joined France, considering the rise of Napoleon to be a guarantee of stability and the end of the revolution. The new alignment pitted Russia against Great Britain. Having lost Malta to the British, Paul, in a fantastic move, sent the Don cossacks to invade distant India over unmapped territory. The emperor's death interfered at this point, and Alexander I promptly recalled the cossacks.

Paul was killed in a palace revolution in March 1801. His rudeness, violent temperament, and unpredictable behavior helped the conspiracy to grow even among the emperor's most trusted associates and indeed within his family. His preference for the troops trained at his own estate of Gatchina antagonized, and seemed to threaten, the guards. The emperor's turning against Great Britain produced new enemies. Count Peter Pahlen, the military governor of St. Petersburg, took an active part in the plot, whereas Grand Duke Alexander, Paul's son and heir, apparently assented to it. It remains uncertain whether murder entered into the plans of the conspirators — Alexander, it seems, had not expected it — or whether it occurred by accident.

THE ECONOMIC AND SOCIAL DEVELOPMENT OF RUSSIA
IN THE EIGHTEENTH CENTURY

> Serfdom in its fullness lasted longer in Russia than in Western coun-
> tries because its economic disadvantages did not earlier outweigh its
> advantages; because the increase of population did not cause suffi-
> ciently acute land shortage among the peasantry until the first half of
> the nineteenth century; because the middle classes were weak in com-
> parison with the serf-owners; because humanitarian and other ideas
> of the value of the individual spirit were little developed; because the
> reaction against the ideas of the French Revolution strengthened the
> *vis inertiae* inherent in any long-established institution; lastly, because
> serfdom was not merely the economic basis of the serf-owners but
> also a main basis of the Russian state in its immense task of somehow
> governing so many raw millions.
>
> SUMNER

> It is significant that none of the contemporary western European
> authors who have written on Russian economics in the late eighteenth
> century and the early nineteenth speaks of Russia as an economically
> backward country. In fact, during some part of the eighteenth century,
> Russian industry, at least in some branches, was ahead not only of all
> the other Continental countries but of England as well. This was par-
> ticularly true of the metal industries. In the middle of the eighteenth
> century Russia was the world's largest producer of both iron and
> copper, and it was not until the 1770's in the case of copper, and the
> very end of the century in the case of iron, that English production
> became equal to that of Russia.
>
> KARPOVICH

As we have already had occasion to observe, the reign of Peter the
Great marked an important divide in the economic and social development
of Russia as well as in the political history of the country. One of the most
significant and least explained changes occurred in the nature of the
population curve, which, it might be added, paralleled the curves in other
European countries: whereas the population of Russia apparently remained
largely stationary for a century and longer prior to the time of Peter the
Great, and whereas it might have decreased during the reformer's hard
reign, it rose rapidly from then on. Within the Russian boundaries of 1725
there lived some thirteen million people in that year, nineteen million in
1762, and twenty-nine million in 1796. Counting approximately seven
million new subjects acquired as a result of Catherine the Great's suc-

cessful foreign policy, the Russian Empire had by the end of the eighteenth century over thirty-six million inhabitants.

In addition to the immediate increase in population, the expansion of Russia in the eighteenth century produced a number of other results important for the economic life of the country. Peter the Great's victory in the Great Northern War gave his state access to the Baltic; and citizens of such ports as Riga, who were more proficient in navigation and commerce than the Russians, were then brought into the empire. "A window into Europe" referred as much to economic affairs as to culture or politics. Catherine the Great's huge gains from the partitions of Poland also brought Russia closer to other European countries and included towns and areas with a relatively more developed economy. Both the German landlords of the Baltic region and the Polish or Polonized gentry of what came to be known as the western provinces were in certain respects more advanced than their Russian counterparts. The acquisitions to the south proved similarly significant. Catherine the Great's success in the two Turkish wars opened vast fertile lands of southern Russia — a further extension of what had been obtained in the preceding decades and centuries — for colonization and development, and established the empire firmly on the Black Sea. Although serfdom restricted mobility, population in the south grew rapidly by means of voluntary migration and the transfer of serfs and state peasants. In the words of one historian concerned with the density of population:

> Prior to the eighteenth century a comparatively dense and settled population in the Russian Empire was to be found only in the center of the plains of European Russia: the region of Moscow with the immediately adjoining provinces, and the upper valley of the Volga. Somewhat less densely settled were the northern part of the Ukraine, and the ancient Smolensk and Novgorod regions, to the west and northwest of Moscow respectively. Finally, starting from central Russia, a narrow strip of fairly dense settlement stretched out toward Archangel, along the river Dvina; and another strip was to be found along the middle course of the Volga and farther east, in the direction of Siberia. To this limited area of comparatively dense settlement now was added a vast territory in the south and the southeast of European Russia.

Agriculture and Other Occupations

Differentiation accompanied expansion. The fertile, mostly "black-earth," agricultural areas of the south became more and more distinct from the more barren regions of the center and north. The system of barshchina, that is, of work for one's master, prevailed in the south, that of obrok, or payments to the landlord in kind or money, in the north. On the rich black earth of the south the serfs tilled their masters' fields as well as their own plots, and they also performed other tasks for the master such as cutting firewood or mowing hay. In addition to the increase in grain and other agricultural products, cattle-raising developed on a large scale. The landlords generally sold the products of their economy on the domestic market, but toward the end of the century export increased.

In the provinces of the center and north, where the earth was not so fertile, the obrok, or quitrent, practice grew. There only modest harvests of rye and other grains suitable to the rigorous climate could be obtained from the soil, so that the peasant population had to find different means to support itself and to discharge its obligations to the landlord and the state. Special crafts developed in various localities. In some places peasants produced iron implements, such as locks, knives, and forks; in others they made wooden utensils, spoons, cups, plates, toys, and the like, or leather goods. Where no such subsidiary local occupations emerged, many peasants left their homes periodically, especially for the winter, to find work elsewhere. Often groups of peasants sought employment together in associations known as *arteli* — singular *artel* — and became carpenters, housepainters, or construction workers. Others earned money in industrial production, transportation, or petty trade. These varied earnings, together with their meager agriculture, made it possible for a large number of peasants to pay their quitrent to the landlord, meet their obligations, and support themselves and their families — although at a very low standard of living. It has been estimated that about one quarter of the peasant population of the less fertile provinces left their villages for winter employment elsewhere.

The great extent and the continuing expansion of agriculture in Russia did not mean that it was modern in technique or very productive. Russian agriculture remained rather primitive, and, because of the backward technique of cultivation, even excellent land gave relatively low yields. Serfdom contributed heavily to the inefficient use of labor and to rural overpopulation. In agriculture Westernization came very slowly indeed. By the end of the century, in spite of the efforts of the Free Economic Society established in 1765 and a few other groups as well as certain individuals, no substantial modernization had occurred. As Marxist historians have repeatedly emphasized, serfdom with its abundant unskilled labor still could effectively satisfy the needs of the rather sluggish and parochial Russian rural economy in the eighteenth century.

Industry and Labor Force

In a sense, the Russians during that period made greater advances in industry. The number of factories grew from 200 or 250 at the time of

Peter the Great's death to 1,200 by the end of the century, to cite one opinion, or possibly even over 3,000, if the smallest manufacturing establishments are included. The total number of workers rose to a considerable figure which has been variously estimated between 100,000 and 225,000. Many factories employed hundreds of hands, with the highest known number in the neighborhood of 3,500. The vitally important mining and metal industries developed so spectacularly as to give Russia a leading position in Europe in this type of production. The Ural area produced at that time some 90 per cent of Russian copper and some 65 per cent of the pig iron. Lesser centers of metal industry existed in Olonets, which is in the north near the Finnish border, and in Tula, south of Moscow. The textile industry flourished in and around Moscow and in some neighboring provinces and, to a lesser extent, in the St. Petersburg area. A number of other industries also developed in eighteenth-century Russia.

However, in the context of Russian society, the acquisition of a suitable labor force often created special problems; Russian manufacturing establishments reflected and in turn affected the social structure of the empire. Thus, in addition to owning and operating some factories outright, the state established in areas of scarce labor supply numerous "possessional factories," which were operated by merchants and to which state peasants were attached as "possessional workers." They were, in fact, industrial serfs, but they belonged to a factory, not to an individual. These possessional factories acquired special prominence in heavy industry. Some landlords, in their turn, set up manorial factories, especially for light industry, where they utilized the bonded labor of their serfs. Nevertheless, free labor also played an increasingly important role in the industrial development of Russia in the eighteenth century. Even when it represented, as it often did, the labor of someone else's serfs out to earn their quitrent, it led to new, more "capitalistic," relationships in the factories. Soviet studies, for example those of Khromov and Poliansky, in contrast to some earlier Marxist works, such as Liashchenko's well-known writings, have emphasized the large scope and vital importance of this free labor and of so-called "merchant" or "capitalist" enterprises based on that labor. For instance, in the middle of the century merchants owned some 70 per cent of textile factories in Russia as well as virtually the entire industry of the Moscow and St. Petersburg regions.

In addition to government managers, merchants, and gentry entrepreneurs, businessmen of a different background, including peasants and even serfs, made their appearance. In a number of instances, peasant crafts were gradually industrialized and some former serfs became factory owners, as, for example, in the case of the textile industry in and around

Industry and Agriculture — 18th Century

- ▲ Iron ore mining
- ■ Iron foundries & smelters
- • Metal processing
- ⬟ Hand-operated blast furnaces
- ◆ Copper smelters
- ◊ Copper processing
- Ⅱ Textiles
- ⚓ Sailcloth & rope
- ⬠ Glassmaking
- ⬡ Leather processing
- ⛵ Shipbuilding
- ◎ Principal fairs

Ivanovo-Voznesensk in central European Russia. Indeed, if we are to follow Poliansky's statistics, peasant participation in industry grew very rapidly and became widespread in the last quarter of the century.

In eighteenth-century Russia the state engaged directly in industrial development but also encouraged private enterprise. This encouragement was plainly evident in such measures as the abolition of various restrictions on entering business — notably making it possible for the gentry to take part in every phase of economic life — and the protective tariffs of 1782 and 1793.

Trade

Trade also grew in eighteenth-century Russia. Domestic commerce was stimulated by the repeal of internal tariffs that culminated in Empress Elizabeth's legislation in 1753, by the building of new canals following the example of Peter the Great, by territorial acquisitions, and especially by the quickened tempo and increasing diversity of economic life. In particular, the fertile south sent its agricultural surplus to the center and the north in exchange for products of industries and crafts, while the countryside as a whole supplied the cities and towns with grain and other foods and raw materials. Moscow was the most important center of internal commerce as well as the main distribution and transit point for foreign trade. Other important domestic markets included St. Petersburg, Riga, Archangel, towns in the heart of the grain-producing area such as Penza, Tambov, and Kaluga, and Volga ports like Iaroslavl, Nizhnii Novgorod, Kazan, and Saratov. In distant Siberia, Tobolsk, Tomsk, and Irkutsk developed as significant commercial as well as administrative centers. Many large fairs and uncounted small ones assisted the trade cycle. The best known among them included the celebrated fair next to the Monastery of St. Macarius on the Volga in the province of Nizhnii Novgorod, the fair near the southern steppe town of Kursk, and the Irbit fair in the Ural area.

Foreign trade developed rapidly, especially in the second half of the century. The annual ruble value of both exports and imports more than tripled in the course of Catherine the Great's reign, an impressive achievement even after we make a certain discount for inflation. After the Russian victory in the Great Northern War, the Baltic ports such as St. Petersburg, Riga, and Libau became the main avenue of trade with Russia, and they maintained this dominant position into the nineteenth century. Russia exported to other European countries timber, hemp, flax, tallow, and some other raw materials, together with iron products and certain textiles, notably canvas for sails. Also, the century saw the beginning of the grain trade which was later to acquire great prominence. This trade became possible on a large scale after Catherine the Great's acquisition of southern Russia and the development of Russian agriculture there as well as the construction of the Black Sea ports, notably Odessa which was won from the Turks in 1792 and transformed into a port in 1794. Russian imports consisted of wine, fruits, coffee, sugar, and fine cloth, as well as manufactured goods. Throughout the eighteenth century exports greatly exceeded imports in value. Great Britain remained the best Russian customer, accounting for

something like half of Russia's total European trade. The Russians continued to be passive in their commercial relations with the West: foreign
businessmen who came to St. Petersburg and other centers in the empire
handled the transactions and carried Russian products away in foreign
ships, especially British and Dutch. Russia also engaged in commerce
with Central Asia, the Middle East, and even India and China, channeling
goods through the St. Macarius Fair, Moscow, Astrakhan, and certain other
locations. A considerable colony of merchants from India lived in Astrakhan
in the eighteenth century.

The Peasants, the Gentry, and Other Classes

Eighteenth-century Russia was overwhelmingly rural. In 1724, 97 per
cent of its population lived in the countryside and 3 per cent in towns; by
1796 the figures had shifted slightly to 95.9 per cent as against 4.1 per
cent. The great bulk of the people were, of course, peasants. They fell into
two categories, roughly equal in size, serfs and state peasants. Toward
the end of the century the serfs constituted 53.1 or even 55 per cent of
the total peasant population. As outlined in earlier chapters, the position
of the serfs deteriorated from the reign of Peter the Great to those of Paul
and Alexander I and reached its nadir around 1800. Increasing economic
exploitation of the serfs acompanied their virtually complete dependence
on the will of their masters, without even the right to petition for redress.
It has been estimated that the obrok increased two and a half times in
money value between 1760 and 1800, while the barshchina grew from
three to four and in some cases even five or more days a week. It was this
striking expansion of the barshchina that Emperor Paul tried to stem with
his ineffectual law of 1797. Perhaps the most unfortunate were the
numerous household serfs who had no land to till, but acted instead as
domestic servants or in some other capacity within the manorial household.
This segment of the population expanded as landlords acquired new tastes
and developed a more elaborate style of life. Indeed, some household serfs
became painters, poets, or musicians, and a few even received education
abroad. But, as can be readily imagined, it was especially the household
serfs who were kept under the constant and complete control of their
masters, and their condition could barely be distinguished from slavery.
State peasants fared better than serfs, although their obligations, too, increased in the course of the century. At best, as in the case of certain areas
in the north, they maintained a reasonable degree of autonomy and prosperity. At worst, as exemplified by possessional workers, their lot could not
be envied even by the serfs. On the whole the misery of the Russian

countryside provides ample explanation for the Pugachev rebellion and for repeated lesser insurrections which occurred throughout the century.

By contrast, the eighteenth century, especially the second half during the reign of Catherine the Great, has been considered the golden age of Russian gentry. Constituting a little over one per cent of the population, this class certainly dominated the life of the country. With the lessening and finally the abolition of their service obligations, the landlords took a greater interest in their estates, and some of them also pursued other lines of economic activity, such as manufacturing. The State Lending Bank, established by Catherine the Great in 1786, had as its main task the support of gentry landholding. Moreover, it was the gentry more than any other social group that experienced Westernization most fully and developed the first modern Russian culture. And, of course, the gentry continued to surround the throne, to supply officers for the army, and to fill administrative posts.

While the gentry prospered, the position of the clergy and their dependents declined. This sizable group of Russians, about one per cent of the total — it should be remembered that Orthodox priests marry and raise families — suffered from the anti-ecclesiastical spirit of the age and especially from the secularization of Church lands in 1764. In return for vast Church holdings populated by serfs, the Church received an annual subsidy of 450,000 rubles, representing about one-third of the revenues from the land and utterly insufficient to support the clergy. With time and inflation the value of the subsidy dropped. Never rich, the Russian priests became poorer and more insecure financially after 1764. They had to depend almost entirely on fees and donations from their usually impoverished parishioners. In the country especially the style of life of the priests and their families differed little from that of peasants. In post-Petrine Russia, in contrast to some other European states, the clergy had little wealth or prestige. Largely neglected by historians, the Russian clerical estate has recently received some valuable attention from Freeze and a few other scholars.

Most of the peasants, the gentry, and the clergy lived in rural areas. The bulk of the town inhabitants were divided into three legal categories: merchants, artisans, and workers. These classes were growing: for instance, peasants who established themselves as manufacturers or otherwise successfully entered business became merchants. Nevertheless, none of these classes were numerous or prominent in eighteenth-century Russia. As usual, it was the government that tried to stimulate initiative, public spirit, and a degree of participation in local affairs among the townsmen by such means as the creation of guilds and the charter of 1785 granting urban self-government. As usual, too, these efforts failed.

Finance. Concluding Remarks

The fiscal policies of the state deserve notice. The successors of Peter the Great, not unlike the reformer himself, ruled in a situation of continuous financial crisis. The state revenue rose from 8.5 million rubles in 1724 to 19.4 million in 1764 and over 40 million in 1794. But expenses tended to grow still more rapidly, amounting to 49.1 million in the last year mentioned. Of that sum, 46 per cent went to the army and the navy, 20 per cent to the state economy, 12 per cent for administration and justice, and 9 per cent to maintain the imperial court. A new item also appeared in the reign of Catherine the Great: this was the state debt, which accounted for 4.5 per cent of the total state expenses in 1794. To make up the difference between revenue and expenses, the government borrowed at home and, beginning in 1769, borrowed abroad too, mainly in Holland. The government also issued paper money on a large scale, especially after the outbreak of the Second Turkish War. By the end of Catherine the Great's reign a paper ruble was worth only 68 per cent of its metallic counterpart. Taxes remained heavy and oppressive.

In effect, the rulers of imperial Russia, perhaps even more than the Muscovite tsars who preceded them, insisted on living beyond their means and thus strained the national economy to the limit. Although a poor, backward, overwhelmingly agricultural, and illiterate country, Russia had a large and glorious army, a complex bureaucracy, and one of the most splendid courts in Europe. With the coming of Westernization, the tragic, and as it turned out fatal, gulf between the small enlightened and privileged segment at the top and the masses at the bottom became wider than ever. We shall consider this again when we deal with Russian culture in the eighteenth century and, indeed, throughout our discussion of imperial Russian history.

XXIV

* * * * * * * * * *

RUSSIAN CULTURE IN THE EIGHTEENTH CENTURY

> The new culture born as a result of the Petrine revolution constituted
> in the beginning nothing but a heterogeneous collection of imported
> articles; but the new elite assimilated them so rapidly that by the
> end of the eighteenth century there already existed a Russian culture,
> more homogeneous and more stable than the old one. That culture
> was Russian in the strictest sense of the word, expressing emotional
> states and creating values that were properly Russian, and if the
> people no more than half understood it, this transpired not because it
> was not sufficiently national, but because the people were not yet a
> nation.
>
> WEIDLE

> . . . A mixture of tongues,
> The language of France with that of Nizhnii Novgorod.
>
> GRIBOEDOV

THE EIGHTEENTH CENTURY constitutes a distinct period in the history of
Russian culture. On one hand it marked a decisive break with the Muscovite past; although, as we know, that break had been foreshadowed
and assisted by certain influences and trends. Peter the Great's violent
activity was perhaps most revolutionary in the domain of culture. All of a
sudden, skipping entire epochs of scholasticism, Renaissance, and Reformation, Russia moved from a parochial, ecclesiastical, quasi-medieval civilization to the Age of Reason. On the other hand, Russian culture of the
eighteenth century also differed significantly from the culture of the following periods. From the beginning of Peter the Great's reforms to the
death of Catherine the Great, the Russians applied themselves to the
huge and fundamental task of learning from the West. They still had much
to learn after 1800, of course; nonetheless, by that time they had acquired and developed a comprehensive and well-integrated modern culture of their own, which later on attracted attention and adaptation
abroad. The eighteenth century in Russia then was an age of apprenticeship
and imitation par excellence. It has been said that Peter the Great, during
the first decades of the century, borrowed Western technology, that Empress
Elizabeth, in the middle of the period, shifted the main interest to Western
fashions and manners, and that Catherine the Great, in the course of the
last third of the century, brought Western ideas into Russia. Although
much too simple, this scheme has some truth. It gives an indication of

the stages in the Russian absorption of Western culture, and it suggests that by 1800 the process had spread to everything from artillery to philosophy.

The Russian Enlightenment

The culture of the Enlightenment, which Russia borrowed, had a number of salient characteristics. It represented notably the triumph of secularism and thus stood in sharp contrast to the Church-centered civilization of Muscovy. To be sure, Orthodoxy remained in imperial Russia and even continued, in a sense, to be linked to the state and occupy a high position. But instead of being central to Russian life and culture, it became, at least as far as the government and the educated public were concerned, a separate and rather neglected compartment. Moreover, within this compartment, to follow Florovsky and other specialists, one could detect little originality or growth. The secular philosophy which dominated the stage in eighteenth-century Europe emphasized reason, education, and the ability of enlightened men to advance the interests of society. The last point applied especially to rulers, so-called enlightened despots, who had the greatest means at their disposal to direct the life of a country. These views fitted imperial Russia remarkably well. Indeed, because of the magnitude and the lasting impact of Peter the Great's efforts to modernize his state, he could be considered the outstanding enlightened despot, although a very early one, while Catherine the Great proved only too eager to claim that title.

In addition to the all-pervasive government sponsorship, Enlightenment came to Russia through the educated gentry. After the pioneer years of Peter the Great, with his motley group of foreign and Russian assistants, the gentry, as we know, increasingly asserted itself to control most phases of the development of the country. Despite some striking individual exceptions, modern Russian culture emerged as gentry culture and maintained that character well into the nineteenth century. It became the civilization of an educated, aristocratic elite, with its salons and its knowledge of French, a civilization which showed more preoccupation with an elegant literary style and proper manners than with philosophy or politics. Nonetheless, this culture constituted the first phase of modern Russian intellectual and cultural history and the foundation for its subsequent development.

Education

The glitter of the age of Catherine the Great was still far away when Peter the Great began his work of educating the Russians. Of necessity,

his efforts were aimed in many directions and dealt with a variety of fundamental matters. As early as 1700 he arranged for publication of Russian
books by a Dutch press; several years later the publishing was transferred
to Russia. Six hundred different books published in the reign of the reformer have come down to our time. In 1702 the first Russian newspaper,
Vedomosti or *News,* began to be published, the monarch himself editing
its first issue. Next Peter the Great took part in reforming the alphabet to
produce what came to be known as the civil Russian alphabet. Composed
of Slavonic, Greek, and Latin letters, the new alphabet represented a considerable simplification of the old Slavonic. The old alphabet was allowed
for Church books, but, following a decree in early 1710, all other works
had to use the new system. Also, Peter the Great introduced Arabic
numerals to replace the cumbersome Slavonic ones.

Peter the Great sent, altogether, hundreds of young Russians to study
abroad, and he opened schools of new types in Russia. For example, as
early as 1701 he established in Moscow a School of Mathematical and
Navigational Sciences. Essentially a secondary school, that institution
stressed the teaching of arithmetic, geometry, trigonometry, astronomy,
and geography. The number of its students reached five hundred by 1715,
and two elementary schools were founded to prepare Russian boys to
enter it. In 1715 a Naval Academy for three hundred pupils opened in St.
Petersburg. Moscow, in turn, received an artillery and an engineering
academy of the same general pattern. Some other special schools, such as
the so-called "admiralty" and "mathematical" ones, also appeared in the
course of the reign. In 1716, in an attempt to develop a broader educational system, the government opened twelve elementary "cypher" schools
in provincial towns. By 1723 their number had increased to forty-two. In
1706 a medical school with a student body of fifty began instruction in
Moscow; in 1709 another medical school, this time with thirty students,
started functioning in St. Petersburg. Peter I also organized small classes
to study such special subjects as Chinese and Japanese and the languages
of some non-Russian peoples within the empire. In addition to establishing state schools, the reformer tried to improve and modernize those of the
Church. Finally, education in Russia expanded by means of private schools
which began to appear in the course of his reign.

Peter the Great's measures to promote enlightenment in Russia also included the founding of a museum of natural science and a large general
library in St. Petersburg. Both were opened free to the public. But the
reformer's most ambitious cultural undertaking was the creation of the
Imperial Academy of Science. Although the Academy came into being
only some months after Peter the Great's death, it represented the realization of a major project of the reformer's last years. The Academy had three
departments, the mathematical, physical, and historical, as well as a sec-

tion for the arts. The academicians gave instruction, and a high school was attached to the Academy to prepare students for this advanced education. Although the Academy operated at first on a small scale and consisted of only seventeen specialists, all of them foreigners, it became before long, as intended, the main directing center of science and scholarship in the Russian Empire. The enormously important Academy of Sciences of the U.S.S.R. stems directly from the Petrine Academy. It has been noted repeatedly, sometimes with an unbecoming derision, that Russia obtained an Academy of Sciences before it acquired elementary schools — a significant comment on the nature of Peter the Great's reforms and the role of the state in eighteenth-century Russian culture.

After the death of Peter the Great, there followed a certain decline in education in Russia. Once the government relaxed its pressure, state schools tended to empty and educational schemes to collapse. Church schools, which were much less dependent on the reformer, survived better. They were to produce many trained Russians, some of whom became prominent in a variety of activities in the eighteenth and subsequent centuries. On the whole, however, Church schools served Church needs, i.e. the training of the clergy, and stood apart from the main course of education in Russia. With the rise of the gentry in the eighteenth century, exclusive gentry schools whose graduates were given certain privileges became increasingly important. Peter the Great's artillery and engineering academies were restricted to members of that class, while new cadet schools were opened under Empress Anne and her successors to prepare sons of the nobility to assume the duties of army officers, in contrast to the first emperor's insistence on rising through the ranks. Home education, often by foreign tutors, also developed among the gentry. Increasing attention was given to good manners and the social etiquette that the Russians began to learn from the West at the time of Peter the Great's reforms: the first emperor had a manual on social etiquette, *A Mirror for Youth,* translated from the German as early as 1717. In the education of the gentry much time and effort were devoted to such subjects as proper bearing in society, fencing, and dancing, as well as to French and sometimes to other foreign languages. As noted in the scheme mentioned earlier, Western manners and fashions came to occupy much of the attention of educated Russians.

While Russian schools showed relatively little vitality or development between the reigns of Peter I and Catherine II, the government did take at least one decisive step forward: in 1755 in Moscow the first Russian university came into existence. Promoted by Ivan Shuvalov and Michael Lomonosov, this first Russian institution of higher learning was to be, all in all, the most important one in the history of the country, as well as a model for other universities. Responsible directly to the Senate and endowed with considerable administrative autonomy, the university possessed three schools: law, medicine, and philosophy. The school of philosophy included both the humanities and sciences, much as reflected in the range of the present-day degree of Doctor of Philosophy. The University of Moscow started with ten professors and some assistants to the professors; of the ten, two were Russians, a mathematician and a rhetorician. In a decade the number of professors about doubled, with Russians constituting approximately half of the total. Originally instruction took place in Latin; but in 1767 Russian began to be used in the university. In 1756 the university started to publish a newspaper, the *Moskovskie Vedomosti* or *Moscow News*. Higher education in Russia, both at the Academy and in the University of Moscow, had a slow and hard beginning, with few qualified students and in general little interest or support. Indeed at one time professors attended one another's lectures! Still, in this field, as in so many others, the eighteenth century bequeathed to its successor the indispensable foundations for further development.

Catherine the Great's reign, or roughly the last third of the century, witnessed a remarkable growth and intensification of Russian cultural life. For instance, we know of 600 different books published in Russia in the reign of Peter the Great, of 2,000 produced between 1725 and 1775, and of 7,500 which came out in the period from 1775 to 1800. Catherine the Great's edict of 1783, licensing private publishing houses, contributed to the trend. The rise of the periodical press proved to be even more striking. Although here too the origins went back to Peter the Great, there was little development until the accession of Catherine II. It was the empress's personal interest in the propagation of her views, together with the interests and needs of the growing layer of educated Russians, that led to the sudden first flowering of Russian journalism. By 1770 some eight periodicals entered the field to comment on the Russian and European scene, criticize the foibles of Russian society, and engage in lively debate with one another, a debate in which Catherine the Great herself took an active part. Societies for the development and promotion of different kinds of knowledge, such as the well-known Free Economic Society, multiplied in Catherine II's reign.

In the sphere of education proper, as in so many other fields, the empress had vast ambitions and plans. Adapting the views of Locke, the Encyclopedists, and Rousseau, Catherine the Great at first hoped to create through education a new, morally superior, and fully civilized breed of people. Education, in this sanguine opinion, could "bestow new existence, and create a new kind of subjects." The empress relied on her close associate and fellow enthusiast of the Enlightenment, Ivan Betsky, to formulate and carry out her educational policy. To be truly transformed, pupils had to be separated from their corrupting environment and educated morally as well as intellectually. Therefore, Catherine the Great and Betsky relied on select boarding schools, including the new Smolny school for girls, the first and the most famous state school for girls in the history of the Russian empire.

Catherine the Great, however, was too intelligent and realistic not to see eventually the glaring weaknesses of her original scheme: boarding schools were very expensive and took care of very few people; moreover, those few, it would seem, could not escape the all-pervasive environment and failed to become paragons of virtue and enlightenment. Therefore, other approaches, broader in scope but more limited in purpose, had to be tried. The empress became especially interested in the system of popular education instituted in the Austrian Empire in 1774 and explained to her by Emperor Joseph II himself. In 1782, following the Austrian monarch's advice, she invited the Serbian educator Theodore Iankovich de Mirievo from Austria and formed a Commission for the Establishment of Popular Schools. The Commission approved Iankovich de Mirievo's plan of a network of schools on three levels and of the programs for the schools. The Serbian educator then concentrated on translating and adapting Austrian textbooks for Russian schools, and also on supervising the training of Russian teachers. A teachers' college was founded in St. Petersburg in 1783. Its first hundred students came from Church schools and were graduated in 1786. In that year a special teachers' seminary began instruction. It was to produce 425 teachers in the course of its fifteen years of existence. Relying on its new teachers, the government opened twenty-six more advanced popular schools in the autumn of 1786 and fourteen more in 1788, all of them in provincial centers. It also proceeded to put popular elementary schools into operation in district towns: 169 such schools, with a total of 11,000 pupils began to function in 1787; at the end of the century the numbers rose to 315 schools and 20,000 students.

Everything considered, Catherine the Great deserves substantial credit in the field of education. Her valuable measures ranged from pioneering in providing education for girls to the institution of the first significant teacher training program in Russia and the spreading of schools to many provincial and district towns. The empress and her advisers, it should be noted, wanted to extend enlightenment to the middle class and hoped to see an educated Third Estate arise in their homeland. Furthermore, as we know, the government's limited efforts did not represent all of Russian education. Church schools continued, and education of the gentry advanced in the last third of the eighteenth century. When not attending exclusive military schools of one kind or another, sons of the nobility received instruction at home by private teachers, augmented, with increasing frequency, by travel abroad. The French Revolution, while it led to the exclusion of France from Russian itineraries, brought a large number of French émigrés to serve as tutors in Russia. But in education, perhaps even more than in other fields, the division of Russian society was glaringly evident. Although the eighteenth century witnessed the rise of modern Russian schools and modern Russian culture, virtually none of this affected the peasants, that is, the great bulk of the people.

Language

The adaptation of the Russian language to new needs in the eighteenth century constituted a major problem for Russian education, literature, and culture in general. It will be remembered that on the eve of Peter the Great's reforms Russian linguistic usage was in a state of transition as everyday Russian began to assert itself in literature at the expense of the archaic, bookish, Slavonicized forms. This basic process continued in the eighteenth century, but it was complicated further by a mass intrusion of foreign words and expressions which came with Westernization and which had to be dealt with somehow. The language used by Peter the Great and his associates was in a chaotic state, and at one time apparently the first emperor wanted to solve the problem by having the educated Russians adopt Dutch as their tongue!

In the course of the century the basic linguistic issues were resolved, and modern literary Russian emerged. The battle of styles, although not entirely over by 1800, resulted in a definitive victory for the contemporary Russian over the Slavonicized, for the fluent over the formal, for the practical and the natural over the stilted and the artificial. Nicholas Karamzin, who wrote in the last decades of the eighteenth and the first of the nineteenth century, contributed heavily to the final decision by effectively using the new style in his own highly popular works. As to foreign words and expressions, they were either rejected or gradually absorbed into the Russian language, leading to a great increase in its vocabulary. The Russian language of 1800 could handle many series of terms and concepts unheard of in Muscovy. That the Russian linguistic evolution of the eighteenth century was remarkably successful can best be seen from the fact that the golden age of Russian literature, still the standard of linguistic and literary excellence in modern Russian, followed shortly after. Indeed Pushkin was born in the last year of the eighteenth century.

The linguistic evolution was linked to a conscious preoccupation with language, to the first Russian grammars, dictionaries, and philological and literary treatises. These efforts, which were an aspect of Westernization, contributed to the establishment of modern Russian literary culture. Lomonosov deserves special praise for the first effective Russian grammar, published in 1755, which proved highly influential. A rich dictionary composed by some fifty authors including almost every writer of note appeared in six volumes in 1789–94. Theoretical discussion and experimentation by Basil Trediakovsky, Lomonosov, and others led to the creation of the now established system of modern Russian versification.

Literature

Modern Russian literature must be dated from Peter the Great's reforms. While, to be sure, the Russian literary tradition goes back to the Kievan age in the *Lay of the Host of Igor* and other works, and even to the prehistoric past in popular song and tale, the reign of the first emperor marked a sharp division. Once Russia turned to the West, it joined the intellectual and literary world of Europe which had little in common with that of Muscovy. In fact, it became the pressing task of educated eighteenth-century Russians to introduce and develop in their homeland such major forms of Western literary expression as poetry, drama, and the novel. Naturally, the emergence of an original and highly creative Russian literature took time, and the slowness of this development was emphasized by the linguistic evolution. The century had to be primarily imitative and in a sense experimental, with only the last decades considerably richer in creative talent. Nevertheless, the pioneer work of eighteenth-century writers made an important contribution to the establishment and development of modern Russian literature.

Antioch Kantemir, 1709–44, a Moldavian prince educated in Russia and employed in Russian diplomatic service, has been called "the originator of modern Russian *belles lettres.*" Kantemir produced original works as well as translations, poetry and prose, satires, songs, lyrical pieces, fables, and essays. Michael Lomonosov, 1711–65, had a greater poetic talent than Kantemir. In literature he is remembered especially for his odes, some of which are still considered classics of their kind, in particular when they touch upon the vastness and glory of the universe. Alexander Sumarokov, 1718–77, a prolific and influential writer, has been honored as the father of Russian drama. In addition to writing tragedies and comedies as well as satires and poetry and publishing a periodical, Sumarokov was the first director of a permanent Russian theater. Sumarokov wrote his plays in the pseudo-classical manner characteristic of the age, and he often treated historical subjects.

The reign of Catherine the Great witnessed not only a remarkable increase in the quantity of Russian literature, but also considerable improvement in its quality. Two writers of the period, not to mention Nicholas Karamzin who belongs to the nineteenth century as well as to the eighteenth, won permanent reputations in Russian letters. The two were Gabriel Derzhavin and Denis Fonvizin. Derzhavin, 1743–1816, can in fairness be called Catherine the Great's official bard: he constantly eulogized the vain empress and such prominent Russians of her reign as Potemkin and

Suvorov. Like most court poets, he wrote too much; yet at his best Derzhavin produced superb poetry, both in his resounding odes, exemplified by the celebrated "God," and in some less-known lyrical pieces. The poet belonged to the courtly world that inspired him and even served as Minister of Justice in the government of Alexander I.

Fonvizin, 1745–92, has received wide acclaim as the first major Russian dramatist, a writer of comedies to be more exact. Fonvizin's lasting fame rests principally on a single work, the comedy whose title has been translated as *The Minor,* or *The Adolescent* — in Russian *Nedorosl.* Pseudo-classical in form and containing a number of artificial characters and contrived situations, the play, nevertheless, achieves a great richness and realism in its depiction of the manners of provincial Russian gentry. The hero of the comedy, the lazy and unresponsive son who, despite his reluctance, in the changing conditions of Russian life has to submit to an elementary education, and his doting, domineering, and obscurantist mother are apparently destined for immortality. In addition to *The Minor,* Fonvizin translated, adapted, or wrote some other plays, including the able comedy *The Brigadier* in which he ridiculed the excessive admiration of France in Russia; he also produced a series of satirical articles and a noteworthy sequence of critical letters dealing with his impressions of foreign countries.

While classicism, or neo-classicism, represented the dominant trend in the European literature of the eighteenth century, other currents also came to the fore toward the end of that period. Again, the Russians eagerly translated, adapted, and assimilated Western originals. Nicholas Karamzin, 1766–1826, can be called the founder of sentimentalism in Russian literature. His sensitive and lacrimose *Letters of a Russian Traveler* and his sensational although now hopelessly dated story *Poor Liza,* both of which appeared at the beginning of the last decade of the century, marked the triumph of the new sensibility in Russia. Karamzin, it might be added, succeeded also as a publisher, and generally helped to raise the stature of the professional writer in Russian life. Other pre-romantic trends in the writings of many authors — rather prominent as Rogger's study indicates — included a new interest in folklore, a concern with the history of the country, and an emphasis on things Russian as opposed to Western.

Social Criticism

The history of ideas cannot be separated from literary history, least of all in the Russian setting. Social criticism constituted the dominant content of both in eighteenth-century Russia. This didactic tendency, highly characteristic of the Age of Reason, found special application in Russia, where so much had to be learned so fast. Kantemir, "the originator of modern

Russian literature," wrote satires by preference, while his translations included Montesquieu's *Persian Letters*. Satire remained a favorite genre among Russian writers of the eighteenth century, ranging from the brilliant comedies of Fonvizin to the pedestrian efforts of Catherine the Great and numerous other aspiring authors. The same satire, the same social criticism inspired journalism; in fact, no clear line divided the two fields. Russian writers and publicists inveighed against the backwardness, boorishness, and corruption of their countrymen, and they neglected no opportunity to turn them toward civilization and light. At the same time they noticed that on occasion "the ungainly beasts" began to admire the West, in particular France, too much and to despise their own country, and that in turn was satirized and denounced throughout the century.

The spirit of criticism developed especially in the reign of Catherine the Great and was aided by the sponsorship and example of the empress herself. Indeed, she gave, so to speak, official endorsement to the far-reaching critiques and views of the *philosophes*. The Free Economic Society even awarded its first prize to a work advocating the abolition of serfdom. A certain kind of Russian Voltairianism emerged, combining admiration for the sage of Ferney with a skeptical attitude toward many aspects of Russian life. Although some historians dismiss this Voltairianism as a superficial fashion, it no doubt served for some Russians as a school of criticism, all the more so because it fitted extremely well the general orientation of the Enlightenment.

Freemasonry became another school of criticism and thought for the Russians, and a more complicated one, for it combined disparate doctrines and trends. It came to Russia, of course, again from the West, from Great Britain, Germany, Sweden, and France. Although the first fraternal lodges appeared at the time of Empress Elizabeth, the movement became prominent only in the reign of Catherine the Great. At that time it consisted of about one hundred lodges located in St. Petersburg, Moscow, and some provincial towns and of approximately 2,500 members, almost entirely from the gentry. In addition to the contribution made by Freemasonry to the life of polite society, which constituted probably its principal attraction to most members, specialists distinguish two main trends within that movement in eighteenth-century Russia: the mystical, and the ethical and social. The first concentrated on such commendable but elusive and essentially individual goals as contemplation and self-perfection. The second reached out to the world and thus constituted the active wing of the movement. Socially oriented Freemasons centered around the University of Moscow. They engaged in education and publishing, establishing a private school and the first large-scale program of publication in Russia outside of the government. They contributed heavily to the periodical literature and its social criticism. Nicholas Novikov, 1744–1818, perhaps the most active publicist of Catherine the Great's reign, led the group which included several other outstanding people.

Of the many things to be criticized in Russia, serfdom loomed largest. Yet that institution was both so well accepted and so fundamental to Russian life that few in the eighteenth century dared challenge it. Catherine the Great herself, after some vague preliminary wavering, came out entirely on the side of the gentry and its power over the peasants. Numerous writers criticized certain individual excesses of serfdom, such as the cruelty of one master or the wastefulness of another, but they did not assail the system itself. Novikov and a very few others went further: their image of serf relations could not be ascribed to individual aberrations, and it cried for reform. Still, it remained to Alexander Radishchev to make the condemnation of serfdom total and unmistakably clear. It was Radishchev's attack on serfdom that broke through the veneer of cultural progressivism and well-being, typical of the reign of Catherine the Great, and served as the occasion for a sharp break between the government and the radical or even just liberal intellectuals.

Radishchev, 1749–1802, was educated at the University of Leipzig as well as in Russia and acquired a wide knowledge of eighteenth-century thought. In particular, he experienced the impact of Rousseau, Mably, and the entire egalitarian, and generally more radical, tendency of the later Enlightenment. A member of the gentry, an official, and a writer of some distinction, Radishchev left his mark on Russian history with the publication in 1790 of his stunning *Journey from Petersburg to Moscow*. Following the first section called "the departure," twenty-odd chapters of that work, named after wayside stations, depicted specific and varied horrors of serfdom. The panorama included such scenes as serfs working on a Sunday, because they could till their own land only on that day, the rest of the week being devoted to the barshchina; the sale at an auction of members of a single family to different buyers; and the forced arrangement of marriages by an overly zealous master. Moreover, Radishchev combined his explicit denunciation of serfdom with a comprehensive philosophical, social, political, and economic outlook, reflected in the *Journey* and in other writings. He assailed Russian despotism and administrative corruption and suggested instead a republic with full liberties for the individual. And he actually drew up a plan for serf emancipation and an accompanying land settlement.

Radishchev's philippic resulted in his being sentenced to death, changed fortunately to ten-year imprisonment in Siberia. Frightened by the French Revolution, Catherine the Great finally turned against the ideas of the Enlightenment, which she had done so much to promote. Novikov and his fellow Masons in Moscow also suffered, and their educational work came to an abrupt end. Edicts against travel and other contacts with the revolutionary West multiplied, reaching absurd proportions in Paul's reign. But the import of the issue proved to be even more profound than a reaction to the French Revolution. Until 1790 the state led Russia on the path of enlightenment. From that year on, it began to apply the brakes. Radishchev's *Journey* meant the appearance of a radical intellectual protest in Russia, a foretaste of the radical intelligentsia.

Science and Scholarship

While secular philosophy, literary debates, and social criticism stood in the center of the Enlightenment, other aspects of culture also developed at that time. Following the West as usual, Russia proceeded to assimilate modern science, scholarship, and the arts. Science took root slowly in Russia, and for a number of decades the Russians had relatively little in this field, except a number of scholars invited from abroad, some of them of great merit. But — to underline the danger of generalizations and schemes — the one great Russian scientist of the eighteenth century appeared quite early on the scene; moreover, his achievements were very rarely if ever to be matched in the entire annals of science in Russia. This extraordinary man was Michael Lomonosov, born in a peasant family in the extreme northern province of Archangel and educated both in Russia and for five years in Germany, most of that time at Marburg University. Lomonosov, 1711–65, who has already been mentioned as a pioneer grammarian, an important literary scholar, and a gifted poet, was also a chemist, a physicist, an astronomer, a meteorologist, a geologist, a mineralogist, a metallurgist, a specialist in navigation, a geographer, an economist, and a historian, as well as a master of various crafts and a tireless inventor. Pushkin was to refer to him, appropriately, as the first Russian university. In considering the work of Lomonosov, we should remember that he lived before the time of extreme scientific specialization, when a single mind still could master many disciplines, and indeed advance them. Lomonosov represented, in other words, the Russian counterpart of the great encyclopedic scholars of the West.

Lomonosov probably did his best work in chemistry, physics, and the border area between these two sciences. In fact, he developed and in 1751 taught the first course of physical chemistry in the world, and in 1752 he published a textbook in that field. The Russian scientist's other most outstanding achievements included the discovery of the law of the preservation of matter and of energy long before Lavoisier, the discovery of atmosphere on Venus, brilliant studies in electricity, the theory of heat, and optics, and the establishment of the nature and composition of crystals, charcoal, and black earth. Lomonosov's scientific work unfortunately proved far ahead of his time, especially in Russia, where it found no followers and was fully rediscovered only by Menshutkin and other twentieth-century scholars.

Although Lomonosov remained essentially an isolated individual, the eighteenth century was also noteworthy in Russian history for large-scale, organized scientific effort. That effort took the form of expeditions to discover, explore, or study distant areas of the empire and sometimes neighboring seas and territories. Geography, geology, mineralogy, botany, zoology, ethnography, and philology, as well as some other disciplines, all profited from these well-thought-out and at times extremely daring undertakings. Begun by Peter the Great, the expeditions led to important results even in the first half of the century. For example, Alaska was discovered in 1732. The so-called First Academic Expedition, which lasted from 1733 to 1742 and included 570 participants, successfully undertook the mammoth task of mapping and exploring the northern shore of Siberia. Numerous expeditions, often of great scholarly value, followed later in the century. Peter-Simon Pallas, a versatile and excellent German scientist in Russian service, deserves special credit for his part in them.

The Russians also applied themselves to what can be called the social sciences and the humanities. Mention has already been made of new Russian scholarship in connection with language and literature. Modern Russian study of economics dates from Peter the Great. Ivan Pososhkov, a wealthy peasant, an extraordinary critic and admirer of the first emperor, and the author of a remarkable treatise, *Books about Poverty and Wealth,* has often been cited as its originator. Pososhkov found his inspiration in Peter the Great's reforms and in the issues facing Russia, not in Western scholarship of which he was ignorant. The study of history too developed quickly in Russia, with the Russians profiting throughout the century from the presence of foreign scholars, such as the outstanding German historian August-Ludwig von Schlözer. Eighteenth-century Russian historians included an important administrator and collaborator of Peter the Great, Basil Tatishchev; Prince Michael Shcherbatov, who argued the case for the rights of the gentry in Catherine the Great's Legislative Commission and produced a number of varied and interesting works; and Major-General Ivan Boltin. From the time of Tatishchev, Russian historians understandably tended to emphasize the role of the monarch and the state.

The Arts. Concluding Remarks

Architecture flourished in eighteenth-century Russia because of the interest and liberality of Peter the Great and his successors. Catherine the Great proved to be a passionate builder, and the same was true of Paul, as well as of Alexander I and Nicholas I in the nineteenth century. St. Petersburg, which rose from the swamps to become one of the truly beautiful and impressive cities of the world, remains the best monument of this imperial devotion to architecture. Baroque at the beginning of the century and the neoclassical style toward the end of the century dominated European and Russian architecture. The builders in the empire of the Romanovs included a number of gifted foreigners, notably Count Bartolomeo Rastrelli, who came as a boy from Italy to Russia, when his sculptor father was invited by Peter the Great, and who erected the Winter Palace and the Smolny Institute in St. Petersburg and the great palace in Tsarskoe Selo — now Pushkin — together with many other buildings. Some excellent Russian architects, such as Basil Bazhenov and Matthew Kazakov, emerged in the second half of the century.

Other arts also grew and developed. In the 1750's the art section of the Academy of Sciences became an independent Academy of Arts. In the field of painting, portrait painting fared best, as exemplified by the work of Dmitrii Levitsky, 1735–1822 — the son, incidentally, of a priest who painted icons. Fedot Shubin, 1740–1805, like Lomonosov a peasant from the extreme north, was the first important Russian sculptor. Having received his initial training in his family of bone carvers, he went on to obtain the best artistic education available in St. Petersburg, Italy, and France, and to win high recognition abroad as well as at home. Shubin's sculptures are characterized by expressiveness and realism.

The eighteenth century also witnessed the appearance in Russia of modern music, notably the opera, as well as ballet and the theater. All of these arts came from the West and gradually, in the course of the century, entrenched themselves in their new locale. All were to be greatly enriched in the future by Russian genius. As to theater, while Peter the Great invited German actors and later sovereigns sponsored French and Italian troupes, a native Russian theater became established only in the 1750's. Its creator was a merchant's son, Theodore Volkov, who organized a successful theater in Iaroslavl on the Volga and was then requested to do the same in the capital. Catherine the Great herself contributed to the new repertoire of Russian plays. By the end of the century Russia possessed several public theaters, a theatrical school, and a periodical, *The Russian Theater,* which began to appear in 1786. Furthermore, theater had won popularity among the great landlords, who maintained some fifteen private theaters in Moscow alone.

Russian culture of 1800 bore little resemblance to that of 1700. In brief, Russia — that is, upper-class, educated Russia — had become Westernized. The huge effort to learn that dominated Russian culture in the eighteenth century was to bear rich fruit. Many Russians, however, from the time of Peter the Great to the present, have worried about this wholesale borrowing from the West. From Pososhkov and Lomonosov to the latest Soviet specialists they have tried to minimize the role of the West and to emphasize native Russian achievements. Unusual among the better pre-revolutionary scholars, this view eventually received a heavy official endorsement in the Soviet Union. As a result, many Soviet discussions of Russia and the West in the eighteenth century became ridiculous. Although common, wounded national pride is an unfortunate and usually unjustified sentiment. To be sure, the Russians not only borrowed from the West, but also assimilated Western culture. For that matter, only two major European countries, England and France, can claim a full continuity of intellectual, literary, and cultural development, and even they, of course, experienced any number of foreign influences. Besides — and to conclude — while the origin of a heritage is important to the historian, its use may well be considered still more significant. We have seen something of that use in this chapter, and shall see much more of it in our subsequent discussions of Russian culture.

THE REIGN OF ALEXANDER I, 1801–25

The book of a brilliant, magnanimous reign opened! Victory is inscribed in it: the conquest of Finland, Bessarabia, Persian territories, the defeat of Napoleon and of the armies of twenty nations, the liberation of Moscow, the capture of Paris twice, the annexation to Russia of the Kingdom of Poland. Magnanimity is inscribed in it: the liberation of Europe, the placing of the Bourbons on the thrones of France, Spain, and the Two Sicilies, the Holy Alliance, the sparing of Paris. There love of learning pointed to the creation of six universities, an academy, a lyceum. There mercy wrote actions worthy of it: rescue of the unfortunate ones, generous pardon of criminals and even of those who insulted His Majesty. There justice marked the affirmation of the rights of the gentry and the law giving the accused full freedom to defend themselves. All the virtues which ennoble man and adorn a tsar mark in this book the reign of Alexander. How many sovereigns of this earth stood impressive in their power and glory, but were there many who, like him, combined humility with power and goodness to enemies with the victories? Alexander of Greece! Caesar of Rome! your laurels are spattered with blood, ambition unsheathed your sword. Our Alexander triumphed virtuously: he wanted to establish in the world the peace of his own soul.

FEDOROV

If, during the two centuries which divide the Russia of Peter the Great from the Bolshevik revolution, there was any period in which the spell of the authoritarian past might have been overcome, the forms of the state liberalized in a constitution, and the course of Russian development merged with the historic currents of the west, it is the earlier part of the reign of Alexander I. Or so, for a moment, one is tempted to think.

CHARQUES

ALEXANDER I WAS twenty-three years old when, following the deposition and assassination of his father, Emperor Paul, he ascended the Russian throne. The new monarch's personality and manner of dealing with other men had thus already been formed, and it is the psychology of the emperor that has fascinated those who became acquainted with him, both his contemporaries and later scholars. Moreover, there seems to be little agreement about Alexander I beyond the assertion that he was "the most complex and most elusive figure among the emperors of Russia." This unusual sovereign has been called "the enigmatic tsar," a sphinx, and "crowned Hamlet," not to mention other similarly mystifying appellations. Striking contradictions or alleged contradictions appear in the autocrat's character and activities. Thus Alexander I was hailed as a liberal by many men, Thomas Jefferson among them, and denounced as a reactionary by numerous others, including Byron. He was glorified as a pacifist, the originator of the Holy Alliance, and in general a man who did the utmost to establish peace and a Christian brotherhood on earth. Yet this "angel" — an epithet frequently applied to Alexander I, especially within the imperial family and in court circles — was also a drill sergeant and a parade-ground enthusiast. Some students of Alexander I's foreign policy have concluded that the tsar was a magnificent and extremely shrewd diplomat, who consistently bested Napoleon. Napoleon himself, it might be added, called him "a cunning Byzantine." But other scholars, again on good evidence, have emphasized the Russian ruler's mysticism and even his growing detachment from reality.

Various elements in the emperor's background have been cited to help account for his baffling character. There was, to begin with, Alexander's difficult childhood and boyhood, in particular his ambiguous relations with his father, Paul, and his grandmother, Catherine the Great, who hated each other. Alexander spent more time with Catherine than with his parents, and he learned early the arts of flattery, dissimulation, and hypocrisy, or at least so his boyhood letters indicate. The empress took a great liking to Alexander from the very beginning and apparently wanted to make him her successor, bypassing Paul. Quite possibly only the suddenness of her death upset this plan. Education also influenced the future emperor's character, views, and activities. Catherine the Great took a personal interest in Alexander's upbringing, which was guided by the ideas of the Enlightenment. A prominent Swiss *philosophe* and liberal, Frédéric-César de LaHarpe, acted as the grand duke's chief tutor and became his close friend. Yet LaHarpe's instruction, full of progressive ideas and humane sentiments, had its disadvantages. LaHarpe, that "very liberal and garrulous French booklet," as Kliuchevsky described the tutor, and his teaching had little in common with Russian reality. The contrast between theory and practice characteristic of Alexander I's reign has been derived by some scholars from this one-sided education. The circumstances of Alexander I's accession to the throne have also been analyzed for their effect on the sovereign's character and rule. Alexander found himself in a precarious position during Paul's reign, especially because Paul thought of divorcing his wife and of disinheriting Alexander and his other sons by her. The young grand duke almost certainly knew of the conspiracy against his father, but the murder of Emperor Paul came to him apparently as a surprise and a shock. Certain critics attribute to the tragedy of his accession Alexander I's strong feeling of guilt and his later mysticism and lack of balance.

Behind Alexander I's reactions to particular incidents and situations of his life there was, of course, his basic character. Alexander I remains a mystery in the sense that human personality has not been and perhaps cannot be fully explained. Yet his psychological type is not especially uncommon, as psychiatrists, psychologists, and observant laymen attest. The emperor belonged with those exceedingly sensitive, charming, and restless men and women whose lives display a constant irritation, search, and disappointment. They lack balance, consistency, and firmness of purpose. They are contradictory. Alexander I's inability to come to terms with himself and pursue a steady course explains his actions much better, on the whole, than do allegations of cynicism or Machiavellianism. As is characteristic for the type, personal problems grew with the passage of time: the emperor became more and more irritable, tired, and suspicious of people, more dissatisfied with life, more frantically in search of a religious or mystical answer; he even lost some of his proverbial charm. The autocrat died in 1825, only forty-eight years old. However, as if to continue the mystery of Alexander I, some specialists insist that he did not die, but escaped from the throne to live in Siberia as a saintly hermit Theodore, or Fedor, Kuzmich. Based on such circumstantial evidence as the emperor's constant longing to shed the burdens of his office, and a court physician's refusal to sign the death warrant, this supposition needs further proof, although it cannot be entirely dismissed. Suicide might offer another explanation for a certain strangeness and confusion associated with the sovereign's death.

Liberalism and Reform

The Russians rejoiced at the accession of Alexander I. In place of an exacting and unpredictable tyrant, Paul, they obtained a young ruler of supreme charm and apparently enormous promise. Alexander I seemed to represent the best of the Enlightenment — that humaneness, progressiveness, affirmation of human dignity, and freedom, which educated Russians, in one way or another, fervently desired. The new emperor's first acts confirmed the expectations. An amnesty restored to their former positions up to twelve thousand men dismissed by Paul; the obnoxious restrictions on travel abroad and on the entry into Russia of foreigners as well as of foreign books and periodicals were abrogated; the censorship was relaxed, and private publishing houses were again allowed to open; torture in investigation was abolished; and the charters granted by Catherine the Great to the gentry and to towns regained their full force. But, of course, these welcome measures marked at best only the beginning of a liberal program. The key issues to be faced included serfdom and autocracy, together with the general backwardness of the country and the inadequacy and corruption of its administrative apparatus. In contrast to Catherine the Great and Paul, Alexander I brought these problems up for consideration, although, as we shall see, the tangible results of his efforts proved to be slight. The reign of Alexander I contained two liberal periods, from 1801 to 1805 and from 1807 to 1812, each, incidentally, followed by war with France.

The first period of reform, following immediately upon Alexander I's acquisition of the crown, was the most far-ranging in purpose and the most hopeful. The new emperor decided to transform Russia with the help of four young, cultivated, intelligent, and liberal friends, the so-called Unofficial Committee. The members of the committee, Nicholas Novosiltsev, Count Paul Stroganov, Count Victor Kochubey, and a Polish patriot Prince Adam Czartoryski, reflected the enlightened opinion of the period, ranging from Anglophilism to Jacobin connections. While they could not be classified as radicals or hotheads, the four did represent a new departure after Paul's administration. The emperor spoke of them jokingly as his "Committee of Public Safety," a reference to the French Revolution which would have made his predecessors shudder. He met with the committee informally and frequently, often daily over coffee.

Our information about the work of the Unofficial Committee — which includes Stroganov's notes on the meetings — suggests that at first Alexander I intended to abolish autocracy and serfdom. However, the dangers and difficulties associated with these issues, as well as the unpreparedness for reform of the administration and the mass of people, quickly became apparent. Serfdom represented, so to speak, the greatest single interest in the empire, and its repeal was bound to affect the entire Russian society, in particular the extremely important gentry class. As to autocracy, the emperor himself, although at one time he had spoken of a republic, hesitated in practice to accept any diminution of his authority. Characteristically, he became disillusioned and impatient with the proceedings and called the Unofficial Committee together less and less frequently. The war of 1805 marked the conclusion of its activities. Russia, thus, went unregenerated and unreformed. Even more limited projects such as the proclamation of a Russian charter of rights failed to be translated into practice.

Although the grand scheme of reform failed, the first years of Alexander's reign witnessed the enactment of some important specific measures. For example, the Senate was restored, or perhaps promoted, to a very high position in the state: it was to be the supreme judicial and administrative institution in the empire, and its decrees were to carry the authority of those of the sovereign, who alone could stop their execution. Peter the Great's colleges, which had a checkered and generally unhappy history in the eighteenth century, were gradually replaced in 1802 and subsequent years by ministries, with a single minister in charge of each. At first there were eight: the ministries of war, navy, foreign affairs, justice, interior, finance, commerce, and education. Later the ministry of commerce was abolished, and the ministry of police appeared.

The government even undertook some limited social legislation. In 1801 the right to own estates was extended from the gentry to other free Russians. In 1803 the so-called "law concerning the free agriculturists" went into effect. It provided for voluntary emancipation of the serfs by their masters, assuring that the emancipated serfs would be given land and establishing regulations and courts to secure the observance of all provisions. The newly emancipated serfs were to receive in many respects the status of state peasants, but, by contrast with the latter, they were to enjoy stronger property rights and exemption from certain obligations. Few landlords, however, proved eager to free their peasants. To be more exact, under the provisions of the law concerning the free agriculturists from the time of its enactment until its suspension more than half a century later on the eve of "the great reforms," 384 masters emancipated 115,734 working male serfs together with their families. It may be added that Druzhinin and other Soviet scholars have disproved the frequently made assertion that Alexander I gave no state peasants, with state lands, into private ownership and serfdom.

Russian backwardness and ignorance became strikingly apparent to the monarch and his Unofficial Committee as they examined the condition of the country. Education, therefore, received a high priority in the official plans and activities of the first years of the reign. Fortunately too this effort did not present quite the dangers and obstacles that were associated with the issues of serfdom and autocracy. Spending large sums of money on education for the first time in Russian history, Alexander I founded several universities to add to the University of Moscow, forty-two secondary schools, and considerable numbers of other schools. While education in Russia during the first half of the nineteenth century will be discussed in a later chapter, it should be noted here that Alexander I's establishment of institutions of learning and his entire school policy were distinctly liberal for his time. Indeed, they have been called the best fruits of the monarch's usually hesitant and brittle liberalism.

The second period of reform in Alexander I's reign, 1807–12, corresponded to the French alliance and was dominated by the emperor's most remarkable assistant, Michael Speransky. Speransky, who lived from 1772 to 1839, was fully a self-made man. In contrast to the members of the Unofficial Committee as well as to most other associates of the sovereign, he came not from the aristocracy but from poor village clergy. It was Speransky's intelligence, ability to work, and outstanding administrative capacity that made him for a time Alexander I's prime minister in fact, if not in name, for no such formal office then existed. As most specialists on Speransky believe, that unusual statesman sought to establish in Russia strong monarchy firmly based on law and legal procedure, and thus free from arbitrariness, corruption, and confusion. In other words, Speransky found his inspiration in the vision of a *Rechtsstaat,* not in advanced liberal or radical schemes. Still, Raeff, the latest major author on the subject, goes too far when he denies that the Russian statesman was at all liberal. In Russian conditions Speransky's views were certainly liberal, as his contemporaries fully realized. Furthermore, they could have been developed more liberally, if the opportunities had presented themselves.

In 1809, at the emperor's request, Speransky submitted to him a thorough proposal for a constitution. In his customary methodical manner, the statesman divided the Russians into three categories: the gentry; people of "the middle condition," that is, merchants, artisans, and peasants or other small proprietors who owned property of a certain value; and, finally, working people, including serfs, servants, and apprentices. The plan also postulated three kinds of rights: general civil rights; special civil rights, such as exemption from service; and political rights, which depended on a property qualification. The members of the gentry were to enjoy all the rights. Those belonging to the middle group received general civil rights and political rights when they could meet the property requirement. The working people too obtained general civil rights, but they clearly did not own enough to participate in politics. Russia was to be reorganized on four administrative levels: the *volost* — a small unit sometimes translated as "canton" or "township" — the district, the province, and the country at large. On each level there were to be the following institutions: legislative assemblies — or *dumy* — culminating in the state duma for all of Russia; a system of courts, with the Senate at the apex; and administrative boards, leading eventually to the ministries and the central executive power. The state duma, the most intriguing part of Speransky's system, showed the statesman's caution, for in addition to the property restriction imposed on its electorate, it depended on a sequence of indirect elections. The assemblies of the volosti elected the district assemblymen, who elected the provincial assemblymen, who elected the members of the state duma, or national assembly. Also the activities of the state duma were apparently to be rather narrowly restricted. But, on the other hand, the state duma did provide for popular participation in the legislative process. That, together with Speransky's insistence on the division of functions, strict legality, and certain other provisions such as the popular election of judges, if successfully applied, would have in time transformed Russia. Indeed, it has been observed that Speransky's fourfold proposal of local self-government and a national legislative assembly represented a farsighted outline of the Russian future. Only that future took extremely long to materialize, offering — in the opinion of many specialists — a classic example of too little and too late. Thus Russia received district and provincial self-government by the so-called zemstvo reform of 1864, a national legislature, the Duma, in 1905–06, and volost self-government in 1917.

In 1809 and the years following, Alexander I failed to implement Speransky's proposal. The statesman's fall from power in 1812 resulted from the opposition of officialdom and the gentry evoked by his measures and projects in administration and finance, from the emperor's fears, suspicions, and vacillations, and also from the break with Napoleon, Speransky having been branded a Francophile. Although Speransky was later to return to public office and accomplish further useful and important work, he never again had the opportunity to suggest fundamental reform on the scale of his plan of 1809. The second liberal period of Alexander I's reign, then, like the first, produced no basic changes in Russia.

Yet, again like the first, the second liberal period led to some significant legislation of a more limited nature. In 1810, on the advice of Speransky — actually this was the only part of the statesman's plan that the monarch translated into practice — Alexander I created the Council of State modeled after Napoleon's *Conseil d'Etat,* with Speransky attached to it as the Secretary of State. This body of experts appointed by the sovereign to help him with the legislative work in no way limited the principle of autocracy; moreover, the Council tended to be extremely conservative. Still, it clearly reflected the emphasis on legality, competence, and correct procedure so dear to Speransky. And, as has been noted for the subsequent history of the Russian Empire, whereas "all the principal reforms were passed by regular procedure through the Council of State, nearly all the most harmful and most mischievous acts of succeeding governments were, where possible, withdrawn from its competence and passed only as executive regulations which were nominally temporary." Speransky also reorganized the ministries and added two special agencies to the executive, one for the supervision of government finance, the other for the development of transport. A system of annual budgets was instituted, and other financial measures were proposed and in part adopted. Perhaps still more importantly, Speransky did yeoman's service in strengthening Russian bureaucracy by introducing something in the nature of a civil service examination and trying in other ways to emphasize merit and efficient organization.

Speransky's constitutional reform project represented the most outstanding but not the only such plan to come out of government circles in the reign of Alexander I. One other should be noted here, that of Novosiltsev. Novosiltsev's *Constitutional Charter of the Russian Empire* emphasized very heavily the position and authority of the sovereign and bore strong resemblance to Speransky's scheme in its stress on legality and rights and its narrowly based and weak legislative assembly. Novosiltsev differed, however, from Speransky's rigorous centralism in allowing something to the federal principle: he wanted the Russian Empire, including Finland and

Russian Poland, to be divided into twelve large groups of provinces which were to enjoy a certain autonomy. The date of Novosiltsev's project deserves attention: its second and definitive version was presented to Alexander I in 1820, late in his reign. Furthermore, the monarch not only graciously accepted the plan, but — it has been argued — proceeded to implement it in small part. Namely, by combining several provinces, he created as a model one of the twelve units proposed by Novosiltsev. Only after Alexander I's death in 1825 was Novosiltsev's scheme completely abandoned, and the old system of administration re-established in the experimental provinces. The story of Novosiltsev's *Charter,* together with certain other developments, introduces qualifications into the usual sharp division of Alexander I's reign into the liberal first half and the reactionary second half, and suggests that a constitution remained a possible alternative for Russia as long as "the enigmatic tsar" presided over its destinies.

Russian Foreign Policy, 1801–12

While the first part of Alexander's rule witnessed some significant developments in internal affairs, it was the emperor's foreign policy that came to occupy the center of the stage. Diplomacy and war in the early years of Alexander I's reign culminated in the cataclysmic events of 1812.

At the beginning of Alexander's reign, peaceful intentions prevailed. After succeeding Paul, who had both fought France and later joined it against Great Britain, the new emperor proclaimed a policy of neutrality. Yet Russia could not long stay out of conflicts raging in Europe. A variety of factors, ranging from the vast and exposed Western frontier of the empire to the psychological involvement of the Russian government and educated public in European affairs, determined Russian participation in the struggle. Not surprisingly, Alexander I joined the opponents of France. Economic ties with Great Britain, and traditional Russian friendship with Austria and Great Britain, together with the equally traditional hostility to France, contributed to the decision. Furthermore, Alexander I apparently came early to consider Napoleon as a menace to Europe, all the more so because the Russian sovereign had his own vision of a new European order. An outline of the subsequent Holy Alliance and concert of Europe, without the religious coloration, can be found in the instructions issued in 1804 to the Russian envoy in Great Britain.

The War of the Third Coalition broke out in 1805 when Austria, Russia, and Sweden joined Great Britain against France and its ally, Spain. The combined Austrian and Russian armies suffered a crushing defeat at the hands of Napoleon on December 2, 1805, at Austerlitz. Although Austria was knocked out of the war, the Russians continued to fight and in 1806 even obtained a new ally, Prussia. But the French armies, in a nineteenth-century version of the *Blitzkrieg,* promptly destroyed the Prussian forces in the battles of Jena and Auerstädt, and, although they could not destroy the Russians, finally succeeded in inflicting a major defeat on them at Friedland. The treaties of Tilsit between France and Russia and France and Prussia followed early in July 1807. The Franco-Prussian settlement reduced Prussia to a second-rate power, saved from complete destruction by the insistence of the Russian sovereign. The agreement between France and Russia was a different matter, for, although Alexander I had to accept Napoleon's redrawing of the map of Europe and even had to support him, notably against Great Britain, Russia emerged as the hegemon of much of eastern Europe and the only major power on the continent other than France.

It was the temporary settlement with France that allowed the Russians to fight several other opponents and expand the boundaries of the empire in the first half of Alexander's reign. In 1801 the eastern part of Georgia, an ancient Orthodox country in Transcaucasia, joined Russia, and Russian sway was extended to western Georgia in 1803–10. Hard-pressed by their powerful Moslem neighbors, the Persians and the Turks, the Georgians had repeatedly asked and occasionally received Russian aid. The annexation of Georgia to Russia thus represented in a sense the culmination of a process, and a logical, if by no means ideal, choice for the little Christian nation. It also marked the permanent establishment of Russian authority and power beyond the great Caucasian mountain range.

As expected, the annexation of Georgia by Russia led to a Russo-Persian war, fought from 1804 to 1813. The Russians proved victorious, and by the Treaty of Gulistan Persia had to recognize Russian rule in Georgia and cede to its northern neighbor the areas of Daghestan and Shemakha in the Caucasus. The annexation of Georgia also served as one of the causes of the Russo-Turkish war which lasted from 1806 to 1812. Again, Russian troops, this time led by Kutuzov, scored a number of successes. The Treaty of Bucharest, hastily concluded by Kutuzov on the eve of Napoleon's invasion of Russia, added Bessarabia and a strip on the eastern coast of the Black Sea to the empire of the Romanovs, and also granted Russia extensive rights in the Danubian principalities of Moldavia and Wallachia. Finally, in 1808–09 Alexander I fought and defeated Sweden, with the result that the Peace of Frederikshamn gave Finland to Russia. Finland became an autonomous grand duchy with the Russian emperor as its grand duke.

The first half of Alexander's reign also witnessed a continuation of Russian expansion in North America, which had started in Alaska in the late eighteenth century. New forts were built not only in Alaska but also in northern California, where Fort Ross was erected in 1812.

Central Europe 1803

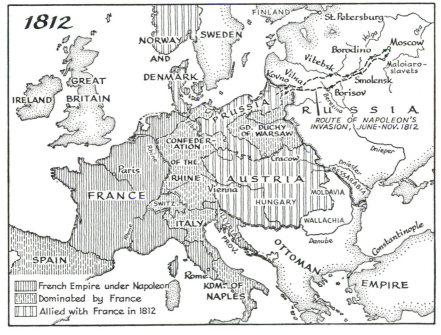

1812

French Empire under Napoleon
Dominated by France
Allied with France in 1812

1812

The days of the Russian alliance with Napoleon were numbered. The agreement that the two emperors reached in Tilsit in 1807, and which was renewed at their meeting in Erfurt in 1808, failed in the long run to satisfy either side. The Russians, who were forced to accept it because of their military defeat, resented Napoleon's domination of the continent, his disregard of Russian interests, and, in particular, the obligation to participate in the so-called continental blockade. That blockade, meant to eliminate all commerce between Great Britain and other European countries and to strangle the British economy, actually helped Russian manufactures, especially in the textile industry, by excluding British competition. But it did hurt Russian exporters and thus the powerful landlord class. Russian military reverses at the hands of the French cried for revenge, especially because they came after a century of almost uninterrupted Russian victories. Also, Napoleon, who had emerged from the fearful French Revolution, who had upset the legitimate order in Europe on an unprecedented scale, and who had even been denounced as Antichrist in some Russian propaganda to the masses, appeared to be a peculiar and undesirable ally. Napoleon and his lieutenants, on their part, came to regard Russia as an utterly unreliable partner and indeed as the last major obstacle to their complete domination of the continent.

Crises and tensions multiplied. The French protested the Russian perfunctory, and in fact feigned, participation in Napoleon's war against Austria in 1809, and Alexander I's failure, from 1810 on, to observe the continental blockade. The Russians expressed bitterness over the development of an active French policy in the Near East and over Napoleon's efforts to curb rather than support the Russian position and aims in the Balkans and the Near East: the French opposed Russian control of the Danubian principalities, objected to Russian bases in the eastern Mediterranean, and would not let the Russians have a free hand in regard to Constantinople and the Straits. Napoleon's political rearrangement of central and eastern Europe also provoked Russian hostility. Notably his deposing the Duke of Oldenburg and annexing the duchy to France, a part of the rearrangement in Germany, offended the Russian sovereign who was a close relative of the duke. Still more ominously for Russia, in 1809 after the French victory over Austria and the Treaty of Schönbrunn, West Galicia was added to the Duchy of Warsaw, a state created by Napoleon from Prussian Poland. This change appeared to threaten in turn the hold of Russia on the vast lands that it had acquired in the partitions of Poland. Even Napoleon's marriage to Marie Louise of Austria added to the tension between Russia and France, because it marked the French emperor's final abandonment of plans to wed a Russian princess, Alexander's sister Anne. Behind specific tensions, complaints, and crises there loomed the fundamental antagonism of two great powers astride a continent and two hostile rulers. In June 1812, having made the necessary diplomatic and military preparations, Napoleon invaded Russia.

France had obtained the support of a number of European states, allies and satellites, including Austria and Prussia: the twelve invading tongues in the popular Russian tradition. Russia had just succeeded in making peace with Turkey, and it had acquired active allies in Sweden and Great Britain. Some 420,000 troops crossed the Russian border with Napoleon to face only about 120,000 Russian soldiers divided into two separate armies, one commanded by Prince Michael Barclay de Tolly and the other by Prince Peter Bagration. Including later reinforcements, an approximate total of 600,000 troops invaded Russia. In addition to its tremendous numbers, Napoleon's army had the reputation of invincibility and what was considered to be an incomparably able leadership. Yet all the advantages were not on one side. Napoleon's *Grande Armée* contained a surprisingly small proportion of veterans. Also, Frenchmen constituted less than half of it, while of the allied troops only the Poles, who fought for a great independent Poland, acquitted themselves with distinction. With the return of the Russian forces from the Turkish front, the arrival of other Russian reinforcements, and the extension of French lines of communication which had to be protected, the invaders gradually lost their numerical superiority. Moreover, on the whole the country rallied solidly behind Alexander I, and the Russian soldiers fought with remarkable tenacity. Indeed, Napoleon's expectations that their early defeats would force the Russians to sue for peace proved groundless. An early and exceptionally cold winter contributed its share to the Russian cause. But, above all, problems of logistics involved in the French campaign turned out to be much more difficult to resolve than Napoleon and his assistants had foreseen.

Napoleon advanced into the heart of Russia along the Vilna-Vitebsk-Smolensk line, just as Charles XII had done a century earlier. The Russians could not stop the invaders and lost several engagements to them, including the bloody battle of Smolensk. However, Russian troops inflicted considerable losses on the enemy, repeatedly escaped encirclement, and continued to oppose French progress. Near Smolensk the two separate Russian armies managed to effect a junction and thus present a united front to the invaders. Under the pressure of public opinion incensed by the continuous French advance, Alexander I put Prince Michael Kutuzov in supreme command of the Russian forces. A disciple of Suvorov, and a veteran of many campaigns, the sixty-seven-year-old Kutuzov did agree in fact with Barclay de Tolly's policy of retreat. Still, he felt it incumbent upon him and his army to fight before surrendering Moscow, and so gave

Napoleon a major battle on the seventh of September near the village of Borodino, seventy-five miles from the great Russian city. The battle of Borodino had few equals in history for the severity of the fighting. Although it lasted but a single day, the Russians suffered 42,000 casualties out of 112,000 combatants, the French and their allies 58,000 out of 130,000. The casualties included scores of generals and thousands of officers, with Prince Bagration and other prominent commanders among the dead or fatally wounded. By nightfall the Russians in the center and on the left flank had been forced to retreat slightly, while they held fast on the right. Kutuzov, however, decided to disengage and to withdraw southeast of Moscow. On the fourteenth of September Napoleon entered the Kremlin.

His expectations of final victory and peace were cruelly deceived. In a rare demonstration of tenacity, Alexander I refused even to consider peace as long as a single French soldier remained on Russian soil, and the country backed its monarch. Far from providing sumptuous accommodations for the French emperor and his army, Moscow, still constructed largely of wood, burned down during the first days of the French occupation. It is possible that Count Theodore Rostopchin, the Russian governor and military commander of the city, deliberately started the conflagration — as most French and some Russian specialists assert — but this remains a disputed issue. Unable to obtain peace from Alexander and largely isolated in the Russian wasteland, Napoleon had to retreat before the onset of winter. The return march of the *Grande Armée,* which started on October 19, gradually became a rout. To begin with, the action of the Russian army at Maloiaroslavets prevented the French from taking a new road through fertile areas untouched by war and forced them to leave the way they had come. As Napoleon's soldiers marched slowly westward winter descended upon them and they were constantly pressed by the pursuing Russian forces — although Kutuzov chose to avoid a major engagement — and harassed by cossacks and other irregulars, including peasant guerrillas. The French and their allies perished in droves, and their discipline began to break down. Late in November, as the remnants of the *Grande Armée* crossed the Berezina River, they were lucky to escape capture through the mistake of a Russian commander. From 30,000 to 50,000 men, out of the total force of perhaps 600,000, finally struggled out of Russia. By the end of the year no foreign soldiers, except prisoners, remained in the country.

The epic of 1812 became a favorite subject for many Russian historians, writers, and publicists, and for some scholars in other lands. Leo Tolstoy's peerless *War and Peace* stands out as the most remarkable, albeit fictionalized, description of the events and human experiences of that cataclysmic year. Other treatments of the subject range from an excellent seven-volume history to some of the best-known poems in Russian literature. While we cannot discuss the poets here, certain conclusions of the historians deserve notice. For example, it has been established that the Russian high command had no over-all "Scythian policy" of retreat with the intention of enticing Napoleon's army deep into a devastated country. The French advance resulted rather from Russian inability to stop the invader and from Napoleon's determination to seize Moscow, which he considered essential for victory. The catastrophic French defeat can be ascribed to a number of factors: the fighting spirit of the Russian army, Kutuzov's wise decisions, Napoleon's crucial mistakes, Alexander's determination to continue the war, the winter, and others. But the breakdown of the transportation and supply of the *Grand Armée* should rank high among the reasons for its collapse. More soldiers of Napoleon died from hunger and epidemics than from cold, for the supply services, handicapped by enormous distances, insecure lines of communication, and bad planning, failed on the whole to sustain the military effort. Finally, it is worth noting that the war of 1812 deserves its reputation in Russian history as a popular, patriotic war. Except for certain small court circles, no defeatism appeared in the midst of the Russian government, educated public, or people. Moreover, the Russian peasants not only fought heroically in the ranks of the regular army but also banded into guerrilla detachments to attack the enemy on their own, an activity unparalleled at the time except in Spain. In fact, as the revising of Tarle's study of the war of 1812 and other works indicate, Soviet historians, while they once neglected it, now tend to overstate the role of the Russian people in the defeat of the French invaders.

Russian Foreign Policy, 1812–25

Alexander I continued the war beyond the boundaries of Russia. Prussia and several months later Austria switched sides to join Russia, Sweden, and Great Britain. The combined forces of Austria, Prussia, and Russia finally scored a decisive victory over Napoleon in the tremendous Battle of Leipzig, known as the "Battle of the Nations," fought from the sixteenth through the nineteenth of October, 1813. Late that year they began to cross the Rhine and invade France. After more desperate fighting and in spite of another display of the French emperor's military genius, the allies entered Paris triumphantly on March 31, 1814. Alexander I referred to that day as the happiest of his life. Napoleon had to abdicate unconditionally and retire to the island of Elba. He returned on March 1, 1815, rapidly won back the French throne, and threatened the allies until his final defeat at Waterloo on the eighteenth of June. The events of the "Hundred Days" moved too quickly for the Russian army to participate in this last war against Napoleon, although, of course, Alexander I was eager to help his allies.

The French emperor's abortive comeback thus failed to undo the new settlement for Europe drawn by the victors at the Congress of Vienna. The Congress, which lasted from September 1814, until the Act was signed on June 8, 1815, constituted one of the most impressive and important diplomatic gatherings in history. Alexander I himself represented Russia and played a leading role at the Congress together with Metternich of Austria, Castlereagh of Great Britain, Hardenberg of Prussia, and, eventually, Talleyrand of France. It must be assumed that the reader has a general knowledge of the redrawing of the political map of Europe and of the colonial settlement that took place in Vienna; however, certain issues in which Russia had a crucial part must be mentioned here. Alexander I wanted to establish a large kingdom of Poland in personal union with Russia, that is, with himself as king; and, by offering to support the Prussian claim to all of Saxony, he obtained Prussian backing for his scheme. Great Britain and Austria, however, strongly opposed the desires of Russia and Prussia. Talleyrand used this opportunity to bring France prominently back into the diplomatic picture, on the side of Great Britain and Austria. The conflict, in the opinion of some specialists, almost provoked a war. Its resolution — which angered the Russian public who expected "gratitude" for "liberating Europe from Napoleon" — constituted a compromise: Alexander I obtained his Kingdom of Poland, but reduced in size, while Prussia acquired about three-fifths of Saxony. More precisely, the Kingdom of Poland contained most of the former Grand Duchy of Warsaw, with Warsaw itself as its capital, but Prussia regained northwestern Poland, and Austria retained most of its earlier share of the country; Cracow became a free city-state under the joint protection of Russia, Austria, and Prussia. New Poland received a liberal constitution from Alexander I. He thus combined the offices of autocratic Russian emperor, constitutional Finnish grand duke, and constitutional Polish king. It might be added that he also favored constitutionalism in France, where the Bourbons returned to the throne as constitutional, not absolute, monarchs.

Alexander I's elated, mystical, and even messianic mood at the time of the Congress of Vienna — a complex sentiment which the Russian sovereign apparently shared in some measure with many other Europeans in the months and years following the shattering fall of Napoleon, and which found a number of spokesmen, such as Baroness Julie de Krüdener, in the tsar's entourage — expressed itself best in a remarkable and peculiar document known as the Holy Alliance. Signed on September 26, 1815, by Russia, Austria, and Prussia, and subsequently by the great majority of European powers, the alliance simply appealed to Christian rulers to live as brothers and preserve peace in Europe. While the Holy Alliance had deep roots in at least two major Western traditions, Christianity and international law, it had singularly little relevance to the international problems of the moment and provided no machinery for the application or enforcement of Christian brotherhood. Indeed, Castlereagh could well describe it as a piece of sublime mysticism and nonsense, while the pope remarked drily that from time immemorial the papacy had been in possession of Christian truth and needed no new interpretation of it.

But, if the Holy Alliance had no practical consequences, the Quadruple Alliance, and the later Quintuple Alliance with which it came to be confused, did. The Quadruple Alliance represented a continuation of the wartime association of the allies and dated from November 20, 1815. At that time Great Britain, Austria, Russia, and Prussia agreed to maintain the settlement with France — that is, the Second Treaty of Paris, which had followed the "Hundred Days" and has superseded the First Treaty of Paris — and in particular to prevent the return of Napoleon or his dynasty to the French throne. The alliance was to last for twenty years. Moreover, its sixth article provided for periodic consultations among the signatory powers and resulted in the so-called "government by conference," also known as the Congress System or Confederation of Europe. Conferences took place at Aix-la-Chapelle in 1818, Troppau and Laibach in 1820–21, and Verona in 1822. At Aix-la-Chapelle, with the payment of the indemnity and the withdrawal of allied occupation troops, France shed its status as a defeated nation and joined the other four great European powers in the Quintuple Alliance. The congresses of Troppau and Laibach considered revolutions in Spain and Italy. Finally, the meeting in Verona dealt again with Spain and also with the Greek struggle against the Turks, to which we shall return in the chapter on the reign of Nicholas I.

After an impressive start, highlighted by the harmony and success of the Aix-la-Chapelle meeting, the Congress System failed to work. A fundamental split developed between Great Britain, which, as the British state paper of May 5, 1820, made plain, opposed intervention in the internal affairs of sovereign states, and Austria, Prussia, and Russia, who, as the Protocol of Troppau spelled out, were determined to suppress revolution, no matter where it raised its head. France occupied something of an intermediate position, although it did invade Spain to crush the liberal regime there. Metternich tended to dominate the joint policies of the eastern European monarchies, especially in the crucial years of 1820–22 when Alexander I, frightened by a mutiny in the elite Semenovskii guard regiment and other events, followed the Austrian chancellor in his eagerness to combat revolution everywhere. The Semenovskii uprising, it might be added, really resulted from the conflict between the regiment and its commanding officer, not from any liberal conspiracy.

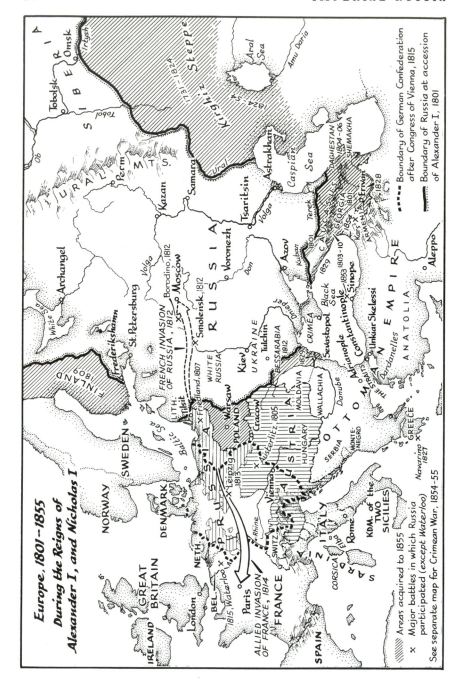

Europe, 1801-1855 During the Reigns of Alexander I, and Nicholas I

The reactionary powers succeeded in defeating liberal revolutions on the continent of Europe, except in Greece, where Christians fought their Moslem masters and the complexity of the issues involved upset the usual diplomatic attitudes and alignments. To be sure, these victories of reaction proved to be short-lived, as the subsequent history of Europe in the nineteenth century was to demonstrate. Also, the British navy prevented their possible extension across the seas, thus barring reactionary Spain and its allies from any attempt to subdue the former Spanish colonies in the new world that had won their independence. The Monroe Doctrine, proclaimed on December 2, 1823, and aimed at preventing European intervention on "the American continents," represented the response of the United States to the potential threat to the countries of the Western hemisphere posed by the reactionary members of the Confederation of Europe, and also, incidentally, a response to the Russian expansion in North America.

The Congress System has been roundly condemned by many historians as a tool of reaction, both noxious and essentially ineffective in maintaining order and stability in Europe. Yet at least one more positive aspect of that unusual political phenomenon and of Alexander I's role in it deserves notice. The architects of the Congress System, including the Russian emperor, created what was at its best more than a diplomatic alliance. In the enthusiastic words of a British scholar writing about the Congress of Aix-la-Chapelle:

> It is clear that at this period the Alliance was looked upon even by British statesmen as something more than a mere union of the Great Powers for preserving peace on the basis of the treaties; and in effect, during its short session the Conference acted, not only as a European representative body, but as a sort of European Supreme Court, which heard appeals and received petitions of all kinds from sovereigns and their subjects alike.

To be sure, this European harmony did not last, and "the Confederation of Europe" seems too ambitious a designation for the alliance following the Congress of Vienna. Yet, if a true Confederation of Europe ever emerges, the Congress System will have to be accepted as its early, in a sense prophetic, predecessor. And it was Alexander I who, more than any other European leader, emphasized the broad construction of the Quadruple and the Quintuple alliances and tried to develop co-operation and unity in Europe. Although Austrian troops intervened in the Italian states and French troops in Spain, the Russian ruler was also eager to contribute his men to enforce the decisions of the powers. In fact, he proposed forming a permanent international army to guarantee the European settlement and offered his troops for that purpose, but the suggestion was speedily rejected by Castlereagh and Metternich. He also proposed, and again unsuccessfully, disarmament.

The Second Half of Alexander's Reign

While "the emperor of Europe" attended international meetings and occupied himself with the affairs of foreign countries, events in Russia took a turn for the worse. The second half of Alexander's reign, that is, the period after 1812, saw virtually no progressive legislation and few plans in that direction; Novosiltsev's constitutional project formed a notable exception. In Poland the constitutional regime, impressive on paper, did not function well, largely because Alexander I proved to be a poor constitutional monarch because he quickly became irritated by criticism or opposition and repeatedly disregarded the law. Serfs were emancipated in the Baltic provinces, but, because they were freed without land, the change turned out to be a doubtful blessing for them. Serfdom remained undiminished and unchallenged in Russia proper, although apparently to the last the sovereign considered emancipating the serfs.

While Speransky was Alexander I's outstanding assistant in the first half of the reign, General Alexis Arakcheev came to occupy that position in the second half — and the difference between the two men tells us much about the course of Russian history in the first quarter of the nineteenth century. Arakcheev, who was once a faithful servant of Paul, was brutal, rude, and a martinet of the worst sort. He became Alexander's minister of war and eventually prime minister, without the title, reporting to the sovereign on almost everything of importance in the internal affairs of Russia and entrusted with every kind of responsibility. Yet the rather common image of the evil genius Arakcheev imposing his will on the emperor badly distorts the relationship. In fact, it was precisely the general's unquestioning and prompt execution of Alexander's orders that made him indispensable to the monarch who was increasingly peremptory and at the same time had lost interest in the complexities of home affairs.

Although Arakcheev left his imprint on many aspects of Russian life during the second half of the reign, his name came to be connected especially with the so-called "military settlements." That project apparently originated with Alexander, but it was executed by Arakcheev. The basic idea of military settlements was suggested perhaps by Turkish practices, a book by a French general, or the wonderful precision and order which reigned on Arakcheev's estates — where, among other regulations, every married woman was commanded to bear a child every year — and it had the appeal of simplicity. The idea was to combine military service with farming and thus reduce drastically the cost of the army and enable its men to lead a normal family life. Indeed, in one of their aspects the military settlements could be considered among the emperor's humanitarian endeavors. The reform began in 1810, was interrupted by war, and attained its greatest impetus and scope between 1816 and 1821, with about one-third of the peacetime Russian army established in military settlements. Troubles and uprisings in the settlements, however, checked their growth. After the rebellion of 1831 Nicholas I turned definitely against the reform, but the last settlements were abolished only much later. Alexander I's and Arakcheev's scheme failed principally because of the extreme regimentation and minute despotism that it entailed, which became unbearable and resulted in revolts and most cruel punishments. In addition — as Pipes has forcefully pointed out — Russian soldiers proved to be very poor material for this venture in state direction and paternalism, resenting even useful sanitary regulations. Arakcheev himself, it may be noted, lost his position with the accession of a new ruler.

Until 1824 two important areas of Russian life, religion and education, remained outside Arakcheev's reach because they formed the domain of another favorite of Alexander's later years, Prince Alexander Golitsyn. Very different from the brutal general, that aristocrat, philanthropist, and president of the important Bible Society in Russia nevertheless had disastrous effects on his country. Like the emperor, Golitsyn was affected by certain mystical and pietistic currents then widespread in Europe — the favorite's eventual fall resulted from allegations of insufficient Orthodoxy. He believed that the Bible contained all essential knowledge and distrusted other kinds of education. It was during Golitsyn's service as minister of education that extreme, aggressive obscurantists, such as Michael Magnitsky and Dmitrii Runich, purged several universities. Magnitsky in particular made of the University of Kazan a peculiar kind of monastic barracks: he purged the faculty and the library of the pernicious influences of the Age of Reason; flooded the university with Bibles; instituted a most severe discipline among the students, with such support as mutual spying and compulsory attendance at religious services; and proclaimed a double system of chronology, the one already in use and a new one dating from the reformation of the university. Magnitsky's fall swiftly followed the change of rulers, for in a secret report he had accused Emperor Nicholas, then a grand duke, of free thinking!

The Decembrist Movement and Rebellion

Disappointment with the course of Alexander I's reign played an important role in the emergence of the first Russian revolutionary group, which came to be known after its unsuccessful uprising in December 1825 as the Decembrists. Most of the Decembrists were army officers, often from

aristocratic families and elite regiments, who had received a good education, learned French and sometimes other foreign languages, and obtained
a first-hand knowledge of the West during and immediately after the
campaigns against Napoleon. Essentially the Decembrists were liberals in
the tradition of the Enlightenment and the French Revolution; they wanted
to establish constitutionalism and basic freedoms in Russia, and to abolish
serfdom. More specifically, the Decembrist plans ranged from those of
Nikita Muraviev, who advocated a rather conservative constitutional
monarchy, to those of Colonel Paul Pestel, the author of the *Russian
Justice,* who favored a strongly centralized republic along Jacobin lines as
well as a peculiar land reform program that would divide land into a public
and a private sector and guarantee every citizen his allotment within the
public sector. While the Decembrists — "our lords who wanted to become
shoemakers," to quote Rostopchin's ironical remark — included some
of the most gifted and prominent Russian youth, and while they enjoyed the
sympathy of many educated Russians, including such literary luminaries
as Pushkin and Griboedov, they had little social backing for their rebellion. That the standard of liberalism had to be carried in the Russia of
Alexander I by aristocratic officers of the guard demonstrates well the
weakness of the movement and above all the feebleness and backwardness of the Russian middle class. Russian liberalism in the early nineteenth
century resembled Spanish liberalism, not English or French.

At first the liberals who later became Decembrists were eager to co-
operate with the government on the road of progress, and their early societies, the Union of Salvation founded in 1816 and the Union of Welfare
which replaced it, were concerned with such issues as the development of
philanthropy, education, and the civic spirit in Russia rather than with
military rebellion. Only gradually, as reaction grew and hopes for a liberal
transformation from above faded away, did the more stubborn liberals
begin to think seriously of change by force and to talk of revolution and
regicide. The movement acquired two centers, St. Petersburg in the
north and Tulchin, the headquarters of the Second Army in southern
Russia. The northern group lacked leadership and accomplished little. In
the south, by contrast, Pestel acted with intelligence and determination.
The Southern Society grew in numbers, developed its organization, discovered and incorporated the Society of the United Slavs, and established
contacts with a Polish revolutionary group. The United Slavs, who pursued aims vaguely similar to those of the Decembrists and had the additional goal of a democratic federation of all Slavic peoples, and who accepted the Decembrist leadership, consisted in particular of poor army
officers, more democratic and closer to the soldiers than were the aristocrats from the guard. Yet, when the hour of rebellion suddenly arrived, the

Southern Society, handicapped by Pestel's arrest, proved to be little better prepared than the Northern.

Alexander I's unexpected death in southern Russia in December 1825 led to a dynastic crisis, which the Decembrists utilized to make their bid for power. The deceased emperor had no sons or grandsons; therefore Grand Duke Constantine, his oldest brother, was his logical successor. But the heir presumptive had married a Polish aristocrat not of royal blood in 1820, and, in connection with the marriage, had renounced his rights to the throne. Nicholas, the third brother, was thus to become the next ruler of Russia, the entire matter having been stated clearly in 1822 in a special manifesto confirmed by Alexander I's signature. The manifesto, however, had remained unpublished, and only a few people had received exact information about it; even the two grand dukes were ignorant of its content. Following Alexander I's death, Constantine and the Polish kingdom where he was commander-in-chief swore allegiance to Nicholas, but Nicholas, the Russian capital, and the Russian army swore allegiance to Constantine. Constantine acted with perfect consistency. Nicholas, however, even after reading Alexander I's manifesto, also felt impelled to behave as he did: Alexander I's decision could be challenged as contrary to Paul's law of succession and also for remaining unpublished during the emperor's own reign, and Nicholas was under pressure to step aside in favor of his elder brother, who was generally expected to follow Alexander I on the throne. Only after Constantine's uncompromising reaffirmation of his position, and a resulting lapse of time, did Nicholas decide to publish Alexander's manifesto and become emperor of Russia.

On December 26, 1825 — December 14, Old Style — when the guard regiments in St. Petersburg were to swear allegiance for the second time within a short while, this time to Nicholas, the Northern Society of the Decembrists staged its rebellion. Realizing that they had a unique chance to act, the conspiring officers used their influence with the soldiers to start a mutiny in several units by entreating them to defend the rightful interests of Constantine against his usurping brother. Altogether about three thousand misled rebels came in military formation to Senate Square in the heart of the capital. Although the government was caught unprepared, the mutineers were soon faced by troops several times their number and strength. The two forces stood opposite each other for several hours. The Decembrists failed to act because of their general confusion and lack of leadership; the new emperor hesitated to start his reign with a massacre of his subjects, hoping that they could be talked into submission. But, as verbal inducements failed and dusk began to gather on the afternoon of that northern winter day, artillery was brought into action. Several canister shots dispersed the rebels, killing sixty or seventy of them. Large-scale arrests followed. In the south too an uprising was easily suppressed. Eventually five Decembrist leaders, including Pestel and the firebrand of the Northern Society, the poet Conrad Ryleev, were executed, while almost three hundred other participants suffered lesser punishment. Nicholas I was firmly in the saddle.

XXVI

* * * * * * * * * *

THE REIGN OF NICHOLAS I, 1825–55

> Here [in the army] there is order, there is a strict unconditional legality, no impertinent claims to know all the answers, no contradiction, all things flow logically one from the other; no one commands before he has himself learned to obey; no one steps in front of anyone else without lawful reason; everything is subordinated to one definite goal, everything has its purpose. That is why I feel so well among these people, and why I shall always hold in honor the calling of a soldier. I consider the entire human life to be merely service, because everybody serves.
>
> NICHOLAS I

> The most consistent of autocrats.
>
> SCHIEMANN

As MAN AND RULER Nicholas I had little in common with his brother Alexander I. By contrast with his predecessor's psychological paradoxes, ambivalence, and vacillation, the new sovereign displayed determination, singleness of purpose, and an iron will. He also possessed an overwhelming sense of duty and a great capacity for work. In character, and even in his striking and powerful appearance, Nicholas I seemed to be the perfect despot. Appropriately, he always remained an army man, a junior officer at heart, devoted to his troops, to military exercises, to the parade ground, down to the last button on a soldier's uniform — in fact, as emperor he ordered alterations of the uniforms, even changing the number of buttons. And in the same spirit, the autocrat insisted on arranging and ordering minutely and precisely everything around him. Engineering, especially the construction of defenses, was Nicholas's other enduring passion. Even as a child "whenever he built a summer house, for his nurse or his governess, out of chairs, earth, or toys, he never forgot to fortify it with guns — for protection." Later, specializing in fortresses, he became head of the army corps of engineers and thus the chief military engineer of his country, perhaps his most important assignment during the reign of his brother; still later, as emperor, he staked all on making the entire land an impregnable fortress.

Nicholas's views fitted his personality to perfection. Born in 1796 and nineteen years younger than Alexander, the new ruler was brought up, not in the atmosphere of the late Enlightenment like his brother, but in that of wars against Napoleon and of reaction. Moreover, Nicholas married a Prussian princess and established particularly close ties with his wife's family, including his father-in-law King Frederick William III and his brother-in-law King Frederick William IV who ruled Prussia in succession. The Russian wing of European reaction, represented by Nicholas I and his government, found its ideological expression in the doctrine of so-called "Official Nationality." Formally proclaimed in 1833 by Count Serge Uvarov, the tsar's minister of education, Official Nationality contained three principles: Orthodoxy, autocracy, and nationality. Autocracy meant the affirmation and maintenance of the absolute power of the sovereign, which was considered the indispensable foundation of the Russian state. Orthodoxy referred to the official Church and its important role in Russia, but also to the ultimate source of ethics and ideals that gave meaning to human life and society. Nationality — *narodnost* in Russian — referred to the particular nature of the Russian people, which, so the official doctrine asserted, made the people a mighty and dedicated supporter of its dynasty and government. However, with some proponents of Official Nationality, especially professors and writers such as Michael Pogodin and Stephen Shevyrev, nationality acquired far-reaching romantic connotations. In particular, the concept for them embraced a longing for a great future for Russia and Slavdom. In sum, in contrast to Alexander I who never entirely gave up his dreams of change, Nicholas I was determined to defend the existing order in his fatherland, and especially to defend autocracy.

Nicholas's "System"

The Decembrist rebellion at the beginning of Nicholas I's reign only hardened the new emperor's basic views as well as his determination to fight revolution to the end. No doubt it also contributed to the emperor's mistrust of the gentry, and indeed of independence and initiative on the part of any of his subjects. Characteristically, Nicholas I showed minute personal interest in the arrest, investigation, trial, and punishment of the Decembrists, and this preoccupation with the dangers of subversion remained with him throughout his reign. The new regime became preeminently one of militarism and bureaucracy. The emperor surrounded himself with military men to the extent that in the later part of his reign there were almost no civilians among his immediate assistants. Also, he relied heavily on special emissaries, most of them generals of his suite, who were sent all over Russia on particular assignments, to execute immediately the will of the sovereign. Operating outside the regular administrative system, they represented an extension, so to speak, of the monarch's own person. In fact, the entire machinery of government came to be permeated by the military spirit of direct orders, absolute obedience, and precision, at least as far as official reports and appearances were concerned. Corruption and confusion, however, lay immediately behind this façade of discipline and smooth functioning.

In his conduct of state affairs Nicholas I often bypassed regular channels, and he generally resented formal deliberation, consultation, or other procedural delay. The importance of the Committee of Ministers, the State Council, and the Senate decreased in the course of his reign. Instead of making full use of them, the emperor depended more and more on special bureaucratic devices meant to carry out his intentions promptly while remaining under his immediate and complete control. As one favorite method, Nicholas I made extensive use of *ad hoc* committees standing outside the usual state machinery. The committees were usually composed of a handful of the most trusted assistants of the emperor, and, because these were very few in number, the same men in different combinations formed these committees throughout Nicholas's reign. As a rule, the committees carried on their work in secret, adding further complication and confusion to the already cumbersome administration of the empire.

The first, and in many ways the most significant, of Nicholas's committees was that established on December 6, 1826, and lasting until 1832. Count Kochubey served as its chairman, and the committee contained five other leading statesmen of the period. In contrast to the restricted assignments of later committees, the Committee of the Sixth of December had to examine the state papers and projects left by Alexander, to reconsider virtually all major aspects of government and social organization in Russia, and to propose improvements. The painstaking work of this select group of officials led to negligible results: entirely conservative in outlook, the committee directed its effort toward hair-splitting distinctions and minor, at times merely verbal, modifications; and it drastically qualified virtually every suggested change. Even its innocuous "law concerning the estates" that received imperial approval was shelved after criticism by Grand Duke Constantine. This laborious futility became the characteristic pattern of most of the subsequent committees during the reign of Nicholas I, in spite of the fact that the emperor himself often took an active part in their proceedings. The failure of one committee to perform its task merely led to the formation of another. For example, some nine committees in the reign of Nicholas tried to deal with the issue of serfdom.

His Majesty's Own Chancery proved to be more effective than the special committees. Organized originally as a bureau to deal with matters that demanded the sovereign's personal participation and to supervise the execution of the emperor's orders, the Chancery grew rapidly in the reign of Nicholas I. As early as 1826, two new departments were added to it: the Second Department was concerned with the codification of law, and the Third with the administration of the newly created corps of gendarmes. In 1828 the Fourth Department was created for the purpose of managing the charitable and educational institutions under the jurisdiction of the Empress Dowager Mary. Eight years later the Fifth Department was created and charged with reforming the condition of the state peasants; after two years of activity it was replaced by the new Ministry of State Domains. Finally, in 1843, the Sixth Department of His Majesty's Own Chancery came into being, a temporary agency assigned the task of drawing up an administrative plan for Transcaucasia. The departments of the Chancery served Nicholas I as a major means of conducting a personal policy which bypassed the regular state channels.

The Third Department of His Majesty's Own Chancery, the political police — which came to symbolize to many Russians the reign of Nicholas I — acted as the autocrat's main weapon against subversion and revolution and as his principal agency for controlling the behavior of his subjects and for distributing punishments and rewards among them. Its assigned fields of activity ranged from "all orders and all reports in every case belonging to the higher police" to "reports about all occurrences without exception"! The new guardians of the state, dressed in sky-blue uniforms, were incessantly active:

> In their effort to embrace the entire life of the people, they intervened actually in every matter in which it was possible to intervene. Family life, commercial transactions, personal quarrels, projects of inventions, escapes of novices from monasteries — everything interested the secret police. At the same time the Third Department received a tremendous number of petitions, complaints, denunciations, and each one resulted in an investigation, each one became a separate case.

The Third Department also prepared detailed, interesting, and remarkably candid reports of all sorts for the emperor, supervised literature — an activity ranging from minute control over Pushkin to ordering various "inspired" articles in defense of Russia and the existing system — and fought every trace of revolutionary infection. The two successive heads of the Third Department, Count Alexander Benckendorff and Prince Alexis Orlov, probably spent more time with Nicholas I than any of his other assistants; they accompanied him, for instance, on his repeated trips of inspection throughout Russia. Yet most of the feverish activity of the gendarmes seemed to be to no purpose. Endless investigations of subversion, stimulated by the monarch's own suspiciousness, revealed very little. Even the most important radical group uncovered during the reign, the Petrashevtsy, fell victim not to the gendarmery but to its great rival, the ordinary police, which continued to be part of the Ministry of the Interior. Short on achievements, the Third Department proved to be long on failings. The gendarmes constantly expanded their pointless work to increase their importance, quarreled with other government agencies, notably the police, and opened the way to some fantastic adventurers as well as to countless run-of-the-mill informers, who flooded the gendarmery with their reports. The false reports turned out to be so numerous that the Third Department proceeded to punish some of their authors and to stage weekly burnings of the denunciations.

The desire to control in detail the lives and thoughts of the people and above all to prevent subversion, which constituted the main aims of the Third Department, guided also the policies of the Ministry of Education — which we shall discuss in a later chapter — specifically in censorship; and, indeed, in a sense they guided the policies of Nicholas's entire regime. As in the building of fortresses, the emphasis was defensive: to hold fast against the enemy and to prevent his penetration. The sovereign himself worked indefatigably at shoring up the defenses. He paid the most painstaking attention to the huge and difficult business of government, did his own inspecting of the country, rushed to meet all kinds of emergencies, from cholera epidemics and riots to rebellion in military settlements, and bestowed special care on the army. Beyond all that, and beyond even the needs of defense, he wanted to follow the sacred principle of autocracy, to be a true father of his people concerned with their daily lives, hopes, and fears.

The Issue of Reform

However, as already indicated, all the efforts of the emperor and his government bore little fruit, and the limitations of Nicholas's approach to reform revealed themselves with special clarity in the crucial issue of serfdom. Nicholas I personally disapproved of that institution: in the army and in the country at large he saw only too well the misery it produced, and he remained constantly apprehensive of the danger of insurrection; also, the autocrat had no sympathy for aristocratic privilege when it clashed with the interests of the state. Yet, as he explained the matter in 1842 in the State Council: "There is no doubt that serfdom, as it exists at present in our land, is an evil, palpable and obvious to all. But to touch it now would be a still more disastrous evil. . . . The Pugachev rebellion proved how far popular rage can go." In fact throughout his reign the emperor feared, at the same time, two different revolutions. There was the danger that the gentry might bid to obtain a constitution if the government decided to deprive the landlords of their serfs. On the other hand, an elemental, popular uprising might also be unleashed by such a major shock to the established order as the coveted emancipation.

In the end, although the government was almost constantly concerned with serfdom, it achieved very little. New laws either left the change in the serfs' status to the discretion of their landlords, thus merely continuing

Alexander's well-meaning but ineffectual efforts, or they prohibited only certain extreme abuses connected with serfdom such as selling members of a single family to different buyers. Even the minor concessions granted to the peasants were sometimes nullified. For instance, in 1847, the government permitted serfs to purchase their freedom if their master's estate was sold for debt. In the next few years, however, the permission was made inoperative without being formally rescinded. Following the European revolutions of 1848, the meager and hesitant government solicitude for the serfs came to an end. Only the bonded peasants of Western Russian provinces obtained substantial advantages in the reign of Nicholas I. As we shall see, they received this preferential treatment because the government wanted to use them in its struggle against the Polish influence which was prevalent among the landlords of that area.

Determined to preserve autocracy, afraid to abolish serfdom, and suspicious of all independent initiative and popular participation, the emperor and his government could not introduce in their country the much-needed fundamental reforms. In practice, as well as in theory, they looked backward. Important developments did nevertheless take place in certain areas where change would not threaten the fundamental political, social, and economic structure of the Russian Empire. Especially significant proved to be the codification of law and the far-reaching reform in the condition of the state peasants. The new code, produced in the late 1820's and the early 1830's by the immense labor of Speransky and his associates, marked, despite defects, a tremendous achievement and a milestone in Russian jurisprudence. In January 1835 it replaced the ancient *Ulozhenie* of Tsar Alexis, dating from 1649, and it was destined to last until 1917.

The reorganization of the state peasants followed several years later after Count Paul Kiselev became head of the new Ministry of State Domains in 1837. Kiselev's reform, which included the shift of taxation from persons to land, additional allotments for poor peasants, some peasant self-government, and the development of financial assistance, schools, and medical care in the villages, has received almost universal praise from prerevolutionary historians. The leading Soviet specialist on the subject, Druzhinin, however, claimed recently, on the basis of impressive evidence, that the positive aspects of Kiselev's reform had a narrow scope and application, while fundamentally it placed an extremely heavy burden on the state peasants, made all the more difficult to bear by the exactions and malpractices of local administration. Finance minister Egor Kankrin's policy, and in particular his measures to stabilize the currency — often cited among the progressive developments in Nicholas I's reign — proved to be less effective and important in the long run than Speransky's and Kiselev's work.

The Last Years

But even limited reforms became impossible after 1848. Frightened by European revolutions, Nicholas I became completely reactionary. Russians were forbidden to travel abroad, an order which hit teachers and students especially hard. The number of students without government scholarships was limited to three hundred per university, except for the school of medicine. Uvarov had to resign as minister of education in favor of an entirely reactionary and subservient functionary, who on one occasion told an assistant of his: "You should know that I have neither a mind nor a will of my own — I am merely a blind tool of the emperor's will." New restrictions further curtailed university autonomy and academic freedom. Constitutional law and philosophy were eliminated from the curricula; logic and psychology were retained, but were to be taught by professors of theology. In fact, in the opinion of some historians, the universities themselves came close to being eliminated and only the timely intervention of certain high officials prevented this disaster. Censorship reached ridiculous proportions, with new agencies appearing, including "a censorship over the censors." The censors, to cite only a few instances of their activities, deleted "forces of nature" from a textbook in physics, probed the hidden meaning of an ellipsis in an arithmetic book, changed "were killed" to "perished" in an account of Roman emperors, demanded that the author of a fortune-telling book explain why in his opinion stars influence the fate of men, and worried about the possible concealment of secret codes in musical notations. Literature and thought were virtually stifled. Even Michael Pogodin, a Right-wing professor of history and a leading exponent of the doctrine of Official Nationality, was impelled in the very last years of the reign to accuse the government of imposing upon Russia "the quiet of a graveyard, rotting and stinking, both physically and morally." It was in this atmosphere of suffocation that Russia experienced its shattering defeat in the Crimean War.

Nicholas I's Foreign Policy

If the Crimean debacle represented, as many scholars insist, the logical termination of Nicholas I's foreign policy and reign, it was a case of historical logic, unique for the occasion and difficult to follow. For, to begin with, the Russian emperor intended least of all to fight other European powers. Indeed, a dedicated supporter of autocracy at home, he became a dauntless champion of legitimism abroad. Nicholas I was determined to maintain and defend the existing order in Europe, just as he considered it his sacred duty to preserve the archaic system in his own country. He saw the two closely related as the whole and its part, and he thought both to be threatened by the same enemy: the many-headed hydra of revolution, which had suffered a major blow with the final defeat of Napoleon but refused to die. Indeed it rose again and again, in 1830, in 1848, and on other occasions, attempting to reverse and undo the settlement of 1815. True to his principles, the resolute tsar set out to engage the enemy. In the course of the struggle, this "policeman of Russia" assumed added responsibilities as the "gendarme of Europe." The emperor's assistants in the field of foreign policy, led by Count Karl Nesselrode who served as foreign minister throughout the reign, on the whole shared the views of their monarch and bent to his will.

Shortly after Nicholas I's accession to the throne, Russia fought a war against Persia that lasted from June 1826 to February 1828. The hostilities, which represented another round in the struggle for Georgia, resulted in the defeat of Persia, General Ivan Paskevich emerging as the hero of the campaigns. While the Treaty of Turkmanchai gave Russia part of Armenia with the city of Erivan, exclusive rights to have a navy on the Caspian sea, commercial concessions, and a large indemnity, Nicholas I characteristically refused to press his victory. In particular, he would not support a native movement to overthrow the shah and destroy his rule.

A few weeks after making peace with Persia, Russia declared war on Turkey. This conflict marked the culmination of an international crisis which had begun with the rebellion of the Greeks against their Turkish masters in 1821, the so-called Greek War of Independence. The Russian government vacillated in its attitude toward the Greek revolution for, on one hand, the Russians sympathized with the Orthodox Greeks and were traditionally hostile to the Turks, while, on the other hand, Russia was committed to the support of the *status quo* in Europe. Moreover, the Greek crisis had unusually complicated diplomatic ramifications and possibilities. Other European powers also found it difficult to maintain a consistent policy toward the struggle of the Greeks against the Turks. Acting more firmly than his brother, Nicholas I tried, first with Great Britain and France, and then on his own, to restrain Turkey and settle the Balkan conflict. On October 20, 1827, in the battle of Navarino, the joint British, French, and Russian squadrons destroyed the Egyptian fleet that had been summoned to help its Turkish overlord. But it was not until April 1828 that the Russo-Turkish hostilities officially began. Although the Porte proved to be more difficult to defeat than the Russian emperor had expected, the second major campaign of the war brought decisive, if costly, victory to the Russian army and forced the Ottoman state to agree to the Treaty of Adrianople in 1829.

That settlement gave Russia the mouth of the Danube as well as considerable territory in the Caucasus; promised autonomous existence, under a Russian protectorate, to the Danubian principalities of Moldavia and Wallachia; imposed a heavy indemnity on Turkey; guaranteed the passage of Russian merchant ships through the Straits; and, incidentally, assured the success of the Greek revolution, which the tsar continued to detest. But in spite of these and other Russian gains embodied in the treaty, it has often and justly been considered an example of moderation in international affairs. The Russian emperor did not try to destroy his former opponent, regarding Turkey as an important and desirable element in the European balance of power. In fact, the decision to preserve the Ottoman state represented the considered judgment of a special committee appointed by Nicholas I in 1829 to deal with the numerous problems raised by the defeat of Turkey and the changing situation in the Balkans. And the committee's report to the effect that "the advantages offered by the preservation of the Ottoman Empire in Europe exceed the inconveniences which it presents," received the Russian sovereign's full endorsement.

The revolution in Paris in July 1830 came as a great shock to the tsar, and its impact was heightened by the Belgian uprising in September and by unrest in Italy and Germany. Nicholas I sent a special emissary to Berlin to co-ordinate action with Prussia and, although the mission failed, assembled an army in Poland prepared to march west. When the regime of Louis-Philippe was promptly accepted by other European governments, the Russian emperor still withheld official recognition for four months and then treated the new French ruler in a grudging and discourteous manner. The revolution of the Belgians against the Dutch similarly provoked the anger of the Russian autocrat who regarded it as another assault on the sacred principle of legitimacy and, in addition, as a clear violation of the territorial provisions of the Treaty of Vienna. Once again failing to obtain diplomatic support from other powers, Nicholas I had to subscribe to the international settlement of the issue, which favored the rebels, although he delayed the ratification of the Treaty of London for several months and did not establish regular diplomatic relations with the new state until 1852. It should be added that the early plans for a Russian military intervention in western Europe might well have been realized, except for the Polish revolution, which broke out late in November 1830, and which took the Russian government approximately a year to suppress.

Patriotic Poles had never accepted the settlement of 1815, which represented to them not a re-establishment of their historic state but the "Fourth Partition of Poland." They resented any link with Russia. And they hoped to regain the vast Lithuanian, White Russian, and Ukrainian lands that Poland had ruled before it was partitioned. Although Nicholas I observed the Polish constitution much better than Alexander I had, tension increased in Warsaw and elsewhere in the kingdom. Finally in 1830, as revolutions spread in Europe, Warsaw rose against the Russians in late November. The commander-in-chief in the kingdom, Grand Duke Constantine, failed in the moment of crisis, and before long Russia lost all control of Poland. Poland, therefore, had to be reconquered in what amounted to a full-fledged war, because the Poles had a standing army of their own that rallied to the national cause. Still, although Paskevich's Russian troops first entered Warsaw nine months later and although more time was required to destroy patriotic detachments and bands in dense Polish forests, the outcome was never in doubt. In addition to their weakness in comparison to the Russians, the Polish nationalists did not keep the strong support of the Polish peasants, and they rashly tried to carry the struggle beyond their ethnic borders where the population would not support them.

The result was another tragedy for Poland. The Polish constitution of 1815 was replaced by the Organic Statute of 1832 that made Poland "an indivisible part" of the Russian Empire. The Statute itself, with its promises of civil liberties, separate systems of law and local government, and widespread use of the Polish language, remained in abeyance while Poland was administered in a brutal and authoritarian manner by its conqueror, the new Prince of Warsaw and Nicholas's viceroy, Marshal Paskevich. The monarch himself carefully directed and supervised his work. The estates of the insurgents were confiscated; Polish institutions of higher learning were closed; the lands of the Catholic Church were secularized and the clergy given fixed salaries. At the same time, Poland was forced more and more into the Russian mold in legal, administrative, educational, and economic matters. The most striking steps in that direction included the subordination of the Warsaw school region to the Russian Ministry of Education in 1839, the abolition of the Polish State Council in 1841, and the abrogation of the customs barrier between Russia and Poland in 1850. The Russian language reigned in the secondary schools as well as in the administration, while a stringent censorship banned the works of most of the leading Polish authors as subversive.

A Russification more thorough than in Poland developed in the western and southwestern provinces, with their White Russian and Ukrainian peasant population and Polonized landlord class. Even prior to the insurrection of 1830–31 the government of Nicholas I had moved toward bringing that territory into closer association with Russia proper, a process connected with the emperor's general penchant for centralization and standardization. After the suppression of the revolution, assimilation proceeded swiftly under the direction of a special committee. Rebels from Lithuanian, White Russian, and Ukrainian provinces were denied the amnesty offered to those from Poland. It was in this territory that the Orthodox Church scored its greatest gain when, in 1839, the Uniates severed their connection with Rome and came into its fold. In 1840 the Lithuanian Statute was repealed in favor of Russian law. Because the landlords represented the Polish element, Nicholas I and his assistants changed the usual policy to legislate against their interests. They went so far as to introduce in some provinces "inventories" which defined and regularized the obligations of the serfs to their masters, and in 1851 to establish compulsory state service for the gentry of the western region. Thousands of poor or destitute families of the petty gentry were reclassified as peasants or townspeople, some of them being transferred to the Caucasus.

But while the Russian government fought against Polish influence, it showed equal hostility to budding Ukrainian nationalism, as indicated by the destruction of the Brotherhood of Cyril and Methodius and the cruel punishment of its members, including the great Ukrainian poet Taras Shevchenko.

Relative stabilization in Europe was followed by new troubles in the Near East. Denied Syria as his reward for help given to the sultan of Turkey in the Greek war, Mohammed Ali of Egypt rebelled against his nominal suzerain and, during the year of 1832, sent an army which conquered Syria and invaded Anatolia, smashing Turkish forces. The sultan's desperate appeals for help produced no tangible results in European capitals, with the exception of St. Petersburg. Nicholas I's eagerness to aid the Porte in its hour of need found ample justification in the political advantages that Russia could derive from this important intervention. But such action also corresponded perfectly to the legitimist convictions of the Russian autocrat, who regarded Mohammed Ali as yet another major rebel, and it supported the Russian decision of 1829 favoring the preservation of Turkey. On February 20, 1833, a Russian naval squadron arrived at Constantinople and, several weeks later, some ten thousand Russian troops were landed on the Asiatic side of the Bosporus — the only appearance of Russian armed forces at the Straits in history. Extremely worried by this unexpected development, the great powers acted in concert to bring Turkey and Egypt together, arranging the Convention of Kutahia between the two combatants and inducing the sultan to agree to its provisions. The Russians withdrew immediately after Orlov had signed a pact with Turkey, the Treaty of Unkiar Skelessi, on July 8, 1833. That agreement, concluded for eight years, contained broad provisions for mutual consultation and aid in case of attack by any third party; a secret article at the same time exempted Turkey from helping Russia in exchange for keeping the Dardanelles closed to all foreign warships. Although, contrary to widespread supposition at the time and since, the Treaty of Unkiar Skelessi did not provide for the passage of Russian men-of-war through the Straits — a point established by Mosely — it did represent a signal victory for

Russia: the empire of the tsars became the special ally and, to a degree, protector of its ancient, decaying enemy, thereby acquiring important means to interfere in its affairs and influence its future.

The events of 1830–31 in Europe, and to a lesser extent recurrent conflicts in the Near East, impressed on Nicholas I the necessity for close co-operation and joint action of the conservative powers. Austria and in a certain measure Prussia felt the same need, with the result that the three eastern European monarchies drew together by the end of 1833. Agreements were concluded at a meeting at Münchengrätz, attended by the emperors of Russia and Austria and the crown prince of Prussia, and at a meeting soon after in Berlin. Russia came to a thorough understanding with the Hapsburg empire, especially regarding their common struggle against nationalism and their desire to maintain Turkish rule in the Near East. Similarly, the Russian agreement with Prussia stressed joint policies in relation to partitioned Poland. More far-reaching in its provisions and its implications was the Convention of Berlin signed by all three powers on October 15, 1833:

> Their Majesties . . . recognize that each independent Sovereign has the right to call to his aid, in case of internal troubles as well as in case of an external threat to his country, every other independent Sovereign. . . .
> In the event that the material help of one of the three Courts, the Austrian, the Prussian, and the Russian, is requested, and if any power would want to oppose this by the force of arms, these three Courts would consider as directed against each one of them every hostile action undertaken with this goal in view.

The agreements of 1833 were thus meant to protect not only the immediate interests of the signatory powers, but also the entire conservative order in Europe. Nicholas I in particular proved eager to police the continent. It was the Russian army that moved quickly in 1846 to occupy the city of Cracow and suppress the uprising there, and it was the Russian emperor who insisted to the somewhat slow and reluctant Austrian government that this remnant of free Poland must become a part of the Hapsburg state, as had been previously arranged among the eastern European monarchies.

The revolution of February 1848 in France opened a new chapter in the struggle between the old order and the rising forces of the modern world in nineteenth-century Europe. While the famous story of Nicholas I telling his guests at a ball to saddle their horses because a republic had just been proclaimed in France is not exact, the Russian autocrat did react immediately and violently to the news from Paris. Although delighted by the fall of Louis-Philippe whom he hated as a usurper and traitor to legitimism, the tsar could not tolerate a revolution, so he broke diplomatic relations with France and assembled three or four hundred thousand troops in western Russia in preparation for a march to the Rhine. But rebellion spread faster than the Russian sovereign's countermeasures: in less than a month Prussia and Austria were engulfed in the conflagration, and the entire established order on the continent began rapidly to crumble into dust.

In the trying months that followed, Nicholas I rose to his full stature as the defender of legitimism in Europe. The remarkable ultimate failures of the initially successful revolutions of 1848 and 1849 can best be explained in terms of the specific political, social, and economic conditions of the different countries involved. Still, the Russian monarch certainly did what he could to tip the balance in favor of reaction. Following a strange and thunderous manifesto against revolution, he proceeded to exercise all his influence to oppose the numerous uprisings that had gripped the continent. For example, the Russian government supplied Austria with a loan of six million rubles and pointed out to Great Britain that, if an outside power were to support an Italian state against the Hapsburgs, Russia would join Austria as a full-fledged combatant. The first Russian military intervention to suppress revolution occurred in July 1848 in the Danubian principalities of Moldavia and Wallachia, where Russia acted for itself and for Turkey to defeat the Rumanian national movement. The most important action took place in the summer of 1849, when Nicholas I heeded the Austrian appeal, on the basis of the agreements of 1833, to help combat the revolt in Hungary, assigning Paskevich and almost two hundred thousand troops for the campaign. The successful Russian intervention in Hungary — which earned the undying hatred of the Hungarians — was directed in part against the Polish danger, as Polish revolutionaries were fighting on the Hungarian side. But its chief rationale lay in the Russian autocrat's determination to preserve the existing order in Europe, for the Austrian empire was one of the main supports of that order. Russia also sided with Austria in the Austrian dispute with Prussia over hegemony in Germany and thus helped the Hapsburgs to score a major diplomatic victory in the Punctation of Olmütz of November 29, 1850, when the Prussians abandoned their attempt to seize the initiative in Germany and accepted a return to the *status quo* and Austrian leadership in that area.

The impressive and in certain ways dominant position which Russia gained with the collapse of the revolutions of 1848–49 on the continent failed to last. In fact, the international standing of the "gendarme of Europe" and the country he ruled was much stronger in appearance than in reality: liberalism and nationalism, although defeated, were by no means dead, and they carried European public opinion from Poland and Hungary to France and England; even the countries usually friendly to the tsar complained of his interference with their interests, as in the case of Prussia, or at least resented his overbearing solicitude, as was true of Austria. On the other hand, Nicholas I himself — in the opinion of some specialists — reacted to his success by becoming more blunt, uncompromising, doctrinaire, and domineering than ever before. The stage was set for a debacle.

The Crimean War

However, when the debacle did come, the accompanying circumstances proved to be exceedingly complex, and they were related especially to issues in the Near East. There the resumption of hostilities between Turkey and Egypt in 1839–40 undid the Treaty of Unkiar Skelessi. European powers acted together to impose a settlement upon the combatants, under terms of the Treaty of London of July 15, 1840, and they also signed the Straits Convention of July 13, 1841. The Convention, in which Great Britain, Austria, Prussia, Russia, and France participated, reaffirmed the closure of the Bosporus and the Dardanelles to all foreign warships in time of peace, substituting an international guarantee of the five signatories for the separate treaty between Russia and Turkey. Nicholas I proved willing to co-operate with the other states, and, in the same spirit, made a particular effort during the years following to come to a thorough understanding with Great Britain. In the summer of 1844 he personally traveled to England and discussed the Near Eastern situation and prospects with Lord Aberdeen, the foreign secretary. The results of these conversations were summarized in an official Russian memorandum, prepared by Nesselrode, which the British government accepted as accurate. According to its provisions, Russia and Great Britain were to maintain the Turkish state as long as possible, and, in case of its impending dissolution, the two parties were to come in advance to an understanding concerning the repartitioning of the territories involved and other problems.

Although the crucial Russo-British relations in the decades preceding the Crimean War have been variously explicated and assessed by different scholars, such as Puryear who saw the picture from the Russian side and Temperley who observed it from the British side, several elements in the situation stand out clearly. Nicholas I's apparently successful agreement with Great Britain had an illusory and indeed a dangerous character. The two main points of the understanding — the preservation and the partitioning of Turkey — were, in a sense, contradictory, and the entire agreement was, therefore, especially dependent on identical, or at least very similar, interpretation by both partners of developments in the Near East, a degree of harmony never to be achieved. Moreover, the form of the agreement also contributed to a certain ambivalence and difference of opinion: while Nicholas I and his associates considered it to be a firm arrangement of fundamental importance, the British apparently thought of it more as a secret exchange of opinions not binding on the subsequent premiers and foreign ministers of Her Majesty's government. The Russian emperor's talks in January and February of 1853 with Sir Hamilton Seymour, the British ambassador, when the tsar dwelt on the imminent collapse of the Ottoman Empire and offered a plan of partition, served only to emphasize the gulf between the two states. The complex and unfortunate entanglement with Great Britain was one of the chief bases for Nicholas I's mistaken belief that his Near Eastern policy had strong backing in Europe.

In 1850 a dispute began in the Holy Land between Catholics and Orthodox in regard to certain rights connected with some of the most sacred shrines of Christendom. Countering Napoleon III's championing of the Catholic cause, Nicholas I acted in his usual direct and forceful manner by sending Prince Alexander Menshikov, in February 1853, with an ultimatum to the Turks: the Holy Land controversy was to be settled in favor of the Orthodox, and the Porte was to recognize explicitly the rights of the vast Orthodox population of its empire. When Turkey accepted the first series of demands, but would not endorse Russian interference on behalf of the Orthodox subjects of the Porte, considering it to be an infringement of Turkish sovereignty, Menshikov terminated the discussion and left Constantinople. Russian occupation of the Danubian principalities as "material guarantees" added fuel to the fire. There is little doubt that the rash actions of Nicholas I precipitated war, although it is probable that he wanted to avoid a conflict. After the first phases of the controversy described above, the Russian government acted in a conciliatory manner, accepting the so-called Vienna Note as a compromise settlement, evacuating the principalities, and repeatedly seeking peace even after the outbreak of hostilities. The war guilt at this later stage should be divided principally among Turkey, France, Great Britain, and even Austria, who pressed increasingly exacting demands on Russia. In any case, after fighting between Russia and Turkey started in October 1853, and the Russians destroyed a Turkish fleet and transports off Sinope on November 30, Great Britain and France joined the Porte in March 1854, and Sardinia intervened the next year. Austria stopped just short of hostilities against Russia, exercising strong diplomatic pressure on the side of the allies. Nicholas I found his country fighting alone against a European coalition.

The Russian emperor's Near Eastern policy, which culminated in the Crimean War, has received various interpretations. Many historians have emphasized Russian aggressiveness toward Turkey, explaining it by the economic requirements of Russia, such as the need to protect grain trade through the Black Sea or to obtain markets in the Near East, by the strategic imperative to control the Straits, or simply by a grand design of political expansion more or less in the footsteps of Catherine the Great. Yet, as we had occasion to observe, the tsar's attitude toward the Ottomans long retained the earmarks of his basic belief in legitimism. Even his ultimate decision to partition the Turkish Empire can be construed as a result of the conviction that the Porte could not survive in the modern world, and that therefore the leading European states had to arrange for a proper redistribution of possessions and power in the Balkans and the Near East in order to avoid anarchy, revolution, and war. In other words, Nicholas's approach to Great Britain can be considered sincere, and the ensuing misunderstanding thus all the more tragic. However, one other factor must also be weighed in an appreciation of Nicholas I's Near Eastern policy: Orthodoxy. Obviously, the Crimean War was provoked partially by religious conflicts. And the tsar himself retained throughout his reign a certain ambivalence toward the sultan. He repeatedly granted the legitimacy of the sultan's rule in the Ottoman Empire, but remained, nevertheless, uneasy about the sprawling Moslem state which believed in the Koran and oppressed its numerous Orthodox subjects. Once the conflict began, Nicholas I readily proclaimed himself the champion of the Cross against the infidels.

Although the Crimean War involved several major states, its front was narrowly restricted. After Austrian troops occupied Moldavia and Wallachia separating the Russians from the Turks in the Balkans, the combatants possessed only one common border, the Russo-Turkish frontier in the Caucasus, and that distant area with its extremely difficult terrain was unsuited for major operations. The allies controlled the sea and staged a number of naval demonstrations and minor attacks on the Russian coasts from the Black, the Baltic, and the White seas to the Bering Sea. Then, in search of a decisive front, they landed in the Crimea in September 1854. The war became centered on the allied effort to capture the Crimean naval base of Sevastopol. Except for the Crimea, the fighting went on only in the Caucasus, where the Russians proved rather successful and even seized the important Turkish fortress of Kars. Sevastopol held out for eleven and a half months against the repeated bombardments and assaults of French, British, Turkish, and Sardinian forces with their superior weapons. While the Russian supply service broke down and the high command showed little initiative, the soldiers and the sailors of the Black Sea fleet, led by such dedicated officers as the admirals Paul Nakhimov and Vladimir Kornilov — both, incidentally, killed in combat — fought desperately for their city. Colonel Count Edward Todtleben, the chief Russian military engineer at Sevastopol, proved to be a great improviser of defenses, who did more than any other man to delay the allied advance. The hell and the heroism of the Crimean War were best related by Leo Tolstoy, himself an artillery officer in the besieged city, in his *Sevastopol Tales.* In English literature the War inspired Tennyson's "Charge of the Light Brigade," a poetic description of an episode in the battle of Balaklava. It might be added that this conflict, which is considered by many scholars as unnecessary and a result of misunderstandings, was the more tragic since typhus

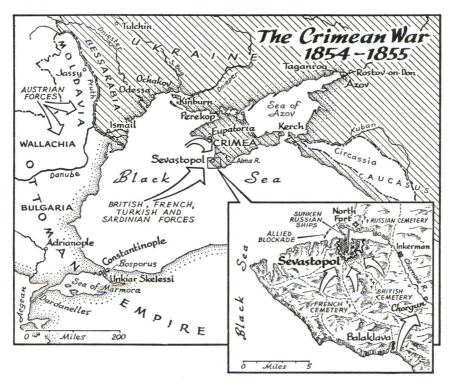

and other epidemics caused even more deaths than did the actual fighting. It was in the Crimean War that Florence Nightingale established a new type of war hospital and worked toward the modernization of nursing, as did French and Russian women.

The Russian forces finally abandoned Sevastopol on September 11, 1855, sinking their remaining ships — others had been sunk earlier to block the harbor — and blowing up fortifications. Nicholas I had died in March, and both his successor, Alexander II, and the allies effectively supported by Austrian diplomacy, were ready early in 1856 to make peace. An impressive international congress met in Paris for a month, from late February until late March. Its work resulted in the Treaty of Paris, signed on the thirtieth of March. By the provisions of the Treaty, Russia ceded to Turkey the mouth of the Danube and a part of Bessarabia and accepted the neutralization of the Black Sea — that is, agreed not to maintain a navy or coastal fortifications there. Further, Russia gave up its claims to a protectorate over the Orthodox in the Ottoman Empire. The Danubian principalities were placed under the joint guarantee of the signatory powers, and an international commission was established to assure safe navigation of the Danube. The Treaty of Paris marked a striking decline of the Russian position in southeastern Europe and the Near East, and indeed in the world at large.

Concluding Remarks

With the major exception of the Marxist scholars, most historians of the reign of Nicholas I — whether they concentrated, like Schilder, on court and government, like Schiemann on foreign policy, like Polievktov on internal developments, or like Lemke on political police and censorship — have noted the importance of the emperor and his firm beliefs for the course of Russian history. Nicholas I, to be sure, gave no new direction to the development of his country. Rather he clung with a desperate determination to the old system and the old ways. The creator of the doctrine of Official Nationality, Count Uvarov, once remarked that he would die with a sense of duty fulfilled if he could succeed in "pushing Russia back some fifty years from what is being prepared for her by the theories." In a sense, Nicholas I and his associates accomplished just that: they froze Russia as best they could for thirty — although not fifty — years, while the rest of Europe was changing. The catastrophe of the Crimean War underlined the pressing need for fundamental reforms in Russia as well as the fact that the hour was late.

However, before we turn to Alexander II and the "great reforms" we shall consider the development of Russian economy, society, and culture in the first half of the nineteenth century. In those fields, as we shall see, by contrast with Nicholas's politics, movement prevailed over stagnation.

XXVII

* * * * * * * * * *

THE ECONOMIC AND SOCIAL DEVELOPMENT OF RUSSIA
IN THE FIRST HALF OF THE NINETEENTH CENTURY

> The development of an exchange or money economy, much more
> rapid and widespread than formerly, must certainly be recognized as
> the main feature of the economic history of Russia in the first half
> of the nineteenth century or — more precisely — until the abolition
> of serfdom. A money economy began perceptibly to develop in
> Russia as early as the middle of the sixteenth century, but at first
> this process went on very slowly and encompassed relatively small
> groups of the population. Only in the nineteenth century did the
> money economy begin to evolve into its second stage of development,
> when a majority of the people becomes engulfed in the trade cycle,
> works for the market, and to satisfy its own needs buys products of
> someone else's labor, also brought to the market as merchandise.
>
> ROZHKOV

THE SECOND HALF of the eighteenth century marked the zenith of manorial
economy and serf agriculture in Russia, but the first decades of the nineteenth witnessed significant changes in the economic picture. Russian estates
sent more and more produce to the market, at home and even abroad, as
southern Russia began to export grain via the Black Sea. New opportunities for marketing, together with a continuing growth of population, led to
a strong and steady rise in land prices. Yet while possibilities beckoned,
Russian agriculture could evolve in the capitalistic direction only to a
limited extent and at great human and economic cost, for it was restricted
by the social structure and the institutions of the country.

Most landlords, entirely unprepared for the task by their education and
outlook, failed to adjust effectively to competition and to establish efficient
production on their estates. In the first half of the nineteenth century, the
proportion of non-gentry landownership grew, despite the fact that only
members of the gentry could own serfs. In addition, the indebtedness of the
gentry to the state increased rapidly, acquiring tremendous proportions by
the middle of the century. It has been estimated that on the eve of the
emancipation of the serfs in 1861 the state held in mortgage two-thirds of
all the serfs. Small estates were especially hard hit. While substantial landlords on the whole adjusted more or less effectively to the new conditions,
their poorer brethren, lacking capital or other sufficient assets, lost out in
the competition. The first half of the century thus saw a concentration of

gentry landholding, and a decline, often pauperization, of small gentry landowners.

Serfdom, of course, lay at the heart of pre-reform Russian agriculture. Considerable evidence indicates that the landlords first responded to the new market opportunities and the generally rising tempo of economic life by trying to obtain a greater yield from their own fields. Barshchina, therefore, increased in scope and became more intensive, a process culminating in the 1840's. But serf labor offered no solution to the problem of achieving efficient, improved production: illiterate, unskilled, and uninterested, the serfs were plainly poor producers. Above all, they lacked incentive and initiative. As a result, in the 1840's and especially in the 1850's obrok increased at the expense of barshchina. Its monetary value rose very markedly; an individual peasant had to pay his master perhaps ten times as much in 1860 as in 1800, while he was encouraged to work hard by the fact that he could retain what remained after the payment. Serfs received additional land in return for obrok, and more of them earned their — and, indeed, their masters' — keep in factories, in transportation, and in other occupations, including agricultural work away from their home. Significantly, more and more free labor came to be hired in agriculture, especially in the Volga region and the Black Sea provinces. Agricultural wages generally rose, although both the amount of rise and the wages themselves remain very difficult to calculate. The increase of free labor in agriculture — even though, of course, that labor frequently represented the work of someone else's serfs hired temporarily — acquires added importance when considered in conjunction with the growth of free labor in industry and, indeed, in virtually all aspects of Russian economy.

While Russian agriculture in the first half of the nineteenth century reacted in a strained and pained manner to new conditions and demands, a certain advance and modernization were achieved. With the use of machinery and fertilizers and improved organization and technique, some estates became successful "capitalistic" producers. In general, too, productivity increased somewhat as Russian agriculture became more intensive. Also, the produce gradually became more diversified. Old staple crops, notably rye and wheat, continued to be grown on a large scale and in fact for the first time attained prominence among Russian exports. But certain new items rose to positions of some importance in the agriculture of the country. These included potatoes and sugar beets, and, in the south, wine, the successful production of which required considerable knowledge and skill. The production of potatoes quintupled in the 1840's, the production of wine tripled between the early 1830's and 1850, and the spread of sugar beets in Russia can be gauged by the number of sugar beet factories: 7 in 1825, 57 in 1836, 206 in 1844, 380 in the early 1850's. The culture of silk and certain vegetable dyes developed in Transcaucasia. Fine wool began to be produced with the introduction into Russia of a new and superior breed of sheep in 1803. With government aid, the number of these sheep increased from 150,000 in 1812 to some 9 million in 1853.

Industry

Industry, no less than agriculture, was affected by the growth of a market economy. Russian manufacturing establishments, counting only those that employed more than fifteen workers, increased in number from some 1,200 at the beginning of the century to 2,818 by 1860. The labor force expanded even faster: from between one and two hundred thousand in 1800 to between five and nine hundred thousand on the eve of the "great reforms." The striking discrepancy in the statistics compiled by various specialists results from both inadequate material and the problem of definition, including definition of the key concepts, "factory" and "worker." Soviet scholars, especially more recently, have on the whole emphasized and exaggerated the industrial development of Russia, but they have also provided some valuable documentation to support certain of their claims.

The relatively new cotton industry grew most rapidly. Its output increased sixteen times over in the course of the half-century, and at the end of the period Russia possessed about one million cotton spindles. The cotton industry required capital, and, in contrast to older woolen and linen manufactures, it was run by free, not serf, labor. On the whole, free labor gained steadily over bonded labor, and "capitalist" factories over both possessional and manorial ones. According to one count, by 1825 "capitalist" factories constituted 54 per cent of all industrial establishments. Wages, although very low to be sure, kept going up.

At the same time, especially after the first quarter of the century, the use of machinery and steam power steadily increased in Russian manufacturing. The Russians imported machinery to the value of 42,500 silver rubles in 1825, 1,164,000 silver rubles in 1845, and 3,103,000 in 1860. Moreover, they began to build their own machines: the country possessed 19 machine-building factories with their annual output valued at 500,000 rubles in 1851, and 99 with an output worth 8,000,000 rubles in 1860. Russian industry, however, remained largely restricted to the Urals, the Moscow area, the rapidly growing St. Petersburg-Baltic region, and several other already well-established centers. In particular, none had as yet arisen in the vast Russian south.

Trade and Transportation

Trade also reflected the quickening tempo of economic life in Russia in the first half of the nineteenth century. Internal trade experienced marked growth. The differentiation of the country into the grain-producing south and the grain-consuming center and north became more pronounced, providing an ever stronger basis for fundamental, large-scale exchange. Thus the north and the center sent the products of their industries and crafts south in return for grain, meat, and butter. Certain areas developed their own specialties. For example, the northwestern region produced flax for virtually all of Russia. A district in the distant Archangel province raised a special breed of northern cows. Several Ukrainian provinces became famous for their horses, while the best sheep were bred in southern Russia, between the Volga and the Don. Even such items as woolen stockings became objects of regional specialization. A number of scholars have noted how, in the first half of the nineteenth century, purchased clothing began gradually to displace the homespun variety among the peasants.

Merchant capital grew and fairs expanded. The famous fair near the Monastery of St. Macarius in the Nizhnii Novgorod province was transferred in 1817 to the town of Nizhnii Novgorod itself and there attained new heights. In 1825 goods worth 12,700,000 rubles were sold at that fair; in 1852 the sum rose to 57,500,000. A number of other fairs also did a very impressive business. The total turnover in Russian internal trade for 1825 has been estimated at the considerable sum of 900,000,000 rubles.

Transportation also developed, if rather slowly. Rivers and lakes continued to play an extremely important role in trade and travel. A number of canals, especially those constructed between 1804 and 1810, added to the usefulness of the water network, by linking, for instance, the Western Dvina to the Dnieper and St. Petersburg to the Volga, thus making it possible to send goods from the upper Volga to the Baltic Sea. The first steamship appeared in Russia in 1815, on the Neva. In 1820 regular steam navigation commenced on the Volga to be extended later to other important rivers and lakes. Following by several years the construction of a small private railroad to serve the needs of a factory, the first public Russian railroad, joining St. Petersburg and the suburban imperial residence of Tsarskoe Selo — present-day Pushkin — was opened to traffic in 1837. In 1851 the first major Russian railroad went into operation, linking St. Petersburg and Moscow on a remarkably straight line as desired by Nicholas I. The Russians even proceeded to establish a railroad industry and build their own locomotives and cars, a development in which Americans, including George Whistler, the father of the painter James McNeill Whistler, played a prominent part. But, considering the size of the country, the systems of transportation remained thoroughly inadequate. In particular, in 1850 Russia possessed only a little over three thousand miles of first-class roads. The Russian army in the Crimea proved to be more isolated from its home bases than the allied forces, which were supplied by sea, from theirs.

Foreign trade — about which we have more precise data than we have concerning domestic commerce — grew swiftly in the first half of the nineteenth century. The annual value of Russian exports on the eve of the "great reforms" has been estimated at 230 million rubles, and of imports at 200 million, compared to only 75 and 52 million respectively at the beginning of the century. Russia continued to export raw materials, such as timber and timber products, hemp, flax, tallow, and increasing quantities of grain. The grain trade resulted from the development of agriculture, notably the raising of wheat, in southern Russia; from the organization of grain export, largely in Greek ships, via the Black Sea; and from the pressing demand for grain in industrializing western Europe. From bare beginnings at the turn of the century, the grain trade rose to 35 per cent of the total value of Russian exports in 1855. It led to the rapid rise of such ports as Odessa and Taganrog and made the Black Sea rival the Baltic as an avenue for commerce with Russia. Russian manufactures, by contrast, found no demand in the West, but — a foretaste of the future — they attracted some customers in Turkey, Central Asia, Mongolia, and China. The Russian imports consisted of tropical produce, such as fruits and coffee, and factory goods, including machinery, as has already been noted.

Social Composition

The population in Russia continued to increase rapidly throughout the period: from 36,000,000 in 1796 to 45,000,000 in 1815 and 67,000,000 in 1851. At the same time its social composition underwent certain changes. While the serfs multiplied in the eighteenth century to constitute, according to Blum, 49 per cent of the total population of Russia in 1796 and as much as 58 per cent in 1811, they failed to keep pace with other social groups after that date. In 1858 they composed 44.5 per cent of the total. Indeed, some scholars have argued that the serfs did not increase in number at all during the decades preceding the emancipation. Semevsky and other students of serfdom have shown what a tremendous and progressively heavier burden of obligations the serfs had to carry, and how hard their life frequently was. These crushing conditions of existence limited the expansion of serfdom and somewhat diminished its relative social weight in Russia in the first half of the nineteenth century.

By contrast, Russian urban population grew both absolutely and as a proportion of the total between 1800 and the "great reforms" — in this case continuing and accelerating an eighteenth-century trend. Townspeople constituted about 4.1 per cent of the inhabitants of the empire at the turn of the century and 7.8 per cent in 1851.

To be sure, the upper class, the gentry, retained its dominant social and economic position and its leadership in most phases of Russian life. Yet, as our brief account of the economic evolution of the country indicated, its problems and difficulties increased. Most landlords failed to adjust effectively to the changing economic conditions, sank gradually deeper into debt, and often slid further toward poverty. The differentiation of the gentry, from successful landed magnates at one extreme to the numerous poor and even destitute gentry at the other, became increasingly prominent. If the reign of Catherine the Great represented the golden age of the Russian gentry, the reigns of Alexander I and Nicholas I witnessed the development of processes leading unmistakably to its decline.

Evaluations of the Russian Economy and Society

There are several ways of looking at Russian economy and society in the first half of the nineteenth century. To many foreign observers, some older Marxist historians, and certain other critics the main characteristics of Russian life in the period preceding the "great reforms" consisted of backwardness, stagnation, and oppression. As a reaction to this extreme view, many historians — ranging from Soviet specialists to such émigré scholars as Karpovich — have stressed the achievements of the Russians during those difficult decades. They have pointed to a wide variety of phenomena in support of their emphasis: the brilliant Russian literature and culture of the period — which we shall discuss in the next chapter — and Kiselev's reform of the condition of the state peasants; the early penetration of capitalism into the country and certain technological improvements made by the Russians; railroads and the cotton industry; the growing middle class and the expanding trade.

Yet this approach, in its turn, must be kept within its proper frame of reference. For, while Russian economy and society certainly did develop in the first half of the nineteenth century, the empire of the tsars failed to keep pace with other European countries. Whereas capitalism began to affect Russia, it was revolutionizing Great Britain, Belgium, and France. Russian industry was less important in the total European and world picture in 1860 than in 1800, and it had to be protected by very high tariffs. Although the Russian urban classes rose rather rapidly during the first half of the nineteenth century, they remained extremely weak compared to the bourgeoisie in different countries of western Europe. Whereas the country obtained some steamships and railroads, its transportation system failed to serve adequately either the peacetime needs or the needs of the Crimean War. The Russians' weapons and military equipment proved inferior to those of their European opponents; the Black Sea fleet, composed of wooden sailing vessels, could not compete with the steam-propelled warships of the allies. And, obviously, in the middle of the nineteenth century Russia could afford even less than at the time of Peter the Great to disregard other states and to live entirely as a world apart. This international dimension of the Russian problem brings into clearer focus Alexander I's vacillations, Nicholas I's stubborn refusal to move, and the urgent need of "great reforms."

XXVIII

* * * * * * * * * *

RUSSIAN CULTURE IN THE FIRST HALF OF THE
NINETEENTH CENTURY

> Pushkin represents an extraordinary and, perhaps, a unique mani-
> festation of the Russian spirit, said Gogol. I shall add on my own:
> also prophetic. . . . His appearance helps greatly to illuminate our
> dark road with a guiding light.
>
> DOSTOEVSKY

> Every age, every nation contains in itself the possibility of original
> art, provided it believes in something, provided it loves something,
> provided it has some religion, some ideal.
>
> KHOMIAKOV

> It has often been noted that the farther east in Europe one goes the
> more abstract and general political ideals become. The English agi-
> tated for the particular and historic rights of Englishmen; the French
> for the universal and timeless rights of man; the Germans sought
> freedom in the realm of the "pure" or "absolute" idea. . . . It is also
> roughly true that the farther east one goes, the more absolute, cen-
> tralized, and bureaucratic governments become, while the middle
> groups between an ignorant peasantry and a military state grow
> smaller and weaker. Moreover, the greater the pressure of the state
> on the individual, the more formidable the obstacles to his inde-
> pendence, and the greater his social loneliness are, the more sweeping,
> general, and abstract are ideologies of protest or compensation.
>
> MALIA

In culture, the eighteenth century in Russia had represented a period of
learning from the West. The learning, to be sure, continued in the nineteenth
century and, in fact, became all the time both broader and deeper. But,
beginning with the reign of Alexander I, Russia developed a glorious
literary culture of its own, which in time became the accepted standard
of excellence in its homeland and a model to be imitated by many writers
in other countries. The "golden age of Russian literature" has been dated
roughly from 1820 to 1880 — from Pushkin's first major poems to
Dostoevsky's last novel — most of it thus falling in the period preceding
the "great reforms." While the arts in Russia did not keep up with Russian
literature, they too advanced in the first half of the nineteenth century.
Music, for example, developed along creative and original lines, leaving
far behind the imitative efforts of the time of Catherine the Great. Russian
science and scholarship also showed noteworthy progress. If the eighteenth

century had its Michael Lomonosov, the reign of Nicholas I witnessed the epoch-making work of Nicholas Lobachevsky. Moreover, whereas Lomonosov had remained something of a paradox in his age, unique, isolated, and misunderstood, learning in Russia in the first half of the nineteenth century gradually acquired a broader and more consecutive character, with its own schools of thought, traditions, and contributions to the total intellectual effort of Western civilization. Even philosophical, political, social, and economic doctrines grew and developed in a remarkable manner in spite of autocracy and strict censorship.

Although people from the lower classes began to acquire prominence on the eve of the "great reforms," Russian culture of the reigns of Alexander I and Nicholas I was essentially gentry culture. Its tone and charm have been best preserved in magnificent works by its representatives, such as Tolstoy's *War and Peace,* Turgenev's *A Gentry Nest,* and Serge Aksakov's family chronicle. Supported by the labor of serfs and confined in a narrow social group — not unlike the culture of the antebellum South in the United States — Russian culture of the first half of the nineteenth century marked, just the same, a great step forward for the country and left many creations of lasting value. The educated gentry, whose numbers grew, continued to enjoy a cosmopolitan, literary upbringing at home, with emphasis on the French language and with the aid of a battery of foreign and Russian tutors. For illustration one can turn to Tolstoy's autobiographical trilogy as well as to a host of other reminiscences of the period. Next, the sons of the gentry often attended select military schools before entering the army as officers, where again the French language and proper social manners were emphasized. Also, members of the gentry often collected valuable libraries on their estates, followed with interest developments in the West, and even frequently traveled abroad to learn about western Europe and its culture first hand. More and more of them attended universities, both at home and in foreign countries.

Education

University education, as well as secondary education in state schools, became more readily available after Alexander I's reforms. With the creation of the Ministry of Education in 1802, the empire was divided into six educational regions, each headed by a curator. The plan called for a university in every region, a secondary school in every provincial center, and an improved primary school in every district. By the end of the reign the projected expansion had been largely completed: Russia then possessed 6 universities, 48 secondary state schools, and 337 improved primary state schools. Alexander I founded universities in Kazan, Kharkov, and St. Petersburg — the latter first being established as a pedagogical institute —

transformed the "main school," or academy, in Vilna into a university, and revived the German university in Dorpat, which with the University of Moscow made a total of six. In addition, a university existed in the Grand Duchy of Finland: originally in Åbo — called Turku in Finnish — and from 1827 in Helsingfors, or Helsinki. Following a traditional European pattern, Russian universities enjoyed a broad measure of autonomy. While university enrollments numbered usually a few hundred or less each, and the total of secondary school students rose only to about 5,500 by 1825, these figures represented undeniable progress for Russia. Moreover, private initiative emerged to supplement the government efforts. It played an important part in the creation of the University of Kharkov, and it established two private institutions of higher education which were eventually to become the Demidov Law School in Iaroslavl and the Historico-Philological Institute of Prince Bezborodko in Nezhin. Finally, it may be noted that the celebrated Imperial Lyceum in Tsarskoe Selo, which Pushkin attended, was also founded during the reign of Alexander I.

The obscurantist purges of the last years of Alexander's rule hurt Russian universities, especially the one in Kazan. But Magnitsky and his associates held power only briefly. The many educational policies under Nicholas I that proved to be noxious rather than beneficial to Russian schools and learning were of greater importance. During the thirty years of Official Nationality, with Uvarov himself serving as minister of education from 1833 to 1849, the government tried to centralize and standardize education; to limit the individual's schooling according to his social background, so that each person would remain in his assigned place in life; to foster the official ideology exclusively; and, above all, to eliminate every trace or possibility of intellectual opposition or subversion.

As to centralization and standardization, Nicholas I and his associates did everything in their power to introduce absolute order and regularity into the educational system of Russia. The state even extended its minute control to private schools and indeed to education in the home. By a series of laws and rules issued in 1833–35, private institutions, which were not to increase in number in the future except where public schooling was not available, received regulations and instructions from central authorities, while inspectors were appointed to assure their compliance. "They had to submit to the law of unity which formed the foundation of the reign." Home education came under state influence through rigid government control of teachers: Russian private tutors began to be considered state employees, subject to appropriate examinations and enjoying the same pensions and awards as other comparable officials; at the same time the government strictly prohibited the hiring of foreign instructors who did not possess the requisite certificates testifying to academic competence and exemplary moral character. Nicholas I himself led the way in supervising and inspecting schools in Russia, and the emperor's assistants followed his example.

The restrictive policies of the Ministry of Education resulted logically from its social views and aims. In order to assure that each class of Russians obtained only "that part which it needs from the general treasury of enlightenment," the government resorted to increased tuition rates and to such requirements as special certificates of leave that pupils belonging to the lower layers of society had to obtain from their village or town before they could attend secondary school. Members of the upper class, by contrast, received inducements to continue their education, many boarding schools for the gentry being created for that purpose. Ideally, in the government's scheme of things — and reality failed to live up to the ideal — children of peasants and of lower classes in general were to attend only parish schools or other schools of similar educational level, students of middle-class origin were to study in the district schools, while secondary schools and universities catered primarily, although not exclusively, to the gentry. Special efforts were made throughout the reign to restrict the education of the serfs to elementary and "useful" subjects. Schools for girls, which were under the patronage of the empress dowager and the jurisdiction of the Fourth Department of His Majesty's Own Chancery, served the same aims as those for boys.

The inculcation of the true doctrine, that of Official Nationality, and a relentless struggle against all pernicious ideas constituted, as we know, essential activities of the Ministry of Education. Only officially approved views received endorsement, and they had to be accepted without question rather than discussed. Teachers and students, lectures and books were generally suspect and required a watchful eye. In 1834 full-time inspectors were introduced into universities to keep vigil over the behavior of students outside the classroom. Education and knowledge, in the estimate of the emperor and his associates, could easily become subversion! As already mentioned, with the revolutionary year of 1848 unrelieved repression set in.

Still, the government of Nicholas I made some significant contributions to the development of education in Russia. Thus, it should be noted that the Ministry of Education spent large sums to provide new buildings, laboratories, and libraries, and other aids to scholarship such as the excellent Pulkovo observatory; that teachers' salaries were substantially increased — extraordinarily increased in the case of professors, according to the University Statute of 1835; that, in general, the government of Nicholas I showed a commendable interest in the physical plant necessary for education and in the material well-being of those engaged in instruction. Nor was quality neglected. Uvarov in particular did much to raise educational and scholarly standards in Russia in the sixteen years during which he headed the ministry. Especially important proved to be the establishment of many new chairs, the corresponding opening up of numerous new fields of learning in the universities of the empire, and the practice of sending promising young Russian scholars abroad for extended training. The Russian educational system, with all its fundamental flaws, came to emphasize academic thoroughness and high standards. Indeed, the government utilized the standards to make education more exclusive at all levels of schooling. Following the Polish rebellion, the Polish University of Vilna was closed; in 1833 a Russian university was opened in Kiev instead. The government of Nicholas I created no other new universities, but it did establish a number of technical and "practical" institutions of higher learning, such as a technological institute, a school of jurisprudence, and a school of architecture, as well as schools of arts and crafts, agriculture, and veterinary medicine.

Science and Scholarship

With the expansion of higher education, science and scholarship grew in Russia. Mathematics led the way. Nicholas Lobachevsky, who lived from 1793 to 1856 and taught at the University of Kazan, was the greatest Russian mathematician of that, or indeed any, period. The "Copernicus of geometry" left his mark in the history of thought by formulating a non-Euclidian geometry. Starting from an attempt to prove the old Euclidian axiom that on a given plane it is possible to draw through a point not on a given line one and only one line parallel to the given line, and proceeding by trying to refute other alternatives, Lobachevsky found his task impossible. He then faced the consequences of his discovery and went on to postulate and develop a non-Euclidian geometry, within which the Euclidian scheme represented but a single instance. While Lobachevsky's revolutionary views received scant recognition from his contemporaries either in Russia or in other countries — although, to be exact, he was not quite alone, for a few Western scholars were approaching similar conclusions at about the same time — they nevertheless represented a major breakthrough in the direction of the modern development of mathematics and the physical sciences. Several other gifted Russian mathematicians of the first half of the nineteenth century also contributed to the growth of their subject.

Astronomy too fared exceptionally well in Russia in the first half of the nineteenth century. In 1839 the celebrated Pulkovo observatory was constructed near St. Petersburg. Directed by one of the leading astronomers of the age who was formerly professor at the University of Dorpat, Frederick William Jacob Struve, and possessing the largest telescope in the world at that time and in general the most up-to-date equipment, Pulkovo quickly became not only a great center of astronomy in Russia, but also a valuable training ground for astronomers from other European countries and the United States. Struve investigated over three thousand double stars, developed methods to calculate the weight of stars and to apply statistics to a study of them, and dealt with such problems as the distribution of stars, the shape of our galaxy, and the absorption of light in interstellar space, a phenomenon which he was the first to establish. Struve's associates and students — in fact, several other members of the Struve family — further expanded the study of astronomy in Russia.

Physics and chemistry also developed in the reigns of Alexander I and Nicholas I. Russian contributors to these branches of knowledge included an early experimental physicist in electricity and other fields, Professor Basil Petrov, who was on the staff of the Medical-Surgical Academy and taught himself physics, and a distinguished chemist, Professor Nicholas Zinin. Zinin worked and taught in Kazan and St. Petersburg and established the first prominent school of Russian chemists. He is perhaps best remembered as a pioneer in the production of aniline dyes.

The natural sciences in Russia grew with the physical, their practitioners including such luminaries as the great Baltic German embryologist Academician Charles Ernest Baer. As in the eighteenth century, the natural sciences were enriched by some remarkable expeditions and discoveries. Russians continued to explore Siberia and traveled repeatedly from the Baltic "around the world" to Alaska. They discovered numerous islands in the Pacific Ocean, which, however, the Russian government did not choose to claim. And in 1821 an expedition led by Thaddeus Belingshausen discovered the antarctic continent.

The humanities and the social sciences progressed similarly in Russia in the first decades of the nineteenth century. Oriental studies, for example, profited both from Russia's proximity to much of Asia and from Uvarov's special patronage. They became established in several universities and made important contributions to knowledge, ranging from pioneer descriptions of some Central Asiatic peoples to Father Iakinf Bichurin's fundamental work on China. Indeed the Russian Orthodox mission in Peking served from the time of Peter the Great to the revolutions of 1917 as an institute of sinology.

The writing of history was developed and gained a new public. Nicholas Karamzin, who must be mentioned more than once in connection with the evolution of the Russian language and literature, also became the first widely popular historian. His richly documented twelve-volume *History of the Russian State,* which began to appear in 1816 and which was left unfinished in the account of the Time of Troubles when the author died in 1826, won the enthusiastic acclaim of the educated public, who enjoyed Karamzin's extremely readable reconstruction of the colorful Russian past. The historian, to be sure, tried to edify as well as entertain: he argued throughout his work that autocracy'and a strong state made Russia great and must remain inviolable. In 1811 Karamzin had expressed similar views more succinctly in his secret *Memoir on Ancient and Modern Russia* given to Alexander I to counteract Speransky's reformist influence. In Russian universities new chairs were founded in history. The hard-working Michael Pogodin, a proponent of Official Nationality, became in 1835 the first professor of Russian history proper at the University of Moscow, to be succeeded in 1845 by a much greater scholar, Serge Soloviev, the bulk of whose work, however, belongs to Alexander II's reign.

Language and Literature

The Russian language evolved further, and so did linguistic and literary studies. If the writings of Karamzin marked the victory of the new style over the old, those of Pushkin already represented the apogee of modern Russian language and literature and became their classic model. The simplicity, precision, grace, and flow of Pushkin's language testify to the enormous development of the Russian literary language since the time of Peter the Great. Such opponents of this process as the reactionary Admiral Alexander Shishkov, who served from 1824 to 1828 as minister of education, fought a losing battle. While writers developed the Russian language, scholars studied it. The first decades of the nineteenth century witnessed the work of the remarkable philologist Alexander Vostokov and the early studies of several other outstanding linguistic scholars. Literary criticism rose to a new prominence. The critics ranged from conservative university professors, typified by Stephen Shevyrev of the University of Moscow, who adhered to the doctrine of Official Nationality, to'the radical firebrand Vissarion Belinsky. Indeed, we shall see that with Belinsky literary criticism in Russia acquired sweeping social, political, and generally ideological significance.

Literature constituted the chief glory of Russian culture in the first half of the nineteenth century, owing to the genius of several writers. It remains the most highly prized legacy from the time of Alexander I and Nicholas I, whether in the Soviet Union, with a virtual cult of Pushkin, or in other countries where such works as *Eugene Onegin* and *Dead Souls* are read.

Karamzin's sentimentalism, mentioned in an earlier chapter, which was popular at the end of the eighteenth and in the first years of the nineteenth century, gradually lost its appeal, while Karamzin himself turned, as we know, to history. New literary trends included what both pre-revolutionary and Soviet scholars have described as romanticism and realism in their various aspects. Romanticism produced no supreme literary figure in Russia except the poet Theodore Tiutchev, 1803–73, who spent much of his life in Germany and had little influence in his native land. It did, however, attract a number of gifted poets and writers and also contributed to the artistic growth of such giants as Lermontov, Pushkin, and Gogol. Of the Russian romanticists proper, Basil Zhukovsky deserves mention. Zhukovsky, who lived from 1783 to 1852, faithfully reflected in his poetry certain widespread romantic moods and traits: sensitivity and concern with subjective feelings, an interest in and idealization of the past, a penchant for the mysterious and the weird. On the whole the poet represented the humane, elegiac, and contemplative, rather than the "demonic" and active, aspects of romanticism. Zhukovsky's value for Russian literature lies in the novel lightness and music of his verse, in the variety of literary forms that he utilized successfully for his poetry, and in his numerous and generally splendid translations. In addition to translating superbly into Russian some works of such contemporary or near-contemporary Western writers as Schiller, Zhukovsky gave his readers an enduring Russian text of Homer's *Odyssey,* translated, characteristically enough, from the German. Incidentally, in 1829 Russians obtained Nicholas Gnedich's excellent translation of the *Iliad* from the Greek.

Realism fared better in Russia than romanticism, a fact which many nineteenth-century and especially Soviet critics have not ceased to point out. They have felt, furthermore, that with realism Russian literature finally achieved true independence and originality and established a firm foundation for lasting greatness. A difficult concept to use, the term *realism* has been applied to a variety of literary developments in Russia in the first half of the nineteenth century. In a sense, the writer of fables, Ivan Krylov, was its best practitioner. Krylov, who lived from 1768 to 1844, but began to write fables only in his late thirties after concentrating unsuccessfully on comedy, tragedy, and satire, achieved something like perfection in his new genre, rivaling such world masters of the fable as Aesop and La Fontaine. Krylov's approximately two hundred fables, which became best sellers as they appeared during the author's lifetime and have remained best sellers ever since, win the reader by the richness and raciness of their popular language, the vividness, precision, and impeccable wording of their succinct narrative, and their author's power of human observation and comment. While animals often act as protagonists, their foibles and predicaments serve as apt illustrations both of Krylov's Russia and of the human condition in general.

Alexander Griboedov's allegiance to realism seems less convincing than Krylov's. That brilliant writer, whose life began in 1795 and ended violently in 1829 when a Persian mob killed him in the Russian legation in Teheran, achieved immortality through one work only: the comedy *Gore ot uma,* translated into English as *Woe from Wit* or as *The Misfortune of Being Clever.* This masterpiece was finished in 1824, but, because of its strong criticism of Russian high society, was put on the stage only in 1831 and then with numerous cuts. *Gore ot uma* is neoclassical in form and contains very little action, but it overflows with wit. It consists almost entirely of sparkling, grotesque, or caustic statements and observations by its many characters, from a saucy maid to the embittered hero Chatsky — all set in the milieu of Muscovite high society. Its sparkle is such that Griboedov's play possesses an eternal freshness and effervescence, while many of its characters' observations — like many lines from Krylov's fables — have become part of the everyday Russian language. Nor, of course, does a comic form exclude serious content. *Gore ot uma* has been praised as the outstanding critique of the leading circles of Russian society in the reign of Alexander I, as a perspicacious early treatment of the subject of the conflict of generations — a theme developed later by Turgenev and other Russian writers — and as providing in its main character, Chatsky, a prototype of the typical "superfluous" hero of Russian literature, at odds with his environment.

Like Griboedov, Alexander Pushkin, the greatest Russian writer of the age, was born near the end of the eighteenth century and became famous in the last years of Alexander I's reign. Again like Griboedov, Pushkin had but a short life to live before meeting violent death. He was born in 1799 and was killed in a duel in 1837. Between 1820, which marked the completion of his first major poem, the whimsical and gently ironic *Ruslan and Liudmila,* and his death, Pushkin established himself permanently as, everything considered, the greatest Russian poet and one of the greatest Russian prose writers, as a master of the lyric, the epic, and the dramatic forms, and even as a literary critic, publicist, and something of a historian and ethnographer. Pushkin's early works, such as *The Fountain of Bakhchisarai* and *The Prisoner of the Caucasus,* magnificent in form, reflected a certain interest in the unusual and the exotic that was characteristic of the age. However, as early as *Eugene Onegin,* written in 1822–31, Pushkin turned to a penetrating and remarkably realistic treatment of Russian educated society and its problems. Onegin became one of the most effective and compelling figures in modern Russian literature, while both he and the heroine of the poem, Tatiana Larina, as well as their simple story, were to appear and reappear in different variations and guises in the works of Lermontov, Turgenev, Goncharov, and many other writers. While *Eugene Onegin* was written in most elegant verse, Pushkin also contributed greatly to the development of Russian prose, especially by such tales as the celebrated *A Captain's Daughter.* In his prose even more than in his poetry Pushkin has been considered a founder of realism in Russia and thus an originator of the main current of modern Russian literature. Pushkin's deeply sensitive and versatile genius ranged from unsurpassed personal lyrics to historical themes — for example, in the tragedy

Boris Godunov and in the long poem, *Poltava*, glorifying his recurrent hero, Peter the Great — and from realistic evocations of the Russia of his day to marvelous fairy tales in verse. He was busily engaged in publishing a leading periodical, *The Contemporary*, and in historical studies when he was killed.

Pushkin's genius has often been described as "classical." Its outstanding characteristic consisted in an astounding sense of form, harmony, and measure, which resulted in perfect works of art. The writer's fundamental outlook reflected something of the same classical balance: it was humane, sane, and essentially affirmative and optimistic. Not that it excluded tragedy. A long poem, *The Bronze Horseman,* perhaps best expressed Pushkin's recognition of tragedy in the world. It depicted a disastrous conflict between an average little man, Eugene, and the bronze statue of the great founder of St. Petersburg, who built his new capital on virtually impassable terrain, where one of the recurrent floods killed Eugene's beloved: a conflict between an individual and the state, human desire and necessity, man and his fate. Yet — although a minority of specialists, including such important critics as Briusov and Lednicki, reject this reading of the poem — *The Bronze Horseman,* too, affirms Peter the Great's work, modern Russia, and life itself.

Pushkin's genius appeared in Russia at the right time. A century of labor since Peter the Great's reforms had fashioned a supple modern language, developed literary forms, and established Russia as a full participant in the intellectual life of Europe. Pushkin, who knew French almost as well as Russian, profited greatly by the riches of Western literature — from Shakespeare to Pushkin's contemporaries — as well as by Russian popular speech and folklore. Yet, while the stage had been set for Pushkin, it was not cluttered. The great writer could thus be the first to realize the potential of modern Russian verse as well as modern Russian prose, of lyric poetry as well as factual narrative, and set the standard. His sweeping influence extended beyond language and literature to the other arts in Russia, and especially to music — where composers, ranging in time from Glinka and Dargomyzhsky through Musorgsky, Rimsky-Korsakov, and Tchaikovsky to Rachmaninov and Stravinsky, created more than twenty operas on the basis of his works. Indeed, he appeared to incarnate the entire glorious spring of Russian literature and culture. Another very great lyric poet, Theodore Tiutchev, expressed this best when he concluded a poem devoted to the tragedy of Pushkin's death: "You, like first love, the heart of Russia will not forget."

If Pushkin is generally regarded as the greatest Russian poet, Michael Lermontov, who also lived and wrote in the first half of the nineteenth century, has often been considered the second greatest. Born in 1814 and killed in a duel in 1841, Lermontov began writing at a very early age and left behind him a literary legacy of considerable size. Very different in temperament and outlook from Pushkin, Lermontov came closest to being the leading romantic genius of Russian letters, the "Russian Byron." His life was a constant protest against his environment, a protest which found expression both in public gestures, such as his stunning poem condemning Russian high society for the death of Pushkin, and in private troubles which resulted in his own death. Lermontov often chose fantastic, exotic, and highly subjective themes, set in the grandeur of the Caucasus, where he spent some time in the army. Throughout most of his life he kept writing and rewriting a magnificent long poem called *A Demon:*

> I am he, whose gaze destroys hope,
> As soon as hope blooms;
> I am he, whom nobody loves,
> And everything that lives curses.

Yet to describe Lermontov as a romantic poet, even a supreme romantic poet, does not do him full justice. For Lermontov's poetic genius had a broad range and kept developing — many critics think it developed toward realism. Also, through his prose writings, particularly his short novel *A Hero of Our Times,* he became one of the founders of the Russian realistic novel, in subject matter as well as in form. Such a discerning critic as Mirsky considers Lermontov's superbly powerful, succinct, and transparent prose superior even to Pushkin's. Lermontov, no doubt, could have done much else had he not been shot dead at the age of twenty-six.

While Pushkin and Lermontov were, in spite of their enormous contribution to Russian prose, primarily poets, Nicholas Gogol's early venture into poetry proved to be an unmitigated disaster. But as a prose writer Gogol had few equals and no superiors, in Russia or anywhere else. Gogol, who lived from 1809 to 1852, came from provincial Ukrainian gentry, and the characteristic society of his stories and plays stood several rungs lower on the social ladder than the world of Chatsky and Onegin. Gogol's first collection of tales, *Evenings on a Farm near Dikanka,* which came out in 1831 and received immediate acclaim, sparkled with a generally gay humor and the bright colors of Ukrainian folklore. The gaiety and the folklore, as well as a certain majestic tone and grand manner — much admired by some critics, but considered affected by others — were to appear in Gogol's later works, for example, the famous cossack prose epic, *Taras Bulba,* which dealt with the struggle of the Ukrainians against the Poles. However, gradually, the real Gogol emerged in literature: the Gogol of the commonplace and the mildly grotesque, which he somehow shaped into an overwhelming psychological world all his own; the Gogol who wrote in an involved, irregular and apparently clumsy style, which proved utterly irresistible. Occasionally, for instance in the stories *Notes of a Madman* and *A Nose,* weird content paralleled these magical literary powers. More frequently, as in the celebrated play, *The Inspector General,* and in Gogol's masterpiece, the novel *Dead Souls,* the subject matter contained nothing out of the ordinary and the plot showed little development.

Dead Souls, published in 1842, demonstrates the scope and might of Gogol's genius and serves as the touchstone for different interpretations of Gogol. That simple story of a scoundrel, Chichikov, who proceeded to visit provincial landlords and buy up their dead serfs — serfs were called "souls" in Russia—to use them in business deals as if they were alive, has been hailed, and not at all unjustly, by critics all the way from Belinsky to the latest Soviet scholars as a devastating, realistic, satirical picture of rural Russia under Nicholas I. But there seems to be much more to Gogol's novel. The landlords of different psychological types whom Chichikov meets, as well as Chichikov himself, appear to grow in vitality with the years, regardless of the passing of that society which they are supposed to mirror faithfully, for, indeed, they are "much more real than life." Russian formalist critics and such writers as Merezhkovsky and Nabokov deserve credit for emphasizing these other "non-realistic" aspects and powers of Gogol. The great novelist himself, it might be added, did not know what he was doing. His withering satire, applauded by the opponents of the existing system in Russia, stemmed directly from his weird genius, not from any ideology of the Left. In fact, in the second volume of *Dead Souls* Gogol tried to reform his characters and save Russia. That project, of course, failed. Still trying to resurrect Russian society, Gogol published in 1847 his unbelievably naïve and reactionary *Selected Passages from Correspondence with Friends,* which suggested, for example, that serfs should remain illiterate and shocked educated Russia. Gogol attempted also to find salvation for himself — and, by extension, for Russia — in religious experience, but to no avail. He died in 1852 after a shattering nervous breakdown when he burned much of the sequel to the first volume of *Dead Souls.*

Karamzin, Zhukovsky, Krylov, Griboedov, Pushkin, Lermontov, and Gogol were by no means the only Russian authors in the reigns of Alexander I and Nicholas I. While no extended discussion of the subject can be offered in a textbook, it should be realized, for instance, that Pushkin did not stand alone, but was the outstanding member of a brilliant generation of poets. Again, the prose writers included, in addition to those already mentioned, the magnificent narrator of provincial gentry life, Serge Aksakov, and other gifted authors. Moreover, pre-reform Russia saw much of the work of another supreme lyric poet who has already been mentioned, Theodore Tiutchev, as well as the first publications of such giants of Russian literature as Turgenev, Dostoevsky, and Tolstoy. It was a golden age.

Ideologies

In spite of the reaction of the last part of Alexander's reign and the steady repression under Nicholas, the first half of the nineteenth century proved to be creative not only in literature but also in Russian political and social thought and in the building of ideologies in general. Herzen could well refer to it as an amazing period of outward political slavery and inward intellectual emancipation. Again Russia profited from its association with the West and from the work performed throughout the eighteenth century in developing education and culture in the country. As we saw earlier, educated Russians shared in the Enlightenment, and indeed, after the outbreak of the French Revolution, produced the first Russian martyrs of the radical ideas of the Age of Reason, such as Novikov and especially Radishchev. Eighteenth-century liberalism or radicalism persisted in the nineteenth century in groups as different as Alexander I's Unofficial Committee and the Decembrists. But on the whole the intellectual scene began to change drastically. Romanticism and German idealistic philosophers replaced the Enlightenment and French *philosophes* as guides for much of European thought. The new intellectual *Zeitgeist* affirmed deep, comprehensive knowledge — often with mystical or religious elements — in opposition to mere rationalism, an organic view of the world as against a mechanistic view, and the historical approach to society in contrast to a utilitarian attitude with its vision limited to the present. It also emphasized such diverse doctrines as struggle and the essential separateness of the component parts of the universe in place of the Enlightenment ideals of harmony, unity, and cosmopolitanism. And it stressed the supreme value of art and culture. In the new world of romanticism such strange problems as the true nature of nations and the character of their missions in history came to the fore.

Romanticism and idealistic philosophy penetrated Russia in a variety of ways. For example, a number of professors, typified by Michael Pavlov who taught physics, mineralogy, and agronomy at the University of Moscow, presented novel German ideas in their lectures in the first decades of the nineteenth century. Educated Russians continued to read voraciously and were strongly influenced by Schiller and other brilliant Western romanticists. Of course, the subjects of the tsar were also Europeans and thus could not help but be part of European intellectual movements. While some Russians showed originality in developing different currents of Western thought, and while in general the Russian response to romantic

Ministries opposite the Winter Palace, Leningrad, 1819-29.

Cathedral of Our Lady of Kazan in Leningrad, 1801–11.

Ivan the Terrible and His Son by Repin.

View of the Admiralty, 1806–10, and St. Isaac's Cathedral, 1768–1858, Leningrad.

Petrodvorets (Peterhof), summer palace built by Peter the Great, 1722–50, near Leningrad.

The Cossacks of the Zaporozhie Writing a Letter to the Turkish Sultan by Repin.

Cathedral of St. Dmitrii, 1194–7, in Vladimır.

A church in ancient Suzdal.

Ivan the Terrible

Catherine the Great

Peter I, the Great

Ivan III, the Great

Leo Tolstoy

Fedor Dostoevsky

Vissarion Belinsky

Ivan Turgenev

ideas can be considered creative rather than merely imitative, there is no convincing reason for dissociating Russian intellectual history of the first half of the nineteenth century from that of the rest of Europe, whether in the name of the alleged uniquely religious nature of the ideological development in Russia or in order to satisfy the peculiar Soviet nationalism.

In particular, two German philosophers, Schelling first and then Hegel, exercised strong influence on the Russians. Schelling affected certain professors and a number of poets — the best Russian expression of some Schellingian views can be found in Tiutchev's unsurpassed poetry of nature — and also groups of intellectuals and even schools of thought, such as the Slavophile. It was largely an interest in Schelling that led to the establishment of the first philosophic "circle" and the first philosophic review in Russia. In 1823 several young men who had been discussing Schelling in a literary group formed a separate society with the study of German idealistic philosophy as its main object. The circle chose the name of "The Lovers of Wisdom" and came to contain a dozen members and associates, many of whom were to achieve prominence in Russian intellectual life. It published four issues of a journal, *Mnemosyne*. The leading Lovers of Wisdom included a gifted poet, Dmitrii Venevitinov, who died in 1827 at the age of twenty-two, and Prince Vladimir Odoevsky, 1803–69, who developed interesting views concerning the decline of the West and the great future of Russia to issue from the combination and fruition of both the pre-Petrine and the Petrine heritages. The Lovers of Wisdom reflected the romantic temper of their generation in a certain kind of poetic spiritualism that pervaded their entire outlook, in their worship of art, in their pantheistic adoration of nature, and in their disregard for the "crude" aspects of life, including politics. The group disbanded after the Decembrist rebellion in order not to attract police attention.

A decade later, the question of the nature and destiny of Russia was powerfully and shockingly presented by Peter Chaadaev. In his *Philosophical Letter,* published in the *Telescope* in 1836, Chaadaev argued, in effect, that Russia had no past, no present, and no future. It had never really belonged to either the West or the East, and it had contributed nothing to culture. In particular, Russia lacked the dynamic social principle of Catholicism, which constituted the basis of the entire Western civilization. Indeed, Russia remained "a gap in the intellectual order of things." Chaadaev, who was officially proclaimed deranged by the incensed authorities after the publication of the letter, later modified his thesis in his *Apology of a Madman.* Russia, he came to believe, did enter history through the work of Peter the Great and could obtain a glorious future by throwing all of its fresh strength into the construction of the common culture of Christendom.

Russian intellectual life grew apace in the 1840's and 1850's. Spurred by Schelling, by an increasing Hegelian influence, and by German romantic thought in general, as well as by the new importance of Russia in Europe ever since the cataclysm of 1812 and by the blossoming of Russian culture, several ideologies emerged to compete for the favor of the educated public. Official Nationality, which we considered in an earlier chapter, represented the point of view of the government and the Right. While it cannot be included in what Herzen called "intellectual emancipation," it did possess influential spokesmen among professors and writers, not to mention censors and other officials, and played a prominent role on the Russian scene. On the one hand, Official Nationality may be regarded as a culmination of reactionary currents in Russia, which found earlier protagonists in such figures as Rostopchin, Shishkov, Magnitsky, and in part Karamzin. On the other hand, it too, in particular its more nationalistic wing that was typified by the Moscow University professors Michael Pogodin and Stephen Shevyrev, testified to the impact of German romanticism on Russia. The Slavophiles and the Westernizers developed the two most important independent, as opposed to government-sponsored, schools of thought. The Petrashevtsy, by contrast, had a briefer and more obscure history. But they did represent yet another intellectual approach to certain key problems of the age.

The Slavophiles were a group of romantic intellectuals who formulated a comprehensive and remarkable ideology centered on their belief in the superior nature and supreme historical mission of Orthodoxy and of Russia. The leading members of the group, all of them landlords and gentlemen-scholars of broad culture and many intellectual interests, included Alexis Khomiakov who applied himself to everything from theology and world history to medicine and technical inventions, Ivan Kireevsky who has been called the philosopher of the movement, his brother Peter who collected folk songs and left very little behind him in writing, Constantine Aksakov, a specialist in Russian history and language, Constantine's brother Ivan, later prominent as a publicist and a Pan-Slav, and George Samarin who was to have a significant part in the emancipation of the serfs and who wrote especially on certain religious and philosophical topics, on the problem of the borderlands of the empire, and on the issue of reform in Russia. This informal group, gathering in the salons and homes of Moscow, flourished in the 1840's and 1850's until the death of the Kireevsky brothers in 1856 and of Khomiakov and Constantine Aksakov in 1860.

Slavophilism expressed a fundamental vision of integration, peace, and harmony among men. On the religious plane it produced Khomiakov's concept of *sobornost,* an association in love, freedom, and truth of believers, which Khomiakov considered the essence of Orthodoxy. Historically, so the Slavophiles asserted, a similar harmonious integration of individuals could be found in the social life of the Slavs, notably in the peasant commune — described as "a moral choir" by Constantine Aksakov — and in such other ancient Russian institutions as the zemskii sobor. Again, the family represented the principle of integration in love, and the same spirit could pervade other associations of men. As against love, freedom, and co-operation stood the world of rationalism, necessity, and compulsion. It too existed on many planes, from the religious and metaphysical to that of everyday life. Thus it manifested itself in the Roman Catholic Church — which had chosen rationalism and authority in preference to love and harmony and had seceded from Orthodox Christendom — and, through the Catholic Church, in Protestantism and in the entire civilization of the West. Moreover, Peter the Great introduced the principles of rationalism, legalism, and compulsion into Russia, where they proceeded to destroy or stunt the harmonious native development and to seduce the educated public. The Russian future lay clearly in a return to native principles, in overcoming the Western disease. After being cured, Russia would take its message of harmony and salvation to the discordant and dying West. It is important to realize that the all-embracing Slavophile dichotomy represented — as pointed out by Stepun and others — the basic romantic contrast between the romantic ideal and the Age of Reason. In particular, as well as in general, Slavophilism fits into the framework of European romanticism, although the Slavophiles showed considerable originality in adapting romantic doctrines to their own situation and needs and although they also experienced the influence of Orthodox religious thought and tradition.

In its application to the Russia of Nicholas I the Slavophile teaching often produced paradoxical results, antagonized the government, and baffled Slavophile friends and foes alike. In a sense, the Slavophiles were religious anarchists, for they condemned all legalism and compulsion in the name of their religious ideal. Yet, given the sinful condition of man, they granted the necessity of government and even expressed a preference for autocracy: in addition to its historical roots in ancient Russia, autocracy possessed the virtue of placing the entire weight of authority and compulsion on a single individual, thus liberating society from that heavy burden; besides, the Slavophiles remained unalterably opposed to Western constitutional and other legalistic and formalistic devices. Yet this justification of autocracy remained historical and functional, therefore relative, never religious and absolute. Furthermore, the Slavophiles desired the emancipation of the serfs and other reforms, and, above all, insisted on the "freedom of the life of the spirit," that is freedom of conscience, speech, and publication. As Constantine Aksakov tried to explain to the government: "Man was created by God as an intelligent and a talking being." Also, Khomiakov and his friends opposed such aspects of the established order as the death penalty, government intrusion into private life, and bureaucracy in general. "Thus the first relationship of the government and the people is the relationship of mutual non-interference. . . ." No wonder Slavophile publications never escaped censorship and prohibition for long.

The Westernizers were much more diverse than the Slavophiles, and their views did not form a single, integrated whole. Besides, they shifted their positions rather rapidly. Even socially the Westernizers consisted of different elements, ranging from Michael Bakunin who came from a gentry home like those of the Slavophiles, to Vissarion Belinsky whose father was an impoverished doctor and grandfather a priest, and Basil Botkin who belonged to a family of merchants. Yet certain generally held opinions and doctrines gave a measure of unity to the movement. The Slavophiles and the Westernizers started from similar assumptions of German idealistic philosophy, and indeed engaged in constant debate with each other, but came to different conclusions. While Khomiakov and his friends affirmed the uniqueness of Russia and the superiority of true Russian principles over those of the West, the other party argued that the Western historical path was the model that Russia had to follow. Russia could accomplish its mission only in the context of Western civilization, not in opposition to it. Naturally, therefore, the Westernizers took a positive view of Western political development and criticized the Russian system. Contrary to the Slavophiles, they praised the work of Peter the Great, but they wanted further Westernization. Also, whereas the Slavophiles anchored their entire ideology in their interpretation and appraisal of Orthodoxy, the Westernizers assigned relatively little importance to religion, while some of them gradually turned to agnosticism and, in the case of Bakunin, even to violent atheism. To be more exact, the moderate Westernizers retained religious faith and an essentially idealistic cast of mind, while their political and social program did not go beyond mild liberalism, with emphasis on gradualism and popular enlightenment. These moderates were typified by Nicholas Stankevich, who brought together a famous early Westernizer circle but died in 1840 at the age of twenty-seven before the movement really developed, and by Professor Timothy Granovsky, who lived from 1813 to 1855 and taught European history very successfully at the University of Moscow. The radical Westernizers, however, largely through Hegelianism and Left Hegelianism, came to challenge religion, society, and the entire Russian and European system, and to call for a revolution. Although few in number, they included such major figures as Vissarion Belinsky, 1811–48, Alexander Herzen, 1812–70, and Michael Bakunin, 1814–76.

Belinsky, the most famous Russian literary critic, exercised a major influence on Russian intellectual life in general. He had the rare good fortune to welcome the works of Pushkin, Lermontov, and Gogol and the debuts of Dostoevsky, Turgenev, and Nekrasov. Belinsky's commentary on the Russian writers became famous for its passion, invective, and eulogy, as well as for its determination to treat works of literature in the broader contexts of society, history, and thought, and to instruct and guide the authors and the reading public. Belinsky's own views underwent important changes and had not achieved cohesiveness and stability at the time of his death. His impact on Russian literature, however, proved remarkably durable and stable: it consisted above all in the establishment of political and social criteria as gauges for evaluating artistic works. As Nekrasov put it later, one did not have to be a poet, but one was under obligation to be a citizen. Following Belinsky's powerful example, political and social ideologies, banned from direct expression in Russia, came to be commonly expounded in literary criticism.

Both Herzen and Bakunin became prominent in the 1830's and 1840's, but lived well beyond the reign of Nicholas I. Moreover, much of their activity, such as Herzen's radical journalistic work abroad and Bakunin's anarchist theorizing and plotting, belonged to the time of Alexander II and will have to be mentioned in a subsequent chapter. Yet their intellectual evolution in the decades preceding the "great reforms" formed a significant part of that seminal period of Russian thought. Herzen, whose autobiographical account *My Past and Thoughts* is one of the most remarkable works of Russian literature, came from a well-established gentry family, like the Slavophiles and Bakunin, but was an illegitimate child. He became a leading opponent of Khomiakov in the Muscovite salons and a progressive Westernizer. Gradually Herzen abandoned the doctrines of idealistic philosophy and became increasingly radical and critical in his position, stressing the dignity and freedom of the individual. In 1847 he left Russia, never to return. Bakunin has been described as "founder of nihilism and apostle of anarchy" — Herzen said he was born not under a star but under a comet — but he began peacefully enough as an enthusiast of German thought, especially Hegel's. Several years earlier than Herzen, Bakunin too left Russia. Before long he turned to Left Hegelianism and moved beyond it to anarchism and a sweeping condemnation of state, society, economy, and culture in Russia and in the world. Bakunin emphasized destruction, proclaiming in a signal early article that the passion for destruction was itself a creative passion. While Herzen bitterly witnessed the defeat of the revolution of 1848 in Paris, Bakunin attended the Pan-Slav Congress in Prague and participated in the revolution in Saxony. After being handed over by the Austrian government to the Russian, he was to spend over a decade in fortresses and in Siberian exile. Both Herzen, disappointed in the

West, and Bakunin, ever in search of new opportunities for revolution and anarchism, came to consider the peasant commune in Russia as a superior institution and as a promise of the future social transformation of Russia — a point made earlier by the Slavophiles, although, of course, from different religious and philosophical positions — thus laying the foundation for subsequent native Russian radicalism.

The Petrashevtsy were another kind of radicals. That informal group of two score or more men, who from late 1845 until their arrest in the spring of 1849 gathered on Fridays at the home of Michael Butashevich-Petrashevsky in St. Petersburg, espoused especially the teaching of the strange French utopian socialist Fourier. Fourier preached the peaceful transformation of society into small, well-integrated, and self-supporting communes, which would also provide for the release and harmony of human passions according to a fantastic scheme of his own invention. Many Petrashevtsy, however, added to Fourierism political protest, demand for reform, and general opposition to the Russia of Nicholas I. The government took such a serious view of the situation that it condemned twenty-one men to death, although it changed their sentence at the place of execution in favor of less drastic punishments. It was as a member of the Petrashevtsy that Dostoevsky faced imminent execution and later went to Siberia. The Petrashevtsy, it might be added, came generally from lower social strata than did the Lovers of Wisdom, the Slavophiles, and the Westernizers, and included mostly minor officials, junior officers, and students.

Several trends in the intellectual history of Russia in the first half of the nineteenth century deserve attention. If we exclude the Decembrists as belonging ideologically to an earlier period, Russian thought moved from the abstract philosophizing and the emphasis on esthetics characteristic of the Lovers of Wisdom, through the system-building of the Slavophiles and, to a lesser extent, the Westernizers to an increasing concern with the pressing issues of the day, as exemplified by the radical Westernizers and, in a different sense, by the Petrashevtsy. At the same time radicalism grew among the educated Russians, especially as German idealistic philosophy and romanticism in general disintegrated. Moreover, socialism entered Russian history, both through such individuals as Herzen and his life-long friend Nicholas Ogarev and through an entire group of neophytes, the Petrashevsty. Also, the intellectual stratum increased in number and changed somewhat in social composition, from being solidly gentry, as the Slavophiles still were, to a more mixed membership characteristic of the Westernizers and the Petrashevsty. All in all, Russian thought in the reigns of Alexander I and Nicholas I, and especially the "intellectual emancipation" of the celebrated forties, was to have a great impact on the intellectual evolution of Russia and indeed on Russian history all the way to 1917 and even beyond.

The Arts

While contemporaries and later many scholars showed special interest in the Russian literature and thought of the first half of the nineteenth century, the fine arts, too, continued to develop in the reigns of Alexander I and Nicholas I. Both emperors were enthusiastic builders in the tradition of Peter and Catherine. At the time of Alexander the neo-classical style, often skillfully adapted to native traditions, reached its height in Russia. It affected not only the appearance of St. Petersburg, Moscow, and other towns, but also the architecture of countless manor houses all over the empire throughout the nineteenth century. The leading architects of Alexander's reign included Hadrian Zakharov, who created the remarkable Admiralty building in St. Petersburg, and Andrew Voronikhin, of serf origin, who constructed the Kazan Cathedral in the capital and certain imperial palaces outside it. Under Nicholas neo-classicism gave way to an eclectic mixture of styles.

Largely guided by the Academy of Arts, painting evolved gradually from neo-classicism to romanticism as exemplified by Karl Briullov's enormous canvas "The Last Day of Pompeii." A few more realistic genre painters also began to appear. Music grew in quantity, quality, and appeal. In particular, Russian opera developed, and it obtained a lasting position in Russia and elsewhere through the genius of Michael Glinka, 1804–57, and the talents of other able composers such as Alexander Dargomyzhsky, 1813–69. As elsewhere in Europe, Russian opera and the Russian musical school generally stressed folk songs, melodies, and motifs. The theater, the ballet, and the opera attracted increasing state support and public interest. The theater profited from the new Russian dramatic literature, which included such masterpieces as *Woe from Wit* and *The Inspector General,* and the emergence of brilliant actors and even traditions of acting. Public theaters existed in many towns, while some landlords continued to establish private theaters on their estates, with serfs as actors. In the ballet too, under the guidance of French and Italian masters, standards improved and a tradition of excellence developed.

On the whole, Chaadaev's claim that Russia had contributed nothing to culture, outrageous in 1836, would have found even less justification in 1855 or 1860. Yet, as the Slavophiles, Herzen, and other thinking Russians realized, not all was well: there remained an enormous gulf between the educated society and the people, between the fortunate few on top and the broad masses. Something had to be done. The future of Russia depended on the "great reforms."

THE REIGN OF ALEXANDER II, 1855–81

> However, sounds of music reached our ears, and we all hurried back
> to the hall. The band of the opera was already playing the hymn,
> which was drowned immediately in enthusiastic hurrahs coming from
> all parts of the hall. I saw Baveri, the conductor of the band, waving
> his stick, but not a sound could be heard from the powerful band.
> Then Baveri stopped, but the hurrahs continued. I saw the stick
> waved again in the air; I saw the fiddle-bows moving, and musicians
> blowing the brass instruments, but again the sound of voices over-
> whelmed the band. . . . The same enthusiasm was in the streets.
> Crowds of peasants and educated men stood in front of the palace,
> shouting hurrahs, and the Tsar could not appear without being fol-
> lowed by demonstrative crowds running after his carriage. . . . I
> was in Nikolskoe in August, 1861, and again in the summer of 1862,
> and I was struck with the quiet, intelligent way in which the peasants
> had accepted the new conditions. They knew perfectly.well how diffi-
> cult it would be to pay the redemption tax for the land, which was in
> reality an indemnity to the nobles in lieu of the obligations of serf-
> dom. But they so much valued the abolition of their personal enslave-
> ment that they accepted the ruinous charges — not without mur-
> muring, but as a hard necessity — the moment that personal freedom
> was obtained. . . . When I saw our Nikolskoe peasants, fifteen
> months after the liberation, I could not but admire them. Their in-
> born good nature and softness remained with them, but all traces of
> servility had disappeared. They talked to their masters as equals talk
> to equals, as if they never had stood in different relations.
>
> KROPOTKIN

> The abolition of serfdom signified the establishment of capitalism as
> the dominant socio-economic formation in Russia.
>
> ZAIONCHKOVSKY

ALEXANDER II succeeded his father, Nicholas I, on the Russian throne at
the age of thirty-seven. He had received a rather good education as well as
considerable practical training in the affairs of state. Alexander's teachers
included the famous poet Zhukovsky, who has often been credited with
developing humane sentiments in his pupil. To be sure, Grand Duke Alexander remained an obedient son of his strong-willed father and showed no
liberal inclinations prior to becoming emperor. Indeed he retained an essentially conservative mentality and attitude throughout his life. Nor can
Alexander II be considered a strong or a talented man. Yet, forced by the
logic of the situation, the new monarch decided to undertake, and actually
carried through, fundamental reforms unparalleled in scope in Russian

history since Peter the Great. These reforms, although extremely important, failed to cure all the ills of Russia and in fact led to new problems and perturbations, which resulted, among other things, in the assassination of the "Tsar-Liberator."

The Emancipation of the Serfs

The last words of Alexander II's manifesto announcing the end of the Crimean War promised reform, and this produced a strong impression on the public. The new emperor's first measures, enacted even before the termination of hostilities, included the repeal of some of the Draconian restrictions of Nicholas I's final years, such as those on travel abroad and on the number of students attending universities. All this represented a promising prologue; the key issue, as it was for Alexander I, the last ruler who wanted to transform Russia, remained serfdom. However, much had changed in regard to serfdom during the intervening fifty or fifty-five years. Human bondage, as indicated in an earlier chapter, satisfied less and less effectively the economic needs of the Russian Empire. With the growth of a money economy and competition for markets, the deficiencies of low-grade serf labor became ever more obvious. Many landlords, especially those with small holdings, could barely feed their serfs; and the gentry accumulated an enormous debt. As we know, free labor, whether really free or merely the contractual labor of someone else's serfs, became more common throughout the Russian economy during the first half of the nineteenth century. Moreover, the serfs perhaps declined in absolute number in the course of that period, while their numerical weight in relation to other classes certainly declined: from 58 per cent of the total population of Russia in 1811 to 44.5 per cent on the eve of the "great reforms," to cite Blum's figures again. Recent interpretations of the Russian economic crisis in mid-nineteenth century range all the way from Kovalchenko's emphatic restatement, with the use of quantitative methods, of the thesis of the extreme and unbearable exploitation of the serfs to Ryndziunsky's stress on the general loosening of the social fabric. In any event, whether the landlords were willing to recognize it or not — and large vested interests seldom obey even economic reason — serfdom was becoming increasingly anachronistic.

Other powerful arguments for emancipation reinforced the economic. Oppressed and exasperated beyond endurance, the serfs kept rising against their masters. While no nineteenth-century peasant insurrection could at all rival the Pugachev rebellion, the uprisings became more frequent and on the whole more serious. Semevsky, using official records, had counted 550 peasant uprisings in the nineteenth century prior to the emancipation. A Soviet historian, Ignatovich, raised the number to 1,467 and gave the following breakdown: 281 peasant rebellions, that is, 19 per cent of the total, in the period from 1801 to 1825; 712 rebellions, 49 per cent, from 1826 to 1854; and 474 uprisings, or 32 per cent, in the six years and two months of Alexander II's reign before the abolition of serfdom. Ignatovich emphasized that the uprisings also increased in length, in bitterness, in the human and material losses involved, and in the military effort necessary to restore order. Still more recently, Okun and other Soviet scholars have further expanded Ignatovich's list of uprisings. Moreover, Soviet scholarship claims that peasant rebellions played the decisive role in the emancipation of the serfs, and that on the eve of the "great reforms" Russia experienced in effect a revolutionary situation. Although exaggerated, this view cannot be entirely dismissed. Interestingly, it was the Third Department, the gendarmery, that had stressed the danger of serfdom during the reign of Nicholas I. Besides rising in rebellion, serfs ran away from their masters, sometimes by the hundreds and even by the thousands. On occasion large military detachments had to be sent to intercept them. Pathetic mass flights of peasants, for example, would follow rumors that freedom could be obtained somewhere in the Caucasus, while crowds of serfs tried to join the army during the Crimean War, because they mistakenly believed that they could thereby gain their liberty.

A growing sentiment for emancipation, based on moral grounds, also contributed to the abolition of serfdom. The Decembrists, the Slavophiles, the Westernizers, the Petrashevtsy, some supporters of Official Nationality, together with other thinking Russians, all wanted the abolition of serfdom. As education developed in Russia, and especially as Russian literature came into its own, humane feelings and attitudes became more widespread. Such leading writers as Pushkin and particularly Turgenev, who in 1852 published in book form his magnificent collection of stories, *Sportsman's Sketches,* where serfs were depicted as full-blown, and indeed unforgettable, human beings, no doubt exercised an influence. In fact, on the eve of the abolition of serfdom in Russia — in contrast to the situation with slavery in the American South — virtually no one defended that institution; the arguments of its proponents were usually limited to pointing out the dangers implicit in such a radical change as emancipation.

Finally, the Crimean War provided additional evidence of the deficiencies and dangers of serfdom which found reflection both in the poor physical condition and listlessness of the recruits and in the general economic and technological backwardness of the country. Besides, as Rieber recently emphasized, Russia had essentially to rely on a standing army without a reserve, because the government was afraid to allow soldiers to return to villages.

At the time of the coronation, about a year after his assumption of power, Alexander II, addressing the gentry of Moscow, made the celebrated statement that it would be better to begin to abolish serfdom from above than to wait until it would begin to abolish itself from below, and asked the gentry to consider the matter. Although the government experienced great difficulty in eliciting any initiative from the landlords on the subject of emancipation, it finally managed to seize upon an offer by the gentry of the three Lithuanian provinces to discuss emancipation without land. The ensuing imperial rescript made it clear that emancipation was indeed official policy and, furthermore, that emancipation would have to be with land. At about the same time restrictions were lifted from the discussion of the abolition of serfdom in the press. In the wave of expectation and enthusiasm that swept the liberals and radicals after the publication of the rescript even Herzen exclaimed to Alexander II: "Thou hast conquered, O Galilean!"

Eventually, in 1858, gentry committees were established in all provinces to consider emancipation, while a bureaucratic Main Committee of nine members was set up in St. Petersburg. Except for a few diehards, the landlords assumed a realistic position and accepted the abolition of serfdom once the government had made its will clear, but they wanted the reform to be carried out as advantageously for themselves as possible. The gentry of southern and south-central Russia, with its valuable, fertile soil, wanted to retain as much land as possible and preferred land to a monetary recompense; the gentry of northern and north-central Russia, by contrast, considered serf labor and the resulting obrok as their main asset and, therefore, while relatively willing to part with much of their land, insisted on a high monetary payment in return for the loss of serf labor. Gentry committees also differed on such important issues as the desirable legal position of the liberated serfs and the administration to be provided for them.

The opinions of provincial committees went to the Editing Commission — actually two commissions that sat together and formed a single body — created at the beginning of 1859 and composed of public figures interested in the peasant question, such as the Slavophiles George Samarin and Prince Vladimir Cherkassky, as well as of high officials. After twenty months of work the Editing Commission submitted its plan of reform to the Main Committee, whence it went eventually to the State Council. After its quick consideration by the State Council, Alexander II signed the emancipation manifesto on March 3, 1861 — February 19, Old Style. Public announcement followed twelve days later.

Throughout its protracted and cumbersome formulation and passage the emancipation reform faced the hostility of conservatives in government and society. That a far-reaching law was finally enacted can be largely credited to the determined efforts of so-called "liberals," including officials such as Nicholas Miliutin, the immediate assistant to the minister of the interior and the leading figure in the Editing Commission, and participants from the public like George Samarin. Two members of the imperial family, the tsar's brother Grand Duke Constantine and the tsar's aunt Grand Duchess Helen, belonged to the "liberals." More important, Alexander II himself repeatedly sided with them, while his will became law for such devoted bureaucrats as Jacob Rostovtsev — a key figure in the emancipation — who cannot be easily classified as either "conservative" or "liberal." The emperor in effect forced the speedy passage of the measure through an antagonistic State Council, which managed to add only one noxious provision to the law, that permitting a "pauper's allotment," which will be mentioned later. Whereas the conservatives defended the interests and rights of the gentry, the "liberals" were motivated by their belief that the interests of the state demanded a thoroughgoing reform and by their views of what would constitute a just settlement.

The law of the nineteenth of February abolished serfdom. Thenceforth human bondage was to disappear from Russian life. It should be noted, however, that, even if we exclude from consideration certain temporary provisions that prolonged various serf obligations for different periods of time, the reform failed to give the peasants a status equal to that of other social classes: they had to pay a head tax, were tied to their communes, and were judged on the basis of customary law. In addition to landowners' serfs, the new freedom was extended to peasants on the lands of the imperial family and to the huge and complex category of state peasants.

Together with their liberty, serfs who had been engaged in farming received land: household serfs did not. While the detailed provisions of the land settlement were extremely complicated and different from area to area, the peasants were to obtain roughly half the land, that part which they had been tilling for themselves, the other half staying with the landlords. They had to repay the landlords for the land they acquired and, because few serfs could pay anything, the government compensated the gentry owners by means of treasury bonds. Former serfs in turn were to reimburse the state through redemption payments spread over a period of forty-nine years. As an alternative, serfs could take one-quarter of their normal parcel of land, the so-called "pauper's allotment," and pay nothing. Except in the Ukraine and a few other areas, land was given not to individual peasants, but to a peasant commune — called an *obshchina* or *mir,* the latter term emphasizing the communal gathering of peasants to settle their affairs — which divided the land among its members and was responsible for taxes, the provision of recruits, and other obligations to the state.

The emancipation of the serfs can be called a great reform, although an American historian probably exaggerated when he proclaimed it to be the greatest legislative act in history. It directly affected the status of some fifty-two million peasants, over twenty million of them serfs of private land owners. That should be compared, for example, with the almost simultaneous liberation of four million black slaves in the United States, obtained as a result of a huge Civil War, not by means of a peaceful legal process. The moral value of the emancipation was no doubt tremendous, if incalculable. It might be added that the arguments of Pokrovsky and some other historians attempting to show that the reform was a clever conspiracy between the landlords and the government at the expense of the peasants lack substance: they are contradicted both by the actual preparation and passage of the emancipation legislation and by its results, for it contributed in a major manner to the decline of the gentry. By contrast, those Soviet specialists and others who emphasize the importance of the abolition of serfdom for the development of capitalism in Russia stand on much firmer ground. The specific provisions of the new settlement have also been defended and even praised, especially on the basis of the understanding that the arrangement had to be a compromise, not a confiscation of everything the gentry owned. Thus, the emancipation of serfs in Russia has been favorably compared to that in Prussia at the beginning of the nineteenth century, and land allotments of Russian peasants, to allotments in several other countries.

And yet the emancipation reform also deserves thorough criticism. The land allotted to the former serfs turned out to be insufficient. While in theory they were to retain the acreage that they had been tilling for themselves prior to 1861, in fact they received 18 per cent less land. Moreover, in the fertile southern provinces their loss exceeded the national average, amounting in some cases to 40 per cent or more of the total. Also, in the course of the partitioning, former serfs often failed to obtain forested areas or access to a river with the result that they had to assume additional obligations toward their onetime landlords to satisfy their needs. Khodsky estimated that 13 per cent of the former serfs received liberal allotments of land; 45 per cent, allotments sufficient to maintain their families and economies; and 42 per cent, insufficient allotments. Liashchenko summarized the settlement as follows: "The owners, numbering 30,000 noblemen, retained ownership over some 95 million dessyatins of the better land immediately after the Reform, compared with 116 million dessyatins of suitable land left to the 20 million 'emancipated' peasants." Other scholars have stressed the overpopulation and underemployment among former serfs who, at least after a period of transition, were no longer obliged to work for the landlord and at the same time had less land to cultivate for themselves. State peasants, although by no means prosperous, received, on the whole, better terms than did the serfs of private owners.

The financial arrangement proved unrealistic and impossible to execute. Although liberated serfs kept meeting as best they could the heavy redemption payments, which were not related to their current income, the arrears kept mounting. By the time the redemption payments were finally abolished in 1905, former serfs paid, counting the interest, one and one half billion rubles for the land initially valued at less than a billion. It should be noted that while officially the serfs were to redeem only the land, not their persons, actually the payments included a concealed recompense for the loss of serf labor. Thus, more had to be paid for the first unit of land, the first desiatina, than for the following units. As a whole the landlords of southern Russia received 340 million rubles for land valued at 280 million; those of northern Russia, where obrok prevailed, 340 million rubles for land worth 180 million rubles. The suspect Polish and Polonized landlords of the western provinces constituted an exception, for they were given slightly less money than the just price of their land.

The transfer of land in most areas to peasant communes rather than to individual peasants probably represented another major error, although this is an extremely complex issue. Arguments in favor of the commune ranged from the Slavophile admiration of the moral aspects of that institution to the desire on the part of the government to have taxes and recruits guaranteed by means of communal responsibility and to the assertion that newly liberated peasants would not be able to maintain themselves but could find protection in the commune. While some of these and other similar claims had a certain validity — indeed, as a practical matter the government could hardly have been expected to break up the commune at the same time the serfs were being freed — the disadvantages of the commune outweighed its advantages. Of most importance was the fact that the commune tended to perpetuate backwardness, stagnation, and overpopulation in the countryside precisely when Russian agriculture drastically needed improvement and modernization.

The emancipation reform disappointed Russian radicals, who considered it inadequate, and it also, apparently, failed to satisfy the peasantry, or at least many peasants, for a rash of agrarian disturbances followed the abolition of serfdom, and the misery, despair, and anger in the countryside remained a powerful threat to imperial Russia until the very end of imperial rule.

Other "Great Reforms"

The emancipation of the serfs made other fundamental changes much more feasible. Alexander II and his assistants turned next to the reform of local government, to the establishment of the so-called zemstvo system. For centuries local government had remained a particularly weak aspect of Russian administration and life. The arrangement that the "Tsar-Liberator" inherited dated from Catherine the Great's legislation and combined bureaucratic management with some participation by the local gentry; the considerable manorial jurisdiction of the landlords on their estates formed another prominent characteristic of the pre-reform countryside. The new law, enacted in January 1864, represented a strong modernization and democratization of local government, as well as a far-reaching effort on the part of the state to meet the many pressing needs of rural Russia and to do this largely by stimulating local initiative and activity. Institutions of self-government, zemstvo assemblies and boards, were created at both the district and provincial levels — the word zemstvo itself connotes land, country, or people, as distinct from the central government. The electorate of the district zemstvo assemblies consisted of three categories: the towns, the peasant communes, and all individual landowners, including those not from the gentry. Representation was proportional to landownership, with some allowance for the possession of real estate in towns. The elections were indirect. Members of district assemblies, in turn, elected from their own midst, regardless of class, delegates to their provincial assembly. Whereas the district and provincial zemstvo assemblies, in which the zemstvo authority resided, met only once a year to deal with such items as the annual budget and basic policies, they elected zemstvo boards to serve continuously as the executive agencies of the system and to employ professional staffs. A variety of local needs fell under the purview of zemstvo institutions: education, medicine, veterinary service, insurance, roads, the establishment of food reserves for emergency, and many others.

The zemstvo system has legitimately been criticized on a number of counts. For example, for a long time it encompassed only the strictly Russian areas of the empire, some thirty-four provinces, not the borderlands. Also, it possessed a limited, many would say insufficient, right to tax. In broader terms, it represented merely a junior partner to the central government, which retained police and much administrative control in the countryside; a governor could in various ways interfere with the work of a zemstvo, but not vice versa. The smallest zemstvo unit, the district, proved too large for effective and prompt response to many popular needs, and the desirability of further zemstvo subdivision soon became apparent. The democracy of the system too had its obvious limitations: because they owned much land, members of the gentry were very heavily represented in the district assemblies, and even more so in the provincial assemblies and the zemstvo boards, where education, leisure, and means to cover the expenses incurred favored gentry delegates. Thus, according to one count, the gentry generally held 42 per cent of the district assembly seats, 74 per cent of the seats in the provincial assemblies, and 62 per cent of the positions on the zemstvo boards. Yet, even such a system constituted a great step toward democracy for autocratic and bureaucratic Russia. It might be added that the zemstvo institutions functioned effectively also in those areas, such as large parts of the Russian north, where there were no landlords and where peasants managed the entire system of local self-government.

Yet, in spite of its deficiencies — and it should be noted that most of the above-mentioned criticisms refer in one way or another to the insufficient extent of the reform and not to substantive defects in it — the zemstvo system accomplished much for rural Russia from its establishment in 1864 until its demise in 1917. Especially valuable were its contributions to public education and health. In effect, Russia obtained a kind of socialized medicine through the zemstvo long before other countries, with medical and surgical treatment available free of charge. As G. Fischer and other scholars have indicated, the zemstvo system also served, contrary to the intentions of the government, as a school for radicalism and especially liberalism which found little opportunity for expression on the national, as distinct from local, scene until the events of 1905 and 1906.

In 1870 a municipal reform reorganized town government and applied to towns many of the principles and practices of the zemstvo administration. The new town government, which was "to take care of and administer urban economy and welfare," consisted of a town council and a town administrative board elected by the town council. The town council was elected by all property owners or taxpayers; but the election was according to a three-class system, which gave the small group on top that paid a third of the total taxes a third of the total number of delegates, the middle taxpayers another third, and the mass at the bottom that accounted for the last third of taxes the remaining third of delegates.

At the end of 1864, the year that saw the beginning of the zemstvo administration, another major change was enacted into law: the reform of the legal system. The Russian judiciary needed reform probably even more than the local government did. Archaic, bureaucratic, cumbersome, corrupt, based on the class system rather than on the principle of equality before the law, and relying entirely on a written and secret procedure, the old system was thoroughly hated by informed and thinking Russians. Butashevich-Petrashevsky and other radicals attached special importance to a reform of the judiciary. A conservative, the Slavophile Ivan Aksakov, reminisced: "The old court! At the mere recollection of it one's hair stands on end and one's flesh begins to creep!" The legislation of 1864 fortunately marked a decisive break with that part of the Russian past.

The most significant single aspect of the reform was the separation of the courts from the administration. Instead of constituting merely a part of the bureaucracy, the judiciary became an independent branch of government. Judges were not to be dismissed or transferred, except by court action. Judicial procedure acquired a largely public and oral character instead of the former bureaucratic secrecy. The contending parties were to present their cases in court and have adequate legal support. In fact, the reform virtually created the class of lawyers in Russia, who began rapidly to acquire great public prominence. Two legal procedures, the general and the abbreviated one, replaced the chaos of twenty-one alternate ways to conduct a case. Trial by jury was introduced for serious criminal offenses, while justices of the peace were established to deal with minor civil and criminal cases. The courts were organized into a single unified system with the Senate at the apex. All Russians were to be equal before the law and receive the same treatment. Exceptions to the general system were the military and ecclesiastical courts, together with special courts for peasants who lived for the most part by customary law.

The reform of the judiciary, which was largely the work of the Minister of Justice Dmitrii Zamiatnin, his extremely important assistant Serge Zarudny, and several other enlightened officials, proved to be the most successful of the "great reforms." Almost overnight it transformed the Russian judiciary from one of the worst to one of the best in the civilized world. Later the government tried on occasion to influence judges for political reasons; and, what is more important, in its struggle against radicalism and revolution it began to withdraw whole categories of legal cases from the normal procedure of 1864 and to subject them to various forms of the courts-martial. But, while the reform of the judiciary could be restricted in application, it could not be undone by the imperial government; and, as far as the reform extended, modern justice replaced arbitrariness and confusion. Russian legal reform followed Western, especially French, models, but, as Kucherov and others have demonstrated, these models were skillfully adapted to Russian needs. It might be added that the courts, as well as the zemstvo institutions, acquired political significance, for they served as centers of public interest and enjoyed a somewhat greater freedom of expression than was generally allowed in Russia.

A reorganization of the military service in 1874 and certain changes within the army have usually been grouped as the last "great reform." Inspired by military needs and technically complex, the reform nevertheless exercised an important general impact on Russian society and contributed to the modernization and democratization of the country. It was executed by Minister of War Dmitrii Miliutin, Nicholas Miliutin's brother, who wanted to profit by the example of the victorious Prussian army. He introduced a variety of significant innovations, of which the most important was the change in military service. The obligation to serve was extended from the lower classes alone to all Russians, while at the same time the length of active service was drastically reduced — from twenty-five years in the beginning of Alexander II's reign to six after the reform of 1874 — and a military reserve was organized. Recruits were to be called up by lot; different exemptions were provided for hardship cases; and, in addition, terms of enlistment were shortened for those with education, a not unwarranted provision in Russian conditions. Miliutin also reformed military law and legal procedure, abolished corporal punishment in the army, strove to improve the professional quality of the officer corps and to make it somewhat more democratic, established specialized military schools, and, a particularly important point, introduced elementary education for all draftees. Measures similar to Miliutin's were carried out in the navy by Grand Duke Constantine.

Other reforms under Alexander II included such financial innovations as Valery Tatarinov's establishment of a single state treasury, publication of the annual budget, and the creation in 1866 of the State Bank to centralize credit and finance, as well as generally liberalizing steps with regard to education and censorship.

The "great reforms" went a long way toward transforming Russia. To be sure, the empire of the tsars remained an autocracy, but it changed in many other respects. Vastly important in themselves, the government's reforms also helped to bring about sweeping economic and social changes, which will be discussed in a later chapter. The growth of capitalism in Russia, the evolution of the peasantry, the decline of the gentry, the rise of the middle class, particularly the professional group, and also of the proletariat — all were affected by Alexander II's legislation. Indeed, Russia began to take long strides on the road to becoming a modern nation. Nor could the changes be undone: there was no return to serfdom or to pre-reform justice.

The Difficult Sixties

However, although the government could not return to the old ways, it could stop advancing on the new road and try to restrict and limit the effectiveness of the changes. And in fact it attempted to do so in the second half of Alexander II's reign, under Alexander III, and under Nicholas II until the Revolution of 1905. While the need for reforms had been apparent, the rationale of reaction proved less obvious and more complicated. For one thing, the reforms, as we know, had their determined opponents in official circles and among the Russian gentry, who did their best to reverse state policy. Special circumstances played their part, such as peasant uprisings, student disturbances, the unexplained fires of 1862, the Polish rebellion of 1863, and Dmitrii Karakozov's attempt to assassinate the emperor in 1866. More important was the fact that the government failed to resolve the fundamental dilemma of change: where to stop. The "great reforms," together with the general development of Russia and the intellectual climate of the time, led to pressure for further reform. Possibly the granting of a constitutional monarchy and certain other concessions would have satisfied most of the demand and provided stability for the empire. But neither Alexander II nor certainly his successors were willing to go that far. Instead they turned against the proponents of more change and fought to preserve the established order. The "great reforms" had come only after the Crimean War had demonstrated the total bankruptcy of the old system, and they owed little to any far-reaching liberalism or vision on the part of Alexander II and his immediate associates. The sequel showed how difficult it was for the imperial government to learn new ways.

After the political stillness and immobility of Nicholas I's reign, and stimulated by the "great reforms," the early 1860's in Russia were loud and active. Peasant riots occurred with great frequency and on a large scale. In 1861 and 1862 disturbances, provoked largely by the clumsy and authoritarian policies of the new minister of education, Count Admiral Evfimii Putiatin, swept Russian universities. In 1862 the provincial assembly of the Tver gentry, led by Alexis Unkovsky, renounced its gentry privileges and demanded the convocation of a constituent assembly representing the entire people to establish a new order in Russia. And in the same year of 1862 a series of mysterious fires broke out in St. Petersburg and in a number of towns along the Volga. Also, in 1861 and 1862 leaflets urging revolution began to appear in different Russian cities. In 1863 Poland erupted in rebellion.

In Poland, too, Alexander II had instituted a liberal policy. Thus in 1862 much of the former Polish autonomy was restored. The change in the Russian attitude found favorable response among Polish moderates, led by Marquis Alexander Wielopolski, but failed to satisfy the nationalists, who wanted complete independence and the historic "greater Poland." Recent successes of Italian unification, the sympathy of Napoleon III and influential French circles, and the general nationalistic spirit of the age encouraged Polish extremists. Following a series of disorders, the government took steps to draft the unruly element, students in particular, into the army. A rebellion followed in January 1863. Although this time, in contrast to the situation in 1831, the Poles possessed no regular army and had to fight for the most part as guerrilla bands, the insurrection spread to Lithuanian and White Russian lands and was not finally suppressed until May 1864. Great Britain, France, and Austria tried to aid the Polish cause by diplomatic interventions, but were rebuffed by Russia. As a result of the rebellion, Poland again lost its autonomous position and became fully subject to Russian administration. Nicholas Miliutin, Samarin, and Cherkassky were dispatched to conquered Poland to study the conditions there and propose appropriate measures. Of their recommendations, however, only those referring to the emancipation of the serfs and land settlement were adopted. Peasants in Poland obtained a more favorable arrangement than those in Russia, while the Polish landlords fared much worse than their Russian counterparts. Otherwise the government chose to rely on centralization, police control, and Russification, with the Russian language made compulsory in Polish schools. A still more intense Russification developed in the western borderlands of Russia, where every effort was made to eradicate the Polish influence. A 10 per cent assessment was imposed there on Polish estates, the use of the Polish language was forbidden, and the property of the Catholic Church was confiscated. In 1875 the Uniates in Poland proper were forcibly reconverted to Orthodoxy.

In spite of the serious troubles of the early 1860's, Alexander II and his associates continued to reform Russia, and the future course of state policy appeared to hang in the balance. For example, while the authorities penalized disaffected Russian students and punished severely — sometimes, as in the case of Nicholas Chernyshevsky, rather clearly on insufficient evidence — those connected with the revolutionary agitation, a considerably more liberal official, Alexander Golovnin, replaced Admiral Putiatin in 1862 as minister of education, and a new and much freer University Statute became law in 1863. Even the Polish rebellion, while it resulted in oppression of the Poles, did not seem to affect the course of reform in Russia. The decisive change away from reform came, in the opinion of many historians, in 1866, following an attempt by an emotionally unbalanced student, Dmitrii Karakozov, to assassinate the emperor. In that year the reactionary Count Dmitrii Tolstoy took charge as minister of education, and the government proceeded gradually to revamp schooling in Russia, intending that stricter controls and heavy emphasis on the classical languages would discipline students and keep their attention away from the issues of the day. Over a period of years reaction also expressed itself in the curbing of the press, in restrictions on the collection of taxes by the zemstvo and on the uses to which these taxes could be put, in the exemption of political and press cases from regular judicial procedure, in continuing Russification, in administrative pressure on magistrates, and the like. On the other hand, despite the reactionary nature of the period, the municipal reform took place in 1870 and the army reform as late as 1874.

New Radicalism and the Revolutionary Movement

Russian history came increasingly to be dominated by a struggle between the government Right and the radical and revolutionary Left, with the moderates and the liberals in the middle powerless to influence the fundamental course of events. The government received unexpected support from the nationalists. It was in 1863, at the time of the Polish rebellion and diplomatic pressure by Great Britain, France, and Austria on behalf of Poland, that the onetime Westernizer, Anglophile, and liberal, journalist Michael Katkov, came out emphatically in support of the government and

Russian national interests. Katkov's stand proved very popular during the Polish war. In a sense Katkov and his fellow patriots who enthusiastically defended the Russian state acted much like the liberals in Prussia and Germany when they swung to the support of Bismarck. Yet, in the long run, it proved more characteristic of the situation in Russia that, although the revolutionaries remained a small minority, they attracted the sympathy of broad layers of the educated public.

While the intellectual history of Russia in the second half of the nineteenth century will be summarized in a later chapter, some aspects of Russian radicalism of the 1860's and 1870's must be mentioned here. Following Turgenev, it has become customary to speak of the generation of the sixties as "sons" and "nihilists" and to contrast these "sons" with the "fathers" of the forties. A powerful contrast does emerge. The transformation in Russia formed part of a broader change in Europe which has been described as a transition from romanticism to realism. In Russian conditions the shift acquired an exaggerated and violent character.

Whereas the "fathers" grew up on German idealistic philosophy and romanticism in general, with its emphasis on the metaphysical, religious, aesthetic, and historical approaches to reality, the "sons," led by such young radicals as Nicholas Chernyshevsky, Nicholas Dobroliubov, and Dmitrii Pisarev, hoisted the banner of utilitarianism, positivism, materialism, and especially "realism." "Nihilism" — and also in large part "realism," particularly "critical realism" — meant above all else a fundamental rebellion against accepted values and standards: against abstract thought and family control, against lyric poetry and school discipline, against religion and rhetoric. The earnest young men and women of the 1860's wanted to cut through every polite veneer, to get rid of all conventional sham, to get to the bottom of things. What they usually considered real and worthwhile included the natural and physical sciences — for that was the age when science came to be greatly admired in the Western world — simple and sincere human relations, and a society based on knowledge and reason rather than ignorance, prejudice, exploitation, and oppression. The casting down of idols — and there surely were many idols in mid-nineteenth-century Russia, as elsewhere — emancipation, and freedom constituted the moral strength of nihilism. Yet few in our age would fail to see the narrowness of its vision, or neglect the fact that it erected cruel idols of its own.

It has been noted that the rebels of the sixties, while they stood poles apart from the Slavophiles and other idealists of the 1830's and 1840's, could be considered disciples of Herzen, Bakunin, and to some extent Belinsky, in their later, radical, phases. True in the very important field of doctrine, this observation disregards the difference in tone and manner: as Samarin said of Herzen, even the most radical Westernizers always retained "a handful of earth from the other shore," the shore of German idealism and romanticism, the shore of their youth; the new critics came out of a simpler and cruder mold. Socially too the radicals of the sixties differed from the "fathers," reflecting the progressive democratization of the educated public in Russia. Many of them belonged to a group known in Russian as *raznochintsy,* that is, people of mixed background below the gentry, such as sons of priests who did not follow the calling of their fathers, offspring of petty officials, or individuals from the masses who made their way up through education and effort. The 1860's and the 1870's with their iconoclastic ideology led also to the emancipation of a considerable number of educated Russian women — quite early compared to other European countries — and to their entry into the arena of radical thought and revolutionary politics. The word and concept "intelligentsia," which came to be associated with a critical approach to the world and a protest against the existing Russian order, acquired currency during that portentous period. Finally, the consecutive history of the Russian revolutionary movement — which, to be sure, had such early and isolated forerunners as the Decembrists — began in the years following the "great reforms."

The Russian revolutionary movement can be traced to the revolutionary propaganda and circles of the 1860's. It first became prominent, however, in the 1870's. By that time the essentially individualistic and anarchic creed of nihilism, with its stress on a total personal emancipation, became combined with and in part replaced by a new faith, populism — *narodnichestvo* — which gave the "critical realists" their political, social, and economic program. While populism also has a broad meaning which could include as adherents Dostoevsky, Tolstoy, certain ideologists of the Right, and other diverse Russian figures, in the narrow sense it came to be associated with the teachings of such intellectuals as Herzen, Bakunin, Nicholas Chernyshevsky, Peter Lavrov, and Nicholas Mikhailovsky — who will be discussed in a later chapter — and the main trend of the Russian radical and revolutionary movement in the last third of the nineteenth century. If nihilists gloried in their emancipation, independence, and superiority to the rotten world around them, populists felt compelled to turn to the masses, which in Russia meant the peasants. They wanted to repay their debt for acquiring education — which had brought the precious emancipation itself — at the expense of the sweat and even the blood of the *muzhik,* and to lead the people to a better future. The intellectuals, it must be added, desired to learn as well as to teach. In particular, following Herzen and Bakunin, they believed in the unique worth and potential of the peasant commune, which could serve as an effective foundation for the just social order of the future. In one way or another most populists hoped to find in the people that moral purity and probity — truth, if you will — which their own environment had denied them. Whether their search stemmed from critical realism or not, represents another matter. Venturi, Itenberg, and others, with all their determined erudition, cannot quite convince the reader that reason ruled the populist movement.

The climax came in 1873, 1874, and the years immediately following. When in 1873 the imperial government ordered Russian students to abandon their studies in Switzerland — where Russians, especially women, could often pursue higher education more easily than in their fatherland — and return home, a considerable number of them, together with numerous other young men and women who had stayed in Russia, decided to "go to the people." And they went to the villages, some two and a half thousand of them, to become rural teachers, scribes, doctors, veterinarians, nurses, or storekeepers. Some meant simply to help the people as best they could. Others nurtured vast radical and revolutionary plans. In particular, the followers of Bakunin put their faith in a spontaneous, elemental, colossal revolution of the people which they had merely to help start, while the disciples of Lavrov believed in the necessity of gradualism, more exactly, in the need for education and propaganda among the masses before they could overturn the old order and establish the new.

The populist crusade failed. The masses did not respond. The only uprising that the populists produced resulted from an impressive but forged manifesto, in which the tsar ordered his loyal peasants to attack his enemies, the landlords. Indeed the muzhiks on occasion handed over the strange newcomers from the cities to the police. The police, in turn, were frantically active, arresting all the crusaders they could find. Mass trials of the 193 and of the 50 in 1877 marked the sad conclusion of the "going to the people" stage of populism. The peasants, to repeat, would not revolt, nor could satisfactory conditions be established to train them for later revolutionary action.

Yet, one more possibility of struggle remained: the one advocated by another populist theoretician, Peter Tkachev, and by an amoral and dedicated revolutionist, Serge Nechaev, and given the name "Jacobin" in memory of the Jacobins who seized power to transform France during the great French Revolution. If the peasants would not act, it remained up to the revolutionaries themselves to fight and defeat the government. Several years of revolutionary conspiracy, terrorism, and assassination ensued. The first instances of violence occurred more or less spontaneously, sometimes as countermeasures against brutal police officials. Thus, early in 1878 Vera Zasulich shot and wounded the military governor of St. Petersburg, General Theodore Trepov, who had ordered a political prisoner to be flogged; a jury failed to convict her, with the result that political cases were withdrawn from regular judicial procedure. But before long an organization emerged which consciously put terrorism at the center of its activity. The conspiratorial revolutionary society "Land and Freedom," founded in 1876, split in 1879 into two groups: the "Black Partition," or "Total Land Repartition," which emphasized gradualism and propaganda, and the "Will of the People" which mounted an all-out terroristic offensive against the government. Members of the "Will of the People" believed that, because of the highly centralized nature of the Russian state, a few assassinations could do tremendous damage to the regime, as well as provide the requisite political instruction for the educated society and the masses. They selected the emperor, Alexander II, as their chief target and condemned him to death. What followed has been described as an "emperor hunt" and in certain ways it defies imagination. The Executive Committee of the "Will of the People" included only about thirty men and women, led by such persons as Andrew Zheliabov who came from the serfs and Sophia Perovskaia who came from Russia's highest administrative class, but it fought the Russian Empire. Although the police made every effort to destroy the revolutionaries and although many terrorists perished, the "Will of the People" made one attempt after another to assassinate the emperor. Time and again Alexander II escaped through sheer luck. Many people were killed when the very dining room of his palace was blown up, while at one time the emperor's security officials refused to let him leave his suburban residence, except by water!

After the explosion in the Winter Palace and after being faced by strikes, student disturbances, and a remarkable lack of sympathy on the part of the educated public, as well as by the dauntless terrorism of the "Will of the People," the emperor finally decided on a more moderate policy which could lead to a *rapprochement* with the public. He appointed General Count Michael Loris-Melikov first as head of a special administrative commission and several months later as minister of the interior. Loris-Melikov was to suppress terrorism, but also to propose reforms. Several moderate or liberal ministers replaced a number of reactionaries. Loris-Melikov's plan called for the participation of representatives of the public, both elected and appointed, in considering administrative and financial reforms — not unlike the pattern followed in the abolition of serfdom. On March 13, 1881, Alexander II indicated his willingness to consider Loris-Melikov's proposal. That same day he was finally killed by the remaining members of the "Will of the People."

Foreign Policy

The foreign policy of Alexander II's reign, while perhaps not quite as dramatic as its internal history, also deserves careful attention. It began with the termination of the Crimean War and the Treaty of Paris, possibly the nadir of the Russian position in Europe in the nineteenth century, and it did much to restore Russian prestige. Notably, the Russians fought a successful war against Turkey and largely redrew the map of the Balkans. Also, in the course of the reign, the empire of the Romanovs made a sweeping expansion in the Caucasus, Central Asia, and the Far East. But not everything went well. Russia experienced important diplomatic setbacks as well as victories. Moreover, the changing pattern of power relations in Europe — fundamentally affected by the unification of Germany, which the tsarist government helped more than hindered — was in many ways less favorable to the state of the Romanovs in 1881 than it had been fifty years earlier.

The Crimean War meant the collapse of the world of Nicholas I, the world of legitimism with himself as its leader. Specifically, it left the Russian government and public bitterly disappointed with Austria, which, in spite of the crucial Russian help in 1849, did everything to aid Russia's enemies short of actually fighting. As Tiutchev insisted, no "Austrian Judas" could be allowed to pay last respects to Nicholas I on behalf of the Hapsburgs! It is worth noting that when the new minister of foreign affairs, Prince Alexander Gorchakov, surveyed the situation, he turned to France as a possible ally, and Napoleon III indicated reciprocal interest. Yet at that time — in contrast to what happened thirty years later — the Franco-Russian *rapprochement* foundered on the Polish rebellion of 1863. As already mentioned, both the French ruler and his people sympathized with the Poles, and, as in the case of Great Britain and Austria, France intervened diplomatically on behalf of the Poles, arguing that from the time of the Congress of Vienna and the creation of the Kingdom of Poland the fate of that country was of international concern and not simply an internal Russian affair. The imperial government could reject the argument of these powers and rebuff their intervention only because of the strong support that it obtained from the Russian public and also from Prussia. Bismarck, who realized the danger of Polish nationalism for Prussia and wanted to secure the goodwill of the tsar, sent Count Constantine Alvensleben to promise the Russians co-operation against the Polish rebels and to sign a convention to that effect. Bismarck's astute handling of the Russians contributed, no doubt, to the rather benevolent attitude on the part of the tsarist government toward the unification of Germany under Prussia, which involved the defeat of Austria in 1866 and of France in 1870. In retrospect, the fact that Russia did nothing to prevent the emergence of Germany as the new continental giant has been called the worst mistake that tsarist diplomacy ever made. To qualify that charge, it should at least be noted that Russian statesmen were not the only ones in that crucial decade totally to misjudge the situation and prospects in Europe. Also, Russia did obtain some compensation through the abrogation of the Black Sea provisions of the Treaty of Paris: at a time when

European attention centered on the Franco-Prussian war, Gorchakov, with Bismarck's backing, repudiated the vexatious obligation not to have a warfleet or coastal fortifications on the Black Sea that Russia had assumed under the Treaty. The British protested and an international conference was held in London in March 1871, but the Russian action was allowed to stand, although the principle of general consent of the signatories as against unilateral action was reaffirmed.

When in the 1870's the tsarist government looked again for allies, it once more found Prussia, or rather Germany, and Austria, which had become Austria-Hungary. For a century the Hohenzollerns had remained, on the whole, the best friends of the Romanovs; as to the Hapsburgs, the Russian rancor against them, generated by their behavior at the time of the Crimean War, had somewhat subsided in the wake of Austrian defeats and other misfortunes. The new alliance, the so-called Three Emperors' League, was formed in 1872 and 1873. Russia's part in it involved a military convention with Germany, according to which each party was to assist with 200,000 troops if its partner were attacked by a European power, and a somewhat looser agreement with Austria-Hungary. The League could be said to represent a restoration of the old association of conservative eastern European monarchies determined to preserve the established order. But, in contrast to earlier decades when Alexander I and Nicholas I led the conservative coalition, the direction of the new alliance belonged to Bismarck. In fact, the Russian government was grateful to be admitted as a partner. Moreover, Russian and German interests did not correspond in some important matters. The lack of harmony became obvious in 1875 when Russia and Great Britain exercised strong pressure on Germany to assure that it would not try a preventive war against France.

The Three Emperors' League finally collapsed over the issue of Turkey and the Balkans, which in the 1870's led to a series of international crises and to war between Russia and the Ottoman Empire. Beginning with the insurrection against Turkish rule in Herzegovina and Bosnia in July 1875, rebellion swept the Balkans. The year 1876 witnessed a brutal Turkish suppression of a Bulgarian uprising, as well as fighting and massacres in other parts of the peninsula, and the declaration of war on the Porte by Serbia and Montenegro. The Russian public reacted strongly to these developments. Pan-Slavism — hitherto no more than a vague sentiment, except for certain small circles of intellectuals — for the first time became an active force. Pan-Slav committees sent up to five thousand volunteers, ranging from prominent members of society to simple peasants and including about eight hundred former Russian army officers, to fight in the Serbian army, which had been entrusted to another Russian volunteer, General Michael Cherniaev. But the Turks defeated the Serbs, hence the last hope of Balkan nationalities in their uneven contest with the Ottomans rested on Russian intervention. The imperial government considered intervention carefully and without enthusiasm. The international situation, with Great Britain and Austria-Hungary hostile to Russia, argued against war; and so did the internal conditions, for reforms were in the process of enactment, notably in the military and financial domains, and there was populist unrest. Besides, Gorchakov and other responsible tsarist officials did not believe at all in Pan-Slavism, the exception being the Russian ambassador to Constantinople, Count Nicholas Ignatiev. However, as the Balkan struggle continued, as international diplomacy failed to bring peace, and as Russia became gradually more deeply involved in the conflict, the tsarist government, having come to an understanding with Austria-Hungary, declared war on Turkey on April 24, 1877.

The difficult, bitter, and costly war, highlighted by such engagements as the Russian defense of the Shipka pass in the Balkan mountains and the Turkish defense of the fortress of Plevna, resulted in a decisive Russian victory. The tsarist troops were approaching Constantinople when the fighting ceased. The Treaty of San Stefano, signed in March 1878, reflected the thorough Ottoman defeat: Russia obtained important border areas in the Caucasus and southern Bessarabia; for the latter, Rumania, which had fought jointly with Russia at Plevna and elsewhere, was to be compensated with Dobrudja; Serbia and Montenegro gained territory and were to be recognized, along with Rumania, as fully independent, while Bosnia and Herzegovina were to receive some autonomy and reform; moreover, the treaty created a large autonomous Bulgaria reaching to the Aegean Sea, which was to be occupied for two years by Russian troops; Turkey was to pay a huge indemnity.

But the Treaty of San Stefano never went into operation. Austria-Hungary and Great Britain forced Russia to reconsider the settlement. Austria-Hungary was particularly incensed by the creation of a large Slavic state in the Balkans, Bulgaria, which Russia had specifically promised not to do. The reconsideration took the form of the Congress of Berlin. Presided over by Bismarck and attended by such senior European statesmen as Disraeli and Gorchakov, who was still the Russian foreign minister, it met for a month in the summer of 1878 and redrew the map of the Balkans. While, according to the arrangements made in Berlin, Serbia, Montenegro, and Rumania retained their independence and Russia held on to southern Bessarabia and most of her Caucasian gains, such as Batum, Kars, and Ardakhan, other provisions of the Treaty of San Stefano were changed beyond recognition. Serbia and Montenegro lost some of their acquisitions. More important, the large Bulgaria created at San Stefano underwent division into three parts: Bulgaria proper, north of the Balkan mountains, which was to be autonomous; Eastern Rumelia, south of the mountains, which was to receive a special organization under

The Balkans
1877-1878

0   Miles   300

••••••• Boundary of Ottoman Empire, 1877
〰〰〰 Proposed boundary of Bulgaria
         by Treaty of San Stefano, 1878
——— Boundaries after Treaty of Berlin, 1878

Turkish rule; and Macedonia, granted merely certain reforms. Also, Austria-Hungary acquired the right to occupy, although not to annex, Bosnia, Herzegovina, and the Sanjak of Novi Bazar, while Great Britain took Cyprus. The diplomatic defeat of Russia reflected in the Berlin decisions made Russian public opinion react bitterly against Great Britain, Austria-Hungary, and, less justifiably, Bismarck, the "honest broker" of the Congress.

Expansion in Asia

Whereas Russian dealings with European powers in the reign of Alexander II brought mixed results, the empire of the tsars continued to expand grandly in Asia. Indeed, many scholars assert the existence of a positive correlation between Russian isolation or rebuffs in the west and the eastward advance. Be this as it may, there can be no doubt that the third quarter of the nineteenth century witnessed enormous Russian gains in Asia, notably in the Caucasus, in Central Asia, and in the Far

East. Also, in 1867, the tsarist government withdrew from the Western hemisphere by selling Alaska to the United States for $7,200,000.

As mentioned earlier, Georgian recognition of Russian rule and successful wars against Persia and Turkey in the first decades of the nineteenth century had brought Transcaucasia and thus all of the Caucasus under the sway of the tsars. But imperial authority remained nominal or nonexistent so far as numerous mountain tribes were concerned. Indeed, Moslem mountaineers reacted to the appearance of Russian troops by mobilizing all their resources to drive the invaders out and by staging a series of desperate "holy wars." The pacification of the Caucasus, therefore, took decades, and military service in that majestic land seemed for a time almost tantamount to a death warrant. Beginning in 1857, however, Russian troops commanded by Prince Alexander Bariatinsky, using a new and superior rifle against the nearly exhausted mountaineers, staged another, this time decisive, offensive. In 1859 Bariatinsky captured the legendary Shamil, who for twenty-five years had been the military, spiritual, and political leader of Caucasian resistance to Russia. That event has usually been considered as the end of the fighting in the Caucasus, although more time had to pass before order could be fully established there. A large number of Moslem mountaineers chose to migrate to Turkey.

The Caucasus needed pacification when Alexander II ascended the throne, but Central Asia had yet to be taken. That was accomplished by a series of daring military expeditions in the period from 1865 to 1876. Led by such able and resourceful commanders as Generals Constantine Kaufmann and Michael Skobelev, Russian troops, in a series of converging movements in the desert, encircled and defeated the enemy. Thus in the course of a decade the Russians conquered the khanates of Kokand, Bokhara, and Khiva, and finally, in 1881, also annexed the Transcaspian region. Russian expansion into Central Asia bears a certain resemblance both to colonial wars elsewhere and to the American westward movement. Central Asia proved attractive for commercial reasons, for the peoples of that area could supply Russia with raw materials, for example cotton, and at the same time provide a market for Russian manufactured goods. Also, Russian settlements had to be defended against predatory neighbors, and that led to further expansion. More important, it would seem that the fluid Russian frontier simply had to advance in one way or another, at least until it came up against more solid obstacles than the khanates of Bokhara and Khiva. In Central Asia, as in the Caucasus, the establishment of Russian rule usually interfered little with the native economy, society, law, or customs.

The Russian Far Eastern boundary remained unchanged from the Treaty of Nerchinsk drawn in 1689 until Alexander II's reign. In the intervening period, however, the Russian population in Siberia had increased considerably, and the Amur river itself had acquired significance as an artery of communication. In 1847 the energetic and ambitious Count Nicholas Muraviev — known later as Muraviev-Amursky, that is, of the Amur — became governor-general of Eastern Siberia. He promoted Russian advance in the Amur area and profited from the desperate plight of China, at war with Great Britain and France and torn by a rebellion, to obtain two extremely advantageous treaties from the Celestial Empire: in 1858, by the Treaty of Aigun, China ceded to Russia the left bank of the Amur river and in 1860, by the Treaty of Peking, the Ussuri region. The Pacific coast of the Russian Empire began gradually to be settled: the town of Nikolaevsk on the Amur was founded in 1853, Khabarovsk in 1858, Vladivostok in 1860. In 1875 Russia yielded its Kurile islands to Japan in return for the southern half of the island of Sakhalin.

XXX

* * * * * * * * * *

THE REIGN OF ALEXANDER III, 1881–94, AND THE

FIRST PART OF THE REIGN OF NICHOLAS II, 1894–1905

> The natural conclusion is that Russians live in a period which Shake-
> speare defined by saying, "The time is out of joint."
>
> M. KOVALEVSKY

THE REIGN OF Alexander III and the reign of Nicholas II until the Revolution of 1905 formed a period of continuous reaction. In fact, as has been
indicated, reaction had started earlier when Alexander II abandoned a
liberal course in 1866 and the years following. But the "Tsar-Liberator"
did enact major reforms early in his rule, and, as the Loris-Melikov episode
indicated, progressive policies constituted a feasible alternative for Russia
as long as he remained on the throne. Alexander III and Nicholas II
saw no such alternative. Narrow-minded and convinced reactionaries, they
not only rejected further reform, but also did their best to limit the
effectiveness of many changes that had already taken place. Thus they
instituted what have come to be known in Russian historiography as
"counterreforms." The official estimate of Russian conditions and needs
became increasingly unreal. The government relied staunchly on the gentry,
although that class was in decline. It held high the banner of "Orthodoxy-
autocracy-nationality," in spite of the fact that Orthodoxy — helped, or
rather hindered, by police and other more direct compulsive measures —
could hardly cement together peoples of many faiths in an increasingly
secular empire, that autocracy was bound to be even more of an anachro-
nism and obstacle to progress in the twentieth than in the nineteenth
century, and that a nationalism which had come to include Russification
could only split a multinational state. Whereas the last two Romanovs to
rule Russia agreed on principles and policies, they differed in character:
Alexander III was a strong man, Nicholas II a weak one; under Nicholas
confusion and indecision complicated further the fundamentally wrong-
headed efforts of the government.

Alexander III, born in 1845, was full of strength and vigor, when he
ascended the Russian throne after the assassination of his father. The new
ruler was determined to suppress revolution and to maintain autocracy,
a point that he made clear in a manifesto of May 11, 1881, which led
to the resignation of Loris-Melikov, Dmitrii Miliutin, Grand Duke Con-

stantine, and the minister of finance, Alexander Abaza. Yet it took a number of months and further changes at the top before the orientation represented by Loris-Melikov was entirely abandoned and the government embarked on a reactionary course. The promoters of reaction included Constantine Pobedonostsev, formerly a noted jurist at the University of Moscow, who had served as tutor to Alexander and had become in 1880 the Ober-Procurator of the Holy Synod; Dmitrii Tolstoy, who returned to the government in 1882 to head the Ministry of the Interior; and Ivan Delianov, who took charge of the Ministry of Education in the same year. Pobedonostsev, the chief theoretician as well as the leading practitioner of reaction in Russia in the last decades of the nineteenth century, characteristically emphasized the weakness and viciousness of man and the fallibility and dangers of human reason, hated the industrial revolution and the growth of cities, and even wanted "to keep people from inventing things." The state, he believed, had as its high purpose the maintenance of law, order, stability, and unity among men. In Russia that aim could be accomplished only by means of autocracy and the Orthodox Church.

"Temporary Regulations" to protect state security and public order, issued late in the summer of 1881, gave officials in designated areas broad authority in dealing with the press and with people who could threaten public order. Summary search, arrest, imprisonment, exile, and trial by courts-martial became common occurrences. The "Temporary Regulations" were aimed primarily at the "Will of the People," which lasted long enough to offer the new ruler peace on conditions of political amnesty and the convocation of a constituent assembly! Although the "Will of the People" had been largely destroyed even before the assassination of the emperor and although most of its remaining members soon fell into the hands of the police, the "Temporary Regulations" were not rescinded, but instead applied, as their vague wording permitted, to virtually anyone whom officials suspected or simply disliked. For many years after the demise of the "Will of the People," terrorism died down in Russia, although occasional individual outbreaks occurred. Yet the "Temporary Regulations," introduced originally for three years, were renewed. Indeed, the tsarist government relied on them during the rest of its existence, with the result that Russians lived under something like a partial state of martial law.

Alexander III's government also enacted "counterreforms" meant to curb the sweeping changes introduced by Alexander II and to buttress the centralized, bureaucratic, and class nature of the Russian system. New press regulations made the existence of radical journals impossible and the life of a mildly liberal press precarious. The University Statute of 1884, which replaced the more liberal statute of 1863, virtually abolished university autonomy and also emphasized that students were to be considered as "individual visitors," who had no right to form organizations or to claim corporate representation. In fact most policies of the Ministry of Education — which will be summarized in a later chapter — whether they concerned the emphasis on classical languages in secondary schools, the drastic curtailment of higher education for women, or the expansion of the role of the Church in elementary teaching, consciously promoted the reactionary aims of the regime.

The tsar and his associates used every opportunity to help the gentry and to stress their leading position in Russia, as, for example, by the creation in 1885 of the State Gentry Land Bank. At the same time they imposed further restrictions on the peasants, whom they considered essentially wards of the state rather than mature citizens. The policies of bureaucratic control of the peasants and of emphasizing the role of the gentry in the countryside found expression in the most outstanding "counter-reform" of the reign, the establishment in 1889 of the office of *zemskii nachalnik,* zemstvo chief, or land captain. That official — who had nothing to do with the zemstvo self-government — was appointed and dismissed by the minister of the interior following the recommendation of the governor of the land captain's province. His assigned task consisted in exercising direct bureaucratic supervision over the peasants and, in effect, in managing them. Thus the land captain confirmed elected peasant officials as well as decisions of peasant meetings, and he could prevent the officials from exercising their office, or even fine, arrest, or imprison them, although the fines imposed by the land captain could not exceed several rubles and the prison sentences, several days. Moreover, land captains received vast judicial powers, thus, contrary to the legislation of 1864, again combining administration and justice. In fact, these appointed officials replaced for the peasants, that is, for the vast majority of the people, elected and independent justices of peace. The law of 1889 stipulated that land captains had to be appointed from members of the local gentry who met a certain property qualification. Each district received several land captains; each land captain administered several volosti, that is, townships or cantons. Russia obtained in this manner a new administrative network, one of land captaincies.

The following year, 1890, the government made certain significant changes in the zemstvo system. The previous classification of landholders, that of 1864, had been based on the form of property, so that members of the gentry and other Russians who happened to hold land in individual ownership were not distinguished. In 1890 the members of the gentry became a distinct group — and their representation was markedly increased. Peasants, on the other hand, could thenceforth elect only candidates for zemstvo seats, the governor making appointments to district zemstvo assemblies from these candidates, as recommended by land captains. In addition, the minister of the interior received the right to confirm chairmen of zemstvo boards in their office, while members of the boards and zemstvo employees were to be confirmed by their respective governors. In 1892 the town government underwent a similar "counterreform," which, among other provisions, sharply raised the property requirement for the right to vote. After its enactment, the electorate in St. Petersburg decreased from 21,000 to 8,000, and that in Moscow from 20,000 to 7,000.

The reign of Alexander III also witnessed increased pressure on non-Orthodox denominations and a growth of the policy of Russification. Even Roman Catholics and Lutherans, who formed majorities in certain western areas of the empire and had unimpeachable international connections and recognition, had to face discrimination: for instance, children of mixed marriages with the Orthodox automatically became Orthodox, and all but the dominant Church were forbidden to engage in proselytizing. Old Believers and Russian sectarians suffered greater hardships. The government also began to oppose non-Christian faiths such as Islam and Buddhism, which had devoted adherents among the many peoples of the empire.

Russification went hand in hand with militant Orthodoxy, although the two were by no means identical, for peoples who were not Great Russians such as the Ukrainians and the Georgians belonged to the Orthodox Church. Although Russification was practiced earlier against the Poles, especially in the western provinces following the rebellions of 1831 and 1863 and to a somewhat lesser extent in Poland proper, and was also apparent in the attempts to suppress the budding Ukrainian nationalism, it became a general policy of the Russian government only late in the nineteenth century. It represented in part a reaction against the growing national sentiments of different peoples of the empire with their implicit threats to the unity of the state and in part a response to the rising nationalism of the Great Russians themselves. Alexander III has often been considered the first nationalist on the Russian throne. Certainly, in his reign measures of Russification began to be extended not only to the rebellious Poles, but, for example, to the Georgians and Armenians in Transcaucasia and even gradually to the loyal Finns.

The Jews, who were very numerous in western Russia as a result of the invitation policy of late medieval Polish kings, were bound to suffer in the new atmosphere of aggressive Orthodoxy and Russification. And indeed old limitations came to be applied to them with a new force, while new legislation was enacted to establish additional curbs on them and their activities. Thus, in contrast to the former lax enforcement of rules, Jews came to be rigorously restricted to residence in the "Pale of Jewish Settlement," that is, the area in western Russia where they had been living for a long time, with the added proviso that even within the Pale they could reside only in towns and smaller settlements inhabited by merchants and craftsmen, but not in the countryside. Educated or otherwise prominent Jews could usually surmount these restrictions, but the great bulk of the poor Jewish population was tied to its location. In 1887 the government established quotas for Jewish students in institutions of higher learning: 10 per cent of the total enrollment within the Pale of Jewish Settlement, 5 per cent in other provinces, and 3 per cent in Moscow and St. Petersburg. In 1881, pogroms — the sad word entered the English language from the Russian — that is, violent popular outbreaks against the Jews, occurred in southwestern Russian towns and settlements, destroying Jewish property and sometimes taking Jewish lives. They were to recur sporadically until the end of imperial Russia. Local authorities often did little to prevent pogroms and on occasion, it is rather clear, even encouraged them. As Pobedonostsev allegedly remarked, the Jewish problem in Russia was to be solved by the conversion to Orthodoxy of one-third of the Russian Jews, the emigration of one-third, and the death of the remaining third. It should be added that the Russian government defined Jews according to their religion; Jews converted to Christianity escaped the disabilities imposed on the others.

Yet even under Alexander III state policies could not be limited to curbing the "great reforms" and generally promoting reaction. Certain more constructive measures were enacted in the domains of finance and national economy where the government had to face a difficult and changing situation, and where it profited from the services of several able ministers. While the development of the Russian economy and of society after the "great reforms" will be discussed in a later chapter, it should be noted here that Nicholas Bunge, who headed the Ministry of Finance from 1881 to 1887, established a Peasant Land Bank, abolished the head tax, introduced the inheritance tax, and also began labor legislation in Russia. His pioneer factory laws included the limitation of the working day to eight hours for children between twelve and fifteen, the prohibition of night work for children and for women in the textile industry, and regulations aimed at assuring the workers proper and regular pay from their employers, without excessive fines or other illegitimate deductions. Factory inspectors were established to supervise the carrying out of the new legislation. It is significant that Bunge had to leave the Ministry of Finance because of the strong opposition to his measures and accusations of socialism. His successors, Ivan Vyshnegradsky, 1887–92, and Serge Witte, 1892–1903, strove especially to develop state railways in Russia and to promote heavy industry through high tariffs, state contracts and subsidies, and other means.

Nicholas II

Nicholas II, Alexander III's eldest son, who was born in 1868, became the autocratic ruler of Russia after his father's death in 1894. The last tsar possessed certain attractive qualities, such as simplicity, modesty, and devotion to his family. But these positive personal traits mattered little in a situation that demanded strength, determination, adaptability, and vision. It may well be argued that another Peter the Great could have saved the Romanovs and imperial Russia. There can be no doubt that Nicholas II did not. In fact, he proved to be both narrow-minded and weak, unable to remove reactionary blinders even when circumstances forced him into entirely new situations with great potentialities, and at the same time unable to manage even reaction effectively. The unfortunate emperor struck many observers as peculiarly automatic in his attitudes and actions, without the power of spontaneous decision, and — as his strangely colorless and un-differentiating diary so clearly indicates — also quite deficient in perspective. Various, often unworthy, ministers made crucial decisions that the sovereign failed to understand fully or to evaluate. Later in the reign the empress, the reactionary, hysterical, and willful German princess Alexandra, became the power behind the throne, and with her even such an incredible person as Rasputin could rise to the position of greatest influence in the state. A good man, but a miserable ruler lost in the moment of crisis — no wonder Nicholas II has often been compared to Louis XVI. As Trotsky and other determinists have insisted, the archaic, rotten Russian system, which was about to collapse, could not logically produce a leader much different from that ineffective relic of the past. Or, as an old saying has it, the gods blind those whom they want to destroy.

Reaction under Nicholas II

Reaction continued unimpeded. The new emperor, who had been a pupil of Pobedonostsev, relied on the Ober-Procurator of the Holy Synod and on other reactionaries such as his ministers of interior Dmitrii Sipiagin and Viacheslav Plehve. The government continued to apply and extend the "Temporary Regulations," to supervise the press with utmost severity, and as best it could to control and often restrict education. The zemstvo and municipal governments experienced further curtailments of their jurisdictions. For example, in 1900 the limits of zemstvo taxation were strictly fixed and the stockpiling of food for emergency was taken away from zemstvo jurisdiction and transferred to that of the bureaucracy. Moreover, the authorities often refused to confirm elections of zemstvo board members or appointments of zemstvo employees, trying to assure that only people of unimpeachable loyalty to the regime would hold public positions of any kind.

Religious persecution grew. Russian sectarians suffered the most, in particular those groups that refused to recognize the state and perform such state obligations as military service. Many of them were exiled from central European Russia to the Caucasus and other distant areas. It was as a result of the policies of the Russian government that the Dukhobory and certain other sects — helped, incidentally, by Leo Tolstoy — began to emigrate in large numbers to Canada and the United States. The state also confiscated the estates and charity funds of the Armenian Church and harassed other denominations in numerous ways. The position of the Jews too underwent further deterioration. Additional restrictions on them included a prohibition from acquiring real estate anywhere in the empire except in the cities and settlements of the Jewish Pale, while new pogroms erupted in southwest Russia, including the horrible one in Kishinev in 1903.

But the case of Finland represented in many respects the most telling instance of the folly of Russification. As an autonomous grand duchy from the time it was won from Sweden in 1809, Finland received more rights from the Russian emperor, who became the Grand Duke of Finland, than it had had under Swedish rule, and remained a perfectly loyal, as well as a relatively prosperous and happy, part of the state until the very end of the nineteenth century and the introduction of a policy of Russification. Finnish soldiers helped suppress the Poles, and in general the Finns participated actively and fruitfully in almost every aspect of the life of the empire. Yet the new nationalism demanded that they too be Russified. While some preliminary measures in that direction had been enacted as early as in the reign of Alexander III, real Russification began with the appointment of General Nicholas Bobrikov as governor-general of Finland and of Plehve as state secretary for Finnish affairs in 1898. Russian authorities argued that Finland could remain different from Russia only so far as local matters were concerned, while it had to accept the general system in what pertained to the entire state. With that end in view, a manifesto concerning laws common to Finland and Russia and a new statute dealing with the military service of the Finns were published in 1899. Almost overnight Finland became bitterly hostile to Russia, and a strong though passive resistance developed: new laws were ignored, draftees failed to show up, and so on. In 1901 freedom of meetings was abrogated in Finland. In 1902 Governor-General Bobrikov received the right to dismiss Finnish officials and judges and to replace them with

Russians. In 1903 he was vested with extraordinary powers to protect state security and public order, which represented a definitive extension of the "Temporary Regulations" of 1881 to Finland. In 1904 Bobrikov was assassinated. The following year the opposition in Finland became part of the revolution that spread throughout the empire.

Witte and the Ministry of Finance

However, under Nicholas II, as in the reign of Alexander III, the Ministry of Finance pursued a more intelligent and far-sighted policy than did the rest of the government; and this affected many aspects of the Russian economy and life. The minister, Serge Witte, was an economic planner and manager of the type common in recent times in the governments of western Europe and the United States, but exceedingly rare in the high officialdom of imperial Russia. Witte devoted his remarkable energy and ability especially to the stabilization of finance, the promotion of heavy industry, and the building of railroads. In 1897, after accumulating a sufficient gold reserve, he established a gold standard in Russia, a measure which did much to add stability and prestige to Russian economic development, and in particular to attract foreign capital. Witte encouraged heavy industry by virtually every means at his command, including government orders, liberal credits, unceasing efforts to obtain investments from abroad, tariff regulations, and improved transportation. As to railroads, the minister, who had risen to prominence as a railroad official, always retained a great interest in them: the Russian railroad network doubled in mileage between 1895 and 1905, and the additions included the enormous Trans-Siberian line, built between 1891 and 1903 — except for a section around Lake Baikal completed later — the importance of which for Siberia can be compared to the importance of the Canadian Pacific Railroad for Canada. Witte's activities, as we shall see presently, affected foreign policy as well as domestic affairs.

Russian Foreign Policy after the Congress of Berlin

Russian foreign policy had been undergoing important changes in the decades that followed the Congress of Berlin. The most significant developments were the final rupture with Austria-Hungary and Germany and the alliance with France. Although the Three Emperors' League had foundered in the Balkan crisis, a new Alliance of the Three Emperors was concluded in June 1881 for three years and renewed in 1884 for another three years. Its most essential provision declared that if one of the contracting powers — Germany, Austria-Hungary, or Russia — engaged in war with a fourth power, except Turkey, the other two were to maintain friendly neutrality.

But, because of their conflicting interests in the Balkans, it proved impossible for Russia and Austria-Hungary to stay in the same alliance. The next major crisis occurred over Bulgaria where — as Jelavich and other specialists have demonstrated — Russia destroyed a great amount of popularity and goodwill by an overbearing and stupid policy. The Russian quarrel with the Bulgarian ruler, Alexander of Battenberg, and the Russian refusal to sanction the unification of Bulgaria and Eastern Rumelia in 1885 failed to stop the unification but resulted in the abdication of Alexander of Battenberg and the election by the Bulgarian Assembly of the pro-Austrian Ferdinand of Saxe-Coburg to the Bulgarian throne. Bulgaria abandoned the Russian sphere of influence and entered the Austrian, leaving the empire of the tsars virtually without Balkan allies. At the same time tension in relations between Russia and Austria-Hungary increased almost to the breaking point. However, Germany, by contrast with Austria-Hungary and despite the fact that in 1879 it had become a close partner of the Hapsburg state, tried at first to retain the Russian connection. Thus when the Alliance of the Three Emperors expired in 1887, Germany and Russia concluded in secret the so-called Reinsurance Treaty, Bismarck's "wire to St. Petersburg" and a veritable *tour de force* of diplomacy: each party was to remain neutral in case the other fought a war, with the exception of an aggressive war of Germany against France or of Russia against Austria-Hungary — the exception making it barely possible for Germany to square the Reinsurance Treaty with its obligations to Austria-Hungary. Nevertheless, following Bismarck's forced resignation in 1890, Germany discontinued the Reinsurance Treaty and thus severed its connection with Russia.

The Russian rupture with the Germanic powers and the general isolation of Russia appeared all the more ominous because of Anglo-Russian tension over the expansion of the Russian Empire in Central Asia, which, the British felt, threatened India. That tension attained its high point in 1885 when the Russians, having reached as far south as the vague Afghan border, clashed with the soldiers of the amir. Although an Anglo-Russian war was avoided and the boundary settled by compromise, Great Britain and Russia remained hostile to each other well past the turn of the century as they competed for influence and control in vast lands south of Russia, especially in Iran.

Political realities pointed to a Franco-Russian alliance — Bismarck's nightmare and the reason behind the Reinsurance Treaty — for France was as isolated as Russia and more threatened. Alexander III, his cautious foreign minister Nicholas Giers, and other tsarist high officials reached that conclusion reluctantly, because they had no liking for the Third Republic and no confidence in it, and because the traditional German orientation in Russian foreign policy died hard. Yet France remained the only possible partner, and it had much to offer. In particular, Paris alone provided a great market for Russian state loans — the Berlin financial market, it might be added, was closed to Russia in 1887 — and thus the main source of foreign financial support much needed by the imperial government. In fact, Frenchmen proved remarkably eager to subscribe to these loans as well as to invest directly in the Russian economy. Economics thus joined politics, although it would be fair to say that politics led the way. The alliance was consolidated in several stages, beginning with the diplomatic understanding of 1891 and ending with the military convention of December 1893–January 1894. B. Nolde, Langer, and other scholars have indicated how through the drawn-out negotiations the French pressed for an ever firmer and more binding agreement, gradually forcing the hand of the hesitant Russians. In its final form the alliance provided that if France were attacked by Germany, or by Italy supported by Germany, Russia would employ all available forces against Germany; and if Russia were attacked by Germany, or by Austria-Hungary supported by Germany, France would employ all available forces against Germany. Additional articles dealt with mobilization, the number of troops to be contributed, and other specific military plans. The Franco-Russian agreement was to remain in force for the duration of the Triple Alliance of Germany, Austria-Hungary, and Italy.

Nicholas II approved Alexander III's foreign policy on the whole and wanted to continue it. However, as we shall see, the new emperor proved to be less steady and more erratic than his father in international relations as in domestic affairs. Also, while Alexander III relied on the careful and experienced Giers throughout his rule, Nicholas II had several foreign ministers whose differences and personal preferences affected imperial diplomacy. In addition, the reign of the last tsar witnessed more than its share of court cliques and cabals which on occasion exercised a strong and at the same time irresponsible influence on the conduct of Russian foreign policy.

Nicholas II appeared prominently on the international scene in 1899, when he called together the first Hague Peace Conference attended by representatives of twenty-six states. Although instigated by Russian financial stringency and in particular by the difficulty of keeping up with Austrian armaments, this initiative was in accord with the emperor's generally peaceful views. While the Conference failed to agree on disarmament or compulsory arbitration of disputes, it did pass certain "laws of war" — later often disregarded in practice, as in the case of the temporary injunction against the use of "projectiles thrown from balloons" — and set up a permanent court of arbitration, the International Court of Justice at the Hague. More important, it became the first of a long series of international conferences on disarmament and peace, on which the hopes of mankind ride today. The Second Hague Peace Conference, in 1907, was also attended by Russian representatives, but again it could not reach agreement on the major issues under discussion.

The Russo-Japanese War

Nicholas II's own policy, however, did not always contribute to peace. Aggressiveness and adventurous involvement characterized Russian behavior in the Far East around the turn of the century, which culminated in the Russo-Japanese War of 1904–5. The construction of the Trans-Siberian railroad between 1891 and 1903, entirely justified in terms of the needs of Siberia, served also to link Russia to Manchuria, China, Korea, and even, indirectly, to Japan. Japan had just gone through a remarkable modernization and in 1894–95 it fought and defeated China, obtaining by the Treaty of Shimonoseki the Chinese territories of Formosa, the Pescadores Islands, and the Liaotung Peninsula, together with other gains, including the recognition of full independence for Korea. Before Japan could profit from the Liaotung Peninsula, Russia, France, and Germany forced her to give it up. Next Russia concluded a secret agreement with China, whereby in return for guaranteeing Chinese territory against outside aggression, it obtained the right to construct a railroad through Manchuria to the coast. Although the new railroad, the East China Railway, belonged nominally to a private company with a large Chinese participation, it marked in effect the establishment of a Russian sphere of influence in northern Manchuria, an influence centered in Harbin and extending along railroad tracks and properties guarded by a special Russian railroad guard.

While Russia had legitimate commercial and other interests in Asia — for one thing, selling the products of its factories in the East when they could not compete in the West — and while up to that point Russian imperialism in the Far East had limited itself to peaceful penetration, the situation became increasingly tense. Moreover, Russia responded to new opportunities more and more aggressively. Thus, when the murder of two German missionaries in November 1897 led to the German acquisition of Kiao-chow through a ninety-nine year lease, Nicholas II demanded and obtained a twenty-five year lease of the southern part of the Liaotung Peninsula with Port Arthur — in spite of Witte's opposition to that move and in flagrant disregard of the Russian treaty with China. Witte in turn proceeded to make the most of the situation and rapidly develop Russian interests in southern Manchuria. Following the so-called Boxer rebellion of the exasperated Chinese against foreigners in 1900–1901, which Russian forces helped to suppress, tsarist troops remained in Manchuria on the pretext that local conditions represented a threat to the railroad. In addition, a group of adventurers with strong connections at the Russian court began to promote a scheme of timber concessions on the Yalu River meant to serve

as a vehicle for Russian penetration into Korea. Witte, who objected energetically to the dangerous new scheme, had to leave the Ministry of Finance; the Foreign Office failed to restrain or control Russian policy in the Far East; and Nicholas II himself sided cheerfully with the adventurers, apparently because he believed in some sort of Russian mission in Asia and, in common with almost everyone else, grossly underestimated Japan. Russian policy could hardly be defended in terms of either justice or wisdom, in spite of the efforts of such able scholars as Malozemoff.

Japan proved to be the more skillful aggressor. Offering partition, which would give the Russians northern Manchuria and the Japanese southern Manchuria and Korea, the Japanese gauged the futility of negotiating, chose their time well, and on February 8, 1904, attacked successfully the unsuspecting Russian fleet in the outer harbor of Port Arthur — thus accomplishing the original Pearl Harbor. What followed turned out to be a humiliating war for the Russians. The Russian colossus suffered defeat after defeat from the Japanese pigmy. This outcome, so surprising at the time, resulted from ample causes: Japan was ready, well-organized, and in effect more modern than Russia, while Russia was unprepared, disorganized, troubled at home, and handicapped by a lack of popular support and even by some defeatism; Japan enjoyed an alliance with Great Britain and the favor of world public opinion, Russia found itself diplomatically isolated; Japan used short lines of communication, Russian forces had to rely on the enormously long single-track Trans-Siberian railroad, with the section around Lake Baikal still unfinished. In any case, although Russian soldiers and sailors fought with their usual courage and tenacity, the Japanese destroyed the Russian navy in the Far East, besieged and eventually captured Port Arthur, and gradually, in spite of bitter engagements near Mukden and elsewhere, pushed the main Russian army north in Manchuria. Finally, on May 27–29, 1905, in the battle of Tsushima Strait, they annihilated Admiral Zinovii Rozhdestvensky's antique fleet which had been sent to the Far East all the way from the Baltic. That fleet, it might be added, had caused a serious international incident when on its journey to the Far East it had fired by mistake at some English fishing vessels on the Dogger Bank, inflicting casualties.

An armistice followed soon after Tsushima. The Russians had suffered numerous defeats, and the government had to cope with revolutionary unrest at home. The Japanese had exhausted their finances and, despite their victories, could not destroy the main Russian army or force a conclusion. In response to a secret Japanese request, President Theodore Roosevelt arranged a peace conference at Portsmouth, New Hampshire, in August 1905. The provisions of the Treaty of Portsmouth reflected the skillful diplomacy of Witte, who headed the Russian delegation, and represented, everything considered, a rather satisfactory settlement for Russia: Russia acknowledged a paramount Japanese interest in Korea and ceded to Japan its lease of the Liaotung Peninsula, the southern part of the railroad up to Chang-chun, and the half of the island of Sakhalin south of the fiftieth degree of latitude; both countries agreed to restore Manchuria to China; in spite of strong Japanese insistence, there was no indemnity.

The Russian government ended the war against Japan none too soon, for, as fighting ceased, the country was already in the grip of what came to be known as the Revolution of 1905.

THE LAST PART OF THE REIGN OF NICHOLAS II: THE REVOLUTION OF 1905 AND THE CONSTITUTIONAL PERIOD, 1905–17

Russia at the dawn of the twentieth century knew no more magic word than "revolution." The idea of revolution was viewed with fear and hatred by the propertied classes of the population, and was loved and revered by all who dreamed of liberty. To the Russians who longed for a new life, there was enchantment in the very sound of the word. Even as they conceived it, even as they pronounced the sacred words, "Long Live the Revolution," Russians felt obscurely that they were already halfway to liberation.

STEINBERG

There is an easier and more convincing explanation for the failure of the constitutional monarchy: it puts the blame primarily on the king himself. Although Louis was well-meaning and showed occasional flashes of insight, his narrow mind had a stubborn and devious quality about it, too. The king did little to consolidate the new system, even though it left him a role of real importance. . . .

. . . The explanation may lie in the constant pressure of the queen and her advisers, which weakened Louis's resolution and changed his flabby mind. Or it may be that this pious king had serious pangs of conscience at some of the reforms built into the new system. . . . Or again, perhaps the course of events brought out his own true character as an irritable, small-minded, stubborn man who built up a neurotic resentment at his loss of initiative after 1789. It is true that even if Louis XVI had been ideally suited to his new role, the system might have broken down nevertheless.

WRIGHT

Whereas actually the main weakness of the Russian monarchy of the imperial period consisted not at all in representing the interests of a *"minority,"* restricted in this or that manner, but in the fact that it represented *no one* whatsoever.*

FLOROVSKY

M. KOVALEVSKY and many other Russians hoped that the period of blind reaction, "the time out of joint," which descended upon Russia in the second half of Alexander II's reign and was certainly present in the reigns of Alexander III and Nicholas II, would give way to a new wave of sweeping liberal reforms. But the government refused to change its course. Instead the country finally exploded into the Revolution of 1905.

* Italics in the original.

The Background of the Revolution of 1905

The Revolution of 1905 could occur because of the social transformation that had been going on in the empire of the tsars and because of the concomitant growth of opposition to the regime. In the decades that followed the "great reforms," capitalism at last became prominent in Russia. In fact, the 1880's and 1890's witnessed rapid industrialization of the country with resulting social changes and tensions. While the Russian society of that period will be discussed in a later chapter, no special exposition is needed to make the point that the growth of capitalism led to the rise of two social groups, the bourgeoisie and the proletariat. The middle class, traditionally weak in Russia at least after the times of Kiev and Novgorod, began finally to come into its own. Even though the Russian commercial and industrial bourgeoisie remained still relatively underdeveloped and inarticulate, professional people seemed eager and ready to participate in politics. These professional groups — whether they should be classified as part of the middle class or as a separate adjoining stratum is of no consequence here — had profited especially from the "great reforms": thus, the judicial reform of 1864 had virtually created the lawyers, while the introduction of the zemstvo system provided numerous openings for doctors, veterinarians, teachers, statisticians, and many other specialists, the so-called "third element" of the zemstva. Liberalism found particularly propitious circumstances for development among the professionals, as well as among some gentry landlords of the zemstva. The rise of the proletariat and the emergence of a labor movement pointed in their turn to a more radical trend in Russian opposition. And, of course, behind dissatisfied bourgeois, critical intellectuals, and bitter workers there spread the human ocean of destitute and desperate peasants — an ocean that had risen in uncounted storms through centuries of Russian history.

The opposition began to organize. The frightful famine of 1891–92 marked the end of a certain lull in Russia and the resumption of social and political activity with emphatic criticism of the regime. The liberals, who could boast of many prominent names in their ranks and who represented at that time the elite of the opposition, eventually formed the Union of Liberation in 1903, with its organ, *The Liberation,* published abroad by the noted economist Peter Struve. In 1905 they organized the Constitutional Democratic party — or "Cadet," a word based on the two initial letters in the Russian name — led by the historian Paul Miliukov and encompassing liberals of different kinds, both constitutional monarchists and republicans.

The radicals formed two important parties around the turn of the century: the Social Democratic, or "SD," and the Socialist Revolutionary, or "SR." The Social Democrats were Marxists, and the creation of their party represented a landmark in the development of Marxism in Russia. Propounded by George Plekhanov and other able intellectuals, Marxism became prominent in the empire of the tsars in the 1880's and especially in the 1890's. Its close association with the labor movement dated at least from 1883, when Plekhanov organized the Emancipation of Labor Group; but a Marxist political party, the Social Democratic, appeared only in 1898. In fact, the convention of 1898 — although commemorated in the U.S.S.R. as the first and founding congress — proved abortive, and most of its few participants were shortly arrested. The party became a reality only after the second convention held in Brussels and London in 1903. At that time the Social Democrats also split into the Bolsheviks, led by Vladimir Ulianov, better known as Lenin, who wanted a tightly knit organization of professional revolutionaries, and the Mensheviks, who preferred a somewhat broader and looser association. In time the ramifications of that relatively slight initial difference acquired great importance. The Socialist Revolutionaries, who engaged in a running debate with the Marxists concerning the nature of Russian society and its future, represented essentially the older populist tradition of Russian radicalism, even though they too were influenced by Marxism. They formed their party in 1901 and had Victor Chernov as their most noted leader.

As the twentieth century opened, Russia was in turmoil. Strikes spread throughout the country. Student protests and disturbances became more frequent, constituting an almost continuous series from 1898 on. Sporadic peasant disturbances kept the tension high in rural areas and offered increased opportunities to the Socialist Revolutionaries, just as the growth of the labor movement encouraged the Social Democrats. In 1902, 1903, and early 1904, committees dealing with the national economy, conferences of teachers and doctors, and other public bodies all demanded reforms. Moreover, the Socialist Revolutionaries resumed the terrorist tactics of their predecessors such as the "Will of the People." Their "Battle Organization" assassinated a number of important officials, including the two especially reactionary ministers of the interior, Sipiagin in 1902 and Plehve in 1904, and early in 1905 Grand Duke Serge, commanding officer of the Moscow military region and Nicholas II's second cousin and brother-in-law. The war against Japan and resulting defeats added fuel to the fire. In November 1904, a zemstvo congress, meeting in St. Petersburg, demanded a representative assembly and civil liberties. The same demands were made with increasing frequency by numerous other public bodies. In particular, professional organizations, such as unions of doctors and teachers, and other associations spread rapidly throughout Russia and made their voices heard. Several months after the zemstvo congress fourteen professional unions united to form a huge Union of Unions led by the Cadets. The government tried both repression and some conciliation, appealing for confidence, but its generally ineffectual efforts only helped to swell the tide of opposition.

The Revolution of 1905

January 22, 1905, came to be known in Russian history as "Bloody Sunday." On that day the police of the capital fired at a huge demonstration of workers led by an adventurer and priest named George Gapon, killing, according to the official estimate, one hundred and thirty persons and wounding several hundred. Ironically, Gapon's union had been essentially a "police union," part of policeman Serge Zubatov's plan to infiltrate the labor movement and direct it into officially desirable channels. Ironically too, the workers were converging on the Winter Palace — ignorant of the fact that Nicholas II was not there — with icons and the tsar's portraits, as faithful subjects, nay, children, of their sovereign, begging him for redress and help. The entire ghastly episode thus testified to official incompetence in more ways than one. The massacre led to a great outburst of indignation in the country and gave another boost to the revolutionary movement. In particular, as many authorities assert, it meant a decisive break between the tsar and those numerous workers who had until that "Bloody Sunday" remained loyal to him.

Under ever-increasing pressure, Nicholas II declared early in March his intention to convoke a "consultative" assembly; in further efforts toward pacification, he proclaimed religious tolerance and repealed some legislation against ethnic minorities; nevertheless, the revolutionary tide kept rising. The summer of 1905 witnessed new strikes, mass peasant uprisings in many provinces, active opposition and revolutionary movements among national minorities, and even occasional rebellions in the armed forces, notably in the celebrated instance of the battleship *Potemkin* in the Black Sea. On August 19 an imperial manifesto created an elective Duma with consultative powers, but that too failed to satisfy the educated public or the masses. The revolutionary movement culminated in a mammoth general strike which lasted from the twentieth to the thirtieth of October and has been described as the greatest, most thoroughly carried out, and most successful strike in history. Russians seemed to act with a single will, as they made perfectly plain their unshakable determination to end autocracy. It was in the course of the strike, and in order to direct it, that workers in St. Petersburg organized a *soviet*, or council — a harbinger of the then unknown future. Paralyzed in their essential activities and forced at last to recognize the immensity of the opposition, Nicholas II and his government finally capitulated. On October 30, the emperor, as advised by Witte, issued the October Manifesto. That brief document guaranteed civil liberties to

the Russians, announced a Duma with the true legislative function of passing or rejecting all proposed laws, and promised a further expansion of the
new order in Russia. In short, the October Manifesto made the empire of
the Romanovs a constitutional monarchy.

Also, it split the opposition. The liberals and moderates of all sorts felt
fundamentally satisfied. The radicals, such as the Social Democrats, on the
contrary, considered the tsar's concession entirely inadequate and wanted
in any case a constituent assembly, not handouts from above. Thus divided,
the opposition lost a great deal of its former power. In the middle of December the government arrested the members of the St. Petersburg Soviet.
The Soviet's appeal for revolution found effective response only in Moscow
where workers and some other radicals fought bitterly against the police
and the soldiers, including a guards' regiment, from the twenty-second of
December until the first of January.

The year 1905 thus ended in Russia in bloody fighting. However, the
revolution had spent itself with that last effort. In the course of the winter,
punitive expeditions and summary courts-martial restored order in many
troubled areas. The extreme Right joined the army and the police; Rightist
active squads, known as the "Black Hundreds," beat and even killed Jews,
liberals, and other intellectuals. Proto-fascist in nature, this newly awakened
Right throve on ethnic and religious hatreds and appealed especially to
wealthy peasants and to members of the lower middle class in towns. More
important, the great bulk of the people was tired of revolution and longed
for peace. It might be added that Witte further strengthened the hand of
the government by obtaining a large loan from France.

The Fundamental Laws

On May 6, 1906, virtually on the eve of the meeting of the First Duma,
the government promulgated the Fundamental Laws. These laws provided
the framework of the new Russian political system; the October Manifesto
had merely indicated some of its guiding lines. According to the Fundamental Laws, the emperor retained huge powers. He continued in complete
control of the executive, the armed forces, foreign policy — specifically
making war and peace — succession to the throne, the imperial court, imperial domains, and so forth. He maintained unchanged his unique domi-
nating position in relation to the Russian Church. And he even retained the
title of autocrat. He was to call together the annual sessions of the Duma
and to disband the Duma, in which case, however, he had to indicate the
time of the election and of the meeting of the new Duma. He had the power
of veto over legislation. Moreover, in case of emergency when the Duma
was not in session, he could issue *ukazes* with the authority of laws, al-

though they had to be submitted for approval to the next session of the Duma no later than two months after its opening.

The Duma, to be sure, received important legislative and budgetary rights and functions by the Fundamental Laws, but these rights were greatly circumscribed. Notably, almost 40 per cent of the state budget, encompassing such items as the army, the navy, the imperial court, and state loans, stayed outside the purview of the Duma, while the remainder, if not passed by the Duma, was re-enacted in the amounts of the preceding year. Ministers and the entire executive branch remained responsible only to the emperor, although the Laws did contain complicated provisions for interpellation, that is, questioning of ministers by the Duma. Furthermore, the State Council, which had functioned since its creation by Alexander I as an advisory body of dignitaries, became rather unexpectedly the upper legislative chamber, equal in rights and prerogatives to the Duma and meant obviously as a conservative counterweight to it. "No more than half" of the membership of the upper house was to be appointed by the emperor — appointed not even for life but by means of annual lists — and the other half elected by the following groups: 56 with very high property standing by the provincial zemstva, 18 by the gentry, 12 by commerce and industry, 6 by the clergy, 6 by the Academy of Sciences and the universities, and 2 by the Finnish Diet.

The First Two Dumas

Whereas the Fundamental Laws introduced numerous restrictions on the position and powers of the Duma, the electoral law emphasized its representative character. The electoral system, despite its complexities and limitations, such as the grouping of the electorate on a social basis, indirect elections, especially in the case of the peasants, and a gross underrepresentation of urban inhabitants, allowed almost all Russian men to participate in the elections to the Duma, thus transforming overnight the empire of the tsars from a country with no popular representation to one which practiced virtually universal manhood suffrage. The relatively democratic nature of the electoral law resulted partly from Witte's decision in December 1905, at the time when the law received its final formulation, to make concessions to the popular mood. More significantly, it reflected the common assumption in government circles that the peasants, the simple Russian people, would vote for their tsar and for the Right. After a free election, the First Duma convened on May 10, 1906.

Contrary to its sanguine expectations, the government had suffered a decisive electoral defeat. According to Walsh, the 497 members of the First Duma could be classified as follows: 45 deputies belonged to parties of the Right; 32 belonged to various national and religious groups, for example, the Poles and the Moslems; 184 were Cadets; 124 were representatives of different groups of the Left; and 112 had no party affiliation. The Cadets with 38 per cent of the deputies thus emerged as the strongest political party in the Duma, and they had the added advantage of an able and articulate leadership well-versed in parliamentary procedure. Those to the Left of the Cadets, on the other hand, lacked unity and organization and wanted mainly to fight against the regime, purely and simply. The cause of the Left in the First Duma had been injured by the fact that both the Socialist Revolutionaries and the Social Democrats had largely boycotted the election to the Duma. The deputies with no political affiliation were mostly peasants who refused to align themselves permanently with any of the political groupings, but belonged in a general sense to the opposition. The government received support only from the relatively few members of the unregenerate Right and also from the more moderate Octobrists. The Octobrists, as their name indicates, split from the Cadets over the October Manifesto, which they accepted as a proper basis for Russian constitutionalism, while the Cadets chose to consider it as the first step on the road to a more democratic system.

Not surprisingly, the government and the Duma could not work together. The emperor and his ministers clearly intended the Duma to occupy a position subordinate to their own, and they further infuriated many deputies by openly favoring the extreme Right. The Duma, in its turn, also proved quite intractable. The Left wanted merely to oppose and obstruct. The Cadets, while much more moderate and constructive, seem to have overplayed their hand: they demanded a constituent assembly, they considered the First Duma to be, in a sense, the Estates-General of 1789, and they objected to the Fundamental Laws, thus in effect telling the government to abdicate. Similarly, while they insisted on a political amnesty, they refused to proclaim their opposition to terrorism, lest their associates to the Left be offended. But the most serious clash came over the issue of land: the Duma wanted to distribute to the peasants the state, imperial family, and Church lands, as well as the estates of landlords in excess of a certain maximum, compensating the landlords; the government proclaimed alienation of private land inadmissible, even with compensation. The imperial regime continued to the last to stand on the side of the landlords. After seventy-three days and forty essentially fruitless sessions, Nicholas II dissolved the First Duma.

The dissolution had a strange sequel. Some two hundred Duma deputies, over half of them Cadets, met in the Finnish town of Viborg and signed a manifesto that denounced the government and called for passive resistance by the people. It urged them not to pay taxes or answer the draft call until the convocation of a new Duma. Although the Viborg Manifesto cited as its justification certain irregularities in the dissolution of the First Duma, in itself it constituted a rash and unconstitutional step. And it turned out to be a blunder as well, for the country failed to respond. The Viborg participants were sentenced to three months in jail. More important, they lost the right to stand for election to the Second Duma which was thus deprived of much of its potential leadership.

In contrast to the first election, the government exerted all possible pressure to obtain favorable results in the election to the Second Duma, and it was assisted by the fact that much of Russia remained in a state of emergency. But the results again disappointed the emperor and his associates. Although — as one authoritative calculation has it — the Duma opposition, including mainly the Cadets and the Left, might have declined from 69 to 68 per cent of the total number of deputies, it also became more extreme. In fact, a polarization of political opinion, with both wings gaining at the expense of the center, constituted the most striking aspect of the election. More specifically, the Cadet representation declined from 184 to 99 deputies, while the Social Democrats and the Socialist Revolutionaries, who this time participated fully in the election, gained respectively 64 and 20 seats. The entire Left membership in the Duma rose from 124 to 216 deputies. Significantly, the Duma personnel underwent a sweeping change, with only 31 members serving both in the First Duma and in the Second, a result not only of the penalties that followed the Viborg Manifesto, but also of a preference for more extreme candidates. Significantly too, the number of unaffiliated deputies declined by about 50 per cent in the Second Duma.

The Second Duma met on March 5, 1907, and lasted for a little more than three months. It also found itself promptly in an impasse with the government. Moreover, its special opponent, the prime minister, was no longer the nonentity Ivan Goremykin — who had replaced the first constitutional prime minister, Witte, early in 1906 — but the able and determined Peter Stolypin. Before it could consider Stolypin's important land reform, he had the Second Duma dissolved on the sixteenth of June, using as a pretext its failure to comply immediately with his request to lift the immunity of fifty-five, and particularly of sixteen, Social Democratic deputies whom he wanted to arrest for treason.

The Change in the Electoral Law and the Last Two Dumas

On the same day, June 16, 1907, Nicholas II and his minister arbitrarily and unconstitutionally changed the electoral law. The tsar mentioned as justification his historic power, his right to abrogate what he had granted, and his intention to answer for the destinies of the Russian state only before the altar of God who had given him his authority! The electoral change was, of course, meant to create a Duma that would co-operate with the government. The peasant representation was cut by more than half and that of the workers was also drastically cut, whereas the gentry gained representation quite out of proportion to its number. Also, Poland, the Caucasus, and some other border areas lost deputies; and the representation of Central Asia was entirely eliminated on the ground of backwardness. At the same time the election procedure became more indirect and more involved, following in part the Prussian model. In addition, the minister of the interior received the right to manipulate electoral districts. It has been calculated that the electoral change of June 1907 produced the following results: the vote of a landlord counted roughly as much as the votes of four members of the upper bourgeoisie, or of sixty-five average middle-class people, or of 260 peasants, or of 540 workers. To put it differently, 200,000 members of the landed gentry were assured of 50 per cent of the seats in the Duma.

The electoral change finally provided the government with a co-operative Duma. And indeed, by contrast with the first two Dumas which lasted but a few months each, the Third Duma served its full legal term of five years, from 1907 to 1912, while the Fourth also continued for five years, from 1912 until the revolution of March 1917, which struck just before the Fourth Duma was to end. In the Third Duma the government had the support of some 310 out of the total of 442 deputies: about 160 representatives of the Right and about 150 Octobrists. The opposition, reduced to 120 seats, encompassed 54 Cadets, smaller numbers of other moderates, and only 33 deputies of the former Left. The Socialist Revolutionaries, it might be noted, boycotted the Third and Fourth Dumas. To indicate another aspect of the change, it has been calculated that whereas non-Great Russians had composed almost half of the membership of the First Duma, in the Third there were 377 Great Russians and 36 representatives of all the other nationalities of the empire.

In the election of 1912 the government made a determined effort to obtain a Right majority that would eliminate its dependence on the Octobrist vote, but it could not quite accomplish its purpose. The Fourth Duma contained approximately 185 representatives of the Right, 98 Octobrists, and 150 deputies to the left of the Octobrists. Because of their crucial central position, the Octobrists continued to play a major role in the Duma, although their number had been drastically diminished. For the rest, the gain of the Right found a certain counterbalance in the gain of the Left.

The Octobrists, who had replaced the Cadets after the electoral change of June 1907 as the most prominent party in the Duma, represented both the less conservative country gentry and business circles. While their Left wing touched the Cadets, Right Octobrists stood close to the old-fashioned

Right. The party enjoyed the advantages of skillful leadership, in particular the leadership of Alexander Guchkov, and operated well in a parliament. The Octobrist deputies, it might be noted, were the wealthiest group in the last two Dumas. The Cadets, who became the loudest voice of the Duma opposition, were, above all, the party of professional people, although their influence extended to large layers of the middle class, especially perhaps of the upper middle class, as well as to some landlords and other groups. The Right, which consisted of more than one party, defended to the limit the interests of the landlords, although it also made demagogic efforts to obtain broader support and paraded some priests and peasants in the Dumas. Bitter dissatisfaction, widespread among the Russian masses, found a modicum of expression in the Duma Left.

Stolypin's Policy

With the Duma under control, the government could develop its own legislative program. The architect of the program, Stolypin, has been described as the last truly effective and important minister of imperial Russia. Stolypin's aim consisted of "pacification" and reform. "Pacification" meant an all-out struggle against the revolutionaries, for, although the mass opposition movements characteristic of 1905 no longer threatened the regime, terrorism continued on a large scale. Practiced especially by the Battle Organization of the Socialist Revolutionaries and by the Socialist Revolutionaries-Maximalists who had split from the main party, terrorism caused some 1,400 deaths in 1906 and as many as 3,000 in 1907. The victims included police officers and agents, various officials, high and low, and numerous innocent bystanders. In August 1906, for example, the Maximalists blew up Stolypin's suburban residence, killing 32 persons and wounding many others, including the prime minister's son and daughter, but not the prime minister himself.

Stolypin acted with directness and severity. By the end of 1906, 82 areas in the Russian Empire had been placed under different categories of special regulations; also, the publication of 206 newspapers had been stopped, and over 200 editors had been brought to court. Moreover, Stolypin introduced summary courts-martial, consisting of officers without juridical training, which tried those accused of terrorism and rebellion. The trials and the execution of sentences were carried out within a matter of some two days or even a few hours. Although the special courts-martial lasted only several months — because Stolypin never submitted the law creating them to the Duma and it expired two months after the Second Duma had met — they led to the execution of well over a thousand persons. "Stolypin's necktie" — the noose — became proverbial in Russia. The policy of "pacification" succeeded on the whole. The Maximalists and many other terrorists were killed or executed, while numerous revolutionaries escaped abroad. A relative quiet settled upon the country.

It should be added that Stolypin continued to sponsor police infiltration of the revolutionary movement and an extremely complex system of agents and informers. Such police practices led, among other things, to the emergence of remarkable double agents, the most notorious of whom, the unbelievable Evno Azeff, successfully combined the roles of the chief informer on the Socialist Revolutionaries and of leader of their Battle Organization. In the latter capacity he arranged the assassination of Plehve and other daring acts of terrorism.

Stolypin intended his "pacification" to constitute a prelude to important changes, especially to a fundamental agrarian reform. That reform, introduced by an imperial legislative order in the autumn of 1906, approved by the Third Duma in the summer of 1910, and developed by further legislative enactments in 1911, aimed at a break-up of the peasant commune and the establishment of a class of strong, independent, individual farmers — Stolypin's so-called wager on the strong and the sober. The emergence of a large group of prosperous and satisfied peasants would, presumably, transform the Russian countryside from a morass of misery and a hotbed of unrest into a conservative bulwark of the regime.

The new legislation divided all peasant communes into two groups: those that did not and those that did engage in land redistribution. In the first type all peasants simply received their landholdings in personal ownership. In the communes with periodic redistribution every householder could at any time request that the land to which he was entitled by redistribution be granted to him in personal ownership. He could also press the commune to give him the land not in scattered strips, but in a single location; the commune had in effect to comply with this request if separation occurred at the time of a general communal redistribution of land, and it had to meet the request "in so far as possible" at other times. Similarly, the commune had to divide its land into consolidated individual plots if requested to do so by not less than one-fifth of the total number of householders. Moreover, separated peasants invariably retained rights to common lands, meadows, forests, and the like. Indeed a partitioning even of pastures and grazing lands was permitted in 1911. Finally, the commune could be entirely abolished: by a majority vote in the case of nonrepartitional communes, and by a two-thirds vote in the case of those that engaged in a redistribution of land. It is significant that the reform made the household elder the sole owner of the land of the household, replacing the former joint family ownership which remained only in the case of households containing members other than the elder's lineal descendants.

Stolypin's major agrarian reform — the impact of which on Russian economy and society will be discussed in a later chapter — received support from a number of related government policies and measures. Notably, the Peasant Land Bank became much more active in helping peasants to buy land, while considerable holdings of the state and the imperial family were put up for sale to them. Also, reversing its earlier attitude, the government began finally to encourage and help peasant migration to new lands in Siberia and elsewhere in the empire. Stolypin's reform, it should be added, made peasants more equal legally to other classes, not only by reducing the power of the commune, but also by limiting that of land captains, and by exempting peasants from some special restrictions. In a different field of action, the ministers and the Dumas worked together to develop education, which made important advances during the last years of the imperial regime. In fact a law of 1908 foresaw schooling for all Russian children by 1922. The government also broadened labor legislation, worked to strengthen the army and national defense, and engaged in a variety of other useful activities.

However, all this fell short of fundamental reform. Only Stolypin's controversial agrarian legislation attempted a sweeping change in the condition of the Russian people, and even that legislation had perhaps too narrow a scope, for Stolypin was determined not to confiscate any gentry land, even with recompense. Moreover, progressive measures remained intertwined with reaction. Thus constitutional Russia witnessed a terrorism of the Right — for example the assassinations in 1906 and 1907 of two Cadet deputies to the First Duma — as well as a terrorism of the Left, and the terrorism of the Right usually went unpunished. Stolypin, himself from the Western borderlands, acted as a nationalist and a Russificator, for one thing reviving the ill-fated policy of trying to Russify Finland. Besides, the government lacked stability. The prime minister, who was after all something of a constitutionalist, antagonized much of the Right in addition to the Left. He managed to have one important piece of legislation enacted only by having the emperor prorogue the legislature for three days and suspend two leading members of the State Council; his high-handed tactics made the Octobrist leader Guchkov resign as chairman of the Third Duma. On September 14, 1911, Stolypin was fatally shot by a police agent associated with a revolutionary group. Stolypin's successor, Count Vladimir Kokovtsov, possessed intelligence and ability, but not his predecessor's determination or influence within the government. After a little more than two years he was replaced by the weak and increasingly senile Goremykin, who thus became prime minister for the second time. Goremykin assumed the leadership of the government in early 1914; in a matter of a few months he and Russia had to face the devastating reality of the First World War.

Russian Foreign Policy, 1905–14

Like the other powers, Russia stumbled into the First World War. The tsarist government contributed its share to international alignments, tensions, and crises, and in the fateful summer of 1914 it decided to support Serbia and thus resort to arms. Yet its part of the celebrated "war guilt" should not be exaggerated or singled out. Russian ambitions and eagerness for war were no greater than those of other countries, while Russian preparedness for an armed conflict proved to be less. The empire of the tsars took no part in the race for colonies overseas which constituted an important aspect of the background of the First World War. Russian interests and schemes in the Balkans and the Near East were paralleled by those of Austria-Hungary and eventually also to some extent by those of Germany. The Pan-Germans were authentic cousins of the Pan-Slavs; and — a point which Fay and many others failed to appreciate — it was the German government, not the Russian, which enjoyed widespread popular support in its own country for a strong national policy. The fatal conflict erupted first between Austria-Hungary and Serbia, and both states can be charged with a responsibility for its tragic outcome which preceded Russia's. Even the early Russian mobilization found its counterpart in the Austrian. Besides, it deserves to be noted that in the summer of 1914 only Austria-Hungary, of all the powers, desired war, although it thought merely of a quick destruction of Serbia, not of a continental conflagration.

In the course of a personal meeting shortly before the opening of the Portsmouth Peace Conference, Emperor William II of Germany talked Nicholas II into signing a defensive alliance, known as the Treaty of Björkö. However, that agreement proved to be stillborn, because leading officials in both governments expressed strong objections to it and especially because France refused co-operation and held Russia to its obligations under the treaty of 1891–94. The years that followed the Russo-Japanese War witnessed an alienation of Russia from Germany, a virtual breakdown of Russo-Austrian relations, and at the same time a further *rapprochement* between Russia and France as well as the establishment of an Anglo-Russian Entente. The agreement with Great Britain, signed on August 31, 1907, was a landmark in Russian foreign policy, for it transformed a relationship of traditional and often bitter hostility into one of cordiality. That result was achieved through compromise in those areas where the interests of the two countries clashed: in Persia, Russia was assigned a large sphere of influence in the northern part of the country, and Great Britain a smaller one in the southeastern section, while the central area was declared neutral; Russia agreed to consider Afghanistan outside its sphere of influence and to deal with the Afghan ruler only through Great

Britain, Great Britain in turn promising not to change the status of that country or interfere in its domestic affairs; both states recognized the suzerainty of China over Tibet. Because Great Britain and France had reached an agreement in 1904, the new accord marked the emergence of the Triple Entente of France, Russia, and Great Britain, poised against the Triple Alliance of Germany, Austria-Hungary, and Italy. On the Russian side, the Entente meant an effective military and political alliance with France and only a vague understanding with Great Britain. Yet, as already indicated, that understanding represented a major reorientation of Russian, as well as British, foreign policy, and it helped to group Europe into two camps. It should be added that the alignment with France and Great Britain gained in popularity in Russia in the years preceding the First World War. It attracted the support of liberals, of many radicals, of business circles closely linked to French and British capital, and also of numerous conservatives who veered toward Pan-Slavism or suffered from tariff wars with Germany and objected to tariff arrangements with that country as detrimental to Russian agriculture.

Alexander Izvolsky, the Russian minister of foreign affairs from 1906 to 1910, not only made an agreement with Great Britain, but also developed an active policy in the Balkans and the Near East. In fact he, his successor Serge Sazonov who headed the ministry from 1910 to 1916, and their various subordinates have been described as a new generation of Russian diplomats eager to advance Russian interests against Turkey and Austria-Hungary after a quarter-century of quiescence. To be sure, as early as 1896 the Russian ambassador in Constantinople, Alexander Nelidov, had proposed to his government that Russia seize the Straits, but that proposal was never implemented. Izvolsky devised a different scheme. In September 1908, in Buchlau, Moravia, he came to an agreement with the Austrian foreign minister, Count Alois von Aehrenthal: Russia would accept the Austrian annexation of Bosnia and Herzegovina, which Austria had been administering according to a decision of the Congress of Berlin; Austria-Hungary in turn would not object to the opening of the Straits to Russian warships. Austria-Hungary proceeded to annex Bosnia and Herzegovina before Russia could prepare diplomatically the desired reconsideration of the status of the Straits — a betrayal of the mutual understanding, according to Izvolsky, but not according to Aehrenthal. Betrayed or not, Russia was left holding the bag, because other powers, especially Great Britain, proved unwilling to see Russian warships in the Straits. The tsarist government experienced further humiliation when it hesitated to endorse the Austrian coup but was finally forced to do so after receiving a near-ultimatum from Germany.

The years following the annexation of Bosnia and Herzegovina witnessed repeated tensions, crises, and conflicts in the Balkans and the Near East.

Like Austria-Hungary and Russia, Germany also pursued a forward policy in that area. William II visited Constantinople and made a point of declaring his friendly feelings for Turkey and the Moslems; German interests pushed the construction of the Berlin-Baghdad railway — a project they had initiated as early as 1898 — and more German military experts came in 1913 to reorganize the Ottoman army. Two important Balkan wars were fought in 1912 and 1913. First Bulgaria, Serbia, Greece, and Montenegro combined to defeat Turkey and expand at Turkish expense. Next, the victors quarreled and the Bulgarians suffered a defeat by the Serbians, the Greeks, and the Montenegrins, as well as by the Rumanians and by the Turks, who resumed hostilities to regain some of their losses. The Balkan wars left a legacy of tensions behind them, in particular making Bulgaria a dissatisfied and revisionist state and further exacerbating the relations between Austria-Hungary and Serbia.

When the heir to the Habsburg throne, Archduke Francis Ferdinand, was assassinated by Serbian patriots on June 28, 1914, and Austria delivered a crushing ultimatum to Serbia, the Russian government decided to support Serbia — the alternative was another, and this time complete, defeat in the Balkans. With the alliances operating almost automatically, Germany backed Austria-Hungary, while France stood by Russia. Austria-Hungary declared war on Serbia on July 28, Germany on Russia on August 1 and on France on August 3. The German attack on Belgium brought Great Britain to the side of France and Russia on August 4. Europe entered the First World War.

Russia in the First World War

From the summer of 1914 until its collapse during the months that followed the overthrow of the imperial regime in 1917, the Russian army fought tenaciously and desperately under most difficult circumstances. The improvised offensive into East Prussia, which opened the hostilities and helped France at the most critical moment, ended in a shattering defeat of the Russians in the battles of Tannenberg and the Masurian Lakes. This offensive, General Michael Alekseev's epic retreat in Poland in 1915, the repeated offensives and counteroffensives in Galicia, and heavy fighting in numerous other sectors of the huge and shifting Eastern front cost the Russians enormous casualties. Quickly the Russian army ran out of its supply of weapons and ammunition, and for a period of time in 1915 up to 25 per cent of Russian soldiers were sent to the front unarmed, with instructions to pick up what they could from the dead. Although later the Russian supply improved, the Russian forces remained vastly inferior to the German and the Austrian in artillery and other weapons.

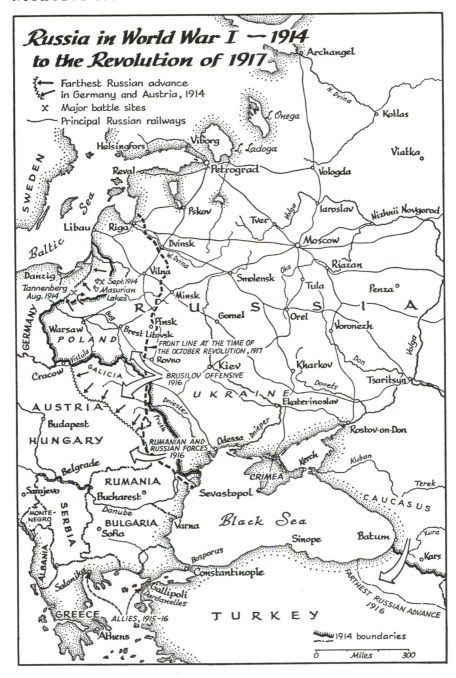

Russia in World War I — 1914 to the Revolution of 1917

- Farthest Russian advance in Germany and Austria, 1914
- x Major battle sites
- Principal Russian railways

Archangel

Kotlas

L. Onega

N. Dvina

Viatka

Viborg

Helsingfors

L. Ladoga

SWEDEN

Reval

Petrograd

Vologda

Baltic Sea

Pskov

Iaroslav

Nizhnii Novgorod

Libau

Riga

Tver

Volga

Dvinsk

Moscow

W. Dvina

Oka

Riazan

Danzig

Vilna

Smolensk

Tula

Penza

x Sept. 1914

Masurian Lakes

Minsk

Gomel

Orel

Voronezh

Volga

Tannenberg Aug. 1914

R   U   S   S   I   A

GERMANY

Warsaw

POLAND

Bug

Pinsk

Brest Litovsk

FRONT LINE AT THE TIME OF THE OCTOBER REVOLUTION, 1917

Kharkov

Cracow

Vistula

Rovno

Kiev

Tsaritsyn

GALICIA

BRUSILOV OFFENSIVE 1916

Don

Donets

AUSTRIA

Dniester

U K R A I N E

Ekaterinoslav

Budapest

Pruth

HUNGARY

RUMANIAN AND RUSSIAN FORCES 1916

Odessa

Dnieper

Rostov-on-Don

Belgrade

Kerch

Kuban

CRIMEA

Terek

Sarajevo

RUMANIA

Sevastopol

CAUCASUS

MONTE-NEGRO

Bucharest

SERBIA

Danube

BULGARIA

Varna

Black Sea

Batum

Kura

Sofia

Sinope

Kars

ALBANIA

Bosporus

FARTHEST RUSSIAN ADVANCE 1916

Saloniki

Constantinople

GREECE

Gallipoli
Dardanelles

ALLIES, 1915–16

T U R K E Y

Athens

1914 boundaries

0   Miles   300

The Allies could help little, for the German navy controlled the Baltic, and approaches through the Black Sea were cut off when Turkey joined the Central Powers in the autumn of 1914. The so-called Gallipoli campaign of the Allies, which aimed to break the Turkish hold on the Straits, failed in 1915. Bulgaria joined the Central Powers in October 1915 to help crush Serbia. The Rumanian entry into the war on the side of the Entente at the end of August 1916 led to a catastrophic defeat of the Rumanians and served to extend the Russian front. Yet the Russian troops went on fighting. In fact, they generally outfought the Austrians, and they also scored successes on the Caucasian front against Turkey. More important, in spite of many defeats and the necessity of retreating, they continued to force Germany to wage a major war on two fronts at the same time. As a present-day British historian has put it: "Despite all defects and difficulties, the Russians fought heroically, and made a decisive contribution to the course of the war." In the field of diplomacy, devoted during those years to the prosecution of the war and the formulation of war aims, the Russian government made a striking gain when in the spring of 1915 Great Britain and France agreed to the Russian acquisition of Constantinople, the Straits, and the adjoining littoral at the peace settlement. Italy, which joined the Entente at the end of August 1916, acquiesced in the arrangement.

While the Russian command made its share of military mistakes, the political mistakes of the Russian government proved to be both greater and more damaging. Nicholas II and his ministers failed to utilize the national rally that followed the outbreak of the war. In fact, they continued to rely on exclusively bureaucratic means to mobilize the resources of the nation, and they proceeded to oppress ethnic and religious minorities in the areas temporarily won from Austria as well as in home provinces. In particular, they failed to make the necessary concessions to the Poles. Russian defeats, the collapse of Russian supply, and the utter incompetence of the war minister, General Vladimir Sukhomlinov, as well as of some other high officials, did lead, to be sure, to certain adjustments. The Duma was finally called together in August 1915 for a short session, Sukhomlinov and three of his colleagues had to resign, and the government began to utilize the efforts of society to support the army. These efforts, it should be added, which were led by public figures and industrialists such as Guchkov, had developed on a large scale, ranging from work in the Red Cross to widespread measures to increase production of military matériel. The Zemstvo Union and the Union of Towns, which joined forces under the chairmanship of Prince George Lvov, and the War Industry Committee, led by Guchkov, became especially prominent.

But the *rapprochement* between the government and the public turned out to be slight and fleeting. Nicholas II would not co-operate with the newly created, moderate Progressive Bloc led by Miliukov, which included the entire membership of the Duma, except the extreme Right and the extreme Left, and which won majority support even in the State Council. Instead he came to rely increasingly on his wife Empress Alexandra and on her extraordinary advisor, the peasant Gregory Rasputin. Moreover, in spite of the protests of ten of his twelve ministers, the sovereign unwisely took personal command of the armed forces, which had been commanded by his relative Grand Duke Nicholas, leaving Alexandra and Rasputin in effective control in the capital. Thus a narrow-minded, reactionary, hysterical woman and an ignorant, weird peasant — who apparently made decisions simply in terms of his personal interest, and whose exalted position depended on the empress's belief that he could protect her son from hemophilia and that he had been sent by God to guide her, her husband, and Russia — had the destinies of an empire in their hands. Ministers changed rapidly in what has been described as a "ministerial leapfrog," and each was more under Rasputin's power than his predecessor. Eventually, after Rasputin's assassination, one of them claimed communion with Rasputin's spirit! That assassination, long and gruesome, took place at the end of December, 1916. It was engineered by a leader of the extreme Right, a member of the imperial family, and another aristocrat related to the imperial family by marriage, who each tried to save the dynasty and Russia. As the year 1917 began, there were rumors of a palace coup that would restore sanity and leadership to the imperial government. But a popular revolution came first.

THE ECONOMIC AND SOCIAL DEVELOPMENT OF RUSSIA FROM THE "GREAT REFORMS" UNTIL THE REVOLUTIONS OF 1917

The last sixty years of Imperial Russia are not only in themselves a period of great historical interest: they are significant for other countries and other periods. The pattern of this period in Russia has repeated and is repeating itself elsewhere. It is not only in Russia, and not only in Europe, that the impact of the nineteenth- or twentieth-century West on a backward country has caused distortions and frustrations, has released revolutionary forces. New countries have been drawn into the world capitalist economy, into the rapid exchange of goods and ideas. The loss of centuries has to be made up in a few years. Improved communications, public order and sanitation increase population faster than output. The impoverished masses become more impoverished. The new ways create a new intelligentsia. The shrieking contrast between the old and the new drive a part of the intelligentsia to revolutionary ideas, and if political conditions make this necessary, to conspiratorial organization. The force which keeps such societies together is the bureaucracy. It holds the power, the privileges and the means of repression. From it and through it come such reforms as are permitted. It is outwardly impressive. It weighs heavily on the backs of the people. But like cast iron, though heavy it is also brittle. A strong blow can shatter it to pieces. When it is destroyed there is anarchy. Then is the moment for a determined group of conspiratorial revolutionary intellectuals to seize power.

H. SETON-WATSON

Whether the general well-being of the peasantry had shown improvement or decline — whether there had been within the peasant mass a tendency to draw together or to draw apart — still, as the day of revolt approached, there was no doubt of the existence in the countryside of a morass of penury sufficiently large, an antithesis between poverty and plenty sufficiently sharp, to give rise to whatever results might legitimately be bred and born of economic misery and economic contrast.

ROBINSON

Who lives joyfully,
Freely in Russia?

NEKRASOV

THE "GREAT REFORMS" made a division in the economic and social development of Russia. Even if we disregard the peculiar Soviet periodization, which considers Russia as feudal from the late Kievan era until the emancipation of the serfs and capitalistic from the emancipation of the serfs until 1917, the crucial significance of the "great reforms" must still be emphasized. In particular, these reforms contributed immensely to the economic changes and the concomitant social shifts which characterized the empire of the Romanovs during its last five or six decades and culminated in its downfall.

Every social class felt the impact of the "great reforms" and of their aftermath. The gentry, to be sure, remained the dominant social group in the country. In fact, as already indicated, both Alexander III and Nicholas II made every effort to strengthen the gentry and to support its interests. Court circles consisted mainly of great landlords. The bureaucracy that ran the empire was closely linked on its upper levels to the landlord class. The ministers, senators, members of the State Council, and other high officials in the capital and the governors, vice-governors, and heads of various departments in the provinces belonged predominantly to the gentry. With the establishment in 1889 of land captains to be appointed from the local gentry, Russia obtained a new network of gentry officials who effectively controlled the peasants. A year later the zemstvo "counterreform" greatly strengthened the role of the gentry in local self-government and emphasized the class principle within that government. In the army most high positions were held by members of the landlord class, while virtually the entire officer corps of the navy belonged to the gentry. The government supported gentry agriculture by such measures as the establishment in 1885 of the State Gentry Land Bank which provided funds for the landlords on highly favorable terms.

Nevertheless, the gentry class declined after the "great reforms." Members of the gentry owned 73.1 million *desiatin* * of land according to the census of 1877, 65.3 million according to the census of 1887, 53.2 in 1905 according to a statistical compilation of that year, and only 43.2 million desiatin in 1911 according to Oganovsky's calculations. At the same time, to quote Robinson: "The average size of their holdings also diminished, from 538.2 *desiatinas* in 1887 to 488 in 1905; and their total possession of work horses from 546,000 in 1888–1891, to 499,000 in 1904–1906 — that is, by 8.5 per cent." Although the emancipation settlement was on the whole generous to the gentry, it should be kept in mind that a very large part of the wealth of that class had been mortgaged to the state before 1861 and that, therefore, much of the compensation that the landlords received as part of the reform went to pay debts, rather little remaining for development and modernization of the gentry economy. Moreover, most landlords failed to make effective use of their resources and opportunities. Deprived of serf labor and forced to adjust to more intense competition and other

* A *desiatina* equals 2.7 acres.

424

harsh realities of the changing world, members of the gentry had little in their education, outlook, or character to make them successful capitalist farmers. A considerable number of landlords, in fact, preferred to live in Paris or Nice, spending whatever they had, rather than to face the new conditions in Russia. Others remained on their estates and waged a struggle for survival, but, as statistics indicate, frequently without success. Uncounted "cherry orchards" left gentry possession. The important fact, much emphasized by Soviet scholars, that a small segment of the gentry did succeed in making the adjustment and proceeded to accumulate great wealth in a few hands does not fundamentally change the picture of the decline of a dominant class.

The Industrialization of Russia

If the "great reforms" helped push the gentry down a steep incline, they also led to the rise of a Russian middle class, and in particular of industrialists, businessmen, and technicians — both results, to be sure, were not at all intentional. It is difficult to conceive of a modern industrial state based on serfdom, although, of course, the elimination of serdom constituted only one prerequisite for the development of capitalism in Russia. Even after the emancipation the overwhelmingly peasant nature of the country convinced many observers that the empire of the tsars could not adopt the Western capitalist model as its own. The populists argued that the Russian peasant was self-sufficient, producing his own food and clothing, and that he, in his egalitarian peasant commune, did not need capitalism and would not respond to it. Perhaps more to the point, the peasant was miserably poor and thus could not provide a sufficient internal market for Russian industry. Also the imperial government, especially the powerful Ministry of the Interior, preoccupied with the maintenance of autocracy and the support of the gentry, for a long time in effect turned its back on industrialization.

Nevertheless, Russian industry continued to grow — a growth traced in detail by Goldsmith and others — and in the 1890's it shot up at an amazing rate, estimated by Gerschenkron at 8 per cent a year on the average. Russian industrialists could finally rely on a better system of transportation, with the railroad network increasing in length by some 40 per cent between 1881 and 1894 and doubling again between 1895 and 1905. In addition to Russian financial resources, foreign capital began to participate on a large scale in the industrial development of the country: foreign investment in Russian industry has been estimated at 100 million rubles in 1880, 200 million in 1890, and over 900 million in 1900. Most important, the Ministry of Finance under Witte, in addition to building railroads and trying to attract capital from abroad, did everything possible to develop heavy industry in Russia. To subsidize that industry Witte increased Russian exports, drastically curtailed imports, balanced the budget, introduced the gold standard, and used heavy indirect taxation on items of everyday consumption to squeeze the necessary funds out of the peasants. Thus, in Russian conditions, the state played the leading role in bringing large-scale capitalist enterprise into existence.

Toward the end of the century Russia possessed eight basic industrial regions, to follow the classification adopted by Liashchenko. The Moscow industrial region, comprising six provinces, contained textile industries of every sort, as well as metal processing and chemical plants. The St. Petersburg region specialized in metal processing, machine building, and textile industries. The Polish region, with such centers as Lodz and Warsaw, had textile, coal, iron, metal processing, and chemical industries. The recently developed south Russian Ukrainian region supplied coal, iron ore, and basic chemical products. The Ural area continued to produce iron, nonferrous metals, and minerals. The Baku sector in Transcaucasia contributed oil. The southwestern region specialized in beet sugar. Finally, the Transcaucasian manganese-coal region supplied substantial amounts of its two products.

The new Russian industry displayed certain striking characteristics. Because Russia industrialized late and rapidly, the Russians borrowed advanced Western technology wholesale, with the result that Russian factories were often more modern than their Western counterparts. Yet this progress in certain segments of the economy went together with appalling backwardness in others. Indeed, the industrial process frequently juxtaposed complicated machinery and primitive manual work performed by a cheap, if unskilled, labor force. For technological reasons, but also because of government policy, Russia acquired huge plants and large-scale industries almost overnight. Before long the capitalists began to organize: a metallurgical syndicate was formed in 1902, a coal syndicate in 1904, and several others in later years. Russian entrepreneurs and employers, it might be added, came from different classes — from gentry to former serfs — with a considerable admixture of foreigners. Their leaders included a number of old merchant and industrialist families who were Old Believers, such as the celebrated Morozovs. As to markets, since the poor Russian people could absorb only a part of the products of Russian factories, the industrialists relied on huge government orders and also began to sell more abroad. In particular, because Russian manufactures were generally unable to compete successfully in the West, export began on a large scale to the adjacent Asiatic countries of Turkey, Persia, Afghanistan, Mongolia, and China. Again Witte and the government helped all they could by such means as the establishment of the Russo-Persian Bank and the Russo-Chinese Bank, and the building of the East China Railway, not to mention the Trans-

Siberian. As already indicated, Russian economic activity in the Far East was part of the background of the Russo-Japanese War.

The great Russian industrial upsurge of the 1890's ended with the depression of 1900, produced by a number of causes, but perhaps especially by the "increasing weakness of the base," the exhaustion of the Russian peasantry. The depression lasted several years and became combined with political unrest and finally with the Revolution of 1905. Still, once order had been restored and the Russians returned to work, industrialization resumed its course. In fact, the last period of the economic development of imperial Russia, from the calling of the First Duma to the outbreak of the First World War, witnessed rapid industrialization, although it was not as rapid as in the 1890's, with an annual industrial growth rate of perhaps 6 per cent compared to the 8 per cent of the earlier period. The output of basic industries again soared, with the exception of the oil industry. Thus, counting in millions of *pudy* * and using 1909 and 1913 as the years to be compared, the Russian production of pig iron rose from 175 to 283, of iron and steel from 163 to 246, of copper from 1.3 to 2.0, and of coal from 1,591 to 2,214.

The new industrial advance followed in many ways the pattern of the previous advance, for instance, in the emphases on heavy industry and on large plants. Yet it exhibited some significant new traits as well. With the departure of Witte, the government stopped forcing the pace of industrialization, decreased the direct support of capitalists, and relaxed somewhat the financial pressure on the masses. Russian industry managed to make the necessary adjustments, for it was already better able to stand on its own feet. Also, the industry often had the help of banks, which began to assume a guiding role in the economic development of the country. But, financial capital aside, the Russian industrialists themselves were gradually gaining strength and independence. Also, it can well be argued that during the years immediately preceding the First World War Russian industry was becoming more diversified, acquiring a larger home market, and spreading its benefits more effectively to workers and consumers.

To be sure, the medal had its reverse side. In spite of increasing production in the twentieth century, imperial Russia was falling further behind the leading states of the West — or so it is claimed by many analysts, especially Marxist analysts. Just as the Russian government relied on foreign loans, Russian industry remained heavily dependent on foreign capital, which rose to almost two and a quarter billion rubles in 1916/17 and formed approximately one-third of the total industrial investment. The French, for example, owned nearly two-thirds of the Russian pig iron and one-half of the Russian coal industries, while the Germans invested heavily in the

* A *pud* equals 36 pounds.

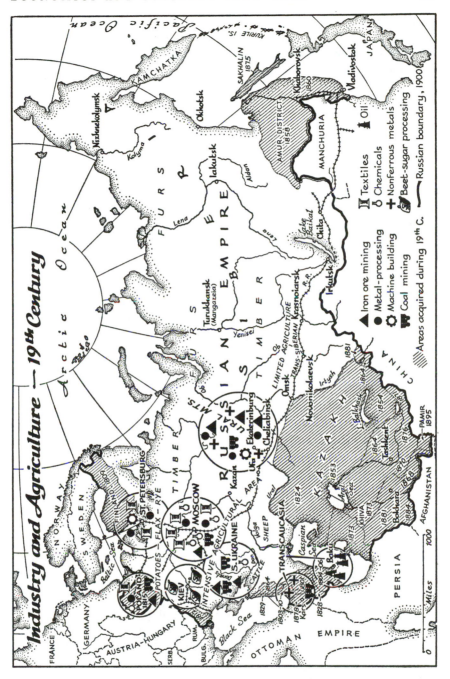

chemical and electrical engineering industries, and the British in oil. On the basis of investment statistics some Marxists even spoke of the "semi-colonial" status of Russia! More ominously, Russian industry rose on top of a bitter and miserable proletariat and a desperately poor peasant mass.

Labor

The industrialization of Russia created, of course, a considerable working class. While Russians began to work in factories in the Urals and elsewhere far back in history, as mentioned in previous chapters, a sizeable industrial proletariat grew in Russia only toward the end of the nineteenth century. Russian industrial workers numbered over 2 million in 1900 and perhaps 3 million out of a population of about 170 million in 1914. Not impressive in quantity in proportion to total population, the proletariat was more densely massed in Russia than in other countries. Because of the heavy concentration of Russian industry, over half the industrial enterprises in Russia employed more than 500 workers each, with many employing more than 1,000 each. The workers thus formed large and closely knit groups in industrial centers, which included St. Petersburg and Moscow.

True, the term "worker" may be too definitive and precise as applied to the Russian situation. Populists, Marxists, and scholars of other persuasions, as well as Western specialists such as Zelnik and Johnson, have debated the extent to which Russian workers remained — or ceased to be — peasants. These workers usually came from the village. Often they belonged to the village commune, left their families behind in the village, and spent a part of every year there, gathering harvest and performing other peasant tasks. For them the village remained their home, while the factory became a novel way to earn obrok, so to speak. When a close relationship with the village ceased, many factory hands still maintained their membership in it and sought to retire to it to end their days in peace. And even after all important ties with the countryside were broken and workers were left entirely and permanently on their own in towns and cities, they could not shed overnight their peasant mentality and outlook. The Russian proletariat tended to be not only the pride but also the despair of the Marxists both before and after 1917. In fact, in the years following the October Revolution much of it vanished into the countryside. Nevertheless, the Marxists were right in their argument with the populists to the extent that they emphasized the continuing growth of capitalism and the proletariat in Russia. With all due qualifications, from the 1880's on, an industrial working class constituted a significant component of Russian population, an essential part of Russian economy, and a factor in Russian politics.

As noted in an earlier chapter, the government initiated modern labor legislation in the 1880's, when Minister of Finance Bunge tried to eliminate or curb certain glaring abuses of the factory system and established factory inspectors to supervise the carrying out of new laws. More legislation followed later, with a law in 1897 applicable to industrial establishments employing more than 20 workers that limited day work of adults to eleven and a half hours and night work to ten hours. The ten-hour day was also to prevail on Saturdays and on the eve of major holidays, while no work was allowed on Sundays or the holidays in question. Adolescents and children were to work no more than ten and nine hours a day respectively. A pioneer labor insurance law, holding the employers responsible for accidents in connection with factory work, came out in 1903, but an improved and effective labor insurance act, covering both accidents and illness, appeared only in 1912. Unions were finally allowed in 1906, and even then exclusively on the local, not the national, level.

However, in spite of labor legislation, and also in spite of the fact that wages probably increased in the years preceding the First World War — a point, incidentally, strongly denied by Soviet scholars — Russian workers remained in general in miserable condition. Poorly paid, desperately overcrowded, and with very little education or other advantages, the proletariat of imperial Russia represented in effect an excellent example of a destitute and exploited labor force, characteristic of the early stages of capitalist development and described so powerfully by Marx in *Capital*.

Not surprisingly, the workers began to organize to better their lot. Indeed, they exercised at times sufficient pressure to further labor legislation, notably in the case of the law of 1897, and they could not be deterred by the fact that unions remained illegal until after the Revolution of 1905 and were still hampered and suspected by the government thereafter. The first significant strikes occurred in St. Petersburg in 1878 and 1879 and at a Morozov textile factory near Moscow in 1885. The short-lived but important Northern Workers' Union, led by a worker and populist, Stephen Khalturin, helped to organize the early labor movement in the capital. Major strikes took place in the '90's, not only in St. Petersburg, but also in Riga, in industrial areas of Russian Poland, and in new plants in the Ukraine. In addition, railwaymen struck in several places. The strike movement again gathered momentum in the first years of the twentieth century, culminating, as we know, in the Revolution of 1905, the creation and the activities of the St. Petersburg Soviet, its arrest by the government, and the unsuccessful armed uprising of workers in Moscow at the very end of the year. A lull of several years followed these events. However, the Russian labor movement revived shortly before the outbreak of the First World War. Strikes became frequent after the massacre of workers in the Lena gold fields in April 1912, when police fired into a crowd of protesting workers killing and wounding more than a hundred of them. In 1912, 725,000 workers went out on strike, 887,000 in 1913, and over a million and a quarter from January to July in 1914. Their demands, it should be noted, were often political, as well as economic, in character. The Social Democrats, both the Mensheviks and the Bolsheviks, developed large-scale activities in the Russian labor movement.

The Peasant Question

Peasants constituted the vast majority of the Russian people, at least three-quarters of the total population according to the census of 1897. In a sense, they were the chief and the most direct beneficiaries of the "great reforms," particularly since the serfs received their freedom and the state peasants escaped some of their bondage to the state. Yet, after the reforms, their condition remained the largest and the gravest problem in Russia. As mentioned, the emancipation provisions proved to be insufficient to develop a healthy peasant economy — whether any provisions would have sufficed is another matter — and some of these provisions were shown to be entirely unrealistic: at the time of the partition former serfs received considerably less than their half of the land, and they simply could not meet the redemption payments. Moreover — a point which we have not discussed in any detail — the emancipation took a long time and followed an uneven course throughout Russia, with periods of transition and other delays to the peasants' full acquisition of their new status. And even that status, when finally attained, did not make the peasants equal to other social groups. Thus they possessed a separate administration and courts and, besides, were tied to the peasant commune in most of European Russia.

The communes, which received the land at the time of the emancipation, were made responsible for taxes and recruits and were in general intended to serve as bulwarks of order and organized life in the countryside. No doubt they helped many peasants keep their bearings in post-reform Russia, and they usually provided at least minimal security for their members. Even industrial workers, as mentioned above, often planned to retire in their villages. But the price of communal services was high. Communes tended to perpetuate backward, indeed archaic, agricultural production: they continued their traditional, ignorant ways, including the partitioning of land into small strips so that each household would receive land of every quality; and they lacked capital, education, and initiative for modernization. Individual householders, even when more progressively inclined, to a large extent had to follow the practices of their neighbors and, besides, acquired little incentive to improve their strips in those communes which periodically redivided the land. At the same time communes greatly hampered peasant mobility and promoted ever-increasing overpopulation in the countryside. Members of a commune frequently found it difficult to obtain permission to leave, because their departure would force the commune to perform its set obligations to the state with fewer men. Also, where communes periodically redivided the land among the households, the head of the household could prevent the departure of one of its members on the ground that that would result in a smaller allotment of land to the household at the next reapportionment. As Gerschenkron commented: "Nothing was more revealing of the irrational way in which the village commune functioned than the fact that the individual household had to retain the abundant factor (labor) as a precondition for obtaining the scarce factor (land)."

Population in Russia grew rapidly after the emancipation: from over 73 million in 1861 to over 125 million according to the census of 1897 and almost 170 million in 1917. Land prices more than doubled between 1860 and 1905, and almost doubled again between 1905 and 1917. In spite of the fact that peasants purchased much of the land sold over a period of time by the gentry, individual peasant allotments kept shrinking. Russian economic historians have calculated that 28 per cent of the peasant population of the country could not support itself from its land allotments immediately after the emancipation, and that by 1900 that figure had risen to 52 per cent. That the allotments still compared reasonably well with the allotments of peasants in other countries proved to be cold comfort, for with the backward conditions of agriculture in Russia they plainly did not suffice. The average peasant ownership of horses also declined sharply, with approximately one-third of peasant households owning no horses by 1901. The peasants, of course, tried a variety of ways to alleviate their desperate plight, from periodic employment in the cities to migration, but with limited success at best. They worked as hard as they could, exhausting themselves and the land, and competing for every bit of it. In this marginal economy droughts became disasters, and the famine of 1891 was a shattering catastrophe. But even without outright famine peasants died rapidly. At the beginning of the twentieth century, the annual death rate for European Russia, with the countryside leading the cities, stood at 31.2 per thousand, compared to 19.6 in France and 16 in England. Naturally, conditions differed in the enormous Russian Empire, with Siberian peasants, for example, reasonably prosperous. On the other hand, perhaps the worst situation prevailed in the thickly populated provinces of central European Russia — caused by the so-called "pauperization of the center." How the peasants themselves felt about their lot became abundantly clear in the massive agrarian disturbances culminating in 1905.

To appreciate the burden that the Russian peasant had to carry, we should take further note of the fiscal system of the empire. Thus, an official inquiry indicated that after the emancipation the peasants paid annually to the state in taxes, counting redemption payments, ten times as much per desiatina of land as did members of the gentry. And even after the head tax was abolished in 1886 and the redemption payments were finally canceled in 1905, the impoverished masses continued to support the state by means of indirect taxes. These taxes, perennially the main source of imperial revenue, were levied on domestic and imported items of everyday consumption such as vodka, sugar, tea, tobacco, cotton, and iron. The tax on alcohol, which Witte made a state monopoly in 1894, proved especially lucrative. While relentless financial pressure forced the peasants to sell all they could, the government, particularly Witte, promoted the export of foodstuffs, notably grain, to obtain a favorable balance of trade and finance the industrialization of Russia. Foodstuffs constituted almost two-thirds in value of all Russian exports in the first years of the twentieth century compared to some two-fifths at the time of the emancipation.

However, the last years of imperial Russia, the period from the Revolution of 1905 to the outbreak of the First World War, brought some hope and improvement — many authorities claim much hope and great improvement — into the lives of the Russian peasants, that is, the bulk of the Russian people. The upswing resulted from a number of factors. As already indicated, the industrialization of Russia no longer demanded or obtained the extreme sacrifices characteristic of the 1890's, and the new Russian industry had more to offer to the consumer. The national income in fifty provinces of European Russia rose, according to Prokopovich's calculation, from 6,579.6 million rubles in 1900 to 11,805.5 million in 1913. In 1913 the per capita income for the whole Russian Empire amounted to 102.2 rubles, a considerable increase even if highly inadequate compared to the figures of 292 rubles for Germany, 355 for France, 463 for England, or 695 for the United States. Luckily, the years preceding the First World War witnessed a series of bountiful harvests. Russian peasants profited, in addition, from a remarkable growth of the co-operative movement, and from government sponsorship of migration to new lands. Co-operatives multiplied from some 2,000 in 1901 and 4,500 in 1905 to 33,000 at the outbreak of the First World War, when their membership extended to 12 million people. Credit and consumers' co-operatives led the way, although some producers' co-operatives, such as Siberian creamery co-operatives, also proved highly successful. As to migration, the government finally began to support it after the Revolution of 1905 by providing the necessary guiding agencies and also by small subsidies to the migrants, suspension of certain taxes for them, and the like. In 1907 over half a million people moved to new lands and in 1908 the annual number of migrants rose to about three-quarters of a million. After that, however, it declined to the immediate pre-war average of about 300,000 a year. Land under cultivation increased from 88.3 million desiatin in 1901–5 to 97.6 million in 1911–13.

Also as mentioned earlier, the Peasant Land Bank became much more active, helping peasants to purchase over 4.3 million desiatin of land in the decade from 1906 to 1915, compared to 0.96 million in the preceding ten years. State and imperial family lands amounting to about a million and a quarter desiatin were offered for sale to the peasants.

Stolypin's land reform could well be considered the most important factor of all in the changing rural situation, because it tried to transform the Russian countryside. Stolypin's legislation of 1906, 1910, and 1911 — outlined in the preceding chapter — aimed at breaking up the peasant commune and at creating a strong class of peasant proprietors. These peasant proprietors were to have their land in consolidated lots, not in strips. To summarize the results of the reform in the words of a hostile critic, Liashchenko:

> By January 1, 1916, requests for acquisition of land in personal ownership were submitted by 2,755,000 householders in European Russia. Among these, some 2,008,000 householders with a total acreage of 14,123,000 *dessyatins* separated from the communes. In addition, 470,000 householders with an aggregate acreage of 2,796,000 *dessyatins* obtained "certified deeds" attesting to their acquisition of personal holdings in communes not practicing any redistribution. Altogether, 2,478,000 householders owning an area of 16,919,000 *dessyatins* left the communes and secured their land in personal ownership. This constituted about 24 per cent of the total number of households in forty provinces of European Russia.

Oganovsky, Robinson, Florinsky, Karpovich and others have arrived at roughly the same figure of about 24 per cent of formerly communal households completing their legal withdrawal from the commune. In contrast to Liashchenko, however, some specialists emphasize a greater spread and potentiality of the reform. Notably they stress the fact that, although only 470,000 households in nonrepartitional communes had time to receive legal confirmation of their new independent status, the law of 1910 made in effect all householders in such communes individual proprietors. Two million would thus be a more realistic figure than 470,000. If we make this adjustment and if we add to the newly established independent households the three million or more hereditary tenure households in areas where communal ownership had never developed, we obtain for European Russia at the beginning of 1916 over seven million individual proprietary households out of the total of thirteen or fourteen million. In other words, peasant households operating within the framework of the peasant commune had declined to somewhat less than half of all peasant households in Russia. Consolidation of strips, a crucial aspect of the reform, proceeded much more slowly than separation from the commune, but it too made some progress. One important set of figures indicates that of the almost two and a half million households that had left communes somewhat more than half had been provided with consolidated farms by 1916.

Still, these impressive statistics do not necessarily indicate the ultimate wisdom and success of Stolypin's reform. True, Stolypin has received much praise from many specialists, including such present-day American scholars as Treadgold, who believe that the determined prime minister was in fact saving the empire and that, given time, his agrarian reform would have achieved its major objective of transforming and stabilizing the countryside. But critics have also been numerous and by no means limited to populists or other defenders of the commune as such. They have pointed, for example, to the limited scope of Stolypin's reform which represented, in a sense, one more effort to save gentry land by making the peasants redivide what they already possessed, and to the element of compulsion in the carrying out of the reform. They argued that the reform had largely spent itself without curing the basic ills of rural Russia. Moreover, it added new problems to the old ones, in particular by helping to stratify the peasant mass and by creating hostility between the stronger and richer peasants whom the government helped to withdraw from the commune on advantageous terms and their poorer and more egalitarian brethren left behind.

Conclusion

To conclude, various evaluations have been given of the development of Russian industry in the last years of the empire, of the development of Russian agriculture, and indeed of the entire economy of the country. Whereas Gerschenkron, Karpovich, Pavlovsky and other scholars have emphasized progress and grounds for optimism, Soviet authorities, as well as such Western specialists as Von Laue, have concluded that in spite of all efforts — perhaps the maximum efforts possible under the old regime — Russia was not solving its problems either in terms of its own requirements or by comparison with other countries. Most close students of the period have come out with the feeling — so pronounced in Robinson's valuable work on rural Russia — that, whether the conditions of life in Russia improved or declined on the eve of the First World War, they remained desperately hard for the bulk of the population.

It has been said that revolutions occur not when the people are utterly destitute, oppressed beyond all measure, and deprived of hope — crushing conditions lead only to blind and fruitless rebellions — but when there is growth, advance, and high expectation, hampered, however, by an archaic and rigid established order. Such a situation existed in Russia in the early twentieth century: in economic and social matters as well as in politics.

XXXIII

* * * * * * * * * *

RUSSIAN CULTURE FROM THE "GREAT REFORMS" UNTIL

THE REVOLUTIONS OF 1917

> There is only one evil among men — ignorance; against this evil
> there is only one medicine — learning; but this medicine must be
> taken not in homeopathic doses, but by the pail and by the forty-
> pail barrel.
>
> PISAREV

> The three points where the new man thought he had made himself
> most secure were: first, his liberation from all the values and
> institutions of the *status quo;* second, his complete faith in human
> reason and the principles it made known to him; and finally, his
> assurance that he was the personal instrument of the historical
> process. . . . They were convinced that they had found the path to
> a state of personal engagement which could sustain them in their
> struggle with the tsarist system, because they believed in the justice
> of their assault and in the inevitability of its ultimate issue. But if
> we view it critically we note that it rested on an "adjustment" that
> was composed in large part of hostility to existing institutions, and
> in equally large part of commitment to a world that had not yet
> come into being. Described so, its precariousness becomes obvious.
>
> MATHEWSON

> Various forces were at work in the 1890's in opposition to the
> Gorky-Andreyev school, and particularly to the dominance of social
> significance and nihilistic thought in literature. There was a definite
> turning away from civic morality to aestheticism, from duty to
> beauty, and cultural and individual values were stressed at the ex-
> pense of political and social values. Most of the participators in this
> movement were brilliant intellectuals, and their efforts represented
> a lofty degree of cultural refinement that had never been achieved
> by any literary group in Russia hitherto.
>
> SIMMONS

THE DECADES that elapsed between the emancipation of the serfs and the
revolutions of 1917 constituted an active, fruitful, and fascinating period
in the history of Russian culture. Education continued to grow at all levels,
in spite of obstacles and even governmental "counterreforms"; in the
twentieth century the rate of growth increased sharply. Russian science
and scholarship, already reasonably well-established at the time of Nicholas
I's death, developed further and blossomed out. In a word, Russia became
a full-fledged contributor to and partner in the intellectual and academic
efforts of the Western world, its new high position in that respect antedating

by decades the October Revolution. Russian literature continued its "golden age," although primarily in prose rather than in poetry and largely through the achievements of several isolated individuals, such as Turgenev, Tolstoy, and Dostoevsky. Later, when the giants died or, as in the case of Tolstoy, stopped writing fiction and the "golden age" came to its end, Chekhov, Gorky, and some other outstanding authors maintained the great tradition of Russian prose. Moreover, the very end of the nineteenth century and the first part of the twentieth witnessed another magnificent literary and artistic revival, designated sometimes as the "silver age." In literature that renaissance meant the appearance once again of superb poetry, especially Alexander Blok's, the introduction of a wide variety of new trends, and the emergence of exceptionally high standards of culture and craftsmanship. The "silver age" also extended to the theater, music, ballet, painting, and sculpture, and in effect to every form of creative expression. It proved especially beneficial to the visual arts, which had produced little of distinction in the age of arid realism, and it scored perhaps its most resounding successes in the ballet and the theater. In the history of ideas, as well as in literature and art, the period can be divided into two uneven parts: from the 1860's to the end of the century and indeed to the revolutions of 1917, the creed of radicalism, utilitarianism, and materialism first proclaimed by left-wing Westernizers dominated student and other active intellectual circles, finding its best expression in nihilism, different forms of populism, and Marxism; yet with the turn of the century and the "silver age" in culture members of the intellectual elite began to return to idealistic metaphysics and religion. The First World War and later the revolutions struck when Russian intellectual and cultural life was exhibiting more vitality, diversity, and sophistication than ever before.

Education

The death of Nicholas I and the coming of the "great reforms" meant liberalization in education as in other fields. The university statute of 1863 reaffirmed the principle of university autonomy, while Nicholas I's special restrictions on universities were among the first regulations to disappear in the new reign. The zemstvo reform of 1864 opened vast opportunities to establish schools in the countryside. In towns or rural areas, the increasing thirst for knowledge on the part of the Russians augured well for education in a liberal age. However, as already mentioned, official liberalism did not last long, and reaction logically, if unfortunately, showed a particular concern for education. As a result, the growth of education in Russia, while it could not be stopped, found itself hampered and to an extent deformed by government action.

After Dmitrii Tolstoy replaced Alexander Golovnin in 1866 as minister of education, the ministry did its best to control education and to direct it into desirable channels. As in the days of Uvarov, high standards were used in universities and secondary schools to keep the number of students down, hindering especially the academic advancement of students of low social background. In secondary education, the emphasis fell on the so-called classical *gymnasia,* which became the only road to universities proper, as distinct from more specialized institutions of higher learning. These gymnasia concentrated on teaching the Latin and Greek languages, to the extent of some 40 per cent of the total class time. Largely because of the rigorous demands, less than one-third of those who had entered the gymnasia were graduated. In addition to the natural obstacles that such a system presented to "socially undesirable" elements, ministers of education made direct appeals in their circulars to subordinates to keep "cook's sons" out of the gymnasia, as did one of Dmitrii Tolstoy's successors, Ivan Delianov, in 1887. In general, the government tried to divide education into airtight compartments that students as a rule could not cross. Under Alexander III and Pobedonostsev, Church schools received special attention. Following the statute of 1884 concerning Church-parish schools, an effort was made to entrust elementary education as much as possible to the Church, the number of Church-parish schools increasing from 4,500 in 1882 to 32,000 in 1894. While inferior in quality, these educational institutions were considered "safe." By contrast, advanced education for women, barely begun in Russia, came to be increasingly restricted. And in all schools and at all levels the Ministry of Education emphasized "conduct" and tried to maintain iron discipline.

Yet, in spite of all the vicissitudes, education continued to grow in Russia. The impact of the zemstva proved especially beneficial. Thus, according to Charnolussky's figures, the sixty provinces of European Russia in 1880 possessed 22,770 elementary schools with 1,141,000 students, 68.5 per cent of the schools having been established after the zemstvo reform of 1864. In addition to the exclusive classical gymnasia, *Realschule,* which taught modern languages and science in place of Greek and Latin, provided a secondary education that could lead to admission to technical institutions of higher learning. Other kinds of schools also developed. In addition to the activities of the ministries of education, war, navy, and of the Holy Synod, Witte promoted commercial schools under the jurisdiction of the Ministry of Finance, establishing some 150 of them between 1896 and 1902, and well over 200 altogether. In 1905 these schools were transferred to the Ministry of Trade and Industry. Moreover, after the Revolution of 1905 schools in Russia profited from a more liberal policy as well as from an increasing interest in education on the part of both the government and the public. As mentioned earlier, plans were drawn to institute schooling for all Russian children by 1922, or, according to a revised estimate following the outbreak of the First World War, by 1925. Educational prospects had never looked brighter in Russia than on the eve of the revolutions of 1917.

The problem, however, remained immense. Russians needed all kinds of training but above all the acquisition of simple literacy. Although by the end of the nineteenth century Russia had 76,914 elementary schools for children and 1,785 for adults with a total of 4.1 million students, and by 1915 the number of students had grown to over 8 million, on the eve of the October Revolution somewhat more than half of the population of the country was illiterate. To be more precise, in 1917 literacy extended in all probability to only about 45 per cent of the people.

At the other end of the educational ladder, universities increased in number, although slowly. The so-called Novorossiiskii University — referring to the name of the area, *Novorossiia,* or New Russia — was founded in Odessa in 1864, the University of Tomsk in Siberia in 1888, the University of Saratov in 1910, of Perm in 1915, and of Rostov-on-Don in 1917. That gave Russia a total of twelve universities, all of them belonging to the state. However, in 1917 the empire also possessed more than a hundred specialized institutions of higher learning: pedagogical, technological, agricultural, and other. Gradually it became possible for women to obtain higher education by attending special "courses" set up in university centers, such as the "Guerrier courses," named after a professor of history, Vladimir Guerrier, which began to function in 1872 in Moscow, and the "Bestuzhev courses," founded in 1878 in St. Petersburg and named after another historian, Constantine Bestuzhev-Riumin. The total number of students in Russian institutions of higher learning in 1917 has been variously estimated between 100,000 and 180,000. It should be noted that while the university statute of 1884 proved to be more restrictive than that of 1863 and over a period of time led to the resignation of a number of noted professors, most of the restrictions disappeared in 1905. In general, and especially after 1905, the freedom and variety of intellectual life in imperial Russian universities invite comparison with the Western universities, certainly not with the Soviet system.

Science and Scholarship

The Academy of Sciences, the universities, and other institutions of higher learning developed, or rather continued to develop, science and scholarship in Russia. In fact, in the period from the emancipation of the serfs until the revolutions of 1917, Russians made significant contributions in almost every area of knowledge. In mathematics, while no one quite rivaled Lobachevsky, a considerable number of outstanding Russian mathematicians made their appearance, including Pafnutii Chebyshev in St.

Petersburg and a remarkable woman, Sophia Kovalevskaia, who taught at
the University of Stockholm. Chemistry in Russia achieved new heights in
the works of many talented scholars, the most celebrated of them being
the great Dmitrii Mendeleev, who lived from 1834 to 1907 and whose
periodic table of elements, formulated in 1869, both organized the known
elements into a system and made an accurate forecast of later discoveries.
Leading Russian physicists included the specialist in magnetism and electricity, Alexander Stoletov, and the brilliant student of the properties of
light, Peter Lebedev, as well as such notable pioneer inventors as Paul
Iablochkov, who worked before Edison in developing electric light, and
Alexander Popov, who invented the radio around 1895, shortly before
Marconi. Russian inventors, even more than Russian scholars in general,
frequently received less than their due recognition in the world both because
of the prevalent ignorance abroad of the Russian language and Russia and
because of the backwardness of Russian technology, which usually failed
to utilize their inventions.

Advances in the biological sciences rivaled those in the physical. Alexander Kovalevsky produced classic works in zoology and embryology,
while his younger brother, Vladimir, the husband of the mathematician,
made important contributions to paleontology — and, incidentally, was
much appreciated by Darwin. The famous embryologist and bacteriologist
Elijah Mechnikov, who did most of his work in the Pasteur Institute in
Paris, concentrated on such problems as the function of the white corpuscles,
immunity, and the process of aging. Medicine developed well in Russia
during the last decades of the empire, both in terms of quality and, after
the zemstvo reform, in terms of accessibility to the masses. Following the
lead of an outstanding anatomist, surgeon, teacher, and public figure,
Nicholas Pirogov, who died in 1881, and others, Russian doctors exhibited
a remarkable civic spirit and devotion to their work and their patients.

Russian contributions to physiology were especially striking and important, and they overlapped into psychology. Ivan Sechenov, who taught
in several universities for about half a century and died in 1905, did
remarkable research on gases in blood, nerve centers, and reflexes and on
other related matters. Ivan Pavlov, who lived from 1849 to 1936 and whose
epoch-making experiments began in the 1880's, established through his
studies of dogs' reactions to food the existence and nature of conditioned
reflexes, and, further developing his approach, contributed enormously to
both theory and experimental work in physiology and to behavioral psychology.

The social sciences and the humanities also prospered. Russian scholars
engaged fruitfully in everything from law to oriental studies and from
economics to folklore. In particular, Russian historiography flourished in
the last decades of the nineteenth and the first of the twentieth century.

Building on the work of Serge Soloviev and other pioneers, Basil Kliu-chevsky, Serge Platonov, Matthew Liubavsky, Paul Miliukov, and their colleagues in effect established Russian history as a rich and many-sided field of learning. Their works have not been surpassed. Other Russians made notable contributions to the histories of other countries and ages, as did the medievalist Paul Vinogradov and the specialist in classical antiquity Michael Rostovtzeff. While Russian historiography profited greatly from the sociological emphasis characteristic of the second half of the nineteenth century, the "silver age" stimulated the history of art, which could claim in Russia such magnificent specialists as Nikodim Kondakov, Alexander Benois, and Igor Grabar, and it led to a revival of philosophy, esthetics, and literary criticism.

Literature

After the "great reforms" as before them, literature continued to be the chief glory of Russian culture, and it also became a major source of Russian influence on the West, and indeed on the world. That happened in spite of the fact that the intellectual climate in Russia changed and became unpropitious for creative expression. Instead of admiring art, poetry, and genius, as had been common in the first half of the nineteenth century, the influential critics of the generation of the sixties and of the following decades emphasized utility and demanded from the authors a clear and simple social message. Logically developed, civic literature led to Chernyshevsky's novel, *What Is To Be Done?*, a worthless literary effort, whatever its intellectual and social significance. With better luck, it produced Nicholas Nekrasov's civic poetry, which showed inspiration and an effective use of language, for Nekrasov was a real poet, although he wrote unevenly and too much. Fortunately for Russian literature, the greatest writers rejected critical advice and proceeded to write in their own manner. That was especially true of the three giants of the age, Ivan Turgenev, Fedor Dostoevsky, and Leo Tolstoy.

Ivan Turgenev lived from 1818 to 1883 and became famous around 1850 with the gradual appearance of his *Sportsman's Sketches*. He responded to the trends of the time and depicted with remarkable sensitivity the intellectual life of Russia, but he failed eventually to satisfy the Left. Six novels, the first of which appeared in 1855 and the last in 1877, described the evolution of Russian educated society and Russia itself as Turgenev, a gentleman of culture, had witnessed it. These novels are, in order of publication, *Rudin, A Gentry Nest, On the Eve,* the celebrated *Fathers and Sons, Smoke,* and *Virgin Soil.* Turgenev depicted Russia from the time of the iron regime of Nicholas I, through the "great reforms," to the return of reaction in the late '60's and the '70's. He concerned himself especially with the idealists of the '40's and the later liberals, nihilists, and populists. Indeed, it was Turgenev's hero, Bazarov, who gave currency to the concept *nihilist* and to the term itself. Although he was a consistent Westernizer and liberal, who was appreciative of the efforts of young radicals to change Russia, Turgenev advocated gradualism, not revolution; in particular he recommended patient work to develop the Russian economy and education. And he refused to be one-sided or dogmatic. In fact, critics debate to this day whether Rudin and Bazarov are essentially sympathetic or unsympathetic characters. Besides, Turgenev's novels were by no means simply *romans à thèse.* The reader remembers not only the author's ideological protagonists, but also his remarkable, strong heroines, the background, the dialogue, and, perhaps above all, the consummate artistry. As writer, Turgenev resembled closely his friend Flaubert, not at all Chernyshevsky. In addition to the famous sequence of novels, Turgenev wrote some plays and a considerable number of stories — he has been described as a better story writer than novelist.

Fedor — that is, Theodore — Dostoevsky, who lived from 1821 to 1881, also became well known before the "great reforms." He was already the author of a novel, *Poor Folk,* which was acclaimed by Belinsky when it was published in 1845, and of other writings, when he became involved, as already mentioned, with the Petrashevtsy and was sentenced to death, the sentence being commuted to Siberian exile only at the place of execution. Next the writer spent four years at hard labor and two more as a soldier in Siberia before returning to European Russia in 1856, following a general amnesty proclaimed by the new emperor. Dostoevsky recorded his Siberian experience in a remarkable book, *Notes from the House of the Dead,* which came out in 1861. Upon his return to literary life, the one-time member of the Petrashevtsy became an aggressive and prolific Right-wing journalist, contributing to a certain Slavophile revival, Pan-Slavism, and even outright chauvinism. His targets included the Jews, the Poles, the Germans, Catholicism, socialism, and the entire West. While Dostoevsky's journalism added to the sound and fury of the period, his immortal fame rests on his late novels, four of which belong among the greatest ever written. These were *Crime and Punishment, The Idiot, The Possessed,* and *The Brothers Karamazov,* published in 1866, 1868, 1870–72, and 1879–80 respectively. In fact, Dostoevsky seemed to go from strength to strength and was apparently at the height of his creative powers in working on a sequel to *The Brothers Karamazov* when he died.

Dostoevsky has often been represented as the most Russian of writers and evaluated in terms of Russian messiahship and the mysteries of the Russian soul — an approach to which he himself richly contributed. Yet, a closer study of the great novelist's so-called special Russian traits demonstrates that they are either of secondary importance at best or even entirely imaginary. To the contrary, Dostoevsky could be called the most international or, better, the most human of writers because of his enormous concern with and penetration into the nature of man. The strange Russian author was a master of depth psychology before depth psychology became known. Moreover, he viewed human nature in the dynamic terms of explosive conflict between freedom and necessity, urge and limitations, faith and despair, good and evil. Of Dostoevsky's several priceless gifts the greatest was to fuse into one his protagonists and the ideas — or rather states of man's soul and entire being — that they expressed, as no other writer has ever done. Therefore, where others are prolix, tedious, didactic, or confusing in mixing different levels of discourse, Dostoevsky is gripping, in places almost unbearably so. As another Russian author, Gleb Uspensky, reportedly once remarked, into a small hole in the wall, where the generality of human beings could put perhaps a pair of shoes, Dostoevsky could put the entire world. One of the greatest anti-rationalists of the second half of the nineteenth century, together with Nietzsche and Kirkegaard, Dostoevsky became with them an acknowledged prophet for the twentieth, inspiring existential philosophy, theological revivals, and scholarly attempts to understand the catastrophes of our time — as well as, of course, modern psychological fiction.

It has been said that, if Dostoevsky was not the world's greatest novelist, then Tolstoy certainly was, and that the choice between the two depends on whether the reader prefers depth or breadth. These are quite defensible views, provided one remembers the range of Dostoevsky, and especially his very numerous secondary and tertiary characters who speak their own language and add their own comment to the tragedy of man, and provided one realizes that Tolstoy too cuts very deep.

Count Leo Tolstoy lived a long, full, and famous life. Born in 1828 and brought up in a manner characteristic of his aristocratic milieu — magnificently described in *Childhood, Boyhood, and Youth* — he received a cosmopolitan, if dilettante, education; engaged in gay social life; served in the army, first in the Caucasus and later in the siege of Sevastopol; and became a happy husband, the father of a large family, and a progressive landowner much concerned with the welfare of his peasants. In addition to these ordinary activities, however, Tolstoy also developed into one of the greatest writers in world literature and later into an angry teacher of mankind, who condemned civilization, including his own part in it, and called for the abandonment of violence and for a simple, moral life. In fact, he died in 1910 at the age of eighty-two as he fled from his family and estate in yet another attempt to sever his ties with all evil and falsehood and to find truth. It is indeed difficult to determine whether Tolstoy acquired more fame and influence in his own country and all over the world as a writer or as a teacher of nonresistance and unmasker of modern civilization, and whether *Anna Karenina* or *A Confession* — an account of the crisis that split his life in two — carries the greater impact. In Russia at least, Tolstoy's position as the voice of criticism that the government dared not silence, as moral conscience, appeared at times even more extraordinary and precious than his literary creations.

But, whatever can be said against Tolstoy as thinker — and much has been justly said about his extraordinary naïveté, his stubborn and at the same time poorly thought-out rationalism, and his absolute insistence on such items as vegetarianism and painless death as parts of his program of salvation — Tolstoy as writer needs no apologies. While a prolific author, the creator of many superb stories and some powerful plays, Tolstoy, like Dostoevsky, is remembered best for his novels, especially *War and Peace,* published in 1869, and *Anna Karenina,* published in 1876. In these novels, as in much else written by Tolstoy, there exists a boundless vitality, a driving, overpowering sense of life and people. And life finds expression on a sweeping scale. *War and Peace* contains sixty heroes and some two hundred distinct characters, not to mention the unforgettable battle and mob scenes and the general background. The war of 1812 is depicted at almost every level: from Alexander I and Napoleon, through commanders and officers, to simple soldiers, and among civilians from court circles to the common people. *Anna Karenina,* while more restricted in scope, has been praised no less for its construction and its supreme art.

The Russian novel, which in the second half of the nineteenth century won a worldwide reputation because of the writings of Turgenev, Dostoevsky, and Tolstoy, had other outstanding practitioners as well. Ivan Goncharov, who lived from 1812 to 1891, produced at least one great novel, *Oblomov,* published two years before the emancipation of the serfs and representing in a sense a farewell, spoken with mixed feelings, to the departing patriarchal Russia, and a welcome, again with mixed feelings, to the painfully evolving new order. Oblomov himself snored his way to fame as one of the most unforgettable as well as most "superfluous" heroes of Russian literature. Other noteworthy novelists of the period included Nicholas Leskov who developed a highly individual language and style and wrote about the provincial clergy and similar topics associated with the Church and the people, and Gleb Uspensky, a populist and a pessimist, deeply concerned with peasant life as well as with the intelligentsia. An able satirist, Michael Saltykov, who wrote under the pseudonym of N. Shchedrin, fitted well into that critical and realistic age and acquired great popularity. A highly talented dramatist, Alexander Ostrovsky, wrote indefatigably from about 1850 until his death in 1886, creating much of the basic repertoire of the Russian theater and contributing especially to the depiction of merchants, minor officials, and the lower middle class in general.

Toward the end of the nineteenth century and in the early twentieth new writers came to the fore to continue the great tradition of Russian prose. One was Vladimir Korolenko, a populist, optimist, and author of charming stories; another was Anton Chekhov; and a third was the restless Alexis Peshkov, better known as Maxim Gorky, who created his own world of tramps and outcasts and went on to become the dean of Soviet writers. Chekhov, who lived from 1860 until 1904, left a lasting imprint on Russian and world literature. A brilliant playwright, he had the good fortune to be writing just as the Moscow Art Theater was rising to its heights. He is even more important as one of the founders and a master craftsman of the modern short story, the literary genre that he usually chose to make his simple, gentle, restrained, and yet wonderfully effective comments on the world.

Poetry fared less well than prose between the "great reforms" and the turn of the century. The very great lyricist Fedor Tiutchev, perhaps the world's outstanding poet of late love and of nature in its romantic, pantheistic, and chaotic aspects, died in 1873, an isolated figure. In the decades following the emancipation neither the small group of poets who championed "art for art's sake," such as the gifted Athanasius Fet-Shenshin, nor the dominant practitioners of "civic poetry," led by Nekrasov, left much of lasting value. The poetic muse had to wait for more propitious circumstances.

These circumstances emerged around 1900 with the dawning of the "silver age." Foreshadowed by certain literary critics and poets in the 1890's, the new period has often been dated from the appearance in 1898 of a seminal periodical, *The World of Art,* put out by Serge Diaghilev and Alexander Benois. What followed was a cultural explosion. Almost overnight there sprung up in Russia a rich variety of literary and artistic creeds, circles, and movements. As Mirsky and other specialists have noted, these different and sometimes hostile groups had little or nothing in common, except their denial of "civic art" and their high standards of culture and craftsmanship. While much of the creative work of the "silver age" tended toward pretentiousness, obscurity, or artificiality, its best products were very good indeed. And even when short of the best, the works of the "silver age" indicated a new refinement, richness, and maturity in Russian culture.

In literature, the new trends resulted in a great revival of poetry and literary criticism, although some remarkable prose was also produced, for example, by Boris Bugaev, known as Andrei Bely. Among the poets, the symbolist Alexander Blok, who lived from 1880 to 1921 and wrote verses of stunning magic and melody to the mysterious Unknown Lady and on other topics, has been justly considered the greatest of the age and one of the greatest in all Russian literature. But Russia suddenly acquired many brilliant poets; other symbolists, for example, Innokentii Annensky, Bely, Valery Briusov, and Constantine Balmont; "acmeists," such as Nicholas

Gumilev and Osip Mandelstam; futurists, such as Velemir Khlebnikov and Vladimir Maiakovsky; or peasant poets, such as Serge Esenin. The poet and novelist Boris Pasternak, who died in 1960, and the poetess Anna Akhmatova, who lived until 1966 as probably the last Russian poet of the first rank, also belong fully to the "silver age." In literary criticism, too, the new trends continued to enrich Russian culture after 1917, producing notably an interesting school of formalist critics, until destroyed by Soviet regimentation and "socialist realism."

The Arts

In art, as in literature, "realism" dominated the second half of the nineteenth century, only to be enriched and in large part replaced by the varied new currents of the "silver age." In painting the decisive turning to realism can even be precisely dated: in 1863 fourteen young painters, led by Ivan Kramskoy and constituting the entire graduating class of the Academy of Arts, refused to paint their examination assignment, "A Feast in Valhalla." Breaking with the stifling academic tradition, they insisted on painting realistic pictures. Several years later they organized popular circulating exhibitions of their works and came to be known as the "itinerants." With new painters joining the movement and its influence spreading, "critical realism" asserted itself in Russian art just as it had in Russian literary criticism and literature. In accord with the spirit of the age, the "itinerants" and their disciples believed that content was more important than form, that art had to serve the higher purpose of educating the masses and championing their interests, and they depicted such topics as the exploitation of the poor, the drunken clergy, and the brutal police. Basil Vereshchiagin, for example, observed wars at firsthand until he went down with the battleship *Petropavlovsk* when it was sunk by the Japanese. He painted numerous and often huge canvases on the glaring inhumanity of wars, characteristically dedicating his "Apotheosis of War," a pyramid of skulls, "to all great conquerors, present, past, and future." To be sure, painting could not be limited to social protest, and realism naturally extended to portraits, genre scenes, landscapes, historical topics — well handled by Basil Surikov — and other subject matter. Still, the Russian artists of the period demonstrated earnestness rather than talent, and added more to the polemics of the age than to art. Even the most famous of them, Elijah Repin, who lived from 1844 to 1930, is less likely to be remembered for his contribution to creative art, than for his active participation in Russian life and culture, and for certain paintings that have become almost inseparable from their subject matter, such as one of the Dnieper cossacks and one of Ivan the Terrible just after he had mortally wounded his son Ivan.

The development of music followed a somewhat different pattern. It, too, responded to the demands of the age, as seen, for example, in Modest

Musorgsky's emphasis on content, realism, and closeness to the masses. Music, however, by its very nature could not be squeezed into the framework of critical realism, and fortunately it attracted much original talent in Russia at the time. The second half of the nineteenth century witnessed a great spread of musical interest and education in the empire, with a conservatory established in St. Petersburg in 1862, headed by the noted composer and magnificent pianist Anton Rubinstein, another one in Moscow in 1866, headed by Anton Rubinstein's younger brother, Nicholas, and still other musical schools in other cities in subsequent years. Moreover, quite a number of outstanding Russian composers came to the fore at that time. The most prominent of them included Peter Tchaikovsky and *dilettante* members of the celebrated "Mighty Bunch," Modest Musorgsky, Nicholas Rimsky-Korsakov, Alexander Borodin, and Caesar Cui. The "Mighty Bunch," or "The Five" — Milii Balakirev, a professional, trained musician, must be added to the four already mentioned — in effect created the national Russian school of music, utilizing folk songs, melodies, tales, and legends, and a romanticized vision of the Russian past to produce such famous operas as Musorgsky's *Boris Godunov,* Borodin's *Prince Igor,* and Rimsky-Korsakov's *Sadko* and *The Tale of the Town of Kitezh.* It hardly needs to be mentioned that much of the instrumental and vocal music of the "Mighty Bunch" has entered the basic musical repertoire all over the world. The same, of course, holds true of Tchaikovsky, who stood apart from "The Five," developing an elegiac, subjective, and psychological approach of his own. Indeed, few pieces in the world of music are better known than Tchaikovsky's *Sixth Symphony* or his ballets, *Swan Lake* and *The Sleeping Beauty.*

The "silver age" brought a renaissance in the fine arts as well as in literature. In music, where Alexander Scriabin initiated the change, it marked the appearance of the genius of Igor Stravinsky and of other brilliant young composers. In a sense, the new ballet masterpieces, for example, Stravinsky's *The Firebird, Petrouchka* — which also belongs to Benois — and *Le Sacre du printemps,* combining as they did superb music, choreography, dancing, and décor, expressed best the cultural refinement, craftsmanship, and many-sidedness of the "silver age." The Russian ballet received overwhelming acclaim when Diaghilev brought it to Paris in 1909, starring such choreographers as Michael Fokine and such dancers as Anna Pavlova and Waslaw Nijinsky. From that time on Russian ballet has exercised a fundamental influence on ballet in other countries. On the eve of 1917 Russia could also boast of leading artists in other musical fields, for instance, the bass Theodore Chaliapin, the conductor Serge Koussevitzky, and the pianist, conductor, and composer, Serge Rachmaninov, to mention three of the best-known names.

Diaghilev's ballets made such a stunning impression in the West in part because of the superb décor and staging. Benois, Constantine Korovin, and other gifted artists of the "silver age" created a school of stage painting that gave Russia world leadership in that field and added immeasurably to operatic and theatrical productions as well as to the ballet. Other Russian artists, notably Marc Chagall and Basil Kandinsky, broke much more radically with the established standards and became leaders of modernism in painting. Still another remarkable development in the "silver age" was the rediscovery of icon painting: both a physical rediscovery, because ancient icons had become dark, been overlaid with metal, or even painted over, and began to be restored to their original condition only around 1900; and an artistic rediscovery, because these icons were newly appreciated, adding to the culture and the creative influences of the period.

Theater, like the ballet a combination of arts, also developed splendidly in the "silver age." In addition to the fine imperial theaters, private ones came into prominence. The Moscow Art Theater, directed by Constantine Stanislavsky who emphasized psychological realism, achieved the greatest and most sustained fame and exercised the strongest influence on acting in Russia and abroad. But it is important to realize that it represented only one current in the theatrical life of a period remarkable for its variety, vitality, and experimentation. Russian art as well as Russian literature in the "silver age" formed an inseparable part of the art and literature of the West, profiting hugely, for example, from literary trends in France or from German thought, and in turn contributing to virtually every form of literary and artistic argument and creative expression. In a sense, Russian culture was never more "Western" than on the eve of 1917.

Ideologies

Russian social, political, and philosophical thought also underwent considerable evolution between the emancipation of the serfs and the First World War. As already mentioned, the radicals of the generation of the sixties, Turgenev's "sons," found their spiritual home first in nihilism, in the denial of all established authorities. As their spokesman, the gifted young literary critic Dmitrii Pisarev, 1840–68, said: "What can be broken, should be broken." The new radical spirit reflected both the general materialistic and realistic character of the age and special Russian conditions, such as a reaction to the stifling of intellectual life under Nicholas I, the autocratic and oppressive nature of the regime, the weak development of the middle class or other elements of moderation and compromise, and a gradual democratization of the educated public.

While nihilism emancipated the young Russian radicals from any allegiance to the established order, it was, to repeat a point, individual rather than social by its very nature and lacked a positive program — both Pisarev and Turgenev's hero Bazarov died young. The social creed came with a vengeance in the form of *narodnichestvo,* or populism, which arose in the 1860's and '70's to dominate much of Russian radicalism until the October Revolution. We have already seen its political impact in such events as the celebrated "going to the people" of 1874, the terrorism of the "Will of the People," and the activities of the Socialist Revolutionary party. Although in a broad sense Russian populism belonged ideologically to the general European radicalism of the age, it also possessed a distinctively Russian character — for Russia was a peasant country *par excellence* — and numerous Russian prophets. The first prophets were the radical Westernizers Herzen and Bakunin, the former surviving until 1870 and the latter until 1876, who both preached that radical intellectuals should turn to the people and proclaimed the virtues of the peasant commune. Bakunin's violent anarchism in particular inspired many of the more impatient populists. Anarchism, it might be added, appealed to a variety of Russian intellectuals, including such outstanding figures as Tolstoy and Prince Peter Kropotkin, a noted geographer, geologist, and radical, who lived from 1842 to 1921 and devoted most of his life to the spreading of anarchism. Kropotkin's activities as a radical included a fantastic escape from the Peter and Paul Fortress, which was described in his celebrated *Memoirs of a Revolutionist* written in English for *The Atlantic Monthly* in 1898–99.

Whereas Herzen and Bakunin were émigrés, populist leaders also arose in Russia after 1855. Nicholas Chernyshevsky, whose views and impact were not limited to populism, but who nevertheless exercised a major influence on Russian populists, deserves special attention. Born in 1828, Chernyshevsky actually enjoyed only a few years of public activity as journalist and writer, especially as editor of a leading periodical, *The Contemporary,* before his arrest in 1862. He returned from Siberian exile only in 1883 and died in 1889. It was probably Chernyshevsky more than anyone else who contributed to the spread of utilitarian, positivist, and in part materialist views in Russia. A man of vast erudition, Chernyshevsky concerned himself with esthetics — developing further Belinsky's ideas on the primacy of life over art — as much as with economics, and wrote on nineteenth-century French history, demonstrating the failure of liberalism, as well as on Russian problems. His extremely popular novel, *What Is To Be Done?,* dealt with the new generation of "critical realists," their ethics and their activities, and sketched both the revolutionary hero and forms of co-operative organization. As to the peasant commune, Chernyshevsky showed more reserve than certain of his contemporaries. Yet he generally believed that it could serve as a direct transition to socialism in Russia, provided socialist revolution first triumphed in Europe. For a time Cherny-shevsky collaborated closely in spreading his ideas with an able radical literary critic, Nicholas Dobroliubov, who died in 1861 at the age of twenty-five.

Chernyshevsky's and Dobroliubov's work was continued, with certain differences, by Peter Lavrov and Nicholas Mikhailovsky. Lavrov, 1823–1900, another erudite adherent of positivism, utilitarianism, and populism, emphasized in his *Historical Letters* of 1870 and in other writings the crucial role of "critically thinking individuals" in the revolutionary struggle and the transformation of Russia. Mikhailovsky, a literary critic who lived from 1842 to 1904, employed the "subjective method" in social analysis to stress moral values rather than mere objective description and to champion the peasant commune, which provided for harmonious development of the individual, by contrast with the industrial order, which led to narrow specialization along certain lines and the atrophy of other aspects of personality. The populist defense of the peasant commune became more desperate with the passage of time, because Russia was in fact developing into a capitalist country and because an articulate Marxist school arose to point that out as proof that history was proceeding according to Marxist predictions. Yet the Socialist Revolutionaries of the twentieth century, led by Victor Chernov, although they borrowed much from the Marxists and had to modify their own views, remained essentially faithful to populism, staking the future of Russia on the peasants and on a "socialization of land."

Marxists proved to be strong competitors and opponents of populists. While Marxism will be discussed in a later chapter, it should be kept in mind that Marxism offered to its followers an "objective knowledge" of history instead of a mere "subjective method" and a quasi-scientific certainty of victory in lieu of, or rather in addition to, moral earnestness and indignation. It claimed to be "tough," where populism was "soft." Moreover, the actual development of Russia seemed to follow the Marxist rather than the populist blueprint. Beginning with the 1890's Marxism made important inroads among Russian intellectuals, gaining adherents both among scholars and in the radical and revolutionary movement. The Social Democrats, divided into the Bolsheviks and the Mensheviks, and their rivals, the Socialist Revolutionaries, gave political expression to the great ideological debate and cleavage of radical Russia.

To be sure, not all thinking and articulate Russians were radicals. But the Right, the conservatives and the reactionaries, had very little to offer. The government did little more than repeat the obsolete formula of Official Nationality, and its ablest theoretician, Constantine Pobedonostsev, determinedly refused to come to terms with the modern world. A few reactionary intellectuals not associated with the government, such as the brilliant writer

Constantine Leontiev, engaged in violent but fruitless criticism of the trends of the time and placed their hopes — desperate hopes indeed — in freezing the social process, in freezing everything!

Perhaps the new-style violent and demagogic Right had brighter prospects than the conservatives did. Its potential might be suggested by the nationalist rally led by Katkov in 1863, by Pan-Slavism in the late 1870's and at certain other times — although Pan-Slavism, especially when it expanded, was by no means limited to the Right — and by the "Black Hundreds" of the twentieth century. Yet all these movements lacked effective organization, continuity, and cohesion, as well as solid ideology. Pan-Slavism, for example, although it had several prophets, including Dostoevsky, and a painstaking theoretician of the quasi-scientific racist variety, Nicholas Danilevsky, whose magnum opus, *Russia and Europe,* was published in 1869, remained an "attitude of mind and feeling" rather than an "organized policy or even a creed." In other words, in times of Balkan crises many Russians sympathized with the Balkan Slavs, but they forgot them once a crisis passed. As a political factor, Pan-Slavism was more a Western bugaboo than a reality. And, in general, whatever racist and fascist possibilities existed in imperial Russia, they failed to develop beyond an incipient stage. Their flowering required a more modern setting than the one offered by the *ancien régime* of the Romanovs.

It can be argued that liberalism, on the other hand, represented a promising alternative for Russia. Moreover, Karpovich, Fischer, and other scholars, as well as a wealth of sources, have demonstrated that Russian liberalism was by no means a negligible quantity. On the contrary, with its bases in the zemstvo system and the professions, it gained strength steadily and it produced able ideologists and leaders such as Paul Miliukov and Basil Maklakov. The important position of the Cadets in the first two Dumas, the only Dumas elected by a rather democratic suffrage, emphasizes the liberal potential. But the government never accepted the liberal viewpoint, nor, of course, did the Russian radical and revolutionary movement accept it. The liberals thus had little opportunity to influence state policies or even to challenge them. Whether liberalism could have satisfied Russian needs will remain an arguable question, because Russian liberalism never received its chance in imperial Russia.

The "silver age" affected Russian thought as well as Russian literature and art. Notably, it marked a return to metaphysics, and often to religion eventually, on the part of a significant sector of Russian intellectuals. Other educated Russians, especially the writers and the artists, tended to become apolitical and asocial, often looking to esthetics for their highest values. The utilitarianism, positivism, and materialism dominant from the time of the '60's, finally had to face a serious challenge.

Philosophy in Russia experienced a revival in the work of Vladimir

Soloviev and his followers. Soloviev, a son of the historian Serge Soloviev, lived from 1853 until 1900 and wrote on a variety of difficult philosophical and theological subjects. A study in ethics, *A Justification of the Good,* is generally considered his masterpiece. A trenchant critic of the radical creed of the age, as well as of chauvinism and reaction, Soloviev remained a rather isolated individual during his lifetime, but came to exercise a profound influence on the intellectual elite of the "silver age." In effect almost everything he had stood for, from imaginative and daring theology to a sweeping critique of the radical intelligentsia, suddenly came into prominence in the early twentieth century.

The new critique of the intelligentsia found its most striking expression in a slim volume entitled *Signposts — Vekhi —* which appeared in 1909. *Signposts* contained essays by seven authors, including such prominent converts from Marxism as Peter Struve, Nicholas Berdiaev, and Serge Bulgakov, and constituted an all-out attack on the radical intelligentsia: Russian radicals were accused of an utter disregard for objective truth, religion, and law, and of an extreme application of the maxim that the end justifies the means, with destruction as their only effective passion. Although *Signposts* represented a minority of Russian intellectuals and attracted strong rebuttals, a new cleavage among educated Russians became apparent — a cleavage all the more revealing because the critics of the intelligentsia could by no means be equated with the Right. Eventually Struve, 1870–1944, became a leading thinker and political figure of the moderate conservatives; Berdiaev, 1874–1948, acquired world fame as a personalist philosopher and champion of "creative freedom"; and Bulgakov, 1871–1944, entered the priesthood and developed into the most controversial Orthodox theologian of the twentieth century. Other prominent intellectuals of the "silver age" included the "biological mystic" Basil Rozanov, who was especially concerned with the problem of sex, the brilliant anti-rationalist Leo Shestov — a pseudonym of Leo Schwartzmann — and the metaphysicians Semen Frank — another contributor to *Signposts* — and Nicholas Lossky. By comparison with the 1860's or even the 1890's, the Russian intellectual scene had indeed changed on the eve of the First World War.

Concluding Remarks

The development of Russian culture in the years preceding 1917 suggests certain significant parallels to the political, economic, and social condition of the country. Most striking was the disparity between the few and the many. In the early twentieth century, Russia possessed a rich variety of poetic schools and the best ballet in the world, but the majority of the people remained illiterate. It was even difficult to communicate across the chasm. One is reminded of Chekhov's story, "The Malefactor," where a peasant brought to court for stealing a bolt from the railroad tracks to weight his fishing tackle fails to see his guilt, explains that enough bolts are left for the train, and in describing his activities constantly refers to "we," meaning the peasants of his village, the people. Again, it can be argued that on the eve of the revolutions Russia exhibited progress and vigorous activity in intellectual as well as in other matters, straining against the confines of the established order. But, contrary to the Soviet view, this intellectual development did not lead ineluctably to Bolshevism. More than that, the cultural climate of the "silver age" indicated that the Russian educated public was finally moving away from the simple materialistic, utilitarian, and activist beliefs professed by Lenin and his devoted followers. It would appear that the Bolsheviks had to succeed soon or not at all. How they did succeed will be told in the next chapter.

XXXIV

* * * * * * * * * *

THE REVOLUTIONS OF 1917

> The collapse of the Romanov autocracy in March 1917 was one of the most leaderless, spontaneous, anonymous revolutions of all time. While almost every thoughtful observer in Russia in the winter of 1916–17 foresaw the likelihood of the crash of the existing regime no one, even among the revolutionary leaders, realized that the strikes and bread riots which broke out in Petrograd on March 8 would culminate in the mutiny of the garrison and the overthrow of the government four days later.
>
> CHAMBERLIN

> The enemies of Bolshevism were numerous, but they were also weak, poorly organized, divided, and apathetic. The strategy of Lenin was calculated to emphasize their divisions, neutralize their opposition, and capitalize on their apathy. In 1902 in *What Is To Be Done?* Lenin had written, "Give us an organization of revolutionaries, and we shall overturn the whole of Russia!" On November 7, 1917, the wish was fulfilled and the deed accomplished.
>
> FAINSOD

As has been indicated in preceding chapters, the constitutional period of Russian imperial history has continued to evoke much controversy, to cite only the contributions by Haimson and other American scholars. Optimistic students of the development of Russia from the Revolution of 1905 to the First World War and the revolutions of 1917 have emphasized that Russia had finally left autocracy behind and was evolving toward liberalism and political freedom. The change in 1907 in the electoral law indicated that the Duma could no longer be abolished. Moreover, the reformed Russian legislature proceeded to play an important part in the affairs of the country and to gain ever-increasing prestige and acceptance at home, among both government officials and the people, as well as abroad. As an Englishman observed, "the atmosphere and instincts of parliamentary life" grew in the empire of the Romanovs. Besides, continue the optimists, Russian society at the time was much more progressive and democratic than the constitutional framework alone would indicate, and was becoming increasingly so every year. Modern education spread rapidly at different levels and was remarkably humanitarian and liberal — as were Russian teachers as a group — not at all likely to serve as a buttress for antiquated ideas or obsolete institutions. Russian universities enjoyed virtually full freedom and a rich creative life. Elsewhere, too, an energetic discussion went on. Even the periodical press, in spite of various restrictions, gave some representation to every point of view, including the Bolshevik. Government prohibitions and penalties could frequently be neutralized by such simple means as a change in the name of a publication or, if necessary, by sending the nominal editor to jail, while important political writers continued their work. To be sure, grave problems remained, in particular, economic backwardness and the poverty of the masses. But, through industrialization on the one hand and Stolypin's land reform on the other, they were on the way to being solved. Above all, Russia needed time and peace.

Pessimistic critics have drawn a different picture of the period. Many of them refused even to call it "constitutional," preferring such terms as *Scheinkonstitutionalismus* — that is, sham constitutionalism — because, both according to the Fundamental Laws and in fact, the executive branch of the government and the ministers in particular were not responsible to the Duma. In any case, the critics asserted, whatever the precise character of the original arrangements, they were destroyed by the arbitrary electoral change of 1907, and by Nicholas II's entire authoritarian and reactionary policy. On the whole, the government refused to honor even its own niggardly concessions to the public. Nonentities, like Goremykin and Sukhomlinov, and the fantastic Rasputin himself, were logical end products of the bankruptcy of the regime. Other aspects of the life of the country, ranging from political terrorism, both of the Left and of the Right, to Russification and interminable "special regulations" to safeguard order, emphasized further the distance that Russia had to travel before it could be considered progressive, liberal, and law-abiding. Social and economic problems were still more threatening, according to the pessimists. Fundamental inequality and widespread destitution could not be remedied by a few large-scale "hothouse" industries and by a redivision of the peasants' inadequate land. Workers in particular, including those concentrated in St. Petersburg and in Moscow, were becoming more radical and apparently more willing to follow the Bolsheviks. Moreover, the government never wanted real reform, because it remained devoted to the interests of the landlords and, secondly, of the great capitalists. Russia was headed for catastrophe.

The optimists, thus, believe that imperial Russia was ruined by the First World War. The pessimists maintain that the war provided merely the last mighty push to bring the whole rotten structure tumbling down. Certainly it added an enormous burden to the load borne by the Russian people. Human losses were staggering. To cite Golovin's figures, in the course of the war the Russian army mobilized 15,500,000 men and suffered greater casualties than did the armed forces of any other country involved in the titanic struggle: 1,650,000 killed, 3,850,000 wounded, and 2,410,000 taken prisoner. The destruction of property and other civilian losses and displacement escaped count. The Russian army tried to evacuate the population as it retreated, adding to the confusion and suffering. It became obvious during the frightful ordeal that the imperial government had again failed in its tasks, as in the Crimean War and the Russo-Japanese War, but on a much larger scale. As mentioned earlier, the Russian minister of war and many other high officials and generals failed miserably in the test of war, Russian weapons turned out to be inferior to the enemies', Russian ammunition in short supply. Transportation was generally bogged down and on numerous occasions it broke down altogether. In addition to the army, the urban population suffered as a result of this, because it experienced serious difficulties obtaining food and fuel. Inflation ran rampant. Worst of all, the government refused to learn any lessons: instead of liberalizing state policies and relying more on the public, which was eager to help, Nicholas II in an anachronistic gesture handed over supreme power to the reactionary empress, and through her to Rasputin, when he assumed command at the front.

The February Revolution and the Provisional Government

The imperial regime died with hardly a whimper. Popular revolution, which came suddenly, was totally unprepared. In the course of the momentous days of March 8 to 11, 1917 (February 23 to 26, Old Style) riots and demonstrations in the capital — renamed "Petrograd" instead of the German "St. Petersburg" during the war — occasioned by a shortage of bread and coal assumed a more serious character. On March 10 reserve battalions sent to suppress the mutineers fraternized with them instead, and there were no other troops in the city. Resolute action, such as promptly bringing in loyal forces from elsewhere, might have saved the imperial government. Instead, with Nicholas II away at the front, authority simply collapsed and many officials went into hiding. Seemingly with one mind, the population of Petrograd turned to the Duma for leadership.

On March 11 members of the Duma sidestepped an imperial dissolution decree, and the next day they created a Provisional Government, composed of a score of prominent Duma leaders and public figures. Prince George Lvov, formerly chairman of the Union of Zemstva and Towns, assumed the positions of chairman of the Council of Ministers, that is, prime minister, and of minister of the interior. His more important colleagues included the Cadet leader Miliukov as minister of foreign affairs, the Octobrist leader Guchkov as minister of war and of the navy, and Alexander Kerensky, the only socialist in the cabinet — associated with the Socialist Revolutionary party — as minister of justice. The new government closely reflected the composition and views of the Progressive Bloc in the Duma, with the Cadets obtaining the greatest single representation.

Nicholas II bowed to the inevitable and on the fifteenth of March abdicated for himself and his only son, Alexis, in favor of his brother, Michael, who in turn abdicated the next day in favor of the decision of the constituent assembly, or in effect in favor of the Provisional Government pending that decision. Nicholas II, on his side, had appointed Lvov prime minister before renouncing the throne. Thus ended the rule of the Romanovs in Russia.

The Provisional Government was quickly recognized, and hailed, by the United States and other Western democracies. But, in spite of its rapid and general acceptance in Russia and abroad, the new government had to deal from the very beginning with a serious rival: the Petrograd Soviet of Workers' and Soldiers' Deputies, which was modeled on the 1905 Soviet. The new Soviet was formed on the twelfth of March, established itself in the Duma building, and proceeded to assert its authority. True, dominated by moderate socialists until the autumn of 1917, it did not try to wrest power from the "bourgeoisie," for it considered Russia unprepared for a socialist revolution, but it made its weight strongly felt nevertheless. In fact, the Provisional Government had been set up by the Duma in consultation with the Soviet and had to take its unofficial partner into account in all its policies and activities. Moreover, the Soviet acted authoritatively on its own, sometimes in direct contradiction to the efforts of the ministers. Notably, as early as March 14 it issued the famous, or notorious, *Order No. 1* to the troops which proclaimed that military units should be run by elected committees, with officers entitled to command only during tactical operations, and which played a role in the demoralization and eventual collapse of the Russian army. Following the Petrograd lead, Soviets began to be formed all over Russia. The first All-Russian Congress of Soviets, which met in the capital on the sixteenth of June, contained representatives from more than 350 local units. The delegates included 285 Socialist Revolutionaries, 245 Mensheviks, and 105 Bolsheviks, as well as some deputies from minor socialist parties. The Congress elected an executive committee which became the supreme Soviet body. Soviets stood much closer to the restless masses than did Lvov and his associates, and thus enjoyed a large and immediate following.

The Provisional Government lasted approximately eight months: from March 12 until November 7, 1917. Its record combined remarkable liberalism with an inability to solve pressing, crucial problems. The new regime promoted democracy and liberty in Russia. All citizens achieved equality before the law. Full freedom of religion, speech, press, assembly, unions, and strikes became a reality. Town and country administration was revamped to make it more democratic, with zemstvo institutions finally introduced at the level of the volost, that is, the township or canton. In addition to equal rights, ethnic minorities received autonomy, while Poland was declared independent. Labor legislation included the introduction of an eight-hour day for some categories of workers.

However, although the Provisional Government demonstrated what liberalism might have done for Russia, it failed to overcome the quite extraordinary difficulties that beset the country, and those who ruled it, in 1917. The new government continued the war in spite of the fact that defeatism spread among the people and that the army became daily less able to fight on. While convinced that all available land should belong to the peasants, it made no definitive land settlement, leaving that to the constituent assembly and thus itself failing to satisfy the peasantry. It proved unable to check inflation, restore transportation, or increase industrial production. In fact, Russian economy continued to run rapidly downhill.

A large part of this failure stemmed from the limited authority and power of the new regime. As already mentioned, it had at all times to contend with the Soviet. It had little in the way of an effective administrative apparatus, the tsarist police in particular having largely gone into hiding. While the high command of the army supported the government, enlisted men remained an uncertain quantity; the Petrograd garrison itself was devoted to the Soviet. What is more, the Provisional Government had to promise the Soviet not to remove or disarm that garrison. Kerensky's derisive appellation, "persuader-in-chief," was in part a reflection of his unenviable position.

The government also made mistakes. It refused to recognize the catastrophic condition of the country and misjudged the mood of the people. Thus, as mentioned, it continued the war, believing that the Russians, like the French at the time of the great French Revolution, would fight better than ever because they were finally free men. In internal affairs, a moderate and liberal position, generally difficult to maintain in times of upheaval, proved quixotic in a country of desperately poor and largely illiterate peasants who wanted the gentry land above all else. The government's temporary, "provisional," nature constituted a special weakness. Its members were deeply conscious of the fact that they had acquired their high authority by chance, that the Duma itself had been elected by the extremely restricted suffrage of 1907, and that the future of Russia must be settled by a fully democratic constituent assembly. Such basic decisions as those involved in the land settlement and in the future status of the national minorities had, therefore, to be left to that assembly. In the suggestive, if controversial, words of a political scientist: "This lack of a representative and responsible parliament helped greatly to distinguish the course of the Russian Revolution from its English, French and American predecessors." Yet, if a constituent assembly meant so much to the members of the Provisional Government, they made perhaps their worst mistake in not calling it together soon enough. While some of the best

Russian jurists tried to draw a perfect electoral law, time slipped by. When a constituent assembly finally did meet, it was much too late, for the Bolsheviks had already gained control of Russia.

The Bolshevik victory in 1917 cannot be separated from the person and activity of Lenin. He arrived, together with some of his associates, at the Finland Station in Petrograd on the sixteenth of April, the Germans having let them through from Switzerland in hopes that they would disorganize the Russian war effort. In contrast to the attitude of satisfaction with the course of the revolution and co-operation with the Provisional Government prevalent even in the Soviet, Lenin assumed an extreme and intransigent position in his "April Theses" and other pronouncements. He declared that the bourgeois revolution had already been accomplished in Russia and that history was moving inexorably to the next stage, the socialist stage, which had to begin with the seizure of power by the proletariat and poor peasants. As immediate goals Lenin proclaimed peace, seizure of gentry land by the peasants, control of factories by committees of workers, and "all power to the Soviets." "War to the palaces, peace to the huts!" shouted Bolshevik placards. "Expropriate the expropriators!"

Although Lenin found himself at first an isolated figure unable to win a majority even in his own party, events moved his way. The crushing burden of the war and increasing economic dislocation made the position of the Provisional Government constantly more precarious. In the middle of May, Miliukov and Guchkov were forced to resign because of popular agitation and pressure, and the cabinet was reorganized under Lvov to include five socialists rather than one, with Kerensky taking the ministries of war and the navy. The government declared itself committed to a strictly defensive war and to a peace "without annexations and indemnities." Yet, to drive the enemy out, Kerensky and General Alexis Brusilov started a major offensive on the southwestern front late in June. Initially successful, it soon collapsed because of confusion and lack of discipline. Entire units simply refused to fight. The Germans and Austrians in turn broke through the Russian lines, and the Provisional Government had to face another disaster. The problem of national minorities became ever more pressing as ethnic and national movements mushroomed in the disorganized former empire of the Romanovs. The government continued its increasingly hazardous policy of postponing political decisions until the meeting of a constituent assembly. Nevertheless, four Cadet ministers resigned in July because they believed that too broad a recognition had been accorded to the Ukrainian movement. Serious tensions and crises in the cabinet were also demonstrated by the resignation of the minister of trade and industry, who opposed the efforts of the new Social Democratic minister of labor to have workers participate in the management of industry, and the clash between Lvov and Victor Chernov, the Socialist

Revolutionary leader who had become minister of agriculture, over the implementation of the land policy. The crucial land problem became more urgent as peasants began to appropriate the land of the gentry on their own, without waiting for the constituent assembly.

The general crisis and unrest in the country and, in particular, the privations and restlessness in the capital led to the so-called "July days," from the sixteenth to the eighteenth of July, 1917, when radical soldiers, sailors, and mobs, together with the Bolsheviks, tried to seize power in Petrograd. Lenin apparently considered the uprising premature, and the Bolsheviks seemed to follow their impatient adherents as much as they led them. Although sizeable and threatening, the rebellion collapsed because the Soviet refused to endorse it, because some military units proved loyal to the Provisional Government, and because the government utilized the German connections of the Bolsheviks to accuse them of treason. Several Bolshevik leaders fled, including Lenin who went to Finland from whence he continued to direct the party; certain others were jailed. But the government did not press its victory and try to eliminate its opponents. On the twentieth of July Prince Lvov resigned and Kerensky took over the position of prime minister; socialists once more gained in the reshuffling of the cabinet.

Ministerial changes helped the regime little. The manifold crisis in the country deepened. In addition to the constant pressure from the Left, the Provisional Government attracted opposition from the Right which objected to its inability to maintain firm control over the army and the people, its lenient treatment of the Bolsheviks, and its increasingly socialist composition. In search of a broader base of understanding and support, the government arranged a State Conference in late August in Moscow, attended by some two thousand former Duma deputies and representatives of various organizations and groups, such as Soviets, unions, and local governments. The Conference produced no tangible results, but underlined the rift between the socialist and the non-socialist approaches to Russian problems. Whereas Kerensky expressed the socialist position and received strong support from socialist deputies, the Constitutional Democrats, army circles, and other "middle-class" groups rallied around the recently appointed commander in chief, General Lavr Kornilov. Of simple cossack origin, Kornilov had no desire to restore the old regime, and he could even be considered a democratic general. But the commander in chief, along with other military men, wanted above all to re-establish discipline in the army and law and order in the country, disapproving especially of the activities of the Soviets.

The "Kornilov affair" remains something of a mystery, although Ukraintsev's testimony and certain other evidence indicate that Kerensky, rather than Kornilov, should be blamed for its peculiar course and its being a fiasco. Apparently the prime minister and the commander in chief had decided that loyal troops should be sent to Petrograd to protect the government. Apparently, too, that "protection" included the destruction of Soviet power in the capital. In any case, when Kornilov dispatched an army corps to execute the plan, Kerensky appealed to the people "to save the revolution" from Kornilov. The break between the prime minister and the general stemmed probably not only from their different views on the exact nature of the strengthened Provisional Government to be established in Russia, and on Kerensky's position in that government, but also from the strange and confusing activities of the man who acted as an intermediary between them.

The revolution was "saved." From the ninth to the fourteenth of September the population of the capital mobilized for defense, while the advancing troops, faced with a railroad strike, encountering general opposition, and short of supplies, became demoralized and bogged down without reaching the destination; their commanding officer committed suicide. Only the Bolsheviks really gained from the episode. Their leaders were let out of jail, and their followers were armed to defend Petrograd. After the Kornilov threat collapsed, they retained the preponderance of military strength in the capital, winning ever more adherents among the increasingly radical masses.

The Provisional Government, on the other hand, came to be bitterly despised by the Right for having betrayed Kornilov — whether the charge was entirely justified is another matter — while many on the Left suspected it of having plotted with him. The cabinet experienced another crisis and was finally able to reconstitute itself — for the third and last time — only on the twenty-fifth of September, with ten socialist and six nonsocialist ministers, Kerensky remaining at the head. It should be added that the Kornilov fiasco, followed by the arrest of Kornilov and several other generals, led to a further deterioration of military discipline, making the position of officers in many units untenable.

The October Revolution

The Bolsheviks finally captured a majority in the Petrograd Soviet on September 13 and in the Moscow Soviet a week later, although the executive committee elected by the first All-Russian Congress of Soviets continued, of course, to be dominated by moderate socialists. Throughout the country the Bolsheviks were on the rise. From his hideout in Finland, Lenin urged the seizure of power. On October 23 he came incognito to Petrograd and managed to convince the executive committee of the party, with some division of opinion, of the soundness of his view. Lenin apparently considered victory a great gamble, not a scientific certainty, but he correctly estimated that the fortunate circumstances had to be exploited, and he did not want to wait until the meeting of the constituent assembly. His opinions prevailed over the judgment of those of his colleagues who, in more orthodox Marxist fashion, considered Russia insufficiently prepared for a Bolshevik revolution and their party lacking adequate support in the country at large. Leon Trotsky — a pseudonym of Leon Bronstein — who first became prominent in the St. Petersburg Soviet of 1905 and who combined oratorical brilliance and outstanding intellectual qualities with energy and organizational ability, proved to be Lenin's ablest and most active assistant in staging the Bolsheviks' seizure of power.

The revolution succeeded with little opposition. On November 7 — October 25, Old Style, hence "the Great October Revolution" — Red troops occupied various strategic points in the capital. In the early night hours of November 8, the Bolshevik-led soldiers of the Petrograd garrison, sailors from Kronstadt, and the workers' Red Guards stormed the Winter Palace, weakly defended by youngsters from military schools and even by a women's battalion, and arrested members of the Provisional Government. Kerensky himself had managed to escape some hours earlier. Soviet government was established in Petrograd and in Russia.

XXXV

* * * * * * * * * *

SOVIET RUSSIA: AN INTRODUCTION

> The philosophers have only *interpreted* the world in various ways;
> the point however is to *change* it.*
>
> MARX

> The conception of a community as an organic growth, which the
> statesman can only affect to a limited extent, is in the main modern,
> and has been greatly strengthened by the theory of evolution. . . .
>   It might, however, be maintained that the evolutionary view of
> society, though true in the past, is no longer applicable, but must, for
> the present and the future, be replaced by a much more mechanistic
> view. In Russia and Germany new societies have been created, in
> much the same way as the mythical Lycurgus was supposed to have
> created the Spartan polity. The ancient law giver was a benevolent
> myth; the modern law giver is a terrifying reality.
>
> RUSSELL

COMMUNIST IDEOLOGY, the Communist party, and Communist direction
have constituted the outstanding characteristics of Soviet Russia, that is,
of the Union of Soviet Socialist Republics. To be sure, other factors,
ranging from the economic backwardness of the country to its position as
a great power in Europe, Asia, and the world, have proved to be of major
importance. Still, it would not be an exaggeration to say that, whereas other
elements in the situation have exercised very significant influences on
Soviet policies, without communism there would have been no Soviet
policies at all and no Soviet Union. Moreover, it is frequently impossible
to draw the line between the communist and the noncommunist aspects
of Soviet Russia and between communist and noncommunist causes of
Soviet behavior because the two modes have influenced and interpenetrated
each other and because Soviet leaders have viewed everything within the
framework of their ideology.

Marxism

The doctrine of communism represents a variant of Marxism, based
on the works of Marx and Engels as developed by Lenin. Working for
several decades, beginning in the 1840's, Marx and Engels constructed

---

* Italics in the original.

a huge and comprehensive, although not entirely consistent, philosophical system. The roots of Marxism include eighteenth-century Enlightenment, classical economics, utopian socialism, and German idealistic philosophy — in other words, some of the main traditions of Western thought. Most important, Marx was "the last of the great system-builders, the successor of Hegel, a believer, like him, in a rational formula summing up the evolution of mankind." While an exposition of Marxism would require another book, certain aspects of the doctrine must be constantly kept in mind by a student of Soviet history.

Marxism postulates dialectical materialism as the key to and the essence of reality. While applicable to philosophy, science, and in fact to everything, dialectical materialism exercised its greatest impact on the study — and later manipulation — of human society, on that combination of sociology, history, and economics that represented Marx's own specialty. "Materialism" asserts that only matter exists; in Marxism it also led to a stress on the priority of the economic factor in man's life, social organization, and history.

> In the social production of their means of existence men enter into definite, necessary relations which are independent of their will, productive relationships which correspond to a definite stage of development of their material productive forces. The aggregate of these productive relationships constitutes the economic structure of society, the real basis on which a juridical and political superstructure arises, and to which definite forms of social consciousness correspond. The mode of production of the material means of existence conditions the whole process of social, political and intellectual life. It is not the consciousness of men that determines their existence, but, on the contrary, it is their social existence that determines their consciousness.

The fundamental division in every society is that between the exploiters and the exploited, between the owners of the means of production and those who have to sell their labor to the owners to earn a living. A given political system, religion, and culture all reflect and support the economic set-up, protecting the interests of the exploiters. The base, to repeat, determines the superstructure.

"Dialectical" adds a dynamic quality to materialism, defining the process of the evolution of reality. For the Marxists insist that everything changes all the time. What is more, that change follows the laws of the dialectic and thus presents a rigorously correct and scientifically established pattern. Following Hegel, Marx and Engels postulated a three-step sequence of change: the thesis, the antithesis, and the synthesis. A given condition, the thesis, leads to opposition within itself, the antithesis, and the tension between the two is resolved by a leap to a new condition, the synthesis. The synthesis in turn becomes a thesis producing a new antithesis, and the dialectic continues. The historical dialectic expresses itself in class struggle: "The history of all hitherto existing society is the history of class struggles." As an antithesis grows within a thesis, "the material productive forces of a society," always developing, "come into contradiction with the existing productive relationships," and social strife ensues. Eventually revolution leads to a transformation of society, only to become itself the new established order producing a new antithesis. In this manner the Italian towns and the urban classes in general revolted successfully against feudalism to inaugurate the modern, bourgeois period of European history. That period in turn ran its prescribed course, culminating in the full flowering of capitalism. But, again inevitably, the capitalists, the bourgeois, evoked their antithesis, their "grave-diggers," the industrial workers or the proletariat. In the words of Marx foretelling the coming revolution:

> The expropriation is brought about by the operation of the immanent laws of capitalist production, by the centralization of capital. . . . The centralization of the means of production and the socialization of labor reach a point where they prove incompatible with their capitalist husk. This husk bursts asunder. The knell of capitalist private property sounds. The expropriators are expropriated.

Interestingly if illogically, the victorious proletarian revolution would mark the end of all exploitation of man by man and the establishment of a just socialist society. In a sense, humanity would return to prehistory, when, according to Marx and the Marxists, primeval communities knew no social differentiation or antagonism.

Leninism

Lenin's theoretical contribution to Marxism could in no sense rival the contributions of the two originators of the doctrine. Still, he did his best to adapt Marxism to the changing conditions in the world as well as to his own experience with the Second International and to Russian circumstances, and he produced certain important additions to and modifications of the basic teaching. More to the point for students of Soviet history is the fact that these amendments became gospel in the Soviet Union, where the entire ideology has frequently been referred to as "Marxism-Leninism."

Among the views developed by Lenin, those on the party, the revolution, and the dictatorship of the proletariat, together with those on the peasantry and on imperialism, deserve special attention. As already mentioned, it was a disagreement on the nature of the party that in 1903 split the Russian Social Democrats into the Lenin-led Bolsheviks and the

Mensheviks. Lenin insisted on a tightly knit body of dedicated professional revolutionaries, with clear lines of command and a military discipline. The Mensheviks, by contrast, preferred a larger and looser organization. With characteristic determination and believing in the imminent worldwide overthrow of the capitalist system, Lenin decided in 1917 that he and his party could then stage a successful revolution in Russia, although at first virtually no one, even among the Bolsheviks, agreed. After the Bolsheviks did seize power in the October Revolution, Lenin proceeded to emphasize the role of the party and the dictatorship of the proletariat.

Lenin's revolutionary optimism stemmed in part from his reconsideration of the role of the peasantry in bringing about the establishment of the new order. Marx, Engels, and Marxists in general have neglected the peasants in their teachings and relegated them, as petty proprietors, to the bourgeois camp. Lenin, however, came to the conclusion that, if properly led by the proletariat and the party, poor peasants could be a revolutionary force: indeed later he proclaimed even the middle peasants to be of some value to the socialist state. The same *April Theses* that urged the transformation of the bourgeois revolution into a socialist one stated that poor peasants were to be part of the new revolutionary wave.

Lenin expanded Marxism in another, even more drastic manner. In his book *Imperialism, the Highest Stage of Capitalism,* written in 1916 and published in the spring of 1917, he tried to bring Marxism up to date to account for such recent developments as intense colonial rivalry, international crises, and finally the First World War. He concluded that in its ultimate form capitalism becomes imperialism, with monopolies and financial capital ruling the world. Cartels replace free competition, and export of capital becomes more important than export of goods. An economic and political partitioning of the world follows in the form of a constant struggle for economic expansion, spheres of influence, colonies, and the like. International alliances and counteralliances arise. The disparity between the development of the productive forces of the participants and their shares of the world is settled among capitalist states by wars. Thus, instead of the original Marxist vision of the victorious socialist revolution as the simple expropriation of a few supercapitalists, Lenin described the dying stage of capitalism as an age of gigantic conflicts, relating it effectively to the twentieth century. Still more important, this externalization, so to speak, of the capitalist crisis brought colonies and underdeveloped areas in general prominently into the picture. The capitalists were opposed not only by their own proletariats, but also by the alien peoples whom they exploited, more or less regardless of the social order and the stage of development of those peoples. Therefore, the proletarians and the colonial peoples were natural allies. Lenin, it is worth noting, paid much more attention to Asia than did Western Marxists. Eventually

— in a dialectical *tour de force* — even the fact that the socialist revolution came to Russia, rather than to such industrial giants as Great Britain, Germany, and the United States, could be explained by the theory of the "weakest link," that is, by the argument that in the empire of the Romanovs various forms of capitalist exploitation, both native and colonial or semi-colonial, combined to make capitalism particularly paradoxical and unstable, so that the Russian link in the capitalist chain snapped first.

Many critics have pointed out that Lenin's special views, while differing from the ideas of Marx and Engels, found their *raison d'être* both in Russian reality and in the Russian radical tradition. A land of peasants, Russia could not afford to rely for its future on the proletariat alone, and at least the poor peasants, if not the wealthier ones, had to be included to bring theory into some correspondence with the facts. Again, in contrast to, for example, Germany, socialism never acquired in imperial Russia a legal standing or a mass following, remaining essentially a conspiracy of intellectuals. If Lenin wanted results, he had to depend on these intellectuals, on a small, dedicated party. Moreover, in doing so he followed the tradition of Chernyshevsky, of Tkachev especially, of the "Will of the People," and even, broadly speaking — though he would have denied it vehemently — of such populists as Lavrov and Mikhailovsky, who emphasized the role of the "critically thinking individuals" as the makers of history. Born in 1870, Lenin grew up admiring Chernyshevsky; and his oldest brother, Alexander, was executed in 1887 for his part in a populistlike plot to assassinate Alexander III. Lenin's later persistent and violent attacks on the populists should not, it has been argued, obscure his basic indebtedness to them.

Yet, although this line of reasoning has some validity and helps to situate the great Bolshevik leader in the history of Russian radicalism — where he certainly belongs as much as in the history of world Marxism — it should not be pursued too far. After all, Lenin dedicated his entire mature life to the theory and practice of Marxism, which he considered to be infallibly true. Besides, while one does not have to subscribe to the official Soviet view that Lenin is the perfect creative Marxist, neither does one have to endorse the view, common among Western Social Democrats, that Lenin and communism betrayed Marxism. In fact, both Lenin's "hard" line, emphasizing the role of the party, revolution, and ruthlessness, and the "soft" approach of Western revisionists can be legitimately deduced from the vast and sometimes inconsistent writings of Marx and Engels.

The Intolerance

Comprehensiveness and ruthless intolerance have been among the most important salient features of Marxism-Leninism. While provoked to an extent by such practical circumstances as the requirements of ruling a state — states, eventually — and the strong and manifold opposition that had to be overcome, these traits nevertheless reside at the heart of the ideology itself. As already explained, Marxism constitutes an all-inclusive view of the world, metaphysical rather than empirical, which omits nothing of importance and possibly — at least so it can be argued in theory — nothing at all. Moreover, its teachings are believed to have the conclusiveness of scientific laws. In other words, to its adherents Marxism-Leninism represents a science, and those who oppose it are regarded by them as absolutely and demonstrably wrong. No matter how sophisticated, these critics ultimately deserve no more consideration than misguided, superstitious peasants who object to inoculation against cholera. More precisely, they are either misguided or class enemies; in the latter case they obviously deserve no favorable consideration at all.

Ruthlessness has also been promoted by the peculiar Marxist ethics, or rather absence of ethics. Ethics, which belongs to the "superstructure" of society, has no independent existence in Marxism. According to that teaching, men behave as they do because of their class nature, because of the fundamental economic and social realities of their lives. Only a change in these realities can and will alter human conduct. Therefore, there will be no moral turpitude and no crime in the ideal society of the future. In the meantime, one is invited to hate the unregenerate world and all its standards and to struggle, with few inhibitions, if any, for the victory of communism.

A pseudo science, Marxism-Leninism also possesses numerous earmarks of a pseudo religion. Berdiaev and other commentators have emphasized the extent to which it proclaims itself to be the truth, the ultimate and entirely comprehensive total, the first and the last, alpha and omega. It determines in effect the right and the wrong and divides the world into white and black. More specifically, it has been suggested that communism has its doctrine of salvation: its Messiah is the proletariat; its paradise is classless society; its church is the party; and its Scriptures are the writings of Marx, Engels, Lenin and, until recently, Stalin. The dialectic of class struggle will suddenly cease when man attains the just society — when man leaps from the kingdom of necessity into the kingdom of freedom. It is probably this pseudoreligious aspect of Marxism-Leninism, even more than its explicit materialism, that makes its frequently fanatical disciples determined enemies of Christianity and of every other religion — for no man can serve two gods.

Needless to say, Marxism-Leninism is not a democratic teaching. While its followers remain convinced that it represents the interests of the masses, the correctness of the ideology and the need to carry it out in practice do not depend in the least on popular approval or disapproval. More than that, Marxism-Lenism has been remarkably exclusive. Where most other major beliefs appeal to all human beings, Marx began with the assumption that the exploiting classes can never have a change of heart, but must be overthrown. Struggle and violence — ruthlessness once more — form the very fabric of the Marxist doctrine. Even among the exploited, Lenin insisted, few could fully comprehend their own situation and the course of history. Left to themselves, workers develop nothing more promising than a trade union mentality. Only the Party, only an elite, can really see the light. And communist parties have invariably continued to be exclusive.

The Appeal

What makes a communist? The ideology itself has no doubt offered numerous attractions to the intellect and helped many people to understand the world. It does represent one of the most impressive systems in the history of Western thought, and it is related to a number of main intellectual currents of the Western tradition. Its greatest strength lies perhaps in its explanation of human exploitation and misery and in its reasoned promise to end both. Those who fail to see the intellectual attractions of communism on either side of the "iron curtain" should consider carefully the testimony of such writers as Milosz who left the Polish "people's democracy," or of the several brilliant ex-Communists who contributed to the book *The God That Failed*. Yet rational persuasiveness stops far short of accounting entirely for the appeal of communism. Of course, both materialism and the dialectic, which are enormously important assumptions, remain unproved. More specific Marxist doctrines, for instance, the crucial labor theory of value, have been very effectively criticized. In addition, Marxist predictions have often been disproved by time. To cite only two of the more important examples, with the middle class growing rather than declining, a polarization into capitalists and workers has failed to take place in capitalist societies; also, in these societies the standard of living of the workers has been improving rather than deteriorating. Marxism possesses no invincible logic, and no scientific certainty; it does provide an elaborate intellectual rationalization and a splendid intellectual façade for those who subscribe to the teaching for nonintellectual reasons.

Especially significant, therefore, might be the link between Marxism-Leninism on the one hand and alienation and protest on the other. Communism has become the vehicle for almost every kind of criticism of the established order, and it has profited from a wide variety of weaknesses and mistakes of noncommunist societies. Indeed, communists seized power not, as predicted, in the advanced industrial countries of the West, but in Russia and in China where relatively backward economic conditions — very different in degree in the two instances — were combined with misery and great tensions and crises. And in both countries the rising class of intellectuals refused to identify itself with the existing system and led the struggle against it. However, even if we allow much for alienation and protest as factors in the rise of communism, we are faced with the question as to why it is communism, rather than some other teaching, that has attracted so many sensitive or dissatisfied people.

To suggest one answer among many, one might mention the four reasons for the appeal of Marxism emphasized by Isaiah Berlin. These include, to begin with, its comprehensiveness and its claim to be the key to knowing everything in the present, the past, and the future. Moreover, the doctrine itself and the knowledge that it gives are allegedly scientific: many social teachings in the nineteenth century, such as Fourier's peculiar utopian socialism or Comte's positivism, claimed scientific validity, but Marxism managed to identify itself more successfully with science than any other. Comprehensiveness and scientific authority become especially attractive with the abandonment of religion and other secure moorings. In the third place, Marxism, in spite of its deterministic aspect, is an activist and optimistic teaching: history is moving in the right direction, and every true believer can have a useful role in furthering its progress. Finally, Marxism possessed from the start a ready-made audience so to speak, the working class, which was invited to take over the world. Later Lenin tried his best to extend the audience to the poor peasants and to colonial peoples.

To move from Berlin's "semirational" reasons for the appeal of Marxism, Lasswell might serve as a representative guide to the slippery area of the irrational appeal. In the language of social psychology and psychoanalysis, he selected such qualities of Marxism-Leninism as its stress on the transitory nature of the present social order, which leads to a redefinition of expectancies about the future and encourages projection. Marxism condemns the capitalist system in clearly moral terms, accusing it in particular of denying affectionate care and attention to the individual and of giving unfair advantage to some over others. The doctrine gains from its prestigious "scientific" form and from its alleged objective quality as well as from specificity, i.e., in analyzing the unjust capitalist society, Marxists point to "surplus value" and "profits" rather than merely to such general factors as human greed or corruption. The extremely vague Marxist utopia, too, serves valuable purposes: it gives free rein to every individual's choice and his craving for omnipotence, and it protects the Marxist ideal from being tied to unpopular or transitory social phenomena.

Doctrines, it should be added, are held no less firmly when they are held irrationally; in fact, it can be argued that they are held more firmly if irrationally.

Concluding Remarks

When Communists seized power in Russia in 1917 they had to face an unforeseen situation: revolution erupted in Russia rather than in the industrial West, and it came to one country only rather than to the entire capitalist world. While Lenin and his associates tried to adjust to these facts, they had also to deal with countless other problems, some of them of utmost urgency. After the first hectic months, and years too, Soviet history has continued to be a story of great pressures, crises, and conflicts. Under these difficult, and at times desperate, circumstances it is remarkable not how little but how much Russian Communist leaders adhered to the pursuit of their ideological goals — from Lenin's determination to build socialism on the morrow of the Revolution, to Stalin's fantastic five-year plans, and to Khrushchev's efforts to speed the establishment of a truly communist society. An account of this pursuit belongs to the following chapters.

XXXVI

* * * * * * * * * *

WAR COMMUNISM, 1917–21, AND THE NEW ECONOMIC POLICY, 1921–28

> You will never be alive again,
> Never rise from the snow:
> Twenty-eight bayonet,
> Five fire wounds.
> A bitter new garment
> I sewed for my friend.
> It does love, does love blood—
> The Russian earth.
>
> AKHMATOVA

> Where are the swans? And the swans have left.
> And the ravens? And the ravens have remained.
>
> TSVETAEVA

> Of all the Governments which were set up in Russia to combat revolutionary rule, only one, that of the Social Revolutionaries at Samara, had the wisdom to assure the peasants that the counterrevolution did not mean the restoration of the land to the landlords. All the rest, in greater or less degree, made plain their policy of re-establishing or compensating them. It was this, and no transcendent virtue in the Bolsheviks, which decided the issue of the three years' struggle, in despite of British tanks and French munitions and Japanese rifles and bayonets.
>
> MAYNARD

ALTHOUGH THE BOLSHEVIKS seized power easily in Russia in November 1917, they managed to consolidate their new position only after several years of bitter struggle. In addition to waging a major and many-faceted civil war, the Soviet government had to fight Poland and deal with the Allied intervention. The Bolsheviks, in a desperate effort to survive, mobilized the population and resources in the area that they controlled and instituted a drastic regime which came to be known as "War Communism." Communist rule did survive, although at a tremendous price. To revive an utterly exhausted, devastated, and starving country, the so-called "New Economic Policy" replaced War Communism and lasted from 1921 to 1928, until the beginning of Stalin's First Five-Year Plan. The period of the New Economic Policy has been rightly contrasted with that of War Communism as a time of relaxation and compromise. Yet, on the whole the Soviet government showed more continuity than change in its policies and pursued its set goals with intelligence and determination — as a brief treatment of the first decade of Communist rule should indicate.

The New Government. Lenin

The Soviet government was organized two days after the October Revolution, on November 9, 1917, under the name of the Council of People's Commissars. Headed by Lenin as chairman, the Council contained such prominent members of the Bolshevik party as Trotsky who became commissar for foreign affairs, Alexis Rykov who became commissar of the interior, and Joseph Dzhugashvili, better known as Stalin, who assumed charge of national minorities. Lenin thus led the government as well as the party and was recognized as by far the most important figure of the new regime in Russia.

Lenin was born in an intellectual family — his father was a school inspector — in 1870 in a town on the Volga named Simbirsk, now Ulianovsk. Vladimir Ulianov proved to be a brilliant student both in secondary school and at the University of Kazan, where he studied law. He early became a radical — the execution of his eldest brother in 1887 for participating in a plot to assassinate Alexander III has sometimes been considered a turning point for him — and then became a Marxist, suffering imprisonment in 1896 and Siberian exile for the three years following. He participated in the publication of a Social Democratic newspaper, *The Spark,* which was printed abroad beginning in 1900, and in other revolutionary activities, often under the pseudonym of *N. Lenin.* At first awed by the "father of Russian Marxism," Plekhanov, Lenin before long struck out on his own, leading the Bolshevik group in the Social Democratic party split in 1903. We have already met Lenin as an important Marxist theoretician. But practice meant more than theory for the Bolshevik leader. Most of his writings in fact were polemical, brief, and to the point: they denounced opponents or deviationists in ideology and charted the right way for the faithful. As Lenin remarked when events in 1917 interrupted his work on a treatise, *The State and Revolution:* "It is more pleasant and more useful to live through the experience of a revolution than to write about it."

The Great October Revolution, masterminded by Lenin, gave him power that he continued to exercise in full until largely incapacitated by a stroke in May 1922. After that he still kept some control until his death on January 21, 1924. Moreover, in contrast to Stalin's later terrorism, Lenin's leadership of the party did not depend at all on the secret police, but rather on his own personality, ability, and achievement. Perhaps appropriately, whereas Stalin's cult experienced some remarkable reversals of fortune shortly after his demise, that of Lenin kept, if anything, gaining in popularity throughout the communist world until its collapse in the late 1980's.

The communist myth of Lenin does not stand far from reality in many respects. For Lenin was a dedicated Bolshevik, who lived and breathed revolution and communism. Moreover, he did so naturally, compulsively to be more exact, rather than as an imposition or a burden. Although not superhumanly clever and virtually infallible, as Soviet propaganda would have it, he did combine high intelligence, an ability for acute theoretical thinking, and practical sense to become a great Marxist "realist." The amalgam proved ideal for communist purposes: Lenin never wavered in his Marxist faith; yet he knew how to adapt it, drastically if need be, to circumstances. Other outstanding qualities of the Bolshevik leader included exceptional will power, persistence, courage, and the ability to work extremely hard. Even Lenin's simple tastes and modest, almost ascetic, way of life were transposed easily and appropriately from the actual man to his mythical image.

To be sure, there is another way to look at this paragon of Communist virtues. Devotion to an exclusive doctrine led to narrow vision. In the opinion of some specialists, the break between Plekhanov and Lenin, between the older Marxist who never lost humanistic standards and culture and the young fanatic confident that the end justified the means, represented a fundamental division in modern Russian history. Ruthlessness followed from fanaticism as well as from Lenin's conviction that he, and sometimes only he, knew the right answer. In the name of a future utopia, horrible things could be sanctioned in the present. Churchill once commented on Lenin: "His aim to save the world. His method to blow it up." The two objectives go ill together.

The First Months

The Second All-Russian Congress of Soviets, which met in Petrograd on the seventh of November, approved the Bolshevik revolution, although moderate socialists walked out of the gathering. In Moscow Soviet authority was established only after a week of fighting, because some military units remained loyal to the Provisional Government. Relying on local Soviets, the Bolsheviks spread their rule to numerous other towns and areas. The first serious challenge to the Bolshevik government occurred in January 1918, when the Constituent Assembly, for which elections had been held in late autumn, finally met. The 707 members who assembled in the capital on January 18 included 370 Socialist Revolutionaries, 40 Left Socialist Revolutionaries who had split from the main party, only 170 Bolsheviks, and 34 Mensheviks, as well as not quite one hundred deputies who belonged to minor parties or had no party affiliation. In other words, the Socialist Revolutionaries possessed an absolute majority. Chernov was elected chairman of the Constituent Assembly. It should be remembered that that Assembly had been awaited for months by almost all political groups in Russia as the truly legitimate and definitive authority in the country. Lenin himself had denounced the Provisional Government for failing to summon it promptly. Yet, in the changed circumstances, he acted in his usual decisive manner and had troops disperse the Constituent Assembly on the morning of the nineteenth of January. No major repercussions followed, and Soviet rule appeared more secure than ever. The lack of response to the disbanding of the assembly resulted in part from the fact that it had no organized force behind it, and in part from the fact that on the very morrow of the revolution the Soviet government had declared its intention to make peace and also had in effect granted the peasants gentry land, thus taking steps to satisfy the two main demands of the people. The Bolsheviks even enjoyed the co-operation of the Left Socialist Revolutionaries who received three cabinet positions, including the ministry of agriculture.

But the making of peace proved both difficult and extremely costly, with the very existence of the Soviet state hanging in the balance. The Allies failed to respond to the Soviet bid for peace and in fact ignored the Soviet government, not expecting it to last. Discipline in the Russian army collapsed entirely, with soldiers often massacring their officers. After the conclusion of an armistice with the Germans in December 1917, the front simply disbanded in chaos, most men trying to return home by whatever means they could find. The Germans proved willing to negotiate, but they offered Draconian conditions of peace. Trotsky, who as commissar for foreign affairs represented the Soviet government, felt compelled to turn them down, proclaiming a new policy: "no war, no peace!" The Germans then proceeded to advance, occupying more territory and seizing an enormous amount of military matériel. In Petrograd many Bolshevik leaders as well as the Left Socialist Revolutionaries agreed with Trotsky that German demands could not be accepted. Only Lenin's authority and determination swung the balance in favor of the humiliating peace. By sacrificing much else, Lenin in all probability saved Communist rule in Russia, for the young Soviet government was in no position whatsoever to fight Germany.

The Soviet-German Treaty of Brest-Litovsk was signed on March 3, 1918. To sum up its results in Vernadsky's words:

The peace conditions were disastrous to Russia. The Ukraine, Poland, Finland, Lithuania, Estonia, and Latvia received their independence. Part of Transcaucasia was ceded to Turkey. Russia lost 26 per cent of her total population; 27 per cent of her arable land; 32 per cent of her average crops; 26 per cent of her railway system; 33 per cent of her manufacturing industries; 73 per cent of her iron industries; 75 per cent of her coal fields. Besides that, Russia had to pay a large war indemnity.

Or to put it in different terms, Russia lost over sixty million people and over five thousand factories, mills, distilleries, and refineries. Puppet states dependent on Germany were set up in the separated border areas. Only the ultimate German defeat in the First World War prevented the Brest-Litovsk settlement from being definitive, and in particular made it possible for the Soviet government to reclaim the Ukraine.

Since Lenin's firm direction in disbanding the Constituent Assembly and capitulating to the Germans had enabled the Soviet government to survive, the great Soviet leader and his associates proceeded rapidly to revamp and even transform Russia politically, socially, and economically. In addition to letting peasants seize land, the government assigned control over the factories to workers' committees and nationalized all banks, confiscating private accounts. Foreign trade became a state monopoly, and a special commissariat was created to handle it. In December 1917, the existing judicial system was declared abolished: the new revolutionary tribunals and people's courts were to be guided by the "socialist legal consciousness." Titles and ranks disappeared. Authorities gradually assumed control over the scarce housing and other material aspects of life. Those who belonged to the upper and middle classes often lost their property, suffered discrimination, and were considered by the new regime to be suspect by definition. Church property was confiscated and religious instruction in schools terminated. The Gregorian or Western calendar — New Style — was adopted on January 31, 1918. The Constitutional Democrats, Socialist Revolutionaries — except for the Left Socialist Revolutionaries until their break with the Bolsheviks — and to an extent Mensheviks, all of whom opposed the new regime, were to be suppressed and hunted down as counterrevolutionaries. As early as December 20, 1917, the government established the Extraordinary Commission to Combat Counterrevolution, Sabotage, and Speculation, the dreaded "Cheka," headed by Felix Dzerzhinsky. From that time on the political police became a fundamental reality of Soviet life.

War Communism and New Problems

With the summer of 1918, War Communism began to acquire a definite shape. The nationalization of industry, which began shortly after the revolution, was extended by the law of June 28, 1918. To cite Carr's listing, the state appropriated "the mining, metallurgical, textile, electrical, timber, tobacco, resin, glass and pottery, leather and cement industries, all steam-driven mills, local utilities and private railways together with a few minor industries." Eventually private industry disappeared almost entirely. Compulsory labor was introduced. Private trade was gradually suppressed to be replaced by rationing and by government distribution of food and other necessities of life. On February 19, 1918, the nationalization of land was proclaimed: all land became state property to be used only by those who would cultivate it themselves. The peasants, however, had little interest in supplying food to the government because, with state priorities and the breakdown of the economy, they could not receive much in return. Therefore, under the pressure of the Civil War and of the desperate need to obtain food for the Red Army and the urban population, the authorities finally decreed a food levy, in effect ordering the peasants to turn in their entire produce, except for a minimal amount to be retained for their own sustenance and for sowing. As the peasants resisted, forcible requisitioning and repression became common. Communism, military and militant, swung into full force.

The rigors of War Communism on the home front largely resulted from and paralleled the bitter struggle the Soviet regime was waging with its external enemies. Beginning with the summer of 1918 the country entered a major, many-faceted, and cruel Civil War, when the so-called Whites — who had rallied initially to continue the war against the Germans — rose to challenge the Red control of Russia. Numerous nationalities, situated as a rule in the border areas of the former empire of the Romanovs, proceeded to assert their independence from Soviet authority. A score of foreign states intervened by sending some armed forces into Russia and supporting certain local movements and governments, as well as by blockading Soviet Russia from October 1919 to January 1920. In 1920, Poland fought a war against the Soviet government to win much of the western Ukraine and White Russia. It appeared that everyone was trying to strike a blow against the Communist regime.

The Civil War

The counterrevolutionary forces, often called vaguely and somewhat misleadingly the White movement, constituted the greatest menace to the Soviet rule, because, in contrast to Poland and various border nationalities, which had aims limited to particular regions, and to the intervening Allied powers, which had no clear aims, the Whites meant to destroy the Reds. The counterrevolutionaries drew their strength from army officers and cossacks, from the "bourgeoisie," including a large number of secondary school students and other educated youth, and from political groups ranging from the far Right to the Socialist Revolutionaries. Such prominent former terrorists as Boris Savinkov fought against the Soviet government, while the crack units of the White Army included a few worker detachments. Most intellectuals joined or sympathized with the White camp.

After the Soviet government came to power, civil servants staged an unsuccessful strike against it. Following their break with the Bolsheviks in March 1918 over the latter's determination to promote class struggle in the villages, the Left Socialist Revolutionaries tried an abortive uprising in Moscow in July. At about the same time and in part in response to the action of the Left SR's, counterrevolutionaries led by the local military commander seized Simbirsk, while Savinkov raised a rebellion in the center of European Russia, capturing and holding for two weeks the town of Iaroslavl on the Volga. These efforts collapsed, however, because of the insufficient strength of the counterrevolutionaries once the Soviet government could concentrate its forces against them. Indeed, it became increasingly clear that the Communist authorities, in particular the Cheka, had a firm grip on the central provinces and ruthlessly suppressed all opponents and suspected opponents. True to their tradition, the Socialist Revolutionaries tried terrorism, assassinating several prominent Bolsheviks, such as the head of the Petrograd Cheka, and seriously wounding Lenin himself in August 1918. Earlier, in July, a Left Socialist Revolutionary had killed the German ambassador, producing a diplomatic crisis. Yet even the terrorist campaign could not shake Soviet control in Moscow — which had again become the capital of the country in March 1918 — Petrograd, or central European Russia. And it provoked frightful reprisals, a veritable reign of terror, during which huge numbers of "class enemies" and others suspected by the regime were killed.

The borderlands, on the other hand, offered numerous opportunities to the counterrevolutionaries. The Don, Kuban, and Terek areas in the south and southeast all gave rise to local anti-Bolshevik cossack governments. Moreover, the White Volunteer Army emerged in southern Russia, led first by Alekseev, next by Kornilov, and after Kornilov's death in combat by an equally prominent general, Anthony Denikin. Other centers of opposition to the Communists sprang up in the east. In Samara, on the Volga, Chernov headed a government composed of members of the Constituent Assembly. Both the Ural and the Orenburg cossacks turned against Red Moscow. The All-Russian Directory of five members was established in Omsk, in western Siberia, in September 1918, as a result of a conference attended by anti-Bolshevik political parties and local governments of eastern Russia. Following a military coup the Directory was replaced by another anti-Red government, that of Admiral Alexander Kolchak. A commander of the cossacks of Transbaikalia, Gregory Semenov, ruled a part of eastern Siberia with the support of the Japanese. New governments

Revolution and Civil War in European Russia 1917–1922

Red Army moves, 1919–21
Deepest penetration of White Army forces, 1919–20
Allied intervention, 1918–20
Boundaries after 1922
1914 boundaries
Areas lost by Russia after World War I

British — Murmansk
Barents Sea
White Sea
Pechora
1920
U.S. — Archangel
Br. + U.S. occupation until late 1919
N. Dvina
Onega
1920
Kotlas
FINLAND Ind. 1917
Kronstadt Revolt Mar. 1921
SCENE OF THE OCTOBER REVOLUTION Nov. 7, 1917 (Oct. 25, O.S.)
Viazma
Helsinki
Stockholm
IUDENICH 1919
Petrograd
Vologda
Allied aid 1919
ESTONIA Ind. 1918
Pskov
KORNILOV Sept. 9–14, 1917
Iaroslavl
U. S. S. R.
(After Dec. 30, 1922)
Nizhnii Novg.
Riga
LATVIA Ind. 1917
Moscow
(Capital after Mar. 1918)
Kazan
KOLCHAK 1919
LITHUANIA Ind. 1917
Danzig
Vilna
FARTHEST GERMAN ADVANCE, MAR.–NOV. 1918
Simbirsk
Samara
Baltic Sea
EAST PRUSSIA
WHITE RUSSIA Ind. 1917
Minsk
Smolensk
Riazan
Oka
Tula
FARTHEST ADVANCE OF DENIKEN Oct. 1919
Penza
Saratov
Volga
CZECHS May–July, 1918
GERMANY
Vistula
Bug
Niemen
Brest-Litovsk
Dnieper
Gomel
Orel
1918
1919
POLAND Ind. 1918
Warsaw
POLES
CURZON LINE 1919–20)
Voronezh
KRASNOV Early 1918
Volga
1919
Cracow
Lvov
JUNE–AUG. 1920
Kiev
Kharkov
Donets
Tsaritsyn
CZECHOSLOVAKIA
UKRAINE Ind. 1918
Dniester
Rostov-on-Don
Don
COSSACKS
Budapest
HUNGARY
BESSARABIA
Pruth
Odessa
Sea of Azov
Kerch
Kuban
Terek
Belgrade
RUMANIA
WRANGEL
1920
Novorossiisk
YUGOSLAVIA
Bucharest
Danube
Sevastopol
French
Black Sea
British
CAUCASUS
GEORGIA Ind. 1918
Batum
Tiflis
BULGARIA
Sofia
Maritsa
ARMENIA Ind. 1918
Ardahan
Kars
ALBANIA
Bosporus
Constantinople
GREECE
Aegean Sea
Dardanelles
T U R K E Y
0 — Miles — 300

emerged also in Vladivostok and elsewhere. Russian anti-Bolshevik forces in the east were augmented by some 40,000 members of the so-called Czech Legion composed largely of Czech prisoners of war who wanted to fight on the side of the Entente. These soldiers were being moved to Vladivostok via the Trans-Siberian Railroad when a series of incidents led to their break with Soviet authorities and their support of the White movement. In the north a prominent anti-Soviet center arose in Archangel, where a former populist, Nicholas Chaikovsky, set up a government supported by the intervening British and French. And in the west, where the non-Russian borderlands produced numerous nationalist movements in opposition to the Soviet government, General Nicholas Iudenich established a White base in Estonia to threaten Petrograd.

The Civil War, which broke out in the summer of 1918, first went favorably for the Whites. In late June and early July the troops of the Samara government captured Simbirsk, Kazan, and Ufa. Although the Red Army managed to eliminate that threat, it immediately had to face a greater menace: the forces of Kolchak, supported by the Czechs, and those of Denikin, aided by cossacks. Kolchak's units, advancing from Siberia, took Perm in the Urals and almost reached the Volga. At this time, on the sixteenth of July, Nicholas II, the empress, their son, and four daughters were killed — apparently in compliance with Lenin's secret order — by local Bolsheviks in Ekaterinburg, where they had been confined, when the Czechs and the Whites approached the town. Denikin's army, after some reversals of fortune, resumed the offensive, and its right wing threatened to link with Kolchak's army in the spring of 1919. While Kolchak's forced retreat eliminated this possibility, Denikin proceeded to occupy virtually all of the Ukraine and to advance on Moscow. In the middle of October his troops took Orel and approached Tula, the last important center south of Moscow. At the same time Iudenich advanced from Estonia on Petrograd, seizing Gatchina, only thirty miles from that city, on October 16, and besieging Pulkovo on its outskirts. As a recent historian of these events has commented: "In the middle of October it appeared that Petrograd and Moscow might fall simultaneously to the Whites."

But the tide turned. Iudenich's offensive collapsed just short of the former capital. Although the Red Army had had to be created from scratch, it had constantly improved in organization, discipline, and leadership under Commissar of War Trotsky, and it managed finally to turn the tables on both Kolchak and Denikin. The admiral, who had assumed the title of "Supreme Ruler of Russia" and had received recognition from some other White leaders, suffered crushing defeat in late 1919 and was executed by the Bolsheviks on February 7, 1920. The general was driven back to the area of the Sea of Azov and the Crimea by the end of March 1920. At that point the Soviet-Polish war gave respite to the southern White Army and even enabled Denikin's successor General Baron Peter Wrangel to recapture a large section of southern Russia. But with the end of the Polish war in the autumn, the Red Army concentrated again on the southern front. After more bitter fighting, Wrangel, his remaining army, and a considerable number of civilians, altogether about 100,000 people, were evacuated on Allied ships to Constantinople in mid-November. Other and weaker counterrevolutionary strongholds, such as that in Archangel, had already fallen. By the end of 1920 the White movement had been effectively defeated.

Allied Intervention

The great Civil War in Russia was complicated by Allied intervention, by the war between the Soviet government and Poland, and by bids for national independence on the part of a number of peoples of the former empire of the Romanovs who were not Great Russians. The intervention began in 1918 and involved fourteen countries; the Japanese in particular sent a sizeable force into Russia — over 60,000 men. Great Britain dispatched altogether some 40,000 troops, France and Greece two divisions each, and the United States about 10,000 men, while Italy and other countries — except for the peculiar case of the Czechs — sent smaller, and often merely token, forces. The Allies originally wanted to prevent the Germans from seizing war matériel in such ports as Archangel and Murmansk, as well as to observe the situation, while the Japanese wanted to exploit the opportunities presented in the Far East by the collapse of Russian power. Japanese troops occupied the Russian part of the island of Sakhalin and much of Siberia east of Lake Baikal. Detachments of American, British, French, and Italian troops followed the Japanese into Siberia, while other Allied troops landed, as already mentioned, in northern European Russia, as well as in southern ports such as Odessa, occupied by the French, and Batum, occupied by the British. Allied forces assumed a hostile attitude toward the Soviet government, blockaded the Soviet coastline from October 1919 to January 1920, and often helped White movements by providing military supplies — such as some British tanks for Denikin's army — and by their very presence and protection. But they often avoided actual fighting. This fruitless intervention ended in 1920 with the departure of Allied troops, except that the Japanese stayed in the Maritime Provinces of the Russian Far East until 1922 and in the Russian part of Sakhalin until 1925.

The War against Poland

The Soviet-Polish war was fought in 1920 from the end of April until mid-October. The government of newly independent Poland opened hostilities to win the western Ukraine and western White Russia, which the Poles considered part of their "historic heritage," although ethnically the areas in question were not Polish. The ancient struggle between the Poles and the Russians resumed its course, with this time the Russians, that is, the Soviet government, in an apparently desperate situation. Actually the war produced more than one reversal of fortune. First, in June and July the Poles overran western Russian areas; next the Red Army, led by Michael Tukhachevsky and others, staged a mighty counteroffensive that reached the very gates of Warsaw; then the Poles, helped by French credits and Allied supplies, defeated the onrushing Reds and gained the upper hand. The Treaty of Riga of March 18, 1921, gave Poland many of the lands it desired, establishing the boundary a considerable distance east of the ethnic line, as well as of the so-called Curzon Line, which approximated the ethnic line and which the Allies had regarded as the just settlement.

National Independence Movements

National independence movements in the former empire of the Romanovs during the years following 1917 defy comprehensive description in a textbook and have to be left to special works, such as Pipes's study. As early as 1917 Finland, Latvia, Lithuania, and White Russia declared their independence. They were followed in 1918 by Estonia, the Ukraine, Poland — once German troops were evacuated — the Transcaucasian Federation — to be dissolved into the separate states of Azerbaijan, Armenia, and Georgia — and certain political formations in the east. The Soviet government had proclaimed the right of self-determination of peoples, but it became quickly apparent that it considered independence movements as bourgeois and counterrevolutionary. Those peoples that were successful in asserting their independence, that is, the Finns, the Estonians, the Latvians, and the Lithuanians, as well as the Poles, did so in spite of the Soviet government which was preoccupied with other urgent matters. Usually they had to suppress their own Communists, sometimes, as in the case of Finland, after a full-fledged civil war. All except Poland and Lithuania became independent states for the first time. In other areas the Red Army and local Communists combined to destroy independence.

Developments in the Ukraine turned out to be perhaps the most complicated of all. There the local government, the Rada or central council, and the General Secretariat, proclaimed a republic of the Ukrainian people after the fall of the Provisional Government in Petrograd. Soviet authorities recognized the new republic, but in February 1918 the Red Army overthrew the Rada. Soviet rule, established in the spring of 1918, was in turn overthrown by the advancing German army. The Germans at first accepted the Rada, but before long they sponsored instead a Right-wing government under Paul Skoropadsky. After the Germans left, the Directory of the Rada deposed Skoropadsky in December 1918, only to be driven out in short order by Denikin's White forces. Following Denikin's withdrawal in the autumn of 1919, Soviet troops restored Soviet authority in the Ukraine. Next the Directory of the Rada made an agreement with the Poles, only to be left out at the peace treaty terminating the Soviet-Polish War which simply divided the Ukraine between Soviet Russia and Poland. Ukrainians supported different movements and fought in different armies as well as in countless anarchic peasant bands. Political divisions survived the collapse of the Ukrainian bid for independence and later divided Ukrainian émigrés. Yet it remains an open question to what extent the young Ukrainian nationalism, nurtured especially among the Ukrainian intellectuals in Austrian Galicia, had penetrated the peasant masses of the Russian Ukraine.

Among the peoples living to the south and southeast of European Russia, many of whom had been joined to the Russian Empire as late as the nineteenth century, numerous independence movements arose and independent states were proclaimed. The new states included the Crimean Tartar republic, the Transcaucasian republics of Georgia, Armenia, and Azerbaijan, the Bashkir, Kirghiz, and Kokand republics, the emirates of Bokhara and Khiva, and others. Time and again local interests clashed and bitter local civil wars developed. In certain instances foreign powers, such as Turkey, Germany, and Great Britain, played important roles. The Menshevik government of Georgia distinguished itself by the relative stability and effectiveness of its rule. But — without going into complicated and varied detail — whether new authorities received much or little popular support, they succumbed eventually to Soviet strength allied with local Communists. The fall of the independent Georgian government in 1921 marked essentially the end of the process, although native partisans in Central Asia, the "Basmachi," were not finally suppressed until 1926.

Reasons for the Red Victory

Few observers believed that the Bolsheviks would survive the ordeal of Civil War, national independence movements, war against Poland, and Allied intervention. Lenin himself, apparently, had serious doubts on that score. The first years of the Soviet regime have justly become a legendary Communist epic, its lustre undimmed even by the titanic events of the Second World War. Yet, a closer look puts the picture into a better focus and helps to explain the Bolshevik victory without recourse to magic in

Marxism or superhuman qualities of Red fighters. To begin with, Allied intervention — the emphatic Soviet view to the contrary notwithstanding — represented anything but a determined and co-ordinated effort to strangle the new Communist regime. Kennan, Ullman, and other scholars have shown how much misunderstanding and confusion went into the Allied policies toward Russia, which never amounted to more than a half-hearted support of White movements. Allied soldiers and sailors, it might be added, saw even less reason for intervening than did their commanders. The French navy mutinied in the Black Sea, while the efficiency of American units was impaired by unrest as well as by a fervent desire to return home. The Labor party in Great Britain and various groups elsewhere exercised what pressure they could against intervention. Ill-conceived and poorly executed, the Allied intervention produced in the end little or no result. The Poles, by contrast, knew what they wanted and obtained it by means of a successful war. Their goals, however, did not include the destruction of the Soviet regime in Russian territory proper. National independence movements also had aims limited to their localities, and were, besides, usually quite weak. The Soviet government could, therefore, defeat many of them one by one and at the time of its own choosing, repudiating its earlier promises when convenient, as in the cases of the Ukraine and the Transcaucasian republics.

The White movement did pose a deadly threat to the Reds. Ultimately there could be no compromise between the two sides. The White armies were many, contained an extremely high proportion of officers, and often fought bravely. The Reds, however, had advantages that in the end proved decisive. The Soviet government controlled the heart of Russia, including both Moscow and Petrograd, most of its population, much of its industry, and the great bulk of military supplies intended for the First World War. The White armies constantly found themselves outnumbered and, in spite of Allied help, more poorly equipped. Also, the Red Army enjoyed the inner lines of communication, while its opponents had to shift around on the periphery. Still more important, the Reds possessed a strict unity of command, whereas the Whites fought, in fact, separate and unco-ordinated wars. Politics, as well as geography, contributed to the White disunity. Anti-Bolshevism represented the only generally accepted tenet in the camp, which encompassed everyone from the monarchists to the Socialist Revolutionaries. Few positive programs were proposed or developed. The Whites' inability to come to terms with non-Russian nationalities constituted a particular political weakness. White generals thought naturally in terms of "Russia one and indivisible" and reacted against separatism; or at least, they felt it quite improper to decide on their own such fundamental questions as those of national independence and boundaries. Thus Denikin antagonized the Ukrainians by his measures to suppress the Ukrainian language and schools, and Iudenich weakened his base in Estonia because he would not promise the Estonians independence.

In the last analysis, the attitude of the population probably determined the outcome of the Civil War in Russia. Whereas the upper and middle classes favored the Whites, and the workers, with some notable exceptions, backed the Reds, the peasants, that is, the great majority of the people, assumed a much more cautious and aloof attitude. Many of them came to hate both sides, for White rule, as well as Red rule, often brought mobilization, requisitions, and terror — as cruel as, if less systematic than, that of the Cheka. In many areas anarchic peasant bands attacked both combatants. Indeed, this so called green resistance proved to be in scope, casualties, and, alas, cruelty quite comparable to the more prominent struggle between the Whites and the Reds, although it was by its very nature local rather than national. Still, on the whole, the peasants apparently preferred the Reds to the Whites. After all, they had obtained the gentry land following the October Revoluiton, while the Whites were associated in their minds — and not entirely unjustly — with some kind of restoration of the old order, a possibility that evoked hatred and fear in the Russian village. *Mutatis mutandis,* one is reminded of the later circumstances of the Communist victory in the civil war in China.

The R.S.F.S.R. and the U.S.S.R.

The first Soviet constitution was adopted by the Fifth All-Russian Congress of Soviets and promulgated on July 10, 1918. It created the Russian Soviet Federated Socialist Republic, or the R.S.F.S.R. Local Soviets elected delegates to a provincial congress of Soviets, and provincial congresses in turn elected the membership of the All-Russian Congress of Soviets. The latter elected the Executive Committee, which acted in the intervals between congressional sessions, and the Council of People's Commissars. Elections were open rather than secret, and they were organized on a class basis, with the industrial workers especially heavily represented. By contrast, the "non-toiling classes" received no vote. In effect, the Communist party, particularly its Central Committee and Political Bureau headed by Lenin, from the beginning dominated the government apparatus and ruled the country. Besides, the same leading Communists occupied the top positions in both party and government, with Lenin at the head of both. On December 30, 1922, the Union of Soviet Socialist Republics came into being as a federation of Russia, the Ukraine, White Russia, and Transcaucasia. Later in the '20's three Central Asiatic republics received "Union Republic" status. Compared to the empire of the Romanovs, the new state had lost Finland, Estonia, Latvia, Lithuania, and the Polish territories, all of which had become independent, and had lost the western Ukraine and western White Russia to Poland, Bessarabia to Rumania, and the Kars-Ardakhan area in Transcaucasia to Turkey. Also, as already mentioned, Japan evacuated all of the Siberian mainland of Russia only in 1922, and the Russian half of the island of Sakhalin in 1925. In spite of these reductions in size, the U.S.S.R. emerged as an enormous country.

The Crisis

At the end of the Civil War Soviet Russia was exhausted and ruined. The droughts of 1920 and 1921 and the frightful famine during that last year added the final, gruesome chapter to the disaster. In the years following the originally "bloodless" October Revolution epidemics, starvation, fighting, executions, and the general breakdown of the economy and society had taken something like twenty million lives. Another two million had left Russia — with Wrangel, through the Far East, or in numerous other ways — rather than accept Communist rule, the émigrés including a high proportion of educated and skilled people. War Communism might have saved the Soviet government in the course of the Civil War, but it also helped greatly to wreck the national economy. With private industry and trade proscribed and the state unable to perform these functions on a sufficient scale, much of the Russian economy ground to a standstill. It has been estimated that the total output of mines and factories fell in 1921 to 20 per cent of the pre-World War level, with many crucial items experiencing an even more drastic decline, for example cotton fell to 5 per cent, iron to 2 per cent, of the prewar level. The peasants responded to requisitioning by refusing to till their land. By 1921 cultivated land had shrunk to some 62 per cent of the prewar acreage, and the harvest yield was only about 37 per cent of normal. The number of horses declined from 35 million in 1916 to 24 million in 1920, and cattle from 58 to 37 million during the same span of time. The exchange rate of an American dollar, which had been two rubles in 1914, rose to 1,200 rubles in 1920.

The unbearable situation led to uprisings in the countryside and to strikes and violent unrest in the factories. Finally, in March 1921, the Kronstadt naval base, celebrated by the Communists as one of the sources of the October Revolution, rose in rebellion against Communist rule. It is worth noting that the sailors and other Kronstadt rebels demanded free Soviets and the summoning of a constituent assembly. Although Red Army units ruthlessly suppressed the uprising, the well-nigh general dissatisfaction with Bolshevik rule could not have been more forcefully expressed. And it was against this background of utter devastation and discontent that Lenin, who, besides, had finally to admit that a world revolution was not imminent, proceeded in the spring of 1921 to inaugurate his New Economic Policy in place of War Communism. Once more Lenin proved to be the realist who had to overcome considerable doctrinaire opposition to have his views prevail in the party and, therefore, in the entire country.

The New Economic Policy

The New Economic Policy was a compromise, a temporary retreat on the road to socialism, in order to give the country an opportunity to recover; and it was so presented by Lenin. The Communist party, of course, retained full political control; the compromise and relaxation never extended to politics. In economics, the state kept its exclusive hold on the "commanding heights," that is, on finance, on large and medium industry, on modern transportation, on foreign trade, and on all wholesale commerce. Private enterprise, however, was allowed in small industry, which meant plants employing fewer than twenty workers each, and in retail trade. The government's change of policy toward the peasants was perhaps still more important. Instead of requisitioning their produce, as had been done during War Communism, it established a definite tax in kind, particularly in grain, replaced later by a money tax. The peasants could keep and sell on the free market what remained after the payment of the tax, and thus they were given an obvious incentive to produce more. Eventually the authorities even permitted a limited use of hired labor in agriculture and a restricted lease of land. The government also revamped and stabilized the financial system, introducing a new monetary unit, the *chervonets;* and it put into operation new legal codes to help stabilize a shattered society.

The New Economic Policy proved to be a great economic success. After the frightful starvation years of 1921 and 1922 — years, incidentally, when many more Russians would have perished, but for the help received from the American Relief Administration headed by Herbert Hoover, from the Quakers, and from certain other groups — the Russian economy revived in a remarkable manner. In 1928 the amount of land under cultivation already slightly exceeded the pre-World War area. Industry on the whole also reached the prewar level. It should be added that during the N.E.P. period, in contrast to the time of War Communism, the government demanded that state industries account for costs and pay for themselves. It was highly characteristic of the N.E.P. that 75 per cent of retail trade fell into private hands. In general, the so-called Nepmen, the small businessmen allowed to operate by the new policy, increased in number in towns, while the *kulaki* — or kulaks, for the term has entered the English language — gained in the villages. *Kulak,* meaning "fist," came to designate a prosperous peasant, a man who held tightly to his own; the prerevolutionary term, used by Soviet sources, also has connotations of exploitation and greed.

These social results of the New Economic Policy naturally worried the

Communists. The Eleventh Party Congress declared as early as 1922 that no further "retreat" could be tolerated. In 1924 and 1925 the government introduced certain measures to restrict the Nepmen, and in 1927 to limit the kulaks. The Party long debated the correct policy to determine the future development of the country. Ideological arguments came to be closely linked to personalities and to the struggle for power that gained momentum after Lenin's death in January 1924.

The Struggle for Power after Lenin's Death

Three main points of view emerged among the Russian Communists during the twenties. The so-called Left position, best developed by Trotsky, maintained that, without world revolution, socialism in Russia was doomed. Therefore, the Bolsheviks had to support revolutionary movements abroad and at the same time pursue a militant and socialist policy at home. An opponent of the N.E.P., Trotsky also came to criticize Stalin for his cooperation with bourgeois forces abroad and for his destruction of democracy within the Party. Such prominent Communist leaders as Gregory Zinoviev — born Radomyslsky — and Leo Kamenev — born Rosenfeld — essentially shared Trotsky's view. The Right faction, led by a prominent theoretician, Nicholas Bukharin, agreed with the Left that socialism in Russia depended on world revolution. But the members of this group concluded that, because such a revolution was not immediately in prospect, the Soviet government should not quixotically force the pace towards socialism, but rather continue the existing compromise and develop the New Economic Policy. Finally, the third faction, the Center headed by Stalin, came to the conclusion that, in spite of the fact that world revolution failed to materialize, socialism could be built within the one country of the Union of Soviet Socialist Republics, with its huge size, large population, and tremendous resources. The Center therefore called for a great effort to transform the Soviet Union. Putting the Right group aside, it should be realized that Trotsky as well as Stalin wanted to build socialism in Russia — Stalin, in fact, has been accused of simply borrowing the Left program — and that Stalin as well as Trotsky aimed at world revolution. The ideological difference between the two was that of emphasis, not of fundamental belief. Yet emphasis can be very important at certain moments in history. Moreover, Stalin's approach for the first time gave Russia, or rather the Soviet Union, the central position in Communist thought and planning.

As has often been described and analyzed, the struggle for power that followed Lenin's death was decided by Stalin's superior control of the Party membership. Acting behind the scenes as the general secretary of the Party, Stalin managed to build up a following strong enough to overcome Trotsky's magnificent rhetoric and great prestige, as well as Kamenev's Party organization in Moscow and Zinoviev's in Petrograd — named Leningrad after Lenin's death. Stalin intrigued skillfully, first allying himself with Kamenev and Zinoviev against Trotsky, whom they envied and considered their rival for Party leadership; then with the Right group against the Left; and eventually, when sufficiently strong, suppressing the Right as well. He kept accusing his opponents of factionalism, of disobeying the established Party line and splitting the Party. Final victory came at the Fifteenth All-Union Congress of the Communist party, which on December 27, 1927, condemned all "deviation from the general Party line" as interpreted by Stalin. The general secretary's rivals and opponents recanted or were exiled; in any case, they lost their former importance. Trotsky himself was expelled from the Soviet Union in January 1929 and was eventually murdered in exile in Mexico in 1940, almost certainly on Stalin's orders.

Still, although Stalin's rise to supreme authority can well be considered an impressive, if gruesome, study in power politics, its ideological aspect should not be forgotten. After all, of the three alternate views present in the Party, the general secretary's possessed the greatest attraction by far for Soviet Communists. The Right, in effect, simply admitted defeat: in spite of the tremendous struggle and all the efforts, socialism could not succeed in the Soviet Union until the uncertain coming of world revolution. Trotsky's Left position, while more sanguine, also tied the Soviet future to world revolution and thus made Bolshevik activity of limited importance and effectiveness at best. Only Stalin offered a sweeping program and a majestic goal to be achieved by Soviet efforts alone. Only he proposed to advance Marxism in the Soviet Union without dependence on problematic developments elsewhere. The same Party congress that condemned all deviations from Stalin's line enthusiastically adopted measures that signified the end of the New Economic Policy and the beginning of the First Five-Year Plan.

THE FIRST THREE FIVE-YEAR PLANS, 1928–41

Enough of living by the law
Given by Adam and Eve.
The jade of history we will ride to death.
Left!
Left!
Left!

MAIAKOVSKY

It [the First Five-Year Plan] asked no less than a complete transformation from backward agricultural individualism to mechanized
collectivism, from hothouse subsidized industry to self-sufficient industry on the greatest, most modern scale, from the mentality of
feudalism, far behind the Western industrial age, to socialism still
ahead of it.

DURANTY

"When a forest is cut down, splinters fly." Of course, it is unfortunate
to be a splinter.

THE REMARK OF A SOVIET CITIZEN
TO THE AUTHOR IN THE SUMMER OF 1958

STALIN'S SWEEPING VICTORY at the Fifteenth All-Union Congress of the
Communist Party in December 1927 marked the inauguration of the era
of Stalin and his five-year plans. The general secretary was to direct the
destinies of the Union of the Soviet Socialist Republics and of world Communism for twenty-five eventful years, becoming in the course of that
quarter of a century perhaps the most totalitarian, powerful, and feared
dictator of all time.

Stalin

Stalin began his life and career humbly enough. In fact, it has often been
mentioned that he was one of the few Bolshevik leaders of more or less
proletarian origin. Born a son of a shoemaker in 1879 in the little town
of Gori near the Georgian capital of Tiflis — or Tbilisi — Joseph Dzhugashvili attended a Church school in Gori until 1894 and then went to the
theological seminary in Tiflis. In 1899, however, he was expelled from the
seminary for reasons that are not entirely clear. By that time, apparently,
Stalin had become acquainted with some radical writers and in particular
with Marx and Lenin. He joined the Social Democratic party and when it

split in 1903 sided firmly with the Bolsheviks. Between 1902 and 1913 Dzhugashvili, or rather Stalin as he came to be known, engaged in a variety of conspiratorial and revolutionary activities, suffering arrest and exile several times. He managed to escape repeatedly from exile, which has suggested police collusion to certain specialists. Stalin's last exile, however, continued from 1913 until the February Revolution. Apparently the Georgian Bolshevik first attracted Lenin's attention when he organized a daring raid to seize funds for the Party. Stalin's revolutionary activity developed in such Transcaucasian centers as Tiflis, Batum, and Baku, as well as in St. Petersburg. In contrast to many other Bolshevik leaders, Stalin never lived abroad, leaving the Russian empire only to attend a few meetings. Because of Stalin's Bolshevik orthodoxy and Georgian origin, the Party welcomed him as an expert on the problem of nationalities, a subject to which he devoted some of his early writings.

One of the first prominent Bolsheviks to arrive in Petrograd, Stalin participated in the historic events of 1917, and after the October Revolution he became the first commissar for national minorities. As a member of the Revolutionary Military Council of the Southern Front he played a role in the Civil War, for example, in the defense of Tsaritsyn against the Whites. Incidentally, Tsaritsyn was renamed Stalingrad in 1925 and Volgograd in 1961. It might be noted that in the course of executing his duties he quarreled repeatedly with Trotsky. But Stalin's real bid for power began in 1922 with his appointment as general secretary of the Party, a position that gave him broad authority in matters of personnel. The long-time official Soviet view of Stalin as Lenin's anointed successor distorts reality, for, in fact, the ailing Bolshevik leader came to resent the general secretary's rigidity and rudeness and in his so-called testament warned the Party leadership against Stalin. But Stalin's rivals failed to heed Lenin's late forebodings, and, before too long, Stalin's Party machine rolled over all opponents. The complete personal dictatorship which began in 1928 was to last until the dictator's death in 1953.

The amount of time that has elapsed since Stalin's death has not been nearly enough for historians to pass a definitive judgment on the Soviet dictator and his historical role. Views of Stalin have ranged from the utterly fantastic eulogy of him as a universal genius, expounded for many years by the propaganda machine of Russian and world communism, to the extremely hostile impression that he was a blood-soaked, man-devouring, oriental monster. Many commentators have made interesting attempts to explain the general secretary, his importance, and his work. Stalin has been credited, for example, with "inflexible will, unwillingness to yield, realistic statesmanship and high organizing abilities." Hardheaded realism and common sense have been mentioned frequently as the dictator's outstanding traits. Deutscher's well-known book presents him as a hard-pressed Marxist realist carrying out a consistent policy and reacting intelligently to the needs of the moment. Yet, as in the case of Ivan the Terrible, there was madness in Stalin's method. That madness, formerly a matter of suspicion and controversy, received fully convincing documentation in Khrushchev's celebrated speech to the Twentieth Party Congress in 1956 and especially during the session of the Twenty-second Party Congress in October 1961, as well as in the memoirs of Stalin's daughter, Svetlana Alliluyeva, published in 1967, and in most recent Soviet material. In addition to fighting real battles and struggling against actual opponents, Stalin lived in the paranoiac world of constant threat and wholesale conspiracy. Fact and fantasy were blended together, making the detection of the dictator's motives extremely difficult. Still, in retrospect, Stalin's proverbial ruthlessness and vindictiveness and his passion for discovering ever-new enemies and plots find their explanation in abnormal psychology rather than in any compelling objective necessity or in any alleged rational advantages of a "permanent purge." That causation is given central attention in Tucker's study of Stalin and it is reflected, in a different sense, in Ulam's recent important book on the general secretary. Paranoiac tendencies joined with Marxism in transforming the Russian scene.

The First Five-Year Plan

The First Five-Year Plan and its successors hit the Soviet Union with tremendous impact. The U.S.S.R. became a great industrial nation: from being the fifth country in production when the plans began, it was eventually second only to the United States. In agriculture individual peasant cultivation gave way to a new system of collective farming. Indeed 1928 and 1929 have been described as the true revolutionary years in Russia: it was then that the mode of life of the peasants, the bulk of the people, underwent a radical change, whereas until the First Five-Year Plan they continued to live much as they had for centuries. A vast social transformation accompanied the economic, while at the same time the entire Soviet system as we have come to know it acquired its definitive form in the difficult decade of the '30's.

Perhaps paradoxically, the five-year plans are not easy to explain. Marxist theory did not specifically provide for them and certainly did not spell out the procedures to be followed. To be sure, the needs of the moment affected the decisions of Soviet leaders. Yet Stalin's and his associates' response to the needs constituted only one alternative line of action, and often not the most obvious. In fact, the leadership rapidly reversed itself on such key subjects as the speed of collectivization.

Certain considerations, however, help to explain the five-year plans. To begin with, although Marxism did not provide for industrialization it insisted on a high level of it. The dictatorship of the proletariat in a land of peasants remained an anomaly. If, contrary to the doctrine, industries and workers were not there in the first place, they had to be created. Marxists in general, and Bolsheviks in particular, thought of socialism entirely in terms of an advanced industrial society. Such authors as Ulam have demonstrated in a rather convincing manner the close and multiple ties between Marxism and industrialization. The collectivization of agriculture, in turn, represented the all-important step from an individual and, therefore, bourgeois system of ownership and production to a collective economy and, therefore, to socialism. As already mentioned, after the October Revolution the Soviet government proceeded to nationalize Russian industry. Lenin showed a special interest in electrification, popularizing the famous slogan: "Electrification plus Soviet power equals communism." In 1921 the State Planning Commission, known as *Gosplan,* was organized to draft an economic plan for the entire country. It studied resources and proposed production figures; eventually it drew up the five-year plans.

Although the New Economic Policy constituted a retreat from socialism, that retreat was undertaken only as a temporary measure and out of sheer necessity. In addition to its social results, unacceptable to most Communists, the N.E.P. raised serious economic problems. While by 1928 Russian industry had regained its pre-World War level, a further rapid advance appeared quite uncertain. With the industrial plant restored and in operation — a relatively easy accomplishment — the Soviet Union needed investment in the producers' goods industries and a new spurt in production. Yet the "socialist sector" of the economy lacked funds, while the "free sector," particularly the peasants, failed to rise to government expectations. The Soviet economy in the 1920's continued to be plagued by pricing problems, beginning with the disparity between the low agricultural prices and the high prices of manufactured consumers' goods, resulting in the unwillingness of peasants to supply grain and other products to the government and the cities — a situation well described as the "scissors crisis." Gerschenkron and other specialists have argued that the Bolsheviks had good reason to fear that a continuation of the N.E.P. would stabilize a peasant society at the point where it was interested in obtaining more consumers' goods, but neither willing nor able to support large-scale industrialization. As already indicated, Stalin's Five-Year Plan proved attractive to the Party because it promised a way out of the impasse: the Soviet Union could abandon the New Economic Policy and become a truly socialist country without waiting for world revolution. "Socialism in one country" gripped many imaginations and became the new Bolshevik battle cry.

Once the Plan went into operation, the economic factors involved in its execution acquired great significance, all the more so because the planners set sail in essentially uncharted waters and often could not foresee the results of their actions. In particular, according to Gerschenkron, A. Erlich, and certain other scholars, the fantastically rapid collectivization of agriculture came about as follows: while the Plan had called for a strictly limited collectivization, set at 14 per cent, the unexpectedly strong resistance on the part of the peasants led to an all-out attack on individual farming; moreover, the government discovered that the collectives, which finally gave it control over the labor and produce of the peasants, enabled it to squeeze from them the necessary funds for industrial investment. It has been estimated that the Soviet state paid to the collectives for their grain only a distinctly minor part of the price of that grain charged the consumer; the remaining major part constituted in effect a tax. That tax, plus the turn-over or sales tax that the Soviet state charged all consumers, together with the ability of the government to keep the real wages down while productivity went up, produced the formula for financing the continuous industrialization of the Soviet Union.

In addition to ideology and economics, other factors entered into the execution of the five-year plans. Many scholars assign major importance to considerations of foreign policy and of internal security and control. Preparation for war, which affected all major aspects of the five-year plans, began in earnest after Hitler came to power in Germany in 1933, and while Japan was further developing its aggressive policies in the Far East. The stress on internal security and control in the five-year plans is more difficult to document. Yet it might well be argued that police considerations were consistently uppermost in the minds of Stalin and his associates. Collectivization, from that point of view, represented a tremendous extension of Communist control over the population of the Soviet Union, and it was buttressed by such additional measures — again combining economics and control — as the new crucial role of the Machine Tractor Stations, the M.T.S., which will be mentioned later.

The First Five-Year Plan lasted from October 1, 1928, to December 31, 1932, that is, four years and three months. The fact that Soviet authorities tried to complete a five-year plan in four years is a significant comment on the enormous speed-up typical of the new socialist offensive. The main goal of the Plan was to develop heavy industry, including machine-building, and that emphasis has remained characteristic of Soviet industrialization from that time on. According to Baykov's calculation, 86 per cent of all industrial investment during the First Five-Year Plan went into heavy industry. Whole new branches of industry, such as the chemical, automobile, agricultural machinery, aviation, machine tool, and electrical, were created from slight beginnings or even from scratch. Over fifteen hundred new factories were built. Gigantic industrial complexes, exemplified by Magnitostroi in the Urals and Kuznetsstroi in western Siberia, began to take shape. Entire cities arose in the wilderness. Magnitogorsk, for instance, acquired in a few years a population of a quarter of a million.

The First Five-Year Plan was proclaimed a great success: officially it was fulfilled in industry to the extent of 93.7 per cent in four years and three months. Furthermore, heavy industry, concerned with means of production, exceeded its quota, registering 103.4 per cent while light or consumers' goods industry produced 84.9 per cent of its assigned total. Of course, Soviet production claims included great exaggerations, difficult to estimate because of the limited and often misleading nature of Soviet statistics for the period. To put it very conservatively and without percentages: "The fact remains beyond dispute that quantitatively, during the years covered by the F.Y.P., industrial production did increase and very substantially." Quality, however, was often sacrificed to quantity, and the production results achieved varied greatly from item to item, with remarkable overfulfillments of the plan in some cases and underfulfillments in others. Besides, the great industrial spurt was accompanied by shortages of consumer goods, rationing, and various other privations and hardships which extended to all of the people, who at the same time were forced to work harder than ever before. The whole country underwent a quasi-military mobilization reminiscent of War Communism.

But the greatest transformation probably occurred in the countryside. As already mentioned, the collectivization of agriculture, planned originally as a gradual advance, became a flood. Tens of thousands of trusted Communists and proletarians — the celebrated "twenty-five thousand" in one instance, actually twenty-seven thousand — were sent from towns into villages to organize kolkhozes and establish socialism. Local authorities and Party organizations, with the police and troops where necessary, forced peasants into collectives. A tremendous resistance developed. About a million of the so-called kulaks, some five million people counting their families, disappeared in the process, often having been sent to concentration camps in far-off Siberia or Central Asia. A frightful famine swept the Ukraine. Peasants slaughtered their cattle and horses rather than bring them into a kolkhoz. Thus from 1929 to 1933 in the Soviet Union the number of horses, in millions, declined from 34 to 16.6, of cattle from 68.1 to 38.6, of sheep and goats from 147.2 to 50.6, and of hogs from 20.9 to 12.2. Droughts in 1931 and 1932 added to the horrors of the transition from private to collectivized farming.

Stalin himself applied the brakes to his own policy after the initial fifteen months. In his remarkable article, "Dizzy with Success," published in March 1930, he criticized the collectivizers for excessive enthusiasm and re-emphasized that collectives were to be formed on the voluntary principle, not by force. At the same time he announced certain concessions to collective farmers, in particular their right to retain a small private plot of land and a limited number of domestic animals and poultry. The new stress on the voluntary principle produced striking results: whereas fourteen million peasant households had joined collective farms by March 1930, only five million remained in collectives in May. But before long their number began to increase again when the authorities resorted to less direct pressure, such as a temporary suspension of taxes and priority in obtaining scarce manufactured goods. By the end of the First Five-Year Plan more than fourteen million peasant households had joined the kolkhoz system. According to one count, at that time 68 per cent of all cultivated land in the Soviet Union was under kolkhoz agriculture, and 10 per cent under sovkhoz agriculture, while only 22 per cent remained for independent farmers. The Plan could well be considered overfulfilled.

A sovkhoz is essentially an agricultural factory owned by the state, with peasants providing hired labor. Although sovkhozes, serving as experimental stations, as enormous grain producers in newly developed regions, and in many other crucial assignments, have been more important for the Soviet economy than their number would indicate, Communist authorities refrained from establishing them as the basic form of agricultural organization in the country. Instead they have relied on the kolkhoz as the norm for the Soviet countryside. A kolkhoz — *kollektivnoe khoziaistvo,* collective economy or farm — is owned by all its members, although it must undertake to deliver the assigned amount of produce to the state and is controlled by the state. Significantly, the produce of a collective farm has generally been allocated as follows: first, the part required by the state, both as taxes and as specified deliveries at set prices; next, the seed for sowing and the part to serve as payment to the Machine Tractor Station that aided the kolkhoz; after that, members of the collective receive their shares calculated on the basis of the "workdays" — a unit of labor to be distinguished from actual days — that they have put in for the kolkhoz; finally, the remainder goes into the indivisible fund of the collective to be used for its social, cultural, and other needs. The members also cultivate their small private plots — and with remarkable intensity and success. The Machine Tractor Stations, finally abolished in 1958, provided indispensable mechanized aid to the collectives, notably at harvest time, helping to co-ordinate the work of different kolkhozes and acting as another control over them. While it might be noted that the Soviet government found it easier to introduce collective farms in those regions where communal agriculture prevailed than in areas of individual proprietors, such as the Ukraine, the kolkhoz bore very little resemblance to the commune. Members of a commune possessed their land in common, but they farmed their assigned lots separately, undisturbed, and in their own traditional way. Organization and regimentation of labor became the very essence of the kolkhoz.

The Second and Third Five-Year Plans

The Second Five-Year Plan, which lasted from 1933 through 1937, and the Third, which began in 1938 and was interrupted by the German invasion in June 1941, continued on the whole the aims and methods of the initial Plan. They stressed the development of heavy industry, completed the collectivization of agriculture, and did their best to mobilize the manpower and other resources of the country to attain the objectives. The Soviet people lived through eight and a half more years of quasi-wartime exertion. Yet these plans also differed in certain ways from the first and from each other. The Second Five-Year Plan, drawn on the basis of acquired knowledge more expertly than the first, tried to balance production to avoid extreme over- or underfulfillment. It emphasized "mastering the technique," including the making of especially complicated machine tools, precision instruments, and the like. Also, it allowed a little more for consumers' goods than the first plan did. However, in the course of the Second Five-Year Plan, and especially during the third, military considerations became paramount. Military considerations linked to ideology had of course always been present in the planning of Soviet leaders. From the beginning of industrialization, Stalin and his associates had insisted that they had to build a powerful socialist state quickly, perhaps in a decade, or be crushed by capitalists. In the 1930's the threat became increasingly real and menacing. Soviet leaders did what they could to arm and equip Red forces, and they accelerated the development of industries inland, east of the Volga, away from the exposed frontiers.

Both the Second Five-Year Plan and the third, as far as it went, were again proclaimed successes, and again the official claims, in spite of their exaggeration, had some sound basis in fact. Industry, especially heavy industry, continued to grow. On the basis of official — and doubtful — figures, the Soviet share in world production amounted to 13.7 per cent in 1937, compared to 3.7 in 1929 and to 2.6 for the Russian Empire in 1913. In the generation of electrical power, for example, the Soviet Union advanced from the fifteenth place among the countries of the world to the third, and it was second only to the United States in machine building, tractors, trucks, and some other lines of production. Moreover, the Soviet Union made its amazing gains while the rest of the world experienced a terrible depression and mass unemployment.

In agriculture collectivization was virtually completed and, except for the wilderness, the Soviet countryside became a land of kolkhozes and

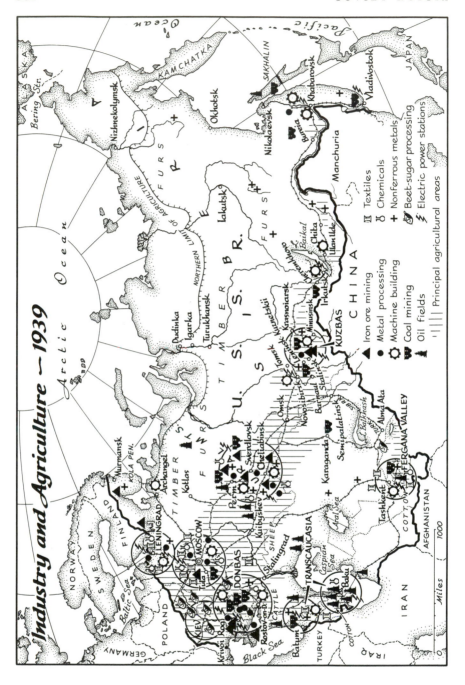

Industry and Agriculture — 1939

▥ Textiles
♀ Chemicals
+ Nonferrous metals
⚘ Beet-sugar processing
⚡ Electric power stations

◆ Iron ore mining
● Metal processing
⚙ Machine building
🗼 Coal mining
🏭 Oil fields
||||| Principal agricultural areas

sovkhozes. Slightly less than 250,000 kolkhozes replaced over 25 million individual farms. The famine and other horrors of the First Five-Year Plan did not recur. In fact, agricultural production increased somewhat, and food rationing was abolished in 1935. Still, the economic success of Soviet agricultural policy remained much more doubtful than the achievements of Soviet industrialization. Peasants regularly failed to meet their production quotas. They showed far greater devotion to their small private plots than to the vast kolkhoz possessions. In other ways, too, they remained particularly unresponsive to the wishes of Communist authorities. A full evaluation of Soviet social engineering should also take account of the costs. As one author summarized the salient human aspects of Soviet agricultural policies during the socialist offensive:

> As a result of collectivization the number of families on the land diminished from 26,000,000 to 21,000,000. This means that 5,000,000 families or approximately 24,000,000 individuals must have left the countryside. Of these the increase in the towns accounts for one half. Twelve millions are not accounted for. A part of them has undoubtedly perished, the other part has found new possibilities in the Far East, in the Arctic, or in Central Asia.

An Evaluation of the Plans

Any over-all judgment of the first three five-year plans is of necessity a complicated and controversial matter, as the writings of Bergson, Grossman, and other economists clearly indicate. The plans did succeed — and succeed strikingly — in developing industry, particularly heavy industry, and in collectivizing agriculture. Skepticism as to the feasibility of the plans, extremely widespread outside the Soviet Union, turned to astonishment and sometimes admiration. To repeat, not only did production greatly increase, but entire new industries appeared, while huge virgin territories, including the distant and difficult far north, began to enter the economic life of the country. Red armed forces, by contrast with the tsarist army, obtained a highly developed industrial and armaments base, a fact which alone justifies the five-year plans, in the opinion of some critics. Moreover, the entire enormous undertaking was carried out almost wholly by internal manpower and financing, except for the very important contribution of several thousand Western specialists in all fields who were invited to help, and some short-term credit extended to the Soviet government by German and other suppliers during the first years of industrialization. Considered by many as Stalin's chimera, the five-year plans proved to be an effective way — if not necessarily the only or the best way — to industrialize a relatively backward country.

Yet the cost was tremendous. Soviet authorities could accomplish their aims only by imposing great hardships on the people and by mobilizing the country in a quasi-military manner for a supreme effort. The very terminology of the five-year plans, with its iron or coal fronts, shock brigades, and constant communiques, spoke of war. Piece work became common and wage differentials grew by leaps and bounds. The new emphasis on "socialist competition" culminated in the Stakhanov movement. In 1935 Alexis Stakhanov, a coal miner in the Donets Basin, was reported to have overfulfilled his daily quota by 1400 per cent in the course of a shift hewing coal. "Stakhanovite" results were soon achieved by other workers in numerous branches of industry. Rewarding the Stakhanovites, whose accomplishments stemmed in different degrees from improved technique, enormous exertion, and co-operation by their fellow workers, the government used their successes to raise general production norms over a period of time. Most workers must have resented this speed-up — some Stakhanovites were actually killed — but they could not reverse it. After the October Revolution, and especially in the '30's, labor unions, to which almost all workers belonged, have served as agencies of the state, to promote its policies and rally the workers behind them, rather than as representatives of labor interests and point of view. Hardships of Soviet life included a desperate shortage of consumers' goods, as well as totally inadequate housing combined with a rigid system of priorities. As a result the black market flourished, and indeed has remained an essential part of the Soviet economic system. Criticisms of the first three five-year plans — in fact, of their successors as well — have also pointed to top-heavy bureaucracy and excessive red tape, to a relatively low productivity per worker and production per inhabitant, to the frequently poor quality of the items produced, and to numerous weaknesses, perhaps outright failure, in agriculture. It can legitimately be asked whether a different regime could have industrialized the country better and with less pain.

For extreme painfulness emerged as a fundamental aspect of the first three five-year plans. While all suffered to some extent, some groups of the population suffered beyond all measure. One such group, as already mentioned, was the kulaks and their families. Another, overlapping but by no means identical with the kulaks, was the inmates of the forced-labor camps. A history of Soviet forced labor remains to be written in full: it has been ignored in Soviet publications as well as by such Soviet sympathizers as Dobb in his interesting and useful studies of the Soviet economy. Scholars who have tried to reconstruct reality have had to do so on the basis of limited and sometimes controversial evidence. Still, after the information provided by numerous former inmates who left the Soviet Union during the Second World War, the researches of such scholars as D. Dallin and Nicolaevsky, the writings of Solzhenitsyn, and other recent material, the basic outlines of the Soviet forced-labor system are reasonably clear. Having begun in the early 'thirties the system encompassed millions of human beings on the eve of the Second World War, in spite of the extremely high mortality rate in the camps. Forced labor was used especially on huge construction projects, such as the Baltic–White Sea and other canals, and for hard work under primitive conditions in distant areas, as in the case of the lumber and gold industries. The political police — from 1922 to 1934 known as the G.P.U. and the O.G.P.U. rather than the Cheka, after 1934 as the N.K.V.D. after the People's Commissariat of Internal Affairs, subsequently as M.V.D. and M.G.B., and since 1954 as K.G.B. — which guarded and administered forced labor, developed veritable concentration-camp empires in the European Russian and Siberian far north, in the Far East, and in certain other areas of the Soviet Union.

The Great Purge

The great purge of the 1930's helped to fill forced-labor camps and formed another major, although perhaps unnecessary, aspect of the five-year plans. It also marked Stalin's extermination of all opposition or suspected opposition and his assumption of complete dictatorial power. Although earlier some engineers and other specialists, including foreigners, had been accused of sabotaging or wrecking the industrialization of the country, the real purge began in December 1934 with the assassination of one of the party leaders who was boss in Leningrad, Serge Kirov, and reached high intensity from 1936 to 1938. The purge eventually became enormous in scope; it was directed primarily against Party members, not against the White Guards or other remnants of the old regime as repressive practices had been before.

The assassin of Kirov, proclaimed to be a member of the Left Opposition, was shot, together with about a hundred alleged accomplices. Revelations at the Twenty-second Party Congress strengthened the suspicions of some specialists that Stalin himself was apparently responsible for Kirov's murder. A Party purge followed. While uncounted people disappeared, the three great public trials featured sixteen Bolshevik leaders, notably Zinoviev and Kamenev, in 1936, another seventeen in 1937, and twenty-one more, including Bukharin and Rykov, in 1938. The accused were charged with association with Trotsky, counterrevolutionary conspiracy, "wrecking," and treasonable alliance with Soviet enemies abroad. Invariably they confessed to the fantastic charges and in all but four cases received the death penalty. Observers and scholars such as Conquest have been trying since to find reasons for the staggering confessions in everything from torture to heroic loyalty to Soviet communism. The purge spread and spread, affecting virtually all Party organizations and government branches, the army, where Marshal Tukhachevsky and seven other top commanders perished at the same time, and almost every other prominent institution, including the political police itself. It reached its height when Nicholas Ezhov — hence *Ezhovshchina* — directed the N.K.V.D. from late September 1936 until the end of July 1938. Fainsod has written the best summary of these events:

> The period of the Yezhovshchina involved a reign of terror without parallel in Soviet history. Among those arrested, imprisoned, and executed were a substantial proportion of the leading figures in the Party and governmental hierarchy. The Bolshevik Old Guard was destroyed. The roll of Yezhov's victims included not only former oppositionists but many of the most stalwart supporters of Stalin in his protracted struggle with the opposition. No sphere of Soviet life, however lofty, was left untouched. Among the purged Stalinists were three former members of the Politburo . . . and three candidate members. . . . An overwhelming majority of the members and candidates of the Party Central Committee disappeared. The senior officer corps of the armed forces suffered severely. According to one sober account "two of five marshals of the Soviet Union escaped arrest, two of fifteen army commanders, twenty-eight of fifty-eight corps commanders, eighty-five of a hundred ninety-five divisional commanders, and a hundred and ninety-five of four hundred and six regimental commanders." The havoc wrought by the purge among naval commanding personnel was equally great. The removal of Yagoda from the NKVD was accompanied by the arrest of his leading collaborators. . . . The Commissariat of Foreign Affairs and the diplomatic service were hard hit. . . . Almost every commissariat was deeply affected.
>
> The purge swept out in ever-widening circles and resulted in wholesale removals and arrests of leading officials in the union republics, secretaries of the Party, Komsomol, and trade-union apparatus, heads of industrial trusts and enterprises, Comintern functionaries and foreign Communists, and leading writers, scholars, engineers and scientists. The arrest of an important figure was followed by the seizure of the entourage which surrounded him. The apprehension of members of the entourage led to the imprisonment of their friends and acquaintances. The endless chain of involvements and associations threatened to encompass entire strata of Soviet society. Fear of arrest, exhortations to vigilance, and perverted ambition unleashed new floods of denunciations, which generated their own avalanche of cumulative interrogations and detentions. Whole categories of Soviet citizens found themselves singled out for arrest because of their "objective characteristics." Old Bolsheviks, Red Partisans, foreign Communists of German, Austrian, and Polish extraction, Soviet citizens who had been abroad or had relations with foreign countries or foreigners, and "repressed elements" were automatically caught up in the NKVD web of wholesale imprisonment. The arrests mounted into the

Michael Lomonosov

Dmitrii Mendeleev

Nicholas Lobachevsky

Ivan Pavlov

Maxim Gorky and Theodore Chaliapin

Anton Chekhov

Nicholas Gogol

Nicholas Chernyshevsky

Michael Lermontov

Alexander Pushkin

Boris Pasternak

Alexander Herzen

Waslaw Nijinsky

Dmitrii Shostakovich

Modest Musorgsky

Peter Tchaikovsky

Ernest Ansermet, Serge Diaghilev, Igor Stravinsky, and Serge Prokofiev

Leon Trotsky

Joseph Stalin

Lenin

Nikita Khrushchev

Stalin's Funeral. *From right:* Khrushchev, Beria, Chou En-lai, Malenkov, Voroshilov, Kaganovitch, Bulganin, Molotov.

Soviet Leaders at Kremlin Meeting of the Supreme Soviet Celebrating the Fiftieth Anniversary of the Bolshevik Revolution, November 4, 1967. *From left:* Brezhnev, Kosygin, Podgorny, Suslov.

millions; the testimony of the survivors is unanimous regarding crowded prison cells and teeming forced labor camps. Most of the prisoners were utterly bewildered by the fate which had befallen them. The vast resources of the NKVD were concentrated on one objective—to document the existence of a huge conspiracy to undermine Soviet power. The extraction of real confessions to imaginary crimes became a major industry. Under the zealous and ruthless ministrations of NKVD examiners, millions of innocents were transformed into traitors, terrorists, and enemies of the people.

Orders were even issued to arrest a certain percentage of the entire population. The total number of those taken by the political police has been estimated at at least eight million. Before the great purge had run its course, Ezhov himself and many of his henchmen fell victim to it after Lavrentii Beria, a Georgian like Stalin, took control of the N.K.V.D.

Stalin's System

The great purge assured Stalin's complete control of the Party, the government, and the country. As frequently pointed out, the Old Bolsheviks, members of the Party before 1917 and thus not creatures of the general secretary, suffered enormous losses. Virtually all of those who had at any time joined any opposition to Stalin perished. But, as already mentioned, some devoted Stalinists also fell victim to the purge; it was on the whole that group, together with the military men, that was posthumously vindicated by Khrushchev. When the Eighteenth All-Union Party Congress gathered in 1939, Old Bolsheviks composed only about 20 per cent of its membership compared to 80 per cent at the Seventeenth Congress in 1934. Moreover, except for a few lieutenants of Stalin, such as Viacheslav Molotov, born Skriabin, almost no leaders of any prominence were left. For example, with the exception of Stalin himself and of Trotsky, who was murdered in 1940, Lenin's entire Politburo had been wiped out.

Absolute personal dictatorship set in. While the Politburo remained by far the most important body in the country, because its fourteen or so members and candidate members were the general secretary's immediate assistants, there is much evidence that they, too, implicitly obeyed their master. Other Party organizations followed the instructions they received as best they could to the letter. Significantly, no Party congress was called between 1939 and 1952. The so-called "democratic centralism" within the Party, that is, the practice of discussing and debating issues from the bottom up, but, once the Party line had been formed, executing orders as issued from the top down, became a dead letter: even within the Communist party framework no free discussion could take place in the Soviet Union, and almost every personal opinion became dangerous.

Through the Communist party apparatus and the several million Party members, as well as through the political police, Stalin supervised the government machine and controlled the people of the country. The peculiar relationship between the Party and the government in the Soviet Union, in which the Party is the leading partner as well as a driving force in carrying out state policies, has been elucidated in such studies as Fainsod's analysis of the Soviet regime in the Smolensk area, based on the Smolensk Party archives which had fallen into Western hands, and Armstrong's investigation of the Communist party in the Ukraine. Not in vain did Article 126 of the Soviet Constitution of 1936, still in operation, declare:

> . . . the most active and most politically conscious citizens in the ranks of the working class and other sections of the working people unite in the Communist Party of the Soviet Union (Bolsheviks), which is the vanguard of the working people in their struggle to strengthen and develop the socialist system and is the leading core of all organizations of the working people, both public and state.

The Party, as will be shown in a later chapter, has in fact dominated the social and cultural, as well as the political and economic life in the Soviet Union.

The Constitution of 1936

The Stalinist Constitution of 1936, which replaced the constitution of 1924 and was officially hailed as marking a great advance in the development of the Union of Soviet Socialist Republics, retained in effect the "dictatorship of the proletariat," exercised by the Communist party and its leadership, specifically Stalin. At the same time it was meant to reflect the new "socialist" stage achieved in the Soviet Union, based on collective ownership of the means of production and summarized in the formula: "From each according to his ability, to each according to his work." It gave the ballot to all Soviet citizens — for no "exploiters" remained in the country — and made elections equal, direct, and secret. In fact, it emphasized democracy and contained in Chapter X a long list of civil rights as well as obligations. Yet, as has often been demonstrated, the permissiveness of the new constitution never extended beyond the Communist framework. Thus Chapter I affirmed that the basic structure of Soviet society could not be challenged. The civil liberty articles began: "In conformity with the interests of the working people, and in order to strengthen the socialist system . . ." — and could be considered dependent on this condition. The Communist party, specifically recognized by the Constitution, was the only political group allowed in the Soviet Union. Still more important, the niceties of the Constitution of 1936 mattered little in a country ruled by an absolute dictator, his party, and his police. Ironically, the height of the great purge followed the introduction of the Constitution.

The Union of Soviet Socialist Republics remained a federal state, its component units being increased to eleven: the Russian Soviet Federated Socialist Republic, and ten Soviet Socialist Republics, namely, the Ukraine, Belorussia or White Russia, Armenia, Georgia, and Azerbaijan in Transcaucasia, and the Kazakh, Kirghiz, Tajik, Turkmen, and Uzbek republics in Central Asia. While the larger nationalities received their own union republics, smaller ones obtained, in descending order, autonomous republics, autonomous regions, and national areas. Altogether, fifty-one nationalities were granted some form of limited statehood. Yet, like much else in the constitution, this arrangement is largely a sham: while important in terms of cultural autonomy — a subject to be discussed in a later chapter — as well as in terms of administration, in fact it gave no political or economic independence to the local units at all. The Soviet Union has been one of the most highly centralized states of modern times.

A bicameral supreme Soviet replaced the congresses of Soviets as the highest legislative body of the land. One chamber, the Union Soviet, represented the entire Soviet people and was to be elected in the proportion of one deputy for every 300,000 inhabitants. The other, the Soviet of Nationalities, represented the component national groups and was to be elected as follows: twenty-five delegates from each union republic, eleven from each autonomous republic, five from each autonomous region, and one from each national area. The two chambers received equal rights and parallel functions, exercising some of them jointly and some separately. Elected for four years — although with the Second World War intervening the second Supreme Soviet was not elected until 1946 — the Supreme Soviet meets twice a year, usually for no more than a week at a time. In the interims between sessions a Presidium elected by the Soviet has full authority. Until very recently, Supreme Soviets unanimously approved all actions taken by their Presidiums. In the words of one commentator: "The brevity of the sessions, already noted, the size of the body, and the complexity of its agendas are all revealing as to the actual power and place of the Supreme Soviet." Still more revealing has been the acquiescence and obsequiousness of the Soviet legislature in its dealings with Soviet rulers.

In the Constitution of 1936 the executive authority continued to be vested in the Council of People's Commissars, which had to be confirmed by the Supreme Soviet. Commissariats were of three kinds: Union — that is, central — Republican, and a combination of the two. Their number exceeded the number of ministries or similar agencies in other countries because many branches of Soviet economy came to be managed by separate commissariats. In general, heavy industry fell under central jurisdiction, while light industry was directed by Union-Republican commissariats.

The Soviet legal system, while extensive and complicated, has served Party and state needs both explicitly and implicitly and has had only an extremely limited independent role in Soviet society. Besides, the political police has generally operated outside even Soviet law. It might be added that the Soviet central government has served as the model for the governments of the union republics, although the latter have established single-chamber, rather than bicameral, legislatures by omitting a chamber for nationalities.

Stalin's Soviet regime, which took its definitive shape in the thirties, was to undergo before long the awesome test of the Second World War. In a sense it passed the test, although it can well be argued that the war raised more questions about the regime than it settled. But, before turning to the Second World War, it is necessary to summarize Soviet foreign policy from the time of Brest-Litovsk and Allied intervention to the summer of 1941.

XXXVIII

* * * * * * * * * *

SOVIET FOREIGN POLICY, 1921–41, AND THE

SECOND WORLD WAR, 1941–45

"Soldiers! isn't Moscow behind us?
Let us then die on the approaches to Moscow,
As our brothers knew how to die!"

LERMONTOV

Our Government committed no few mistakes; at times our position
was desperate, as in 1941–42, when our army was retreating,
abandoning our native villages and towns in the Ukraine, Byelorussia,
Moldavia, the Leningrad Region, the Baltic Region, and the Karelo-
Finnish Republic, abandoning them because there was no other
alternative. Another people might have said to the government: You
have not come up to our expectations. Get out. We shall appoint
another government, which will conclude peace with Germany and
ensure tranquillity for us. But the Russian people did not do that,
for they were confident that the policy their Government was
pursuing was correct; and they made sacrifices in order to ensure
the defeat of Germany. And this confidence which the Russian
people displayed in the Soviet Government proved to be the decisive
factor which ensured our historic victory over the enemy of mankind,
over fascism.
 I thank the Russian people for this confidence!
 To the health of the Russian people!

STALIN

Soviet foreign policy — still, with its nuclear capability, one of the
concerns of the world today — can be considered in several contexts. To
begin with, there is the Marxist-Leninist ideology. True, Marxism did not
provide any explicit guidance for the foreign relations of a Communist state.
In fact, it preached a world revolution that would eliminate foreign policy
altogether. That Lenin and his associates had to conduct international relations after their advent to power represented, in Marxist terms, one of several major paradoxes of their position. Not surprisingly, they assumed for
months and even for a few years the imminence of a revolution that would
destroy the entire capitalist world system. The alternative appeared to
be their own immediate destruction by the capitalists. When neither happened, the Soviet leadership, in foreign relations as in home affairs,
proceeded to adapt ideology to circumstances. Marxism supplied the
goal of world revolution, although the time of that revolution could no
longer be predicted with exactitude. Marxism, especially as developed by

509

Lenin, with such key concepts as finance capitalism and imperialism, provided also the framework within which the Soviet leadership sought to understand and interpret the world.

However, when Lenin and his associates seized power in Petrograd, they inherited an international position and interests that had nothing in common with Marxism. The Bolsheviks did their best to break the ties with tsarist Russia, repudiating treaties and debts and publishing secret diplomatic documents. Still, they could not entirely divest the country of its past or separate the communist from the noncommunist aspects of their new role in the world. In fact, as the Soviet regime developed and after Soviet Russia explicitly became the center of Communist interest following the inauguration of the First Five-Year Plan, Soviet foreign policy evolved, in the opinion of many observers, in the direction of traditionalism and nationalism, acquiring a pronounced "Russian" character. Or, to make a different emphasis and suggest yet another context for Soviet foreign relations, the U.S.S.R. can be analyzed simply as a gigantic modern state, and its foreign policy as a product of such considerations of *Realpolitik* as security, rather than considerations of Marxist ideology or of national tradition.

The dichotomy between revolutionary idealism and conservative nationalism, between the Communist International and the Commissariat for Foreign Affairs, and even between Party and state interests has been most frequently used to explain Soviet foreign policy. Often the issue is presented also in terms of a gradual shift from a revolutionary to a more traditional position, whether the change is said to have occurred in the twenties, the thirties, or during the Second World War. This approach, it should be pointed out, although at times enlightening, has its grave dangers. We have no documentary knowledge of how Soviet foreign policy was made: this fact has caused some cautious scholars, such as Beloff, to omit this subject from their discussion of Soviet foreign relations, while it has helped other writers to ascribe Soviet actions on the international scene to everything from an all-embracing Communist conspiracy to a return of alleged long-term Russian imperialism. On the basis of our limited information, it seems best not to emphasize dichotomies and splits in a remarkably monolithic system, which tolerated no deviation. It seems best also to allow generously for the importance of Communist ideology in Soviet foreign policy. The crux of the matter may be not whether the Soviet leaders were fanatics or realists, idealists or cynics, but the fact that, whatever they were they thought naturally in Marxist categories. For example, the argument that Soviet foreign policy should be understood primarily in terms of defense, as a reaction to the threat from the outside, has some validity. But it should be realized that this outside threat, the menace of the capitalist world, loomed large to the men in the Kremlin not only because of the facts of the case, but also as a fundamental tenet of their ideology. Perhaps we have become too accustomed to thinking of international relations as intercourse among a considerable number of rather similar states, a situation that prevailed in the eighteenth and nineteenth centuries. In certain respects the world scene after 1917 appeared to bear a greater resemblance to the struggle between Christian and Moslem in the Middle Ages or to the age of the Reformation and the Counter-Reformation — those epochal rivalries eventually bogged down, to be sure, in attenuation, particularism, and compromise. The "new thinking" in Soviet foreign policy may well have represented such an attenuation or shift from Marxist orthodoxy. According to Edward Shevardnadze, it started from the realization that the old approach put almost the entire world into the enemy camp. Moreover, the ongoing collapse of the Soviet Union makes Soviet foreign policy, as well as other Soviet activities, increasingly a matter of the past, with the immediate attention of the world moving to the nature and behavior of the emerging successor states.

Soviet Foreign Policy in the Twenties

When Trotsky became commissar of war in 1918, his assistant, George Chicherin, replaced him as commissar of foreign affairs. Chicherin was to occupy that position until 1930; because of Chicherin's ill health, however, his eventual successor, Maxim Litvinov, directed the commissariat from 1928. Chicherin was of gentry origin and for many years of Menshevik, rather than Bolshevik, affiliation. In fact, he never entered the narrow circle of Communist leaders. Nevertheless, because of his ability and special qualifications for the post — Chicherin had originally begun his career in the tsarist diplomatic service and was a fine linguist with an excellent knowledge of the international scene — he was entrusted for over a decade with the handling of Soviet foreign policy, although, to be sure, he worked under the close supervision of Lenin, Stalin, and the Politburo. As mentioned previously, positions of real power in the Soviet system have been at the top of the Party hierarchy, not in any of the commissariats.

One of Chicherin's main tasks was to obtain recognition for the Soviet Union and to stabilize its position in the world. In spite of transitory successes in Hungary and Bavaria, Communist revolutions had failed outside Soviet borders. On the other hand, with the defeat of the White movement and the end of Allied intervention, the Bolshevik regime appeared to be firmly entrenched in Russia. "Coexistence" became a reality, and both sides sought a suitable modus vivendi. Yet the Soviet Union supported the Third or Communist International — called the Comintern — established in 1919 with Zinoviev as chairman, and it refused to pay tsarist debts or compensate foreigners for their confiscated property, demanding in its turn huge reparations for Allied intervention. In particular the Comintern, composed of Communist parties scattered throughout the world, who were bent on subversion and revolution and were clearly directed from the Soviet Union in Soviet interests, constituted a persistent obstacle to normal diplomatic relations. Most other states, on their side, looked at Soviet Russia with undisguised hostility and suspicion.

The Soviet Union managed to break out of isolation in the spring of 1922. A Soviet delegation attended then for the first time an international economic conference, held in Genoa. Although the conference itself produced no important results, bogging down on the above-mentioned issues of debts and reparations, among others, Soviet representatives used the occasion to reach an agreement with Germany. The Treaty of Rapallo of April 16, 1922, supplemented later by a commercial agreement, established economic co-operation between the Soviet Union and Germany and even led to some political and military ties. It lasted until after Hitler's advent to power. While the Treaty of Rapallo produced surprise and indignation in many quarters, its rationale was clear enough and, as in the case of most other Soviet agreements, it had nothing to do with the mutual sympathy or antipathy of the signatories: both Soviet Russia and Germany were outcasts in the post-Versailles world, and they joined hands naturally for mutual advantage.

Early in 1924 Great Britain formally recognized the Soviet Union; it was followed by France, Italy, Austria, Sweden, Norway, Denmark, Greece, Mexico, and China before the end of the year. In 1925 Japan established normal relations with the U.S.S.R., evacuating at last the Russian part of the island of Sakhalin, although retaining certain oil, coal, and timber concessions there. The recognition of Soviet Russia by many states marked simply their acceptance of the existence of the Bolshevik regime, accompanied sometimes by hopes of improving trade relations, rather than any real change in their attitude toward the U.S.S.R. Lloyd George's remark on trading even with cannibals has often been quoted. Moreover, other countries, including notably the United States and most Slavic states of eastern Europe, continued to ignore the Soviet Union and refuse it recognition. Still, all in all, Chicherin succeeded in bringing Soviet Russia into the diplomatic community of nations.

That the course of Soviet foreign policy could be tortuous and even paradoxical became clear in the case of China. There Stalin chose to support the Kuomintang, the nationalist movement of Sun Yat-sen and Sun's successor Chiang Kai-shek, sending hundreds of military specialists to help the Nationalists and directing the Chinese Communists to follow "united front" tactics. For a time Communist infiltration appeared successful, and Soviet position and prestige stood high in China. But in 1927 as soon as Chiang Kai-shek had assured himself of victory in the struggle for the control of the country, he turned against the Communists, massacring them in Shanghai and evicting Soviet advisers. When the Chinese Communists, on orders from Moscow, retaliated with a rebellion in Canton, they were bloodily crushed. Yet, although defeated in China, the Soviet Union managed to establish control over Outer Mongolia after several changes of fortune. Also, in the mid-twenties it concluded useful treaties of neutrality and friendship with Turkey, Persia, and Afghanistan. It should be added that the Bolshevik regime renounced the concessions and special rights obtained by the tsarist government in such Asiatic countries as China and Persia. But it held on to the Chinese Eastern Railway, weathering a conflict over it with the Chinese in 1929.

Soviet Foreign Policy in the Thirties

Chicherin's efforts in the '20's to obtain recognition for his country and to stabilize Soviet diplomatic relations developed into a more ambitious policy in the '30's. Devised apparently by Stalin and the Politburo and executed by Maxim Litvinov, who served as commissar for foreign affairs from 1930 until 1939, the new approach aimed at closer alliances with *status quo* powers in an effort to check the mounting aggression of the "have-nots." It culminated in the Soviet entrance into the League of Nations and Litvinov's emphasis on disarmament and collective security. To appreciate the shift in Soviet tactics, it should be realized that the Bolshevik leadership had for a long time regarded Great Britain and France as their main enemies and the League of Nations as the chief international agency of militant imperialism. Indeed, the Politburo placed its hopes, it would seem, in the expected quarrels among leading capitalist powers, and in particular in a war between Great Britain and the United States! Under the circumstances, the Japanese aggression that began on the Chinese mainland in 1931 and especially the rise of Hitler to power in Germany in January 1933, together with his subsequent policies, came as rude shocks. The Soviet government, caught quite unprepared by the appearance of Hitler, was slow to appreciate the new danger — in all fairness it should be added that other governments, although not handicapped by Marxist blinkers, were equally surprised and slow. Yet, once the handwriting on the wall became clear, the Bolshevik leadership did what it could to counteract the Fascist enemy, for that purpose mobilizing Communist parties all over the world as well as using orthodox diplomatic means. Hence the celebrated "popular fronts" of the 1930's and the strange *rapprochement* between the U.S.S.R. and Western democracies as well as a new cordiality between the U.S.S.R. and Chiang Kai-shek. Based on dire expediency rather than on understanding or trust and vitiated by mistakes of judgment on all sides, the *rapprochement* with the West collapsed in a catastrophic manner in 1938 and 1939 to set the stage for the Second World War.

As early as 1929 the Soviet Union used the occasion of the making of the Kellogg-Briand Pact outlawing war to formulate the Litvinov Protocol, applying the pact on a regional basis. Poland, Rumania, Latvia, Estonia, Lithuania, Turkey, Persia, and the Free City of Danzig proved willing to sign the Protocol with the U.S.S.R. In 1932 the Soviet Union concluded treaties of nonaggression with Poland, Estonia, Latvia, and Finland, as well as with France. In 1933 the United States finally recognized the Soviet Union, obtaining from the Soviets the usual unreliable promise to desist from Communist propaganda in the U.S. In the spring of 1934 the nonaggression pacts with Poland and the Baltic states were expanded into ten-year agreements. In the summer of that year the Soviet government signed treaties with Czechoslovakia and Rumania — the establishment of diplomatic relations with the latter country marked the long delayed, temporary Soviet reconciliation to the loss of Bessarabia. And in the autumn of 1934 the U.S.S.R. joined the League of Nations.

The following year witnessed the conclusion of the Soviet-French and the Soviet-Czech alliances. Both called for military aid in case of an unprovoked attack by a European state. The Soviet-Czech treaty, however, added the qualification that the U.S.S.R. was obliged to help Czechoslovakia only if France, which had concluded a mutual aid treaty with the Czechs, would come to their assistance. France, it is worth noting, failed to respond to Soviet pressure for a precise military convention, while neither Poland nor Rumania wanted to allow the passage of the Red Army to help the Czechs in case of need.

Also in 1935 the Third International, which had become somewhat less active as a revolutionary force in the course of the preceding years, at its Seventh Congress proclaimed the new policy of popular fronts: Communist parties, reversing themselves, were to co-operate in their respective countries with other political groups interested in checking Fascist aggression, and they were to support rearmament. In its turn the Soviet government demanded in the League of Nations and elsewhere that severe sanctions be applied to aggressors and that forces of peace be urgently mobilized to stop them. Yet both the League and the great powers individually accomplished little or nothing. Italy completed its conquest of Ethiopia, while Japan developed its aggression on the Asiatic mainland. In the summer of 1936 a great civil war broke out in Spain, pitting Franco's Fascist rebels and their allies against the democratic and Leftwing republican government. Once more, the Soviet Union proved eager to stop Fascism, while France and Great Britain hesitated, compromised, emphasized nonintervention, and let the Spanish republic go down.

Whereas Italian divisions and German airmen and tankmen aided Franco, none but Soviet officers and technicians were sent to assist the Loyalists, while the international Communist movement mobilized its resources to obtain and ship volunteers who fought in the celebrated "international brigades." Although much in the Soviet intervention in Spain remains obscure and controversial, studies by Cattell and others demonstrate both the seriousness of the Soviet effort to defeat Franco and the remarkable way in which the Communists, including the secret police, proceeded to extend their hold on republican Spain and to dispose of their rivals. But, with massive Italian and German backing, the insurgents won the bitter civil war in Spain, hostilities ending in the spring of 1939.

The position and prospects of the Soviet Union became graver and graver in the course of the '30's. In November 1936, Germany and Japan concluded the so-called Anti-Comintern Pact aimed specifically against the U.S.S.R. Italy joined the Pact in 1937 and Spain in 1939. In the Far East in 1935 the Soviet Union sold its dominant interest in the Chinese Eastern Railway to the Japanese puppet state of Manchukuo, thus eliminating one major source of conflict. But relations between Japan and the U.S.S.R. remained tense, as Japanese expansion and ambitions grew, while the Soviet leaders continued to send supplies to Chiang Kai-shek as well as to direct and support Communist movements in Asia. In fact, in 1938 and again in 1939 Japanese and Soviet troops fought actual battles on the Manchurian and Mongolian borders, the Red Army better than holding its own and hostilities being terminated as abruptly as they had begun. Hitler's Germany represented an even greater menace to the Soviet Union than Japan. The Führer preached the destruction of communism and pointed to the lands east as the natural area of German expansion, its legitimate *Lebensraum*. Again, as in the cases of Japan and Italy, the Western powers failed to check the aggressor. Following the remilitarization of the Rhineland in 1936, Hitler annexed Austria to the Third Reich in March 1938, making a shambles of the Treaty of Versailles.

Soviet Foreign Policy from September 1938 until June 1941

The climax of appeasement came in September 1938 at Munich. Great Britain and France capitulated to Hitler's demand for Germany's annexation of the Sudetenland, a largely ethnically German area of Czechoslovakia; Chamberlain and Daladier flew to Munich and sealed the arrangement with Hitler and Mussolini. The unpreparedness and unwillingness of the Western democracies to fight, rather than any collusion of the West with Hitler against the U.S.S.R., motivated the Munich surrender. Still, the extreme Soviet suspicion of the settlement can well be understood, especially since the Soviet government was not invited to participate in it. Although it had expressed its readiness to defend Czechoslovakia, the Soviet Union had been forced to remain a helpless bystander when France failed to come to the aid of the Czechs and Prague had to accept its betrayal by the great powers. Moreover, after Munich the Franco-Russian alliance appeared to mean very little, and the U.S.S.R. found itself, in spite of all its efforts to promote collective security, in highly dangerous isolation.

His appetite whetted by appeasement, Hitler in the meantime developed further aggressive designs in eastern Europe. In March 1939 he disposed of what remained of Czechoslovakia, establishing the German protectorate of Bohemia and Moravia and another one of Slovakia. This step both destroyed the Munich arrangement and made plain Nazi determination to expand beyond ethnic German boundaries. Next Hitler turned to Poland, demanding the cession of Danzig to Germany and the right of extraterritorial German transit across the Polish "corridor" to East Prussia. The alternative was war.

Poland, however, did not stand alone against Germany in the summer of 1939. France and Great Britain finally saw the folly of appeasement after Germany had seized the remainder of Czechoslovakia. At the end of March they made clear their determination to fight if Poland were attacked. As war clouds gathered, the position of the Soviet Union became all the more significant. In May Molotov replaced Litvinov as commissar for foreign affairs, retaining at the same time his office of Chairman of the Council of People's Commissars, equivalent to prime minister, as well as his membership in the Politburo. Thus for the first time since Trotsky in 1918 a Communist leader of the first rank took charge of Soviet foreign policy. Moreover, in contrast to his predecessor Litvinov, Molotov had not been personally committed to collective security and, therefore, could more easily undertake a fresh start. In retrospect commentators have also noted the fact that Molotov, again in contrast to Litvinov, was not Jewish. After an exchange of notes in the spring of 1939, Great Britain and France began in the summer to negotiate with the U.S.S.R concerning the formation of a joint front against aggression. But the Western powers failed to come to terms with the Soviet Union, or even to press the negotiations, sending a weak and low-ranking mission to Moscow. The Soviet government, on its side, remained extremely suspicious of the West, especially after the Munich settlement, and eagerly sought ways of diverting impending hostilities away from its borders. On August 23 a German-Russian agreement of strict neutrality was signed in Moscow — secret talks had begun as early as May — an event which produced surprise and shock in the world. Fortified by the pact, Hitler attacked Poland on the first of September. On the third, Great Britain and France declared war on Germany. The Second World War became a reality.

The Bolsheviks and the Nazis hated each other and considered themselves to be irreconcilable enemies. That no illusions were involved in their agreement is indicated, among many other things, by the fact that Molotov, who signed the treaty for the Soviet Union and thus represented the "pro-German orientation," retained his position and Stalin's favor after Hitler attacked the U.S.S.R. Yet both parties to the pact expected to gain major temporary advantages by means of it. Germany would be free to fight Western powers. The Soviet Union would escape war, at least for the time being. Besides, the agreement was accompanied by a secret protocol dividing the spheres of influence and enabling the Soviet Union to expand in eastern Europe.

The Red Army occupied eastern Poland, incorporating its White Russian and Ukrainian areas into the corresponding Soviet republics. Next the Soviet government signed mutual assistance pacts with Estonia, Latvia, and Lithuania, obtaining a lease of Baltic bases. But in July 1940 these states were occupied by Soviet troops and, following a vote of their beleaguered parliaments, they were incorporated into the U.S.S.R. as union republics — a procedure that the Western democracies have with excellent reasons failed to recognize to this day. Finland was more troublesome: the Finnish government turned down the Soviet demand that they move the Finnish boundary some twenty miles further away from Leningrad, abandoning a Finnish defense line, in exchange for a strip of Karelia; a war between the two countries resulted and lasted from the end of November 1939 until mid-March 1940. In spite of the heroic Finnish defense and the surprising early reverses of the Red Army, the Soviet Union eventually imposed its will on Finland. Finally, in the summer of 1940 the U.S.S.R. utilized its agreement with Germany to obtain from Rumania, by means of an ultimatum, the disputed region of Bessarabia as well as northern Bukovina. The new Moldavian Soviet Socialist Republic was formed from the territory acquired from Rumania. In April 1941 the Soviet Union signed a five-year nonaggression treaty with Japan, which had chosen to expand south rather than into Siberia.

But, although the Soviet government did not know it, time was running short for its efforts to strengthen its position on the European and Asiatic continents. Following his stunning victory in the west in the summer of 1940, Hitler decided to invade the Soviet Union. In December he issued precise instructions for an attack in May 1941. The defeat that Germany suffered in the autumn in the aerial Battle of Britain apparently only helped convince the Nazi dictator that he should strike his next major blow in the east. The schedule, however, could not quite be kept. A change of government in Yugoslavia made the Germans invade Yugoslavia as well as Greece, which had stopped an earlier Italian offensive. While brilliantly successful, the German campaign in the Balkans, together with a certain delay in supplying the German striking force with tanks and other vehicles, postponed by perhaps three weeks the invasion of Soviet Russia. The new date was June 22, and on that day German troops aided by Finnish, Rumanian, and other units attacked the U.S.S.R. along an enormous front from the Baltic to the Black Sea.

The Soviet Union in the Second World War

The blow was indeed staggering. Hitler threw into the offensive some 175 divisions, including numerous armored formations. A huge and power- ful air force closely supported the attack. Moreover, perhaps surprisingly, the German blow caught the Red Army off guard. Apparently, although Stalin and the Politburo were preparing for war, they had ignored Western warnings as well as their own intelligence and did not expect such an early, sudden, and powerful offensive. The Germans aimed at another *Blitzkrieg*, intending to defeat the Russians within two or three months or in any case before winter. Although it encountered some determined resistance, the German war machine rolled along the entire front, particularly in the north towards Leningrad, in the center towards Moscow, and in the south towards Kiev and Rostov-on-Don. Entire Soviet armies were smashed and taken prisoner at Bialystok, Minsk, and Kiev, which fell in September. The south- ern wing of the invasion swept across the Ukraine. In the north, Finnish troops pushed to the Murmansk railroad, and German troops reached, but could not capture, Leningrad. The city underwent a two-and-a-half-year siege, virtually cut off from the rest of the country; its population was de- creased by starvation, disease, and war from four to two and a half million. Yet the city would not surrender, and it blocked further German advance north.

The central front proved decisive. There the Germans aimed their main blow directly at Moscow. But they were delayed in fierce fighting near Smolensk. The summer *Blitzkrieg* became a fall campaign. Hitler in- creased the number of his and his allies' divisions in Russia to 240 and pushed an all-out effort to capture the Soviet capital. In the middle of October German tanks broke through the Russian lines near Mozhaisk, some sixty miles from Moscow. Stalin and the government left the city for Kuibyshev, formerly Samara, on the Volga. Yet, instead of abandon- ing Moscow as in 1812, its defender, Marshal George Zhukov, had his troops fall slowly back on the capital, reducing the German advance to a crawl. The Germans proceeded to encircle the city on three sides, and

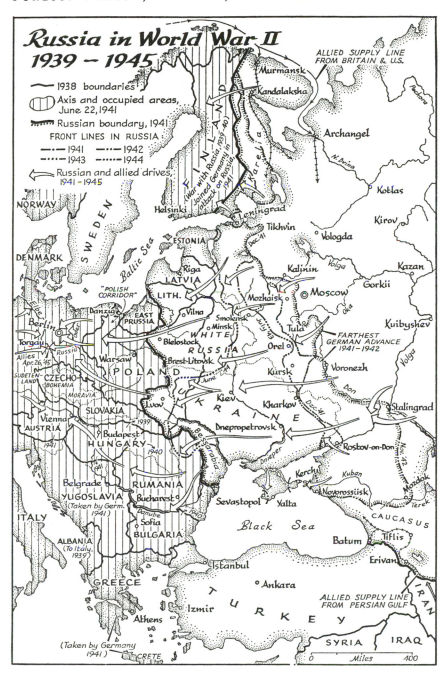

Russia in World War II
1939 – 1945

—— 1938 boundaries
Axis and occupied areas,
June 22, 1941
Russian boundary, 1941
FRONT LINES IN RUSSIA:
—·— 1941   —··— 1942
—···— 1943   —···— 1944
Russian and allied drives,
1941 – 1945

NORWAY
SWEDEN
DENMARK
Baltic Sea
"POLISH CORRIDOR"
Berlin
Elbe
Torgau
Allies Apr. 26, 45
SUDETEN-LAND
CZECHO-BOHEMIA
MORAVIA
SLOVAKIA
Vienna
AUSTRIA
1941
HUNGARY
Budapest
1940
Belgrade
YUGOSLAVIA
(Taken by Germ. 1941)
ITALY
ALBANIA
(To Italy, 1939)
GREECE
Athens
(Taken by Germany 1941)   CRETE
Izmir

FINLAND
(War with Russia 1939-40)
(Joined Germany in attack on Russia 1941)
Karelia
Helsinki
Leningrad
ESTONIA
Riga
LATVIA
LITH.
Vilna
Danzig
EAST PRUSSIA
Warsaw
POLAND
Brest-Litovsk
Bielostock
WHITE RUSSIA
Minsk
Smolensk
Lvov
Kiev
U K R A I N E
Dnepropetrovsk
Bessarabia
Dnieper
1939
1940
RUMANIA
Bucharest
Sofia
BULGARIA
Istanbul
Ankara
T U R K E Y
SYRIA
Sevastopol
Yalta
Kerch
Black   Sea

Murmansk
Kandalaksha
ALLIED SUPPLY LINE
FROM BRITAIN & U.S.
Pechora
Archangel
N. Dvina
Kotlas
Kirov
Vologda
Tikhvin
Dec. 41
Kalinin
Volga
Kazan
Gorkii
Mozhaisk
Moscow
Oka
Kuibyshev
Tula
Orel
July 42
FARTHEST
GERMAN ADVANCE
1941–1942
Kursk
Voronezh
Don
Volga
Kharkov
Dec. 42
Stalingrad
Rostov-on-Don
Nov. 42
Kuban
Novorossiisk
Terek
Mosdok
CAUCASUS
Batum
Tiflis
Erivan
IRAN
IRAQ
ALLIED SUPPLY LINE
FROM PERSIAN GULF

0   Miles   400

they came to within twenty miles of it, but no further. Late in November the Red Army started a counteroffensive against the extremely extended German lines on the southern front, recapturing Rostov-on-Don at the end of the month. In early December it struck on the central front, attacking both north and south of Moscow as well as in the Moscow area itself. The Germans suffered enormous losses and had to retreat. Winter came to play havoc with unprepared German troops and to assist the Russians. On January 20 the Red Army recaptured Mozhaisk, thus eliminating any immediate threat to Moscow. But German troops had to retreat much further west before they could stabilize the front. In fact, its lines overextended, its troops unequipped for cold weather and exhausted, the German army probably came near complete collapse in the winter of 1941/42. Some specialists believe that only Hitler's frantic determination to hold on prevented a catastrophic withdrawal. As it was, the German army gave up about one hundred thousand square miles of Soviet territory, but retained five hundred thousand when fighting finally quieted down.

In retrospect it seems clear that, in spite of its many splendid victories, the great German campaign of 1941 in Russia failed. The Red Army remained very much in the field, and the *Blitzkrieg* turned into a long war on an enormous front. Quite possibly Hitler came close to crushing the Soviet Union in 1941, but he did not come close again. Taking into account Soviet resources and the determination to resist, the Nazis had to win quickly or not at all. German losses in their initial eastern campaign, large in quantity, were still more damaging in quality: the cream of German youth lined the approaches to Moscow.

Furthermore, although the Soviet Union bore the brunt of Nazi armed might from the summer of 1941 until the end of the Second World War in Europe, it certainly did not fight alone. Churchill welcomed Soviet Russia as an ally the day of the German attack — although shortly before he had been ready to wage war against the U.S.S.R. in defense of Finland. Great Britain and the United States arranged to send sorely needed supplies to the Soviet Union; and after the Japanese strike at Pearl Harbor on December 7, 1941, the United States became a full-fledged combatant. In spite of German submarines and aircraft and the heavy losses they inflicted, British convoys began to reach Murmansk and Archangel in the autumn of 1941, while American aid through Persia started to arrive in large quantity in the spring of 1942. More important, the Axis powers had major enemies to fight in Africa, and eventually in southern and western Europe, as well as in the east.

The second great German offensive in Russia, unleashed in the summer of 1942, was an operation of vast scope and power, even though it was more limited in its sweep and resources than the original attack of 1941: in 1942 the Germans and their allies used about 100 divisions and perhaps a million men in an attack along the southern half of the front, from Voronezh to the Black Sea. Having occupied the Kerch area and captured Sevastopol after a month of bitter fighting, the Germans opened their main offensive early in July. They struck in two directions: east toward the Volga, and south toward the Caucasus. Blocked on the approaches to Voronezh, the German commander, Marshal Fedor von Bock had his main army of over 300,000 men cross the Don farther south and drive to the Volga. At the end of August the Nazis and their allies reached Stalingrad.

That industrial city of half a million people, strung along the right bank of the Volga, had no fortifications or other defensive advantages. Yet General Basil Chuikov's 62nd Army, supported by artillery massed on the other bank, fought for every house and every foot of ground. Reduced to rubble, the city became only more impassable to the invaders in spite of all their weapons and aircraft. Both sides suffered great losses. Hitler, who had assumed personal command of the German army in December 1941 and possibly saved his troops from catastrophe in the winter of 1941/42, began to make disastrous strategic errors. He kept pounding at Stalingrad for fruitless weeks and even months and, disregarding professional opinion, would not let his troops retreat even when a Soviet counteroffensive began to envelop them. Eventually, at the end of January 1943, Marshal Friedrich Paulus and some 120,000 German and Rumanian troops surrendered to the Red Army, their attempt to break through to the Volga thus ending in a complete fiasco. The German offensive southward had captured Rostov-on-Don once more and had swept across the northern Caucasus, the attackers seizing such important points as the port of Novorossiisk and the oil center of Mozdok. But again the extended German lines crumbled under Zhukov's counteroffensive in December. The invaders had to retreat fast into the southern Ukraine and the Crimea and were fortunate to extricate themselves at all.

After some further retreats and counterattacks in the winter of 1942/43, the Germans tried one more major offensive in Russia the following summer. They struck early in July in the strategic watershed area of Kursk, Orel, and Voronezh with some forty divisions, half of them armored or motorized, totaling approximately half a million men. But after initial successes and a week or ten days of tremendous fighting of massed armor and artillery the German drive was spent, and the Red Army in its turn opened an offensive. Before very long the Red drive gathered enough momentum to hurl the invaders out of the Soviet Union and eventually to capture Budapest, Vienna, Prague, and Berlin, stopping only with the end of the war. The smashing Soviet victory was made possible by the fact that the German forces had exhausted themselves. Their quality began to decline probably about the end of 1941, while the increasing numbers of satellite troops pressed into service, notably Rumanians, could not at all measure up to the German standard. Hitler continued to make mistakes. Time and again, as in the case of Stalingrad, he would not allow his troops to retreat until too late. The Red Army, on the other hand, in spite of its staggering losses, improved in quality and effectiveness. Its battle-tested commanders showed initiative and ability; its weapons and equipment rolled in plentiful supply both from Soviet factories, many of which had been transported eastward and reassembled there, and through Allied aid, while the German forces suffered from all kinds of shortages. As long as they fought on Soviet soil, the Germans had to contend with a large and daring partisan movement in their rear as well as with the Red Army. And they began to experience increasing pressure and defeat on other fronts, as well as from the air, where the Americans and the British mounted a staggering offensive against German cities and industries. The battle of Stalingrad coincided with Montgomery's victory over Rommel in Egypt and Allied landings in Morocco and Algeria. Allied troops invaded Sicily in the summer of 1943 and the Italian mainland that autumn. Finally, on June 6, 1944, the Americans, the British, and the Canadians landed in Normandy to establish the coveted "second front." As the Russians began to invade the Third Reich from the east, the Allies were pushing into it from the west.

The Red Army recovered much of occupied Soviet territory in the autumn of 1943 and in the winter of 1943/44. On April 8, 1944, Marshal Ivan Konev crossed the Pruth into Rumania. In the following months Soviet armies advanced rapidly in eastern and central Europe, while other armies continued to wipe out the remaining German pockets on Soviet soil. Rumania and Bulgaria quickly changed sides and joined the anti-German coalition. The Red Army was joined by Tito partisans in Yugoslavia and in September 1944 entered Belgrade. After some bitter fighting, Red forces took Budapest in February 1945 and Vienna in mid-April. In the north, Finland had to accept an armistice in September 1944. The great offensive into Germany proper began in the autumn of 1944 when Red forces, after capturing Vilna, penetrated East Prussia. It gained momentum in January 1945 when large armies commanded by Konev in the south, Zhukov in the center, and Marshal Constantine Rokossovsky in the north invaded Germany on a broad front. On April 25, 1945, advanced Russian units met American troops at Torgau, on the Elbe, near Leipzig. On the second of May, Berlin fell to Zhukov's forces after heavy fighting. Hitler had already committed suicide. The Red Army entered Dresden on the eighth of May and Prague on the ninth. On that day, May 9, 1945, fighting ceased: the Third Reich had finally surrendered unconditionally to the Allies, first in Rheims on the seventh of May and then formally in Berlin on the eighth.

Urged by its allies and apparently itself eager to participate, the Soviet Union entered the war against Japan on August 8, 1945, three months after the German surrender. By that time Japan had already in fact been defeated by the United States and other powers. The American dropping of an atomic bomb on Hiroshima on the sixth of August and on Nagasaki on the ninth eliminated the need to invade the Japanese mainland, convincing the Japanese government that further resistance was useless. In spite of subsequent claims of Soviet historians and propagandists, the role of the U.S.S.R. in the conflict in the Far East and the Pacific was, therefore, fleeting and secondary at best. Yet it enabled the Red forces to occupy Manchuria, the Japanese part of the island of Sakhalin, and the Kurile islands, and to capture many prisoners — all at the price of considerable casualties, for the Japanese did resist. The formal Japanese surrender to the Allies took place on board the U.S. battleship *Missouri* in Tokyo Bay on September 2, 1945. It marked the end of the Second World War.

Wartime Diplomacy

Diplomacy accompanied military operations. In the course of the war the Soviet Union established close contacts with its allies, in particular with Great Britain and the United States. It accepted the Atlantic Charter formulated by Roosevelt and Churchill in August 1941, which promised freedom, self-determination, and equality of economic opportunity to all countries, and it participated fully in the preparation and the eventual creation of the United Nations Organization. It concluded a twenty-year agreement with Great Britain "for the joint achievement both of victory and of a permanent peace settlement" in June 1942 and later made a treaty with France also.

Of the various high-level conferences of the Allies during the war, the three meetings of the heads of state were the most impressive and important. They took place at Teheran in December 1943, at Yalta in the Crimea in February 1945, and at Potsdam near Berlin in July and August 1945. Stalin, who had assumed the position of prime minister and generalissimo, that is, chief military commander, while remaining the general secretary of the Party, represented the Soviet Union on all three occasions. Roosevelt headed the American delegation at Teheran and Yalta, and Truman, after Roosevelt's death, at Potsdam. Churchill and later Attlee spoke for Great Britain. The heads of the three world powers devoted large parts of their conferences to a discussion of such major issues of the Second World War as the establishment of the "second front" and the eventual entry of the Soviet Union into the struggle against Japan. But, especially as victory came nearer, they also made important provisions for the time when peace would be achieved. These included

among others: the division of Germany into zones of occupation, with
Berlin receiving special status; the acceptance of the incorporation of the
Königsberg district of East Prussia into the Soviet Union; the determination of the Polish eastern frontier, which was to follow roughly the Curzon
Line, Poland being granted an indefinite compensation in the west; the
decision to promote the establishment of democratic governments based
on free elections in all restored European countries; and provisions concerning the formation of the United Nations. Considerable, if largely
deceptive, harmony was achieved. Roosevelt in particular exuded optimism.

Yet even during the war years important disagreements developed among
the Allies. The Soviet Union was bitterly disappointed that the Western
powers did not invade France in 1942 or in 1943. In spite of the importance of contacts with the West and the enormous aid received from there,
Soviet authorities continued to supervise closely all relations with the
outside world and to restrict the movement and activities of foreigners
in the Soviet Union. Perhaps more important, early difficulties and disagreements concerning the nature of postwar Europe became apparent.
Poland served as a striking case in point. After Germany attacked the
Soviet Union, Soviet authorities established relations with the Polish government in exile in London. But the co-operation between the two broke
down before long. The Polish army formed in the Soviet Union was transferred to Iran and British auspices, while the Soviet leadership proceeded
to rely on a smaller group of Left-Wing Poles who eventually organized
the so-called Lublin government in liberated Poland. The historic bitterness between the Poles and the Russians, the problem of the frontier,
and other controversial issues were exacerbated by the events of the war
years. In April 1943, the German radio announced to the world the
massacre by the Reds of thousands of Polish officers in the Katyn Forest
near Smolensk before the capture of that area by German troops. This
charge, which led to the break in relations between Moscow and the
Polish government in London, has now been confirmed. Again, when the
Red Army reached the Vistula in August 1944, it failed to assist a desperate Polish rebellion against the Germans in Warsaw, which was finally
crushed in October. In this manner it witnessed the annihilation of the anti-
German, but also anti-Soviet, Polish underground. The official assertion
that Red troops could not advance because they had exhausted their supplies and needed to rest and regroup had its grounds. But Soviet authorities
would not even provide airstrips for Allied planes to help the Poles. Under
the circumstances the Yalta decision to recognize the Lublin government
expanded by several representatives of the London Poles and to hold free
elections and establish a democratic regime in Poland proved unrealistic
and amounted in the end to a Western surrender to Soviet wishes. This and

other grave problems of postwar eastern Europe are treated in the next chapter.

The Soviet Union in the Second World War: An Evaluation

The Soviet performance in the Second World War presents a fascinating picture of contrasts. Seldom did a country and a regime do both so poorly and so well in the same conflict. Far from purposely enticing the Germans into the interior of the country or executing successfully any other strategic plan, the Red Army suffered catastrophic defeat in the first months of the war. Indeed, the Russians were smashed as badly as the French had been a year earlier, except that they had more territory to retreat to and more men in reserve. Moreover, while the German army was at the time the best in the world, Soviet forces did not at all make the most of their admittedly difficult position. Some top Red commanders, such as the Civil War cavalry hero Marshal Semen Budenny, proved to be as incompetent as the worst tsarist generals. The fighting spirit of Soviet troops varied greatly: certain units fought heroically, while others hastened to surrender. The enormous number of prisoners taken by the Germans testified not only to their great military victory, but also in part to the Soviet unwillingness to fight. Even more significantly, the Soviet population often welcomed the Germans. This was strikingly true in the recently acquired Baltic countries and in large areas of the Ukraine and White Russia, but it also occurred in Great Russian regions near Smolensk and elsewhere. After a quarter of a century of Communist rule many inhabitants of the U.S.S.R. greeted invaders, any invaders, as liberators. In addition to Red partisans there developed anti-Soviet guerrilla movements, which were at the same time anti-German. In the Ukraine, nationalist bands continued resisting Red rule even long after the end of the Second World War. To the great surprise of the Western democracies, tens of thousands of Soviet citizens liberated by Allied armies in Europe did all they could not to return to their homeland.

But the Soviet regime survived. In spite of its staggering losses, the Red Army did finally hold the Germans and then gradually push them back until their defeat became a rout. Red infantry, artillery, cavalry, and tanks all repeatedly distinguished themselves in the Second World War. Uncounted soldiers acted with supreme heroism. The names of such commanders as Zhukov and Rokossovsky became synonymous with victory. In addition to the regular army, daring and determined partisans also fought the invader to the death. The government managed under most difficult conditions to organize the supply of the armed forces. It should be stressed that while Soviet military transportation depended heavily on vehicles from Lend-Lease, the Red Army was armed with Soviet weapons.

Although many people died of starvation in Leningrad and elsewhere, government control remained effective and morale did not break on the home front. Eventually the Soviet Union won, at an enormous cost, it is true, a total victory.

Much has been written to explain the initial Soviet collapse and the great subsequent rally. For example, it has been argued that the Germans defeated themselves. Their beastly treatment of the Soviet population — documented in A. Dallin's study and in other works — turned friends into enemies. It has even been claimed that to win the war the Nazis had merely to arm Soviet citizens and let them fight against their own government, but Hitler was extremely reluctant to try that. The Russian Liberation Army of Andrew Vlasov, a Soviet general who had been taken prisoner by the Germans and had proceeded to organize an anti-Communist movement, received no chance to develop and prove itself in combat until it was too late. Commentators have also rightly stressed the importance of the Soviet appeal to patriotism and other traditional values. The Communist government consciously utilized the prestige of Russian military heroes of the past and the manifold attractions of nationalism. It emphasized discipline and rank in the army, reducing the power of the commissars. Concurrently it made concessions to the practice of religion and spoke of a new and better life which would follow the end of the war. The Russians, it has been maintained, proved ready to die for their country and for that new life, while they felt only hostility to the Soviet regime.

These and other similar explanations of the Soviet turnabout and of the German defeat appear to contain much truth. Yet, in the last analysis, they might give as one-sided a picture of the Soviet scene as the wholesale admiration of the Communist regime and its virtues popular during and immediately after the war in less critical Western circles. The salient fact remains that in one way or another Stalin and his system prevailed over extreme adversity. Besides, whatever its wartime appeals and promises, the regime did not change at all in essence — as subsequent years were to demonstrate to the again astonished world.

XXXIX

* * * * * * * * * *

STALIN'S LAST DECADE, 1945–53

> We demand that our comrades, both as leaders in literary affairs
> and as writers, be guided by the vital force of the Soviet order —
> its politics. Only thus can our youth be reared, not in a devil-may-
> care attitude and a spirit of ideological indifference, but in a strong
> and vigorous revolutionary spirit.
>
> ZHDANOV

> When the immediate passions of the war recede into the background
> and it becomes possible to view the decade after 1939 in greater
> perspective, the statesmanship exhibited during World War II by
> Roosevelt, Churchill, and Stalin will doubtless be more fully under-
> stood. What is remarkable is not that the Western democracies
> and the Soviet Union failed to reach any general agreement as to
> the postwar organization of Europe, but rather that they were able
> to maintain their coalition until the end of the war with so few
> alarms and disagreements. It is now clear that the success of the
> coalition must be attributed more to the immediacy and gravity of
> the common danger represented by the military might of Germany
> and Japan, than to any harmony of opinion among the Allies
> regarding the political bases of a stable peace. During the long
> period since the winter of 1917–18, when the Bolsheviks had
> negotiated a separate peace with the Central Powers, agreement
> between Russia and the West had been the exception rather than
> the rule. Close co-operation had been achieved almost as a last
> resort in the face of an immediate threat to their security, and once
> the enemy was defeated the differences in political outlook which
> had been temporarily overlooked inevitably reappeared.
>
> BLACK AND HELMREICH

THE SECOND WORLD WAR brought tremendous human losses and material destruction to the Soviet Union. In addition to the millions of soldiers who died, millions of civilians perished in the shifting battle zone and in German-occupied territory. Of the hundreds of thousands of Soviet citizens who went west, either as Nazi slave labor or of their own will, only a part ever returned to their homeland. The brutality of the invaders defied description. Red Army prisoners starved to death in very large numbers in German camps; whole categories of people, such as Jews, Communists, government officials and gypsies were exterminated wherever they could be found. Partisan warfare led to horrible reprisals against the population. In contrast to the First World War, most atrocity stories of the Second World War were true. The total number of Soviet military and civilian dead in the dreadful conflict remains quite uncertain. In 1946 the Soviet government set the figure at seven million. A similiar total has been reached by a few specialists outside the Soviet Union, such as Mironenko. Most foreign scholars, however, have arrived at much higher figures, for instance, Prokopovich estimates fourteen million and Schuman twenty million. Latest calculations based on some newly available material raised the figure even to twenty-seven million. It is generally believed that the losses were about evenly divided between the military and civilian. To the dead must be added perhaps another twenty million for the children that were not born in the decade of the forties. Population figures announced by the Soviet Union in the spring of 1959 tend to support high rather than low estimates of the Second World War losses. Significantly, the ratio of males to females among the peoples of the U.S.S.R. in 1959 stood at 45 per cent males to 55 per cent females.

Material losses were similarly enormous. In addition to the destruction suffered in the fighting, huge areas of the country were devastated — frequently more than once — at the hands of the retreating Red Army or the withdrawing Germans. The Red Army followed the scorched-earth policy, trying to destroy all that could be of military value to the enemy. The Nazis, when they were forced to abandon Soviet territory, attempted to demolish everything, and often did so with remarkable thoroughness. For example, they both flooded and wrecked mines and developed special devices to blow up railroad tracks. Much of the Soviet Union became an utter wasteland. According to official figures — probably somewhat exaggerated as all such Soviet figures tend to be — Soviet material losses in the war included the total or partial destruction of 1,700 towns, 70,000 villages, 6,000,000 buildings, 84,000 schools, 43,000 libraries, 31,000 factories, and 1,300 bridges. Also demolished were 98,000 kolkhozes and 1,876 sovkhozes. The Soviet economy lost 137,000 tractors and 49,000 combine-harvesters, as well as 7,000,000 horses, 17,000,000 head of cattle, 20,000,000 hogs, and 27,000,000 sheep and goats. Soviet authorities estimated the destruction in the U.S.S.R. at half the total material devastation in Europe during the Second World War. It may have also amounted to two-thirds of the reproducible wealth of occupied Soviet areas and one-quarter of the reproducible wealth of the Soviet Union.

The war affected Soviet Russia in other ways as well. It led to a strong upsurge of patriotism and nationalism, promoted by the Communist government itself which did all it could to mobilize the people for supreme effort and sacrifice. The army acquired new prominence and prestige, whereas from the time of the Civil War it had been kept in the background in the Soviet state. Religion, as already mentioned, profited from a more tolerant attitude on the part of authorities. In addition, a striking religious revival developed in German-occupied territory. While the Soviet government maintained control over the people, in certain respects it relaxed somewhat its iron grip. Many Soviet citizens apparently felt more free than before the war. In particular, some kolkhozes simply collapsed, the peasants dividing the land and farming it in private. On the whole, because of lessened controls and a great demand for food, many peasants improved their position during the war years. In the German zone of occupation the people immediately disbanded the collectives. The Nazis, however, later in part reintroduced them as useful devices to control peasants and obtain their produce. The war also led to closer and friendlier relations with Western allies and made widespread contacts of the Soviet and the non-Soviet world inevitable. Moreover, millions of Soviet citizens, prisoners of war, deportees, escapees, and victorious Red Army soldiers, had their first look at life outside Soviet borders. Other millions, the inhabitants of the Baltic countries, eastern Poland, Bessarabia, and northern Bukovina, brought up under alien systems and in different circumstances, were joined to the Soviet Union.

Another obvious result of the Second World War was the great rise in the Soviet position and importance in the world. The U.S.S.R. came to dominate eastern Europe, except for Greece, and much of central Europe. Barring the Allied expeditionary forces, it had no military rival on the entire continent. The international Communist movement, which had reached its nadir with the Soviet-German treaty and Hitler's victory in the west, was experiencing a veritable renaissance. After the German attack on the U.S.S.R., Communists had played major roles in numerous resistance movements, and they emerged as a great political force in many European countries, including such important Western states as France and Italy. With the total defeat and unconditional surrender of Germany and Japan, the earlier defeat of Italy, and the collapse of France, only exhausted Great Britain and the United States remained as major obstacles to Soviet ambition in the whole wide world.

In a sense, Stalin and the Politburo had their postwar policy cut out for them. They had to rebuild the Soviet Union and to continue the industrial and general economic advance. They had to reimpose a full measure of socialism on the recalcitrant peasant, and to supervise and control closely such non-Marxist sources of inspiration and belief as religion and nationalism. They had to combat the "contamination" that had come to their country from the non-Soviet world, and they had to make all their people, including the inhabitants of the newly acquired territories, into good Soviet citizens. They had to maintain complete control over the army. They had to exploit the new position of the U.S.S.R. and the new, sweeping opportunities open to the Soviet Union and international communism in the postwar world. Those numerous observers who were surprised by the course of Soviet politics at home and abroad from 1945 until Stalin's death in the spring of 1953 for the most part either had altogether failed to understand the nature of the Soviet system or believed that it had undergone a fundamental change during the Second World War.

Reconstruction and Economic Development

To repair war damage and resume the economic advance, Stalin and the Politburo resorted, characteristically, to a five-year plan, and indeed to a sequence of such plans. The Fourth Five-Year Plan, which lasted from 1946 to 1950 and was proclaimed overfulfilled in four years and three months, was cut out of the same cloth as its predecessors. It stressed heavy industry, which absorbed some 85 per cent of the total investment, particularly emphasizing the production of coal, electrical power, iron, steel, timber, cement, agricultural machinery, and trucks. The demobilization of more than ten million men provided the needed additional manpower, for the total number of workers and employees had declined from 31 million in 1940 to 19 million in 1943. The rebuilding of devastated towns and villages, which had begun as soon as the Germans had left, gathered momentum after the inauguration of the Plan. But the Fourth Five-Year Plan aimed at more than restoration: Russian industry, especially heavy industry, was supposed to achieve new heights of production, while labor productivity was to rise 36 per cent, based on an increase in the amount of capital per worker of about 50 per cent. As usual, every effort was made to force the Soviet people to work hard. A financial reform of December 1947 virtually wiped out wartime savings by requiring Soviet citizens to exchange the money they had for a new currency at the rate of ten to one. Piece work and bonuses received added emphasis. Official retail prices went up, although the concurrent abolition of rationing and of certain other forms of distribution alleviated somewhat the hardships of the consumer. Foreign economists noted a certain improvement in the urban standard of living as well as a redistribution of real income within the urban population, primarily against the poorer groups.

The Fourth Five-Year Plan obtained a great boost from reparations and other payments collected from defeated Germany and its allies. In 1947, for example, three-fourths of Soviet imports came from eastern Europe and the Soviet zone of Germany, that is, from the area dominated by Red military might. The total value of Soviet "political" imports, including reparations, especially favorable trade provisions, and other economic arrangements, as well as resources spent by different countries for the support of Red Army troops stationed in those countries, has been estimated at the extraordinary figure of over twenty billion dollars. Some reparations were made in the form of complete factories that were dismantled, transported to the Soviet Union, and reassembled there.

In the end the Plan could well be considered a success in industry, much like its predecessors, in spite of the frequently inferior quality of products and uneven results, which included large overfulfillments and underfulfillments. While industry was rebuilt and even expanded in the Ukraine and other western areas, the Plan marked a further industrial shift east, which grew in relative economic importance compared to the prewar period. By mobilizing resources the Soviet Union managed to maintain during the Fourth and Fifth Five-Year Plans the very high annual industrial growth rate characteristic of the first three plans and estimated by Western economists at some 12 to 14 per cent on the average — a figure composed of much higher rates in the late forties and much lower in the early fifties. The Fifth Five-Year Plan lasted from 1951 to 1955 and thus continued beyond Stalin's rule. Similar to all the others in nature and accomplishments, it apparently made great advances in such complex fields as aviation and armament industries and atomic energy. Its completed projects included the Volga-Don canal.

Agriculture, as usual, formed an essential aspect of the plans and, again as usual, proved particularly difficult to manage successfully. The war, to repeat, produced sweeping destruction, a further sharp decline in the already insufficient supply of domestic animals, and at the same time a breakdown of discipline in many kolkhozes, where members proceeded to divide the land and farm it individually or at least to expand their private plots at the expense of the collective. Discipline was soon restored. By September 1, 1947, about fourteen million acres had been taken away from the private holdings of members of collectives as exceeding the permissible norm. Moreover, the Politburo and the government mounted a new offensive aimed at turning the peasants at long last into good socialists. This was to be done by greatly increasing the size of the collectives — thereby decreasing their number — and at the same time increasing the size of working units in a collective, in the interests of further mechanization and division of labor. Nikita Khrushchev, who emerged as one of the leaders in postwar Soviet agriculture, spoke even of grouping peasants in *agrogoroda,* veritable agricultural towns, which would do away once and for all with the diffusion of labor, the isolation, and the backwardness characteristic of the countryside. The agrogoroda proved unrealistic, or at least premature, but authorities did move to consolidate some 250,000 kolkhozes into fewer than 100,000 larger units. In spite of all the efforts — some hostile critics believe largely because of them — peasants failed to satisfy the demands of Soviet leaders and insufficient agricultural production remained a major weakness of the Soviet economy, as Khrushchev in effect admitted after Stalin's death.

Politics and Administration

The postwar period also brought some political changes. As already mentioned, the Soviet Union acquired five new republics during the time of the Russo-German agreement. They were lost, together with other large territories, when Germany and its allies invaded the U.S.S.R. and reacquired when the Red Army advanced west. The five Soviet Socialist Republics, the Estonian, Latvian, Lithuanian, Karelo-Finnish, and Moldavian, raised the total number of component units of the U.S.S.R. to sixteen. In July 1956, however, the Karelo-Finnish S.S.R. was downgraded to its prewar status of an autonomous republic within the R.S.F.S.R., reducing the number of union republics to the present total of fifteen. The Karelo-Finnish Republic, consisting both of some older Soviet lands and of territory acquired from Finland in 1940 and again in 1944, largely failed as an expression of Finnish culture and nationality; in particular, because the inhabitants had a choice of staying or moving to Finland, virtually no people remained in the area that the Soviet Union annexed from Finland. The downgrading, therefore, seemed logical, although it might have been connected with the desire to Russify that strategic area still more effectively. While the number of union republics increased as a result of the Second World War, the number of autonomous republics was reduced: five of the latter, the Volga-German Autonomous Republic and four in the Crimea, the northern Caucasus, and adjacent areas were disbanded for sympathizing with or assisting the Germans, their populations being transported to distant regions. In the case of the Volga Germans, the N.K.V.D. apparently staged a fake parachute raid, pretending to be a Nazi spearhead in order to uncover the sympathies of the people. Mass deportations also took place in the newly acquired areas that were rapidly and ruthlessly incorporated into the Soviet system. For example, most of the members of the upper and middle classes, including a great many intellectuals, disappeared from the Baltic republics. The concentration-camp empire of Stalin and Beria bulged at the seams.

By contrast, although the Union expanded and rigorous measures were applied to bring all parts of it into conformity with the established order, the Soviet political system itself changed little. Union-wide elections were held in 1946 for the first time since 1937, and again in 1950. The new Supreme Soviets acted, of course, as no more than rubber stamps for Stalin and the government. Republican and other local elections also took place. The minimum age for office holders was raised from eighteen to twenty-three. In 1946 people's commissariats became ministries. More important, their number was reduced in the postwar years and they were more strongly centralized in Moscow. Shortly before his death, Stalin carried out a potentially important change in the top Party administration: the Politburo as well as the Organizational Bureau were abolished and replaced by the Presidium to consist of ten Politburo members, the eleventh being dropped, plus another fifteen high Soviet leaders. But Stalin died without calling together the Presidium. After his death its announced membership was reduced to ten, so that as an institution it differed from the Politburo in nothing but name, and even the name was restored after Khrushchev's fall.

The postwar years witnessed also a militant reaffirmation of Communist orthodoxy in ideology and culture. While more will be said about this subject in a later chapter, it might be noted here that scholarship, literature, and the arts all suffered from the imposition of a Party strait jacket. Moreover, Andrew Zhdanov, a member of the Politburo and the Party boss of Leningrad during the frightful siege, who led the campaign to restore orthodoxy, emerged as Stalin's most prominent lieutenant from 1946 until Zhdanov's sudden death in August 1948. That death — engineered by Stalin in the opinion of some specialists — again left the problem of succession wide open. The aging dictator was surrounded during his last years by a few surviving old leaders, his long-time associates, such as Molotov, Marshal Clement Voroshilov, Lazarus Kaganovich, and Anastasius Mikoyan, as well as by some younger men who had become prominent after the great purge, notably Beria, Khrushchev, and George Malenkov. Malenkov in particular appeared to gain consistently in importance and to loom as Stalin's most likely successor.

Foreign Policy

Stalin's last decade saw extremely important developments in Soviet foreign policy. Crucial events of the postwar years included the expansion of Soviet power in eastern Europe, the breakdown of the wartime cooperation between the U.S.S.R. and its Western allies, and the polarization of the world into the Communist and the anti-Communist blocs, headed by the Soviet Union and the United States respectively. That the Soviet Union proved intractable in its dealings with the West, that it did what it could to expand its own bloc, and that it received support from the Communist movement all over the world, followed logically from the nature and new opportunities of Soviet communism. A persistent refusal on the part of many circles in the West to face reality testified simply to their wishful thinking or ignorance. Yet it does not follow that every Soviet move was a cleverly calculated step of a prearranged conspiracy. It appears more likely that the Soviet leaders, too, had prepared little for the postwar period, and that in their preparation they had concentrated on such objectives as rendering Germany permanently harmless. The sweeping Soviet expansion in eastern Europe occurred at least in part because of special circumstances: the rapid Western withdrawal of forces and demobilization, the fact that it became apparent that free elections in most eastern European countries would result in anti-Soviet governments, and the pressure of local Communists as well as, possibly, the urging of the more activist group within the Soviet leadership. In the opinion of Mosely and certain other observers, Stalin embarked on a policy of intransigence and expansion shortly after Yalta.

The Soviet Union and the Allies co-operated long enough to put into operation their arrangement for dividing and ruling Germany and to bring top Nazi leaders to trial before an international tribunal at Nuremberg in 1946. Also, in February 1947, the victorious powers signed peace treaties with Italy, Rumania, Bulgaria, Hungary, and Finland. The Soviet Union confirmed its territorial gains from Rumania and Finland, including a lease of the Finnish base of Porkkala, and obtained extensive reparations. Rounding out its acquisitions, the U.S.S.R. obtained the so-called Carpatho-Ruthenian area from friendly Czechoslovakia in 1945. While most inhabitants of that region spoke Ukrainian, they had not been connected with any Russian state since the days of Kievan Russia.

But on the whole co-operation between the U.S.S.R. and the Western powers broke down quickly and decisively. No agreement on the international control of atomic energy could be reached, the Soviet Union refusing to participate in the Atomic Energy Commission created by the United Nations in 1946. In the same year a grave crisis developed over the efforts of the Soviet government to obtain significant concessions from Persia, or Iran, and its refusal to follow the example of Great Britain and the United States and withdraw its troops from that country after the end of the war. Although, as a result of Western pressure and the airing of the question in the United Nations, Soviet forces did finally leave Iran, the hostility between former allies became increasingly apparent.

The Communist seizure of power in eastern Europe contributed very heavily to the division of the world into two opposed blocs. While many details of the process varied from country to country, the end result in each case, that is, in Yugoslavia, Albania, Bulgaria, Rumania, Hungary, and Poland, was the firm entrenchment of a Communist regime co-operating with and dominated by the Soviet Union. The same happened in eastern Germany. Only Greece and Finland managed to escape the Communist grasp. Liberated Greece fell into the British rather than the Soviet sphere, and its government, supported by Great Britain and the United States, managed to win a bitter civil war from the Communist-led Left. The fact that Finland survived as a free nation remains puzzling. It could be that Moscow first overestimated the strength of Finnish Communists, who did play a prominent part in the government of the country immediately after the war, and then decided not to force the issue in a changing international situation after the Finnish Communists failed to seize power. In particular, the Soviet Union probably wanted to avoid driving Sweden into the camp of Soviet enemies. Similarly — at a greater distance from the U.S.S.R. — the large and strong Communist and allied parties in France and Italy, very prominent in the first years following the war, were forced out of coalition governments and had to limit themselves to the role of an opposition bent largely on obstruction.

It has frequently been said that communism won in Europe only in countries occupied by the Red Army, and that point deserves to be kept in mind. Yet it does not tell the whole story. Whereas in Poland, for example, native Communists were extremely weak, in Yugoslavia and Albania they had led resistance movements against the Axis powers and had attained dominant positions at the end of the war. Perhaps more important, the Soviet Union preferred to rely in all cases on local Party members, while holding the Red Army in readiness as the ultimate argument. Usually, the "reactionary" elements, including monarchs where such were present and the upper classes in general as well as Fascists, would be forced out of political life and a "united front" of "progressive" elements formed to govern the country. Next the Communists destroyed or at least weakened and neutralized their partners in the front to establish in effect, if not always in form, their single-party dictatorship even though the party might be known as the "workers'" or "socialist unity" party rather than simply "Communist." It is worth noting that the eastern European Communists had the most trouble with agrarian parties, just as the Bolsheviks had met their most dangerous rivals in the Socialist Revolutionaries. In Roman Catholic countries, such as Poland and Hungary, they also experienced strong and persistent opposition from the Church. The Communist seizure of power in Czechoslovakia proved particularly disturbing to the non-Communist world, because it occurred as late as 1948 and disposed of a regime headed by President Beneš which had enjoyed popular support and maintained friendly relations with the Soviet Union. The new totalitarian governments in eastern Europe proclaimed themselves to be "popular democracies." They followed the Soviet lead in introducing economic plans, industrializing, collectivizing agriculture — sometimes gradually, however — and establishing minute regulation of all phases of life, including culture. As in the U.S.S.R., the political police played a key role in social transformation and control. An "iron curtain" came to separate the Communist world from the non-Communist.

Churchill, at the time out of office, in a speech in Fulton, Missouri, in March 1946, stressed the danger to the democratic world of the Communist expansion. He was one of the first Western statesmen to point out this danger. When another year of negotiations with the U.S.S.R. produced no results, President Truman appealed to Congress for funds to provide military and economic aid to the neighbors of the U.S.S.R. — Greece and Turkey — the independence of which was threatened directly or indirectly by the Communist state; this policy came to be known as the Truman Doctrine. In June 1947 the Marshall Plan was introduced to help rebuild the economies of European countries devastated by war. Because the Soviet Union and its satellites would not participate, the plan became a powerful bond for the Western bloc. Next, in 1949, twelve Western countries, the United States, Great Britain, Canada, France, Belgium, the Netherlands, Luxembourg, Norway, Denmark, Iceland, Italy, and Portugal signed the Atlantic Defense Pact of mutual aid against aggression. A permanent North Atlantic Treaty Organization and armed force were subsequently created, under General Eisenhower's command. Also in 1949, the U.S. Congress passed a broad Mutual Defense Assistance Program to aid American allies all over the world. With these agreements and with numerous bases girding the U.S.S.R., the United States and other countries were finally organized to meet the Soviet threat.

The Communist bloc also organized. In 1947 the Communist Information Bureau, known as Cominform, replaced the Communist International which had been disbanded in 1943. Bringing together the Communist parties of the U.S.S.R., eastern Europe, France, and Italy, the Cominform aimed at better co-ordination of Communist efforts in Europe. Zhdanov, who represented the Soviet party, set the unmistakably militant tone of the organization. But Communist co-operation was dealt a major blow by the break between Yugoslavia and the U.S.S.R., backed by its satellites, in the summer of 1948. Tito chose to defy Stalin because he wanted to retain full effective control of his own country and resented the role assigned to Yugoslavia in the economic plans and other plans of the Soviet bloc. He succeeded in his bold undertaking because he had a strong organization and support at home in contrast to other eastern European Communist leaders, many of whom were simply Soviet puppets, and because the Soviet Union did not dare invade Yugoslavia, apparently from fear of the probable international complications. Tito's unprecedented defection created the new phenomenon of "national" communism, independent of the Soviet bloc. It led to major purges of potential heretics in other eastern European Communist parties, which took the lives of some of the most important Communists of eastern Europe and resembled in many respects the great Soviet purge of the thirties.

The Western world confronted the Soviet in many places and on many issues. Continuous confrontation in the United Nations resulted in little more than Soviet Russia's constant use of its veto power in the Security Council. Thus, of the eighty vetoes cast there in the decade from 1945 to 1955, seventy-seven belonged to the Soviet Union. The two sides also faced each other in Germany. Because of the new enmity of the wartime allies, the

Allied Control Council in Germany failed to function almost from the beginning, and no agreement could be reached concerning the unification of Germany or the peace treaty with that country. Finally, the Federal Republic of Germany with its government in Bonn was established in the Western-occupied zones in May 1949, while the German Democratic Republic was created in the Soviet-held area in October of the same year. The first naturally sided with the West and eventually joined NATO. The second formed an integral part of the Soviet bloc. Cold war in Germany reached its height in the summer of 1948 when Soviet authorities stopped the overland supply of the American, British, and French sectors of Berlin. Since that city, located 110 miles within the Soviet zone, was under the jurisdiction of the four powers, three of them Western, it, or rather West Berlin, remained a highly provocative and disturbing "window of freedom" in rapidly Communized eastern Germany and eastern Europe. But Soviet hopes to force the Western powers to abandon their part of the city failed: a mammoth airlift was maintained for months by American and British planes to keep West Berlin supplied until the Soviet Union discontinued its blockade.

Postwar events in Asia were as important as the developments in Europe. Communists made bids to seize power in such different areas as Indonesia, Malaya, and Burma. They succeeded in China. The great Chinese civil war ended in 1949 with Chiang Kai-shek's evacuation to Formosa — or Taiwan — and the proclamation of the Communist Chinese People's Republic, with Mao Zedong at its head, on the mainland. While the Soviet Union took no direct part in the Chinese war and at first apparently even tried to restrain Mao, it helped Chinese Communists with supplies and backed fully Mao's new regime. And indeed Communist victory in a country of great size inhabited by some half a billion people meant an enormous accretion of strength to the Soviet bloc, although it also created serious problems: China could not be expected to occupy the role of a satellite, such as Bulgaria or Czechoslovakia, and the Communist world acquired in effect a second center of leadership. By an agreement concluded in 1950, the U.S.S.R. ceded to Communist China its railroad possessions in Manchuria, although briefly retaining a naval base at Port Arthur.

In Korea cold war turned to actual hostilities. There, as in Germany, no agreement could be reached by the victorious powers, and eventually two governments were formed, one in American-occupied southern Korea and the other in the Soviet north, the thirty-eighth parallel dividing the two. At the end of June 1950, North Korea attacked South Korea. In the ensuing years of fighting, which resulted in the two sides occupying approximately the same positions when the military action stopped as they had in the beginning, U.S. forces and some contingents from other countries came to the assistance of South Korea in execution of a mandate of the United Nations, whereas tens and even hundreds of thousands of Chinese "volunteers" intervened on the North Korean side. The Soviet army itself did not participate in the war, although the North Koreans and the Chinese used Soviet-made aircraft and weapons, and although it is more than likely that Soviet advisers, as well as Soviet pilots and other technicians, were in North Korea. Although the front became stabilized in the summer of 1951, no armistice could be concluded until the summer of 1953, after Stalin's death.

The End of Stalin

Stalin's final months had a certain weird quality to them. It could be that the madness that kept peering through the method during his entire rule asserted itself with new vigor. In any case, events which then transpired will have to be elucidated by future historians. With international tension high, dark clouds gathered at home. In January 1953, nine doctors were accused of having assassinated a number of Soviet leaders, including Zhdanov. Beria's police were charged with insufficient vigilance. The press whipped up a campaign against traitors. Everything pointed to another great purge. Then on March 4 it was announced that Stalin had suffered a stroke on the first of the month, and on the morning of the sixth the news came that he had died the previous night. Some of the dictator's entourage especially close to him disappeared at the same time.

The Stalin Revolution I: Collectivization and Industrialization
   Lecture (April 13). Hellie, "The Stalin Revolution"
   April 15. Daniels, 1-12, 63-130. Sholokhov, *Virgin Soil Upturned* in *Readings*.
      Stalin, in *Readings*. Riasanovsky, 492-508.
   April 17. Scott, 3-116. Kotkin, "Stalinism as a Civilization" or "Introduction" to Scott.

The Stalin Revolution II: Culture and Politics. The Purges
   Lecture (April 20). Hellie, "The Purges"   Read that section in *Readings*.
   April 22. Daniels, 131-70.
   April 24. Chukovskaia. Scott, 173-208.

World War II and the End of the Stalin Era
   Lecture (April 27). Burton, "The Great Patriotic War and Late Stalinism"
   April 29. Sholokhov, "Fate" in *Readings*. Salisbury. Rokossovsky, A. Tolstoy in
      Readings. Riasanovsky, 509-38.
   May 1. Rapoport, Iakov. *The Doctors' Plot of 1953*, 23-55, 86-158. Daniels, 183-263.

Khrushchev and the Thaw
   Lecture (May 4). Hellie, "Soviet Literature"
   May 6. N. S. Khrushchev, "The Crimes of the Stalin Era," in *Readings*.
      Riasanovsky, 539-89.
   May 8. Rasputin, "Money for Malia"; Soloukhin, "Varvara Ivanova"; and Breslauer in
      Readings.

The Brezhnev Era of Stagnation
   Lecture (May 11). Burton, "The Late Soviet Economy"
   May 13. *Chronicle of Current Events*; Aleksandrov; Pliushch; Camp 35 - all in *Readings*.
   May 19-20. Zaslavskaia and Trifonov in *Readings*.

Gorbachëv, Perestroika, &
   Lecture (May 18). Hellie, "Perestroika"
   May 20. Tarasulo, 1-62, 169-215, 238-53, 277-352. *Glasnost'*, Andreeva, Zhurkin in
      Readings. Riasanovsky, 590-612.
   May 22. Tarasulo, 67-125.

The Collapse of the Soviet System
   Lecture (May 27 NB!). Hellie, "The Collapse of the USSR"
   May 29. Gaddy.

The New Russia
   Lecture (June 1) Hellie, "What Has 1991 Given the Russian People?"
   June 3. Goltz, Holmes (in *Readings*).

Course Requirements:

Option A

1) A two-hour written examination (as scheduled by the Registrar) on the reading lectures, and class discussions. 50% of the grade.

2) A twelve-page (roughly) term paper on a specific problem or general essay on a topic the student chooses in consultation with his/her instructor within the time period cover by the course this quarter. Students are advised to consult useful reference works, such as: *T American Bibliography of Russian and East European Studies*, 1956-- [Z 2483 .A5 RR2S]; Dav Lewis Jones, *Books in English on the Soviet Union 1917-73. A Bibliography* [Z 2491 .J76 Harp Reserve]; *Widener Shelf List* [Z 7044.5 .H34 RR2S]. Also useful may be: *The Grand Sov Encyclopaedia* [AE 55 .B75304 RR]; *The Modern Encyclopaedia of Russian and Soviet Histo* [DK 14 .M68]; *The Modern Encyclopaedia of Russian and Soviet Literature* [PG 3007 .M RR2S]. Having consulted these reference works, the student is advised to submit to his/h instructor before the mid-point of the quarter a one-page proposal: a tentative/working tit several sentences about the proposed topic itself; a list of the sources he/she intends to consu The completed paper must be submitted by the end of the reading period if the student desir a qualitative grade on this quarter's transcript. **The proposal must be appended to the fin draft of the submitted paper.** {Students are **strongly** urged to retain **all** their notes and oth work on the paper as well as a copy of the final draft at least until they are **sure** that the fir grade has been recorded by the Registrar.} 50% of the grade.

Option B

1) A fifteen-minute spot-check oral examination on the readings, lectures, and cl discussions at a pre-arranged time prior to the scheduled examination. Key words and topics w be typed on cards which the student will draw; he/she will be expected to be able to identify a discuss them reality and coherently. (This is only possible when the student has completed reading assignments and attended the lectures and class discussions.) 25% of the grade.

2) The same as (2) above under Option A. 75% of the grade.

Office Hours: Richard Hellie, SS 204, TuTh 1:30-2:30. x 2-8377
Christopher Burton. SS 204 MW 10-11.

Readings

CONTENTS

THE COLLEGE OF THE UNIVERSITY OF CHICAGO
Introduction to Russian Civilization - 3

Social Sciences 242   Spring 1998

To be Purchased at the Seminary Coop Bookstore:

1. Chukovskaia, Lydia. *Sofia Petrovna*. Northwestern.
2. Daniels, Robert, ed. *The Stalin Revolution*. Heath.
3. Gaddy, Clifford G. *The Price of the Past. Russia's Struggle with the Legacy of a Militarized Economy.*
4. Gray, Camilla. *The Russian Experiment in Art 1863-1922*. Thames and Hudson.
5. Gladkov, Fëdor. *Cement*.
6. Reed, John. *Ten Days That Shook the World*. Penguin
7. Riasanovsky, Nicholas V. *A History of Russia*. Oxford.
8. Scott, John. *Behind the Urals*. Indiana.
9. Tarasulo, Isaac J., ed. *Gorbachev and Glasnost. Viewpoints from the Soviet Press*. Scholarly Resources.

Available from the Social Sciences Duplicating Service, SS 103:

10. Hellie & Burton, eds. *Selected Readings for Russian Civilization*. $13.50 cash or check.

Available on Harper Reserve:

Blok, Aleksandr. "The Intelligentsia and the Revolution." in Raeff, *Russian Intellectual History* 364-71. [DK 32.7 .R17.]
Kotkin, Steve. "Afterword: Stalinism as a Civilization," in his *Magnetic Mountain*. [DK 65 .M159K6750 1995.]
Rapoport, Iakov. *The Doctors' Plot of 1953*. [DS 135.R95R37130 1991.]
Salisbury, Harrison. *The 900 Days. The Siege of Leningrad.* [D 764.3 .L5S19.]

Reading and Lecture Schedule:

1. The Revolutionary Year 1917
   Lecture (March 30). Hellie, "The Russian Revolution"
   April 1. Reed, 1-116. Lenin, first 4 selections in *Readings*. Riasanovsky, 453-61.
   April 3. Reed, 117-271. Blok, "Intelligentsia & the Revolution," in Raeff. Blok in *Readings*.

2. The Civil War & NEP
   Lecture (April 6). Hellie, "Literary Developments 1917-1933"
   April 8. Gladkov, *Cement*. Riasanovsky, 465-91. Lenin's Testament in *Readings*.
   April 10. Il'f & Petrov in *Readings*. Gray, 219-76. Daniels, 13-61.

X L

* * * * * * * * * *

THE SOVIET UNION AFTER STALIN, 1953–85

> One of the fundamental principles of party leadership is collectivity in deciding all important problems of party work. It is impossible to provide genuine leadership if inner party democracy is violated in the party organization, if genuine collective leadership and widely developed criticism and self-criticism are lacking. Collectiveness and the collegium principle represent a very great force in party leadership. . . .
>
> SLEPOV

> As long as we confine ourselves, in substance, to denouncing the personal faults of Stalin as the cause of everything we remain within the realm of the "personality cult." First, all that was good was attributed to the superhuman, positive qualities of one man: now all that is evil is attributed to his equally exceptional and even astonishing faults. In the one case, as well as in the other, we are outside the criterion of judgment intrinsic in Marxism. The true problems are evaded, which are why and how Soviet society could reach and did reach certain forms alien to the democratic way and to the legality which it had set for itself, even to the point of degeneration. . . .
>
> TOGLIATTI

> It is difficult to exaggerate the historical significance of the Sino-Soviet conflict. It has influenced every facet of international life, not to speak of the Soviet block itself. No analysis of the relationship between Washington and Moscow, of the problem of nuclear proliferation, or the orientation of Indian nationalism, of the thrust of revolutionary movements in the Third World would be complete without taking into account the impact of the increasingly bitter dispute between the two onetime seemingly close allies. For the international Communist movement, it has been a tragic disaster, comparable in some respects to the split in Christianity several centuries ago. The Communist and Christian experience both showed that in theologically or ideologically oriented movements disagreements even only about means and immediate tactical concerns can escalate into basic organizational and doctrinal, indeed, even into national conflicts, fundamentally destructive of the movement's unity.
>
> BRZEZINSKI

STALIN'S STROKE — if its official date is to be believed — was followed by three days of silence from the Kremlin and, in all probability, by hard bargaining among top Soviet leaders. When the dictator's demise was announced, the new leadership proclaimed itself ready to govern the country, emphasizing the solidarity of its members as well as its unity with the people. The shrill tone and the constant repetition of both assertions must have covered many suspicions and fears. Malenkov emerged clearly in the chief role, for he became presumably both the senior Party secretary, which had been Stalin's most important office, and prime minister. Beria and Molotov stood next to Malenkov, forming a triumvirate of successors to the dictator. The three, in that order, were the key living figures during the burial of Stalin in the Lenin Mausoleum in Red Square on the ninth of March, making appropriate speeches on the occasion.

The Rise, Rule, and Fall of Nikita Khrushchev

As early as the middle of March, however, it was announced that Malenkov had resigned as the Party secretary, although he remained prime minister and continued to be treated as the top personage in the Soviet Union. The new Presidium of the Party was reduced to ten members. Later it was announced that Khrushchev had been promoted to the position of first Party secretary, the title used instead of that of general secretary associated with Stalin. In the summer of 1953, Beria was arrested and then executed in secret, with a number of his followers, on charges of treason and conspiracy; or, as Khrushchev related to some visitors, Beria was killed at the Presidium meeting at which he had expected to assume full power. In any case, it would seem that in the race to dispose of one another Beria had narrowly lost out. Beria's fall marked a certain weakening in the power of the political police. In February 1955, Malenkov resigned as prime minister, saying that he was guilty of mistakes made in the management of Soviet agriculture and of having incorrectly emphasized the production of consumer goods at the expense of heavy industry. Nicholas Bulganin, a prominent Communist leader who had been a member of the Politburo since 1948, replaced Malenkov as head of the government. Bulganin and Khrushchev, the chief of the government and the chief of the Party, then occupied the center of the Soviet stage and also held the limelight in international affairs, suggesting to some observers the existence of something resembling a diarchy in the U.S.S.R. Marshal Zhukov, a great hero of the Second World War who had been reduced by Stalin to provincial commands and had returned to prominence after Stalin's death, took over Bulganin's former office of minister of defense. Zhukov's rise marked the first appearance of an essentially military, rather than Party, figure in high governing circles in Soviet Russia.

The struggle in the Kremlin continued. Probably its most astounding event was Khrushchev's speech to a closed session of the Twentieth Party Congress in February 1956, in which the new first secretary denounced his predecessor, Stalin, as a cruel, irrational, and bloodthirsty tyrant, who had destroyed many innocent with the guilty in his great purge of the Party and the army in the thirties and at other times. In fact, Stalin and the "cult of personality" he had fostered were blamed also for military unpreparedness and defeats in the Second World War as well as for other Soviet mistakes and weaknesses. At the same time, paradoxically, Khrushchev presented Stalin's colossal aberrations as mere deviations of an essentially correct policy, entirely rectified by the collective leadership that replaced the despot. Khrushchev's explosive speech remains difficult to explain: after all, it was certain to produce an enormous shock among Communists and do great damage to the Communist cause — to say the least, the transition from years of endless adulation of Stalin to Khrushchev's revelations was bound to be breathtaking; besides, Khrushchev could not help but implicate himself and his associates, at least indirectly, in Stalin's crimes and errors. The answer to the riddle of the speech lies probably in the exigencies of the struggle for power among Soviet leaders. Khrushchev's sensational denunciation of Stalin struck apparently at some "old Stalinists," his main competitors. Besides, it would seem that Khrushchev tried both to put the blame for many of the worst aspects of the Soviet past on Stalin, implying that these evils could not happen again, and to set the correct line of policy for the future.

The conflict at the top reached its culmination in the spring and early summer of 1957, after the Hungarian rebellion of the preceding autumn and certain other events at home and abroad had raised grave questions concerning the orientation and activities of the new Soviet administration and indeed concerning the stability of the whole Soviet system. Defeated in the Presidium of the Party, Khrushchev took his case to its entire Central Committee, successfully reversing the unfavorable decision and obtaining the ouster from the Presidium and other positions of power of the "anti-Party group" of Malenkov, Molotov, Kaganovich, and Dmitrii Shepilov, a recent addition to the Soviet front ranks. While Khrushchev's enemies were dropped from the Presidium, its membership was increased to fifteen, giving the general secretary further opportunities to bring his supporters into that extremely important body. Marshal Zhukov, who, it would seem, had provided valuable assistance to Khrushchev in the latter's bid for power, again fell into disgrace several months later. Finally in March 1958, Bulganin, who had been disloyal to Khrushchev the preceding year, resigned as head of the government. Khrushchev himself replaced Bulganin, thus combining the supreme effective authority of the Party and of the state. Clearly that self-made man of peasant background and limited education no longer had any equals within the collective leadership or elsewhere in the U.S.S.R.

The remarkable Twenty-second Party Congress held in the second half of October, 1961, confirmed on the whole Khrushchev's dominant position. As expected, it gave ready approval to the new leader's twenty-year program of "building communism" and denounced his enemies at home and abroad. Another old leader, Voroshilov, was linked to the "anti-Party group." In a much more unexpected development, however, Khrushchev and the Congress returned to the grizzly issue of Stalinism, detailing and documenting many of its atrocities. The removal of Stalin's body from the mausoleum in Red Square, the renaming of the cities named after Stalin, with Stalingrad becoming Volgograd, and the publicity given for the first time to certain aspects of the great purge must have had a powerful impact on many Soviet minds.

Yet, although Khrushchev managed to assert his will at the Twenty-second Party Congress and even evict Stalin from the mausoleum, it can be seen in retrospect that by 1961 his fortunes were on the decline. In fact, 1958 probably marked Khrushchev's zenith. The year followed the new leader's decisive victory over the "anti-Party group," and the sensational Soviet inauguration of the space age the preceding autumn. It was blessed with a bounteous harvest. In spite of serious problems, industrial production continued to grow at a high rate. The ebullient Khrushchev could readily believe that all roads led to a communism that was bound to bury capitalism in the not-too-distant future.

Disillusionments followed in rapid succession. Economic development went sour; Khrushchev's exhortations, and his economic, administrative, and party reorganizations, together with his hectic campaigns to remedy particular deficiencies — all to be discussed later in this chapter — were increasingly ineffective in resolving the crisis. In his last years and months in office Khrushchev saw the rate of industrial growth decline sharply while he had to resort to an unprecedented purchase of Canadian wheat to forestall hunger at home. De-Stalinization or, more broadly, a certain "liberalization" of Soviet life seemed to produce as many problems as it resolved. It led in effect to soul-searching and instability rather than to any outburst of creative communist energy. The world situation — also to be discussed later — deteriorated even more sharply from the Soviet point of view. In 1960 the conflict with China, which dated back at least to Khrushchev's original de-Stalinization of 1956, burst into the open, and from about 1963 the break between the former allies seemed irreparable. In the relations with the West, Khrushchev's aggressive enthusiasm, spurred by the successes of Soviet space technology, received repeated checks in Germany and finally suffered a smashing defeat in October 1962 in the crucial confrontation with the United States over the Soviet missiles in Cuba. Khrushchev's survival of the catastrophe of his apparently largely personal foreign policy might be considered a tribute to Soviet totalitarianism. Yet totalitarianism too was deteriorating in the Soviet Union. Observers noted that although the Twenty-second Party Congress confirmed and extended Khrushchev's victory over the "anti-Party group" these enemies of the leader were not even expelled from the Party. New fissures and problems appeared in the ensuing months and years. It would seem that during this time Khrushchev made the mistake of acting in an increasingly autocratic and arbitrary manner even though his power was not nearly as great as Stalin's had been.

On October 15, 1964, it was announced in Moscow that Nikita Sergeevich Khrushchev had been "released" from both his Party and his government positions, because of "advanced age and deterioration of his health."

Brezhnev and Kosygin

The ten years or so of Khrushchev's rule of the Soviet Union have often been described as a transitional period, but they also marked a culmination. When Khrushchev assumed power in the Kremlin he became both the head of the U.S.S.R. and the leader of an essentially united, ever-victorious, and ever-expanding world communism. He could still believe in the identity of interests of the state and the movement. Indeed he delighted in counting the years, twenty or fifteen, at the end of which the Soviet Union would enter full communism, and additional years, perhaps to the time of "our grandchildren," which would establish communism all over the globe. Khrushchev's own rags-to-riches story was about to be repeated on a universal scale. By the time the enthusiastic leader "retired," communism was hopelessly split between the antagonistic centers of Moscow and Peking, while the Cuban confrontation and defeat spelled out to the Soviet leaders in an unforgettable manner the realities of the atomic age, of which Marx and Engels and even Lenin had had no inkling. At home de-Stalinization kept releasing new furies, and the economic situation called for emergency measures to improve productivity, distribution, and services rather than for blueprints of a communist utopia. All these, and many other problems, fell upon the shoulders of Khrushchev's successors, and in particular on Leonid Brezhnev, who obtained the top position in the Party, and Alexis Kosygin, who as prime minister became the effective head of the government.

The new leaders had the usual record of Party and government service, and their views could not be easily distinguished from those of Khrushchev. In fact, Khrushchev apparently had thought of Brezhnev as his eventual successor. The Chinese and certain others who expected Soviet policy to be transformed by the fall of Khrushchev were quickly disappointed. Instead of challenging Khrushchev on fundamentals, the new leadership assailed his personal performance and style of work, accusing him of "subjectivism," authoritarianism, ignorance, "hare-brained schemes," and "mad improvisations." Khrushchev's sweeping reorganizational reforms were repealed, some promptly and some after a period of time. It was a certain businesslike, low-key quality of the new administration that presented a striking contrast to the flamboyancy and bombast of the deposed leader. The overturn of October 1964 could also be considered a reassertion of collective leadership, eliminating as it did the latest cult of personality.

Although the fall of Khrushchev strengthened the forces opposed to de-Stalinization, they proved unable to gain the upper hand. Instead the leadership resorted to compromise which found its characteristic expression in the mammoth Twenty-third Party Congress held in the spring of 1966. There were 4,942 delegates and additional representatives of 86 foreign communist parties and sympathizing organizations — the number announced was 86, but there were actually somewhat fewer. The Congress avoided mention of such crucial issues as China, Stalin, or, for that matter, Khrushchev. With the Vietnam war in full swing, it adopted doctrinaire, anti-imperialistic planks in foreign policy and expounded a hard line in matters of ideology and culture. On the other hand, it upheld a certain economic "liberalization" and took a more realistic view of the economic and social development and potentialities of the Soviet Union than had been customary under Khrushchev. In the words of critical commentators, the Soviet authorities opted for economic development without its consequences. It was by the decision of the Twenty-third Congress that the Presidium became again, as in the days of Stalin, the Politburo, and the first secretary of the Party became again the general secretary.

Brezhnev was also the central figure and delivered the main address at the Twenty-fourth Party Congress in March and April 1971, the Twenty-fifth Party Congress in late February 1976, and the Twenty-sixth Party Congress in February and March 1981. His authority grew with the years, and one could speak even of a cult of Brezhnev, especially after the general secretary of the Party also became, in 1976, a Marshal of the Soviet Union and his autobiographical writings were given tremendous prominence. Yet he continued, apparently, to work closely with other leaders of the Politburo; besides, Brezhnev, born in 1906, was becoming increasingly an old and sick man. His name and efforts came to be associated with the policy of détente, which the Soviet propaganda machine preferred to call "irreversible," with the great strengthening of the Soviet military might vis-à-vis the United States, and also with economic policies emphasizing such crucial sectors as agriculture and energy. Living standards rose markedly, and some commentators began to write of a Soviet version of consumer attitudes and a consumer society. The Party and government elite in particular came to enjoy high living, although, not surprisingly, it has remained touchy on that subject. In the anecdotal words of Brezhnev's uneducated mother, at the sight of her son's splendid collection of motor cars: "That is fine, my dear son, but what if the Bolsheviks return?" Yet, as the Soviet economic situation became more difficult in the late 1970's and early 1980's, the Brezhnev government attempted essentially palliatives rather than fundamental reform: "the decision has been made 'to settle for short-run solutions to long-term problems.' " Certain observers concluded that, although Brezhnev did not have complete control of the Soviet Union, he had full veto power of any reforms and that, therefore, no real change could be expected as long as he remained at the helm.

When Brezhnev died, finally, on November 10, 1982, at the age of seventy-five, he had outlived such near-peers as Kosygin, by about two years, and the chief Party ideologist, Michael Suslov, by less than a year; Nicholas Podgorny had been ousted from the leadership in 1977; Brezhnev's long-time lieutenant, Andrew Kirilenko, slightly older than his patron, lost his Politburo position in 1982, whether for political or medical reasons. Yet the remaining leaders still belonged to the same well-established group and were of comparable age and, as far as one could tell, orientation. Nicholas Tikhonov, who replaced Kosygin as prime minister, was born, like Brezhnev, in 1906; Constantine Chernenko, probably closest to Brezhnev at the time of the latter's death, was only five years younger; Dmitrii Ustinov, the man in charge of what may be described as the Soviet military-industrial complex, was born in 1908. That the General Secretaryship of the Party went to Iurii Vladimirovich Andropov, sixty-eight, was not unexpected, although some observers were surprised by the rapidity and smoothness of the transition. Credited with uncommon intelligence and general ability, as well as a certain sophistication, Andropov became well-known as the head of the K.G.B., the political police, for the fifteen years before he switched in May 1982 to work in the Party secretariat. He had also been a prominent Politburo member from 1973. Andropov's earlier service included the position of ambassador to Hungary in 1954–57, when he became linked, apparently, both to the brutal suppression of the Hungarian revolt and to the institution of a liberal economic policy in Hungary in its wake. A sharp critic of the stagnation and corruption under Brezhnev and, apparently, a determined reformer of sorts, Andropov addressed himself immediately to purging the administrative apparatus and to strengthening labor discipline by such spectacular measures as police searches in public places for absentee workers. But his activity was cut short by kidney failure, and he died after only about a year and three months in office. Andropov was replaced by Chernenko, Brezhnev's intended heir and already a sick man, who lived for barely another year. Then, on March 11, 1985, Mikhail Sergeevich Gorbachev, Andropov's fifty-four-year-old protégé, was elected by the Politburo to the general secretaryship of the Party.

Economic Development

When Stalin died, the Fifth Five-Year Plan was in full swing. It was duly completed in 1955, yielding the usual result of accomplishment in industry — checkered, to be sure, with huge overfulfillments and under-fulfillments — based on great exertion and privation. The Sixth Five-Year Plan, scheduled to run from 1956 to 1960, promptly succeeded the fifth. Its period was truncated in 1958, however, and a Seven-Year Plan to last from 1959 through 1965 was proclaimed instead. The official explanation for the change, which stressed the discovery of vast new natural resources that altered Soviet economic prospects, was not convincing. Apparently, the Sixth Five-Year Plan had fallen considerably behind its assigned norms of production and the Soviet leadership decided to try a fresh start.

Another change in Soviet economic life occurred in 1957, when Khrushchev, in a move aimed at a geographic dispersion, or deconcentration — although not organizational decentralization — of authority, transferred the direction of a good proportion of industry from the ministries in Moscow to regional Economic Councils. Reflecting the constant Soviet search for the most effective and efficient economic organization, this reform was nevertheless considered by many observers as primarily political in motivation: it removed from Moscow large economic managerial staffs which, it would seem, had supported Malenkov in the struggle for power within the Kremlin. Another aim might have been to give the local Party bosses more authority in economic matters.

The industrial goals of the Seven-Year Plan were pronounced realistic by such Western economists as Campbell and Jasny. While concentrating as usual on heavy industry, with special attention paid to, for example, further electrification and development of the chemical industry, the plan called for a rate of industrial growth approximately 20 per cent slower than that achieved during the Fifth Five-Year Plan. In this sense it was also less ambitious than the abortive Sixth Five-Year Plan. In evaluating the Seven-Year Plan, Campbell made the following comparison between the Soviet and the U.S. economies:

> If it is assumed that industry in the United States will continue to expand at the rate of about 4 per cent that has characterized the postwar period, and that the rate of growth planned by the Russians for their industry is actually achieved, their industrial output will rise from about 45 per cent of ours at the beginning of the seven-year period to about 61 per cent at the end. In other words they will still be a long way behind us (and even further behind in terms of *per capita* output, it may be added), though they will certainly have made a remarkable gain on us.

In fact, Campbell's prediction proved to be generally intelligent, although impossible to evaluate definitively with any degree of precision — incidentally, such comparisons give vastly different results depending on whether they are made in rubles or in dollars. After Khrushchev too, it might be added, the Soviet economy continued to gain in relative output on the American economy, helped by such developments as the recession in the United States and the Western world in general in the 1970's and 1980's.

Although concentrating on capital goods, the Seven-Year Plan allowed somewhat more for the everyday needs of the people than had generally been true of previous Soviet industrialization. Especially interesting was the ambitious housing and general building program of the plan, which aimed to increase the total Soviet building investment by 83 per cent. Even when executed not in its entirety and with buildings of inferior quality, this aspect of the Seven-Year Plan constituted a major contribution to the improvement of the Soviet standard of living. Superior quality and unflagging attention were devoted, by contrast, to such advanced technical fields as atomic energy, rockets, missiles, and space travel. From the launching of the first artificial earth satellite, Sputnik I, in October 1957, the U.S.S.R. has achieved a remarkable series of pioneer successes in rockets and space travel.

Important developments took place in Soviet agriculture during the Khrushchev years. Indeed, frantic efforts to raise agricultural production constituted, together with certain concessions to the consumer, the salient new features of Soviet economic policy. The magnitude of the Soviet farm problem can be seen from the fact that, by contrast with industrial achievements, the gross output of agriculture in 1952 was only some 6 per cent above 1928. In 1954 Khrushchev set into full operation his sweeping "virgin lands" project: huge areas of arid lands in Asiatic Russia, eventually totaling some seventy million acres, were to be brought under cultivation. The undertaking, supported by great exertion as well as by a mighty propaganda effort, gave remarkably mixed results from year to year, depending in large part on weather conditions, but did not live up to expectations. The new first secretary also started a huge corn-planting program. He further decided to boost drastically the production of such foods as meat, milk, and butter. These items came to rival electric power and steel in Soviet propaganda and to serve as significant gauges in "surpassing America."

Yet the condition of Soviet agriculture remained bad. Official claims and promises, especially the latter, differed sharply from reality. Indeed, the mass planting of corn, often in unsuitable conditions, and even the huge gamble on the virgin lands, which are difficult to cultivate, might have been unwise. To increase production Soviet authorities resorted to the old method of further socialization. Between 1953 and 1957 the number of sovkhozes increased from 4,857 to 6,000, while the number of kolkhozes declined at the same time from 91,000 to 78,900, reducing the kolkhoz share of land under cultivation from 84 to 72 per cent. By 1961 the number of collective farms had fallen to 44,000 — and by the end of the decade they were to be reduced further, through amalgamation, and absorptions by the sovkhozes, to under 35,000; by 1974, there was in the Soviet Union slightly more land under sovkhoz than under kolkhoz cultivation, with only some 30,000 collective farms still in operation. As late as September 1958, Khrushchev, other leaders, and the propaganda machine still spoke of the more truly socialist nature, as well as of the technical superiority, of the sovkhoz system of agriculture over that of the kolkhoz. Yet, apparently because of the strength of peasant resistance, especially of the passive kind, the first secretary stopped the attack on kolkhozes in early 1959 at the Twenty-first Party Congress.

The official policy toward the collective farms continued to be ambivalent. There is a consensus among experts that the income of the members of the kolkhozes, extremely low at the time of Stalin's death, increased markedly in subsequent years. The set prices paid by the state for compulsory deliveries of collective farm produce were raised to more realistic levels in 1956 and immediately afterward, enlarging the income of individual kolkhoz members by as much as 75 per cent, according to Marchenko's calculations. The collectives themselves also gained in strength. In 1958, in an abrupt reversal of previous policy, the government enacted measures to disband the Machine Tractor Stations, enabling the kolkhozes to obtain in ownership all the agricultural equipment which they needed. And, as already mentioned, early in 1959 attacks on the collectives ceased and they were again recognized as the proper form of agricultural organization at the given stage of development of the Soviet economy and society.

But, on the other hand, state and Party pressure on the kolkhozes continued and in certain respects even gained momentum. The years witnessed a great stress on increasing the "indivisible fund" of a collective — that is, that part of its revenue which belongs to the entire kolkhoz and is not parceled out among individual members — and on using this fund for such "socially valuable" undertakings as building schools and roads in the locality. The purchase of M.T.S. machinery by the collectives in itself necessitated heavy expenditure. Also, Khrushchev and other leaders returned to the theme that the private plots of the members of a kolkhoz are meant merely to augment a family's food supply rather than to produce for the market and that they should become entirely unnecessary with further successes of socialist agriculture.

Moreover, the Seven-Year Plan goals of increasing agricultural production by 70 per cent and raising labor productivity in the kolkhozes by 100 per cent and in the sovkhozes by 60 to 65 per cent proved to be impossible to attain. Perhaps they had been predicated on a further drastic socialization of Soviet agriculture, and in particular the elimination or near elimination of the twenty million small private plots of the members of the collectives, which the leadership did not dare carry out.

Again, in the opinion of Bergson and other Western observers, the agricultural goals adopted by the Twenty-second Party Congress as part of the program of creating a "material basis" for communism by 1980 seemed fantastically optimistic and quite unreal — an estimate that did not apply to nearly the same extent to the industrial goals. Khrushchev's frantic efforts after the Congress to bolster farm production — this time demanding the abolition of the grass rotation system in favor of planting feed crops such as sugar beets, corn, peas, and beans — served to emphasize further the crisis of Soviet agriculture. It is also probably in connection with the economic, especially the agricultural, crisis that Khrushchev enacted, in 1962, his strangest reorganizational measure: the across-the-board division of the hitherto monolithic Communist party into two party hierarchies, one to deal with industry and the other with agriculture.

Khrushchev's enthusiasm and ambition in the economic and other fields found characteristic expression in his insistence on the early building of communism, which was to replace socialism as the culminating phase in the evolution of Soviet society. The Twenty-second Party Congress, in October 1961, paid much attention to this issue, proclaiming that the preconditions for communism should be established in the U.S.S.R. by 1980. Although the concept of communism remained fundamentally vague and lacked substantiating detail, Feldmesser and other Western scholars have been able to draw a generally convincing picture of the projected Soviet utopia.

Communism would be based on an economy of abundance which would satisfy all the needs of the population. These needs, however, were to be defined by the authorities. In the words of Khrushchev, "Of course, when we speak of satisfying people's needs, we have in mind not whims or claims to luxuries, but the healthy needs of a culturally developed person." Presumably, the authorities could also determine that some people had more needs than others. Nevertheless, the main thrust of communism would be toward equality. Income differentials would be drastically reduced. More than that, communism would finally eliminate the distinction between town and country, industrial and agricultural work, mental and manual labor, and thus the differences in the styles of life. Members of the new society would be "broad-profile workers," that is, persons trained in two or three related skills who would, in addition, engage without pay in one or more other socially useful occupations in their leisure hours.

The collective would obviously dominate. Even some of the abundant consumer goods would be available in the form of "appliance pools" of refrigerators, washing machines, or vacuum cleaners. Apparently, Khrushchev objected to the last to private automobile ownership and projected instead public car pools. On a still broader scale, life would become increasingly socialized. Free public health services and transportation would be followed, for example, by free public meals which would virtually eliminate kitchen drudgery for women. The Academician Stanislav Strumilin and others constructed models of communal cities of the future, with parents allowed a daily visit to their children, who would live separately under the care of a professional staff. Indeed communism would seem to imply a great diminution in the role of the family, if not its abolition, although most Soviet commentators have refused to face this conclusion. By contrast, the role of the school would expand, and so would the roles of labor brigades, comrades' courts, and other public organizations. Lenin's, or Khrushchev's, authoritarian Marxist system would in no sense be diluted, or even diversified, in communism, but only strengthened and more effectively "socialized," so to speak, and internalized. In the end, only mentally deranged persons would seriously object to it.

According to a bitter Chinese remark, largely applicable in the economic as in other fields, the fall of Khrushchev resulted simply in Khrushchevism without Khrushchev. Yet, as already indicated, it brought at least a striking change in the manner of execution and in tone, if not in fundamental policy. The new leaders abolished Khrushchev's reorganizational reforms, such as the division of the Party in two and the creation of the sovnarkhozy, and discontinued some of his pet projects. They stopped the discussion of the imminent building of communism and the propaganda concerning the early surpassing of the United States in the production of consumer goods. Instead they revealed grave economic shortcomings and failures of the past administration and took a more realistic view of the potentialities of the Soviet economy.

It was in the middle and late 1960's especially that fundamental measures were enacted to bolster Soviet agriculture. Collective farmers finally received a guaranteed wage, which made their position comparable to that of the sovkhoz workers, whereas earlier they had the last claim in the distribution of gain, frequently rendering their very existence marginal, a point emphasized by Lewin and other scholars. Also, pensions and social services were extended to the kolkhoz members. Over a period of years the state greatly increased the amount of resources devoted to agriculture so that investment in agriculture came to constitute over a third in the allocation of the total national investment. Another 4½ per cent of the national income was assigned to subsidize retail food prices to consumers, to keep these prices down in spite of heavy production costs. Still other large sums went into agricultural research. If one adds to these huge expenses some five billion dollars spent by the Soviet Union in 1975–76 alone to buy grain abroad, more money to buy meat and butter, as well as similar huge purchases later, one can get an idea of the enormous effort mounted by the Soviet leadership in the last two and a half decades or so to develop the agricultural sector and to supply the Soviet public with increasing amounts of food at more or less stable prices. Indeed it has been said that instead of being the most depressed social group in their country, the Soviet peasants have become the most pampered, at least in relative terms, that is, by comparison with the treatment they formerly received. This makes their poor condition and the poor condition of Soviet agriculture today all the more remarkable.

The new Five-Year Plan, 1966–70 — eventually designated as the Eighth — presented by Kosygin to the Twenty-third Party Congress in the spring of 1966, reset a number of Khrushchev's economic goals from 1965 to 1970. The economy was to strive for a 49–52 per cent increase in the output of heavy industry and a 43–46 per cent increase in consumer goods, with the annual growth rate of 8.5 per cent and 7.7 per cent respectively — a strikingly high figure for consumer goods in relation to heavy industry, although in line with Khrushchev's thought on the matter on the eve of his fall. Subsequently the Soviet government signed contracts with Italian and French companies to help develop the automobile industry in the Soviet Union.

The Eighth Five-Year Plan was followed by the Ninth, 1971–75, then the Tenth was promoted as the "Five-Year Plan of Quality," the "Basic Directions" for which came out in mid-December 1975, some two months before the Twenty-fifth Party Congress. Yet rather than recapturing the earlier drive the new Plans seemed to testify to a slowdown of the Soviet economy, accentuated by the disastrous crop failures of 1972 and especially of 1975, which necessitated massive purchase of grain abroad — supplied, particularly by the United States in 1972, on terms remarkably advantageous, to be sure, to the Soviet Union. To quote an expert evaluation of the economic position of the U.S.S.R. at the time of the Twenty-fifth Party Congress:

> Soviet economic growth slowed down significantly during the Ninth FYP period (1971–75), and the Plan's ambitious targets were generally — sometimes widely — missed, affected as the USSR has been by declining reserves of labor and other retardational forces and under the blows of two major crop failures. Particularly hard hit was agricultural production, and consumer goods output and consumption levels rose much less than planned. Civilian equipment production and capital formation also fell short of expectations. Nonetheless, Soviet heavy industries expanded

at high rates, and (presumably) military production did well too. The recession in the West made Soviet industrial performance look particularly good.

The just-announced Tenth FYP (1976–80) provides for further retardation in growth throughout the economy. The advance in consumption levels is expected to slow down even further, as are fixed investment and capital formation. Labor productivity will also rise more slowly. Despite the relative moderation of the Plan's goals, they may still turn out to be rather ambitious in relation to resources. No liberalizing reforms seem to be in the wings; rather, there is strong emphasis on centralism in planning and management, with considerable hope placed on mergers of enterprises into rather large units and on computerization. Yet withal the industrial basis of Soviet power — including military might — will certainly continue to grow at a pace that would be creditable for any advanced industrial power.

Since the fall of Khrushchev and in general since the death of Stalin, the standard of living of the urban, and especially of the poverty-stricken rural, population apparently continued to improve, at least until very recently. At the same time, the Soviet Union has been bearing very heavy military expenditures, exemplified by the deployment of anti-missile ballistic systems and by the tremendous growth of the Soviet navy. Economic activities in the U.S.S.R. have spread out, and the economic map of the country is undergoing constant change. Recent illustrations of this change include the rise of Novosibirsk as a great scientific and technological center in Siberia, the Bratsk Dam, the Baikal-Amur mainline railway, the new problem of the industrial pollution of Lake Baikal, and the shift in the center of oil production since the Second World War from its long-time location in the Caucasus to new fields between the Volga and the Urals, and, most recently, also to oil and natural gas fields beyond the Urals.

The new leadership also resorted to economic reform, described generally as an economic "liberalization" and associated with the name of a Kharkov economist, Evsei Liberman. Faced with an economic slow-down, characterized by a drop in the growth rate of the gross national product and by a marked decline in the return from investment and in the growth of productivity of labor as well as by a great loss accruing from an under-utilization of capital and labor resources, the government decided to shift the emphasis and the incentives from the sheer volume of production, where they had been from the inauguration of the First Five-Year Plan, to sales and profits. Under the new system managerial bonuses were to depend not on the output as such, but on sales and profits, the latter factor finally giving serious recognition to the element of cost in Soviet production. In January 1966 forty-three enterprises from seventeen industries, with a total of 300,000 workers, were switched to the new system. Others followed in subsequent months and years. Some economic reform was realized in industry, transportation, and retail trade, and it spread to the sovkhozes and to the construction sector. Yet, ambivalent and probably insufficient to begin with, it was emasculated in the process of implementation, with the result that there proved to be very little difference between the new system and the old system before 1965. More prominent was the new emphasis on material incentives, provisions of more and more differentiated rewards. However, although widely applied, these incentives did not lead to an important improvement in performance.

Indeed, the Tenth Five-Year Plan, 1976–80, and the Eleventh which succeeded it, although on the whole less ambitious than their predecessors, witnessed repeated inability of the Soviet economy to meet set goals, a decline in the increase of labor productivity, and other signs of stagnation. Some specialists considered 1979, the first of the unprecedented four successive years of bad grain harvests, a disastrous turning point. Then and in the years immediately following, seemingly everything, from transportation bottlenecks and difficulty in maintaining the supply of energy to ever-increasing alcoholism and inflation, combined to retard Soviet economic development and to emphasize the seriousness of Soviet economic problems. Other observers wrote more generally of the first successful period of the Brezhnev regime, when the growth of Soviet military and industrial might went hand in hand with a sharp rise in living standards, and of the last stagnant and disappointing years with their ubiquitous shortages of food and consumer goods. At the time of Brezhnev's death perhaps the best evaluation of his eighteen-year stewardship of the Soviet economy, from 1964 to 1982, went as follows (accompanied by a telling comparison with the United States). On the one hand, there was

Steady growth of aggregate output over the eighteen-year period, averaging 3.8 per cent per year, with industrial output growing at an average annual rate of 4.9 per cent.

Steady increase in living standards of the Soviet population, with per capita consumption rising at an average annual rate of 2.7 per cent.

Significant growth in Soviet military power in absolute terms — achieved through a steady increase in real Soviet defense expenditures averaging 4 to 5 per cent per year — as well as in relative terms vis-à-vis the United States.

Reduction of the gap in aggregate and per capita output (GNP) between the Soviet Union and the United States. Whereas in 1965 Soviet GNP was only about 46 per cent that of the United States (38 per cent on a per capita basis), by 1982 it was 55 per cent (47 per cent on a per capita basis).

Reduction of the gap in productivity between the Soviet Union and the United States. While in 1965 the productivity of an average Soviet worker was only 30 per cent that in the United States, by 1982 it was 41 per cent.

Increase in the output of major industrial commodities to the point where, at the beginning of the 1980's, the physical output of many key commodities in the Soviet Union equaled or exceeded that of the United States.

On the other hand, there also was

Steady deceleration in the growth of the Soviet economy. The average annual growth of GNP declined from the peak of 5.2 per cent during 1966–70 to 3.7 per cent during 1971–75, to 2.7 per cent during 1976–80, and to an estimated 2.0 per cent during 1981–82.

Steady deceleration in the growth of living standards, with the average annual growth of per capita consumption declining from a peak of 4.3 per cent during 1966–70 to 2.6 per cent during 1971–75, to 1.7 per cent during 1976–80, and to an estimated 1.2 per cent during 1981–82.

Failure to achieve satisfactory growth in Soviet agriculture. Over the eighteen-year period the average growth rate of GNP originating in agriculture amounted to only 1.7 per cent.

Lack of growth of agricultural productivity both in absolute terms and in relative terms vis-à-vis the United States. While in 1965 the productivity of an average Soviet farm worker was only 14 per cent that in the United States (in the Soviet Union one worker supplied six persons; in the United States one worker supplied forty-three persons), by 1981 it actually declined to a mere 12 per cent (in the Soviet Union one worker supplied eight people; in the United States the corresponding figure was sixty-five).

Although a significant effect of *long-term* weather cycles on grain output in the Soviet Union cannot be ruled out, the most significant failure of the Brezhnev era appears to be grain harvests, which after 1972 repeatedly fell far short of expectations and needs. There were six of these poor harvests over the eleven years: 1972, 1975, 1979, 1980, 1981, and 1982. Whereas the Soviets appeared to be closing the gap in aggregate output with respect to the United States through the mid-1970's, the dramatic slowdown that has taken place in the Soviet Union since 1976 has resulted in some widening of the output gap. The Brezhnev reign was characterized by the highest priority being given to the growth of investment and defense spending except during the period 1964–70. As a result, the per capita consumption of an average Soviet citizen today is still not much more than one-third that in the United States — in fact, over the eighteen-year period under Brezhnev's rule the relative gap remained almost constant.

But while the facts and the statistics seemed reasonably reliable, explanations of them differed. Possibly the most important issue was to what extent Soviet economic difficulties were of a temporary and relatively remediable character and to what extent they were intrinsic to the system.

"The Thaw"

Soviet economic policies from the death of Stalin to the advent of Gorbachev thus demonstrated both the continuation of the main course of development pursued by the deceased dictator, and certain hesitations, reversals, and changes. Also, they indicated somewhat more attention to the immediate needs and wishes of the population than had hitherto been the case. *Mutatis mutandis*, the same or similar generalizations can be made in regard to other aspects of the evolution of the Soviet Union in those years. Stalin's death and especially Beria's fall in the summer of 1953 resulted in a considerable diminution in the role and power of the political police. Khrushchev's denunciation of Stalin gave another shock to the state security apparatus, for it emphasized its horrible past crimes and mistakes and led to a vindication, usually posthumous, of some of its prominent victims. Two developments in relation to the police after Stalin's death deserve special notice: the numbers of forced-labor camps and their inmates was drastically reduced; also, it seems that Soviet citizens gradually lost the immediate and all-pervasive dread of the political police which they had acquired under Stalin. But, although milder, the Soviet Union remained a police state.

As we shall see in a later chapter, Stalin's death was also followed by some relaxation of Party control in the field of culture. Khrushchev's denunciation of the late dictator in itself suggested the need of thorough revaluation of a great many former assumptions and assertions. It also created much confusion. For a number of months in 1956 some Soviet writers exercised remarkable freedom in their approach to Soviet reality and their criticism of it. But, after the Polish crisis and the Hungarian uprising in the autumn of that year, severe restrictions reappeared. After 1956 and until the proclamation of *glasnost*, in the "quiet" years between Stalin and Gorbachev, actual Soviet culture, although not as much hampered and badgered as in the worst days of Stalin and Zhdanov, on the whole faithfully reflected totalitarian Party control. Khrushchev's fall made little difference in this respect. In fact, it can be argued that his successors generally assumed a harder line against dissent, as illustrated by the arrest, trial, and sentencing of Andrei Siniavsky and Julius Daniel in 1965–66 and numerous other instances of cultural suppression since that trial.

The amount of covert opposition and bitterness that this control and the Soviet system in general created can only be surmised. Yet it should be noted that uprisings against Communist regimes took place not only in East Germany, Poland, Czechoslovakia, and Hungary, but also in the

U.S.S.R. itself: notably in the Vorkuta forced-labor camps in the north of European Russia in 1953; in Tbilisi — or Tiflis — the capital of Georgia, in 1956; in Temir-Tau in Kazakhstan among young Russian construction workers, most of them members of the Union of Communist Youth — or Komsomol — in 1959; and in Novocherkassk in 1962. Sporadic riots, strikes, and student demonstrations against the government also occurred in the Soviet Union in more recent years, as in Dneprodzherzhinsk in 1973.

Short of physical violence, the thawing of Soviet society and the emerging opposition views gave rise to the blossoming of a striking and varied *samizdat*, that is, self-published, illegally produced, reproduced, and distributed literature, and to the appearance of dissenting intellectuals and even groups of intellectuals on the fringes of official cultural life. Harassed and suppressed in many ways, including on occasion incarceration in dreadful mental hospitals, the opposition kept nevertheless delivering its message, or rather messages, ranging from a kind of conservative nationalism and neo-Slavophilism to former hydrogen-bomb physicist the late Andrei Sakharov's progressive, generally Westernizer, views and the late Andrei Amalrik's personal, catastrophic, almost Chaadaev-like vision. And it produced the phenomenon of Alexander Solzhenitsyn. Whatever one thinks of that writer, now an exile in the West, in terms of literary stature, ideological acumen, or scholarly precision, most of his works, especially the *Gulag* volumes, are likely to be linked as indissolubly to the Russia of Stalin as Pushkin's *Eugene Onegin* and Turgenev's *Gentry Nest* have been linked to the Russia of the landed gentry — probably unto the ages of ages. Isolated, weak, armed only with a belief in individual moral regeneration, so prominent in Solzhenitsyn, the intellectual opposition has nevertheless been a highly troublesome element in Soviet society and a forerunner of glasnost.

Jewish self-affirmation, protest, and massive migration to Israel (about 235,000 emigrants up to 1985, some 10 per cent of the total Jewish population of the U.S.S.R., with many more applying) — together with the permitted emigration of some non-Jews — represented another development to disturb the post-Stalin Soviet scene, a development closely linked to the intellectual opposition, although also quite distinct. One suspects that the decision to let numerous dissatisfied Soviet citizens leave, while solving the immediate problem of dealing with those people as well as responding in a conciliatory way to world public opinion, potentially raised more questions for the Soviet system than it settled. It is apparently among many Soviet Jews that the alienation from the established order was especially thoroughgoing, as in the anecdotal story of the Moscow Jew who was accused of receiving a letter from a brother in Tel Aviv, although he had claimed that he had no relatives abroad. He explained: "You don't understand: my brother is at home; I am abroad."

The post-Stalin relaxation of restrictions appeared especially striking in an area that spans domestic and foreign policies: foreign travel and international contacts in general. Modifying the former Draconian regulations, which had made a virtually impenetrable "iron curtain" between the Soviet people and the outside world, Soviet authorities began to welcome tourists, including Americans, and allow increasing numbers of their citizens to travel abroad. Always strong on organization, they proceeded to arrange numerous "cultural exchanges," ranging from advanced study in many fields of learning to motion pictures and books for children. Soviet scientists, scholars, athletes, dancers, and musicians, not to mention the astronauts, drew deserved attention in many countries of the world. At the same time Soviet citizens welcomed distinguished visitors from the West and vigorously applauded their performances. In 1976, following the Helsinki agreements of the preceding year, foreign travel and cultural exchange gained further strength, supplying the U.S.S.R. with more international contacts than had been the case at any time since the discontinuation of the N.E.P. Bit by bit, the Soviet Union was becoming better acquainted with the West and the world.

Foreign Relations

Soviet foreign policy after Stalin's death also continued to follow the established pattern in many respects as the U.S.S.R. and the Communist bloc faced the United States and its allies. No conclusive agreements on such decisive issues as control of atomic weapons, general disarmament, or Germany were reached between the two sides. Crises in widely scattered areas appeared in rapid succession. The Soviet Union made a special effort to profit by the emancipation of former Asian and African colonies from Western rule. Yet the post-Stalin policy, especially as developed by Khrushchev, also had its more conciliatory side. The new party secretary elevated the fact of coexistence of the two worlds into a dogma and asserted that all problems would be solved without war. The apparent contradiction of the two approaches probably stemmed from a real inconsistency in Khrushchev's thinking rather than from tactical considerations. It reflected further the dilemma faced by aggressive communism in an age of hydrogen warfare. Brezhnev was to pursue the substance, if not the flamboyant style, of his predecessor's foreign policy, engaging in an enormous arms race and pushing hard Soviet influence and interests in Europe, Asia, the Near East, Africa, and elsewhere, while emphasizing at the same time détente with the United States and the march of history towards peaceful evolution and international cooperation.

Stalin's death and Malenkov's assumption of the leading role in the Soviet Union marked some lessening of international tensions as well as some relaxation at home. The new prime minister asserted that all disputed questions in foreign relations could be settled peacefully, singling out the United States as a country with which an understanding could be reached. In the summer of 1953 an armistice was finally agreed upon in Korea. In the spring of 1954 an international conference ended the war in Indo-China by partitioning it between the Communist Vietminh in the north and the independent state of Vietnam in the south. Although the Soviet Union had not participated directly in the Indo-Chinese conflict, that local war had threatened to become a wider conflagration, and its termination enhanced the chances of world peace. In January 1954, the Council of Foreign Ministers of the four powers, inoperative for a long time, met in Berlin to discuss the German and Austrian treaties, but without result. The Soviet Union joined the United Nations Educational, Scientific, and Cultural Organization, or UNESCO, and the International Labor Organization, or ILO, that April. Malenkov spoke of a further improvement of international relations and of a summit meeting.

That a policy of even moderate relaxation had its dangers for the Soviet bloc became, however, quickly apparent. In early June 1953, demonstrations and strikes erupted in Czechoslovakia, assuming a dangerous form in Pilsen — or Plzeň — where rioters seized the city hall and demanded free elections. In the middle of the month East Berlin and other centers in East Germany rose in a rebellion spearheaded by workers who proclaimed a general strike. Soviet troops re-established order after some bitter fighting. Beria's fall that summer might have been affected by these developments, for the police chief had stressed relaxation and legality since the death of Stalin. Malenkov's resignation from the premiership in February 1955 ended the role of that former favorite of Stalin on the world scene.

Bulganin, who replaced Malenkov as head of the government, became the most prominent Soviet figure in international affairs, although he usually traveled in the company of and acted jointly with the Party chief, Khrushchev. Molotov, in the meantime, continued in charge of the foreign office. "B. and K." diplomacy, as it came to be known, included much showy journeying on goodwill missions in both Europe and Asia. The Soviet Union paid special court to India and other neutralist countries, which had formerly been condemned as lackeys of imperialism. At the same time the two Soviet leaders claimed to be ready to settle the points at issue with the United States and the West. And, indeed, in May 1955 the great powers managed to come to an agreement and conclude a peace treaty with Austria, which included the permanent neutralization of that state as well as certain Austrian payments and deliveries to the U.S.S.R. in recompense for the Soviet return of German property in Austria to the Austrian government. The height of the détente was reached at the summit conference in Geneva in July 1955. While no concrete problems were solved at that meeting, the discussion took place in a remarkably cordial atmosphere, with both Bulganin and Eisenhower insisting that their countries would never engage in aggressive action. The following month Soviet authorities announced a reduction of their armed forces by 640,000 men. In September the U.S.S.R. returned the Porkkala base to Finland and concluded a treaty of friendship with the Finns for twenty years. Yet in the autumn of 1955, as soon as the ministers of foreign affairs tried to apply the attitude of accommodation and understanding expressed by their chiefs to the settlement of specific issues, a deadlock resulted, with Molotov not budging an inch from the previous Soviet positions and demands. The "spirit of Geneva" proved to be an enticing dream rather than a reality.

Since the *rapprochement* between the U.S.S.R. and the West failed to last, the polarization of the world continued. Following the Communist victory in northern Indo-China, the Manila pact of September 1954 created the Southeast Asia Treaty Organization, or SEATO. Great Britain, France, Pakistan, and Thailand joined the four countries already allied, the United States, Australia, New Zealand, and the Philippines, to establish a new barrier to Communist expansion in Asia. In Europe, West Germany rose steadily in importance as an American ally and a member of the Western coalition. The Soviet Union in its turn concluded the so-called Warsaw Treaty with its satellites in May 1955 to unify the Communist military command in Europe.

The year 1956 was a memorable one in Soviet foreign policy. Khrushchev's February speech denouncing Stalin further shook the discipline in the Communist world. On the other hand, the improvement in Soviet-Yugoslav relations, which had begun with Bulganin's and Khrushchev's visit to Belgrade in 1955, received a boost, the break between the two states now being blamed on Stalin himself as well as on Beria. In April 1956 the Cominform was dissolved, and in June Shepilov replaced Molotov as foreign minister. The ferment in the Soviet satellite empire finally led to explosions in Poland and in Hungary. In late June 1956, workers in Poznan clashed with the police and scores of people were killed. Polish intellectuals and even many Polish Communists clamored for a relaxation of controls and a generally milder regime. On October 19, Wladyslaw Gomulka, who had been imprisoned as a Titoist and had been reinstated in

August, became the Party secretary. That same day Khrushchev and other Soviet leaders flew to Warsaw to settle the crisis. In spite of extreme tension, an understanding was reached: the U.S.S.R. accepted Gomulka and a liberalization of the Communist system in Poland and agreed to withdraw Soviet troops from that country.

Events in Hungary took a graver turn. There, under the influence of the happenings in Poland, a full-scale revolution took place in late October, during which the political police were massacred. The army sided on the whole with the revolutionaries. The overturn was spearheaded by young people, especially students and workers. The new government of a revisionist Communist, Imre Nagy, constituted a political coalition rather than single-party rule and withdrew Hungary from the Warsaw Treaty. But on November 4, after only a few days of freedom, Soviet troops began storming Budapest and crushed the revolution. The imprudent attack on Egypt staged at that time by Great Britain, France, and Israel over the issue of the Suez Canal helped the Soviet move by diverting the attention of the world, splitting the Western camp, and engaging some of its forces. While crushing the Hungarians, the U.S.S.R. championed the cause of Egypt and threatened its assailants. But the moral shock of the Hungarian intervention proved hard to live down: it led to the greatest popular condemnation of the Communist cause and the most widespread desertions from Communist party ranks in the free world since the Second World War. There were strikes, demonstrations, and protests even in the Soviet Union.

As already suggested, Khrushchev might have been lucky to survive these grave perturbations in the Communist world. Yet he did defeat and dismiss Malenkov, Molotov, and Kaganovich, together with Shepilov who sided with them, in the spring and summer of 1957. After Bulganin's fall in March 1958, the first secretary, now also prime minister, became the undisputed chief of Soviet foreign policy, while Andrei Gromyko headed the foreign office. Khrushchev's behavior on the international scene showed a certain pattern. He remained essentially intransigent, pushing every advantage he had, be it troubles in newly independent states, such as the Congo, or Soviet achievements in armaments and space technology. Nevertheless, he talked incessantly in favor of coexistence and summit conferences to settle outstanding issues. Also, he paid friendly visits to many countries, including the U.S.A. in 1959. The summit conference in the summer of 1960 was never held, for two weeks before it was scheduled to begin Khrushchev announced that an American U-2 spy plane had been brought down deep in Soviet territory. But in 1961 Khrushchev met the new American president, John F. Kennedy, in Vienna. In the summer of 1962 both aspects of Soviet foreign policy stood in bold relief: fanned by the U.S.S.R., a new Berlin crisis continued to threaten world peace; yet, on the other hand, Khrushchev emphasized more than ever coexistence abroad and peaceful progress at home, having made that his signal theoretical contribution to the program that was enunciated at the Twenty-second Party Congress. To be sure, as officially defined in the Soviet Union, coexistence meant economic, political, and ideological competition with the capitalist world until the final fall of capitalism. But that fall, Soviet authorities came to assert, would occur without a world war.

However, in the autumn of that same year, Khrushchev overreached himself and brought the world to the brink of a thermonuclear war. The confrontation between the United States and the U.S.S.R. in October 1962 over the Soviet missiles in Cuba, which resulted in a stunning Soviet defeat, can be explained, at least in part, by the Soviet leader's enthusiasm and his conviction that the United States and capitalism in general were on the decline and would retreat when hard pressed. The outcome, no doubt, strengthened the argument for peaceful coexistence and emphasized caution and consultation in foreign policy, symbolized by the celebrated "hot line" between Washington and Moscow. The Soviet Union proceeded to measure carefully its reactions and its involvement even in such complicated and entangling crises as the Israeli-Arab wars of 1967 and 1973 and the Vietnam War. In the latter conflict, the Soviet Union denounced of course "American imperialism" and provided extremely valuable matériel to North Vietnam, but it avoided escalation. Yet, following the complete victory of communism in Indo-China in 1975 and the shattering impact of the catastrophic American policy in Vietnam on the American public, the Soviet Union might have felt that it had a freer hand on the international stage, in Angola or elsewhere.

With the Soviet Union as well as the United States acquiring a second-strike capability, that is, the ability to retaliate and inflict "an unacceptable damage" on the enemy after absorbing a nuclear blow, a true balance of terror settled on the world. Ever-improving technology made virtually all established strategic concepts obsolete. Numerous bases and indeed whole sections of the globe lost their importance in terms of the possible ultimate showdown between the two nuclear giants.

From the mid-seventies it was authoritatively estimated in the West — and apparently realistically in contrast to earlier alarms about alleged "missile gaps" and the like — that the U.S.S.R. had caught up with the United States in overall nuclear military strength, and indeed had perhaps moved slightly ahead. Even the Soviet navy, insignificant compared with its American rival at the end of the Second World War, had risen to be, according to many indices, the strongest fleet in the world, although still behind the Americans in aircraft carriers and perhaps in such intangibles as naval tradition and the expertise and spirit of its personnel. Yet the enormous economic burden, terror, and inconclusiveness of the arms race have not so far led to a full negotiated settlement. Important results were achieved, to be sure. Following the earlier banning of nuclear tests in the atmosphere, the nuclear non-proliferation agreement was signed by the two superpowers and other states in early 1968. Other agreements were reached concerning outer space, where 1975 witnessed the celebrated joint effort of the Russians and the Americans. The crucial issue of military limitations itself was tackled in numerous negotiations, including the so-called S.A.L.T. II talks and President Ford's discussions with Brezhnev in Vladivostok in 1974. Still, in spite of a considerable measure of agreement, the S.A.L.T. II talks remained inconclusive, primarily because of the problems of the Backfire bombers on the Soviet side and of the cruise missile on the American. Moreover, as Edward Teller and other scientists have pointed out, the difficulty in the negotiations resides not only in the entire complex of aims, attitudes, and policies of the two superpowers, but also in the very nature of scientific and technological advance, which rapidly makes pre-arranged schemes of limitation obsolete.

The very closely related but even larger issue of détente between the Soviet Union and the United States also sailed to an uncertain future. With explicit "cold war" a thing of the past, détente scored a resounding success at the Helsinki conference in the summer of 1975, where the United States and other Western countries accepted in effect the communist redrawing of the map of central and eastern Europe following the Second World War in exchange for unsubstantiated promises of greater contacts between the two worlds and a greater degree of freedom in those contacts. But a comprehensive economic agreement between the U.S.S.R. and the U.S.A. failed over the questions of the most favored nation clause, credits, and the American concern with the fate of Soviet Jews. Furthermore, before long détente was again swamped by new international developments, to be detailed later in this chapter.

Ironically, while Soviet-American relations improved and became more stable after the Cuban confrontation, and while the Soviet leaders found welcome in Gaullist France and other capitalist countries, their standing in the communist world deteriorated. The conflict with China broke out into the open around 1960 and has widened and deepened since. After the abrupt withdrawal of Soviet personnel from China in August of that year and the discontinuance of assistance, relations between the two countries quickly became those of extreme antagonism. To the sound of violent mutual denunciations the two states and parties competed with each other for the leadership of world communism, the Chinese usually championing the revolutionary position against Soviet "revisionism." Moreover, China became an atomic power and formulated large claims on Soviet Asiatic territory. Observers noted that international crises such as the war in Vietnam only intensified the hostility between the two great communist states. Although China remained far behind the U.S.S.R. in industrial and technological development and although it was fully preoccupied with a "cultural revolution," its aftermath, and other internal problems, it could pose a major threat to the Soviet Union in the future, if not in the immediate present.

Problems in eastern Europe have proved to be more pressing. The twelve years which followed the suppression of the Hungarian revolution witnessed Soviet attempts to adjust to changing times, to allow for a communist pluralism with a considerable measure of institutional and eventually even ideological diversity. In Brzezinski's phrase, satellites were to become junior allies. Even Tito usually received a kind of fraternal recognition, and he spoke with authority. Yet tensions persisted and indeed increased, both between the different east European countries and the Soviet Union and within those countries as most of them proceeded with de-Stalinization, economic liberalization, and other important changes. The break with China led in 1961 to the unexpected departure of Albania into "the Chinese camp." Rumania under its new leader, Nicholas Ceausescu, showed a remarkable, even stunning, independence from the Soviet Union, although it remained barely within the communist bloc and continued a hard-line policy at home. Poland, belying the promise of 1956, had its progress toward freedom arrested, and concentrated its energy on trying to contain, by petty and persistent persecution, the Catholic church, liberal intellectuals and students, and other forces favoring change. Its problems and plight might be particularly relevant to the future evolution of Soviet society itself.

The developments in Czechoslovakia led to a catastrophe. That highly Western country with a democratic tradition remained long under a form of Stalinism practiced by Antonin Novotny and his clique. But when in the early months of 1968 Novotny was finally deposed, the new Party leadership, of Alexander Dubček and others, championed an extremely liberal course which included the abolition of censorship. The sweeping liberal victory in Czechoslovakia which was to be confirmed and extended at a forthcoming Party congress led to consternation in the governing circles of the Soviet Union, East Germany, Poland, Bulgaria, and possibly Hungary. Exchanges of opinion and an unprecedented face-to-face discussion between the members of the Politburos of the Soviet Union and of Czechoslovakia seemed momentarily to resolve the conflict. Then on the twentieth of August, Soviet troops, assisted by the troops of the four allies, invaded Czechoslovakia and quickly occupied the country. There was very little bloodshed, because the Czechoslovak armed forces had been instructed not to resist. Soviet intervention was probably caused, in no certain order of priority, by fear for the Warsaw Pact which the Czechs wanted to modify although not abandon, by the hatred of Czech liberalization with its critique of the U.S.S.R., by the concern lest liberalism at home be too much encouraged, and by the need to respond to the pleas of the Soviet allies, especially East Germany, who saw the developments in Czechoslovakia as an immediate threat to their own regimes. The repercussions of the intervention lasted long after the summer of 1968.

In Poland, the 1970 replacement of Gomulka by Gierek as Party secretary was followed by the introduction of an ambitious scheme to modernize and expand Polish industry and trade with the aid of Western capital and technology. By 1976, it was evident that Gierek's loudly hailed economic "acceleration" had begun to fail. Continuing world economic crisis and mismanagement and corruption at all levels of Party and government apparatus, as well as the ever-increasing cost of participating in the Soviet-directed Council of Mutual Economic Assistance (COMECON) and the Warsaw Pact, all contributed to Poland's difficulties. In 1976, workers' protests and strikes over drastic increases in food prices were followed by the rapid formation and activation of dissident organizations and clandestine printing establishments. The Catholic Church, its traditional prestige fortified by the election of the Archbishop of Cracow, Cardinal Karol Wojtyla, to the papal throne (Pope John Paul II), also spoke out strongly against many of the Communist government's policies. The Gierek regime was unable to suppress the opposition effectively, in part at least, it would seem, because of its heavy dependence on continuing Western loans, required to keep the economy solvent, and the consequent need to avoid drastic action which could lead to the cutting off of Western funds.

The summer of 1980, with continuing labor unrest and economic near-collapse, led to the change of Party leadership and to a formal agreement between the Polish government and the great majority of Polish workers, now mostly represented by independent "Solidarity" trade unions and led by a charismatic veteran of the struggle for workers' rights in Poland, electrician Lech Walesa. The agreement, accepted by the workers as a foundation for a dialogue with the government, appears to have been a tactical maneuver of the Communist authorities. No regular contacts with the Solidarity leadership and the Catholic hierarchy aimed at creating a constructive and meaningful national consensus were initiated by the government. By exploiting its monopoly over the mass media and over the distribution of increasingly scarce food supplies and consumer goods, the government attempted to undermine the position of the opposition while at the same time strongly seconding Moscow's accusations that Solidarity was attempting to subvert the political structure and international position of People's Poland. The rise to prominence of General Wojciech Jaruzelski, who progressively combined the posts of Minister of Defense, Premier, and

First Secretary of the Party, coincided with a gradual militarization of the administration of important branches of government and industry.

All this was done in preparation for the military coup which was executed in close cooperation with the Soviet authorities on December 13, 1981. Active resistance against the overwhelming forces of the regime was quite limited, and, from a military standpoint, the operation was carried out rather effectively. Nevertheless, the "success" of General Jaruzelski's junta was very dubious. Although thousands of Solidarity activists, including Lech Walesa, and other dissidents were arrested and placed in internment camps, some leaders of the movement escaped arrest and an underground opposition began to form. Western economic sanctions and continuing passive resistance to the regime in the factories, offices, schools, and universities were making the task of running the country extremely difficult for the Jaruzelski regime. By the end of 1982, there appeared to be two clear choices before the military government of Poland: either to continue with the martial law administration, further alienating the population and risking a total economic collapse of the country, or to end martial law and attempt to open the few remaining channels of contact with the great majority of the Polish population in an effort to reduce tensions and improve the performance of the economy. The choice was not an easy one for the Polish Communist authorities — and their Soviet sponsors.

The Soviet invasion of Afghanistan in late December 1979 produced a strong impression in the world. The impression was exacerbated by the fact that, although the so-called Afghan rebels could not match the Red Army in open fighting, they could not be entirely destroyed either. More than five years after the original invasion, when Gorbachev came to power, the Soviet Union was still employing perhaps 100,000 of its troops in the Moslem country, and it was not clear how much of that country, outside the main cities, was under Soviet control. Critics pointed out that the Afghan invasion represented the first direct Soviet use of military force outside "its own" east European empire since the Second World War. The massive intervention was also interpreted as the first step in a bid for the oil of the Middle East and a general takeover of that region. It can well be argued, on the other hand, that the decisive Soviet move was essentially defensive: communism had actually come to Afghanistan some two years earlier in a peculiar internal struggle which pitted two communist factions against each other as well as against other groups; the Soviet choice in late 1979 was that between intervention and witnessing a neighboring communist state, which it had already welcomed and supported as part of the communist world, go down to popular opposition. But, defensive or not, the Soviet step was certainly a grave and disturbing one.

As of 1985, tension between the Soviet Union and the United States, the East and the West, was not confined to the crucial problems of Afghanistan and Poland. Rather, the two sides opposed each other all over the world, from Central America to southern Africa, Lebanon, and Cambodia. To be sure, western European countries, in spite of strong United States objections and even sanctions against particular companies, continued to support the building of a natural-gas pipeline from western Siberia to western Europe. But they were also apparently prepared to proceed with the installation of United States middle-range missiles to counteract the already established Soviet ones, an installation most especially opposed for years by Brezhnev. The virtually all-important Soviet-American disarmament negotiations remained deadlocked. S.A.L.T. II was not ratified by the United States Senate, and its future chances appeared slim, especially after the departure of Carter from the Presidency. In fact, numerous critics accused the tougher anti-Soviet tone of the Reagan administration as largely precluding adjustment and agreement. Yet the administration itself and others claimed that it was precisely this firmer approach, and especially the concurrent building up of the United States nuclear and military might, that would force the U.S.S.R. to negotiate effectively for disarmament.

X L I

* * * * * * * * * *

SOVIET SOCIETY AND CULTURE

The Soviet Union is a contradictory society halfway between capitalism and socialism, in which: (a) the productive forces are still far from adequate to give the state property a socialist character; (b) the tendency toward primitive accumulation created by want breaks out through innumerable pores of the planned economy; (c) norms of distribution preserving a bourgeois character lie at the basis of a new differentiation of society; (d) the economic growth, while slowly bettering the situation of the toilers, promotes a swift formation of privileged strata; (e) exploiting the social antagonisms, a bureaucracy has converted itself into an uncontrolled caste alien to socialism; (f) the social revolution, betrayed by the ruling party, still exists in property relations and in the consciousness of the toiling masses; (g) a further development of the accumulating contradictions can as well lead to socialism as back to capitalism; (h) on the road to capitalism the counterrevolution would have to break the resistance of the workers; (i) on the road to socialism the workers would have to overthrow the bureaucracy. In the last analysis, the question will be decided by a struggle of living social forces, both on the national and the world arena.

TROTSKY

The party leadership of literature must be thoroughly purged of all philistine influences. Party members active in literature must not only be the teachers of ideas which will muster the energy of the proletariat in all countries for the last battle for its freedom; the party leadership must, in all its conduct, show a morally authoritative force. This force must imbue literary workers first and foremost with a consciousness of their collective responsibility for all that happens in their midst. Soviet literature, with all its diversity of talents, and the steadily growing number of new and gifted writers, should be organized as an integral collective body, as a potent instrument of socialist culture.

GORKY

THE BOLSHEVIKS' SEIZURE of power in Russia in November 1917 meant a social as well as a political revolution. The decades that followed "Great October" witnessed a transformation of Russian society into Soviet society. They also saw the emergence and development of an unmistakably Soviet style of culture. In spite of its enormous size, huge population, and tremendous variety of ethnic and cultural strains, the U.S.S.R. is a remarkably homogeneous land, for it reflects throughout its length and breadth — "from Kronstadt and to Vladivostok," to quote a Soviet song — some seventy-five years of Communist engineering, social and cultural as well as political and economic.

The Communist Party of the Soviet Union

The Communist party has played in fact, as well as in theory, the leading role in Soviet society. Its membership, estimated at the surprisingly low figure of less than twenty-five thousand in 1917, passed the half million mark in 1921 and the million mark in the late twenties. The number of Soviet Communists continued to rise, in spite of repeated purges which included the frightful great purge of the thirties, and reached the total of almost four million full members and candidates when Germany invaded the U.S.S.R. While many Communists perished in the war, numerous new members were admitted into the Party, especially from front-line units. Postwar recruitment drives further augmented Party membership to seven to nine million in the immediate postwar years, as much as thirteen million in 1967, 16,380,000 in 1978, and almost 20 million in the 1980s. More recent developments, however, led to the loss of some 2.7 million members, resulting in the figure of 16.5 million members in April 1991. (The Komsomol, mentioned below, was reduced from 42 to 23.6 million members.)

These figures, of course, by no means tell the entire story of Communist penetration into Soviet life. As already emphasized, the party, in the Leninist view which served to differentiate the Bolsheviks from the Mensheviks, comprises a fully conscious and dedicated elite, exclusive by definition, but also educating and guiding other organizations and, indeed, the broad masses. In addition to the Party proper, there exist huge youth organizations: Little Octobrists for young children, Pioneers for those aged from nine to fifteen, and the Union of Communist Youth, or Komsomol, with members in the fourteen to twenty-six age range. The first two organizations, and, of late, even the Komsomol, have acted as Party agencies for the general education of the younger Soviet generations, opening their doors wide to members. The Party has also worked with and directed uncounted institutions and groups: professional, social, cultural, athletic, and others. In fact, from the official standpoint, Soviet society has only one ideology and only one outlook, the Communist; citizens and groups of citizens differ solely in the degree to which they incarnate it. That sweeping assumption, it might be added, expresses especially well the monolithic and totalitarian nature of the Soviet system, and its modification or abolition, indeed the proscription of the party itself, is one of the crucial issues today.

The Party demands the entire man or woman. Lenin's example illustrates the ideal of absolute and constant dedication to Party purposes. The word *partiinost,* translated sometimes as "Party-mindedness," summarizes the essential quality of a Communist's life and work. While the early emphasis on austerity has been greatly relaxed since the thirties, especially in the upper circles, the requirements of implicit obedience and hard work generally remain. In particular, Party members are expected throughout their lives both to continue their own education in Marxism-Leninism and to utilize their knowledge in all their activities, carrying out Party directives to the letter and influencing those with whom they come in contact. While exacting, the "Party ticket" opens many doors. It constitutes in effect the greatest single mark of status, importance, and, above all, of being an "insider" in the Soviet Union. Although, to be sure, many Soviet Communists are people of no special significance, virtually all prominent figures in the country are members of the Party. Since the Second World War special efforts have been made to assure that such fields as university teaching and scientific research are largely in the hands of Communists. Conversely, it has become much easier for outstanding people to join the Party.

The social composition of the Communist party of the Soviet Union has shown some fluctuation. Ostensibly the true party of the proletariat, prior to 1917 it had a largely bourgeois leadership and no mass following of any kind. The workers as a group, however, did support it in November 1917 and during the hard years that followed the Bolshevik seizure of power. The Party naturally welcomed them, while at the same time displaying extreme suspiciousness toward those of "hostile" class origin. With the stabilization of the Soviet system and the inauguration of the five-year plans, "Soviet intellectuals," in particular technical and administrative personnel of all sorts, became prominent. On the eve of the Second World War the Party was described as composed 50 per cent of workers, 20 per cent of peasants, and 30 per cent of Soviet intellectuals, with the last group on the increase. That increase has continued since the war, as social origin became less significant with time and the authorities tried to bring all prominent people into the Party. It might be noted that, in relation to their numbers, peasants have been poorly represented, indicating the difficulty the Communists have experienced in permeating the countryside. The proportion of women has increased, and they constitute at present about one-quarter of the membership of the Party.

The Communist party of the Soviet Union is very thoroughly organized. Starting with primary units, or cells, which are established where three or more Communists can be found, that is, in factories, collective farms, schools, military units, and so forth, the structure rises from level to level to culminate in periodic Party congresses, which constitute important events in Soviet history, and in the permanently active Central Committee, Secretariat, and Politburo. At every step, from an individual factory or collective farm to the ministries and other superior governing agencies, Communists are supposed to provide supervision and inspiration, making it their business to see that no undesirable trends develop and that production goals

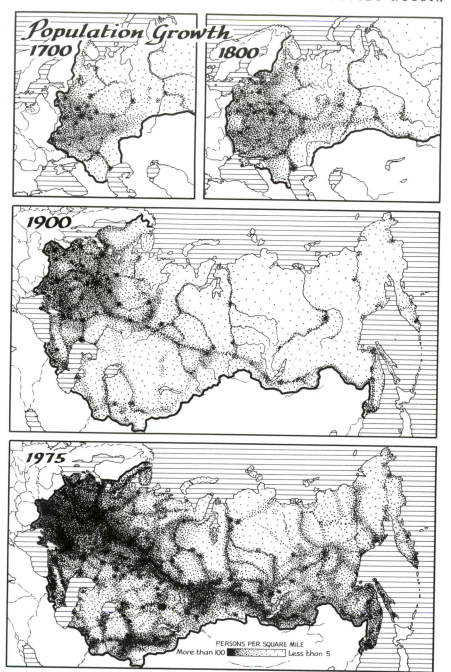

Population Growth
1700
1800
1900
1975

PERSONS PER SQUARE MILE
More than 100   Less than 5

are overfulfilled. At higher government levels, as already indicated, the entire personnel consists of Communists, a fact which nevertheless does not eliminate Party vigilance and control. In general, rotation between full-time government positions and Party administrative positions is common. It should be noted that the guiding role of the Party has asserted itself with increased force since Stalin's death, for — as L. Schapiro and other close students of Soviet communism have indicated — the late general secretary's dictatorial power had grown to such enormous proportions that it had put even the Party into the shade.

It is this enormous and enormously important Communist party and apparatus that the Soviet successor states have to incorporate or replace.

The Destruction of the Old Society

Whereas the Great October Revolution catapulted the Communist party to power, it led to the destruction of entire social classes. Indeed, its initial impact resulted in a sweeping leveling of traditional Russian society. The landowning gentry, for centuries the top social group in Russia, disappeared rapidly in 1917 and 1918 as peasants seized their land. The upper bourgeoisie, financial, industrial, and commercial, was similarly eliminated when the Bolsheviks nationalized finance, industry, and trade. The middle and especially the lower bourgeoisie, to be sure, staged a remarkable comeback during the years of the New Economic Policy. Their final destruction, however, came with the implementation of the five-year plans. If the gentry occupied the stage in Russia too long, the bourgeoisie was cut down before it came into its own. The clergy, the monks and nuns, and other people associated with the Church, constituted yet another group to suffer harsh persecution, although in their case it stopped short of complete annihilation. The great majority of the intellectuals, too, found themselves in opposition to the new regime. Many of them emigrated. Many others perished in the frightful years of civil war and famine. In fact, although some of its members remained, the intelligentsia as a cohesive, articulate, and independent group was no more.

The Peasants

Whereas the Bolsheviks regarded the upper and middle classes as enemies by definition, they believed themselves to be acting in the interests of the masses, that is, of the workers and of the peasants. As it turned out, however, the peasants have borne the brunt of the privations and sacrifices imposed by the Soviet "builders of socialism." The total population of the U.S.S.R. was officially given in the spring of 1959 as only 208,826,000 — and as 262,400,000 according to the census of 1979 — a low figure which

testifies to two demographic catastrophes: the one associated with the
First Five-Year Plan, more especially the collectivization of agriculture,
and the other resulting from the Second World War. In both cases
peasants — and peasants as soldiers — suffered the most, dying by the
millions. The extent to which the Soviet Union has been a land of peasants
is indicated by the fact that the rural population constituted 82 per cent
of the total in 1928 and that, after some sixty-five years of industrialization
and urbanization, it still constitutes one-third of it today.

Of course, peasants have carried such a heavy burden in the U.S.S.R.
not only because of their vast numbers, but also because of the policies
pursued by the government. Lenin's original endorsement of the peasant
seizure of gentry land had great appeal in the countryside. Influenced by
the Bolshevik land policy and by revolutionary soldiers returning home —
a point effectively emphasized by Radkey — the rural masses proved
reasonably well inclined toward the new regime and on the whole apparently preferred it to the Whites during the great civil war. But War
Communism antagonized many of them. Besides, the Bolsheviks tried to
split the peasants, inciting the poor against the better-off and later attempting to utilize the poor and the middle peasant against the so-called
kulak. While some social differentiation did exist in the villages, the authorities, applying abstract Marxist formulas where they did not fit, exaggerated it beyond all measure and ended by, in effect, condemning and
punishing all peasants who did not behave in the prescribed manner.

The respite during the N.E.P., in the course of which rural Russia
recovered and in part even began to experience something akin to prosperity, was followed by the all-out offensive of the First Five-Year Plan.
Five million kulaks and members of their families disappeared. Countless
peasants, recalcitrant or relatively prosperous or simply unlucky, populated
forced-labor camps. Other uncounted peasants starved to death. Scenes
of horror in the once bounteous Ukraine defied description. But, as we
know, the peasants, in spite of their resistance, were finally pushed and
pulled into collectives. The typical member of a kolkhoz is a new phenomenon in Russian history. The novelty lies not in his wretched poverty,
not even in the extremely heavy exactions imposed upon him, but in the
minute state organization and control of his work and life. While peasants
profited from certain Soviet policies, notably the spread of education, and
while some of them rose to higher stations in society, on the whole the
condition of the rural masses, the bulk of the Soviet people, remained
miserable and at times desperate. Largely supporting the five-year plans
by their labor, as already explained, Soviet peasants received very little
in return. After Stalin's death, Khrushchev and other leaders admitted the
grave condition of the Soviet countryside, while writers presented some
unforgettable pictures of it during the relative freedom of expression that

prevailed for several months in 1956. Recent years, to be sure, have witnessed an improvement. Yet rural Russia remains poor. Moreover, the party and the government continued their social engineering, as clearly indicated in such postwar measures and projects as the increase in the size of the collective farms, the abortive agrogoroda, the temporary emphasis on the sovkhoz form of agriculture, and the periodic campaigns against the private plots of kolkhoz members. Indeed — logically, from their point of view — Communists were not likely to relax until peasants disappeared as a separate group, having been integrated into a completely socialized, mechanized, and urbanized economy. No wonder that the coming of *perestroika* made peasant landownership a central issue and one very difficult to handle.

The Workers

Industrial workers in many ways profited most from the Bolshevik revolution. That revolution was made in their name, and they gave the new regime its greatest social support. Because of this, perhaps a million and a half workers and their children rose to new importance. They became Party functionaries, Red Army officers, and even organizers of collective farms. Many received rapid training to be graduated as technologists. Persons of a proletarian background enjoyed priorities in institutions of higher learning and elsewhere. The upward social mobility of workers was all the more remarkable because their total number was not very large, and it contrasted sharply with the relatively static nature of tsarist society. Many prominent Soviet people in all walks of life today owe their positions to that rise.

But, of course, while many workers went up the social ladder, new men and women entered the factories. After the inauguration of the five-year plans the influx turned into a deluge. Peasants of yesterday became workers of today. Russia finally acquired vast crowds of proletarians characteristic of the industrial revolution. Whether the condition of the workers in the Soviet Union improved compared to tsarist times remains an open question. That it continued to be miserable cannot be reasonably doubted. Soviet workers profited from increased educational and cultural opportunities, but their pitiful real wages probably remained below the prerevolutionary level as late as the early 'fifties. After all, the huge industrialization was made possible by keeping industrial wages down as well as by squeezing the peasants. In addition, workers suffered from the totally inadequate and deteriorating urban housing, and, together with other Soviet citizens, they had to contribute their efforts and their scarce time to various "voluntary" projects, to their own and others' political education, and to other prescribed activities. In contrast to tsarist days, they could not strike or otherwise openly express their discontent. The material condition of the Soviet proletariat has improved, however, since the death of Stalin. How much it can improve further is indissolubly linked to the present crisis of the Soviet economy and its future resolution.

The "New Class"

Whereas the initial impact of the Bolshevik revolution, coupled with famine and other catastrophes, did much to level Russian society, smashing the rigid class structure of imperial Russia and even destroying entire classes, before long social differentiation began to grow again. In particular, the five-year plans produced a tremendous expansion of administrative and technical personnel, which, together with the already existing Party and government bureaucracies, became, broadly speaking, the leading class in the country. One author estimated that the Soviet economy employed 1,700,000 bookkeepers alone! Scientists, writers, artists, professors, and other intellectuals, purged and integrated into the new system, became prominent members of the privileged group. Army and naval officers and their families provided additional members. Altogether, the privileged, distinguished primarily by their education and nonmanual occupations, came to compose about 15 per cent of the total population. Relatively speaking — paradoxically, if you will — they have enjoyed greater advantages compared to the masses than have their counterparts in Western capitalist societies. It is also of interest that material differences within the educated class and within the worker and peasant classes, who are often paid according to some form of the piece rate, have been very marked in Soviet Russia. Paid vacations and other rewards supplied by the regime have been distributed in a similarly uneven manner. In fact, wages and salaries have tended to show a greater differentiation in the U.S.S.R. than in the West, although, of course, Soviet citizens could not accumulate fortunes based on profits, rent, or interest.

The "Great Retreat"

As the new Soviet elite advanced to the fore, Soviet society lost many of its revolutionary traits and began to acquire in certain respects a strikingly conservative character. The transformation occurred essentially during the thirties, but on the whole it continued and developed further during the Second World War and in the postwar years. While state laws and regulations were crucial in this process, they reflected, as well as contributed to, basic social and economic changes.

Initially the Bolshevik regime took a disdainful and even negative view of the family. Marriages became matters of little importance in the eyes of the state, while divorce could be obtained simply by declaration of one of the parties involved. Abortions were legal and extremely common. In the thirties, all that changed. Authorities declared themselves in favor of a strong Soviet family. Particular emphasis was placed on having many children. Mothers with five or six living offspring received the Motherhood Medal, those with seven or eight were awarded a decoration known as Motherhood Glory, while those with ten achieved the status of Heroine Mother. Financial grants to large families helped further the implementation of the new policy. At the same time abortions lost their legal sanction, while divorce became much more difficult to obtain in the U.S.S.R. than in most countries in the West. The family — the proper, Marxist, Soviet family, to be sure — was hailed as a mainstay of the socialist order.

Discipline improved in the army, and it made an effective reappearance in schools and elsewhere. Ranks, titles, decorations, and other distinctions, whether bureaucratic, military, or academic, were restored and acquired vast importance. Even social manners made a comeback. Pomp and circumstance re-entered the stage. Uniforms blossomed everywhere, reminding observers of tsarist Russia. Generalissimo Stalin toasting his marshals at a gargantuan Kremlin reception presented a far different picture from Lenin in his worn-out coat haranguing workers in squares and factory yards. In a sense, the Soviet regime had arrived. Equally important changes took place, as we shall see, in education and culture, where the avantgarde and experimental approach of the early years gave place to rockribbed conservatism. Patriotism and historical tradition emerged again, although in a minor key and as aids, rather than rivals, to the fundamentally Marxist ideology.

Women and Feminism

Women have constituted half, actually considerably more than half, the population of the Soviet Union, and they have certainly contributed their share to its history. In a very real sense they have carried half, or more, of the burden of that history on their shoulders. The communist program included liberating women from oppression, discrimination, and drudgery as part of the liberation of humankind. The first decade or more after the October Revolution was full of promise for Soviet feminists, as well as of new departures in the position and activities of Soviet women, perhaps most notably and permanently so among the Islamic peoples of the country. But, for the Soviet leaders, feminist ideals were always ancillary to the fundamental Marxist vision of class struggle and the building of socialism. And they were crushed, together with other autonomous views, once the U.S.S.R. was set in the firm Stalinist mold. There has been some relaxation but no basic change in the situation since the death of the crucial dictator.

Lapidus and other scholars have done much recently to present and interpret the position of Soviet women in both its positive and its negative aspects. The former include, notably, the great increase in education, to where women are now proportionately better represented as students in Soviet institutions of higher learning than men. Concurrently women have risen remarkably in the professions, so that today, for example, the great majority of the doctors of medicine in the U.S.S.R. are women. Yet, as it has been repeatedly pointed out, few women reach the top rungs of their profession, medicine included, and they are strikingly absent at the highest levels of both Party and government. Moreover, Soviet women both hold full-time jobs — to the point that, broadly speaking, there is no reserve of employable women left at present in the Soviet economy — and perform the great bulk of the work at home, a task made all the more difficult by the hard conditions of life in the Soviet Union. It might be added that feminism in the Western sense is at best in its incipient stage in the U.S.S.R. today. Nor are all its emphases — as a student of Soviet society will readily understand — particularly relevant to the Soviet scene.

The Nationalities

Its multinational composition has been a major problem for the Soviet Union as it was for the Russian Empire. While Great Russians form about half of the population of the U.S.S.R., and Ukrainians and White Russians, or Belorussians, approximately another quarter, the remaining quarter consists of a staggering variety of ethnic and linguistic groups. The Caucasus alone contains a fantastically complicated mixture of peoples. More than a hundred and fifty languages and dialects are spoken in the Soviet Union. Soviet nationalities range from ancient civilized peoples, such as the Armenians and the Georgians, to primitive Siberian tribes. They include Lutherans and Catholics as well as Orthodox, and Moslems and Buddhists together with shamanists. Moreover, many of these peoples showed nationalist tendencies in the years of revolution and civil war, which corresponded only too well to the generally nationalist atmosphere of the twentieth century.

Soviet authorities developed several basic policies in dealing with national groups. They allowed them no independence in ideological, political, economic, or social matters, and even no deviation from the established official line. The U.S.S.R. remained essentially a most highly centralized state. The single Communist party of the Soviet Union acted as an especially important foundation and guarantee of that unity. At the same time, Soviet rulers granted a kind of cultural autonomy to the nationalities in the Soviet Union, stating that their cultures should be "national in form, and socialist in content." The form included the language and the cultural tradition of a given people, which, however, had to be fitted, as in the case of the Russians proper, into the Soviet-Marxist framework. Thus, the government tried to destroy Islam as well as Orthodoxy and interpreted Georgian history as well as Russian in the simple terms of a class struggle.

But this dual approach to nationalities proved difficult to maintain in practice. Cultural autonomy could easily become cultural nationalism, and that in turn would lead to separatism. Always suspicious, the Soviet leadership kept uncovering "bourgeois nationalists" in union republics and lesser subdivisions of the U.S.S.R. In the crucially important case of the Ukraine, for example, the Party apparatus itself suffered several sweeping purges because of its "deviations." Moreover, after a controlled measure of Great Russian patriotism and nationalism became respectable in the Soviet Union, Stalin and the Politburo began to stress the Russian language and the historical role of the Great Russian people as binding cement of their multinational state. This trend continued during the Second World War and in the postwar years. Eastern peoples of the U.S.S.R. were made to use the Cyrillic in place of the Latin alphabet for their native tongues, while the Russian language received emphasis in all Soviet schools. Histories had to be rewritten again to demonstrate that the incorporation of minority nationalities into the Russian state was a positive good rather than merely the lesser evil as compared to other alternatives. Basically contrary to Marxism, the new interpretation was fitted into Marxist dress by such means as stress on the progressive nature of the Russian proletariat and the advanced character of the Russian revolutionary movement, which benefited all the peoples fortunate enough to be associated with the Russians. But Stalin, and some other Soviet leaders as well, went further, giving violent expression to some of the worst kinds of prejudices. Notably the quite un-Marxist vice of anti-Semitism found fertile soil in the Soviet Union. Yiddish intellectuals were among the groups virtually wiped out by the purges. All Jews were apparently excluded from the Soviet diplomatic service. Stalin's and Zhdanov's fierce attack on "cosmopolitanism" after the Second World War seemed particularly difficult to reconcile with the international character of Marxism or with the legacy of Lenin. The present Jewish emigration from the U.S.S.R. has more than one sound reason behind it.

Education

Education has played an extremely important role in the development of the Soviet Union. Educational advances were a most important part of state planning and made the striking Soviet economic and technological progress possible. As already indicated, education also stood at the heart of the evolution of Soviet society. In the future, too, it is likely to exercise a crucial influence on the fortunes of the U.S.S.R.

Somewhat less than half of the Russian people were literate at the time of the Bolshevik revolution. Furthermore, the years of civil war, famine, epidemics, and general disorganization that followed the establishment of the Soviet regime resulted in a decline of literacy and in a general lowering of the educational level in the country. Beginning in 1922, however, the authorities began to implement a large-scale educational program, aiming not only at establishing schools for all children, but also at eliminating illiteracy among adults. By the end of the Second Five-Year Plan, that is, by 1938, a network of four-year elementary schools covered the U.S.S.R., while more advanced seven-year schools had been organized for urban children. The total elimination of illiteracy proved more difficult, although the government created more than 19,000 "centers for liquidating illiteracy" by 1925 and persevered in its efforts. The census of 1926 registered 51 per cent of Soviet citizens, aged ten and above, as literate; that of 1939 81.1 per cent. Projecting the increase, 85 per cent of the Soviet people must have been literate at the time of the German invasion. Almost all are today.

The four-year and the seven-year schools became basic to the Soviet system. But ten-year schools also appeared in quantity. This type of school, for boys and girls from seven to seventeen, provides more class hours in its ten years than does the American educational system in twelve. At present, while the compulsory seven years of schooling have not yet in practice been extended to all Soviet children, there is emphasis on the new eight-year school. Although in 1940 tuition was introduced in the last three years of the ten-year school, as well as in the institutions of higher learning — and has been repealed and restored since — an extremely widespread system of scholarships and stipends has been used at all times to make advanced education available to those with ability.

After initial experimentation with some progressive education and certain quite radical methods of teaching the young and combining school and life, Soviet education returned to entirely traditional, disciplinarian, and academic practices. The emphasis is on memorization and recitation, with a tremendous amount of homework. It has been estimated that, if Soviet schoolchildren were to do all their assignments conscientiously and to the full, they would be reading 280 printed pages a day! Soviet schools are especially strong in mathematics and science, that is, in physics, chemistry, biology, and astronomy, as well as in geography, and drafting. But they also stress language, literature, foreign languages, and history, together with certain other academic subjects. For instance, six years of a foreign language are taught in a ten-year school. There are no electives.

Before he lost power Khrushchev emphasized the need to bring schools closer to life and to combine education in the upper grades with some apprenticeship work in factories and farms. But educational reforms along these lines proved to be abortive. Many students, however, are forced to spend at least two years "in production," that is, in factory or agricultural work, before proceeding from secondary to higher education. The Soviet Union also has special schools for children with musical and artistic gifts, military schools, and the like. In addition, many boarding schools for the general education of Soviet children have been established. They numbered 2,000, with 500,000 pupils, in the autumn of 1961, and were described as the "new school of Communist society" at the Twenty-second Party Congress. Probably because of their great expense and the generally more modest tone of the new leadership, they have been given less prominence in recent years.

Beyond secondary schools, there are technical and other special schools, as well as full-fledged institutions of higher learning. The number of these higher schools is constantly growing. While Soviet authorities have developed the old university system, they have placed much more emphasis in higher education on institutes that concentrate on a particular field, such as technology, agriculture, medicine, pedagogy, or economics. Study in the institutes ranges from four to six years; a university course usually takes five years. Applicants to universities and institutes must take competitive entrance examinations, and it has been estimated that frequently as many as two out of three qualified candidates have been rejected because of lack of space in recent years. The older Soviet students, as well as the schoolchildren, must attend all their classes, are in general subject to strict discipline, and follow a rigidly prescribed course of study.

The educational effort of the Party and the government extends beyond schools to libraries, museums, clubs, the theater, the cinema, radio, television, and even circuses. All of these, of course, are owned by the state, are constantly augmented, and are closely co-ordinated to serve the same purposes. More peculiarly Soviet has been the practice of constant oral propaganda in squares and at street corners, with more than two million propagandists sponsored by the Party. Bereday has written authoritatively on the spread of education in the Soviet Union, and has compared this spread with the situation in the United States:

. . . [In 1958] there were in the Soviet Union approximately 110,000 elementary four-year schools, 60,000 seven-year schools, and 25,000 ten-year schools, a total of nearly 200,000 regular schools of general education. There are, in addition, some 7,000 auxiliary special and part-time schools, 3,750 technikums and professional schools, 730 institutes of higher education, and 39 universities. The countryside is dotted by

150,000 libraries, 850 museums, 500 theaters, 2,700 Pioneer palaces, 500 stations for young technicians and naturalists, 240,000 movie theaters, and 70 circuses. A task force of 1,625,000 teachers and other personnel mans this extensive enterprise. . . . Population and school-attendance figures substantiate the ambitions of the Soviet educational plan to reach all the people. The figures now available estimate the situation as follows: 2,500 out of each 10,000 people were in some type of school in 1955–1956; 814 of these were in grades five to ten of the general secondary school, 100 were in professional secondary schools, and 93 were in institutions of higher learning. These figures, which account for one-fourth of the total population, expand as we single out for consideration only the present younger generation. Approximately 10 per cent of the appropriate age group attend institutions of higher education, second largest proportion in the world after the United States, with 33 per cent of its youth in colleges. About 30 per cent of the appropriate age group complete secondary education, a close second after the United States, with 45 per cent. At the age of fourteen 80 per cent of the age group are still in school, in the United States some 90 per cent.

Education on the job and by correspondence is also extremely widespread in the U.S.S.R. Moreover, a further expansion and diffusion of education have constituted an essential part of the recent five-year plans, although the rate of educational advance has slowed down compared to the earlier period, while comparison with the United States has been affected by the great expansion of American higher education in the 1960's.

Soviet education, and indeed Soviet culture in general, has greatly profited from the prerevolutionary legacy. The high standards, the serious academic character, and even the discipline of Soviet schools date from tsarist days. The main Communist contribution has been the dissemination of education at all levels and on a vast scale, although it should be remembered that imperial Russia was, on the whole, moving in the same direction and that given a little more time it would have established universal schooling. Many observers have noted that Soviet students study with remarkable diligence and determination. That probably stems both from the old tradition, which holds education in high esteem, and from contemporary conditions of life: education provides for Soviet citizens the only generally available escape from the poverty and drabness of the kolkhoz and the factory. If generous subsidization and energetic promotion have constituted the main Soviet virtues in education, the allpervasive emphasis on Marxism has been the chief vice. While a detailed criticism of the Soviet school system must be left to DeWitt, Lilge, Kline, and other specialists, it is important to realize that Soviet Marxism distorts whatever it touches and that, therefore, the quality of Soviet education and culture frequently deteriorates in direct proportion to its proximity to doctrine. For this reason Soviet mathematics, in schools or universities, is vastly preferable to Soviet history, and Soviet chemistry to Soviet philosophy.

Soviet Culture

Soviet science, scholarship, literature, and arts reflect the same traits of the Soviet regime as does education, from which, in any case, they cannot be entirely separated. The Soviet performance in all these fields is noteworthy for its vast scope, liberal expenditure of funds, extremely thorough organization, co-ordination, and planning, and ubiquitous party control. All Soviet intellectuals are in effect employed by the state. Even when their income depends primarily on royalties, their books cannot be published nor their music played without official authorization. The quality of Soviet creative work has ranged from some brilliant developments in science and excellent compositions in music to the dreary poverty of "socialist realism" in literature and its virtually unrelieved worthlessness in painting and sculpture. But in almost all fields, fruitful as well as barren, the stifling grip of the Party and its ideology has left its mark.

Science and Scholarship

For a variety of reasons, science has been a privileged area of Soviet culture. It was obviously and immediately useful and, indeed, indispensable if the U.S.S.R. were to become the military, technological, and economic leader of the world. It was fully endorsed by Marxism, which prided itself on its own allegedly scientific character. In fact, some writers have commented on an almost religious admiration of science and technology in the Soviet Union, an expression in part of the old revolutionary titanism and determination to transform the world. Yet science, while subject to the dialectic, lies on the whole outside Marxist doctrines, which concentrate on human society, and thus constitutes a "safer" field in the Soviet Union than, for example, sociology or literature. Not that it escaped the Party and the ideology altogether. Communist interference with science included such important instances as Soviet difficulties in accepting Einstein's "petty bourgeois" theories, as well as Trofim Lysenko's virtual destruction of Soviet biology, particularly genetics, together with the elimination of a number of leading Soviet biologists, notably Nicholas Vavilov. Lysenko, an agricultural expert and a dangerous quack and fanatic, claimed to have disproved the basic laws of heredity and obtained Party support for his claims: Lysenko's theories gave Marxist environmentalism a new dimension and made a Communist transformation of the world seem more feasible than ever — the only trouble was that Lysenko's theories were false. But Einstein's views had to be accepted, at least for practical purposes; and even Soviet biology staged a comeback, although it took many years and several turns of fortune finally to dispose of Lysenko's authority. Moreover, thousands of scientists, in contrast, for example, to writers, could continue working in their fields more or less undisturbed. And science especially profited from the large-scale financing and organization of effort provided by the state.

The Sputniks, the shot at the moon, the photographing of the far side of the moon, and Soviet astronauts' orbiting of the earth, together with atomic and hydrogen explosions, have emphasized the achievements of Soviet applied science, and in particular Soviet rockets, missiles, and atomic and space technology.* In these fields, as in others, the Soviet Union profited from the prerevolutionary legacy, especially from the continuing work of such scholars as the pioneer in space travel Constantine Tsiolkovsky, 1857–1935. The contributions of espionage and of German scientists brought to the U.S.S.R. after the Second World War are more difficult to assess. The state, of course, financed and promoted to the full all the extremely expensive technological programs referred to above. It also organized, in connection with the five-year plans, a great search for new natural resources, vast geographic expeditions, and other, similar projects. The work of Soviet scientists in the far north acquired special prominence. The Academy of Sciences continues to direct Soviet science as well as other branches of Soviet scholarship.

* Soviet "firsts" in space include: first earth satellite, Sputnik I, launched October 4, 1957; first satellite with animal aboard, Sputnik II, November 3, 1957; first moon rocket, Lunik I, January 2, 1959; first photographs of hidden side of moon, October 18, 1959; first retrieval of animal from orbit, August 20, 1960; first launching from orbit, Venus probe, February 12, 1961; first man in space, Lieut. Col. Iurii A. Gagarin, April 12, 1961; first double launching with humans, Major Andrian Nikolaev, August 11, 1961, Lieut Col. Pavel Popovich, August 12, 1962; first woman in space, Valentina Tereshkova, June 16, 1963; first triple-manned launching, Col. Vladimir Komarov, space commander, Konstantin Feoktistov, scientist, Dr. Boris Egorov, physiologist, October 12, 1964; first man to walk in cosmic space, Lieut. Col. Aleksei A. Leonov from Voskhod II (flight commander, Col. Pavel Beliaev) March 19, 1965; first flight around the moon and return of an automatic space craft, Zond 5, September 15–22, 1968; establishment of first orbital experimental station during flight of Soyuz 4 and Soyuz 5 spaceships, January 1969; first self-propelled automatic laboratory on the surface of the moon, Lunokhod-1, November 17, 1970; first manned research station, Salyut, in circumterrestrial orbit, June 7, 1971; first soft landing on the surface of Mars and transmission of video signal to Earth by Mars-3 probe, December 2, 1971; first soft landing on the sunward surface of Venus by Venera-8 probe and transmission to Earth of atmospheric and surface measurements for 50 minutes, July 22, 1972. The Soviet Union also announced the first loss of a man in actual space flight, Col. Vladimir Komarov, Soyuz 1, April 24, 1967.

While Soviet applied science has now received perhaps too much praise in the press of the world, the over-all excellence of Soviet science has on the whole not yet been sufficiently appreciated. With theoretical physicists like Leo Landau, experimental physicists like Abraham Joffe and Peter Kapitza, chemists like Nicholas Semenov, mathematicians like Ivan Vinogradov, astronomers like Victor Ambartsumian, geochemists like Vladimir Vernadsky, and botanists like Vladimir Komarov — to select only a very few out of many names — the Soviet Union has had outstanding scientific talent, while the scope of its scientific effort has exceeded that of all other countries except the United States.

Soviet social sciences and humanities do not compare with the sciences. The dead hand of Soviet Marxism has stifled virtually all growth in such fields as philosophy and sociology. Moreover, the official ideology has proved to be remarkably barren with the result that even Marxist thought in the U.S.S.R. has been crude and undeveloped compared to certain Western and satellite varieties. Clearly, and for a number of good reasons, the best talent has gone into science. To be sure, there have been important variations in the Soviet control of scholarship and the results achieved.

Thus, until the early and middle thirties, Mikhail Pokrovsky's negativistic school held sway in history. A convinced Marxist, Pokrovsky took an extremely critical and bitter view of the Russian past, in effect declaring it of no importance. With the Soviet consolidation and turn to cultural conservatism in the thirties, Pokrovsky and his school were denounced, and the authorities began to promote intense work in the field of history and in such related disciplines as archaeology. In particular, Soviet historians have turned to collecting and editing sources. Some valuable work has also been done in social and economic history, with at least one Soviet historian, Boris Grekov, originally a prerevolutionary specialist, making contributions of the first rank. Yet in general, in spite of the change in the thirties and a certain further liberalization following Stalin's death, Soviet historiography has suffered enormously from the Party strait jacket, most especially in such fields as intellectual history and international relations.

Linguistic studies followed a somewhat different pattern. There Nicholas Marr, 1864–1934, an outstanding scholar of Caucasian languages who apparently fell prisoner to some weird theories of his own invention, played the same sad role that Trofim Lysenko had played in biology. Endorsed by the Party, Marr's strange views almost destroyed philology and linguistics in the Soviet Union, denying as they did the established families of languages in favor of a ubiquitous and multiform evolution of four basic sounds. The new doctrine seemed Marxist because it related, or at least could relate, different families of languages to different stages of the material development of a people, but its implications proved so confusing and even dangerous that Stalin himself turned against the Marr school in 1950, much to the relief and benefit of Soviet scholarship.

Most areas of Soviet scholarship, however, profited much more by Stalin's death than by his dicta. From the spring of 1953, Soviet scholars enjoyed more contact with the outside world and somewhat greater freedom in their own work. In particular, they no longer had to praise Stalin at every turn, prove that most things were invented first by Russians, or deny Western influences in Russia — as they had had to do in the worst days of Zhdanov. Entire disciplines or sub-disciplines, such as cybernetics and certain kinds of economic analysis, were eventually permitted and even promoted. Yet, while some of the excesses of Stalinism were gone, compulsory Marxism-Leninism and partiinost remained. Soviet assertions that their scholars were free men retained a hollow — and indeed tragic — ring. Glasnost, to be sure, has represented a real breakthrough into honest scholarship, but it could not immediately eliminate all institutional, psychological, and material obstacles to it, while its very existence is bound to depend on the future course of Soviet society.

Literature

Literature in Soviet Russia in the twenties continued in certain ways the trends of the "silver age," in spite of the heavy losses of the revolutionary and civil war years and the large-scale emigration of intellectuals. Some poets went on publishing excellent poetry, and writers created numerous groups and movements. The formalist critics rose and flourished. All that, of course, could not last under the new system. First, the R.A.P.P. — a Russian abbreviation for the Russian Association of Proletarian Writers — came to dominate the scene, preaching, somewhat along Pokrovsky's lines, the discarding of all culture except the proletarian. In 1932 the government disbanded the largely nihilistic R.A.P.P. and proceded to organize all the writers into the Union of Soviet Writers and to impose on them the new official doctrine that came to be known as "socialist realism." Guided by the correct principles, Soviet writers were to participate fully and prominently in the "building of socialism" as, to quote Stalin, "engineers of human souls." "Socialist realism" became synonymous with literature in the Soviet Union, other approaches being proscribed. Most of the prominent figures of the "silver age" disappeared early in the Soviet period: Blok died in 1921, Briusov in 1924, Bely in 1934, Gumilev was shot as a counterrevolutionary in 1921, Esenin committed suicide in 1925, and Maiakovsky, whose futurist verses rang the praises of the revolution and whose "Left March" had become almost its un-official poetic manifesto, took his own life in 1930. The few outstanding figures who remained, such as Akhmatova and Pasternak, either fell into silence — at best writing for themselves and their friends — or concentrated on translating from foreign languages. Akhmatova was expelled from the Union of Soviet Writers by Zhdanov in 1946, following the publication of some poems where she had displayed an unsocialist loneliness among other vices, and Pasternak was ejected in 1958 after the appearance abroad of his celebrated novel, *Doctor Zhivago*, for which he was offered the Nobel prize.

Although the concept of socialist realism sponsored by Stalin and the Politburo was never made entirely clear, it referred ostensibly to a realistic depiction of life in its full revolutionary social dimension, in part in the tradition of Pushkin and Tolstoy and indeed of the main stream of modern Russian literature. But because the Party had its own view of life, based essentially on Marxist clichés misapplied to Russian reality, socialist realism turned into crude and lifeless propaganda. Writers had to depict the achievements of the five-year plans and other "significant" subjects or at least write realistic historical novels. More important, they had to do it in a prescribed manner. Black was to be made black and white white with no shades in between. The Soviet hero had to be essentially a paragon of virtue, with no fundamental inner conflicts and no psychological ambiguities. Instead of the grim world around them, authors were urged to see things as they should appear and will appear in the future. Pessimism was banned.

Not surprisingly, in terms of quality the results of "socialist realism" have been appalling. After Gorky's death in 1936 — a death arranged by Stalin, according to some specialists — no writer of comparable stature rose in Soviet letters. A few gifted men, such as Alexis N. Tolstoy, 1883–1945, the author of popular historical and contemporary novels and Michael Sholokhov, 1905–, who wrote the novels *The Quiet Don* and *Virgin Soil Upturned*, describing Don cossacks in civil war and collectivization, managed to produce good works more or less in line with the requirements of the regime, although they too had to revise their writings from edition to edition to meet changing Party demands. Other talented writers, for instance, Iurii Olesha, failed on the whole to adjust to "socialist realism." More typical Soviet practitioners have turned out simple, topical, and at times interesting, but unmistakably third-rate, pieces. An example is Constantine Simonov, a writer of stories, novelist, playwright, and poet, as well as an editor and war correspondent, who drew international attention by his novel about the defense of Stalingrad, *Days and Nights*, and his play concerning American attitudes toward the

U.S.S.R., *The Russian Question*. The battle of Stalingrad, however, was depicted also in a great work, Vasily Grossman's long novel *Life and Fate*, which, together with a few short poems by Anna Akhmatova, is likely to remain as the imperishable literary tribute to the Soviet people in the Second World War. The overwhelming bulk of Soviet literature became extremely monotonous, drab, and lacking in artistry or in any kind of ability. Soviet poetry, especially hampered by the injunction to be simple and easy to understand, as well as socialist and realist, proved to be inferior even to Soviet prose. The government no doubt contributed more to the enjoyment of its readers by publishing on a large scale the Russian classics and world classics in translation. As a matter of fact, most of the best Russian literature during the last few decades has been written abroad. Some of the outstanding expatriate authors were the novelist, story writer, and poet Ivan Bunin and the highly original prose writer with a unique style, Alexis Remizov, who both died in Paris, in 1953 and 1957, respectively. In all fairness, however, one should note a certain revival in Soviet literature since Stalin's death. Yet that revival, too, has a tragic ring. Its leading figures, such as the poet Joseph Brodsky and indeed Alexander Solzhenitsyn himself, were hounded in their native land until their exile abroad, their works and thought forbidden to Soviet readers. Ironically, Solzhenitsyn's writings may well be considered to represent the long-delayed success of socialist realism: they focus on central problems and situations of Soviet life; they deal with common people, in fact all kinds of people; they are meant for the masses; and they are certainly realistic. G. Struve's two books, one on Soviet and the other on émigré writers, should be read together to obtain the best picture available of Russian literature since 1917 and of its hard lot.

The Arts

The Soviet record in the arts has paralleled that in literature. Again, the twenties, linked closely to the "silver age" and to contemporary trends in the West, were an interesting and vital period. Notably, in architecture functionalism flourished, producing some remarkable buildings, while new and experimental approaches added vigor and excitement to other arts also. However, once "socialist realism" established its hold on Soviet culture, arts in the Soviet Union acquired a most conservative and indeed antiquated character. Impressive in quantity, Soviet realistic painting and sculpture are essentially worthless in quality, being in general poor imitations of a bygone style. While Soviet architecture has on the whole had more to offer, it too traveled the sad road from inspired and novel creations in the earlier period of Communist rule to the utterly tasteless and contrived Moscow skyscrapers of Stalin's declining years, exemplified by the much-publicized new university building. Although certain stirrings have been detected since the late dictator's death, and indeed although many modern utilitarian buildings are now being erected, there has so far been no basic change of orientation in Soviet arts. Music, it is true, has been somewhat more fortunate throughout the period, both because of its greater distance from Marxist and "realistic" injunctions — which nevertheless did not prevent the Party from attacking "formalism" and modernism in music and from tyrannizing in that field — and, perhaps, because of the accident of talent. In any case, the one-time figure of the "silver age" and émigré Serge Prokofiev, 1891–1953, the creator of such well-known pieces as the *Classical Symphony* and *Peter and the Wolf,* and Dmitrii Shostakovich, 1906–75, together with a few other composers, managed to produce works of lasting value in spite of ideological obstacles.

Short on creativity and development, certain Soviet arts have been long on execution and performance. Again, the high standards were continuations from tsarist days, aided by increased state subsidies and by the fact that schooling and culture spread to more people. Soviet musicians have performed brilliantly on many instruments, both at international competitions of the thirties, and again in more recent years when the best among them, such as the violinist David Oistrakh and the pianists Sviatoslav Richter and Emil Gilels, were allowed to tour the world. The ballet, while in a sense stagnant — the clock having stopped for most purposes in 1917 — continued to do its dances beautifully, and was apparently backed by more funds and a better system of schools and selection than in any other country. The Moscow Art Theater is still one of the most remarkable centers of acting anywhere, although unfortunately its school of acting had for a long time a monopoly in the Soviet Union, all other approaches to acting and the theater having been proscribed. Good acting has also characterized many Soviet films. In fact the Soviet cinema continued to be creative longer than other Soviet arts — in part probably because it had no nineteenth-century tradition in the image of which it could be conveniently frozen. Soviet film directors included at least one great figure, Serge Eisenstein, 1898–1948, as well as other men of outstanding ability.

In the arts as in literature, the years of glasnost have brought great promise as well as dislocation and worry.

Religion

Religion in the Soviet Union constituted an anomaly, a threat, and a challenge from the Communist point of view. Marxist theory considers it an "opiate of the masses" and finds its *raison d'être* in the efforts of the exploiting classes to keep the people obedient and docile. Russian practice seemed to add weight to the theory, for the Orthodox Church in Russia was closely linked to the imperial regime, and it naturally sided with the Whites in the Civil War. Clearly, its social basis gone, religion would cease to exist in a socialist society. But this has not occurred. Therefore, the Soviet leadership had to compromise and allow religion a highly restricted position in the U.S.S.R., while looking forward to its eventual, much delayed, disappearance. Religion, it might be added, also proved to be one of the main obstacles to the Communist transformation of man and society in other eastern European countries.

Outright persecution lasted well into the thirties. In addition to executing and exiling many clerics, monks, and Orthodox laymen, confiscating church implements "for victims of famine," and closing churches and converting them into antireligious museums, the authorities tried to break up the Church from the inside by assisting a modernist "Living Church" group within it — fruitlessly, for the people would not follow that group. After the death in 1925 of Patriarch Tikhon — elected by a Church council in 1918 to resume the patriarchal form of ecclesiastical organization which had been discontinued by Peter the Great — the government prevented any patriarch being elected in his stead, and Church leadership fell to provisional appointees. Yet, according to an official report based on the never-published census of 1936, 55 per cent of Soviet citizens still identified themselves as religious — while many others presumably concealed their belief.

That stubborn fact in conjunction with the general social stabilization of the thirties made Stalin and the Politburo assume a more tolerant attitude toward religion. The war and the patriotic behavior of the Church in the war added to its acceptance and standing. In 1943 the Church was permitted to elect a patriarch, the statesmanlike Metropolitan Sergius obtaining that position. After his death in 1945, Sergius was succeeded by Alexis, who continued as "Patriarch of Moscow and All Russia" for a quarter of a century. In 1971, following Alexis's death, Pimen was elected patriarch to head the Church, followed in 1990 by Alexis II. The ecclesiastical authorities were also allowed to establish a few theological schools, required to prepare students for the priesthood, and to open a limited number of new churches. The activities of the Union of the Godless and anti-religious propaganda in general were curtailed. In return the patriarchal Church declared complete loyalty to the regime, and supported, for example, its international peace campaigns and its attempts to influence the Balkan Orthodox. More unfortunately, the two co-operated in bringing the two or three million Uniates of former eastern Poland into Orthodoxy. The Church in the U.S.S.R., however, remained restricted to strictly religious, rather than more general social and educational, functions — even the constitution proclaims merely the freedom of religious confession, as against the freedom of anti-religious propaganda — and, while temporarily tolerated within limits, it remained a designated enemy of Marxist ideology and Communist society. In fact, Khrushchev especially, as well as his successors, increased the pressures against religion even when "liberalizing" other aspects of Soviet life. A remarkably tenacious relic of the past, the Church has no future in the Communist view.

Yet one of the most striking aspects of glasnost and perestroika was to be an improvement in the position of the Church, now allowed a much greater range of activity and apparently considered even by Communist leaders as a source of morality and stability.

It should be added that other Soviet Christians, such as Baptists, and other religious groups, such as the numerous Moslems, shared their histories with the Orthodox. They, too, led a constricted and precarious existence within a fundamentally hostile system, profiting from relaxations when they occurred and entering a new life as a result of the policy of glasnost.

X L I I

* * * * * * * * * *

THE GORBACHEV PERIOD AND BEYOND, 1985–92

The river of time in its flow
Carries away all the works of human beings . . .
DERZHAVIN

AT PRESENT the peoples of the Soviet Union, and to some extent even peoples outside its borders, are in the midst of the Gorbachev-initiated maelstrom, which makes an objective judgment of the situation exceptionally difficult. To be sure, the Soviet leader was not in control of his country and its citizens, and, indeed, he had been repeatedly obtaining results opposite to those intended — after all, Nicholas II also made important contributions to the revolutions of 1917. And as many of our best specialists tell us, it is dangerous to personalize major historical issues, and another Gorbachev or still other lines of development would have produced similar results. But as long as history is an account of what happened and is happening rather than of the logical alternatives, the period of glasnost and perestroika will remain linked to its extraordinary protagonist, whose main assets appear to have been optimism, glibness, and marvelous political agility and adroitness, which enabled him to dance on top of and around historical developments, if not to preside over their course — an obvious transitory figure who refused to transit.

Gorbachev's Early Years

There is no doubt that Gorbachev started the ball rolling. Exactly what he and his original associates, such as Eduard Shevardnadze and Alexander Yakovlev, had in mind when they began reforming the Soviet Union may never become clear, even to them. Suppositions and explanations of their intent abound, but the overwhelming factors in what transpired appear to be the absence of correspondence between plans and reality and the dizzying power of contradictory forces unchained by even slight reform. Recent events in the U.S.S.R. and eastern Europe stunned everyone, especially those who had any regard for the communist system, and that includes by definition the entire Soviet leadership. There may well be, however, one quite major exception to this almost total disjunction between purpose and

590

accomplishment. Gorbachev, Shevardnadze especially, and other prominent Soviet figures insisted that one of the pillars of their new thinking was the absolute realization of the inadmissibility of nuclear war in human affairs and, therefore, of the necessity for at least a minimum of international cooperation, in particular between the Soviet Union and the United States. With all qualifications, it can be argued that Soviet foreign policy came to reflect that realization. If so, the gain to the world was incalculable, although the realization itself is elementary and its roots even in the Soviet Union largely preceded Gorbachev. Otherwise, it hardly needs reminding that in his book *Perestroika* — published in English as well as in Russian in October 1987 and a good way to become acquainted with its author — and even later Gorbachev emphasized the supreme importance of Lenin and the Communist Party in the Soviet Union, rejected privatization and political pluralism in communist states, and praised Soviet solutions to social and nationality problems. And it should be remembered that the Warsaw Pact was renewed and extended for twenty years on April 26, 1985; it was abolished following the complete collapse of communism in eastern Europe on February 25, 1991. The river of time does carry away the works of human beings.

At the foundation of Gorbachev's reforming lay the need to escape the economic cul de sac, which had become increasingly and unmistakably apparent by the end of the Brezhnev regime. Although some specialists claim that the Soviet state and society had already lost their forward motion with the fall of Khrushchev or shortly after, the long years of Brezhnev's rule became incontrovertibly a time of stagnation and corruption. Economic indicators generally pointed downward, although thus far only the rate of increase of productivity and product rather than productivity and product themselves declined. In spite of great expenditures and extensive efforts, the condition of Soviet agriculture remained dismal. In industry, as before, quality lagged behind quantity. Even more important, the entire industrial establishment, a direct inheritance from the initial five-year plans, failed to respond competitively to the new age of computers and electronics. Indeed, falling behind in science and technology became one of the main Soviet worries. Military needs continued to devour huge chunks of the gross national product — percentagewise more than twice that devoured in the United States. Stagnation and economic crisis found their natural counterpoints in pessimism and low morale, which pervaded the country.

The first two or three years of the Gorbachev regime, inaugurated on March 11, 1985, displayed a fairly "traditional" cast. The new Party secretary had to concentrate on strengthening his position, and, indeed, over a period of time he effected a major turnover of ruling and high administrative personnel. Thus on July 1, 1985, Shevardnadze became a member of the Politburo, and on the following day he was appointed foreign minister, replacing Andrei Gromyko, who was moved to a more ceremonial high office. Other new men entered the Politburo, while Victor Grishin, Gorbachev's original rival for the position of Party secretary, retired.

Perestroika, Gorbachev's proposed rebuilding of the Soviet country and system, was loud in promise but, everything considered, initially quite similar to the proposals and exhortations of earlier Soviet reformers. The draft plan, as presented by Gorbachev in October 1985, called for doubling the national income in fifteen years, with special emphasis on the modernization of equipment and an increase in labor productivity. It was all-important to overcome stagnation, to get the Soviet Union moving. But although the leader spoke of a "radical transformation of all spheres of life" and although such concepts as profits and profitability, decentralization, initiative, and even market economy and private enterprise increasingly entered national discourse, in practice the effort was limited mainly to an attempt at a speed-up, in particular by eliminating such evils as absenteeism and drunkenness — witness the major anti-alcohol campaign mounted in May 1985 and in subsequent months and years.

Gorbachev showed more originality in gradually promoting the concept of glasnost, or free discussion in speech and print. In its full, or at least rich, development a stunning novelty for Soviet society, it largely won for Gorbachev the initial support of the intellectuals and, even more broadly, of the educated public at home and great acclaim abroad. Foreign praise was powerfully augmented, of course, by the increasingly accommodating and peaceful foreign policy of the Soviet Union. Yet it is worth noting that whereas at the time the deficiencies of the Brezhnev regime were endlessly excoriated and whereas Gorbachev himself resumed de-Stalinization in July 1987 by condemning Stalinist terror, Lenin remained the lodestar of the new course. As of that time perestroika and glasnost, too, were expected to fit neatly into a gloriously reformed, somewhat humanized, and professedly Leninist Soviet Union.

Unfortunately, almost nothing worked during the first years of the Gorbachev regime. The economy would not respond to mere exhortations. Indeed, the government's own economic, especially financial, policies led to budget deficits and inflation and thus made matters worse. Even the anti-alcohol campaign proved to be a disaster, its only incontrovertible result a great increase in the illegal production of spirits, to the extent that sugar disappeared from stores in parts of the U.S.S.R. Before long under the new administration and its vacillating and confusing direction, the economy began to lose what cohesion it had had under Brezhnev without gaining anything to replace it. The war in Afghanistan, exceedingly painful to the population of the Soviet Union, continued to take its toll. On April 28, 1986, a nuclear reactor exploded in Chernobyl; the resulting medical and environmental catastrophe threw a glaring light on multiple Soviet deficiencies, from those in engineering to those in the news media. Indeed, that tragic episode, treated at first in the firm tradition of Stalinist secrecy, eventually became both an opening into glasnost and a strong argument in its favor. But then glasnost itself, a valuable and undeniable achievement of the Gorbachev years, was becoming increasingly dangerous to the Soviet regime and all its plans. Gorbachev's worst miscalculation might well have been his belief that glasnost would strengthen rather than destroy communism. Freedom of speech meant freedom to ask questions, and there were so many questions the Soviet government would rather not answer. Freedom of speech also meant freedom of different political and other opinions, and consequently the legitimacy of different political and other parties, an obvious conclusion which Gorbachev tried for a time to deny by upholding glasnost but rejecting political pluralism. Glasnost and related measures of liberalization would lead to the appearance of diverse groups — from monarchists to Fascists and from Orthodox clergy to the champions of homosexuals — in the streets and squares of Moscow and other Soviet cities. Perhaps most important, they led to the revival of numerous nationalisms, suppressed but still alive in the Marxist superstate. The new time of troubles, like the original one at the end of the sixteenth century and the beginning of the seventeenth, was to have its national phase.

The Rise of Nationalisms and the Breakup of the Soviet Union

Because of the number, richness, variety, and specificity of ensuing developments, it is impossible to present in a brief general account an adequate summary of the rise of nationalism, or rather nationalisms, in the Soviet Union after centralized control was removed or even merely weakened. All fifteen constituent republics were radically affected. Moreover, many ethnic subdivisions within these republics and still other ethnic minorities also entered the fray. In line with the nature of nationalism, the relations of the participants were usually antagonistic, sometimes to the point of physical combat. It was illustrative of the many-sided struggle that Tskhinvali, the main town of the South Ossetian Autonomous Region within the Georgian Soviet Socialist Republic, came to be held one-third by the local Ossetian militia, one-third by Georgian nationalist forces, and one-third by the Soviet army. This treatment of the issue of nationalism in the Soviet Union is limited to mentioning a few highlights and suggesting certain emerging patterns.

In many respects, the three Baltic republics — Estonia, Latvia, and

Lithuania — led the way. Independent states between the two world wars (and in the case of Lithuania, of course, in the much longer, richer, and more complex historical past), forced to join the Soviet Union only about fifty years ago, and possessed of their own languages and, on the whole, of a skilled and well-educated citizenry, the three republics, once self-expression became possible, made no doubt of their desire for independence. It was in Estonia that the first large-scale noncommunist political coalition, the People's Front, received recognition, in June 1988, and it was Estonia that proclaimed on November 17, 1988, the right to reject Soviet laws when they infringed on its autonomy. On January 18, 1989, Estonian became the official language of the republic; legislation was enacted in an even more rigorous form a week later for the Lithuanian language in Lithuania, and still later, after mass demonstrations, for the Latvian language in Latvia. In May 1989, the Lithuanian legislature adopted a resolution seeking independence, and on August 22, 1989, it declared null and void the Soviet occupation and annexation of Lithuania in 1940. In early December 1989, Lithuania became the first republic to abolish the Communist Party's guaranteed monopoly of power, while later that month the Communist Party in Lithuania voted to break away from Moscow, thus becoming the first local and independent Communist Party in the U.S.S.R., and to endorse political separation. On March 11, 1990, Lithuania, led by its president, Vytautas Landsbergis, proclaimed full independence. Events in Estonia and Latvia followed a similar course. It is worth noting that whereas Lithuanians constituted at least three-quarters of the total population of their republic, Latvians and Estonians composed only a little more than half of theirs, and that all three new states tended toward rather exclusive policies that mandated a single official language and, for citizenship, a residential or familial connection with the pre-Soviet period to eliminate Russian newcomers. Yet in spite of the resulting built-in opposition, which claimed discrimination, in February 1991, 91 per cent of the voters in Lithuania approved independence; in March, referendums in Estonia and Latvia gave independence a three to one majority — clearly, not only the Balts, but also many Russians and people of still other ethnic backgrounds wanted above all to escape the Soviet system.

Gorbachev drastically underestimated the power of nationalism in the Baltic area, as well as elsewhere, and at first tried to ignore or dismiss the demands for recognition and independence. Once the crisis became obvious, he attempted persuasion, political maneuvering with the many elements involved, including different kinds of communists, and coercion, although never to the extent of mass military repression. Thus on January 11, 1990, he went to Vilnius, the capital of Lithuania, hoping to convince both leaders and milling crowds to check the nationalist course of development, but his trip was in vain. More successful was the oil blockade, a great reduction in the supply of oil to Lithuania, which began in mid-April 1990 and forced the republic to suspend, although not repeal, its declaration of independence on May 16. More violent coercion consisted of such incidents as army intervention in Vilnius, resulting in the death of fourteen people, and the assault by Black Berets on a Latvian government ministry building in Riga, both in January 1991 — aborted coups d'état in the opinion of some — as well as repeated attacks on border posts and customs personnel of the nationalist republics, the signs of their new independence. The perpetrators included special army forces, such as the Black Berets, and perhaps some paramilitary groups, as in the case of seven Lithuanian customs and police officers killed on July 31, 1991. Officially, all these violent acts were labeled local incidents or transgressions; Gorbachev, in particular, denied any complicity. In fact, he emphasized that he objected to the manner of procedure of the Baltic republics, not to their goal of independence, which could be legitimately obtained in time, although personally he retained the hope that they would decide to remain in the new Soviet Union.

While nationalisms developed in a parallel and even co-operative way in the Baltic area, they were from the start on a collision course in Transcaucasia. Of the three Soviet republics beyond the great Caucasian mountain range — Armenia, Georgia, and Azerbaijan — the first two represent two of the oldest yet entirely distinct peoples and cultures of the world. With their histories antedating Christianity by far, the Georgians and the Armenians also built Christian states and cultures centuries before the vaunted conversion of the Rus in A.D. 988. The Georgians are Orthodox; the Armenians are Eastern Christians but not Orthodox. Becoming part of the Russian Empire when it finally reached them in the early nineteenth century may have been important, even essential, for their survival in the face of hostile Moslem Turks and Persians, now Iranians. (Armenians who remained in Turkey did not survive.) Yet if the Baltic republics kept referring to 1940 as the crucial year when Soviet power crushed their independence, the Georgians focused on 1918, when, following the Russian revolutions of 1917, they created an independent Menshevik-led state, only to be overcome after three years by the Red Army. Azerbaijan, not a distinct historical entity, represented the Turkic element so prominent in the past and present life of the area; its inhabitants are historically Moslem and are closely related to other Turkic speakers in the U.S.S.R. from the Volga to the Chinese border, especially in four of the five Soviet Central Asian republics, as well as to Turks abroad.

The Georgian revolution had its central event, a Georgian "Bloody Sunday." On April 9, 1989, the particularly brutal suppression of a nationalist demonstration in Tbilisi led to the death of 20 participants and the injury of more than 200. Although authorities in Moscow blamed local officials and started an investigation, communist control could not in effect be restored. The local Party, which, as in Lithuania, tried to play an independent role, lost the crucial ensuing election, and Georgia emerged with a noncommunist government headed by Zviad Gamsakhurdia. On April 1, 1991, Georgians responded to the question of whether they agreed "that the state independence of Georgia should be restored on the basis of the independence act of May 26, 1918," with a turnout, according to official sources, of 90.53 per cent of the 3.4 million Georgian voters and the affirmative reply of 98.93 per cent of them. Whatever their exact political future, Georgians, like the Baltic peoples, definitely want to live outside the Soviet Union. In the summer of 1991, there even existed widespread interest in restoring the ancient Georgian monarchy, although in a modern constitutional form, in the person of Georgii II Bagration, presently living in Spain but invited by President Gamsakhurdia and the parliament to visit Georgia. Yet in Georgia, too, nationalism brought no easy solutions. In particular, while asserting their own rights, Georgians did their best to limit and control those of the constituent minority groups in their state — the Adzharians, the Abkhazians, and perhaps especially the Ossetians — sometimes to the point of fighting on a considerable scale.

But the most extensive fighting in Transcaucasia, and indeed in the entire Soviet Union, has taken place between the Armenians and the Azerbaijanis and their respective republics. The historical hostility of the two peoples came to center on Nagorny Karabakh, an Armenian-populated area within the republic of Azerbaijan. The Armenians claimed it for themselves on grounds of nationality and of alleged mistreatment of its inhabitants. The Azerbaijanis responded by attacking Armenians wherever they could be found. During the last four months of 1989, they also blockaded railroads leading into Armenia and carrying supplies vital to that republic. Especially traumatic were assaults in Baku, the capital of Azerbaijan, on Armenians and some Russians in January 1990 which resulted in at least twenty-five deaths. On January 20, the Soviet army intervened against the Azerbaijani rioters, in effect recapturing Baku from them. The central government was blamed both for intervening and for intervening late and was even accused by both sides of inflaming hostility among nationalities for its own nefarious purposes. More likely, it was trying to make the best of a very bad situation, breaking the railroad blockade, and attempting to prevent massacres, while rejecting Armenian claims to Nagorny Karabakh. The Armenian-Azerbaijani border was transformed into front lines, with the opponents remarkably well provided with weapons and matériel stolen or otherwise obtained from the Soviet army. Although there is a lull at the front at present, the situation is anything but peaceful. Masses of people migrated between the two republics and even to Moscow and other distant points. Some Armenians have been brutally moved by the Soviet army into the Armenian republic from their native villages in Azerbaijan. Armenia has declared its independence and broken with the Soviet Union, although because of its especially difficult predicament and the volatile nature of events, its decision appears perhaps less definitive than that of the Baltic republics.

More co-operative, it would appear, are the other five "Moslem" republics of the U.S.S.R., all located in Central Asia: the Turkic Kazakh, Kirghiz, Turkmen, and Uzbek republics and the Iranian Tajik republic. Deeply affected by the present political and nationalist turmoil, affirming in the train of other republics their "rights" and their "sovereignty," and in constant conflict with central authorities, their minorities, and at times one another, they have proved nevertheless to be so far among the less self-assertive major components of the Soviet Union. The Party and the administration have been relatively successful in maintaining their positions in Soviet Central Asia. The explanation for that success may well lie in the comparative underdevelopment of the area, with its extreme reliance on a single crop (cotton), its poverty, its population explosion, and especially its dependence on huge government subsidies vital to the economy and even to the existence of its peoples. Kazakhstan, by far the largest republic of the five, represents a special case: it is no more than half Kazakh, the southern half, while the north is predominantly Russian and therefore claimed even by such Russian nationalists as Solzhenitsyn who are eager to separate Russians from alien peoples.

Whereas all the republics discussed thus far can be considered peripheral from the standpoint of Russian geography as well as Russian history and, typically, entered that history relatively recently, this judgment in no sense applies to the Ukraine, as readers of this book or of any other book treating Russian history in the large must know. Correspondingly, the historic future of the Ukraine will be of immense importance to that of Russia proper. The nationalist tide brought to power in Kiev, after elections, a coalition government led by rather nationalistically minded Ukrainian communists and joined by a noncommunist nationalist movement known as Rukh. In contrast to more exclusive Baltic nationalists, Ukrainian politicians appealed to all the inhabitants of the republic. As to its relation to Soviet and, later, Russian governments, the Ukraine gave some indication of willingness to participate in certain kinds of associations but always with reservations and conflicting problems. At present the problems include Ukrainian sovereignty over the Crimea, the management and disposal of atomic weapons, and the division and control of the armed forces. The eastern and the smaller western parts of the Ukraine are sharply different from each other. It is especially in the latter, Soviet only since 1939 or 1945, that the many-sided religious revival includes the restoration, at times a militant restoration, of the formerly prohibited Uniate Church, a Catholic jurisdiction, while anti-communism and anti-Russian nationalism ride high.

Adjacent and closely related to Ukrainians, as well as involved in the general course of Russian history from its very inception, Belorussians have been slow in developing a nationalism of their own, perhaps a generation or two behind the Ukrainians. Also, the Party proved to be stronger in the Belorussian republic than in some others. Still, the new nationalist wave had its effect. Thus in late July 1990, Belorussia issued a resounding declaration of its "sovereignty." And while it apparently constitutes one of the more cooperative members of the commonwealth, the future is difficult to predict.

The Moldavian Soviet Socialist Republic, bordering the Ukraine on the southwest, exemplifies well some of the conundrums and miseries of contemporary nationalisms in the Soviet Union. In a sense a fake nationalism to begin with — for the Moldavian language is really Rumanian and Moldavians are part of the Rumanian people, the postulation and promotion of differences between the two being a deliberate Soviet policy — it has nevertheless gripped the titular ethnic group, to the detriment of numerous minorities, such as the Turkic-speaking Gagauz people, the Ukrainians, and the Russians. In August and September 1989, large demonstrations and counterdemonstrations erupted over the introduction of Moldavian as the only official language of the republic. Tensions exploded sporadically into fighting and led to Soviet army intervention, along the lines of its peace-keeping efforts in Transcaucasia. Moldavian authorities announced their break with the Soviet Union and their refusal to take part in any new federal or other arrangements. The future of the area is far from clear.

Gorbachev and his government received no support from the Russian republic, the gigantic R.S.F.S.R., as they were trying to control the non-Russian nationalities of the Soviet Union. To the contrary, before long the Russian republic, too, was making declarations and demands aimed at the central authorities and frequently co-operating with other discontented entities — all for good reason. To be sure, Russians enjoyed certain advantages within the Soviet Union, such as the privileged position of their language and a greater acceptance of their cultural and historical past, albeit in a Marxist-Leninist interpretation, but they remained poor, even poorer than the inhabitants of a number of other republics, and, all in all, they bore their full share of the deprivation, suffering, and oppression characteristic of the Soviet system. They were even denied such "local" institutions, granted to other republics, as their own branch of the Communist Party and their own

Leaders of the communist world in Moscow, 1986. *From left*: Kadar of Hungary, Ceausescu of Rumania, Honecker of East Germany, Gorbachev of the Soviet Union, Chinh of Vietnam, Jaruzelski of Poland, Castro of Cuba, Zhivkov of Bulgaria, Husak of Czechoslovakia, Tsedenbal of Mongolia.

Muscovites attending an Eastern Orthodox Christmas procession in Red Square, 1991.

Patriarch Aleksy II of the Russian Orthodox Church blessing Yeltsin, the first freely elected president of the Russian Soviet Federated Socialist Republic.

Yeltsin being inaugurated as president of the Russian republic, July 10, 1991. Speaker of the legislature Khasbulatov stands to the left.

Ethiopian youths standing on the toppled statue of Lenin, May 23, 1991, two days after the end of the Soviet-backed Ethiopian regime.

Demonstrators pulling down the statue of Dzerzhinsky in front of the KGB headquarters, August 1991.

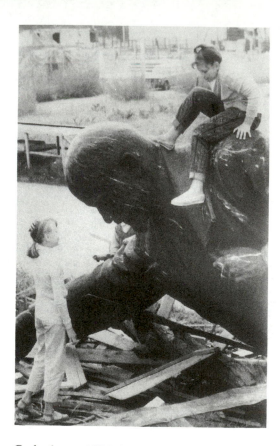

Children playing on a toppled statue of Lenin in Lithuania following the failed Kremlin coup, August 1991.

Gorbachev and Yeltsin at the Extraordinary Congress of People's Deputies, September 3, 1991.

academy of sciences, apparently, at least in part, because of the fear that these organizations might become too powerful and compete with the central Soviet ones.

The Russian republic acquired a remarkable, idiosyncratic leader in the person of Boris Yeltsin, an associate of Gorbachev, similar to him in his Party background and career, but much different in his extravagant manner and his populist and charismatic appeal. Dismissed on November 11, 1987, from his position as head of the Moscow Party organization for his criticism of the slowness of Gorbachev's reforming activity, Yeltsin made an unprecedented career in the Russian republic, culminating in his landslide election as its president on June 12, 1991, a stunning display of democratic procedure and popular support which neither Gorbachev nor any other leader in the central government could claim. Elections had already brought other liberals to office in the Russian republic, especially in its great cities, with Anatolii Sobchak becoming mayor of Leningrad and Gavriil Popov, of Moscow. It is worth noting that Yeltsin resigned from the Communist Party on July 12, 1990, and Sobchak and Popov on the following day. Liberalism was combined with nationalism and a religious revival as historic towns, places, and streets regained their old names and as religious services, including public religious services, multiplied. The day Yeltsin was elected president, the Leningrad voters also decided that their city should again become St. Petersburg. Immense problems, of course, continued; indeed, the entire dazzling change acquired a certain operatic quality, while the basic processes of economic and social life were grinding down. The mere administration of the R.S.F.S.R. became a near impossibility, with everyone from the Tartars on the Volga to the Yakuts in eastern Siberia and the nomadic tribes of the far north laying claim to their historic rights, their diamonds, or their reindeer. The excruciating interplay between Gorbachev and Yeltsin, with its repeated reversals of positions, ranging from close collaboration to determined attempts by each to drive the other out of politics, came to occupy center stage on the Soviet scene. In late summer 1991, the two were again together, in spite of major differences on such crucial issues as the role of the Communist Party in Soviet society.

Eastern Europe and the World

Just when everything was beginning to unglue at home, Gorbachev and the Soviet Union lost eastern Europe, which, in turn, contributed mightily to a further ungluing. In retrospect, there appear to be two main explanations for the stunning events of the miraculous year of 1989: the enormous extent of the opposition — indeed, hatred — of the peoples of the satellite states to their communist system and regimes, and Gorbachev's decision against any Soviet army intervention in defense of his communist allies. It was the extent to which communism was bankrupt and despised in eastern Europe that most outside observers failed to take into account. As to Gorbachev's decision, the Soviet leader apparently initially naively believed that there should be perestroika in the satellite countries, as well as in the Soviet Union, and that restructuring would only strengthen the system. But once the system unraveled, and at a terrifying speed, he concluded that nothing could be done to save the old order. As he thundered against Soviet hardliners who accused him of betrayal, only tanks could block change in eastern Europe, and tanks could not be used forever.

Thus 1989 witnessed the collapse of communism in Poland, Czechoslovakia, Hungary, Rumania, Bulgaria, and, of course, East Germany, which was to disappear entirely through absorption into the Federal Republic of Germany. Masses of refugees crossing newly opened frontiers, the once-formidable Berlin Wall acquiring souvenir status as it was being disassembled piece by piece, the corpses of Ceausescu and his wife, executed immediately after the overturn in Rumania, and so many other episodes and details will be enshrined in history books and human memory for ages to come. While each national case had its own peculiarities, such as the tremendous importance of West Germany for what happened in and to East Germany or the unique role of Solidarity in Poland, there were also common characteristics. Above all, communist regimes proved unable to survive intellectual and political freedom — glasnost, if you will — and, especially, free elections, beginning with the election in Poland on June 5, 1989. Even in the controversial cases of Rumania and Bulgaria — perhaps especially relevant for the Russian and Soviet future — where much of the establishment seems to have been preserved so far intact, the issues are the continuation of privilege and the brakes that old personnel may put on the democratic development of these countries, not the fear of a return to the days of Ceausescu and Zhivkov.

Although very hard hit, Gorbachev reacted to the events rapidly and imaginatively. Instead of mounting any kind of rear-guard action, especially on the central issue of the unification of Germany, Gorbachev fully accepted the unification, earning German gratitude — in particular, that of Chancellor Helmut Kohl — as well as advantageous financial provisions for the withdrawal and relocation at home of Soviet troops and some other German aid. Moreover, the solution of the German problem and the Soviet abandonment of troublesome eastern Europe meshed well with Gorbachev's policy of peace and international co-operation. Soviet army troops finally left Afghanistan, although the Soviet Union continued to provide massive military aid to the government forces in the seemingly endless civil war.

Extremely complex and long-drawn-out negotiations with the United States resulted, at the end of July 1991, in an agreement to reduce certain kinds of armaments. Commentators noted at the time that the new spirit of co-operation was even more significant than the particular provisions of the treaty. Although some disagreements and tensions remained, for example, in connection with the Japanese determination to regain some small islands in the Kurile chain seized by the Soviet Union toward the end of the Second World War or the American pressure to have the U.S.S.R. dump Castro and Cuba altogether, Gorbachev and his country were rapidly becoming respected supporters of world order. They played that role successfully in 1990 in the crisis and war following the Iraqi occupation of Kuwait, al-though the Soviet Union did not intervene militarily, and in 1991 in the aftermath of that war when international attention shifted to the continuous Arab-Israeli conflict. It should be added that in October 1990, Gorbachev was awarded the Nobel Peace Prize. Gorbachev's foreign policy could thus be considered a catastrophe, a great success, or both, depending largely on the point of view.

Domestic Developments

But at home, catastrophe prevailed. The optimism and confidence of the early Gorbachev years were gone. The economy kept deteriorating, leading to enormous shortages of consumer goods and even fear of famine. Major strikes, especially of miners, erupted in the Ural region, in Siberia, and in the Ukraine. Specific measures, such as the decree of January 23, 1991, withdrawing 50- and 100-ruble notes from circulation and compelling the exchange of these notes under highly restrictive conditions, turned into disasters. Very poorly managed, that decree failed to check inflation or limit crime, while it hit hard the average working citizens and pensioners. In fact, proliferating decrees and directives only led to utter confusion. With the new self-assertion of the union republics and of lesser jurisdictions, it was not at all clear who owned or managed what. The same piece of property or sphere of economic activity could be claimed by the central government, a union republic, a regional administration, or a municipality. Reforming measures by all kinds of authorities were at best partial, hap-hazard, and difficult, if not impossible, to implement. Major general eco-nomic reform, while repeatedly promised, kept being postponed.

Natural and man-made catastrophes together with their aftermaths, whether in the case of the earthquake in Armenia in December 1988, which killed some 25,000 people and left another 500,000 homeless, or in that of the train collision and gas explosion near Asha in the Urals in June 1989, with 190 persons listed as dead, 270 as missing, and 720 as hospitalized, served to underline the manifold deficiencies, including the incompetence, of the Soviet system. Ecological issues loomed ever larger as the nature and extent of the ecological damage in the country became better known. Perhaps even more damaging to the government and system were repeated discoveries of mass graves: some 102,000 bodies found near Minsk in Belorussia in October 1988; between 200,000 and 300,000 burials outside Kiev which a special commission determined in March 1989 to contain victims of Stalin, not of the Nazis; about 300,000 more bodies in mass graves near Cheliabinsk and Sverdlovsk in the Urals, uncovered on October 2, 1989; and still others. Glasnost not only provided information about all these matters and contributed to the rehabilitation of many communists executed in the purges of the 1930's as well as of Russian cultural figures abroad, such as the distinguished musician Mstislav Rastropovich and the brilliant satirist Vladimir Voinovich, but also led to a great diversity of opinion and variety of criticism. Gorbachev and his policies were attacked from the right, from the left, and from every direction.

Like his predecessors, Gorbachev began his career as the supreme Soviet leader when he attained the post of Party secretary as voted by the Politburo. Characteristically, as already mentioned, his initial main concern was to strengthen his position in the Politburo and, more broadly, in the upper echelons of the Party, a personnel policy that was on the whole successful, although never to the extent of giving the new leader complete control. But before long, great complications arose. The policies of perestroika and glasnost produced strong and continual opposition both by the Right, which felt fundamentally threatened by them, and by the Left, as in the case of Yeltsin, which complained that the reforming activity was not sufficiently efficacious or expeditious.

Furthermore, the position of the Party itself was changing. Gorbachev introduced competitive elections within the Communist Party to fight stagnation and obtain supporters against the entrenched traditionalists. On January 13, 1990, he reversed his earlier stand by declaring his willingness to accept the existence of other political parties in the U.S.S.R. Following a demonstration in February of some 250,000 people in Moscow and other gatherings elsewhere, on March 13, 1990, the legislature repealed Article 6 of the constitution, which had guaranteed "the leading role" of the Communist Party, that is, its monopoly of political power in the Soviet Union. Although the reformers have been slow in creating a comprehensive political organization outside the Party, there emerged in July 1991 the Democratic Reform Movement, led by such prominent former associates of Gorbachev as Shevardnadze and Yakovlev as well as other notable liberals. And the Party itself, once absolutely monolithic, came to be under the constant threat of a major split, and thus pluralism, in the summer of 1991, most obviously in its huge Russian branch.

The same day that Article 6 was abolished, Gorbachev, as president of the country, received the right to rule when necessary by executive decree. Combining Party and state offices was nothing new for Soviet leaders; the novelty of the latest arrangement consisted in the fact that the state position could now be used against the Party. Gorbachev had prepared his state base of power well, succeeding Gromyko to the title of president in October 1988, obtaining election to that office by the 2,250-member Congress of People's Deputies on April 25, 1989, and greatly enhancing its prominence. As the sway of the Politburo and the Party declined, close advisory bodies to the president, such as the eighteen-member Presidential Council, which lasted from March to December 1990, and especially the eight-member Security Council of the U.S.S.R., which succeeded it, acquired greater significance. The latter was composed mainly of the more important ministers of state. In the summer of 1991, speculation was rife that Gorbachev might abandon the Party altogether and stake everything on the state administration and reform. Actually, he turned in the opposite direction, winning once more sufficient Party support and apparently determined to carry it with him on his wayward way.

It is not easy to evaluate or even simply present Gorbachev's policy. Often it seems impossible to distinguish his own projects, plans, and aims from the political and other tactical concessions and compromises he had to make, and even from extraneous elements imposed on him by other political forces in the Soviet Union. The net result was a tortuous course most notable for its meandering between reform and restraint. To mention only the latest turnings, in October 1990 Gorbachev endorsed the so-called Shatalin plan, associated with the economist Stanislav Shatalin and meant to establish within five hundred days a market economy in the U.S.S.R. But at the last moment he held back and in the subsequent weeks and months adopted instead a conservative and even reactionary policy, characterized by hard-line key governmental appointments and the granting of new powers to the police and the army acting as police, and brought about the above-mentioned instances of military violence in Latvia and Lithuania. It was at that time that Shevardnadze resigned as foreign minister in protest and warning and was replaced by the former ambassador to the United States, Alexander Bessmertnykh, apparently without any change in the nature of Soviet foreign policy. Yet spring and summer brought another turning, with Gorbachev more enthusiastic than ever in the cause of economic and general reform, ready to condemn even Marxism as such on occasion and full of promising plans for the Soviet Union, although still without specifics or a timetable.

However, it may be most appropriate to end this brief discussion of perestroika where it began, that is, with the economic crisis, and for that to turn to Gregory Grossman's compelling presentation of the nature and the problem of the Soviet economic collapse in his testimony to congressional committees on June 25, 1991.*

One can hardly recall an instance in modern history in which — major war or its effects apart — the economic condition of an important country plunged so deep so fast as has that of the Soviet Union in the last few years. Less than a decade ago, serious Western observers could still seriously consider whether the global economic competition would eventually be "won" by the East, with all that implied for the world's future. Today, equally serious people equally seriously advocate Marshall-like assistance from the West in the hundreds of billions of dollars lest the Soviet economy (and polity and society) fall even deeper into destitution and disorder, with all *that* would imply for the world's future.

Although the present economic condition is indeed catastrophic, it has not been quite as unexpected as one might have assumed from appearances alone. In fact, the underlying forces of rot and ruin have been at work for decades, albeit concealed by the secretiveness of the dictatorial regime and the silence of an intimidated population (but for a relatively few dissidents). Among such long-term, corrosive trends one might mention the huge diversion of national resources to military and imperial ends; heedlessly wasteful depletion of natural and human reserves for economic growth and progress, combined with lags in civilian technological advance and improvement in quality; inability to feed the population without massive imports; enormous physical degradation and contamination of the environment with major effects on human health; growing sclerosis of the centralized system of economic planning and governance, aggravated by rigid price-wage controls and monetary mismanagement; steady growth of a large underground economy intimately linked with widespread official corruption and (with time) major organized crime; deterioration of work incentives and work morale, not to say initiative, enterprise (except in the underground), and sense of responsibility. And consequent steady retardation of economic growth, and actual decline.

One could extend this dismal list of the underlying economic factors (not to mention the political, social, and ethnic ones) that have been propelling the Soviet economy for decades towards its historic moment of deep crisis. That moment arrived under Gorbachev, not because Gorbachev is the most skilled economic reformer the USSR could have sooner or later produced — very likely he is not — but because it is difficult to imagine another communist leader, and it would have to be one, who could have more quickly and thoroughly discredited the shams of the past.

But Gorbachev has not yet destroyed, if indeed he intends to fully de-

---

* Slightly abridged statement of Gregory Grossman (Berkeley) submitted at the Joint Hearing of the Subcommittee on Europe and the Middle East of the Committee on Foreign Affairs of the U.S. House of Representatives and of the Joint Economic Committee of the U.S. Congress.

stroy, either the old ways of running the economy or the social groups that have traditionally run it, namely, the Party apparat (himself being its titular head) and the state's nomenklatura, or the idea of socialism itself. Which is one reason why the economy is currently in such deep trouble.

But first we should note the economic agenda that has informed and inflamed Soviet politics since Gorbachev took over in March 1985. (A) The first urgent problem that Gorbachev addressed was reducing the military-economic-technological gap vis-à-vis the West. He attacked it in the traditional Stalinist way with the same institutions and the same people . . . and failed, wasting resources and 1–2 years' time in the process. (B) Then he turned to the idea that apparently had been germinating in his mind for some time, namely, the boldest institutional reform yet, perestroika (as well as its concomitant, glasnost) to attract the educated to his side to help him discredit the past.

Economic reform (later, in its more radical, system-transforming, privatizing form usually termed "transition") imposed an agenda of economic measures which — not at once but with time, and not always in the same order of urgency — came to consist of the following main components. (1) At this point most urgently, stopping the rapid downward slide of the "real" economy. (2) Stabilization: moving to a condition in which — in starkest contrast to the current situation — domestic money is a functioning common denominator of value and an effective medium of exchange, and prices (and foreign exchange rates and interest rates) are market-clearing, reasonably stable, and sustainable.

By 1990–91 stabilization became the foremost issue, overshadowing all the others on the agenda. (3) Marketization of the command economy, a task of no mean order. (4) Legalization of the widest range of private business activity, plus privatization of hitherto state-owned businesses. (5) Resolving the explosive issue of new formulas for union (federal) to republic relations, inter-republic relations, secession or extreme autonomy of some of the republics (now usually calling themselves sovereign states), etc. (6) Integration of the USSR (or what is left of it) and of its individual constituent parts into the world economy. (7) A social safety net adequate to the economic dislocations and the political upheavals that can be fairly expected. (8) Fiscal and banking structures corresponding to the solutions taken under (1)–(7). And of course drafting and enacting a vast amount of legislation. A more daunting agenda would be hard to imagine even for a country not riven by nearly every kind of strife and feud as is the USSR.

Now, in mid-1991, important developments are moving at once on several planes and in contradictory ways. First and foremost, the economy's downward slide has accelerated markedly. Above, I used the word "catastrophic" in this regard; it is not overdrawn. For the first four months of this year GNP fell by over 9 percent in relation to the corresponding period the year before (but see the remarks regarding Soviet statistics immediately below). The record for the whole year may well be substantially worse, as is generally believed likely by well-informed Soviet economists with whom I met in Moscow and in the West during the past 30 days. Hence, I find myself in broad agreement with the "grim" conclusions in the paper entitled "Beyond Perestroyka: The Soviet Economy in Crisis" (16 May 1991, presented by the CIA and the DIA to the Joint Economic Committee of the U.S. Congress). To wit:

> There is no doubt that 1991 will be a worse year for the Soviet economy than 1990, and it is likely to be radically worse. . . . [If the standoff between the center and the republics continues], real gross national product likely would decline 10 to 15 percent and the annual inflation rate could easily exceed 100 percent. (p. iv)

The last may have already happened, largely by virtue of the sharp and mostly administrative increases in official retail prices on 2 April 1991. (It seems that the producers and ministries "ran away" with the prices and raised them significantly above permitted levels.) The monetary "compensation" for the April retail price increase that is being paid to the public, differentiated by social groups, offsets perhaps about half of the actual price rise. Further price increases can be expected before the end of this year.

In the event that the standoff between the center and the republics moderates, the results for 1991 could be less "grim" — but hardly very much less.

This said, in truth it is extremely difficult to establish with any numerical exactness just what is currently happening in the Soviet economy. Although, under its current leadership, the official statistical establishment is trying to improve the quality and comparability of its data, as well as increase the volume of statistical publication, its efforts are thwarted by the fractious realities of the moment. Great ranges of multiple prices for individual goods in a given place and time; physical shortages of goods; wide use of barter (for lack of faith in money); corruption, ubiquitous black markets, evasion of taxes and of administrative regulations; the fragmentation of the country into scores if not hundreds of what amounts to semi-independent principalities (due to erosion of central control), each with its own rationing norms, administrative rules, trade barriers, price controls, etc. — combine to make the statistician's lot not a happy one in the USSR. And yet, the broad trends in production and consumption are probably what we think they are; in other words, close to catastrophic.

No less important is the picture in regard to the distribution of personal income and wealth. Here, our data are even more opaque; yet it is difficult to escape the impression that 1991 has hit a new high (or low) in terms of the differentiation of personal income and wealth. The losers are that large portion of the population whose livelihood depends chiefly on official sources of income (wages, salaries, pensions, etc.), which have generally fallen in real terms this year; and/or whose *informal* incomes have risen less than inflation; and/or those with few lucrative personal connections. But many others have gained considerably riding the crest of the many new opportunities both for profitable production and commerce and for quick arbitrage and black-market dealing — by dint of growing shortages and soaring prices, substantial liberalization of private business activity in various new legal forms, and the chaos itself.

Thanks to such rising private opportunities in the midst of general confusion and chaos in the economy, the Soviet Union may today present some of the best prospects for quick personal enrichment. Nor need one keep one's wealth in rubles; there are innumerable ways of various degrees of legality or illegality for transferring private money abroad.

But note that until money and prices are stabilized, the federal issue is resolved, and the legal underpinnings of private business are further secured, private money will continue to shun long-term, illiquid investment and to seek the quick and easy deals. We must not be misled by stories of successful new Soviet entrepreneurs. Few of them are long-term investors.

Needless to say, large personal gain frequently derives from *de jure* or *de facto* "spontaneous appropriation" at the state's expense. And since those closest to the state's assets, the traditional elite (including the old management), obviously have the best possibilities of appropriating it, the phenomenon of the "propriation of the nomenklatura" has acquired if not mass dimensions then at least mass attention (as in Eastern Europe) with definite political implications.

There may be nothing wrong with the propriation of the nomenklatura or the use of "dirty money" to buy the state's assets — and indeed this may be the quickest road to privatization — but is this the kind of privatization we have in mind when we list the conditions for Western assistance, such as marketization and privatization of the economy and democratization of the polity? Do we want to see a market economy of sorts run by cartels of the same old communist bosses and the same lords of corruption and organized crime in a new guise?

If privatization by quick and dubious appropriation be ruled out, how many decades will it take to marketize and privatize the Soviet economy? In the meantime, will democracy develop and survive? Some other economic arrangement would have to be in place. Would it be largely the old command system — thus foreclosing hope of both economic progress and democratic revival? Or will there be a protracted interim phase of state-owned firms operating in a market context, i.e. a socialist market economy (to use an old term)? Indeed, for the reasons just suggested, and despite its disadvantages, the latter may yet have a lease on life in the USSR and in Eastern Europe as an alternative to something less attractive.

What will the USSR be like five years hence? Indeed, what will it be like five months hence? In 1986 who could have foreseen the USSR of 1991, the world of 1991? I suspect that it will be more decomposed (in various ways) than democratic, with an economy still as much administered as marketized, an economy still more chaotic than capitalistic, with great private-profit opportunities precisely for these reasons rather than despite them.

The reasons are that the economic problems are too deep, the political and national problems too acute, the legacy of the old days too tenacious. The process of privatization of state assets will be too slow (if not too threatening, in the way described above). And most of all, after everything else is set aright, the physical capital stock will have to be rebuilt almost from scratch. This rebuilding and re-equipping, its technical upgrading, and most of all, its re-orientation toward new social objectives and toward the outside world, will be protracted and enormously costly. As will be the management of the accumulated environmental destruction and toxic contamination.

The country will remain poor for a long time; and it will be in need of an expensive social safety net to avert or contain political disasters. Consequently, for some time the role of the state (or most successor states) will remain large, the tax burden heavy, and the rate of investment (private and public) in GNP will continue to be high. The day of a Soviet Thatcher may yet come, but not very soon. The United States should have no illusions on these scores.

It was in this situation of economic, social, and political flux that it was announced on August 19, 1991, that Gorbachev had relinquished, for reasons of health, his high government and Party posts. The rightist coup appeared to be, at least initially, successful. Yet it collapsed within three days because of the popular opposition led by Yeltsin, poor organization, and, apparently, the refusal of key military and police units to execute the orders of the leaders of the coup. The results of the collapse of the coup included the return of Gorbachev, a great enhancement of Yeltsin's position, the arrest of the leaders of the coup, and the discrediting of many more government, Party, and military leaders — indeed of the Party and the police themselves — with Gorbachev leaving the Party and Party activities being "suspended" pending investigation. Also, the process of the dissolution of the U.S.S.R. gathered great momentum. Lithuania, Latvia, and Estonia declared immediate independence, which received international and even Soviet recognition. Most other republics, including the Ukraine, also proclaimed independence, although, on the other hand, plans and negotiations continued for a new confederation or perhaps commonwealth, if not an effective federation. In fact, after a negative vote and subsequent modifications, the All-Union Congress of People's Deputies approved on September 6 Gorbachev's "transitional" plan of governing the Soviet Union, which gave more power and authority to the republics, but still provided for some central institutions, including the State Council to deal with foreign affairs, military matters, law enforcement, and security.

On December 25, 1991, Gorbachev resigned. All the Union republics declared their independence, although most of them indicated their willingness to form an undefined loose association which came to be known as the Commonwealth of Independent States. Boris Yeltsin stood out as the central figure in a new and highly unsettled situation. Finally, under Yeltsin, crucial economic reforms began to be introduced in Russia, and substantial help for these reforms was promised by the West.

X L I I I

* * * * * * * * * *

SOVIET RUSSIA: CONCLUDING REMARKS

> The history of mankind is executing in our days one of the greatest,
> most difficult turnings, of an unlimited, one can without the slightest
> exaggeration say: of a world-liberating importance. . . . from an
> abyss of suffering, torment, famine and barbarization to the bright
> future of a communist society, general well-being, and firm peace.
>
> LENIN

> Human sacrifices
> Does a canal redeem?
> He is godless, your engineer,
> And what power he acquired! *
>
> PASTERNAK

> Still one believes — suffering is not eternal,
> Though there is little justice, it exists in the world.
> Tenacious as the old hag is, just look,
> Children will nevertheless outlive her!
>
> KLENOVSKY

THE SOVIET UNION has been described and analyzed in a flood of books
and articles. Unfortunately — although perhaps inevitably, considering the
contemporary and critical nature of the subject — the great bulk of this
writing leaves much to be desired. Omitting from discussion the Communists' own propaganda, one can designate the next most biased approach as the school of extreme and blind hatred. Authors of this tendency,
usually fortified by formidable and even invincible ignorance, typically
lump Russia and the Soviet Union together and denounce them un-
reservedly from top to bottom. Not satisfied with the actual Soviet record,
they invent additional sins and crimes, such as the alleged brotherly feeling and even identity of the Communists and the Nazis. Regarding both
the activities of Communist parties everywhere and the pronouncements
of Marxists from Marx to Khrushchev on the eventual triumph of communism as too vague, they produce more specific blueprints of world
conquest, ascribing them to Lenin, or Stalin, or even Peter the Great. As
Brumberg and others have demonstrated, they have an easy way with quo-
tations and attributions. But on the whole, because of its crude and out-

---

* Professor Vladimir Markov of the University of California in Los Angeles first
pointed out this inexact and extremely suggestive translation by Pasternak from
Goethe, which was changed in a later edition of Pasternak's translation.

609

spoken character, the school of blind hatred is not likely to mislead people who are at all informed, unless they want to be misled.

More complicated and controversial are the efforts of genuine scholars and other intellectuals who have really tried to understand the Soviet Union but have nevertheless failed to form a realistic image of it. Many, for example, have confused the Soviet system with democratic socialism and even with liberalism and the general progress of mankind. Thus the U.S.S.R. has been intrepreted by various persons primarily in terms of economic democracy, a scientific outlook on life, or the emancipation of women. Prominent commentators, such as Sidney and Beatrice Webb, or even Schuman, to take a more recent example, have not been immune to this approach. Sometimes, as in the case of Gide, a look at Soviet reality set the perspective straight. Others have persisted in their rosy beliefs.

More skeptical scholars have at times been affected by the sheer weight of Soviet material and the nearly exclusive prominence of the Soviet point of view. Many events of Soviet history are known only in the official version, while the great bulk of archival and other primary sources has remained in the U.S.S.R., to be seen or published only by permission of the authorities. As a result, even Carr — perhaps the best as well as the most productive historian of the Soviet Union to date — has probably relied too much on Lenin's and on the government's versions of developments. It should be emphasized that an extremely reserved and cautious general attitude toward Soviet reports and claims can easily be combined in particular instances with an excessive dependence on Soviet sources, frequently the only ones available.

A different kind of error has threatened other interpreters of the Soviet system, especially among social scientists, who have used their analytical techniques to demonstrate the logicality, if not the inevitability, of the Bolshevik regime. Almost everything, from the swaddling of Russian babies to the particular stage and configuration of the economic development of the country, has been brought forth to explain "why they behave like Russians." Usually it has been argued that whereas the particular Soviet regime did not necessarily have to come about, any possible substitute would have looked and acted much the same. Russians, it has been asserted, could not make democracy work in any case. Also, their attitudes toward civil liberties, political police, and government control were basically different from those prevalent in the West. The danger of this approach lies in underestimating the magnitude and novelty of Soviet social engineering and in treating it as essentially rational and suited to the Russian situation. In its extreme form, the approach tends to deny that the Russian people have any human qualities.

Many other observers of the Soviet development have recognized the radical nature of the Bolshevik experiment, but have concluded that at a certain point the Soviet regime either "betrayed the revolution" or at least congealed into a more conservative, established system. This group includes, for example, Trotsky and his followers, as well as certain other Leftists. It might be noted, however, that Trotsky could never take a completely negative view of the Soviet system, but preferred to regard it as a mixture of genuine socialist revolution and Stalinist reaction.

By contrast, numerous proponents of the thesis of the Soviet retreat from revolutionary positions to conservatism look at the U.S.S.R. from the right rather than from the left. Most frequently they insist that at a given time nationalism and need for social stability asserted themselves with sufficient force in the Soviet Union to check the revolutionary current and consolidate the new system, conservatism replacing revolutionary dynamism. Examples of other revolutions, especially the great French Revolution, are often adduced to support the argument. Thus the inauguration of the New Economic Policy was interpreted by many as the end of the revolution and a return to the traditional Russian pattern. Therefore, a number of Russian émigré intellectuals, for instance, the gifted Nicholas Ustrialov, reversed their former position and proclaimed their willingness to support the new regime in protecting and developing Russian national interests. The group came to be known as *smenovekhovtsy*, a name suggested by the title of their collection of articles and meaning "those who changed the sign posts." The Soviet leaders, however, abolished the N.E.P. and switched to the all-out socialist offensive of the five-year plans. Ustrialov later returned to the U.S.S.R. and disappeared in the great purge.

Again, the conservative turning in social and cultural matters characteristic of the Soviet Union in the 1930's led some observers to believe that the decisive break with the revolutionary past had arrived. Such well-informed scholars as Timasheff and Sorokin expounded, respectively, the thesis of the "great retreat" and the view that the democratic provisions of the constitution of 1936 were sincerely meant and that the U.S.S.R. was on its way to becoming a democracy. Other Russians, together with some non-Russians, rejoiced simply at the return of Pushkin, of Russian historical heroes, or of discipline in Soviet schools. Yet the retreat proved to be highly limited and rigidly controlled. Most important, Stalin and the Politburo retained full power and freedom of action, and, needless to say, they never renounced their Marxist ideology or Communist intentions. Stalin, if you will, assumed the roles of Mirabeau, Danton, Robespierre, and even Napoleon, but he always remained the same ruthless Georgian Bolshevik, relentless in his hatred and determined to pursue what he considered to be correct Party goals.

The Second World War, of course, suggested more turning points as nationalism rose in Soviet Russia, the army acquired a new prominence, and religion a new respectability. Moreover, the U.S.S.R. was allied to Western democracies in a titanic common struggle. The immediate postwar years, therefore, proved to be the greatest disappointment of all to the democratic world and possibly as well to Soviet citizens themselves: bitter hostility and the cold war, punctuated by actual heavy fighting in such restricted areas as Korea and Indo-China, replaced the wartime co-operation between communism and the West, while at home Stalin and Zhdanov moved to isolate their people from foreign influences, restore Marxist orthodoxy, and punish the deviationists. Communism proceeded to expand across the maps of Europe and Asia.

Finally, the death of Stalin in the spring of 1953 and the events that followed it convinced some observers that the Soviet system was at last breaking down or undergoing a radical change. With the brief rise of Zhukov, much was written about the new power of the army and the possibilities of a military regime in the U.S.S.R. Other authors argued that the Party would have to give way to technicians and indeed a managerial society. Actually nothing of the sort occurred between 1953 and 1985.

What happened since was sketched in the immediately preceding chapter. The old Soviet regime and the expanding Soviet empire are now irretrievably gone, at last. Yet the historiographical argument about the nature of the Soviet communist system and its change may be more apposite than ever. Would Gorbachev's survival have meant a definitive end of the Soviet period and the dawn of a new age? Will Yeltsin's? How much transformation can one expect, and within what time frame? Some historical regimes conditioned their successors for generations and even centuries. Others, such as Nazism, after a total defeat to be sure, seemed to melt into thin air. There is ground for optimism for the peoples of the area to the extent that for them any future, except for such unlikely extremes as a many-sided and murderous civil war, a neo-Stalinist restoration, or a Fascist victory, is likely to be happier than their Soviet past. Beyond that, one hesitates to predict and can only cite, as a counterpoint to Derzhavin's, Heraclites' even more famous statement: "$\tau\acute{\alpha}\ \pi\acute{\alpha}\nu\tau\alpha\ \acute{\rho}\epsilon\hat{\iota}$," "Everything flows."

BIBLIOGRAPHY

Alliluyeva, Svetlana Iosifovna (1926–   ) Stalin's daughter. Works include *Twenty Letters to a Friend.*

Armstrong, John A. (1922–   ) American political scientist. Works include *The Politics of Totalitarianism: The Communist Party of the Soviet Union from 1934 to the Present; The Soviet Bureaucratic Elite: A Case Study of the Ukrainian Apparatus; Ukrainian Nationalism, 1939–1945; The European Administrative Elite; Ideology, Politics and Government in the Soviet Union.*

Baumgarten, Nicolas Pierre Serge von (1867–1939) Russian historian. Works include "Aux origines de la Russie," *Orientalia Christiana Analecta* 119 (1939); "Chronologie ecclésiastique des terres russes du Xe au XIIIe siècle," *Orientalia Christiana* (January 1930); "Généalogies des branches régnantes des Rurikides russes du XIIIe au XVIe siècle," *Orientalia Christiana* (June 1934); "Généalogies et mariages occidentaux des Rurikides russes du Xe au XIIIe siècle," *Orientalia Christiana* (May 1927).

Bayer, Gottlieb Siegfried (1694–1738) German historian who worked in Russia under Empress Anne. Works include *De Russorum prima expeditione Constantinopolitana; De Varagis; Geographia Russiae . . . ex Constantino Porphyrogenneta; Geographia Russiae ex Scriptoribus Septentrionalibus; Origines Russicae.*

Baykov, Alexander (1899–   ) Russian-British economist. Works include *The Development of the Soviet Economic System: An Essay on the Experience of Planning in the USSR; Soviet Foreign Trade.*

Beloff, Max (1913–   ) British political scientist. Works include *The Foreign Policy of Soviet Russia, 1929–1941* (2 vols.); *Soviet Policy in the Far East, 1944–1951.*

Berdiaev, Nikolai Aleksandrovich (1878–1948) Russian cultural philosopher. Works include *Istoki i smysl russkogo kommunizma (The Origin of Russian Communism); The Russian Revolution: Two Essays on its Implications in Religion and Psychology; Russkaia ideia: Osnovnye problemy russkoi mysli XIX veka i nachala XX veka (The Russian Idea).*

Bereday, George Z. F. (1920–   ) American specialist on Soviet education. Works include *The Changing Soviet School: The Comparative Education Society Field Study in the U.S.S.R.* (ed. with William W. Brickman and Gerald H. Read); *The Politics of Soviet Education* (ed. with Jaan Pennar); "Education: Organization and Values since 1917," in *Transformation of Russian Society* (ed. C. E. Black).

Bergson, Abram (1914–   ) American economist. Works include *The Real National Income of Soviet Russia since 1928; The Structure of Soviet Wages: A Study in Socialist Economics; Soviet Economic Growth: Conditions and Perspectives* (ed.).

Berlin, Sir Isaiah (1909–   ) British intellectual historian. Works include *The Hedgehog and the Fox, An Essay on Tolstoy's View of History; Karl Marx: His Life and Environment;* "Russia and 1848," *Slavonic and East European Review* (April 1948); "The Marvelous Decade," *Encounter* (June, November, December 1955; May 1956); "The Silence in Russian Culture," *Foreign Affairs* (October 1957); "Tolstoy and Enlightenment," *Encounter* (February 1961); *Russian Thinkers.*

Black, Cyril E. (1915–1989) American historian. Works include *Twentieth-Century Europe: A History* (with E. C. Helmreich); *Rewriting Russian History: Soviet Interpretations of Russia's Past* (ed.); *The Transformation of Russian Society: Aspects of Social Change since 1861* (ed.); *The Modernization of Japan and Russia* (with others); "The Nature of Imperial Russian Society,"

Slavic Review (December 1961); *Understanding Soviet Politics: The Perspective of Russian History.*

Blum, Jerome (1913–   ) American historian. Works include *Lord and Peasant in Russia from the Ninth to the Nineteenth Century;* "The Rise of Serfdom in Eastern Europe," *American Historical Review* (July 1957); "Russian Agriculture in the Last 150 Years of Serfdom," *Agricultural History* (January 1960).

Bogoslovsky, Mikhail Mikhailovich (1867–1929) Russian historian. Works include *Oblastnaia reforma Petra Velikogo, provintsiia 1719–1727 gg.; Petr I, materialy dlia biografii* (5 vols.).

Boltin, Ivan Nikitich (1735–1792) Russian historian. Works include *Kriticheskie primechaniia gen.-maiora Boltina na pervyi-vtoroi tom istorii kniazia Shcherbatova* (2 vols.); *Otvet gen.-maiora Boltina na pismo Kn. Shcherbatova; Primechaniia na istoriiu gospodina Leklerka* (2 vols.).

Briusov, Valerii Iakovlevich (1873–1924) Russian poet, writer, and literary scholar. Works include "Mednyi Vsadnik," in *Biblioteka velikikh pisatelei pod redaktsiei S. A. Vengerova: Pushkin* (vol. 3).

Brumberg, Abraham (1926–   ) American specialist on communism and editor of *Problems of Communism.* Works include "Apropos of Quotation Mongering," *New Republic* (August 29, 1960); *Chronicle of a Revolution: A Western-Soviet Inquiry into Perestroika* (ed.).

Brzezinski, Zbigniew K. (1928–   ) American political scientist. Works include *The Soviet Bloc: Unity and Conflict; The Permanent Purge; Ideology and Power in Soviet Politics; Alternative to Partition: For a Broader Conception of America's Role in Europe; The Grand Failure: The Birth and Death of Communism in the Twentieth Century.*

Campbell, Robert W. (1926–   ) American economist. Works include *Soviet Economic Power: Its Organization, Growth, and Challenge.*

Carr, Edward H. (1892–1982) British historian. Works include *A History of Soviet Russia: The Bolshevik Revolution, 1917–1922* (vols. 1–3), *The Interregnum, 1923–1924* (vol. 4), *Socialism in One Country, 1924–1926* (vols. 5–7), *Foundations of a Planned Economy, 1926–1929* (vol. 8, in two parts; with R. W. Davies, vol. 9); *Michael Bakunin; The Romantic Exiles: A Nineteenth-Century Portrait Gallery; The Soviet Impact on the Western World; The October Revolution: Before and After.*

Cattell, David T. (1923–   ) American political scientist. Works include *Communism and the Spanish Civil War; Soviet Diplomacy and the Spanish Civil War.*

Chamberlin, William Henry (1897–1969) American journalist and specialist on the Soviet Union. Works include *The Russian Revolution, 1917–1921* (2 vols.); *Russia's Iron Age.*

Charnolussky, Vladimir Ivanovich (1865–1941) Russian-Soviet specialist on education. Works include "Nachalnoe obrazovanie vo vtoroi polovine XIX stoletiia," in *Istoriia Rossii v XIX veke* (vol. 7); "Narodnoe obrazovanie v pervoi polovine XIX veka," in *Istoriia Rossii v XIX veke* (vol. 4).

Charques, Richard Denis (1899–   ) British writer, literary scholar, and historian. Works include *A Short History of Russia; The Twilight of Imperial Russia.*

Cherepnin, Lev Vladimirovich (1905–1977) Soviet historian. Works include *Obrazovanie russkogo tsentralizovannogo gosudarstva: Ocherki sotsialno-ekonomicheskoi i politicheskoi istorii Rusi; Osnovnye etapy razvitiia feodalizma v Rossii; Russkaia istoriografiia do XIX veka: Kurs lektsii; Russkie feodalnye arkhivy XIV–XV vekov; Knigi moskovskikh prikazov v fondakh TsGADA: opis, 1495–1718 gg.; Novgorodskie berestianye gramoty kak istoricheskii istochnik; Krestianskie voiny v Rossii semnadstatogo-vosemnadtsatogo vekov:*

problemy, poiski, resheniia. Sbornik statei (ed.); *Feodalnaia Rossiia vo vsemirno-istoricheskom protsesse; Puti razvitiia feodalizma* (with A. P. Novoseltsev and V. T. Pashuto); *Zemskie sobory russkogo gosudarstva v XVI–XVII vv; Otechestvennye istoriki XVIII–XX vv.*

Chizhevsky (Chyzhevskyi), Dmitrii (1894–1977) Ukrainian-German specialist in Russian and Slavic literature and thought. Works include *Gegel v Rossii* (Ger. *Hegel in Russland*); *Geschichte der altrussischen Literatur im 11., 12., und 13. Jahrhundert, Kiever Epoche; Hegel bei den Slaven* (ed.); *Das heilige Russland: Russische Geistesgeschichte I, 10.–17. Jahrhundert; History of Russian Literature from the Eleventh Century to the End of the Baroque; Russische Literaturgeschichte des 19. Jahrhunderts* (2 vols.); *Outline of Comparative Slavic Literatures.*

Churchill, Sir Winston S. (1874–1965) British statesman and historian. Works include *The Aftermath; The Second World War* (6 vols.).

Conquest, Robert (1917–   ) British specialist in the Soviet Union. Works include *The Great Terror: Stalin's Purge of the Thirties; The Great Terror: A Reassessment; The Harvest of Sorrow; Agricultural Workers in the USSR* (ed.); *Industrial Workers in the USSR* (ed.); *The Nation Killers: The Soviet Deportation of Nationalities; V. I. Lenin; Stalin and the Kirov Murder; Stalin: Breaker of Nations.*

Cross, Samuel H. (1891–1946) American specialist in Slavic languages, literatures, and cultures. Works include *Mediaeval Russian Churches* (ed. Kenneth J. Conant); *The Russian Primary Chronicle, Laurentian Text* (trans. and ed. with O. P. Sherbowitz-Wetzor); "The Lay of the Host of Igor" (trans.), in *La Geste du Prince Igor* (ed. Henri Grégoire, Roman Jakobson, and Marc Szeftel).

Crossman, R. H. S. (1907–1974) British intellectual and politician. Works include *The God that Failed: Six Essays on Communism* (ed.).

Dallin, Alexander (1921–   ) American political scientist. Works include *German Rule in Russia, 1941–1945: A Study of Occupation Policies; The Soviet Union and Disarmament, an Appraisal of Soviet Attitudes and Intentions; The Soviet Union at the United Nations; The Black Box.*

Dallin, David J. (1889–1962) Russian-American historian. Works include *The Changing World of Soviet Russia; Forced Labor in the Soviet Union* (with B. I. Nicolaevsky); *The New Soviet Empire; Russia and the Far East; Soviet Foreign Policy after Stalin.*

Deutscher, Isaac (1907–1967) Polish-British historian. Works include *The Prophet Armed: Trotsky, 1879–1921; The Prophet Unarmed: Trotsky, 1921–1929; The Prophet Outcast: Trotsky, 1929–1940; Stalin: A Political Biography; The Unfinished Revolution, 1917–1967.*

Dewitt, Nicholas (1923–   ) American specialist in Soviet education and economics. Works include *Education and Professional Employment in the U.S.S.R.; Soviet Professional Manpower: Its Training and Supply.*

Diakonov, Mikhail Aleksandrovich (1856–1919) Russian historian. Works include *Izbranie Mikhaila Fedorovicha na tsarstvo; Ocherki obshchestvennogo i gosudarstvennogo stroia drevnei Rusi* (Ger. *Skizzen zur Gesellschaft und Staatsordnung des alten Russlands*); *Vlast moskovskikh gosudarei: Ocherki po istorii politicheskikh idei drevnei Rusi do kontsa XVI veka.*

Dobb, Maurice Herbert (1900–1976) British economist. Works include *Soviet Economic Development since 1917.*

Druzhinin, Nikolai Mikhailovich (1886–   ) Soviet historian. Works include *Gosudarstvennye krestiane i reforma P. D. Kiseleva* (2 vols.); *Krestianskoe dvizhenie v 1861 godu posle otmeny krepostnogo prava; Russkaia derevnia na perelome, 1861–1880 gg.*

Dunlop, Douglas Morton (1909–1987) British historian. Works include *The History of the Jewish Khazars.*

Duranty, Walter (1884–1957) American journalist. Works include *Duranty Reports Russia; I Write as I Please; Stalin & Co.: The Politburo, the Men Who Run Russia.*

Dvornik, Francis (1893–1975) Czech-American historian. Works include *The Slavs in European History and Civilization;* "Byzantine Influences in Russia," *Geographical Magazine* (1947); "The Kiev State and Its Relations with Western Europe," *Transactions of the Royal Historical Society* (1947); "Byzantine Political Ideas in Kievan Russia," *Dumbarton Oaks Papers* (1956).

Erlich, Alexander (1912–1985) American economic historian. Works include *The Soviet Industrialization Debate, 1924–1928.*

Fainsod, Merle (1907–1972) American political scientist. Works include *How Russia Is Ruled; Smolensk under Soviet Rule.*

Fay, Sidney B. (1876–1967) American historian of Europe. Works include *The Origins of the World War* (2 vols.).

Fedotov, Georgii Petrovich (1886–1951) Russian-American historian of religion and culture. Works include *The Russian Religious Mind: Kievan Christianity, the Tenth to the Thirteenth Centuries* (vol. 1), *The Middle Ages, the Thirteenth to the Fifteenth Centuries* (vol. 2); *Sviatye Drevnei Rusi (X–XVII st.);* *A Treasury of Russian Spirituality.*

Feldmesser, Robert A. (1925–   ) American sociologist. Works include "Stratification and Communism," in *Prospects for Soviet Society* (ed. A. Kassof).

Fischer, George (1923–   ) American historian and political scientist. Works include *Russian Liberalism, from Gentry to Intelligenntsia; Soviet Opposition to Stalin: A Case Study in World War II.*

Florinsky, Michael T. (1894–1981) Russian-American economist and historian. Works include *The End of the Russian Empire; Russia: A History and Interpretation* (2 vols.).

Florovsky, George Vasilevich (1893–1979) Russian-American Orthodox theologian and intellectual historian. Works include *Puti russkogo bogosloviia (Ways of Russian Theology,* part 1); "O patriotizme pravednom i grekhovnom," in the Eurasian book *Na putiakh;* "The Problem of Old Russian Culture," *Slavic Review* (March 1962).

Freeze, Gregory (1945–   ) American historian. Works include *Description of the Clergy in Rural Russia: The Memoir of a Nineteenth Century Parish Priest* (ed. and trans.); *The Parish Clergy in Nineteenth Century Russia: Crisis, Reform, Counter-reform; The Russian Levites: Parish Clergy in the Eighteenth Century.*

Gerschenkron, Alexander (1904–1978) Russian-American economist. Works include *Economic Backwardness in Historical Perspective;* "Agrarian Policies and Industrialization, Russia 1861–1917," in *Cambridge Economic History of Europe* (vol. 6, pt. 2); *Continuity in History and Other Essays; Europe in the Russian Mirror: Four Lectures in Economic Theory.*

Gide, André (1869–1951) French writer. Works include *Retour de l'U.R.S.S. (Back from the USSR); Retouches à "Retour de l'U.R.S.S." (Afterthoughts, A Sequel to "Back from the USSR").*

Goldsmith, Raymond W. (1904–1988) American economist. Works include "The Economic Growth of Tsarist Russia 1860–1913," *Economic Development and Cultural Exchange* (April 1961).

Golikov, Ivan Ivanovich (1735–1801) Russian historian who collected source material on Peter the Great. Works include *Deianiia Petra Velikogo, mudrogo preobrazitelia Rossii: Sobrannye iz dostovernykh istochnikov i raspolozhennye po godam* (12 vols.); *Dopolnenie* (18 vols.).

Golovin, Nikolai Nikolaevich (1875–1944) Russian general and author. Works include *Rossiiskaia kontrrevoliutsiia v 1917–1918 gg.* (5 vols.); *The Russian Army in the World War.*

Golubinsky, Evgenii Evsigneevich (1834–1912) Russian Church historian. Works include *Istoriia russkoi tserkvi* (2 vols.).

Gooch, George Peabody (1873–1968) British historian of Europe. Works include *Catherine the Great and Other Studies.*

Grabar, Igor Emmanuilovich (1871–1960) Russian-Soviet specialist in art and art history. Works include *Istoriia russkogo iskusstva* (6 vols.).

Grekov, Boris Dmitrievich (1882–1953) Soviet historian. Works include *Feodalnye otnosheniia v kievskom gosudarstve; Kievskaia Rus (Kiev Rus); Krestiane na Rusi s drevneishikh vremen do XVII veka* (Ger. *Die Bauern in der Rus von den altesten Zeiten biz zum 17. Jahrhundert* [2 vols.]); *Zolotaia Orda i ee padenie* (with A. Iu Iakubovskii) (Fr. *La Horde d'or*).

Grossman, Gregory (1921–   ) American economist. Works include *Value and Plan: Economic Calculation and Organization in Eastern Europe* (ed.); "National Income," in *Soviet Economic Growth* (ed. A. Bergson); "A Note on the Fulfillment of the Fifth Five-Year Plan in Industry," *Soviet Studies* (April 1957); "The Structure and Organization of the Soviet Economy," *Slavic Review* (June 1962); "Thirty Years of Soviet Industrialization," *Soviet Survey* (October–December 1958); "Notes for a Theory of the Command Economy," *Soviet Studies* (October 1963); "The Soviet Economy and the Waning of the Cold War," in *Beyond the Cold War* (ed. R. Goldwin); "Innovation and Information in the Soviet Economy," *The American Economic Review* (May 1966); "Economic Reforms: A Balance Sheet," *Problems of Communism* (November–December 1966); "Gold and the Sword: Money in the Soviet Command Economy," in *Industrialization in Two Systems: Essays in Honor of Alexander Gerschenkron* (ed. H. Rosovsky); "The Solidary Society: A Philosophical Issue in Communist Economic Reforms," in *Essays in Socialism and Planning in Honor of Carl Landauer;* "The Economy at Middle Age," *Problems of Communism* (March–April 1976); "Economics of Virtuous Haste: A View of Soviet Industrialization and Institutions," in *Marxism, Central Planning, and the Soviet Economy: Economic Essays in Honor of Alexander Erlich* (ed. Padma Desai); "A Note on Soviet Inflation," in U.S. Congress, Joint Economic Committee, *Soviet Economy in the 1980s, Problems and Prospects, Part I: Selected Papers;* "The Party as Manager and Entrepreneur," in *Entrepreneurship in Imperial Russia and the Soviet Union* (ed. Gregory Guroff and Fred V. Carstensen); "The Second Economy: Boon or Bane for the Reform of the First Economy?" in *Economic Reforms in the Socialist World* (ed. Stanislaw Gomulka, Yong-Chool Ha, and Cae-One Kim).

Grunwald, Constantine de (   ) Russian-French historian. Works include *Alexandre Ier: Le tsar mystique; La Russie de Pierre le Grand (Peter the Great); Trois siècles de diplomatie russe; La Vie de Nicolas Ier (Tsar Nicholas I).*

Haimson, Leopold (1927–   ) American historian. Works include "The Problem of Social Stability in Urban Russia, 1905–1917," *Slavic Review* (December 1964–March 1965); *The Russian Marxists and the Origins of Bolshevism; The Politics of Rural Russia, 1905–1914* (ed.).

Halecki, Oscar (1891–1973) Polish-American historian. Works include *Borderlands of Western Civilization: A History of East Central Europe; From Florence to Brest (1459–1596); A History of Poland; The Limits and Divisions of European History; Cambridge History of Poland* (ed. with W. F. Reddaway, J. H. Penson, and R. Dyboski [2 vols.]); "Imperialism in Slavic and East European History," *American Slavic and East European Review* (February 1952).

Hellie, Richard 1937–   ) American historian. Works include *Enserfment and Military Change in Muscovy;* "Recent Soviet Historiography on Medieval and Early Modern Russian Slavery," *Russian Review* (January 1976); *Slavery in Russia; The Muscovite Law Code (Ulozhenie) of 1649 Part 1: Text and Translation* (ed. and trans.).

Hook, Sidney (1902–1989) American political philosopher. Works include *From Hegel to Marx; The Hero in History: A Study in Limitation and Possibility; Towards an Understanding of Karl Marx.*

Hrushevsky (Grushevsky), Mikhail Sergeevich (1866–1934) Ukrainian historian. Works include *Istoriia Ukrajiny-Rusy* (10 vols.) (Eng. trans. of a different, much briefer study, *A History of the Ukraine*).

Ignatovich, Inna Ivanovna (1879–1967) Russian-Soviet historian. Works include *Borba krestian za osvobozhdenie; Pomeshchichi krestiane nakanune osvobozhdeniia;* "Krestianskie volneniia pervoi chetverti XIX veka," *Voprosy istorii* (1950).

Inkeles, Alex (1920–   ) American sociologist. Works include "Models and Issues in the Analysis of Soviet Society," *Survey* (July 1966); *How the Soviet System Works* (with R. Bauer and C. Kluckhohn); *The Soviet Citizen* (with R. Bauer); *Public Opinion in the Soviet Union.*

Itenberg, Boris Samuilovich (1921–   ) Soviet historian. Works include *Dvizhenie revolutsionnogo narodnichestva; Pervyi Internatsional i revoliutsionnaia Rossiia; Iuzhno-rossiiskii soiuz rabochikh: vozniknovenie i deiatelnost; P. L. Lavrov v russkom revoliutsionnom dvizhenii.*

Jakobson, Roman (1896–1982) Russian-American philologist and historian of literature. Works include *Remarques sur l'évolution phonologique du russe comparée à celle des autres langues slaves; Russian Epic Studies* (with E. J. Simmons); *Slovo o Polku Igoreve v perevodakh kontsa vosemnadtsatogo veka; La Geste du Prince Igor* (ed. with Henri Grégoire and Marc Szeftel).

Jasny, Naum (1883–1967) Russian-American economist. Works include *The Socialized Agriculture of the U.S.S.R.: Plans and Performance; Soviet Industrialization, 1928–1952; The Soviet 1956 Statistical Handbook: A Commentary;* "The Soviet Seven-Year Plan: Is It Realistic?" *Bulletin of the Institute for the Study of the USSR* (May 1959); "The Soviet Statistical Yearbooks, 1955–1960," *Slavic Review* (March 1962).

Jelavich, Charles (1922–   ) American historian of Eastern Europe. Works include *Tsarist Russia and Balkan Nationalism, 1879–1886.*

Johnson, Robert E. (1943–   ) American historian. *Peasant and Proletarian: The Working Class of Moscow at the End of the Nineteenth Century.*

Karamzin, Nikolai Mikhailovich (1766–1826) Russian writer and historian. Works include *Istoriia Gosudarstva Rossiiskogo* (12 vols.); *Karamzin's Memoir on Ancient and Modern Russia: The Russian Text* (ed. R. E. Pipes) (*Karamzin's Memoir on Ancient and Modern Russia: A Translation and Analysis,* ed. R. E. Pipes).

Karpovich, Michael (1888–1959) Russian-American historian. Works include *Economic History of Europe since 1750* (with Witt Bowden and Abbott P. Usher); *Imperial Russia, 1801–1917;* Russian sections of *An Encyclopedia of World History* (ed. W. L. Langer); "A Forerunner of Lenin: P. N. Tkachev," *Review of Politics* (1944); "Two Types of Russian Liberalism: Maklakov and Miliukov," in *Continuity and Change in Russian and Soviet Thought* (ed. E. J. Simmons); "Vladimir Soloviev on Nationalism," *Review of Politics* (1946).

Keep, J. L. H. (1926–   ) British historian. Works include "The Decline of the Zemsky Sobor," *Slavonic and East European Review* (December 1957); "The Regime of Filaret," *Slavonic and East European Review* (June 1960); *The Russian Revolution: A Study in Mass Mobilization; Soldiers of the Tsar: Army and Society in Russia, 1462–1874.*

Kennan, George F. (1904–   ) American diplomat and historian. Works include *Soviet-American Relations, 1917–1920: Russia Leaves the War* (vol. 1), *The Decision to Intervene* (vol. 2); *Soviet Foreign Policy, 1917–1941;* "Russia and the Versailles Conference," *American Scholar* (Winter 1960–61); "Soviet Historiography and America's Role in the Intervention," *American Historical Review* (January 1960); *The Decline of Bismarck's European Order: Franco-Russian Relations, 1875–1890.*

Kerner, Robert J. (1887–1956) American historian. Works include *Northeastern Asia: A Selected Bibliography; Slavic Europe: A Selected Bibliography; The Urge to the Sea: The Course of Russian History.*

Khodsky, Leonid Vladimirovich (1854–1918) Russian economist. Works include *Osnovy gosudarstvennogo khoziaistva; Politicheskaia ekonomiia v sviazi s finansami; Pozemelnyi kredit v Rossii i otnoshenie ego k krestianskomu zemlevladeniiu.*

Khromov, Pavel Alekseevich (1907–   ) Soviet economic historian. Works include *Ekonomicheskoe razvitie Rossii v XIX–XX vekakh; Ocherki ekonomiki feodalizma v Rossii; Ocherki ekonomiki Rossii perioda monopolisticheskogo kapitalizma; Ocherki ekonomiki tekstilnoi promyshlennosti SSSR.*

Kirchner, Walther (1905–   ) German-American historian. Works include *The History of Russia; Eine Reise durch Siberien im achtzehnten Jahrhundert: Die Fahrt des schweizer Doktors Jakob Fries; The Rise of the Baltic Question.*

Kizevetter, Aleksandr Aleksandrovich (1866–1933) Russian historian. Works include *Gorodovoe polozhenie Ekateriny II; Istoricheskie ocherki; Istoricheskie otkliki; Na rubezhe dvukh stoletii: Vospominaniia 1881–1914; Posadskaia obshchina v Rossii v XVIII st.;* chapters in *Histoire de Russie* (ed. P. N. Miliukov, C. Seignobos, and L. Eisenmann [3 vols.]); "Vnutrenniaia politika v tsarstvovanie Nikolaia Pavlovicha," in *Istoriia Rossii iv XIX veke* (vol. 1).

Kline, George L. (1921–   ) American specialist in Russian philosophy and culture. Works include *Soviet Education* (ed.); *Spinoza in Russian Philosophy* (ed.); "Recent Soviet Philosophy," *Annals of the American Academy of Political and Social Science* (January 1956); "Russia's Lagging School System," *New Leader* (March 16, 1959); "Philosophy, Ideology, and Policy in the Soviet Union," *Review of Politics* (April 1964); "Economic Crime and Punishment," *Survey* (October 1965).

Kliuchevsky (Klyuchevsky), Vasilii Osipovich (1841–1911) Russian historian. Works include *Boiarskaia duma drevnei Rusi; Istoriia soslovii v Rossii; Kurs russkoi istorii* (5 vols.) (*A History of Russia*); *Opyty i issledovaniia* (3 vols.).

Konovalov, Sergei (1899–1982) Russian-British historian, former editor and frequent contributor to the *Oxford Slavonic Papers.*

Kostomarov, Nikolai Ivanovich (1817–1888) Ukrainian historian. Works include *Deiateli russkoi tserkvi v starinu; Deux nationalités russes; Istoricheskie monografii i issledovaniia; O znachenii Velikogo Novgoroda; Russkaia istoriia v zhizneopisaniiakh ee glavneishikh deiatelei* (3 vols.).

Kovalchenko, Ivan Dmitrievich (1923–   ) Soviet historian. Works include *Russkoe krepostnoe krestianstvo v pervoi polovine XIX veka.*

Kovalevsky, Maksim Maksimovich (1851–1916) Russian sociologist, political scientist, and historian. Works include *Istoriia nashego vremeni; Modern Customs and Ancient Laws of Russia; Ocherk proiskhozhdeniia i razvitiia semi i sobstvennosti; Le Régime économique de la Russie; Russian Political Institutions; La Russie sociale.*

Kucherov, Samuel (1892–1972) Russian-American specialist in legal history and Soviet affairs. Works include *Courts, Lawyers, and Trials under the Last Three Tsars; The Organs of Soviet Administration of Justice: Their History and Operation.*

Langer, William L. (1896–1977) American historian of Europe. Works include *The*

Diplomacy of Imperialism, 1890–1902 (2 vols.); *European Alliances and Alignments, 1870–1890; The Franco-Russian Alliance, 1890–1894; An Encyclopedia of World History* (ed.).

Lantzeff, George V. (1892–1955) Russian-American historian. Works include *Siberia in the Seventeenth Century: A Study in Colonial Administration; Eastward to Empire* (with R. A. Pierce).

Lapidus, Gail Warshofsky (1939–   ) American political scientist. Works include *Women in Soviet Society: Equality, Development, and Social Change; Women in Russia* (ed. with Dorothy Atkinson and Alexander Dallin).

Lasswell, Harold D. (1902–1978) American sociologist and psychologist. Works include *World Politics and Personal Insecurity.*

Lednicki, Waclaw (1891–1967) Polish-American specialist in Slavic and European literature. Works include *Pushkin's Bronze Horseman: The Story of a Masterpiece; Russia, Poland and the West: Essays in Literary and Cultural History; Russian-Polish Relations: Their Historical, Cultural and Political Background.*

Lemke, Mikhail Konstantinovich (1872–1923) Russian historian. Works include *Epokha tsenzurnykh reform, 1859–1865; Politicheskie protsesy; Nikolaevskie zhandarmy i literatura.*

Leontovich (Leontovitsch), Victor (1922–1960) German historian. Works include *Geschichte des Liberalismus in Russland* (*A History of Liberalism in Russia*).

Lewin, Moshe (1921–   ) British historian. Works include *Russian Peasants and Soviet Power: A Study of Collectivization* (trans.); 'Soviet Policies of Agricultural Procurements before the War," in *Essays in Honor of E. H. Carr* (ed. Ch. Abramsky); *Lenin's Last Struggle* (trans. A. M. Sheridan Smith); *The Gorbachev Phenomenon: A Historical Interpretation.*

Liashchenko, Petr Ivanovich (1876–1955) Russian-Soviet economic historian. Works include *Istoriia narodnogo khoziaistva SSSR* (*History of the National Economy of Russia to the 1917 Revolution*); *Krestianskoe delo i poreformennaia zemleustroitelnaia politika; Ocherki agrarnoi evoliutsii Rossii; Russkoe zernovoe khoziaistvo v sisteme mirovogo khoziaistva; Sotsialnaia ekonomiia selskogo khoziaistva* (2 vols.).

Lilge, Frederic (1911–   ) German-American specialist in Soviet education. Works include *Anton Semyonovitch Makarenko: An Analysis of His Educational Ideas in the Context of Soviet Society;* "Impressions of Soviet Education," *International Review of Education* (1959); "The Soviet School Today," *Survey* (July 1963); "Lenin and the Politics of Education," *Slavic Review* (June 1968).

Liubavsky, Matvei Kuzmich (1860–1937) Russian historian. Works include *Lektsii po drevnei russkoi istorii do konsta shestnadtsatogo veka; Obrazovanie osnovnoi gosudarstvennoi territorii velikorusskoi narodnosti; Ocherk istorii Litovsko-Russkogo gosudarstva.*

Lord, Robert Howard (1885–1954) American historian of European diplomacy. Works include *The Second Partition of Poland: A Study in Diplomatic History;* "The Third Partition of Poland," *Slavonic and East European Review* (1925).

Madariaga, Isabel de (1919–   ) British historian. Works include *Britain, Russia, and the Armed Neutrality of 1780; Russia in the Age of Catherine the Great.*

Makovsky, D. P. (1899–1970) Soviet historian. Works include *Razvitie tovarno-denezhnykh otnoshenii v selskom khoziastve russkogo gosudarstva v XVI veke.*

Malia, Martin E. (1924–   ) American historian. Works include *Alexander Herzen and the Birth of Russian Socialism, 1812–1855;* "Schiller and the Early Russian Left," in *Harvard Slavic Studies* (vol. 4); "What Is the Intelligentsia?" in *The Russian Intelligentsia* (ed. R. Pipes); *Comprendre la Révolution russe.*

Malozemoff, Andrew Alexander (1910–1954) American historian. Works include *Russian Far Eastern Policy, 1881–1904, with Special Emphasis on the Causes of the Russo-Japanese War.*

Marchenko (Martschenko), Vasilii Pavlovich (1900–   ) Soviet-Canadian economist. Works include *Osnovnye cherty khoziaistva poslestalinskoi epokhi,* in *Issledovaniia i materialy* of the Institute for the Study of the USSR.

Markov, Vladimir (1920–   ) American specialist on Russian language and literature. Works include "Unnoticed Aspect of Pasternak's Translations," *Slavic Review* (October 1961).

Mathewson, Rufus W., Jr. (1918–1978) American specialist in Russian literature. Works include *The Positive Hero in Russian Literature;* "The Hero and Society: The Literary Definition, 1855–1865, 1934–1939," in *Continuity and Change in Russian and Soviet Thought* (ed. E. J. Simmons); "The Soviet Hero as the Literary Heritage," *American Slavic and East European Review* (December 1953).

Maynard, Sir John Herbert (1865–1943) British historian. Works include *Russia in Flux: Before October* (abridged as *Russia in Flux*); *The Russian Peasant and Other Studies.*

Mazon, André (1881–1967) French specialist in Russian language and literature. Works include *Le Slovo d'Igor.*

Menshutkin, Boris Nikolaevich (1874–1938) Russian-Soviet historian of science. Works include *Mikhail Vasilevich Lomonosov* (*Russia's Lomonosov: Chemist, Courtier, Physicist, Poet*).

Merezhkovsky, Dmitri Sergeevich (1865–1941) Russian writer and critic. Works include *Gogol i chort, issledovanie.*

Meyendorff, John (1926–   ) Russian-American theologian and historian. Works include *The Byzantine Legacy in the Orthodox Church; Byzantium and the Rise of Russia: A Study of Byzantino-Russia Relations in the Fourteenth Century.*

Miakotin, Venedikt Aleksandrovich (1867–1937) Russian historian. Works include *Ocherki sotsialnoi istorii Ukrainy v XVII–XVIII vv.* (3 vols. in 1); *Protopop Avvakum, ego zhizn i deiatelnost: Biograficheskii ocherk;* chapters in *Histoire de Russie* (ed. P. N. Miliukov, C. Seignobos, and L. Eisenmann [3 vols.]).

Miliukov, Pavel Nikolaevich (1859–1943) Russian historian and statesman. Works include *Glavnye techeniia russkoi istoricheskoi mysli* (Eng. summary "The Chief Currents of Russian Historical Thought," in *The American Historical Association Annual Report for 1904*); *Gosudarstyennoe khoziaistvo Rossii v pervoi chetverti XVIII stoletiia i reforma Petra Velikogo; Histoire de Russie* (with C. Seignobos, L. Eisenmann, and others [3 vols.]); *Iz istorii russkoi intelligentsii* (Fr. *Le Mouvement intellectuel russe*); *Ocherki po istorii russkoi kultury* (4 vols.) (abridged Eng. trans. *Outlines of Russian Culture;* abridged Fr. trans. *Essais sur l'histoire de la civilisation russe*); *Russia and Its Crisis; Spornye voprosy finansovoi istorii moskovskogo gosudarstva.*

Milosz, Czeslaw (1911–   ) Polish-American poet, writer, and specialist in Slavic literature. Works include *The Captive Mind.*

Mironenko, Iurii Pavlovich (1909–   ) Soviet-German specialist on the Soviet Union. Works include "K voprosu o dinamike naseleniia Sovetskogo Soiuza s 1939 po 1956 god," *Vestnik Instituta po Izucheniiu SSSR* (1956).

Mirsky (Sviatopolk-Mirsky), Dmitrii Petrovich (1890–1938) Russian-British historian of Russian literature. Works include *Contemporay Russian Literature, 1881–1925; A History of Russian Literature from Its Beginnings to 1900* (ed. Francis J. Whitfield); *A History of Russian Literature from the Earliest Times to the Death of Dostoevsky; Russia: A Social History.*

Mosely, Philip Edward (1905–1972) American historian and political scientist. Works include *The Kremlin and World Politics; Russian Diplomacy and the Opening of the Eastern Question in 1838 and 1839; Russia since Stalin* (ed.).

Muratov, Pavel Pavlovich (1881–1950). Russian-French art historian. Works include *L'Ancienne Peinture russe; Les Icones russes.*

Nabokov, Vladimir V. (1899–1977) Russian-American writer. Works include *Nikolai Gogol.*

Nicolaevsky, Boris I. (1889–1966) Russian-American specialist in the Russian revolutionary movement and the U.S.S.R. Works include *Azeff, the Russian Judas; Forced Labor in the Soviet Union* (with D. J. Dallin).

Nolde, Boris E. (1876–1948) Russian-French historian and legal scholar. Works include *L'Alliance franco-russe: Les Origines du système diplomatique d'avant guerre; L'Ancien Régime et la révolution russe; La Formation de l'empire russe* (2 vols.); *Ocherki russkogo gosudarstvennogo prava; Russia in the Economic War; Vneshniaia politika.*

Nosov, Nikolai Evgenievich (1925–   ) Soviet historian. Works include *Ocherki po istorii mestnogo upravleniia russkogo gosudarstva pervoi poloviny XVI veka.*

Obnorsky, Sergei Petrovich (1888–1962) Soviet philologist and historian of literature. Works include *Khrestomatiia po istorii russkogo iazyka; Kultura russkogo iazyka; Ocherki po istorii russkogo literaturnogo iazyka starshego perioda.*

Obolensky, Dimitrii (1918–   ) British historian. Works include "Russia's Byzantine Heritage," in *Oxford Slavonic Papers* (vol. 1); "Byzantium, Kiev and Moscow: A Study in Ecclesiastical Relations," *Dumbarton Oaks Papers* (1957); *The Byzantine Commonwealth.*

Oganovsky, Nikolai Petrovich (1874–   ) Russian economist. Works include *Narodnoe khoziaistvo SSSR v sviazi s mirovym; Selskoe khoziaistvo Rossii v dvadtsatom veke.*

Okun, Semen Bentsionovich (1908–1972) Soviet historian. Works include *Ocherki istorii SSSR: Konets XVIII-pervaia chetvert XIX veka; Ocherki istorii SSSR: Vtoraia chetvert XIX veka; Rossiisko-Amerikanskaia Kompaniia (The Russian-American Company).*

Pares, Sir Bernard (1867–1949) British historian. Works include *The Fall of the Russian Monarchy; A History of Russia; My Russian Memoirs; Russia.*

Pavlov-Silvansky, Nikolai Pavlovich (1869–1908) Russian historian. Works include *Feodalizm v drevnei Rusi; Feodalizm v udelnoi Rusi; Gosudarevy sluzhilye liudi: Proiskhozdenie russkogo dvorianstva; Ocherki po russkoi istorii XVIII–XIX vv.*

Pavlovsky, Georgii Alekseevich (1887–   ) Russian-British agrarian historian. Works include *Agricultural Russia on the Eve of the Revolution.*

Pipes, Richard E. (1923–   ) American historian. Works include *The Formation of the Soviet Union: Communism and Nationalism, 1917–1923; Karamzin's Memoir on Ancient and Modern Russia: A Translation and Analysis; Social Democracy and the St. Petersburg Labor Movement, 1885–1897;* "Karamzin's Conception of the Monarchy," in *Harvard Slavic Studies* (vol. 4); "The Russian Military Colonies," *Journal of Modern History* (1950); *The Russian Intelligentsia* (ed.); *Revolutionary Russia* (ed.); *Struve: Liberal on the Left, 1870–1905; Russia under the Old Regime; Struve: Liberal on the Right, 1905–1944; U.S.-Soviet Relations in the Era of Détente; Russia Observed: Collected Essays on Russian and Soviet History; The Russian Revolution.*

Platonov, Sergei Feodorovich (1860–1933) Russian historian. Works include *Boris Godunov* (Fr. *Boris Godounov, tsar de Russie, 1598–1605*); *Lektsii po russkoi istorii* (Fr. *Histoire de Russie*); *Moskva i zapad v XVI–XVII vekakh* (Mos-

cow and the West); *Ocherki po istorii smuty v moskovskom gosudarstve XVI–XVII vv*; *Petr Velikii, lichnost i deiatelnost; Smutnoe vremia: Sotsialnyi krizis smutnogo vremeni* (*Time of Troubles: A Historical Study of the Internal Crisis and Social Struggle in 16th and 17th Century Muscovy*); "Ivan Groznyi v russkoi istoriografii," in *Russkoe proshloe* (vol. 1).

Pogodin, Mikhail Petrovich (1800–1875) Russian historian and right-wing intellectual. Works include *Issledovaniia, zamechaniia i lektsii o russkoi istorii* (7 vols.).

Pokrovsky, Mikail Nikolaevich (1868–1932) Russian-Soviet historian. Works include *Dekabristy: Sbornik statei; Diplomatiia i voiny tsarskoi Rossii v XIX stoletii; Istoricheskaia nauka i borba klassov* (2 vols.); *Ocherk istorii russkoi kultury; Russkaia istoricheskaia literatura v klassovom osveshchenii* (2 vols.) (ed.); *Russkaia istoriia s drevneishikh vremen* (5 vols.) (abridged Eng. trans. *History of Russia from the Earliest Times to the Rise of Commercial Capitalism* [2 vols.]); *Russkaia istoriia v samom szhatom ocherke* (2 vols.) (*Brief History of Russia* [2 vols.]).

Poliansky, Fedor Iakovlevich (1907–   ) Soviet historian. Works include *Ekonomicheskii stroi manufaktury v Rossii XVIII veka; Istoriia narodnogo khoziaistva SSSR; Pervonachalnoe nakoplenie kapitala v Rossii; Remeslo v Rossii XVIII veka.*

Polievktov, Mikhail Aleksandrovich (1872–1946) Russian historian. Works include *Baltiiskii vopros v russkoi politike; Nikolai I: Biografiia i obzor tsarstvovaniia.*

Presniakov, Aleksandr Evgenievich (1870–1929) Russian historian. Works include *Kniazhoe pravo v drevnei Rusi, ocherki po istorii X–XII stoletiia; Lektsii po russkoi istorii: Kievskaia Rus; Moskovskoe tsarstvo; Obrazovanie velikorusskogo gosudarstva, ocherki po istorii XII–XV stoletti* (Eng. trans. without notes, *The Formation of the Great Russian State*).

Priselkov, Mikhail Dmitrievich (1881–1941) Russian-Soviet historian. Works include *Ocherki po tserkovno-politicheskoi istorii kievskoi Rusi X–XII vv.*

Prokopovich, Sergei Nikolaevich (1871–1955) Russian-American economic historian. Works include *Krestianskoe khoziaistvo; Narodnoe khoziaistvo SSSR* (Fr. *Histoire économique de l'URSS*).

Puryear, Vernon J. (1901–1970) American historian of European diplomacy. Works include *England, Russia, and the Straits Question, 1844–1856; International Economics and Diplomacy in the Near East: A Study of British Commercial Policy in the Levant, 1834–1853; Napoleon and the Dardanelles.*

Radkey, Oliver Henry (1909–   ) American historian. Works include *The Agrarian Foes of Bolshevism: The Promise and Default of the Russian Socialist Revolutionaries; The Sickle under the Hammer: The Russian Socialist Revolutionaries in the Early Months of Soviet Rule; The Elections to the Russian Constituent Assembly of 1917;* "Chernov and Agrarian Socialism before 1918," in *Continuity and Change in Russian and Soviet Thought* (ed. E. J. Simmons); *The Unknown Civil War in Soviet Russia: A Study of the Green Movement in the Tambov Region, 1920–1921.*

Raeff, Marc (1923–   ) American historian. Works include *Michael Speransky: Statesman of Imperial Russia; Siberia and the Reform of 1822; Origins of the Russian Intelligentsia: The Eighteenth-Century Nobility; Imperial Russia, 1682–1825: The Coming of Age of Modern Russia; Russia Abroad: A Cultural History of the Russian Emigration, 1919–1939.*

Rieber, Alfred J. (1931–   ) American historian. Works include *The Politics of Autocracy: Letters of Alexander II to Prince A. T. Bariatinsky, 1857–1864* (ed.); *Merchants and Entrepreneurs in Imperial Russia.*

Robinson, Geroid Tanquary (1892–1971) American historian. Works include *Rural Russia under the Old Regime.*

Rogger, Hans (1923–   ) American historian. Works include *National Consciousness in Eighteenth-Century Russia;* "Russian Ministers and the Jewish Question, 1881–1917," *California Slavic Studies* (1975); *Jewish Policies and Right-Wing Politics in Imperial Russia.*

Rostovtzeff, Mikhail I. (1870–1952) Russian-American historian of the ancient world. Works include *Iranians and Greeks in South Russia;* "South Russia in the Prehistoric and Classical Period," *American Historical Review* (January 1921).

Rozhkov, Nikolai Aleksandrovich (1868–1927) Russian-Soviet historian. Works include *Agrarnyi vopros v Rossii i ego reshenie v programmakh razlichnykh partii; Gorod i derevnia v russkoi istorii; Obzor russkoi istorii s sotsiologicheskoi tochki zreniia; Russkaia istoriia* (12 vols.); "Ekonomicheskoe razvitie Rossii v pervoi polovine XIX veka," and "Finansovaia reforma Kankrina," in *Istoriia Rossii v XIX veke* (vol. 1).

Russell, Bertrand, Earl (1872–1970) British philosopher. Works include *Bolshevism: Practice and Theory; A History of Western Philosophy.*

Rybakov, Boris Aleksandrovich (1908–   ) Soviet historian. Works include *Istoriia kultury drevnei Rusi; Obrazovanie drevnerusskogo godsudarstva; Remeslo drevnei Rusi;* "Predposylki obrazovaniia drevnerusskogo gosudarstva," in *Ocherki istorii SSSR III–IX vv; Gerodotova Skifiia.*

Ryndziunsky, Pavel Grigorievich (1909–   ) Soviet historian. Works include *Gorodskoe grazhdanstvo doreformennoi Rossii; Utverzhdenie kapitalizma v Rossii, 1850–1880 g.*

Savelev (Saveliev) Pavel Stepanovich (1814–1859) Russian numismatist and historian. Works include *Mukhammedanskaia numizmatika v otnoshenii k russkoi istorii.*

Schapiro, Leonard (1908–1983) British historian and political scientist. Works include *The Communist Party of the Soviet Union; The Origins of the Communist Autocracy: Political Opposition in the Soviet State, First Phase, 1917–1922; Rationalism and Nationalism in Russian Nineteenht-Century Political Thought; 1917: The Russian Revolutions and the Origins of Modern Communism; Russian Studies.*

Schiemann, Theodor (1847–1921) German historian. Works include *Geschichte Russlands unter Kaiser Nikolaus I* (4 vols.).

Schilder (Shilder), Nikolai Karlovich (1842–1902) Russian historian. Works include *Imperator Aleksandr Pervyi, ego zhizn i tsarstvovanie* (4 vols.); *Imperator Nikolai Pervyi, ego zhizn i tsarstvovanie* (2 vols.).

Schlözer, August Ludwig von (1735–1809) German historian who worked in Russia. Works include *Nestor: Russische Annalen in ihrer slavonischen Grundsprache verglichen, übersetzt und erklärt von A. L. Schlözer; Tableaux de l'histoire de Russie; Probe russischer Annalen.*

Schuman, Frederick L. (1904–1981) American political scientist. Works include *Russia since 1917: Four Decades of Soviet Politics; Soviet Politics at Home and Abroad.*

Semevsky, Vasilii Ivanovich (1848–1916) Russian historian. Works include *Krestiane v tsarstvovanie imperatritsy Ekateriny II; Krestianskii vopros v Rossii v XVIII i pervoi polovine XIX veka* (2 vols.); *Obshchestvennye dvizheniia v Rossii v pervuiu polovinu XIX veka* (with V. I. Bogucharsky and P. E. Shchegolev); *Politicheskie i obshchestvennye idei dekabroistov.*

Seton-Watson, Hugh (1916–1984) British historian. Works include *The Decline of Imperial Russia, 1855–1914; The East European Revolutions; Eastern Europe between the Wars, 1918–1941; From Lenin to Khrushchev; The Russian Empire, 1801–1917.*

Shakhmatov, Aleksei Aleksandrovich (1864–1920) Russian specialist in Slavic languages and literature. Works include *Drevneishiia sudby russkogo plemeni; Razyskaniia o drevneishikh russkikh letopisnykh svodakh; Povest vremennykh let: Vvodnaia chast, Tekst, Primechaniia; "Povest vremennykh let" i ee istochniki.*

Shchapov, Afanasii Prokofevich (1830–1876) Russian historian. Works include *Russkii raskol staroobriadstva, rassmatrivaemyi v sviazi s vnutrennim sostoianiem russkoi tserkvi i grazhdanstvennosti v XVII veke i pervoi polovine XVIII; Sotsialno-pedagogicheskie usloviia umstvennogo razvitiia russkogo naroda.*

Shcherbatov, Mikhail Mikhailovich (1733–1790) Russian historian. Works include *Istoriia Rossiiskaia ot drevneishikh vremen* (7 vols.); *Kratkaia povest o byvshikh v Rossii samozvantakh; O povrezhdenii nravov v Rossii.*

Simmons, Ernest Joseph (1903–1972) American historian and specialist in Russian literature. Works include *Continuity and Change in Russian and Soviet Thought* (ed.); *Dostoevsky: The Making of a Novelist; Leo Tolstoy; An Outline of Modern Russian Literature; Pushkin; Through the Glass of Soviet Literature; Views of Russian Society* (ed.).

Slepov, Lazar Andreevich (1905–   ) Soviet journalist. Works include "Collectivity Is the Highest Principle of Party Leadership," *Pravda,* reprinted in *A Documentary History of Communism* (ed. R. V. Daniels).

Solovev (Soloviev), Sergei Mikhailovich (1820–1879) Russian historian. Works include *Istoriia otnoshenii mezhdu russkimi kniaziami Riurikova doma; Istoriia Rossii s drevneishikh vremen* (29 vols.) (certain volumes in Eng.).

Soloveytchik (Soloveichik), George M. de (1902–1982) Russian-British author. Works include *Potemkin: Soldier, Statesman, Lover, and Consort of Catherine of Russia.*

Sorokin, Pitirim A. (1889–1968) Russian-American sociologist. Works include *Russia and the United States; Sovremennoe sostoianie Rossii.*

Steinberg, Isaac Nachman (1888–1957) Russian political figure and intellectual. Works include *Ot fevralia po oktiabr 1917 g.* (*In the Workshop of the Revolution*).

Stender-Petersen, Adolph (1893–1963) Danish philologist and historian of literature. Works include *Geschichte der russischen Literatur; Slavisch-germanische Lehnwortkunder: Eine Studie über die ältesten germanischen Lehnwörter im Slavischen in sprach- und kulturgeschichtlicher Beleuchtung;* "Die Varägersage als Quelle der altrussischen Chronik," *Acta Jutlandica* 6, no. 1; *Varangica.*

Stepun, Fedor Avgustovich (1884–1965) Russian-German intellectual historian. Works include "Die deutsche Romantik und die Geschichtsphilosophie der Slavophilen," *Logos* (1927); "Nemetskii romantism i russkoe slavianofilstvo," *Russkaia Mysl* (March 1910).

Stokes, Antony Derek (1927–   ) British historian. Works include "The Status of the Russian Church, 988–1037," *Slavonic and East European Review* (June 1959); "Tmutarakan," *Slavonic and East European Review* (June 1960).

Struve, Gleb (1898–1985) Russian-American specialist in Russian literature. Works include *Russkaia literatura v izgnanii; Russian Literature Under Lenin and Stalin, 1917–1953.*

Sumner, Benedict Humphrey (1893–1951) British historian. Works include *Peter the Great and the Emergence of Russia; Peter the Great and the Ottoman Empire; Russia and the Balkans, 1870–1880; A Short History of Russia.*

Tarle, Evgenii Viktorovich (1874–1955) Rsusian-Soviet historian. Works include *Evropa v epokhu imperializma, 1871–1918 gg.; Kontinentalnaia blokada; Krymskaia voina* (2 vols.) *Nashestvie Napoleona na Rossiiu v 1812 godu* (*Napoleon's Invasion of Russia in 1812*); *Ocherki i kharakteristiki iz istorii evropeiskago obshchestvennago dvizheniia v XIX veke; Severnaia voina i shvedskoe nahestvie na Rossiiu.* Concerning the revising of Tarle's study of Napoleon's invasion of Russia mentioned in the text, see Ann K. Erickson, "E. V. Tarle, the Career of a Historian under the Soviet Regime," *American Slavic and East European Review* 14 (April 1960): 202–16.

Tatishchev, Vasilii Nikitich (1686–1750) Russian historian. Works include *Istoriia Rossiiskaia s samykh drevneishikh vremen* (7 vols.).

Temperley, Harold W. V. (1879–1939) British diplomatic historian. Works include *England and the Near East, the Crimea; The Foreign Policy of Canning.*

Thomsen, Vilhem Ludvig Peter (1842–1927) Danish philologist. Works include *The Relations between Ancient Russia and Scandinavia and the Origins of the Russian State.*

Tikhomirov, Mikhail Nikolaevich (1893–1965) Soviet historian. Works include *Drevnerusskie goroda* (*The Towns of Ancient Rus*); *Issledovanie o russkoi pravde: Proiskhozhdenie tekstov; Istochnikovedenie istorii SSSR* (with S. A. Nikitin); *Krestianskie i gorodskie vosstaniia na Rusi XI–XIII vv.; Ocherki istorii istoricheskoi nauki v SSSR* (ed. with others [2 vols.]); "Soslovnopredstavitelnye uchrezhdeniia (zemskie sobory) v Rossii XVI veka," *Voprosy istorii* (1958).

Timasheff, Nikolai S. (1886–1970) Russian-American sociologist. Works include *The Great Retreat: The Growth and Decline of Communism in Russia; Religion in Soviet Russia, 1917–1942.*

Treadgold, Donald W. (1922–   ) American historian. Works include *The Great Siberian Migration: Government and Peasant in Resettlement from Emancipation to the First World War; Lenin and His Rivals: The Struggle for Russia's Future; Twentieth-Century Russia;* "Was Stolypin in Favor of the Kulaks?" *American Slavic and East European Review* (February 1955); *The West in Russia and China: Religion and Secular Thought in Modern Times: Russia, 1472–1917* (vol. 1).

Trotsky, Leon (Lev Davidovich Bronstein) (1879–1940) Russian revolutionary figure, Soviet leader, and historian. Works available in many languages include in English *The History of the Russian Revolution* (3 vols.); *My Life; The Permanent Revolution; The Revolution Betrayed: What Is the Soviet Union and Where Is It Going?; Stalin: An Appraisal of the Man and His Influence.*

Tucker, Robert C. (1918–   ) American political scientist. Works include *Stalin as Revolutionary, 1879–1929: A Study in History and Personality; Philosophy and Myth in Karl Marx; The Soviet Political Mind; The Marxian Revolutionary Idea; The Great Purge Trial* (ed. with S. F. Cohen); *Stalin in Power: The Revolution from Above, 1928–1941.*

Ukraintsev, N. (   ) Russian military jurist. Works include "Delo Kornilova," *Novoe russkoe slovo* (August 12, October 21, and October 28, 1956) ("A Document in the Kornilov Affair," *Soviet Studies* [October 1973]).

Ulam, Adam Bruno (1922–   ) American political scientist. Works include *The Unfinished Revolution: An Essay on the Sources of Influence of Marxism and Communism; The New Face of Soviet Totalitarianism; The Bolsheviks; Stalin: The Man and His Era; Expansion and Coexistence: The History of Soviet Foreign Policy, 1917–1967; In the Name of the People; Russia's Failed Revolutions: From the Decembrists to the Dissidents.*

Ullman, Richard H. (1933–   ) American historian. Works include *Anglo-Soviet Relations, 1917–1921: Intervention and the War* (vol. 1), *Britain and the Russian Civil War, November 1918–February 1920* (vol. 2), *Anglo-Soviet Accord* (vol. 3).

Ustrialov, Nikolai Vasilevich (1890–1930's?) Russian legal scholar. Works include *Na novom etape; Pod znakom revoliutsii;* "Patriotica," *Smena Vekh* (July 1921).

Vasilev (Vasiliev), Aleksandr Aleksandrovich (1867–1953) Russian-American historian of Byzantium. Works include *The Goths in the Crimea; the Russian Attack on Constantinople in 860.*

Venturi, Franco (1914–   ) Italian historian. Works include *Il moto decabrista e i fratelli Poggio; Il populismo russo* (2 vols.) (*Roots of Revolution: A History of the Populist and Socialist Movements in Nineteenth-Century Russia*).

Vernadsky, George (1887–1973) Russian-American historian. Works include *A History of Russia* (1 vol.); *A History of Russia: Ancient Russia* (vol. 1), *Kievan Russia* (vol. 2), *The Mongols and Russia* (vol. 3), *Russia at the Dawn of the Modern Age* (vol. 4), *The Tsardom of Moscow, 1547–1682* (vol. 5, 2 books); *The Origins of Russia;* "The Death of the Tsarevich Dmitry: A Reconsideration of the Case," in *Oxford Slavonic Papers* (vol. 5).

Vladimirsky-Budanov, Mikhail Flegontovich (1838–1916) Russian legal historian. Works include *Gosudarstvo i narodnoe obrazovanie v Rossii s XVII veka do uchrezhdeniia ministerstv; Gosudarstvo i narodnoe obrazovanie v Rossii XVIII veka; Obzor istorii russkogo prava.*

Von Laue, T. H. (1916–   ) American historian. Works include *Sergei Witte and the Industrialization of Russia; Why Lenin? Why Stalin?; The Global City.*

Voyce, Arthur (1889–   ) American art historian. Works include *The Moscow Kremlin: Its History, Architecture, and Art Treasures; Russian Architecture: Trends in Nationalism and Modernism; The Art and Architecture of Medieval Russia.*

Walsh, Warren B. (1909–1979) American historian. Works include *Russia and the Soviet Union; Readings in Russian History* (ed.); "Political Parties in the Russian Dumas," *Journal of Modern History* (June 1950).

Webb, Sidney (1859–1947) and Beatrice (1858–1943) Lord and Lady Passfield. British writers and Fabian socialists. Works include *Soviet Communism: A New Civilisation?; The Truth about Soviet Russia.*

Weidle, Wladimir (1895–1979) Russian-French historian of art and culture. Works include *La Russie absente et presente* (*Russia Absent and Present*); "Some Common Traits in Early Russian and Western Art," in *Oxford Slavonic Papers* (vol. 4).

Wipper (Vipper), Robert Iurevich (1859–1954) Russian-Soviet historian. Works include *Ivan Groznyi* (*Ivan Grozny*).

Zaionchkovsky, Petr Andreevich (1904–1983) Soviet historian. Works include *Otmena krepostnogo prava a Rossii* (*The Abolition of Serfdom in Russia*); *Provedenie v zhizn krestianskoi reformy 1861 g.; Voennye reformy 1860–1870 godov v Rossii; Krizis samoderzhaviia na rubezhe 1870–1880 godov* (*The Russian Autocracy in Crisis, 1878–1882*); *Rossiiskoe samoderzhavie v kontse XIX stoletiia* (*The Russian Autocracy under Alexander III*); *Samoderzhavie i russkaia armiia na rubezhe XIX–XX stoletii, 1881–1903.*

Zelnik, Reginald E. (1936–   ) American historian. Works include *Labor and Society in Tsarist Russia: The Factory Workers of St. Petersburg, 1855–1870; A Radical Worker in Tsarist Russia: The Autobiography of Semen Ivanovich Kanatchikov* (ed. and trans.); "Russian Bebels," in *Russian Review* (July and October 1976); " 'To the Unaccustomed Eye': Religion and Irreligion in the Experience of St. Petersburg Workers in the 1870's," in *Russian History* (1989).

Zenkovsky, Serge A. (1907–1990) Russian-American historian. Works include *Pan-Turkism and Islam in Russia;* "The Ideological World of the Denisov Brothers," in *Harvard Slavic Studies* (vol. 3); "The Russian Church Schism: Its Background and Repercussions," *Russian Review* (October 1957); *Russkoe staroobriadchestvo: dukhovnye dvizheniia semnadtsatgogo veka.*

Zimin, Aleksandr Aleksandrovich (1920–1980) Soviet historian. Works include *I.S.*

Peresvetov i ego sovremenniki, ocherki po istorii russkoi obshchestvenno-politicheskoi mysli serediny XVI veka; Reformy Ivana Groznogo, ocherki sotsialno-ekonomicheskoi i politicheskoi istorii Rossii serediny XVI veka; Oprichnina Ivana Groznogo; Rossiia na poroge novogo vremeni: ocherki politicheskoi istorii Rossii pervoi treti XVI veka; Kholopy na Rusi (s drevneishikh vremen do kontsa XV veka); "Pripiska k Pskovskomu Apostolu 1307 Goda i 'Slovo o polku Igoreve,' " *Russkaia Literatura* (1966); "Spornye voprosy tekstologii 'Zadonshchiny,' " *Russkaia Literatura* (1967).

APPENDIX: TABLE 1

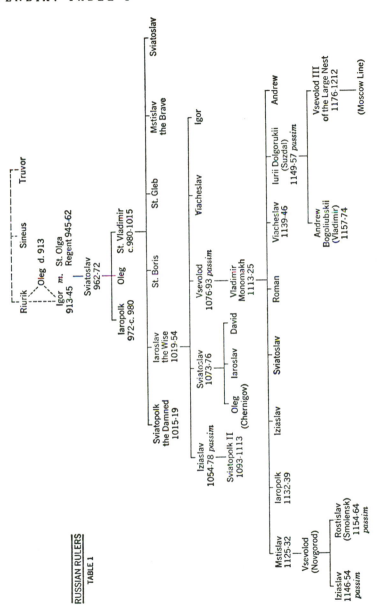

RUSSIAN RULERS
TABLE 1

629

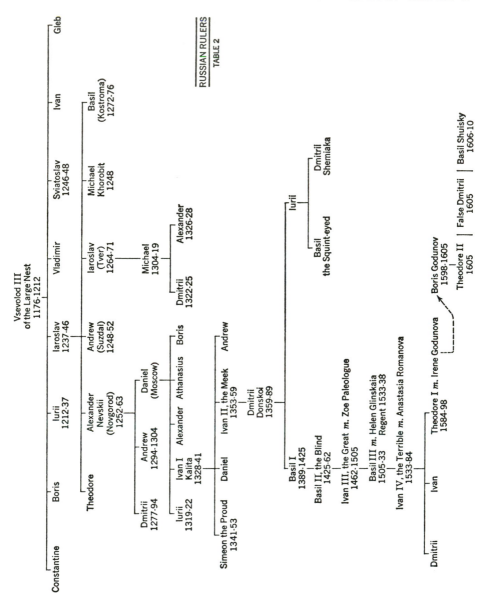

RUSSIAN RULERS

TABLE 2

631

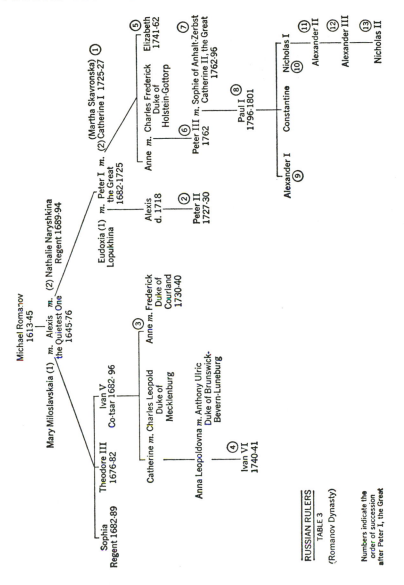

Michael Romanov
1613-45

Mary Miloslavskaia (1) *m.* Alexis *m.* (2) Nathalie Naryshkina
the Quietest One   Regent 1689-94
1645-76

Sophia
Regent 1682-89

Theodore III
1676-82

Ivan V
Co-tsar 1682-96

Anne *m.* Frederick
Duke of
Courland
1730-40 ③

Catherine *m.* Charles Leopold
Duke of
Mecklenburg

Anna Leopoldovna *m.* Anthony Ulric
Duke of Brunswick-
Bevern-Luneburg

Ivan VI
1740-41 ④

Eudoxia (1) *m.* Peter I *m.* (Martha Skavronska) ①
Lopukhina   the Great   (2) Catherine I 1725-27
1682-1725

Alexis
d. 1718

Peter II
1727-30 ②

Anne *m.* Charles Frederick
Duke of
Holstein-Gottorp

Elizabeth
1741-62 ⑤

Peter III *m.* Sophie of Anhalt-Zerbst
1762   Catherine II, the Great
1762-96 ⑦ ⑥

Paul I
1796-1801 ⑧

Alexander I ⑨

Constantine

Nicholas I ⑩

Alexander II ⑪

Alexander III ⑫

Nicholas II ⑬

RUSSIAN RULERS
TABLE 3
(Romanov Dynasty)

Numbers indicate the
order of succession
after Peter I, the Great

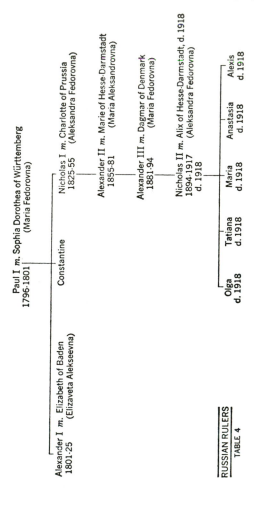

Paul I *m.* Sophia Dorothea of Württemberg
1796-1801
(Maria Fedorovna)

Alexander I *m.* Elizabeth of Baden
1801-25   (Elizaveta Alekseevna)

Constantine

Nicholas I *m.* Charlotte of Prussia
1825-55   (Aleksandra Fedorovna)

Alexander II *m.* Marie of Hesse-Darmstadt
1855-81   (Maria Aleksandrovna)

Alexander III *m.* Dagmar of Denmark
1881-94   (Maria Fedorovna)

Nicholas II *m.* Alix of Hesse-Darmstadt, d. 1918
1894-1917   (Aleksandra Fedorovna)
d. 1918

Olga   Tatiana   Maria   Anastasia   Alexis
d. 1918   d. 1918   d. 1918   d. 1918   d. 1918

RUSSIAN RULERS
    TABLE 4

U.S.S.R.: INDEXES OF GROWTH OF POPULATION AND GROSS NATIONAL PRODUCT BY USE, 1928-82

OUTLAY	1937 = 100					1955 = 100			1970 = 100			
	1928	1937	1940	1950	1955	1955	1968	1970	1970	1975	1980	1982
Population	91.3	100	118.3	108.8	118.5	100	122	123	100	105	109	111
Household consumption[a]	93	100	114	130	197	100	198	223	100	119	136	n.a.
Per capita[a]	102	100	96	119	166	100	162	181	100	113	125	n.a.
Communal services[a]	27	100	119	145	178							
Government administration, including security forces	40	100	146	200	142							
Defense (as recorded in the budget)	10	100	266	245	358							
Gross investment	30	100	90	155	234	100			100	130	160	n.a.
Gross national product	62	100	121	150	216	100	208	231	100	120	137	n.a.
Per capita	67	100	102	138	182	100	171	187	100	114	126	n.a.

Sources:

1928-55:

Population estimates are those of Warren W. Eason, in *Prospects of Soviet Society*, ed. Allen Kassof (New York, 1968), Table 4, p. 200. Gross national product and its components, 1928-55, extracted from Abram Bergson, *The Real National Income of Soviet Russia since 1928* (Cambridge, Mass., 1961), Table 51, p. 210. Per capita series were computed by the present author. Estimates are at ruble factor cost of 1937.

1955-70:

Population from official Soviet statistics.

Gross national product from Stanley H. Cohn, "General Growth Performance of the Soviet Economy," in U.S. Congress, Joint Economic Committee, *Economic Performance and the Military Burden in the Soviet Union* (Washington, D.C., 1970), p. 17 (estimates at 1959 factor cost), and John P. Hardt, "Introduction," in U.S. Congress, Joint Economic Committee, *Soviet Economic Prospects for the Seventies* (Washington, D.C., 1973), p. ix (estimates at 1968 factor cost).

Household consumption from David W. Bronson and Barbara S. Severin, "Soviet Consumer Welfare: The Brezhnev Era," in *Soviet Economic Prospects for the Seventies*, pp. 398-401. Estimates are in 1968 prices.

1970-82:

Official Soviet statistics 2, 6, 7.—U.S. C.I.A.; National Foreign Assessment Center, *Handbook of Economic Statistics 1981* (Washington, D.C., 1981).

a For 1955-80, communal services (i.e, free services to consumers, such as health and education) are included in household consumption.

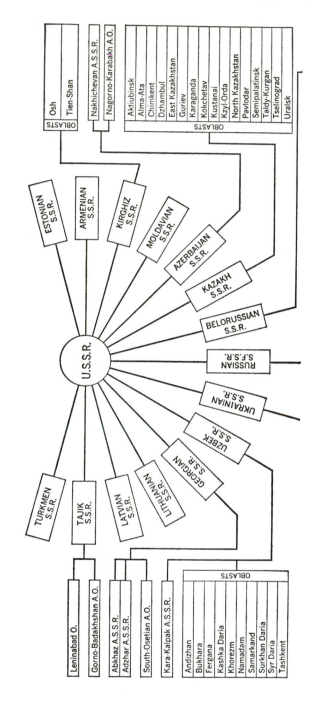

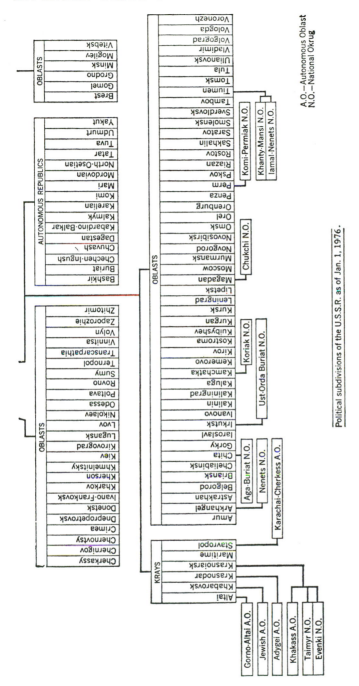

Political subdivisions of the U.S.S.R. as of Jan. 1, 1976.

A.O.—Autonomous Oblast
N.O.—National Okrug

A SELECT LIST OF READINGS IN ENGLISH ON RUSSIAN HISTORY

* * * * * * * * * *

FROM EARLIEST TIMES TO 1983

A. Bibliography and Historiography

American Bibliography of Russian and East European Studies. Bloomington, Ind. Published annually since 1957.

Fisher, H. H., ed. *American Research on Russia.* Bloomington, Ind., 1959.

Horecky, P., ed. *Basic Russian Publications: A Selected and Annotated Bibliography on Russia and the Soviet Union.* Chicago and London, 1962.

Horecky, P., ed. *Russia and the Soviet Union: A Bibliographic Guide to Western-Language Publications.* Chicago and London, 1965.

Maichel, K. *Guidse to Russian Reference Books.* Stanford, 1962.

Martianov, N. N. *Books Available in English by Russians and on Russia.* New York, 1960.

Mazour, A. G. *Modern Russian Historiography.* Princeton, N.J., 1958.

Miller, W. W. *USSR.* Cambridge, 1961.

Morley, C. *Guide to Research in Russian History.* Columbus, Ohio, 1951.

Pierce, R. A. *Soviet Central Asia. A Bibliography: Part 1: 1558–1866. Part 2: 1867–1917. Part 3: 1917–1966.* Berkeley, 1966.

Pushkarev, S. G. *A Source Book for Russian History from Early Times to 1917.* Edited by A. Ferguson et al. 3 vols. New Haven, Conn., 1972.

Pushkarev, S. G., comp. *Dictionary of Russian Historical Terms from the Eleventh Century to 1917.* Edited by G. Vernadsky and R. Fisher, Jr., New Haven, Conn., 1970.

Shapiro, D. *A Selected Bibliography of Works in English on Russian History, 1801–1917.* New York and London, 1962.

Szeftel, M. *Russia before 1917,* in *Bibliographical Introduction to Legal History and Ethnology.* Edited by J. Glissen. Brussels, 1966.

B. Encyclopedias

Florinsky, M. T., ed. *McGraw-Hill Encyclopedia of Russia and the Soviet Union.* New York, 1961.

Kubijovyč, V., et al., eds. *Ukraine: A Concise Encyclopedia.* Vol. 1. Toronto, 1963.

Utechin, S. V. *Everyman's Concise Encyclopedia of Russia.* New York, 1961.

Wieczynski, J. L., ed. *The Modern Encyclopedia of Russian and Soviet History.* 54 vols. Gulf Breeze, Fla., 1976.

C. Geography and Demography

Chew, A. F. *An Atlas of Russian History: Eleven Centuries of Changing Borders.* New Haven, Conn., 1970.

Gilbert, M. *Russian History Atlas.* New York, 1972.

Hooson, D. J. *The Soviet Union: People and Regions.* Belmont, Calif., 1966.

Jorré, G. *The Soviet Union: The Land and Its People.* Translated by E. D. Laborde. London, 1950. 3rd ed., 1967.

D. Nationality Question

Allen, W. E. D. *A History of the Georgian People.* New York, 1971.

Allworth, E. A. *The Modern Uzbeks from the Fourteenth Century to the Present: A Cultural History.* Stanford, 1990.

Baron, S. W. *The Russian Jew under Tsars and Soviet.* New York, 1964.

Becker, S. *Russia's Protectorates in Central Asia: Bukhara and Khiva, 1865–1924.* Cambridge, Mass., 1968.

Chase, T. *The Story of Lithuania.* New York, 1946.

Chirovsky, N. L. *Old Ukraine: Its Socio-Economic History Prior to 1781.* Madison, N.J., 1963.

Dmytryshyn, B. *Moscow and the Ukraine, 1918–1953.* New York, 1956.

Doroshenko, D. *History of the Ukraine.* Edmonton, Alberta, 1941.

Dubnow, S. M. *History of the Jews in Russia and Poland.* 3 vols. Philadelphia, 1916.

d'Encausse, H. C. *Islam and the Russian Empire: Reform and Revolution in Central Asia.* Berkeley, 1988.

Greenberg, L. S. *The Jews in Russia,* 2 vols. New Haven, Conn., 1944, 1951.

Grousset, R. *The Empire of the Steppes: A History of Central Asia.* Translated by N. Walford. New Brunswick, N.J., 1970.

Halecki, O. *A History of Poland.* New York, 1943.

Hrushevskyi, M. *A History of the Ukraine.* New Haven, Conn., 1941.

Kazemzadeh, F. *The Struggle for Transcaucasia, 1917–1921.* New York and Oxford, 1951.

Kubijovyč, V., et al., eds. *Ukraine: A Concise Encyclopedia.* Vol. 1. Toronto, 1963.

Lang, D. M. *A Modern History of Georgia.* London, 1962.

Lewis, R. A., R. H. Rowland, and R. S. Clem, eds. *Nationality and Population Change in Russia and the USSR: An Evaluation of Census Data, 1870–1970.* New York, 1976.

Nelbandian, L. *The Armenian Revolutionary Movement.* Berkeley, 1967.

Pierce, R. A. *Russian Central Asia, 1867–1917: A Study in Colonial Rule.* Berkeley, 1960.

Potichnyj, P. J., and H. Aster, eds. *Ukrainian-Jewish Relations in Historical Perspective.* Edmonton, Alberta, 1988.

Raun, T. U. *Estonia and the Estonians.* Stanford, 1987.

Senn, A. E. *The Emergence of Modern Lithuania.* New York, 1959.

Senn, A. E. *Lithuania Awakening.* Berkeley, 1990.

Sullivant, R. S. *Soviet Politics and the Ukraine, 1917–1957.* New York, 1962.

Suny, R. G. *The Making of the Georgian Nation.* Bloomington, Ind., 1988.

Suny, R. G., ed. *Transcaucasia: Nationalism and Social Change. Essays in the History of Armenia, Azerbaijan, and Georgia.* Ann Arbor, Mich., 1983.

Swietochowski, T. *Russian Azerbaijan, 1905–1920: The Shaping of National Destiny in a Muslim Community.* Cambridge, 1985.

Tarulis, A. N. *Soviet Policy toward the Baltic States, 1918–1940.* Notre Dame, Ind., 1959.

Thaden, E. C., ed. *Russification in the Baltic Provinces and Finland.* Princeton, N.J., 1981.

Vakar, N. *Belorussia: The Making of a Nation.* Cambridge, Mass., 1956.

Wheeler, G. *The Modern History of Soviet Central Asia.* London, 1964.

Zenkovsky, S. A. *Pan-Turkism and Islam in Russia.* Cambridge, Mass., 1960.

E. General Histories

Auty, R., and D. Obolensky, eds. *An Introduction to Russian History.* Vol 1, *Companion to Russian Studies.* Cambridge, 1976.

Charques, R. *A Short History of Russia.* London, 1959.

Clarkson, J. *A History of Russia.* New York, 1961.

Florinsky, M. T. *Russia: A History and an Interpretation.* 2 vols. New York, 1953.

Harcave, S. *Russia: A History.* Chicago, 1956.

Klyuchevsky (Kliuchevsky), V. O. *Course of Russian History.* Translated by C. J. Hogarth. 5 vols. New York, 1911–31.

Miliukov, P., C. Seignobos, L. Eisenmann, et al. *History of Russia.* Translated by C. L. Markmann. 3 vols. New York, 1968.

Pares, B. *A History of Russia.* London, 1926.

Pipes, R. E. *Russia Under the Old Regime.* New York, 1974.

Pokrovsky, M. N. *Brief History of Russia.* 2 vols. London, 1933.

Sumner, B. H. *Survey of Russian History.* London, 1944.

Vernadsky, G. *A History of Russia.* New Haven, Conn., 1929. 5th rev. ed., 1961.

Vernadsky, G., and M. Karpovich. *A History of Russia.* Vol. 1, *Ancient Russia.* Vol. 2, *Kievan Russia.* Vol. 3, *The Mongols and Russia.* Vol. 4, *Russia at the Dawn of the Modern Age.* Vol. 5, *Tsardom of Moscow, 1547–1682.* 2 books. New Haven, Conn., 1943, 1948, 1953, 1959, 1968.

F. Specialized Histories and Interpretative Essays

Avrich, P. *Russian Rebels, 1600–1800.* New York, 1972.

Billington, J. *The Icon and the Axe: An Interpretative History of Russian Culture.* New York, 1966.

Black, C., et al. *The Modernization of Japan and Russia.* New York, 1975.

Blum, J. *Lord and Peasant in Russia from the Ninth to the Nineteenth Century.* Princeton, N.J., 1961.

Bulgakov, S.N. *The Orthodox Church.* New York and London, 1935.

Cherniavsky, M. *Tsar and People: Studies in Russian Myths.* New Haven, Conn., 1961.

Chyzhevskyi, D. *History of Russian Literature from the Eleventh Century to the End of the Baroque.* New York and The Hague, 1960.

Fennell, J., and A. Stokes. *Early Russian Literature.* Berkeley, 1974.

Gasiorowska, X. *The Image of Peter the Great in Russian Fiction.* Madison, Wis., 1979.

Gerschenkron, A. *Continuity in History and Other Essays.* Cambridge, Mass., 1968.

Gerschenkron, A. *Europe in the Russian Mirror: Four Lectures in Economic History.* Cambridge, Mass., 1970.

Grey, I. *The Romanovs. The Rise and Fall of a Dynasty.* Garden City, N.Y., 1970.

Hans, N. *History of Russian Educational Policy, 1701–1917.* London, 1931.

Hans, N. *The Russian Tradition in Education.* London, 1963.

Haxthausen, A. von. *The Russian Empire, Its People, Institutions, and Resources.* Translated by R. Farie. New York, 1970.

Hingley, R. *The Russian Secret Police: Muscovite, Imperial Russian and Soviet Political Security Operations, 1565–1970.* New York, 1970.

Hunczak, T., ed. *Russian Imperialism from Ivan the Great to the Revolution.* New Brunswick, N.J., 1974.

Iswolsky, H. *Christ in Russia: The History, Tradition and Life of the Russian Church.* Milwaukee, 1960.

Kerner, R. J. *The Urge to the Sea: The Course of Russian History.* New York, 1971.

Lantzeff, G. V., and R. A. Pierce. *Eastward to Empire.* Montreal and London, 1973.

Lappo-Danilevsky, A. S. "The Development of Science and Learning in Russia." In *Russian Realities and Problems,* edited by J. D. Duff. Cambridge, 1917.

Leonard, R. *A History of Russian Music.* London, 1956.

Liashchenko, P. I. *A History of the National Economy of Russia to the 1917 Revolution.* Translated from Russian. New York, 1949.

Lincoln, W. B. *The Romanovs: Autocrats of All the Russias.* New York, 1981.

Longworth, P. *The Cossacks: Five Centuries of Turbulent Life on the Russian Steppes.* New York, 1970.

Lossky, N. O. *History of Russian Philosophy.* New York, 1951.

Masaryk, T. G. *The Spirit of Russia.* Translated from German. 3 vols. New York, 1955–67.

Miliukov, P. N. *Outlines of Russian Culture.* Edited by M. Karpovich. Translated and abridged from Russian. 4 vols. Philadelphia, 1942–75.

Miliukov, P. N. *Russia and Its Crisis.* Chicago, 1905.

Mirsky, D. S. *A History of Russian Literature.* New York, 1927.

Mirsky, D. S. *Russia: A Social History.* London, 1931.

Obolensky, D. *The Byzantine Commonwealth.* New York, 1971.

Pokrovsky, M. N. *Russia in World History: Selected Essays.* Edited by R. Szporluk. Translated by R. Szporluk and M. Szporluk. Ann Arbor, Mich., 1970.

Rice, T. T. *Russian Art.* London, 1949.

Schmemann, A. *The Historical Road of Eastern Orthodoxy.* New York, 1963.

Soloviev, A. V. *Holy Russia: The History of a Religious-Social Idea.* New York, 1959.

Stephan, J. J. *Sakhalin. A History.* New York, 1971.

Stokes, A. D. (with John Fennell). *Early Russian Literature.* Berkeley, 1974.

Thaden, E. C. *Russia since 1801: The Making of a New Society.* New York, 1971.

Treadgold, D. W. *The West in Russia and China: Religion and Secular Thought in Modern Times.* Vol. 1, *Russia, 1472–1917.* Cambridge, Mass., 1973.

Utechin, S. V. *Russian Political Thought: A Concise History.* New York and London, 1964.

Volin, L. *A Century of Russian Agriculture: From Alexander II to Khrushchev.* Cambridge, Mass., 1970.

Vucinich, A. S. *Science in Russian Culture: A History to 1860.* Stanford, 1963.

Vucinich, A. S. *Science in Russian Culture (1861–1917).* Stanford, 1970.

Weidle, V. *Russia Absent and Present.* New York, 1952.

Williams, R. C. *Culture in Exile: Russian Emigrés in Germany, 1881–1941.* Ithaca, N.Y., 1972.

Wren, M. C. *The Western Impact upon Tsarist Russia.* Chicago, 1971.

Zenkovsky, V. V. *A History of Russian Philosophy.* 2 vols. New York, 1953.

G. Collected Essays

Atkinson, D., A. Dallin, and G. W. Lapidus, eds. *Women in Russia.* Stanford, 1977.

Curtiss, J. S., ed. *Essays in Russian and Soviet History in Honor of G. T. Robinson.* Leiden, 1963.

Ferguson, A. D., and A. Levin, eds. *Essays in Russian History: A Collection Dedicated to George Vernadsky.* Hamden, Conn., 1964.

McLean, H., M. Malia, and G. Fischer, eds. *Russian Thought and Politics.* Harvard Slavic Studies, vol. 5. Cambridge, Mass., 1957.

Mendelsohn, E., and M. S. Shatz, eds. *Imperial Russia, 1700–1917: Essays in Honor of Marc Raeff.* DeKalb, Ill., 1988.

Oberlander, E., et al., eds. *Russia Enters the Twentieth Century, 1894–1917.* Translated by G. Onn. New York, 1971.

Oliva, L. Jay, ed. *Russia and the West from Peter to Khrushchev.* Boston, 1965.

Pipes, R., ed. *Revolutionary Russia.* Cambridge, Mass., 1968.

Pipes, R., ed. *The Russian Intelligentsia.* New York, 1961.

Rabinowitch, A., J. Rabinowitch, and L. Kristof, eds. *Revolution and Politics in Russia: Essays in Memory of B. I. Nicolaevsky.* Bloomington, Ind., 1972.

Treadgold, D. W., ed. *Soviet and Chinese Communism: Similarities and Differences.* Seattle, 1967.

Vucinich, W. S., ed. *The Peasant in Nineteenth Century Russia.* Stanford, 1968.

Vucinich, W. S., ed. *Russia and Asia: Essays on the Influence of Russia on the Asian Peoples.* Stanford, 1972.

H. Readings, Anthologies

Harcave, S., ed. *Readings in Russian History.* 2 vols. New York, 1962.

Page, S. W., ed. *Russia in Revolution: Selected Readings in Russian Domestic History since 1855.* Princeton, N.J., 1965.

Raeff, M., ed. *Russian Intellectual History: An Anthology.* New York, 1966.

Riha, T., ed. *Readings in Russian Civilization.* 3 vols. Chicago, 1964.

Rubinstein, A. Z., ed. *The Foreign Policy of the Soviet Union.* New York, 1960.

Schmemann, A., ed. *Ultimate Questions: An Anthology of Modern Russian Religious Thought.* New York, 1965.

Senn, A. E., ed. *Readings in Russain Political and Diplomatic History.* 2 vols. Homewood, Ill., 1966.

Walsh, W. B., ed. *Readings in Russian History.* Syracuse, N.Y., 1959.

PRE-PETRINE RUSSIA (TO 1682)

A. Source Materials

Avvakum. *The Life of the Archpriest Avvakum by Himself.* Translated by V. Nabokov. New York, 1960.

Baron, S. H., ed. and trans. *The Travels of Olearius in Seventeenth-Century Russia.* Stanford, 1967.

Cross, S. H., and O. P. Sherbovitz-Wetzor, trans. and eds. *The Russian Primary Chronicle, Laurentian Text.* Cambridge, Mass., 1953.

Dewey, H. W. "The White Lake Chapter: A Medieval Russian Administrative Statute." *Speculum* 32 (1957).

Dmytryshyn, B., ed. *Medieval Russia: A Source Book, 900–1700.* New York, 1967.

Esper, T., ed. and trans. *Heinrich von Staden: The Land and Government of Muscovy: A Sixteenth-Century Account.* Stanford, 1967.

Fedotov, G. P. *A Treasury of Russian Spirituality.* New York, 1948.

Fennell, J., ed. and trans. *The Correspondence between Prince A. M. Kurbsky and Tsar Ivan IV of Russia, 1564–1579, with Russian Text.* New York, 1955.

Fletcher, G. *Of the Russe Commonwealth, 1591.* Facsimile ed. Introduction by R. Pipes. Cambridge, Mass., 1966.

Howes, R. C., ed. and trans. *The Testaments of the Grand Princes of Moscow.* Ithaca, N.Y., 1967.

Michell, R., and N. Forbes, trans. *The Chronicle of Novgorod, 1016–1471.* In *Royal Historical Society Publications,* Camden 3rd ser., vol. 25. London, 1914.

Palmer, W. *The Patriarch and the Tsar.* 6 vols. London, 1871–76.

Vernadsky, G., trans. *Medieval Russian Laws.* In *Records of Civilization,* no. 41, edited by A. P. Evans. New York, 1947.

Zenkovsky, S. A. *Medieval Russian Epics, Chronicles and Tales.* New York, 1963.

B. Specialized Studies

Anderson, M. S. *Britain's Discovery of Russia, 1553–1815.* London, 1958.

Andreyev, N. "Filofey and His Epistle to Ivan Vasilyevich." *Slavonic and East European Review* 38 (1959).

Andreyev, N. "Kurbsky's Letters to Vasyan Muromtsev." *Slavonic and East European Review* 33 (1955).

Baron, S. H. *Muscovite Russia.* London, 1980.

Bond, Sir E. A. *Russia at the Close of the Sixteenth Century.* London, 1856.

Cherniavsky, M. "Old Believers and the New Religion." *Slavic Review* 25 (March 1966).

Cherniavsky, M. *Tsar and People: Studies in Russian Myths.* New Haven, Conn., 1961.

Chyzhevskyi, D. *History of Russian Literature from the Eleventh Century to the End of the Baroque.* New York and The Hague, 1960.

Conybeare, F. C. *Russian Dissenters.* Cambridge, Mass., 1921.

Crummey, R. O. *The Formation of Muscovy, 1304–1613.* London and New York, 1987.

Dewey, H. W. "The 1497 Sudebnik: Muscovite Russia's First National Law Code." *American Slavic and East European Review* 29 (1951).

Dukes, P. *The Making of Russian Absolutism, 1613–1801.* London and New York, 1990.

Dunlop, D. M. *The History of the Jewish Khazars.* Princeton, N.J., 1954.

Dvornik, F. "Byzantine Influences in Russia." *Geographical Magazine* 20 (May 1947).

Dvornik, F. "Byzantine Political Ideas in Kievan Russia." *Dumbarton Oaks Papers,* nos. 9 and 10 (1956).

Dvornik, F. "The Kiev State and Its Relations with Western Europe." *Transactions of the Royal Historical Society* 29 (1947).

Dvornik, F. *The Slavs, Their Early History and Civilization.* Boston, 1956.

Fedotov, G. *The Russian Religious Mind.* Vol 1, *Kievan Christianity: The Tenth to the Thirteenth Centuries.* Vol. 2, *The Middle Ages: The Thirteenth to the Fifteenth Centuries.* Edited by J. Meyendorff. Cambridge, Mass., 1946, 1966.

Fennell, J. "The Attitude of the Josephians and the Trans-Volga Elders to the Heresy of the Judaisers." *Slavonic and East European Review* 29 (1951).

Fennell, J. *The Crisis of Medieval Russia, 1200–1304.* London and New York, 1983.

Fennell, J. *The Emergence of Moscow, 1304–1359.* Berkeley, 1968.

Fennell, J. *Ivan the Great of Moscow,* London, 1961.

Florovsky, G. "The Problem of Old Russian Culture." *Slavic Review* 21 (March 1962).

Fuhrmann, J. T. *The Origins of Capitalism in Russia: Industry and Progress in the Sixteenth and Seventeenth Centuries.* Chicago, 1972.

Graham, S. *Boris Godunov.* London, 1933.

Grekov, B. *Kiev Rus.* Translated from Russian. Moscow, 1959.

Grey, I. *Ivan III and the Unification of Russia.* New York, 1964.

Hammer, D. P. "Russia and the Roman Law." *American Slavic and East European Review* 16 (1957).

Hellie, R. *Enserfment and Military Change in Muscovy.* Chicago, 1971.

Honigmann, E. "Studies in Slavic Church History." *Byzantion* 17 (1945).

Keep, J. "The Decline of the Zemsky Sobor." *Slavonic and East European Review* 36 (1957).

Keep, J. "The Regime of Filaret." *Slavonic and East European Review* 38 (1960).

Kliuchevsky, V. O. *A Course in Russian History: The Seventeenth Century* (translation of Vol. 3 of the 1957 Soviet edition of his *Collected Works*). Translated by N. Duddington. Introduction by A. Rieber. Chicago, 1968.

Kliuchevsky, V. O. *Peter the Great.* New York, 1959 (part of Vol. 4 of his *Course of Russian History*).

Kliuchevsky, V. O. *The Rise of the Romanovs.* Edited and translated by L. Archibald. New York, 1970.

Kliuchevsky, V. O. "St. Sergius: The Importance of His Life and Work." *Russian Review* (London) 2 (1913).

Kondakov, N. P. *The Russian Icon.* Translated from Russian. Oxford, 1927.

Lantzeff, G. *Siberia in the Seventeenth Century: A Study of Colonial Administration.* Berkeley, 1943.

Leatherbarrow, W. J., and D. C. Offord, eds. *A Documentary History of Russian Thought: From the Enlightenment to Marxism.* Ann Arbor, Mich., 1987.

Medlin, W. K. *Moscow and East Rome: A Political Study of the Relation of Church and State in Muscovite Russia.* New York and Geneva, 1952.

Medlin, W. K., and C. G. Patrinelis. *Renaissance Influences and Religious Reforms in Russia: Western and Post-Byzantine Impacts on Culture and Education (16th–17th Centuries).* Geneva, 1971.

Norretranders, B. *The Shaping of Tsardom under Ivan Grozny.* Copenhagen, 1964.

Nowak, F. *Medieval Slavdom and the Rise of Russia.* New York, 1970.

Obolensky, D. "Byzantium, Kiev and Moscow: A Study in Ecclesiastical Relations." *Dumbarton Oaks Papers,* no. 11 (1957).

Obolensky, D. "Russia's Byzantine Heritage." *Oxford Slavonic Papers* 1 (1950).

O'Brien, C. B. *Muscovy and the Ukraine: From the Pereiaslavl Agreement to the Truce of Andrusovo.* Berkeley, 1963.

Paszkiewicz, H. *The Making of the Russian Nation.* London, 1963.

Paszkiewicz, H. *The Origin of Russia.* London, 1954.

Payne, R., and N. Romanoff. *Ivan the Terrible.* New York, 1975.

Pelenski, J. *Russia and Kazan: Conquest and Imperial Ideology (1438–1560's).* The Hague and Paris, 1974.

Platonov, S. F. *Moscow and the West.* Edited and translated by J. Wieczynski. Hattiesburg, Miss., 1972.

Platonov, S. F. *The Time of Troubles: A Historical Study of the Internal Crisis and Social Struggle in Sixteenth- and Seventeenth-Century Muscovy.* Translated by J. Alexander. Lawrence, Kans., 1970.

Pokrovsky, M. N. *History of Russia from the Earliest Times to the Rise of Commercial Capitalism.* Translated and edited by J. D. Clarkson and M. R. Griffiths. New York, 1931.

Prawdin, M. *The Mongol Empire: Its Rise and Legacy.* Translated by E. Paul and C. Paul. New York and London, 1940. 2nd ed., 1967.

Presniakov, A. E. *The Formation of the Great Russian State: A Study of Russian History in the Thirteenth to Fifteenth Centuries.* Translated by A. E. Moorhouse. Chicago, 1970.

Raeff, M. "An Early Theorist of Absolutism: Joseph of Volokolamsk." *American Slavic and East European Review* 8 (1949).

Riasanovsky, N. V. "The Norman Theory of the Origin of the Russian State." *Russian Review* (Autumn 1947).

Runciman, S. "Byzantium, Russia and Caesaropapism." *Canadian Slavonic Papers* 2 (1957).

Ševčenko, I. "Byzantine Cultural Influences." In *Rewriting Russian History,* edited by C. Black. Princeton, N.J., 1962.

Ševčenko, I. "A Neglected Byzantine Source of Muscovite Ideology." *Harvard Slavic Studies* 2 (1945).

Soloviev, A. V. *Holy Russia: The History of a Religious-Social Idea.* New York, 1959.

Spinka, M. "Patriarch Nikon and the Subjection of the Russian Church to the State." *Church History* 10 (1941).

Stremoukhoff, D. "Moscow, the Third Rome: Sources of the Doctrine." *Speculum* (1953).

Szeftel, M. *Russian Institutions and Culture up to Peter the Great.* London, 1975.

Thomsen, V. *Relations Between Ancient Russia and Scandinavia and the Origins of the Russian State.* Oxford, 1877.

Tikhomirov, M. N. *The Towns of Ancient Rus.* Translated from Russian. Moscow, 1959.

Vasiliev, A. A. *The Goths in the Crimea.* Cambridge, Mass., 1936.

Vernadsky, G. *Ancient Russia.* New Haven, Conn., 1943.

Vernadsky, G. *Bohdan, Hetman of Ukraine.* New Haven, Conn., 1941.

Vernadsky, G. *Kievan Russia.* New Haven, Conn., 1948.

Vernadsky, G. *The Mongols and Russia.* New Haven, Conn., 1953.

Vernadsky, G. *The Origins of Russia.* Oxford, 1959.

Vernadsky, G. *Russia at the Dawn of the Modern Age.* New Haven, Conn., 1959.

Vernadsky, G. *The Tsardom of Moscow, 1547–1682.* New Haven, Conn., 1959.

Voyce, A. *The Art and Architecture of Medieval Russia.* Norman, Okla., 1967.

Voyce, A. *Moscow and the Roots of Russian Culture.* Norman, Okla., 1964.

Voyce, A. *The Moscow Kremlin.* Berkeley, 1954.

Wolff, R. L. "The Three Romes: The Migration of an Ideology and the Making of an Autocrat." *Daedalus* (Spring 1959).

Zernov, N. *St. Sergius, Builder of Russia.* London, 1938.

Zernov, N. "Vladimir and the Origin of the Russian Church." *Slavonic and East European Review* 28 (1949–50).

IMPERIAL RUSSIA, 1682–1917

A. Source Materials

Annenkov, P. *The Extraordinary Decade: Literary Memoirs by P. V. Annenkov.* Edited by A. P. Mendel. Ann Arbor, Mich., 1968.

Bakunin, M. *Selected Writings.* Edited by A. Lehning. New York, 1974.

Barratt, G. R. V. *Voices in Exile: The Decembrist Memoirs.* Montreal and London, 1974.

Bing, E. J., ed. *The Letters of Tsar Nicholas and Empress Marie.* London, 1937.

Bock, M. P. von. *Reminiscences of My Father, Peter A. Stolypin.* Edited and translated by M. Patoski. Metuchen, N.J., 1970.

Buchanan, G. *My Mission to Russia.* 2 vols. Boston, 1923.

Bulgakov, S. N. "Heroism and Service: Thoughts on the Religious Character of the Russian Intelligentsia." *Russian Review* (London) 3 (1914).

Bulgakov, S. N. "The Russian Public and Religion." *Russian Review* (London) 1 (1912).

Catherine the Great. *The Memoirs of Catherine the Great.* Edited by D. Maroger. Translated by M. Budberg. New York, 1961.

Chaadaev, P. *The Major Works of Peter Chaadaev.* Translated and with commentary by R. T. McNally. Notre Dame, Ind., 1969.

Chernyshevsky, N. G. *Selected Philosophical Essays.* Moscow, 1953.

Chernyshevsky, N. G. *What Is to Be Done?* New York, 1961.

Dostoevsky, F. M. *The Diary of a Writer.* Translated by R. Brasol. New York, 1954.

Giers, N. K. *The Education of a Russian Statesman: The Memoirs of N. K. Giers.* Edited by C. Jelavich and B. Jelavich. Berkeley, 1962.

Golder, F. A., ed. *Documents of Russian History, 1914–1917.* New York and London, 1927.

Gurko, V. I. *Features and Figures of the Past. Government and Opinion in the Reign of Nicholas II.* Stanford, 1939.

Herzen, A. I. *My Past and Thoughts.* Translated by C. Garnett. 6 vols. New York, 1924–28, 1973.

Izvolsky, A. P. *Recollections of a Foreign Minister.* Garden City, N.Y., 1921.

Karamzin, N. M. *Letters of a Russian Traveler, 1789–1790: An Account of a Young Russian Gentleman's Tour through Germany, Switzerland, France and England.* Translated by F. Jonas. Edited by E. Simmons. New York, 1957.

Karamzin, N. M. *Memoir on Ancient and Modern Russia.* Translated and analysis by R. Pipes. Edited by R. Pipes. Cambridge, Mass., 1959.

Kohn, H., ed. *The Mind of Modern Russia: Historical and Political Thought of Russia's Great Age.* New Brunswick, N.J., 1955.

Kokovtsov, V. N. *Out of My Past.* Edited by H. H. Fisher. Stanford, 1935.

Kravchinsky, S. M. (Stepniak) *Underground Russia: Revolutionary Profiles and Sketches from Life.* Preface by P. L. Lavrov. New York, 1883.

Kropotkin, P. A. *Memoirs of a Revolutionist.* New York, 1899.

Kropotkin, P. A. *Modern Science and Anarchism.* London, 1913.

Kropotkin, P. A. *The State: Its Part in History.* London, 1898, 1943.

Lavrov, P. *Historical Letters.* Edited and translated by J. P. Scanlan. Berkeley, 1967.

Maklakov, V. A. *Memoirs of V. A. Maklakov: The First State Duma: Contemporary Reminiscences.* Edited by M. Belkin. Bloomington, Ind., 1964.

Maximoff, G. P., ed. *The Political Philosophy of Bakunin: Scientific Anarchism.* Chicago, 1953.

Miliukov, P. *Political Memoirs, 1905–1917.* Edited by A. P. Mendel. Ann Arbor, Mich., 1967.

Paleologue, G. *An Ambassador's Memoirs.* 3 vols. London, 1925.

Pares, B. "Conversations with Mr. Stolypin." *Russian Review* (London) 2 (1913).

Pares, B. *My Russian Memoirs.* London, 1931.

Pares, B., ed. *Letters of the Tsaritsa to the Tsar, 1914–1916.* London, 1923.

Pobedonostsev, K. P. *Reflections of a Russian Statesman.* London, 1898.

Radishchev, A. N. *A Journey from St. Petersburg to Moscow.* Edited with an introduction and notes by R. P. Thaler. Cambridge, Mass., 1958.

Raeff, M., ed. *The Decembrist Movement.* Englewood Cliffs, N.J., 1966.

Raeff, M., ed. *Plans for Political Reform in Russia, 1730–1905.* Englewood Cliffs, N.J., 1966.

Read, H., ed. *Kropotkin: Selections from His Writings.* London, 1942.

Reddaway, W. F., ed. *Documents of Catherine the Great.* Cambridge, 1931.

Rieber, A. *The Politics of Autocracy: Letters of Alexander II to Prince A. I. Bariatinskii, 1857–1864.* Paris, 1966.

Rosen, R. R. *Forty Years of Diplomacy.* 2 vols. New York, 1922.

Rozanov, V. V. *Fallen Leaves.* Translated by S. S. Koletiansky. London, 1920.

Rozanov, V. V. *Selected Works.* Edited by G. Ivask. New York, 1956.

Sazonov, S. D. *Fateful Years, 1909–1916.* New York, 1928.

Signposts. See *Vekhi.*

Soloviev, V. S. *Lectures on Godmanhood.* Poughkeepsie, N.Y., 1944. London, 1948.

Soloviev, V. S. *Russia and the Universal Church.* Translated by H. Rees. London, 1948.

Soloviev, V. S. *A Soloviev Anthology.* Edited by S. L. Frank. London, 1950.

Soloviev, V. S. *War, Progress and the End of History.* Translated by A. Bakstry. London, 1915.

Tikhomiroff, L. *Russia: Political and Social.* Translated by E. Aveling. 2 vols. London, 1888.

Tolstoy, L. *Works.* Translated by L. Maude and A. Maude. 21 vols. London and New York, 1928–37.

Vekhi:

Signposts: A Collection of Articles on the Russian Intelligentsia. Moscow, 1909. [*Vekhi: Sbornik Statei o Russkoi Intelligentsii*]. Translated and edited by M. Shatz and J. Zimmerman.

Canadian Slavic Studies, beginning with vol. 2, no. 2 (Summer 1968).

Wallace, D. M. *Russia.* New York, 1880.

Watrous, S. D., ed. *John Ledyard's Journey through Russia and Siberia, 1787–1788: The Journal and Selected Letters.* Madison, Wis., 1966.

Witte, S. *The Memoirs of Count Witte.* Edited by A. Yarmolinsky. Garden City, N.Y., 1921.

B. General Studies

Benois, A. *The Russian School of Painting.* London, 1916.

Bird, A. *A History of Russian Painting.* Boston, 1987.

Crummey, R. O. *The Old Believers and the World of Antichrist: The Vyg Community and the Russian State, 1694–1855.* Madison, Wis., 1970.

Dmytryshyn, B., ed. *Modernization of Russia under Peter I and Catherine II.* New York and Toronto, 1974.

Florinsky, M. T. *The End of the Russian Empire.* New Haven, Conn., 1931.

Harcave, S. *Years of the Golden Cockerel: The Last Romanov Tsars, 1814–1917.* New York, 1968.

Ivanits, L. J. *Russian Folk Belief.* Armonk, N.Y., 1989.

Karpovich, M. *Imperial Russia, 1801–1917*. New York, 1932.
Kornilov, A. *Modern Russian History from the Age of Catherine the Great to the End of the Nineteenth Century*. Translated by A. Kaun. Bibliography by J. Curtiss. 2 vols. New York, 1970.
Maynard, J. *Russia in Flux*. New York, 1948.
Miliukov, P. *Russia and Its Crisis*. Chicago, 1905; New York, 1962.
Mirsky, D. S. *History of Russian Literature*. New York, 1927.
Pares, B. *The Fall of the Russian Monarchy: A Study of the Evidence*. New York, 1939.
Pares, B. *Russia: Between Reform and Revolution*. Edited by F. B. Randall. New York, 1962.
Pavlovsky, G. *Agricultural Russia on the Eve of the Revolution*. London, 1930.
Pushkarev, S. *The Emergence of Modern Russia*. Translated from Russian. New York, 1963.
Raeff, M. *Imperial Russia, 1682–1825: The Coming of Age of Modern Russia*. New York, 1971.
Robinson, G. T. *Rural Russia under the Old Regime: A History of the Landlord-Peasant World and a Prologue to the Peasant Revolution of 1917*. New York, 1932.
Seton-Watson, H. *The Russian Empire, 1801–1917*. Oxford, 1967.
Treadgold, D. W. *Twentieth Century Russia*. Chicago, 1959.
Vucinich, A. S. *Science in Russian Culture: A History to 1860*. Stanford, 1963.
Vucinich, A. S. *Science in Russian Culture (1861–1917)*. Stanford, 1970.
Vucinich, W. S., ed. *The Peasant in Nineteenth Century Russia*. Stanford, 1968.
Wallace, D. M. *Russia on the Eve of War and Revolution*. New York, 1961.
Westwood, J. N. *Endurance and Endeavour: Russian History, 1812–1980*. 3rd ed. The Short Oxford History of the Modern World. Oxford, 1987.

C. Specialized Studies: Government, Society, and Institutions
Alexander, J. *Emperor of the Cossacks: Pugachev and the Frontier Jacquerie of 1773–1775*. Lawrence, Kans., 1973.
Alexander, J. T. *Catherine the Great: Life and Legend*. New York, 1989.
Anderson, B. *Internal Migration during Modernization in Late Nineteenth-Century Russia*. Princeton, N.J., 1980.
Becker, S. *Nobility and Privilege in Late Imperial Russia*. De Kalb, Ill., 1985.
Black, C., ed. *Aspects of Social Change since 1861: The Transformation of Russian Society*. Cambridge, Mass., 1960.
Clausewitz, Carl von. *The Campaign of 1812 in Russia*. London, 1843.
Crisp, O., and L. Edmonson, eds. *Civil Rights in Imperial Russia*. New York, 1989.
Curtiss, J. S. *Church and State in Russia, 1900–1917*. New York, 1940.
Curtiss, J. S. *The Russian Army under Nicholas I, 1825–1855*. Durham, N.C., 1965.
Dukes, P. *Catherine the Great and the Russian Nobility: A Study Based on the Materials of the Legislative Commission of 1767*. Cambridge, 1968.
Emmons, T. *The Russian Landed Gentry and the Peasant Emancipation of 1861*. Cambridge, 1967.
Emmons, T., ed. *Emancipation of the Russian Serfs*. New York, 1970.
Field, D. *The End of Serfdom: Nobility and Bureaucracy in Russia, 1855–1861*. Cambridge, Mass., 1976.
Glickman, R. L. *Russian Factory Women: Workplace and Society, 1880–1914*. Berkeley, 1984.
Golovin, N. N. *The Russian Army in the World War*. New Haven, Conn., 1931.
Gooch, G. P. *Catherine the Great and Other Studies*. London, 1954.
Gronsky, P., and N. Astrov. *The War and the Russian Government*. New Haven, Conn., 1929.
Haimson, L. H., ed. *The Politics of Rural Russia, 1905–1914*. Bloomington, Ind., 1979.
Hamm, M. F. *The City in Russian History*. Lexington, Ky., 1976.

Harcave, S. *Years of the Golden Cockerel: The Last Romanov Tsars, 1814–1917.* New York, 1968.

Jones, R. E. *The Emancipation of the Russian Nobility, 1762–85.* Princeton, N.J., 1973.

Kassow, S. D. *Students, Professors, and the State in Tsarist Russia.* Berkeley, 1989.

Kennan, G. *Siberia and the Exile System.* 2 vols. New York, 1891.

Kovalevsky, M. M. *Russian Political Institutions.* Chicago, 1902.

Kucherov, S. *Courts, Lawyers and Trials under the Last Three Tsars.* New York, 1953.

LeDonne, J. P. *Ruling Russia: Politics and Administration in the Age of Absolutism, 1762–1796.* Princeton, N.J., 1984.

Levin, A. *The Second Duma: A Study of the Social-Democratic Party and the Russian Constitutional Experiment.* New Haven, Conn., 1940.

Lincoln, W. B. *The Great Reforms: Autocracy, Bureaucracy and the Politics of Change in Imperial Russia.* De Kalb, Ill., 1990.

Lincoln, W. B. *Nicholas I: Emperor and Autocrat of All the Russias.* Bloomington, Ind., 1978.

Lincoln, W. B. *The Romanovs: Autocrats of All the Russias.* New York, 1981.

MacKenzie, D. *The Lion of Tashkent: The Career of General M. G. Cherniaev.* Athens, Ga., 1974.

Madariaga, I. de. *Russia in the Age of Catherine the Great.* New Haven, Conn., 1981.

Massie, R. K. *Nicholas and Alexandra.* New York, 1967.

McClelland, J. C. *Autocrats and Academics: Education, Culture, and Society in Tsarist Russia.* Chicago and London, 1979.

McGrew, R. E. *Russia and the Cholera, 1823–1832.* Madison, Wis., 1965.

Miller, F. *Dimitrii Miliutin and the Reform Era in Russia.* Nashville, Tenn., 1968.

Monas, S. *The Third Section: Police and Society under Nicholas I.* Cambridge, Mass., 1961.

Nichols, R. L. and T. G. Stavrou, eds. *Russian Orthodoxy under the Old Regime.* Minneapolis, 1978.

O'Brien, C. B. *Russia under Two Tsars, 1682–1689.* Berkeley, 1952.

Orlovsky, D. T. *The Limits of Reform: The Ministry of Internal Affairs in Imperial Russia, 1802–1881.* Cambridge, Mass., 1981.

Paleologue, G. M. *The Enigmatic Tsar: The Life of Alexander I of Russia.* London, 1938.

Papmehl, K. A. *Freedom of Expression in Eighteenth Century Russia.* The Hague, 1971.

Pearson, T. S. *Russian Officialdom in Crisis: Autocracy and Local Self-Government, 1861–1900.* Cambridge, 1989.

Pintner, W. M., and D. K. Rowney, eds. *Russian Officialdom: The Bureaucratization of Russian Society from the Seventeenth to the Twentieth Century.* Chapel Hill, N.C., 1980.

Raeff, M. *Michael Speransky: Statesman of Imperial Russia.* The Hague, 1957.

Raeff, M. *Siberia and the Reforms of 1822.* Seattle, 1956.

Raeff, M., ed. *Peter the Great: Reformer or Revolutionary?* Boston, 1963.

Ransel, D. L. *Mothers of Misery: Child Abandonment in Russia.* Princeton, N.J., 1988.

Rodzianko, M. V. *Reign of Rasputin: An Empire's Collapse.* New York, 1927.

Rogger, H. "Russian Ministers and the Jewish Question 1881–1917." *California Slavic Studies* 8 (1975).

Ruud, C. A. *Fighting Words: Imperial Censorship and the Russian Press, 1804–1906.* Toronto, 1982.

Sinel, A. *The Classroom and the Chancellery: State Education Reform in Russia under Count Dimitrii Tolstoy.* Cambridge, Mass., 1973.

Soloveytchik, G. *Potemkin: Soldier, Statesman, Lover and Consort of Catherine of Russia.* New York, 1947.

Starr, S. F. *Decentralization and Self-Government in Russia, 1830–1870.* Princeton, N.J., 1972.

Sumner, B. H. *Peter the Great and the Emergence of Russia.* New York, 1962.

Treadgold, D. W. *The Great Siberian Migration: Government and Peasant in Resettlement from Emancipation to the First World War.* Princeton, N.J., 1957.

Troyat, H. *Catherine the Great.* Translated by J. Pinkham. New York, 1980.

Wade, R. A., and S. J. Seregny, eds. *Politics and Society in Provincial Russia: Saratov, 1590–1917.* Columbus, Ohio, 1989.

Walkin, J. *The Rise of Democracy in Pre-Revolutionary Russia: Political and Social Institutions under the Last Three Tsars.* New York, 1962.

Wirtschafter, E. K. *From Serf to Russian Soldier.* Princeton, N.J., 1990.

Wortman, R. S. *The Development of a Russian Legal Consciousness.* Chicago and London, 1976.

Yaney, G. *Systematization of Russian Government: Social Evolution in the Domestic Administration of Imperial Russia, 1711–1905.* Urbana, Ill., 1973.

Zaionchkovskii, P. A. *The Abolition of Serfdom in Russia.* Gulf Breeze, Fla., 1978.

Zaionchkovskii, P. A. *The Russian Autocracy in Crisis, 1878–1882.* Gulf Breeze, Fla., 1979.

Zaionchkovskii, P. A. *The Russian Autocracy under Alexander III.* Gulf Breeze, Fla., 1976.

Zaitsev, P. *Taras Shevchenko: A Life.* Edited, abridged, and translated with an introduction by G. S. N. Luckyj. Toronto, 1988.

Zelnik, R. E. *Labor and Society in Tsarist Russia: The Factory Workers of St. Petersburg, 1855–1870.* Stanford, 1971.

D. Specialized Studies: Foreign Affairs

Barker, A. J. *The War Against Russia, 1854–1856.* New York, 1971.

Bromley, J. S., ed. *The Rise of Great Britain and Russia, 1688–1715/25.* Vol. 6 of *The New Cambridge Modern History.* Cambridge, Mass., 1970.

Curtiss, J. S. *Russia's Crimean War.* Durham, N.C., 1979.

Dallin, D. J. *The Rise of Russia in Asia.* New Haven, Conn., 1949.

Donnelly, A. S. *The Russian Conquest of Bashkiria: A Case Study in Imperiailsm, 1552–1740.* New Haven, Conn., 1968.

Fay, S. B. *The Origins of the World War.* 2 vols. New York, 1928.

Fisher, A. W. *The Russian Annexation of the Crimea, 1772–1783.* Cambridge, 1970.

Golder, F. A. *Russian Expansion on the Pacific, 1641–1850.* Cleveland, 1914.

Jelavich, B. *A Century of Russian Foreign Policy, 1814–1914.* New York, 1964.

Jelavich, B. *Russia and Greece During the Regency of King Otton, 1832–1835.* Thessalonika, 1962.

Jelavich, B. *Russia and the Greek Revolution of 1843.* Munich, 1966.

Jelavich, B. *Russia and the Rumanian National Cause, 1858–1859.* Bloomington, Ind., 1959.

Jelavich, B., and C. Jelavich. *Russia in the East, 1876–1880.* Leiden, 1959.

Jelavich, C. *Tsarist Russia and Balkan Nationalism: Russian Influence in the Internal Affairs of Bulgaria and Serbia, 1879–1886.* Berkeley, 1958.

Kaplan, H. H. *The First Partition of Poland.* New York, 1962.

Kennan, G. F. *The Decline of Bismarck's European Order: Franco-Russian Relations, 1875–1890.* Princeton, N.J., 1979.

Langer, W. L. *The Diplomacy of Imperialism, 1890–1902.* 2 vols. New York, 1935.

Langer, W. L. *European Alliances and Alignments, 1871–1890.* New York, 1950.

Langer, W. L. *The Franco-Russian Alliance, 1890–1894.* Cambridge, Mass., 1929.

Lederer, I., ed. *Russian Foreign Policy: Essays in Historical Perspective.* New Haven, Conn., 1962.

Lensen, G. A. *The Russian Push toward Japan: Russo-Japanese Relations, 1697–1875.* Princeton, N.J., 1959.

Lobanov-Rostovsky, A. *Russia and Asia.* Ann Arbor, Mich., 1951.

Lobanov-Rostovsky, A. *Russia and Europe, 1789–1825.* Durham, N.C., 1947.

Lobanov-Rostovsky, A. *Russia and Europe, 1825–1878.* Ann Arbor, Mich., 1954.

Lord, R. H. *The Second Partition of Poland.* Cambridge, Mass., 1915.

Madariaga, I. de. *Britain, Russia, and the Armed Neutrality of 1780: Sir James Harris's Mission to St. Petersburg During the American Revolution.* New Haven, Conn., 1962.

Malozemoff, A. *Russian Far Eastern Policy, 1881–1904.* Berkeley, 1958.

Montesquiou-Fezensac, R. *The Russian Campaign, 1812.* Translated by L. Kennett. Athens, Ga., 1970.

Mosely, P. *Russian Diplomacy and the Opening of the Eastern Question in 1838 and 1839.* Cambridge, Mass., 1934.

Mosse, W. E. *The European Powers and the German Question, 1848–1871.* Cambridge, 1958.

Okun, S. B. *The Russian-American Company.* Translated from Russian. Cambridge, Mass., 1951.

Pierce, R. A. *Russian Central Asia, 1867–1917: A Study in Colonial Rule.* Berkeley, 1960.

Pierce, R. A. *Russia's Hawaiian Adventure, 1815–1817.* Berkeley, 1965.

Puryear, V. J. *England, Russia and the Straits Question, 1844–1856.* Berkeley, 1931.

Ragsdale, H. *Détente in the Napoleonic Era: Bonaparte and the Russians.* Lawrence, Kans., 1980.

Romanov, B. *Russia in Manchuria, 1892–1906.* Translated by S. Jones. Ann Arbor, Mich., 1952.

Smith, C. J. *The Russian Struggle for Power, 1914–1917: A Study of Russian Foreign Policy During the First World War.* New York, 1956.

Sumner, B. H. *Peter the Great and the Ottoman Empire.* Oxford, 1949.

Sumner, B. H. *Russia and the Balkans, 1870–1880.* Oxford, 1937.

Sumner, B. H. *Tsardom and Imperialism in the Far East and Middle East, 1880–1914.* London, 1940.

Tarle, E. V. *Napoleon's Invasion of Russia, 1812.* Translated from Russian first edition. New York, 1942.

Taylor, A. J. P. *The Struggle for Mastery in Europe, 1848–1918.* Berkeley, 1931.

Thaden, E. C. *Russia and the Balkan Alliance of 1912.* University Park, Pa., 1965.

Thomson, G. S. *Catherine the Great and the Expansion of Russia.* London, 1947.

Warner, D., and P. Warner. *The Tide at Sunrise: A History of the Russo-Japanese War, 1904–1905.* New York, 1974.

White, J. A. *The Diplomacy of the Russo-Japanese War.* Princeton, N.J., 1964.

E. Specialized Studies: Economic Development

Ames, E. "A Century of Russian Railway Construction, 1837–1936." *American Slavic and East European Review* 6 (1947).

Blackwell, W. L. *The Beginnings of Russian Industrialization, 1800–1860.* Princeton, N.J., 1968.

Gerschenkron, A. "Agrarian Policies and Industrialization, Russia 1861–1917." In *The Cambridge Economic History.* Vol. 6, part 2. Cambridge, 1966.

Gerschenkron, A. "The Early Phases of Industrialization in Russia." In *The Economics of Take-Off into Sustained Growth,* edited by W. Rostow. London, 1963.

Gerschenkron, A. *Economic Backwardness in Historical Perspective.* Cambridge, Mass., 1962.

Lih, L. *Bread and Authority in Russia, 1914–1921.* Berkeley, 1990.

McKay, J. P. *Pioneers for Profit: Foreign Entrepreneurship and Russian Industrialization, 1885–1913.* Chicago, 1970.

Pintner, W. W. *Russian Economic Policy under Nicholas I.* Ithaca, N.Y., 1967.

Portal, R. "The Industrialization of Russia." In *The Cambridge Economic History.* Vol. 6, part 2. Cambridge, 1966.

Rieber, A. J. *Merchants and Entrepreneurs in Imperial Russia.* Chapel Hill, N.C., 1982.

Rozman, G. *Urban Networks in Russia, 1750–1800, and Premodern Periodization.* Princeton, N.J., 1976.

Smith, R. E. F., and D. Christian. *Bread and Salt: A Social and Economic History of Food and Drink in Russia.* New York, 1981.

Tugan-Baranovsky, M. I. *The Russian Factory in the 19th Century.* Translated by A. Levin, C. Levin, and G. Grossman. Homewood, Ill., 1970.

Von Laue, T. H. "Russian Labor between Field and Factory, 1892–1903." *California Slavic Studies* 3 (1964).

Von Laue, T. H. *Sergei Witte and the Industrialization of Russia.* New York, 1963.

F. Specialized Studies: Intellectual History and the Revolutionary Movement

Ambler, E. *Russian Journalism and Politics 1861–1881: The Career of Aleksei S. Suvorin.* Detroit, 1972.

Andrew, J. *Women in Russian Literature, 1780–1863.* New York, 1988.

Ascher, A. *The Revolution of 1905: Russia in Disarray.* Stanford, 1988.

Avrich, P. *The Russian Anarchists.* Princeton, N.J., 1967.

Baron, S. H. *Plekhanov: The Father of Russian Marxism.* Stanford, 1963.

Berdiaev, N. *Constantin Leontieff.* Translated by H. Iswolsky. Paris, 1937.

Berdiaev, N. *The Russian Idea.* Translated by R. French. London, 1947.

Berlin, I. *The Hedgehog and the Fox: An Essay on Tolstoy's View of History.* New York, 1953.

Berlin, I. "The Marvelous Decade." *Encounter* (1955, 1956).

Berlin, I. *Russian Thinkers.* Edited by H. Hardy and A. Kelly. New York, 1978.

Billington, J. *Mikhailovsky and Russian Populism.* New York, 1958.

Bonnell, V. E. *Roots of Rebellion: Workers' Politics and Organizations in St. Petersburg and Moscow, 1900–1914.* Berkeley, 1983.

Bowman, H. *Vissarion Belinskii, 1811–1848: A Study in the Origins of Social Criticism in Russia.* Cambridge, Mass., 1954.

Broido, E. *Memoirs of a Revolutionary.* Edited and translated by V. Broido. New York, 1967.

Brower, D. R. *Training the Nihilists: Education and Radicalism in Tsarist Russia.* Ithaca, N.Y., 1975.

Brown, E. J. *Stankevich and His Moscow Circle, 1830–1840.* Stanford, 1966.

Byrnes, R. F. *Pobedonostsev: His Life and Thought.* Bloomington, Ind., 1968.

Carr, E. H. *Michael Bakunin,* New York, 1961.

Carr, E. H. *The Romantic Exiles: A Nineteenth Century Portrait Gallery.* London, 1933.

Chmielewski, E. *Tribune of the Slavophiles: Konstantin Aksakov.* University of Florida Monograph. Social Sciences, no. 12. Gainesville, Fla., 1961.

Christoff, P. K. *An Introduction to Nineteenth-Century Russian Slavophilism: A Study in Ideas.* Vol. 1, *A. S. Xomjakov.* The Hague, 1961.

Christoff, P. K. *An Introduction to Nineteenth-Century Russian Slavophilism: A Study in Ideas.* Vol. 2, *I. V. Kireevskij.* The Hague, 1972.

Christoff, P. K. *An Introduction to Nineteenth-Century Russian Slavophilism.* Vol. 3, *K. S. Aksakov: A Study in Ideas.* Princeton, N.J., 1982.

Christoff, P. K. *An Introduction to Nineteenth-Century Russian Slavophilism.* Vol. 4, *Iu. F. Samarin.* Boulder, Colo., 1991.

Christoff, P. K. *Third Heart: Some Intellectual-Ideological Currents in Russia, 1800–1830.* The Hague, 1970.

Edelman, R. *Gentry Politics on the Eve of the Russian Revolution: The Nationalist Party, 1907–1917.* New Brunswick, N.J., 1980.

Edelman, R. *Proletarian Peasants: The Revolution of 1905 in Russia's Southwest.* Ithaca, N.Y., 1987.

Fadner, F. *Seventy Years of Pan-Slavism: Karazin to Danilevskii, 1800–1870.* Washington, D.C., 1962.

Field, D. *Rebels in the Name of the Tsar.* Boston, 1976.

Fischer, G. *Russian Liberalism, from Gentry to Intelligentsia.* Cambridge, Mass., 1958.

Frank, J. *Dostoevsky: The Seeds of Revolt, 1821–1849.* Princeton, N.J., 1976.

Frank, J. *Dostoevsky: The Stir of Liberation, 1860–1865.* Princeton, N.J., 1986.

Frank, J. *Dostoevsky: The Years of Ordeal, 1850–1859.* Princeton, N.J., 1983.

Gerstein, L. *Nikolai Strakhov.* Cambridge, Mass., 1971.

Getzler, I. *Martov: A Political Biography of a Russian Social Democrat.* New York, 1967.

Haimson, L. H. *The Russian Marxists and the Origins of Bolshevism.* Cambridge, Mass., 1955.

Hardy, D. *Land and Freedom: The Origins of Russian Terrorism, 1876–1879.* Westport, Conn. 1987.

Hare, R. *Portraits of Russian Personalities between Reform and Revolution.* New York, 1959.

Karpovich, M. "P. L. Lavrov and Russian Socialism." *California Slavic Studies* 2 (1963).

Katz, M. *Mikhail N. Katkov: A Political Biography, 1818–1887.* The Hague, 1966.

Keep, J. L. H. *The Rise of Social Democracy in Russia.* Oxford, 1963.

Kindersley, R. *The First Russian Revisionists: A Study of Legal Marxism in Russia.* Oxford, 1962.

Kohn, H. *Pan Slavism: Its History and Ideology.* Notre Dame, Ind., 1953; New York, 1960.

Lampert, E. *Sons against Fathers: Studies in Russian Radicalism and Revolution.* London, 1965.

Lampert, E. *Studies in Rebellion.* London, 1957.

Lang, D. M. *The First Russian Radical: Alexander Radishchev, 1749–1802.* New York, 1960.

Lednicki, W. *Russia, Poland and the West: Essays in Literary and Cultural History.* London and New York, 1954.

Leslie, R. F. *Reform and Insurrection in Russian Poland, 1856–1865.* London, 1963.

Lincoln, W. B. *In War's Dark Shadow: The Russians Before the Great War.* New York, 1983.

Lukashevich, S. *Ivan Aksakov, 1823–1866; A Study in Russian Thought and Politics.* Cambridge, Mass., 1965.

Lukashevich, S. *Konstantin Leontey, 1831–1891: A Study in Russian "Heroic Vitalism."* New York, 1967.

Lukashevich, S. *N. F. Fedorov (1828–1903): A Study in Roman Eupsychian and Utopian Thought.* Newark, N.J., 1977.

Malia, M. *Alexander Herzen and the Birth of Russian Socialism, 1812–1855.* Cambridge, Mass., 1961.

Malia, M. "What Is the Intelligentsia?" In *The Russian Intelligentsia,* edited by R. Pipes, New York, 1961.

Mazour, A. *The First Russian Revolution, 1825: The Decembrist Movement.* Stanford, 1937.

Mazour, A. G. *Women in Exile: Wives of the Decembrists.* Tallahassee, Fla., 1975.

McDaniel, T. *Autocracy, Capitalism, and Revolution in Russia*. Berkeley, 1988.

McNally, R. T. *Chaadaev and His Friends: An Intellectual History of Peter Chaadaev and His Russian Contemporaries*. Tallahassee, Fla., 1971.

McNeal, R. H., ed. *Russia in Transition, 1905–1914. Evolution or Revolution?* New York, 1970.

Mehlinger, H. D., and J. M. Thompson. *Count Witte and the Tsarist Government in the 1905 Revolution*. Bloomington, Ind., 1972.

Mendel, A. P. *Dilemmas of Progress in Tsarist Russia: Legal Marxism and Legal Populism*. Cambridge, Mass., 1961.

Mendel, A. P. *Michael Bakunin: Roots of Apocalypse*. New York, 1981.

Mochulsky, K. *Dostoevsky: His Life and Work*. Translated by M. Minihan. Princeton, N.J., 1967.

Mohrenschildt, D. von. *Russia in the Intellectual Life of 18th Century France*. New York, 1936.

Paperno, I. *Chernyshevsky and the Age of Russian Realism*. Stanford, 1988.

Petrovich, M. B. *The Emergence of Russian Pan-Slavism, 1856–1870*. New York, 1956.

Pipes, R. *Social Democracy and the St. Petersburg Labor Movement, 1885–1897*. Cambridge, Mass., 1963.

Pipes, R. *Struve: Liberal on the Left, 1870–1905*. Cambridge, Mass., 1970.

Pipes, R. *Struve: Liberal on the Right, 1905–1944*. Cambridge, Mass., 1980.

Pipes, R., ed. *Revolutionary Russia*. Cambridge, Mass., 1968.

Pipes, R., ed. *The Russian Intelligentsia*. New York, 1961.

Plamenatz, J. *German Marxism and Russian Communism*. London, 1954.

Pollard, A. "The Russian Intelligentsia: The Mind of Russia." *California Slavic Studies* 3 (1964).

Pomper, P. *Peter Lavrov and the Russian Revolutionary Movement*. Chicago, 1972.

Putnam, G. F. *Russian Alternatives to Marxism: Christian Socialism and Idealistic Liberalism in Twentieth-Century Russia*. Knoxville, Tenn., 1977.

Rabinowitch, A. *Prelude to Revolution: The Petrograd Bolsheviks and the July 1917 Uprising*. Bloomington, Ind., 1968.

Raeff, M. *Origins of the Russian Intelligentsia: The Eighteenth Century Nobility*. New York, 1966.

Randall, F. *N. G. Chernyshevskii*. New York, 1967.

Riasanovsky, N. V. *Nicholas I and Official Nationality in Russia, 1825–1855*. Berkeley, 1959.

Riasanovsky, N. V. *A Parting of Ways: Government and the Educated Public in Russia, 1801–1855*. Oxford, 1976.

Riasanovsky, N. V. *Russia and the West in the Teachings of the Slavophiles*. Cambridge, Mass., 1952.

Rogger, H. *National Consciousness in 18th Century Russia*. Cambridge, Mass., 1960.

Sablinsky, W. *The Road to Bloody Sunday. The Role of Father Gapon and the Assembly in the Petersburg Massacre of 1905*. Princeton, N.J., 1976.

Schapiro, L. *Rationalism and Nationalism in Russian Nineteenth Century Political Thought*. New Haven, Conn., 1967.

Schneiderman, J. *Sergei Zubatov and Revolutionary Marxism. The Struggle for the Working Class in Tsarist Russia*. Ithaca, N.Y., 1976.

Schwarz, S. M. *The Russian Revolution of 1905: The Workers' Movement and the Formation of Bolshevism and Menshevism*. Translated by G. Vakar. Inter-University Project on the History of the Menshevik Movement. Chicago and London, 1967.

Seregny, S. J. *Russian Teachers and Peasant Revolution: The Politics of Education in 1905*. Bloomington, Ind., 1989.

Serge, V. *Memoirs of a Revolutionary, 1901–1941*. Translated by P. Sedgwick. London, 1963.

Sites, R. *The Women's Liberation Movement in Russia: Feminism, Nihilism, and Bolshevism, 1860–1930*. Princeton, N.J., 1978.

Surh, G. D. *1905 in St. Petersburg: Labor, Society and Revolution*. Stanford, 1989.

Thaden, E. C. *Conservative Nationalism in 19th Century Russia*. Seattle, 1964.

Treadgold, D. W. *Lenin and His Rivals: The Struggle for Russia's Future, 1898–1906*. New York, 1955.

Trotsky, L. *1905*. Translated by A. Bostock. New York, 1971.

Ulam, A. B. *The Bolsheviks*. New York, 1965.

Ulam, A. B. *In the Name of the People: Prophets and Conspirators in Prerevolutionary Russia*. New York, 1977.

Venturi, F. *Roots of Revolution: A History of the Populist and Socialist Movements in Nineteenth Century Russia*. Translated by F. Haskell. New York, 1960.

Vucinich, A. *Social Thought in Tsarist Russia: The Quest for a General Science of Society 1861–1917*. Chicago and London, 1976.

Walicki, A. *The Controversy Over Capitalism*. Oxford, 1969.

Walicki, A. "Personality and Society in the Ideology of Russian Slavophiles: A Study in the Sociology of Knowledge." *California Slavic Studies* 2 (1963).

Walicki, A. *The Slavophile Controversy. History of a Conservative Utopia in Nineteenth Century Russian Thought*. Translated by H. Andrews-Rusiecka. Oxford, 1975.

Weeks, A. L. *The First Bolshevik: A Political Biography of Peter Tkachev*. New York, 1968.

Wildman, A. *The Making of a Workers' Revolution: Russian Social Democracy, 1891–1903*. Inter-University Project on the History of the Menshevik Movement. Chicago and London, 1967.

Wilson, E. *To the Finland Station*. New York, 1940.

Woehrlin, W. F. *Chernyshevsky: The Man and the Journalist*. Cambridge, Mass., 1971.

Wolfe, B. E. *Three Who Made a Revolution: A Biographical History*. New York, 1948.

Yarmolinsky, A. *Road to Revolution: A Century of Russian Radicalism*. London, 1957.

Zenkovsky, V. V. *Russian Thinkers and Europe*. Translated from Russian. Ann Arbor, Mich., 1953.

Zetlin, M. *The Decembrists*. Translated by G. Panin. New York, 1958.

SOVIET RUSSIA, 1917–1983

A. Revolution and Civil War

(i) GENERAL STUDIES, INCLUDING WORKS BY PARTICIPANTS

Carr, E. H. *A History of Soviet Russia*. Vols. 1–3, *The Bolshevik Revolution, 1917–1923*. Vol. 4, *The Interregnum, 1923–1924*. Vols. 5–7, *Socialism in One Country, 1924–1926*. Vols. 8–9, *Foundations of a Planned Economy, 1926–1929* (Vol. 8 with R. W. Davies). New York, 1951–53, 1954, 1958, 1971–72.

Chamberlin, W. N. *The Russian Revolution, 1917–1921*. 2 vols. New York, 1935.

Chernov, V. *The Great Russian Revolution*. New Haven, Conn., 1936.

Curtiss, J. S. *The Russian Revolutions of 1917*. Princeton, N.J., 1957.

Denikin, A. I. *The Russian Turmoil*. London, 1922.

Fitzpatrick, S. *The Russian Revolution*. New York, 1982.

Footman, D. *Civil War in Russia*. New York, 1961.

Footman, D. *The Russian Revolution*. New York, 1962.

Katkov, G. *Russia, 1917: The February Revolution*. New York, 1967.

Kerensky, A. *The Catastrophe: Kerensky's Own Story of the Russian Revolution*. New York, 1927.

Kerensky, A. *The Crucifixion of Liberty*. New York, 1934.

Kerensky, A. *Russia and History's Turning Point*. London, 1965.

Liebman, M. *The Russian Revolution*. Translated by A. Pomperans. New York, 1970.

Lincoln, W. B. *Passage Through Armageddon: The Russians in War and Revolution, 1914–1918*. New York, 1986.

Stites, R. *Revolutionary Dreams: Utopian Vision and Experimental Life in the Russian Revolution*. New York, 1989.

Sukhanov, N. N. *The Russian Revolution of 1917*. New York, 1955.

Trotsky, L. *The History of the Russian Revolution*, 3 vols. New York, 1932–57.

Tucker, R. C., ed. *The Lenin Anthology*. New York, 1975.

Von Mohrenschildt, D., ed. *The Russian Revolution of 1917: Contemporary Accounts*. New York, 1971.

Woytinsky, W. S. *Stormy Passage. A Personal History through Two Russian Revolutions to Democracy and Freedom: 1905–1960*. New York, 1961.

(ii) READINGS ON SPECIAL TOPICS

Adams, A. E. *Bolsheviks in the Ukraine: The Second Campaign*. New Haven, Conn., 1963.

Adams, A. E., ed. *The Russian Revolution and Bolshevik Victory: Why and How?* Boston, 1960.

Anweiler, O. *The Soviets: The Russian Workers, Peasants, and Soldiers Councils, 1905–1921*. Translated by R. Hein. New York, 1974.

Avrich, P. *Kronstadt 1921*. Princeton, N.J., 1970.

Browder, R., and F. Kerensky, eds. *The Russian Provisional Government, 1917*. 3 vols. Stanford, 1961.

Bunyan, J., ed. *Intervention, Civil War and Communism in Russia, April–December, 1918: Documents*. Baltimore, 1936.

Bunyan, J., and H. H. Fisher, eds. *The Bolshevik Revolution, 1917–1918: Documents*. Stanford, 1934.

Burdzhalov, E. N. *Russia's Second Revolution: The February Uprising in Petrograd*. Translated and edited by D. J. Raleigh. Bloomington, Ind., 1987.

Carr, E. H. *The October Revolution: Before and After*. New York, 1969.

Dan, Th. *The Origins of Bolshevism*. London, 1964.

Daniels, R. V. *Red October. The Bolshevik Revolution of 1917*. London, 1968.

Deutscher, I., ed. *The Age of Permanent Revolution: A Trotsky Anthology*. New York, 1964.

Farnsworth, B. *Alexandra Kollontai: Socialism, Feminism, and the Bolshevik Revolution*. Stanford, 1980.

Ferro, M. *The Russian Revolution of February 1917*. Translated by J. L. Richards. Englewood Cliffs, N.J., 1972.

Fischer, Louis. *The Life of Lenin*. New York, 1964.

Galili, Z. *The Menshevik Leaders in the Russian Revolution: Social Realities and Political Struggles*. Princeton, N.J., 1989.

Gankin, O. H., and H. H. Fisher, eds. *The Bolsheviks and the World War: Documents*. Stanford, 1940.

Haimson, L. H., ed. *The Mensheviks: From the Revolution of 1917 to the Second World War*. Chicago, 1974.

Hasegawa, T. *The February Revolution: Petrograd, 1917*. Seattle, 1981.

Hunczak, T., ed. *The Ukraine, 1917–1921: A Study in Revolution*. Cambridge, Mass., 1977.

Keep, J. *The Russian Revolution: A Study in Mass Mobilization*. New York, 1976.

Kennan, G. F. *Soviet-American Relations, 1917–1920*. Vol. 1, *Russia Leaves the War*. Vol. 2, *The Decision to Intervene*. Princeton, N.J., 1956, 1958.

Laqueur, W. *The Fate of the Revolution: Interpretations of Soviet History*. New York, 1967.

Lehovich, D. V. *White Against Red: The Life of General Anton Denikin*. New York, 1974.

Lenin, V. I. *Collected Works*. New York, 1927–42.

Lenin, V. I. *Imperialism: The Highest Stage of Capitalism*. New York, 1927.

Lenin, V. I. *The State and Revolution*. New York, 1927.

Lenin, V. I. *What Is to Be Done?* Moscow, 1947.

Lewin, M. *Lenin's Last Struggle*. New York, 1968.

Luckett, R. *The White Generals: An Account of the White Movement and the Russian Civil War*. New York, 1987.

Mally, L. *Culture of the Future: The Proletkult Movement in Revolutionary Russia*. Berkeley, 1990.

Medvedev, R. A. *The October Revolution*. Translated by G. Saunders. Foreword by H. G. Salisbury. New York, 1979.

Melgunov, S. P. *The Bolshevik Seizure of Power*. Edited by S. G. Pushkarev and B. S. Pushkarev. Translated by J. Beaver. Santa Barbara, Calif., 1972.

Payne, R. *The Life and Death of Lenin*. New York, 1964.

Pipes, R. *The Formation of the Soviet Union: Communism and Nationalism, 1917–1923*. Cambridge, Mass., 1954.

Pipes, R., ed. *Revolutionary Russia: A Symposium*. 2nd ed. rev. Cambridge, Mass., 1968.

Possony, S. T. *Lenin: The Compulsive Revolutionary*. Chicago, 1964.

Rabinowitch, A. *The Bolsheviks Come to Power: The Revolution of 1917 in Petrograd*. New York, 1976.

Radkey, O. H. *The Agrarian Foes of Bolshevism: Promise and Default of the Russian Socialist Revolutionaries, February to October, 1917*. New York, 1958.

Radkey, O. H. *The Election to the Russian Constituent Assembly of 1917*. Cambridge, Mass., 1950.

Radkey, O. H. *The Sickle under the Hammer: The Russian Socialist Revolutionaries in the Early Months of Soviet Rule*. New York, 1963.

Raleigh, D. J. *Revoluiton on the Volga: 1917 in Saratov*. Ithaca, N.Y., 1986.

Raskolnikov, F. F. *Kronstadt and Petrograd in 1917*. Translated and annotated by B. Pearce. London, 1982.

Reed, J. *Ten Days That Shook the World*. New York, 1960.

Reshetar, J. S. *The Ukrainian Revolution, 1917–1920*. Princeton, N.J., 1952.

Rosenberg, A. *A History of Bolshevism*. Garden City, N.Y., 1967.

Rosenberg, W. G. *A. I. Denikin and the Anti-Bolshevik Movement in South Russia*. Amherst, Mass., 1961.

Rosenberg, W. G. *Liberals in the Russian Revolution: The Constitutional Democratic Party, 1917–1921*. Princeton, N.J., 1974.

Rosmer, A. *Moscow under Lenin*. Translated by I. Birchall. New York, 1972.

Serge, V. *Year One of the Russian Revolution*. Translated by P. Sedgwick. New York, 1972.

Shapiro, L. *The Origins of the Communist Autocracy: Political Opposition in the Soviet State: First Phase, 1917–1922*. Cambridge, Mass., 1955.

Smith, E. E. *The Young Stalin*. New York, 1967.

Stewart, G. *The White Armies of Russia*. New York, 1933.

Suny, R. G. *The Baku Commune, 1917–1918: Class and Nationality in the Russian Revolution*. Princeton, N.J., 1972.

Tirado, I. A. *Young Guard! The Communist Youth League, Petrograd 1917–1920*. New York, 1988.

Ukraintsev, N. "A Document in the Kornilov Affair." *Soviet Studies* (October 1973).

Varnock, E., and H. H. Fisher. *The Testimony of Kolchak and Other Siberian Materials and Documents*. Stanford, 1935.

Wheeler-Bennett, J. W. *The Forgotten Peace: Brest-Litovsk*. New York, 1939.

B. Soviet Period: General

Abramovitch, R. *The Soviet Revolution, 1917–1939.* New York, 1962.

Alliluyeva, S. *Twenty Letters to a Friend.* New York, 1967.

Amalrik, A. *Involuntary Journey to Siberia.* New York, 1970.

Amalrik, A. *Notes of a Revolutionary.* Translated by G. Daniels. New York, 1982.

Amalrik, A. *Will the Soviet Union Survive until 1984?* New York, 1970.

Benet, S., ed. and trans. *The Village of Viriatino: An Ethnographic Study of a Russian Village from before the Revolution to the Present.* Garden City, N.Y., 1970.

Berdiaev, N. *The Origins of Russian Communism.* London, 1948.

Berdiaev, N. *The Russian Revolution: Two Essays on Its Implications in Religion and Psychology.* London, 1931.

Bialer, S. *Stalin's Successors: Leadership, Stability, and Change in the Soviet Union.* Cambridge, 1980.

Breslauer, G. W. *Khrushchev and Brezhnev as Leaders: Building Authority in Soviet Politics.* London, 1982.

Chalidze, V. *To Defend These Rights: Human Rights and the Soviet Union.* Translated by G. Daniels. New York, 1974.

Cohen, S. *Rethinking the Soviet Experience: Politics and History Since 1917.* New York, 1985.

Conquest, R. *Stalin and the Kirov Murder.* New York, 1989.

Crossman, R., ed. *The God That Failed.* London, 1950.

Daniels, R. V. *A Documentary History of Communism.* New York, 1960.

Daniels, R. V. *Russia: The Roots of Confrontation.* Cambridge, Mass., 1985.

Daniels, R. V., ed. *The Stalin Revolution: Fulfillment or Betrayal of Communism?* Boston, 1965.

Deutscher, I. *The Prophet Armed: Trotsky, 1879–1921.* New York, 1954.

Deutscher, I. *The Prophet Outcast: Trotsky, 1929–1940.* New York, 1963.

Deutscher, I. *The Prophet Unarmed: Trotsky, 1921–1929.* New York, 1959.

Deutscher, I. *Stalin: A Political Biography.* New York, 1949. 2nd ed., 1966.

Dmytryshyn, B. *USSR: A Concise History.* New York, 1971.

Dornberg, J. *Brezhnev: The Masks of Power.* New York, 1974.

Dunn, S., and E. Dunn. *The Peasants of Central Russia.* New York, 1967.

Fischer, G. *Soviet Opposition to Stalin: A Case Study in World War II.* Westport, Conn., 1970.

Fitzpatrick, S. *Cultural Revolution in Russia, 1928–1931.* Bloomington, Ind., 1978.

Friedrich, C. J., and Z. K. Brzezinski. *Totalitarian Dictatorship and Autocracy.* Cambridge, Mass., 1956.

Galanskov, I., et al. *The Trial of the Four: A Collection of Materials on the Case of Galanskov, Ginzberg, Dobrovolsky and Lashkova, 1967–68.* Edited by P. Reddaway. Translated by J. Saprets, H. Sternberg, and D. Weissbort. New York, 1972.

Geiger, H. K. *The Family in Soviet Russia.* Cambridge, Mass., 1968.

Gerstenmaier, C. *The Voices of the Silent.* Translated by S. Hecker. New York, 1972.

Hahn, W. G. *Postwar Soviet Politics: The Fall of Zhdanov and the Defeat of Moderation, 1946–53.* Ithaca, N.Y., 1982.

Hendel, S., and R. Braham, eds. *The USSR after 50 Years: Promise and Reality.* New York, 1967.

Hindus, M. *Red Bread: Collectivization in a Russian Village.* Foreword by R. G. Suny. Bloomington, Ind., 1988.

Hosking, G. *The First Socialist Society: A History of the Soviet Union from Within.* Cambridge, Mass., 1985.

Khrushchev, N. S. *Khrushchev Remembers.* Vol 1, edited and translated by S. Talbott. Introduction and Commentary by E. Crankshaw. Vol 2, *The Last Testament.*

Edited and translated by S. Talbott. Foreword by E. Crankshaw. Introduction by J. L. Schecter. Boston and Toronto, 1970, 1974.

Lapidus, G. W. *Women in Soviet Society: Equality, Development, and Social Change.* Berkeley, 1978.

Lapidus, G. W., ed. *Women, Work and Family in the Soviet Union.* Armonk, N.Y., 1982.

Mandelshtam, N. *Hope Abandoned.* Translated by M. Hayward. New York, 1974.

Mandelshtam, N. *Hope Against Hope.* Translated by M. Hayward. New York, 1970.

Marcuse, H. *Soviet Marxism.* New York, 1958.

McNeal, R. *Stalin: Man and Ruler.* New York, 1988.

McNeal, R. H. *The Bolshevik Tradition: Lenin, Stalin, Khrushchev.* Englewood Cliffs, N.J., 1963.

McNeal, R. H., ed. *Lenin, Stalin, Khrushchev: Voices of Bolshevism.* Englewood Cliffs, N.J., 1963.

Medvedev, R. *Let History Judge. The Origins and Consequences of Stalinism.* Edited by D. Joravsky and G. Haupt. Translated by C. Taylor. New York, 1971.

Medvedev, R. *On Socialist Democracy.* Edited and translated by E. de Kadt. New York, 1975.

Medvedev, Zh. A., and R. A. Medvedev. *A Question of Madness.* New York, 1971.

Meyer, A. *Communism.* New York, 1960.

Meyer, A. *Leninism.* Cambridge, Mass., 1957.

Meyer, A. *Marxism.* Cambridge, Mass., 1954.

Nettl, J. P. *The Soviet Achievement.* London, 1967.

On Trial: The Case of Sinyavsky (Tertz) and Daniel (Arzhak). Documents edited by L. Labedz and M. Hayward. Russian text translated by M. Harari and M. Hayward. French texts translated by M. Villiers. London, 1967.

Pankratova, A. M., ed. *A History of the USSR.* Compiled by K. V. Bazilevich et al. 3 vols. New York, 1970.

Rabinowitch, A., and J. Rabinowitch, eds. *Revolution and Politics in Russia. Essays in Memory of B. I. Nicolaevsky.* Bloomington, Ind., 1972.

Ratushinskaya, I. *In the Beginning.* Translated by A. Kojevnikov. New York, 1991.

Rauch, B. von. *A History of Soviet Russia.* New York, 1957.

Reddaway, P., ed. and trans. *Uncensored Russia: Protest and Dissent in the Soviet Union. The unofficial Moscow journal A Chronicle of Current Events.* London, 1972.

Reve, K. van het, ed. *Dear Comrade: Pavel Litvinov and the Voices of Soviet Citizens in Dissent.* New York, 1969.

Rieber, A. J., and R. C. Nelson. *A Study of the USSR and Communism: An Historical Approach.* Chicago, 1962.

Rostow, W. W. *The Dynamics of Soviet Society.* New York, 1963.

Rothberg, A. *The Heirs of Stalin: Dissidence and the Soviet Regime, 1953–1970.* Ithaca, N.Y., 1972.

Schlesinger, R. *Changing Attitudes in Soviet Russia: The Family.* London, 1949.

Shatz, M. S. *Soviet Dissent in Historical Perspective.* Cambridge, 1981.

Souvarine, B. *Stalin, A Critical Survey of Bolshevism.* New York, 1939.

Treadgold, D. W. *Twentieth-Century Russia.* Chicago, 1959. 6th ed., 1987.

Trotsky, L. *The Revolution Betrayed.* New York, 1937.

Trotsky, L. *Stalin: An Appraisal of the Man and His Influence.* New York, 1941.

Tucker, R. C. *Stalin as Revolutionary, 1879–1929: A Study in History and Personality.* New York, 1973.

Ulam, A. B. *Stalin: The Man and His Era.* New York, 1973.

Von Laue, T. H. *The Global City.* Philadelphia and New York, 1969.

Von Laue, T. H. *Why Lenin? Why Stalin?* Philadelphia and New York, 1964.

Webb, S., and B. Webb. *Soviet Communism. A New Civilization?* 2 vols. New York, 1936.

Werth, A. *Russia: The Post-War Years.* New York, 1972.

Westwood, J. N. *Endurance and Endeavour: Russian Hisotry, 1812–1980.* 3rd ed. The Short Oxford History of the Modern World. Oxford, 1987.

C. *Ideology, Government, Administration, and Law*

Armstrong, J. A. *The Politics of Totalitarianism: The Communist Party of the Soviet Union.* New York, 1961.

Armstrong, J. A. *The Soviet Bureaucratic Elite: A Case Study of the Ukrainian Apparatus.* New York, 1961.

Avtorkhanov, A. *Stalin and the Soviet Communist Party.* Munich, 1959.

Azrael, J. R. *Managerial Power and Soviet Politics.* Cambridge, Mass., 1966.

Barghoorn, F. C. *Soviet Russian Nationalism.* New York, 1956.

Barron, J. *KGB: The Secret Work of Soviet Agents.* New York, 1974.

Berman, H. J. *Justice in Russia.* Cambridge, Mass., 1950. Rev. ed., 1963.

Berman, H. J., and M. Kerner. *Soviet Military Law and Administration.* Cambridge, Mass., 1955.

Berman, H. J., and P. B. Maggs. *Disarmament Inspection under Soviet Law.* Dobbs Ferry, N.Y., 1967.

Berman, H. J., and J. W. Spindler, trans. *Soviet Criminal Law and Procedure: The RSFSR Code.* Introduction and analysis by H. J. Berman. Cambridge, Mass., 1965.

Brzezinski, Z. K. *The Permanent Purge.* Cambridge, Mass., 1956.

Conquest, R. *The Great Terror: A Reassessment.* New York, 1990.

Conquest, R. *The Great Terror: Stalin's Purge of the Thirties.* New York, 1971.

Dallin, A., and T. B. Larson, eds. *Soviet Politics since Khrushchev.* Englewood Cliffs, N.J., 1968.

Daniels, R. V. *The Conscience of the Revolution: Communist Opposition in Soviet Russia.* Cambridge, Mass., 1960.

Deacon, R. *A History of the Russian Secret Service.* New York, 1972.

Dinerstein, H. S., and L. Goure. *Communism and the Russian Peasant.* Glencoe, Ill., 1955.

Dinerstein, H. S., and L. Goure. *War and the Soviet Union.* New York, 1962.

Erickson, J. *The Soviet High Command: A Military-Political History, 1918–1941.* New York, 1962.

Fainsod, M. *How Russia Is Ruled.* Cambridge, Mass., 1953.

Fainsod, M. *Smolensk under Soviet Rule.* Cambridge, Mass., 1958.

Fischer, G. *Soviet Opposition to Stalin.* Cambridge, Mass., 1952.

Graham, L. R. *The Soviet Academy of Sciences and the Communist Party, 1927–1932.* Princeton, N.J., 1967.

Gsovsky, V. *Soviet Civil Law.* 2 vols. Ann Arbor, Mich., 1948, 1949.

Gsovsky, V., and K. Grybowski. *Government, Law and Courts in the Soviet Union and Eastern Europe.* New York, 1959.

Hahn, W. G. *Postwar Soviet Politics: The Fall of Zhdanov and the Defeat of Moderation, 1946–1953.* Ithaca, N.Y., 1982.

Hammer, D. P. *USSR: The Politics of Oligarchy.* Hinsdale, Ill., 1974.

Harcave, S. *The Structure and Functioning of the Lower Party Organizations in the Soviet Union.* University, Ala., 1954.

Harvey, M. L., L. Goure, and V. Prokofieff. *Science and Technology as an Instrument of Soviet Policy.* Washington, D.C., 1972.

Hazard, J. N. *The Soviet System of Government.* Chicago, 1957.

Hendel, S., ed. *The Soviet Crucible: Soviet Government in Theory and Practice.* Princeton, N.J., 1960.

Inkeles, A., and R. A. Bauer. *The Soviet Citizen*. Cambridge, Mass., 1959.

Inkeles, A., R. Bauer, and C. Kluckhohn. *How the Soviet System Works*. Cambridge, Mass., 1956.

Karcz, J., ed. *Soviet and East European Agriculture*. Berkeley, 1967.

Kassof, A., ed. *Prospects for Soviet Society*. New York, 1968.

Kucherov, S. *The Organs of Soviet Administration of Justice: Their History and Operation*. London, 1970.

Leites, N., and E. Bernaut. *Ritual of Liquidation: The Case of the Moscow Trials*. Glencoe, Ill., 1954.

Leonard, W. *The Kremlin since Stalin*. New York, 1962.

Levytsky, B. *The Uses of Terror. The Soviet Secret Police, 1917–1970*. Translated by H. Piehler. New York, 1972.

Matthews, M. *Class and Society in Soviet Russia*. New York, 1972.

Matthews, M., ed. *Soviet Government: A Selection of Official Documents on Internal Policies*. New York, 1974.

Meissner, B. *The Communist Party of the Soviet Union: Party Leadership*. New York, 1956.

Moore, B. *Soviet Politics: The Dilemma of Power*. Cambridge, Mass., 1950.

Odom, W. E. *The Soviet Volunteers: Modernization and Bureaucracy in a Public Mass Organization*. Princeton, N.J., 1973.

Reshetar, J. S. *A Concise History of the Communist Party of the Soviet Union*. New York, 1960.

Rigby, T. H. *Communist Party Membership in the USSR, 1917–1967*. Princeton, N.J., 1968.

Schapiro, L. B. *The Communist Party of the Soviet Union*. New York, 1971.

Schlesinger, R. *Soviet Legal Theory*. London, 1945.

Schuman, F. *Government in the Soviet Union*. New York, 1961.

Scott, D. J. R. *Russian Political Institutions*. New York, 1958.

Sorenson, R. *The Life and Death of Soviet Trade Unionism*. New York, 1969.

Swearer, H. R., and M. Rush. *The Politics of Succession in the USSR: Materials on Khrushchev's Rise to Leadership*. Boston, 1964.

Towster, J. *Political Power in the USSR, 1917–1947*. New York, 1948.

Tucker, R. C. *The Soviet Political Mind*. New York, 1963.

Ulam, A. *The Bolsheviks*. New York, 1965.

Ulam, A. *The New Face of Soviet Totalitarianism*. Cambridge, Mass., 1963.

Ulam, A. *The Unfinished Revolution: An Essay on the Sources of Influence of Marxism and Communism*. New York, 1960.

Vyshinsky, A. *The Law of the Soviet State*. New York, 1948.

Weinberg, E. A. *The Development of Sociology in the Soviet Union*. Boston, 1974.

Wolin, S., and R. Slusser, eds. *The Soviet Secret Police*. New York, 1957.

D. Economic Development

Ball, A. M. *Russia's Last Capitalists: The Nepmen, 1921–1929*. Berkeley, 1987.

Baykov, A. *The Development of the Soviet Economic System*. Cambridge, 1947.

Bergson, A. *The Economics of Soviet Planning*. New Haven, Conn., 1964.

Bergson, A. *Planning and Productivity under Soviet Socialism*. New York, 1968.

Bergson, A. *Real National Income of Soviet Russia since 1928*. Cambridge, Mass., 1961.

Bergson, A. *Soviet Economic Growth*. Evanston, Ill., 1953.

Bergson, A. *The Structure of Soviet Wages*. Cambridge, Mass., 1944.

Bergson, A., and S. Kuznets, eds. *Economic Trends in the Soviet Union*. Cambridge, Mass., 1963.

Campbell, R. *Soviet Economic Power: Its Organization, Growth and Challenge*. Boston, 1960.

Dallin, D. J., and B. I. Nicolaevsky. *Forced Labor in Soviet Russia.* New Haven, Conn., 1947.

Davies, R. W. *The Industrialization of Soviet Russia.* 2 vols. Cambridge, Mass., 1980.

Deutscher, I. *Soviet Trade Unions.* London, 1950.

De Witt, N. *Education and Professional Employment in the USSR.* Washington, D.C., 1961.

Dobb, M. *Soviet Economic Development Since the 1917 Revolution.* London, 1948.

Erlich, A. *The Soviet Industrialization Debate, 1924–28.* Cambridge, Mass., 1960.

Gregory, P. R., and R. C. Stuart. *Soviet Economic Structure and Performance.* New York, 1974.

Grossman, G. "The Economy at Middle Age." *Problems of Communism* (March–April 1976).

Grossman, G. "The Solidary Society: A Philosophical Issue in Communist Economic Reforms." In *Essays in Socialism and Planning in Honor of Carl Landauer,* edited by G. Grossman. Englewood Cliffs, N.J., 1970.

Grossman, G., ed. *Money and Plan: Financial Aspects of East European Economic Reforms.* Berkeley, 1968.

Holzman, F. *Soviet Taxation.* Cambridge, Mass., 1955.

Holzman, F., ed. *Readings on the Soviet Economy.* Chicago, 1962.

Jasny, N. *The Socialized Agriculture of the USSR.* Stanford, 1949.

Jasny, N. *Soviet Economy during the Plan Era.* Stanford, 1951.

Jasny, N. *Soviet Industrialization, 1928–1932.* Chicago, 1961.

Laird, R. D., ed. *Soviet Agricultural and Peasant Affairs.* Lawrence, Kans., 1963.

Lewin, M. *Russian Peasants and Soviet Power: A Study of Collectivization.* Translated by I. Nove. Evanston, Ill., 1968.

Lewin, M. "Soviet Policies of Agricultural Procurements before the War." In *Essays in Honor of E. H. Carr,* edited by Ch. Abramsky, and B. I. Williams. Hamden, Conn., 1974.

Moskoff, W. *The Bread of Affliction: The Food Supply in the USSR During World War II.* Cambridge, 1990.

Nove, A. *An Economic History of the USSR.* London, 1969.

Pryde, P. R. *Conservation in the Soviet Union.* Cambridge, 1972.

Quigley, J. *The Soviet Foreign Trade Monopoly: Institutions and Laws.* Columbus, Ohio, 1974.

Schwartz, H. *Russia's Soviet Economy.* 2nd ed. New York, 1954.

Schwartz, H. *The Soviet Economy since Stalin.* Philadelphia, 1965.

Schwarz, S. *Labor in the Soviet Union.* New York, 1952.

E. Foreign Affairs

Adams, A. E. *Readings in Soviet Foreign Policy: Theory and Practice.* Boston, 1961.

Barghoorn, F. C. *The Soviet Cultural Offensive: The Role of Cultural Diplomacy in Soviet Foreign Policy.* Princeton, N.J., 1960.

Barghoorn, F. C. *Soviet Foreign Propaganda.* Princeton, N.J., 1964.

Beloff, M. *The Foreign Policy of Soviet Russia, 1929–1941.* 2 vols. New York, 1947, 1949.

Beloff, M. *Soviet Foreign Policy in the Far East, 1944–1951.* London, 1953.

Bishop, D. G., ed. *Soviet Foreign Relations: Documents and Readings.* Syracuse, N.Y., 1952.

Brandt, C. *Stalin's Failure in China, 1924–1927.* Cambridge, Mass., 1958.

Brzezinski, Z. K. *The Soviet Bloc: Unity and Conflict.* Cambridge, Mass., 1960. Rev. ed., 1967.

Carell, P. *Scorched Earth: The Russian-German War, 1943–1944.* Translated by E. Osers. Boston, 1970.

Carr, E. H. *The Soviet Impact on the Western World*. London, 1946.

Carr, E. H. *Twilight of the Comintern, 1930–1935*. New York, 1982.

Cattell, D. T. *Communism and the Spanish Civil War*. Berkeley, 1955.

Chew, A. F. *The White Death: The Epic of the Soviet-Finnish Winter War*. East Lansing, Mich., 1971.

Dallin, A. *German Rule in Russia, 1941–1945: A Study of Occupation Policies*. New York, 1957.

Dallin, D. J. *Soviet Russia and the Far East*. New Haven, Conn., 1948.

Dallin, D. J. *Soviet Russia's Foreign Policy, 1939–1942*. New Haven, Conn., 1942.

Degras, J. *The Communist International, 1919–1943*. New York, 1956.

Degras, J. *Soviet Documents on Foreign Policy, 1917–1941*. 3 vols. New York, 1951–53.

Degras, J., ed. *Calendar of Soviet Documents on Foreign Policy, 1917–1941*. London, 1948.

Eudin, X., and H. H. Fisher. *Soviet Russia and the West, 1920–1927*. Stanford, 1957.

Eudin, X., and R. C. North, eds. *Soviet Russia and the East, 1920–1927*. Stanford, 1957.

Farnsworth, B. *William C. Bullitt and the Soviet Union*. Bloomington, Ind., 1967.

Fischer, L. *The Soviets in World Affairs, 1917–1929*. Princeton, N.J., 1951.

Fischer, R. *Stalin and German Communism*. Cambridge, Mass., 1948.

Floyd, D. *Mao against Khrushchev: A Short History of the Sino-Soviet Conflict*. New York, 1963.

Garthoff, R. L. *Soviet Strategy in the Nuclear Age*. New York, 1962.

Griffith, W. E. *Communism in Europe: Continuity, Change and the Sino-Soviet Dispute*. Cambridge, Mass., 1966.

Gruber, H. *International Communism in the Era of Lenin: A Documentary History*. Greenwich, Conn., 1967.

Harvey, D. L., and L. C. Ciccoritti. *U.S.-Soviet Cooperation in Space*. Washington, D.C., 1974.

Impact of the Russian Revolution, 1917–1967: The Influence of Bolshevism on the World Outside Russia, The. London, 1967.

Jamgotch, N., Jr. *Soviet-East European Dialogue: International Relations of a New Type*. Stanford, 1968.

Kapur, H. *Soviet Russia and Asia, 1917–1927: A Study of Soviet Policy towards Turkey, Iran and Afghanistan*. London, 1966; New York, 1967.

Kennan, G. F. *Russia and the West under Lenin and Stalin*. Boston, 1961.

Kennan, G. F. *Soviet Foreign Policy, 1917–1941*. Princeton, N.J., 1960.

Laqueur, W. Z. *The Soviet Union and the Middle East*. New York, 1959.

Laserson, M. M., ed. *The Development of Soviet Foreign Policy in Europe, 1917–1942: A Selection of Documents*. International Conciliation, no. 386 (January 1943).

Leach, B. A. *German Strategy Against Russia, 1939–1941*. Oxford, 1973.

Mackintosh, J. *Strategy and Tactics of Soviet Foreign Policy*. New York, 1962.

Maisky, I. *Memoirs of a Soviet Ambassador: The War 1939–1943*. New York, 1968.

Menon, R. *Soviet Power and the Third World*. New Haven, Connecticut, and London, 1986.

Moore, H. L. *Soviet Far Eastern Policy, 1931–1945*. Princeton, N.J., 1945.

Mosely, P. E. *The Kremlin and World Politics*. New York, 1961.

Mosely, P. E. *Russia after Stalin*. New York, 1955.

Mosely, P. E., ed. *The Soviet Union, 1922–1962: A Foreign Affairs Reader*. New York and London, 1963.

North, R. C. *Moscow and Chinese Communists*. 2nd ed. Stanford, 1962.

O'Connor, T. *Diplomacy and Revolution: G. V. Chicherin and Soviet Foreign Affairs, 1918–1930*. Ames, Ia., 1988.

Pethybridge, R., ed. *The Development of the Communist Bloc*. Boston, 1965.

Ro'i, Y. *From Encroachment to Involvement: A Documentary Study of Soviet Policy in the Middle East, 1945–1973*. New York, 1974.

Rubinstein, A. Z. *Soviet Foreign Policy Since World War II: Imperial and Global*. Cambridge, Mass., 1981.

Rubinstein, A. Z., ed. *The Foreign Policy of the Soviet Union*. New York, 1960.

Rubinstein, A. Z., ed. *Soviet and Chinese Influence in the Third World*. New York, 1975.

Seaton, A. *The Battle for Moscow, 1941–1942*. New York, 1971.

Seaton, A. *The Russo-German War, 1941–45*. New York, 1971.

Seton-Watson, H. *The East European Revolution*. 3rd ed. New York, 1956.

Seton-Watson, H. *From Lenin to Khrushchev: The History of World Communism*. New York, 1951.

Shotwell, J. T., and M. M. Laserson. *Poland and Russia, 1919–1945*. New York, 1945.

Shulman, M. D. *Beyond the Cold War*. New Haven, Conn., 1966.

Shulman, M. D. *Stalin's Foreign Policy Reappraised*. Cambridge, Mass., 1963.

Sontag, R. J., and J. S. Beddie, eds. *Nazi-Soviet Relations, 1939–1941: Documents from the Archives of the German Foreign Office*. Washington, D.C., 1948.

Swearingen, A. R., and P. Langer. *Red Flag in Japan*. Cambridge, Mass., 1952.

Ulam, A. B. *Expansion and Coexistence: A History of Soviet Foreign Policy, 1917–1967*. New York, 1968.

Ulam, A. B. *Titoism and the Cominform*. Cambridge, Mass., 1952.

Ullman, R. H. *Anglo-Soviet Relations, 1917–21*. Vol. 1, *Intervention and the War*. Vol. 2, *Britain and the Russian Civil War, November 1918–February 1920*. Vol. 3, *Anglo-Soviet Accord*. Princeton, N.J., 1961, 1968, 1973.

Villmow, J. R. *The Soviet Union and Eastern Europe*. Englewood Cliffs, N.J., 1965.

Warth, R. *Soviet Russia in World Politics*. New York, 1963.

Weinberg, G. *Germany and the Soviet Union, 1939–1941*. Leiden, 1954.

Zinner, Paul E. *Communist Strategy and Tactics in Czechoslovakia, 1918–1948*. New York, 1963.

F. Nationalities

Armstrong, J. A. *Ukrainian Nationalism, 1939–1945*. New York, 1955.

Browne, M., ed. *Ferment in the Ukraine: Documents by V. Chornovil, I. Kandyba, L. Lukyanenko, V. Moroz and Others*. New York, 1971.

Carve, Sir O. *Soviet Empire: The Turks of Central Asia and Stalinism*. London, 1967.

Comrie, B. *The Languages of the Soviet Union*. Cambridge, 1981.

Conquest, R. *The Nation Killers: The Soviet Deportation of Nationalities*. London, 1970.

Dmytryshyn, B. *Moscow and the Ukraine, 1918–1952: A Study of Russian Bolshevik Nationality Policy*. New York, 1956.

Dunlop, J. B. *The Faces of Contemporary Russian Nationalism*. Princeton, N.J., 1983.

Dunn, S. P. *Cultural Processes in the Baltic Area under Soviet Rule*. Berkeley, 1967.

Fedyshyn, O. S. *Germany's Drive to the East and the Ukrainian Revolution, 1917–1918*. New Brunswick, N.J., 1971.

Goldhagen, E., ed. *Ethnic Minorities in the Soviet Union*. New York, 1967.

Hovannisian, R. G. *The Republic of Armenia*. Vol. 1, *The First Year, 1918–1919*. Berkeley, 1971.

Israel, G. *The Jews in Russia*. Translated by S. L. Chernoff. New York, 1975.

Katz, Z., R. Rogers, and F. Harned, eds. *Handbook of Major Soviet Nationalities*. New York and London, 1975.

Kirchner, W. *The Rise of the Baltic Question*. Westport, Conn., 1970.

Kolarz, W. *Russia and Her Colonies*. New York, 1953.

Kolasky, J. *Education in the Soviet Ukraine*. Toronto, 1968.

Kolasky, J. *Two Years in Soviet Ukraine: A Canadian's Personal Account of Russian Oppression and the Growing Opposition.* Toronto, 1970.

Levin, N. *The Jews in the Soviet Union Since 1917: Paradox of Survival.* 2 vols. New York and London, 1990.

Lubachko, I. S. *Belorussia under Soviet Rule, 1917–1957.* Lexington, Ky., 1972.

Nahaylo, B., and V. Swoboda. *Soviet Disunion: A History of the Nationalities Problem in the USSR.* New York, 1991.

Nove, A., and J. A. Newth. *The Soviet Middle East.* London, 1967.

Peters, V. *Nestor Makhno: The Life of an Anarchist.* Winnipeg, 1970.

Rumer, B. Z. *Soviet Central Asia: "A Tragic Experiment."* Boston, 1989.

Schwarz, S. *The Jews in the Soviet Union.* Syracuse, N.Y., 1951.

von Rauch, G. *The Baltic States: The Years of Independence: Estonia, Latvia, Lithuania, 1917–1940.* Translated by G. Onn. Berkeley, 1974.

G. Religion, Education, and Culture

Bereday, G., and J. Pennar, eds. *The Politics of Soviet Education.* New York, 1960.

Brown, E. J. *The Proletarian Episode in Soviet Literature, 1929–1932.* New York, 1953.

Brown, E. J. *Russian Literature since the Revolution.* New York, 1963.

Churchward, L. G. *The Soviet Intelligentsia: An Essay on the Social Structure and Roles of Soviet Intellectuals during the 1960's.* London and Boston, 1973.

Clark, K. *The Soviet Novel: History as Ritual.* Chicago, 1981.

Curtiss, J. S. *The Russian Church and the Soviet State, 1917–1950.* Boston, 1953.

Dunham, V. S. *In Stalin's Time: Middle-Class Values in Soviet Fiction.* New York, 1976.

Dunlop, J. B., R. Haugh, and A. Klimoff, eds. *Aleksandr Solzhenitsyn: Critical Essays and Documentary Materials.* Belmont, Mass., 1973.

Ehrenburg, I. *Memoirs: 1921–1941.* Translated by T. Shebunina. New York, 1963.

Ellis, J. *The Russian Orthodox Church: A Contemporary History.* Bloomington, Ind., 1986.

Enteen, G. M. *The Soviet Scholar-Bureaucrat: M. N. Pokrovskii and the Society of Marxist Historians.* University Park, Pa., 1978.

Evtushenko, E. *A Precocious Autobiography.* New York, 1963.

Fitzpatrick, S. *Education and Social Mobility in the Soviet Union, 1921–1934.* Cambridge, 1979.

Fletcher, W. C. *A Study in Survival: The Church in Russia, 1927–1943.* New York, 1965.

Gleason, A., P. Kenez, and R. Stites, eds. *Bolshevik Culture: Experiment and Order in the Russian Revolution.* Bloomington, Ind., 1985.

Hansson, C., and K. Linden. *Moscow Women: Thirteen Interviews by Carola Hansson and Karin Linden.* Translated by G. Bothmer, G. Blechler, and L. Blechler. Introduction by G. W. Lapidus. New York, 1983.

Hayward, M., and W. C. Fletcher, eds. *Religion and the Soviet State: A Dilemma of Power.* New York, 1969.

Hosking, G. *Beyond Socialist Realism: Soviet Fiction Since "Ivan Denisovich."* New York, 1980.

Jacoby, S. *Inside Soviet Schools.* New York, 1974.

Johnson, P. *Khrushchev and the Arts: The Politics of Soviet Culture, 1962–1964.* Cambridge, Mass., 1965.

Keep, J., ed. *Contemporary History in the Soviet Mirror.* New York, 1964.

Lowe, D. *Russian Writing Since 1953: A Critical Survey.* Cambridge, 1987.

Luckyj, G. *Literary Politics in the Soviet Ukraine, 1917–1934.* New York, 1955.

McLean, H., and W. N. Vickery. *The Year of Protest, 1956: An Anthology of Soviet Literary Materials.* New York, 1961.

Medvedev, Zh. A. *The Rise and Fall of T. D. Lysenko.* Translated by I. M. Lerner. New York, 1969.

Mihailov, M. *Moscow Summer*. Translated from Serbo-Croatian. New York, 1965.

Pospielovsky, D. *The Russian Church Under the Soviet Regime, 1917–1982*. 2 vols. Crestwood, N.Y., 1984.

Sakharov, A. D. *Progress, Coexistence and Intellectual Freedom*. Translated and introduction by H. Salisbury, New York, 1968.

Shlapentokh, V. *Public and Private Life of the Soviet People: Changing Values in Post-Stalin Russia*. New York, 1989.

Simon, G. *Church, State and Opposition in the USSR*. Translated by K. Matett. Berkeley, 1974.

Slonim, M. *Soviet Russian Literature: Writers and Problems*. New York, 1964.

Solzhenitsyn, A. *The Cancer Ward*. Translated by R. Frank. New York, 1968.

Solzhenitsyn, A. *The First Circle*. Translated by M. Guydon. London, 1968.

Solzhenitsyn, A. *The Gulag Archipelago*. 3 vols. Vols. 1 and 2 translated by T. Whitney; vol. 3 translated by H. Willetts, New York, 1974–1978.

Starr, S. F. *Red and Hot: The Fate of Jazz in the Soviet Union, 1917–1980*. New York, 1983.

Stroyen, W. *Communist Russia and the Russian Orthodox Church, 1943–1962*. Washington, D.C., 1967.

Struve, G. *Russian Literature under Lenin and Stalin, 1917–1953*. Norman, Okla., 1971.

Struve, N. *Christians in Contemporary Russia*. Translated by L. Sheppard and A. Manson. New York, 967.

Swayze, H. *Political Control of Literature in the USSR, 1946–1959*. Cambridge, Mass., 1962.

Timasheff, N. *Religion in Soviet Russia*. New York, 1942.

Vucinich, A. *Empire of Knowledge: The Academy of Sciences of the USSR, 1917–1970*. Berkeley, 1984.

Weiner, D. R. *Models of Nature: Ecology, Conservation, and Cultural Revolution in Soviet Russia*. Bloomington, Ind., 1988.

INDEX

* * * * * * * * * *

Abaza, Alexander, 392

Aberdeen, Earl of, 336

Abkhazians, and Georgian nationalism, 596

Abo (Turku): Treaty of, 252; university in, 350

Abraham of Smolensk, Saint, 126

Absolutism, *see* Autocracy; Centralization

Academy of Arts, Imperial, 298, 367, 445

Academy of Sciences, Imperial, 237, 287, 298

Academy of Sciences, Russian, 599

Academy of Sciences of the U.S.S.R., 287; direction of Soviet scholarship, 582

"Acmeists," 444

Adashev, Alexis, 145, 149

Admiralty, 367

Adrianople: threatened by Sviatoslav, 33; Treaty of, 330–1

Adriatic Sea, reached by Mongols, 71

Adzharians, and Georgian nationalism, 596

Aehrenthal, Count Alois von, and Buchlau agreement, 417

Aesop, 355

Afghanistan: conflict with Great Britain, 399, settlement, 416–7; Soviet relations with, 512, 566; trade with, 425; war in, 566, 592–3, 600–1

Agapius, Saint, 86

Age of Reason, 230, 256, 285, 293, 319; and ideologies, 360ff; *see also* Enlightenment, the

Agrarian parties, and Communists, 535

Agriculture: in Kievan Russia, 45–8; in appanage Russia, 114; in Muscovite Russia, 183–4; in 18th century, 277–8; 1800–1861, 341–3; and peasant communes, 373–4; in post-reform period, 430–4; Soviet policy, 501; in post-war period, 531; in post-Stalin period, 547–9, 550–1; and private

plots, 548–9, 573, 591; *see also* Collectivization; Peasantry; Serfdom

Agrogoroda, 531, 573

Ahmad, Khan: campaigns against Moscow (1451, 1455, 1461), 102; campaign against Moscow (1480), 106

Aigun, Treaty of, 390

Aix-la-Chapelle, conference at, 315

Akhmatova, Anna, 445, 474, 584, 585

Aksakov, Constantine, 362–3; quoted, 43

Aksakov, Ivan, 362, 376

Aksakov, Serge, 359; *Family Chronicle*, 349

Alans, 14

Alaska, 297, 308; sold to U.S., 389

Albania, Communist regime in, 534; defiance of Soviet Union, 563, 564

Aleksandrov (town), 150

Alekseev, General Michael, 418, 480

Alesha Popovich, 56

Alevisio, 128–9

Alexander I, Emperor, 229, 231, 273, 275, 347; and architecture, 297; personality, 300–2; reign of, 300–22, first period of reform (1801–5), 302–4, second period of reform (1807–12), 304–6; in 1812, 310–3; and foreign policy (1812–25); 313–8; and succession, 321; and establishment of universities, 349–50; and Karamzin, 353–4

Alexander II, Emperor: and Crimean War, 339; reign of, 368–90; reforms, 369–78, 380; and emancipation, 371–4; reaction, 379–80; *rapprochement* with public, 384; foreign policy, 384–90; assassination, 384

Alexander III, Emperor: reign of, 391–6; support for Church school, 437

Alexander of Battenberg (Prince of Bulgaria), 399

Alexander Nevskii: Prince of Novgorod, Grand Prince, Saint, 92, 110, 121, 126; defeat of Swedes, 80; reign of, 79–80; submission to Mongols, 80–81

665

by Alexander II, 370; Jews in restricted, 395; in December (1905) uprising and labor unrest, 408, 429; State Conference in, 459, Soviet (1917), majority captured by Bolsheviks, 460, 476; uprising of Left SR's, 480–81; becomes capital (March 1918), 481; Denikin advances to, 482; controlled by Reds in Civil War, 486; party organization and Kamenev, 490–91; in Second World War, 518; defense of, 518, 519; and election of Popov, 599; demonstration, of February 1990, 602; *see also* Muscovite Russia

Moscow Art Theater, 444, 447, 587

Mosely, Philip E.: on Treaty of Unkiar Skelessi, 332; on post-Yalta policies, 534

Moskovskie Vedomosti, 289

Moskvitianin, Ivan (cossack, explorer), 194

Moses, Archbishop of Novgorod, 85

Moslems, *see* Minorities; Religion

Mount Athos, 53, 198

Mozhaisk, annexed to Moscow, 97

Mstislav (Prince of Galicia), 91

Mstislav the Brave of Tmutorokan, 37

Mstislav of Toropets, 79

Mstislav (Vladimirovich), Grand Prince, 40, 78

Mstislavsky, Prince Theodore, 167

M.T.S., *see* Machine Tractor Stations

Mukden, battle of, 403

Münchengrätz, meeting at, 334

Munich agreement, 515

Municipal government: Petrine, 231; reform of 1870, 376; and "counter-reforms," 394; *see also* Local administration

Münnich, Count Burkhard C., 218; in reign of Empress Anne, 245; in wars with Turkey, 252

Muratov, Pavel P., on influence of St. Sergius on iconography, 130

Muraviev, Nikita, 320

Muraviev-Amursky, Count Nicholas, 390

Murmansk: and Civil War, 483; convoys to, 520

Muscovite, or Naryshkin, baroque, 204

Muscovite Russia, 63, 81, 95–209 *passim;* autocracy and Mongol legacy, 74–5; and Mongols, 74–5, 102–107; 111–12; and Novgorod, 85, 104–5; rise of principality, 95–113; struggle with Tver, 97–105 *passim;* reaction to centralization, 102; legends of princely descent from Roman emperors and regalia of Constantine Monomakh,

107, 124–25; struggle with Lithuania, 108, 134, 152; reasons for success, 109–13, 138–9, geographic location, 109–10, economic factors, 110, policies, 111–12, role of Church, 112, role of rulers, 110; institutions and society, 144–5, 175–7, 188–92; union with the Ukraine, 180; economy, 185–8; expansion, eastward, 193–5; views on, 112–13; culture, 196–209; and schism, 198–201; and Petrine Russia, 232–3; *see also* Appanage Russia *and individual rulers*

Muscovite Russians, *see* Great Russians

Music: in Imperial Russia, 357, 367, 436, 445–6; in Soviet Russia, 581, 586–7

Musorgsky, Modest, 357; *Boris Godunov,* 156, 446

Mussolini, Benito, and Munich, 515

Mutual Defense Assistance Program (U.S.), 536

Muzhi (Kievan upper classes), 49

Mysticism, absence in Kievan Russia, 53

Nabokov, Vladimir V., on Gogol, 359

Nagorny Karabakh, war over, 596

Nagy, Imre, 560

Nakaz, see Instruction to the Legislative Commission

Nakhimov, Admiral Paul, 338

Namestnik, 192

Naples, 274

Napoleon I, Emperor, 271, 275, 306, 611; on Alexander I, 301; and Russian foreign policy (1801–12), 307–8; in 1812, 310–14; and Russian foreign policy, 313–14 *et passim*

Napoleon III, Emperor: and Crimean War, 337; and sympathy with Poles, 379, 385; Russian relations with, after Crimean War, 385

Narodnichestvo, see Populism

Narva, fortress captured by Russians, 222; battle of, 222, 224, 229

Naryshkin, Leo (boyar), 216

Naryshkin baroque, or Muscovite baroque, 204

Naryshkina, Nathalie (second wife of Tsar Alexis and mother of Peter the Great), 213–16, 238

Naryshkins and their party, 213–14

"National" communism, and Yugoslavia, 536

National income, and Gorbachev's reforms, 592

National independence movements, 479, 484–5; *see also* Minorities *and individual countries*

694

Ossetians (*Cont.*)
Alans), 14; and Georgian nationalism, 593, 596
Ostermann, Andrew, 218, 245, 251
Ostrogoths, 15
Ostromirovo Gospel, 86
Ostrovsky, Alexander, 443
Otrepiev, Gregory, 161f
Ottoman Empire, *see* Turkey
Outer Mongolia, Soviet control in, 512
Oxford Slavonic Papers, 207

Pacific Ocean: coast reached, 194; explorations in, 353; Soviet Russia and the war in, 523
"Pacification" (post-1905 policy), 413–14
Paganism, in Russia, 52–3
Pahlen, Count Peter, 275
Painting, *see* Arts
Pakistan, in SEATO, 559
Palaea, 57
Paleologue, Zoe or Sophia, marriage to Ivan III, 107
Pallas, Peter-Simon, 297
Pan-Germanism, and World War I, 416
Pan-Slav Congress in Prague, 365
Pan-Slavism, 450; support for Balkan insurrection against Turkey, 386–7; popularity before World War I, 416
Panin, Count Nikita (Ivanovich), and Russian foreign policy, 264
Panin, Count Nikita (Petrovich), 275
Pares, Barnard, quoted on Ivan IV, 143
Paris: visited by Peter I, 225; entered by the allies (1814), 313; First and Second Treaties of (1814–15), 315; Congress, 339; Treaty of (1856), 339, 385; Black Sea provisions abrogated, 386
Parties, emergence of, 405–6
Partitions of Poland, 251, 256, 267–72, 277; First, 264, 268; Second, 269–70, 272; Third, 264, 270, 272; and Duchy of Warsaw, 310; and Congress of Vienna, 314, 385
Party, role of, as envisaged by Lenin, 468, 471; *see also* Communist party; Social Democratic party
"Party-mindedness" (*partiinost*), 568, 584
Pashkov, Philip, 165
Paskevich, General Prince Ivan, 331; and Polish revolt, 332; and Hungarian revolution (1849), 335
Pasternak, Boris, 445, 585; *Doctor Zhivago,* 585; quoted, 609; translation of Goethe, 609n
Paterikon, 57

Patkul, Johann R., role in the Great Northern War, 221, 223
Patriarchate in Russia: established, 155, 198; suspended, 232; reinstituted in Soviet Russia, 588
Patriotism, Great Russian, 528, 577
Patzinaks, *see* Pechenegs
Paul I, Emperor, 257f, 282, 297, 318; bypassed by Catherine II, 248; and Catherine, 257f; and Law of 1797, 274, 282; personality, and reign of, 274–5; relationship with Alexander, 301
Paulus, Field Marshal Friedrich von, 521
"Pauper's allotment," 372
Pavlov, Ivan, 439
Pavlov, Michael, 360
Pavlov-Silvansky, Nikolai P., on feudalism, 49, 115f; quoted, 114
Pavlova, Anna, 446
Pavlovsky, Georgii A., on economic development of Russia, 434
Peasant commune (*mir* or *obshchina*): in views of, Slavophiles, 363, 365–6, radicals, 365–6, populists, 382–3, 424, 448, 449; land obtained by the Emancipation Edict, 372; role of, 374, 430–34; and Stolypin's reforms, 414–15; and collectivization, 499
Peasant Land Bank, 395, 415, 433
Peasantry: in Kievan Russia, 49; in appanage Russia, 118; in Muscovite Russia, preceding Time of Troubles, 159; in Time of Troubles, 165 *passim;* represented in zemskii sobor of 1613, 171–2; treatment of, by Poles, 179f; in Muscovite Russia, 185–7; at end of 18th century, 278–80, 282–3; in struggle against Napoleon, 313; unrest of, 369–70, 379, 406–7; in ideology of populism, 382; and revolutionary movement, 383; and reaction, 393–4; and SR's, 406; and elections to Dumas, 409–10; representations in Dumas, 411–12; and Stolypin reforms, 414–15; after emancipation, 430–34; question, following "great reforms," 430–34; death rate among, 431; and views of Lenin, 468; and October Revolution, 474, 571, 572–3; and War Communism, 479, 488–9; support Reds, 487; opposition to five-year plans, 495–6; during Second World War, 529; and increased control over, 531; and Khrushchev's policies, 547–9; *see also* Collectivization; "Great reforms"; *Kulaks;* Land; Serfdom
Peasants, state; represented in Legisla-

Zhivkov, Todor, 600

Zhukov, Marshal George, 518, 521, 522, 525, 612; and Khrushchev, 540

Zhukovsky, Basil, 359; and *Odyssey* translation, 355; tutor to Alexander II, 368

Zimin, A. A.: on *The Lay of the Host of Igor*, 57

Zinin, Nicholas, 353

Zinoviev (Radomyslsky), Gregory, 490f, 503, 511

Zolkiewski, Stanislaw, 168

Zorndorf, battle of, 252

Zosima (monk), travels, 126

Zubatov, Serge, 407

Zyriane, 122